Grundlehren der mathematischen Wissenschaften

A Series of Comprehensive Studies in Mathematics

Volume 357

Grundlehren der mathematischen Wissenschaften (subtitled Comprehensive Studies in Mathematics), Springer's first series in higher mathematics, was founded by Richard Courant in 1920. It was conceived as a series of modern textbooks. A number of significant changes appear after World War II. Outwardly, the change was in language: whereas most of the first 100 volumes were published in German, the following volumes are almost all in English. A more important change concerns the contents of the books. The original objective of the Grundlehren had been to lead readers to the principal results and to recent research questions in a single relatively elementary and accessible book. Good examples are van der Waerden's 2-volume Introduction to Algebra or the two famous volumes of Courant and Hilbert on Methods of Mathematical Physics.

Today, it is seldom possible to start at the basics and, in one volume or even two, reach the frontiers of current research. Thus many later volumes are both more specialized and more advanced. Nevertheless, most books in the series are meant to be textbooks of a kind, with occasional reference works or pure research monographs. Each book should lead up to current research, without over-emphasizing the author's own interests. There should be proofs of the major statements enunciated, however, the presentation should remain expository. Examples of books that fit this description are Maclane's Homology, Siegel & Moser on Celestial Mechanics, Gilbarg & Trudinger on Elliptic PDE of Second Order, Dafermos's Hyperbolic Conservation Laws in Continuum Physics ... Longevity is an important criterion: a GL volume should continue to have an impact over many years. Topics should be of current mathematical relevance, and not too narrow.

The tastes of the editors play a pivotal role in the selection of topics.

Authors are encouraged to follow their individual style, but keep the interests of the reader in mind when presenting their subject. The inclusion of exercises and historical background is encouraged.

The GL series does not strive for systematic coverage of all of mathematics. There are both overlaps between books and gaps. However, a systematic effort is made to cover important areas of current interest in a GL volume when they become ripe for GL-type treatment.

As far as the development of mathematics permits, the direction of GL remains true to the original spirit of Courant. Many of the oldest volumes are popular to this day and some have not been superseded. One should perhaps never advertise a contemporary book as a classic but many recent volumes and many forthcoming volumes will surely earn this attribute through their use by generations of mathematicians.

More information about this series at http://www.springer.com/series/138

David Mond • Juan J. Nuño-Ballesteros

Singularities of Mappings

The Local Behaviour of Smooth and Complex Analytic Mappings

Springer

David Mond
Mathematics Institute
University of Warwick
Coventry, UK

Juan J. Nuño-Ballesteros
Departament de Matemàtiques
Universitat de València
Burjassot, Spain

ISSN 0072-7830 ISSN 2196-9701 (electronic)
Grundlehren der mathematischen Wissenschaften
ISBN 978-3-030-34439-9 ISBN 978-3-030-34440-5 (eBook)
https://doi.org/10.1007/978-3-030-34440-5

Mathematics Subject Classification (2010): 14A10, 14B05, 14B07, 14J17, 14B10, 13H10

This Springer imprint is published by the registered company Springer Nature Switzerland AG.
The registered company address is: Gewerbestrasse 11, 6330 Cham, Switzerland

To Gabi, to my sons Gabriel and Isaac, and to Maíra and Ruben who were entrusted to me

D.M.

To Conxín, and our children Carme and Joan

J.J.N.B.

Preface

This book is concerned with the local behaviour of mappings. The local behaviour at an immersive point, where the derivative is injective, and at a submersive point, where the derivative is surjective, is uninteresting; in the first case, the mapping is equivalent, under right-left equivalence (smooth changes of coordinates in source and target), to a linear inclusion. In the second, it is equivalent to a linear projection. Singularity theory begins where neither of these two conditions holds.

Singularities of algebraic varieties have been studied by algebraic geometers since the beginning of the subject and continue to play a central role in the minimal model programme for the birational classification of varieties of dimension ≥ 3. This book has a different focus. It is not concerned with singularities of varieties though they appear and play an important role, so much as with singularities of mappings, and its focus is not algebraic geometry, but smooth (C^∞) and complex analytic geometry though the gap is not all that wide.

There are many sources for this part of the subject.

- The earliest is probably the Morse Lemma (Example 3.1 below), proved by Marston Morse in the 1920s (see [Mor96]): any smooth function of n variables with a non-degenerate critical point at x_0 can be transformed, by a change of coordinates in some neighbourhood of x_0, into a quadratic form. The Morse Lemma is an example of "finite determinacy" (*cf* Chap. 6), in which a finite portion of the Taylor series of a smooth or analytic map determines it up to some kind of equivalence—in this case, equivalence under change of coordinates in the domain, which we refer to from now on as *right equivalence*. Finite determinacy allows the algebraisation of the spaces and maps it covers, and this can have important consequences. The simple normal form for a function in the neighbourhood of a non-degenerate critical point is used in Morse Theory to prove important statements about the global topology of manifolds.
- In the 1940s and 1950s, Whitney was concerned with immersing manifolds in Euclidean spaces. Immersions are dense in the space of proper maps $M^n \to \mathbb{R}^{2n}$ (see below, Exercise A.7). An n-manifold can also be immersed in $\mathbb{R}^{2n-1}$, even though immersions are no longer dense: some non-immersive points cannot be

removed by small deformations. To remove them (by large deformations), one needs to understand their geometry. In [Whi44], Whitney found and described the singularities which persist, up to right-left equivalence—see Exercise 7.2.4, and in [Whi55] he did the same in the case of maps from surfaces to the plane—see Exercises 3.3.6 and 3.3.7. These are examples of what we now call *stable* singularities.

- Knot theorists defined invariants by means of "generic planar projections": planar projections with only nodes as singularities. K. Reidemeister, seeking to understand how one generic planar projection could be deformed into another, described the three other singularities which were unavoidable in the process (see [Rei27b]).They are now known as the three Reidemeister moves; in the language of singularity theory they are the codimension 1 singularities of parameterised plane curves—see Example 3.4 below.
- Moving on from questions about the representability of homology classes by embedded submanifolds, René Thom developed a general theory of smooth maps, and in the process proved his celebrated transversality theorem (see Theorem A.1 in Appendix A), which puts the "generic" into every subsequent theorem about the generic behaviour of maps. Maps $f, g : N \to P$ are *left-right equivalent* if there are diffeomorphisms φ on N and ψ of P such that $f = \psi \circ g \circ \varphi$. A smooth mapping $f : N \to P$ is *stable* if its orbit under left-right equivalence is open in the space $C^\infty(N, P)$ of smooth mappings with its Whitney topology. Thom asked whether the set of stable mappings is dense in $C^\infty(N, P)$ and quickly found a counterexample (See Sect. 5.2 below). His question became *when* are stable mappings dense. The answer was provided by John Mather, in a remarkable series of six papers collectively called "Stability of C^∞ mappings", [Mat68a, Mat68b, Mat69a, Mat69b, Mat70, Mat71].

Part I of this book is concerned with these ideas. In particular, we cover most of the material in [Mat68a, Mat68b, Mat69a, Mat69b, Mat70, Mat71] though our focus is always local rather than global. Mather went on to develop a theory of topological stability, in which φ and ψ are required merely to be homeomorphisms; we comment only very briefly on this, in Sect. 5.3.

John Milnor started another strand in the subject with his 1968 book "Singular points of complex hypersurfaces", [Mil68], by studying the level sets of a complex polynomial $f(x_1, \ldots, x_{n+1})$ in the neighbourhood of an isolated critical point. Intersected with a suitably small ball around the critical point, these level sets have the homotopy type of a wedge of n-spheres. The resulting fibration over the complement of the critical value, the so-called Milnor fibration, has been the object of intensive study since then, notably by the Russian school led by Vladimir Arnold and the German school led by Egbert Brieskorn. One of the salient features of the subject is the fact that the Milnor number of the singularity, the number of spheres in the wedge, is equal to its deformation-theoretic codimension with respect to right equivalence (see Definition 3.3 and Sect. 8.2 below). If we regard the Milnor fibre as the stable object near to the (unstable) singular hypersurface, then Milnor's study sets the theme for Part II of the book, where we study the stable mapping (on a suitably defined domain) which results from perturbing an unstable map-germ.

It is important to point out a bifurcation in the subject here. Milnor was concerned with preimages, and the largest part of singularity theory in this direction, studying singularities of *varieties*—see e.g. [GLS07, Dim92, Loo84, AGZV12a, AGZV12b]. As mentioned earlier, there is also a large literature on singularities of varieties in the context of birational geometry and the minimal model programme in algebraic geometry. In this book, we are interested in singularities of *mappings*, and those properties which are preserved by right-left equivalence. As Felix Klein remarked, the shape of the subject is determined by the equivalence relation one uses, and for right-left equivalence, it turns out that the topological counterpart to the deformation theory of an unstable singularity is found not in the fibres of the map, but in the discriminant of a stable perturbation, in case the dimension of the source, n, is greater than or equal to the dimension of the target, p, and the image otherwise. When $n \geq p - 1$, the discriminants and images of stable perturbations have the homotopy type of wedges of spheres, just like Milnor fibres, and the number of spheres in the wedge is closely related to the deformation-theoretic codimension of the germ itself, echoing Milnor's findings unexpectedly closely. Part II of the book is concerned with discriminants and images and with the relation between their topology and the deformation theory.

The book is a monograph and not a textbook—its shape reflects the subject, or rather our knowledge of it, rather than the structure of a course. Even so we hope that it could be used as the basis for a graduate course. We have included many exercises and some preliminary material that will make it approachable for a beginning graduate student. A graduate course consists of

- Chapter 3: Sects. 3.2, 3.3, 3.4, 3.5
- Chapter 4: Sects. 4.1, 4.2, 4.3, 4.4
- Chapter 5: Sects. 5.1, 5.5
- Chapter 6: Statement of finite determinacy theorems
- Chapter 7: Sects. 7.1, 7.2
- Chapter 8: Sects. 8.1, 8.2, 8.3, 8.4, 8.5.

Many people have contributed to our understanding of the subject. DM is especially grateful to Terry Wall, Jim Damon, Terry Gaffney, Duco van Straten, Vladimir Arnol'd and Victor Goryunov for explanations, conversations and collaborations. He also thanks Jairo Charris and Carlos Ruiz for encouragement when it most counted. JJNB thanks Lê Dũng Tráng for helpful conversations and Raúl Oset Sinha, Guillermo Peñafort Sanchis and Roberto Giménez Conejero for their help with the revision of some parts of this book. And we both apologise to the people—surely there are many—whose important work we have not been able to cover.

Coventry, UK | David Mond
Burjassot, Spain | Juan-José Nuño-Ballesteros
July 2019

Contents

Chapter 1
Introduction

What is singularity theory? The crucial notion is the *derivative* of a smooth or analytic map: if $f : X \to Y$ is a map of manifolds and $x \in X$ then $d_x f : T_x X \to T_{f(x)} Y$ is the derivative, defined by

$$d_x f(\hat{x}) = \lim_{h \to 0} \frac{f(x + h\hat{x}) - f(x)}{h}$$

if X and Y are open sets in linear spaces. If X and Y are contained, but not open, in linear spaces, $d_x f$ can be defined by restricting to $T_x X$ the derivative of a suitable extension of f to an open set in the ambient space; otherwise one uses charts. It is also worth recalling that every tangent vector $\hat{x} \in T_x X$ is the tangent vector $\gamma'(0)$ to a parameterised curve $\gamma : (\mathbb{R}, 0) \to (X, x)$ (or $\gamma : (\mathbb{C}, 0) \to (X, x)$ in the complex analytic category), and that $d_x f$ satisfies

$$d_x f(\gamma'(0)) = (f \circ \gamma)'(0). \tag{1.1}$$

This may be taken as the definition. It gives a useful heuristic in infinite-dimensional cases, such as where X is a group of diffeomorphisms.

A point $x_0 \in X$ is a *regular point* of f if $d_{x_0} f$ is surjective, in which case f is a *submersion* at x_0, and a *critical point* otherwise, and f is an *immersion* at x_0 if $d_{x_0} f$ is injective. The image of a critical point is a *critical value* of f; any point in Y which is not a critical value is a *regular value* (even if it has no preimages). The set of all critical values is often called the *discriminant* of the map f. If f is a submersion at x_0, then from the inverse function theorem follows

Theorem 1.1 (Normal Form for Submersions) *Suppose that* $\dim X = n \geq k = \dim Y$ *and* x_0 *is a regular point of* $f : X \to Y$*. Then one can choose coordinates* $x_1, \ldots, x_n$ *on* X *around* x_0*, and* $y_1, \ldots, y_k$ *on* Y *around* $f(x_0)$*, such that* f *takes the form* $f(x_1, \ldots, x_n) = (x_1, \ldots, x_k)$.

D. Mond, J. J. Nuño-Ballesteros, *Singularities of Mappings*,
Grundlehren der mathematischen Wissenschaften 357,
https://doi.org/10.1007/978-3-030-34440-5_1

These notions are only of interest when $\dim X \geq \dim Y$; when $\dim X < \dim Y$, all points of X are critical points, and the set of critical values of f is the whole image of f. In this case one is interested in whether or not $d_x f$ is injective. If it is, one has

Theorem 1.2 (Normal Form for Immersions) *Suppose that* $\dim X = n \leq k = \dim Y$ *and that* $f : X \to Y$ *is an immersion at* x_0*. Then one can choose coordinates around* x_0 *and* $f(x_0)$ *such that* f *takes the form* $f(x_1, \ldots, x_n) = (x_1, \ldots, x_n, 0, \ldots, 0)$.

Both statements follow from the inverse function theorem, by incorporating f into a suitable auxiliary mapping whose derivative is invertible; we give the standard proof, as Corollary 2.1, in the introductory chapter on smooth manifolds and maps which follows this introduction.

Both statements can be summarised by saying that any smooth mapping can be "linearised" in a neighbourhood of a regular point (that is, the local behaviour of a mapping near a regular point is like a linear mapping). Both are examples of the phenomenon of *finite determinacy*, which we study in Chaps. 3 and 6, in which the behaviour of a mapping on some neighbourhood of a point x_0 is determined up to diffeomorphisms in source and/or target by a finite portion of its Taylor series at x_0. A finitely determined mapping is thus equivalent to the polynomial mapping defined by this portion of its Taylor series. In the case of immersions and submersions, first order information alone is enough, and thus they are equivalent to linear injections and projections, respectively.

Singularity theory begins where f is neither a submersion nor an immersion—though it is also concerned with the way that multiple immersions can interact.

In this book we concentrate on the local behaviour of mappings. This is best described with the notion *germ of mapping*, which we describe in Sect. 2.1.

1.1 Real or Complex?

A large part of the theory we will cover holds both in the complex analytic category, where one considers complex analytic (holomorphic) maps between complex manifolds, and in the C^∞ category, where one considers real C^∞ manifolds and mappings. In order to give a single account, where possible we will refer to both real affine space $\mathbb{R}^n$ and complex affine space $\mathbb{C}^n$ as $\mathbb{F}^n$. In the parts where the two are treated jointly, we will refer to both C^∞ manifolds and mappings, and complex analytic manifolds and mappings, simply as *smooth*. However, we are not in any sense trying to work over a general unspecified field. Our $\mathbb{F}$ is at all times either $\mathbb{R}$ or $\mathbb{C}$ or both. Both $\mathbb{R}$ and $\mathbb{C}$ have a Euclidean metric (which for $\mathbb{C}$ renders it the same as $\mathbb{R}^2$) and we make free use of this.

1.2 Structure of the Book

This book is divided into two parts.

Part I is concerned primarily with the purely local theory of germs of mappings, and of their unfoldings. All but a few minor details here is the work of other authors, beginning with René Thom and especially with John Mather, whose six papers jointly called Stability of C^∞ mappings [Mat68a, Mat68b, Mat69a, Mat69b, Mat70, Mat71] set the groundwork for the subject. It also contains later work of Bruce, Damon, Gaffney, Looijenga, Martinet, du Plessis, K. Saito, Wall and others.

Part II is of more recent origin, though it takes its inspiration from Milnor's book [Mil68]. It is concerned with what we can call the semi-local theory: what happens in a suitably small neighbourhood of an unstable point of a mapping when the mapping is deformed. That this makes sense at all is a slightly subtle point, though our intuition easily furnishes a working definition of what it means, for example, for a repeated root of a complex polynomial to split into a certain number of ordinary roots. The formal definitions here rely on the locally conical structure of analytic sets. Locally conical structure is easy to prove in spaces with isolated singularities: one integrates a radial vector field, which is not hard to construct. In spaces with non-isolated singularities, considerably more technical sophistication is needed, in particular the theory of Whitney stratifications and controlled vector fields. Incorporating every technical development as it is needed would give a very uneven narrative, so we consign a significant number, including those just mentioned, to appendices, where they are summarised. We hope that this may also make the book useful as a reference.

Three aspects to the local theory are covered in Part I. The fundamental invariant in the theory is the left-right *codimension*. For now one can use the following relatively non-technical working definition: the codimension is the minimal number of parameters for a family of mappings in which a singularity left-right equivalent to the one in question occurs 'stably' or "irremovably". In other words, it is the smallest numbers of parameters for a parameterised family in which one can expect to see singularities of the given type. The three key notions are, first, left-right stability itself, studied in Chap. 3, which then extends to the second notion, of (left-right) versal unfolding, studied in Chap. 5. We prove Martinet's theorem [Mar82] that every germ of finite left-right codimension has a versal unfolding, by means of his infinitesimal criterion for versality. We introduce the bifurcation set in the base of a versal unfolding, and discuss Mather's nice dimensions, originally defined as those dimension-pairs (n, p) for which globally stable mappings are dense in $C^\infty_{\text{pr}}(N, P)$, whenever $\dim N = n$ and $\dim P = p$. We show that this property is equivalent to the property that every germ of finite left-right codimension $(\mathbb{F}^n, 0) \to (\mathbb{F}^p, 0)$ has a stable perturbation. In Chap. 6 we study the third of the key notions, finite determinacy. We prove Mather's theorem [Mat68b] that germs of finite left-right codimension are finitely determined, with the improvements to Mather's original estimates for the determinacy degree due to Gaffney and du Plessis, and then discuss the work of Bruce, du Plessis and Wall in [BdPW87] on unipotent groups and their role in finite determinacy.

In almost every aspect of the local theory it is unnecessary to distinguish between $\mathbb{R}$ and $\mathbb{C}$. One of the reasons for this is finite determinacy mentioned above. Every real polynomial, and every convergent real power series, has a complexification, defined by the same formula. Of course there are C^∞ functions which do not have a complexification, such as the smooth functions one uses to construct partitions of unity, where there are points where all derivatives vanish. But in a sense which will become clear later, such functions and singularities are "infinitely degenerate". And a finitely determined mapping is equivalent to a polynomial mapping, which does have a natural complexification. All of the singularities of finite codimension, and in particular, all stable ones, have this property. And the codimension of a real singularity is the same as the codimension of its complexification: the algebraic object with which one calculates the codimension is in general a vector space, real in the C^∞ case and complex in the complex analytic case, and in every case the complex vector space corresponding to the complexification of a real singularity is the tensor product with $\mathbb{C}$ of the real vector space for the real singularity (Proposition 3.8, below). *Thus the calculations involved in determining it are formally the same.*

Note that behind the proof of finite determinacy of germs of finite codimension for left-right equivalence is the Weierstrass preparation theorem, proved in the late nineteenth century, in the complex analytic case, and, in the C^∞ case, the preparation theorem proved by Malgrange in the early 1960s. Malgrange's remarkable result is an essential underpinning for the close link between the C^∞ and the complex analytic theory.

However, in general classification over $\mathbb{R}$ is more complicated than over $\mathbb{C}$. A single complex equivalence class may have many inequivalent real forms, as witness the non-degenerate critical point (Morse singularity) of a function of n variables. Here, in the complex case, the Morse Lemma tells us that with respect to suitable coordinates centred on the critical point, f is equal to $f(0)+\sum_j x_j^2$. In the real case there are $n+1$ different possibilities, if we allow coordinate changes only in the source: f is equivalent to $f(0)-\sum_{j=1}^k x_j^2+\sum_{j=k+1}^n x_j^2$ for some $k=0,\ldots,n$ (the index of the critical point). Over $\mathbb{C}$, the substitution of a new complex variable ix_j for x_j changes x_j^2 to its negative, and so as complex germs all these are equivalent. Nevertheless, we do not attempt a systematic discussion of the relation between real and complex classification.

Part II is concerned with the geometrical theory. Classical singularity theory is interested in *preimages* $f^{-1}(y_0)$, where f is a germ of map $(\mathbb{F}^n,0)\to(\mathbb{F}^p,0)$ with $n>p$. *We do not cover this part of the theory*; it is already the subject of several excellent books, in particular [Loo84, Dim92, GLS07]. There, the link between the semi-local theory and the local is the "Milnor–Tjurina relation": the Milnor number of an isolated complete intersection singularity (the rank of the middle homology of a nearby non-singular fibre) is greater than or equal to its Tjurina number, its codimension for $\mathscr{K}$-equivalence, with equality in the weighted homogeneous case (see [Gre80, LS85, Vos02]). For the theory of singularities of mappings, one is interested in left-right equivalence, and the Milnor number bears no relation to the left-right codimension. Instead, it is in the topology of images and discriminants of stable perturbations that one finds invariants which do relate to the left-right codimension. In Chap. 8 we introduce image and discriminant Milnor numbers, due to Jim Damon and the first author, and explore their relation with left-right codimension. Part of this relation is still obstinately conjectural, though well-supported by examples.

Images and discriminants of germs of maps $\mathbb{F}^n \to \mathbb{F}^p$ are in general highly singular, and this makes calculating their topology rather difficult. It turns out that a useful handle on this topology and, indeed, other useful invariants of singularities of maps come from the study of their multiple points. Here we take a rather naive definition: if $f : X \to Y$, then $D^k(f)$ is the closure in X^k of the set

$$\{(x_1, \ldots, x_k) \in X^k : x_i \neq x_j \text{ if } i \neq j,\ f(x_i) = f(x_j) \text{ for all } i, j\}.$$

In Chap. 9 we show that for germs $(\mathbb{F}^n, 0) \to (\mathbb{F}^p, 0)$ with $n \leq p$ of corank 1 and with finite left-right codimension, the $D^k(f)$ are isolated complete intersection singularities, of which the multiple point spaces of a stable perturbation f_t of f are smoothings. The Milnor numbers of the $D^k(f)$ are left-right invariants, but there is more: from certain invariants of the symmetric group action on $D^k(f_t)$, permuting the copies of X (the "alternating homology"), one can calculate the homology of the image of f_t by means of the "image computing spectral sequence" ICSS. The ICSS was introduced by Goryunov and the first author in [GM93] for rational homology, and then further developed by Goryunov in [Gor95] and by Houston in [Hou97] and other papers. In Chap. 10 we give a new construction of the ICSS, due to José Luis Cisneros and the first author, using the triangulability of stratified mappings proved by Hardt in [Har77] for light mappings and more generally by Verona in [Ver84]. With this approach, convergence of the ICSS to the homology of the image is easily proved. The fact that for a corank 1 map-germ the $D^k(f)$ are isolated complete intersection singularities means that in this case the ICSS for a stable perturbation collapses at E^1, making the homology of the image easy to read off from the alternating homology of the $D^k(f_t)$. We also discuss (though we do not prove) the remarkable results of Houston in [Hou97] which show that the same collapse occurs even in the case of germs of corank > 1, where the $D^k(f)$ are neither complete intersection singularities nor necessarily isolated.

Chapter 11 focuses on the case of germs of maps $(\mathbb{C}^n, 0) \to (\mathbb{C}^{n+1}, 0)$ and studies the information which can be obtained from the matrix of a presentation of $\mathscr{O}_{\mathbb{C}^n,0}$ as $\mathscr{O}_{\mathbb{C}^{n+1},0}$-module. In particular we are interested in the Fitting ideals, which, it turns out, give a natural analytic structure to the multiple point sets

$$M_k(f) = \{y \in Y : |f^{-1}(y)| \geq k,\ \text{(counting multiplicity)}\}.$$

We reprove results of Catanese, Mond–Pellikaan, de Jong–van Straten, Kleiman-Lipman–Ulrich and Sharland.

The book ends with a number of appendices on technical aspects of the theory which are needed in the main body of the text.

1.3 The Nearby Stable Object

Much of Part II centres on the notion of the *nearby stable object*. This term will be made more precise in Chap. 8; for now, we make do with three examples. The first comes from classical singularity theory, concerned with preimages. The second comes from the singularity theory of images, an important part of the subject of this book. The third comes from the singularity theory of the discriminant, the set of critical values of a map.

The first example is the nearby level set of a non-degenerate critical point x_0 of a complex analytic function. The critical level set is unstable—by means of an arbitrarily small deformation (or choice of other level) it is smoothed, and in particular it is qualitatively changed. The nearby stable object is a nearby non-singular level set, "localised" by being intersected with a small ball around the critical point. Being smooth, it does not change qualitatively under small changes in the function or level.

By the Morse lemma, f is equal to $f(x_0) + \sum_j x_j^2$ in some local coordinate system centred on the point x_0. For simplicity let us assume $f(0) = 0$. The critical level set, $f^{-1}(0)$, is contractible within a sufficiently small ball $B_\varepsilon(x_0)$. For $t \neq 0$, $f^{-1}(t) \cap B_\varepsilon(x_0)$ has the homotopy-type of an $n-1$-sphere. The extra homology acquired by $f^{-1}(t) \cap B_\varepsilon(x_0)$ as t leaves 0 is known as the *vanishing homology* of the singularity.[1]

The second example of nearby stable object and vanishing homology is provided by the third Reidemeister move of knot theory, where three immersed pieces of parameterised plane curve meet, two-by-two transversely, at a point. This configuration is evidently unstable: one can move any one of the three to form a triangle (see Fig. 1.1). Since now all intersections are transverse, the altered configuration is stable. It is the "nearby stable object" for this example, and its vanishing homology, generated by the 1-cycle highlighted in the drawing on the right, once again has rank 1. In this example, the inclusion of real in complex is a homotopy equivalence; the drawing on the right is a "good real picture" (cf [Mon96, MM96]).

The left-right codimension here is equal to 1; that is why it is important in knot theory. Given two plane projections of the same knot, one can be deformed to the other in such a way that during the deformation, only three types of qualitative change occur. These are the three Reidemeister moves of classical knot theory

[1]The term "emerging homology" might be more appropriate, but because algebraic geometers saw singularities as degenerations of the objects they really wanted to study, we are stuck with "vanishing homology".

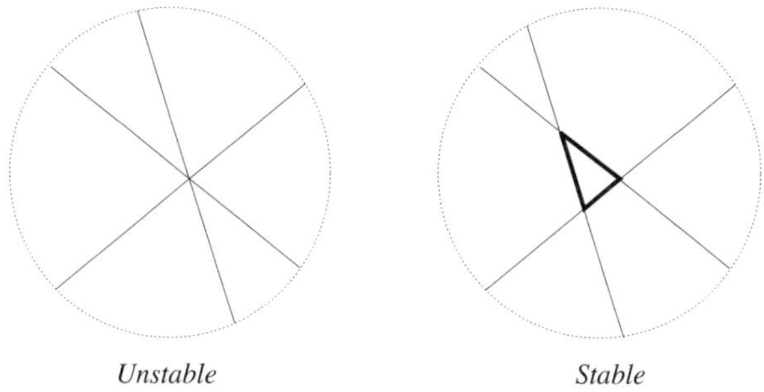

Fig. 1.1 Stable perturbation of the triple point

[Rei27a], each of which is passage through a codimension 1 singularity. Our example shows the third of these. They cannot be avoided in a 1-parameter family of projections, other more complicated singularities can be.

The third example is familiar to every schoolchild. When the real cubic x^3 is deformed to $x^3 - ux$, the unstable critical point at 0 splits a non-degenerate local maximum at $c_- := -\sqrt{u/3}$ and a non-degenerate local minimum at $c_+ := \sqrt{u/3}$, with critical values $v_{\max}$ and $v_{\min}$. For $u > 0$, the discriminant consists of the two points $v_{\max}$ and $v_{\min}$. Its relative 0th homology is generated by the class $[v_{\max}] - [v_{\min}]$, which vanishes as u returns to 0.

Example 1.1 (Looking at Bent Wires) A number of the themes of the book can be appreciated by considering plane projections of knots in 3-space. We return to this several times. The topic was first studied from this point of view by J.M. Soares David (published after a long delay) in [Dav83] and by C.T.C. Wall in [Wal77], and subsequently in [Mon95, Wal09, DNnB08, Dia15] among others. Take a thin copper wire, easy to bend but thick enough to support its own weight, and join the two ends after bending it to form a knot—which (making allowances for the fact that the wire is not infinitely thin) should be a C^∞ embedding of the circle in $\mathbb{R}^3$. You should obtain something looking like Fig. 1.2.

Fig. 1.2 A knot in 3-space

Regarding this picture as a subset of the plane, the view shown here is "a generic projection" of the knot to the plane: the only singular points on the image are nodes—transverse crossings of two branches. From certain points of view one can see other singularities than just nodes. If one spends a little time experimenting, looking at the knot from different points of view, then provided the knot is generic, one can convince oneself that there is only a rather short list of topologically distinct local pictures. The intuitive term "local picture" can be understood with the help of the fact that every analytic set $X \subset \mathbb{R}^N$ is locally conical, in the sense that for every point $x_0 \in X$ there is an $\varepsilon > 0$ such that $X \cap B_\varepsilon(x_0)$ is homeomorphic to the cone on its boundary $X \cap \partial B_\varepsilon(x_0)$ (see Sect. B.3 below). By a local picture of a set X at x_0 we mean such a neighbourhood of a point x_0 in X. Of course, the planar pictures of our knot in 3-space may not be analytic subsets of the plane; but for a smooth generic knot, all of the germs of projection to the plane are finitely determined, and thus equivalent to the polynomial map-germs defined by finite segments of their Taylor series. So in some neighbourhood of each point, the image is analytic with respect to suitable local coordinates, and thus locally conical.

In the case of plane projections of a knot, by slightly changing our viewpoint we can deform any unstable view so that it becomes stable. Each local picture then turns into a "nearby stable picture": the portion of the deformed curve inside $B_\varepsilon(x_0)$. To capture this notion more precisely, the deformation must be so small that the neighbourhood boundary $X \cap \partial B_\varepsilon(x_0)$ does not change during the deformation. Note that here the nearby stable picture associated to each unstable local picture is not unique: depending on the deformation we choose, we may end up with one of several different possibilities. This can be seen in the case of the first two Reidemeister moves, whose nearby stable pictures are shown below.

To Do Experiment with a copper knot, as described, and make a list of topologically distinct local pictures in its planar projections. For each one, determine its nearby stable pictures by gently moving the viewpoint (or the knot).

For each type X the following two numbers can be determined experimentally:

1. the codimension in $\mathbb{R}^3$ of the set View(X) of centres of projection (viewpoints) for which a singularity of type X appears, and
2. the maximum number of nodes, n, into which the singularity X splits when the centre of projection is moved.

For example, one sees a first order cusp by looking at the curve along one of its tangent lines at a point of non-vanishing torsion; so when X is a cusp, View(X) is the tangent developable surface of the curve (minus the tangent lines at torsion zero points), and has codimension 1.

To Do For each singularity in your list, determine the codimension of the viewing set and the maximum number of nodes into which it splits.

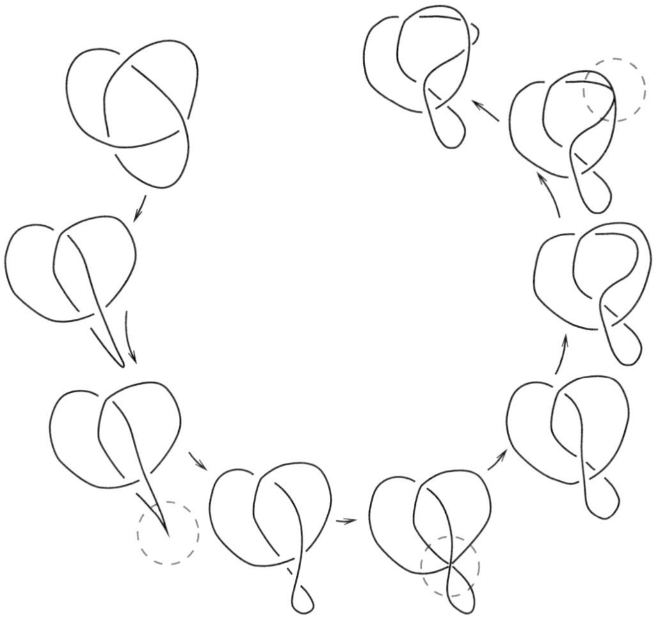

Fig. 1.3 Sequence of projections of the trefoil knot

You should find two other singularities whose viewing set has codimension 1: the tacnode and the triple point. The cusp, tacnode and triple point are the Reidemeister moves mentioned above. Each one appears in the sequence of images of Fig. 1.3.

In the neighbourhood of the cusp, tacnode and triple point in the sequence here, or in any other generic one-parameter family, the local picture changes in an entirely predictable way, shown in the diagram of Fig. 1.4. These changes can be observed with a copper knot, by moving one's eye along a path which crosses any of the codimension 1 viewing sets transversely.

For a "generic" knot, the deformations of the local pictures obtained by moving the viewpoint are, with one exception, universal, in the sense that, up to equivalence, they contain all possible deformations: they are *versal deformations*. The exception is the quadruple point, where four smooth branches of the projected curve cross at a point, pairwise transversely. Here there is a modulus in the smooth classification, namely the cross ratio of the four tangent lines. This cannot be altered merely by moving the viewpoint: the only movement retaining a view of the quadruple crossing is along the line of sight, and this does not change the cross ratio.

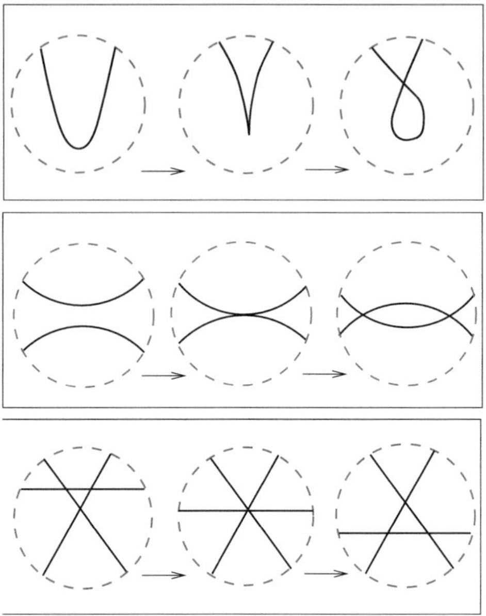

Fig. 1.4 The three Reidemeister moves

1.4 Exercises and Open Questions

We suppose that the book falls somewhere between a textbook and a monograph. Nevertheless every reader will no doubt want to practice their skills, and we have tried to provide exercises for every section. We hope that enough of them are easy and enough are hard. One of the pleasures of the subject is the unlimited number of examples for which one can make interesting calculations, which range from easy and gratifying to stubbornly difficult. At the end of some of the sections of Part II of the book we have listed some open questions. We would welcome questions and answers from anyone interested in tackling them.

1.5 Notation

Proofs are ended with □ and Remarks and Examples are ended with ◇.

Part I
Thom-Mather Theory: Right-Left Equivalence, Stability, Versal Unfoldings, Finite Determinacy

Chapter 2
Manifolds and Smooth Mappings

This chapter gives a brief account of the definitions and results we need from elementary differential topology, ranging from the notion of manifold to transversality. We assume the reader is already familiar with these concepts, so we do not give many details. The only novelty is that we consider real C^∞ and complex analytic manifolds together, highlighting the differences between the two cases.

2.1 Germs

In order to study local properties of mappings, we introduce the notion of germ of a mapping $f: X \to Y$ between topological spaces at a subset $S \subset X$.

Definition 2.1 Let X, Y be topological spaces, and let $S \subset X$.

1. Let $f: U \to Y$ and $g: V \to Y$ be maps, where U and V are open neighbourhoods of S in X. We say that f and g *have the same germ at* S, if there is a neighbourhood $W \subset U \cap V$ of S in X such that f and g coincide on W. This is evidently an equivalence relation, and a *germ of mapping at* S or *map-germ at* S is an equivalence class under this relation.
2. Two subsets X_1 and X_2 of X *have the same germ at* S if there is a neighbourhood U of S in X such that $X_1 \cap U = X_2 \cap U$. A *germ of subset of* X or *set-germ* at S is an equivalence class of subset under this relation.
3. We denote a germ at S of a mapping $f: U \to Y$ by $f: (X, S) \to Y$, or $f: (X, S) \to (Y, T)$ if $f(S) \subset T \subset Y$. Given a map-germ $f: (X, S) \to Y$, each member $f: U \to Y$ of the corresponding equivalence class is called a *representative*. Analogously, the germ at S of a subset X_1 of X is denoted by (X_1, S) and each member X_1 of the corresponding equivalence class is called a *representative* of the set-germ.

D. Mond, J. J. Nuño-Ballesteros, *Singularities of Mappings*,
Grundlehren der mathematischen Wissenschaften 357,
https://doi.org/10.1007/978-3-030-34440-5_2

Terminology A germ at a one point set is sometimes called a *mono-germ*, to distinguish it from *multi-germ*, a germ at a finite point set with more than one point. When S is not a finite point set, it is common to use the expression *germ along* S instead of germ at S. In fact we will not be concerned with cases where S is infinite.

We say that a map-germ $f\colon (X, S) \to Y$ is *continuous* if there exists a continuous representative $f\colon U \to Y$. If a germ $f\colon (X, S) \to (Y, T)$ is continuous and $g\colon (Y, T) \to Z$ is another germ, then we can define the composition in the obvious way: take $f\colon U \to Y$ and $g\colon V \to Z$ representatives with $f(U) \subset V$ and define the *composition* $g \circ f\colon (X, S) \to Z$ as the germ of $g \circ f\colon U \to Z$ at S. The continuity of f is necessary to ensure the existence of such representatives and it is easy to prove that the definition does not depend on the choice of representatives.

A map-germ $\phi\colon (X, S) \to (Y, T)$ is called a *homeomorphism* if there exists a representative $\phi\colon U \to V$ which is a homeomorphism, where U and V are open neighbourhoods of S and T in X and Y, respectively. Equivalently, ϕ is invertible as a continuous map-germ, that is, there exists another continuous map-germ $\phi^{-1}\colon (Y, T) \to (X, S)$ such that

$$\phi \circ \phi^{-1} = \mathrm{id}_{(Y,T)}, \quad \phi^{-1} \circ \phi = \mathrm{id}_{(X,S)} .$$

Let $f\colon (X, S) \to (Y, T)$ be a continuous map-germ and let (Y_1, T) be a set-germ in Y. We take representatives $f\colon U \to Y$ and $Y_1 \subset Y$, respectively. The *inverse image* $f^{-1}(Y_1, T)$ is defined as the set-germ $(f^{-1}(Y_1), S)$. Again it is not difficult to prove that the definition does not depend on the choice of representatives.

Example 2.1 In general, the image of a continuous map-germ is not well defined, since it may depend on the choice of the representative. For instance, consider the continuous mapping $f\colon [0, 2\pi) \to \mathbb{R}^2$ given by $f(x) = (\cos x, \sin x)$. For each ϵ with $0 < \epsilon < 2\pi$, f and the restriction $f|_{[0,\epsilon)}\colon [0, \epsilon) \to \mathbb{R}^2$ have the same germ at 0, but their images S^1 and $f([0, \epsilon))$ do not have the same germ at $(1, 0)$. ◊

A mapping $f\colon X \to Y$ is called *open onto its image at* $S \subset X$ if for every W neighbourhood of S in X, $f(W)$ is a neighbourhood of $f(S)$ in $f(X)$. We say that a map-germ $f\colon (X, S) \to Y$ is *open onto its image* if there is representative $f\colon U \to Y$ which is open onto its image at S. If $f\colon (X, S) \to Y$ is open onto its image, the *image* $f(X, S)$ is defined as the set-germ $(f(U), f(S))$ in Y, where $f\colon U \to Y$ is any representative which is open onto its image at S. The fact that the definition does not depend on the choice of the representative is shown in the next lemma.

Lemma 2.1 *Assume that $f\colon U \to Y$ and $g\colon V \to Y$ have the same germ at $S \subset X$ and that each mapping is open onto its image at S. Then $f(U)$ and $g(V)$ have the same germ at $f(S)$ in Y.*

Proof On the one hand, there exists an open neighbourhood $W \subset U \cap V$ of S such that f and g coincide on W. On the other hand, there exist $A, B \subset Y$, neighbourhoods of $f(S)$, in Y such that $f(W) = A \cap f(U)$ and $g(W) = B \cap g(V)$. Thus,

$$f(U) \cap A \cap B = f(W) \cap A \cap B = g(W) \cap A \cap B = g(V) \cap A \cap B,$$

which implies that $f(U)$ and $g(V)$ have the same germ at $f(S)$ in Y. □

The following result provides a simple and useful criterion to guarantee that a continuous map-germ is open onto its image.

Proposition 2.1 *Assume that X and Y are locally compact, Hausdorff and first countable spaces. Then every continuous map-germ $f: (X, S) \to (Y, y)$ with S finite and such that $(f^{-1}(\{y\}), S) = (S, S)$ is open onto its image.*

Proof We first take a continuous representative $f: U' \to Y$ such that $f^{-1}(y) = S$. Since X is locally compact and S is finite, there exists another open neighbourhood $U \subset X$ of S whose closure $\overline{U}$ is compact and such that $\overline{U} \subset U'$. We will show that the restriction $f: U \to Y$ is open onto its image at S.

In fact, if this was false, there would be a neighbourhood $W \subset U$ of S whose image $f(W)$ is not a neighbourhood of y in $f(U)$. Since Y is first countable, this means that there exists a sequence $\{y_n\}$ in $f(U) \setminus f(W)$ which converges to y. For each $n \geq 1$, there is an $x_n \in U \setminus W$ such that $f(x_n) = y_n$. By the compactness of $\overline{U}$, we can take a subsequence $\{x_{n_k}\}$ which converges to some $x' \in \overline{U}$. Now f is continuous, so $\{y_{n_k}\}$ converges to $f(x')$ and hence $f(x') = y$, since Y is Hausdorff. This implies that $x' \in S$, but this is in contradiction with the fact that W is a neighbourhood of S. □

The notion of germ is particularly natural in the (real or complex) analytic category, because of uniqueness of analytic continuation: if U_1 and U_2 are open sets in $\mathbb{F}^n$ with $U_1 \cap U_2$ connected, and $f_i: U_i \to \mathbb{C}^p$ are analytic maps which coincide on some open $V \subset U_1 \cap U_2$, they coincide on all of $U_1 \cap U_2$. Thus, if the f_i have the same germ at some point of $U_1 \cap U_2$, then they coincide on all of $U_1 \cap U_2$. The same is not true of C^∞ maps, of course.

2.2 Manifolds and Their Tangent Spaces

We consider the affine space $\mathbb{F}^n$, where $\mathbb{F}$ is either $\mathbb{R}$ or $\mathbb{C}$. By convention, a smooth mapping will mean any mapping $f: A \to \mathbb{F}^p$, where $A \subset \mathbb{F}^n$ is an open subset, which is differentiable of class C^∞ in the case $\mathbb{F} = \mathbb{R}$ or holomorphic (complex analytic) in the case $\mathbb{F} = \mathbb{C}$.

Definition 2.2 Let X be a topological space and let $n \geq 1$.

- A *smooth atlas* in X over $\mathbb{F}^n$ is a family $\mathcal{A} = \{\phi_i : U_i \to A_i\}_{i \in I}$ of homeomorphisms called *charts*, where $U_i \subset X$ and $A_i \subset \mathbb{F}^n$ are open subsets for all $i \in I$, such that:

  1. $X = \bigcup_{i \in I} U_i$,
  2. for each $i, j \in I$, the restriction

  $$\phi_j \circ \phi_i^{-1} : \phi_i(U_i \cap U_j) \longrightarrow \phi_j(U_i \cap U_j)$$

  is a smooth map (between open subsets of $\mathbb{F}^n$).

- Two smooth atlases $\mathcal{A}$ and $\mathcal{A}'$ are called *compatible* if $\mathcal{A} \cup \mathcal{A}'$ is again a smooth atlas. This defines an equivalence relation. Each equivalence class $[\mathcal{A}]$ is called a *smooth structure* in X over $\mathbb{F}^n$.
- A *manifold* of dimension n is a pair $(X, [\mathcal{A}])$, where X is a Hausdorff and second countable topological space and $[\mathcal{A}]$ is a smooth structure in X over $\mathbb{R}^n$.
- A *complex manifold* of dimension n is a pair $(X, [\mathcal{A}])$, where X is a Hausdorff and second countable topological space and $[\mathcal{A}]$ is a smooth structure in X over $\mathbb{C}^n$.
- A *manifold* of dimension 0 is any discrete and countable topological space.

Observe that a complex manifold of dimension n is a C^∞ manifold of dimension $2n$. We will use the word "manifold" to refer either to a C^∞ manifold or to a complex manifold, whenever there is no risk of confusion. It is also common to use the notation X for a manifold, instead of $(X, [\mathcal{A}])$, as long as only one smooth structure is in play.

Definition 2.3 Let $f : X \to Y$ be a continuous mapping between manifolds.

- We say that f is *smooth* if for every $x \in X$, there exist charts $\phi : U \to A$ in X and $\psi : V \to B$ in Y such that $x \in U \subset f^{-1}(V)$ and the mapping $\psi \circ f \circ \phi^{-1} : A \to B$ is smooth.
- We say that f is a *diffeomorphism* if it is bijective and f and f^{-1} are both smooth.
- A germ $f : (X, S) \to Y$ is called *smooth* if there exists a smooth representative $f : U \to Y$, where U is an open neighbourhood of S in X.

Next, we introduce the notion of tangent space of a manifold X at a point $x_0 \in X$. We introduce the following notation: C_{X,x_0} is the set of smooth curve germs $\alpha : (\mathbb{F}, 0) \to (X, x_0)$ and $\mathcal{O}_{X,x_0}$ is the set of smooth function germs $h : (X, x_0) \to \mathbb{F}$. We define an equivalence relation in C_{X,x_0} as follows:

$$\alpha \sim \beta \iff (h \circ \alpha)'(0) = (h \circ \beta)'(0), \ \forall h \in \mathcal{O}_{X,x_0}.$$

Definition 2.4 The *tangent space* of X at x_0 is defined as

$$T_{x_0} X := \frac{C_{X,x_0}}{\sim}.$$

The elements of $T_{x_0}X$ are called *tangent vectors* of X at x_0. Given a tangent vector $v = [\alpha] \in T_{x_0}X$ and a function $h \in \mathscr{O}_{X,x_0}$, the *derivative* of h in the direction v is defined as

$$v(h) = D_v(h) := (h \circ \alpha)'(0).$$

To endow $T_{x_0}X$ with the structure of $\mathbb{F}$-vector space, we make use of the action of tangent vectors as derivations on $\mathscr{O}_{X,x_0}$.

Germs of smooth functions can be added and multiplied, so $\mathscr{O}_{X,x_0}$ has the structure of a commutative $\mathbb{F}$-algebra. In fact, it is a local algebra whose maximal ideal, $\mathfrak{m}_{X,x_0}$, is the subset of germs $h \in \mathscr{O}_{X,x_0}$ such that $h(x_0) = 0$. Let $\phi \colon U \to A$ be a chart in X with $x_0 \in U$. After composing ϕ with a translation of $\mathbb{F}^n$, we can assume that $\phi(x_0) = 0$. The associated germ $\phi \colon (X, x_0) \to (\mathbb{F}^n, 0)$ is a germ of diffeomorphism which induces an isomorphism of $\mathbb{F}$-algebras $\phi^* \colon \mathscr{O}_{\mathbb{F}^n,0} \to \mathscr{O}_{X,x_0}$ given by $h \mapsto h \circ \phi$. The maximal ideal $\mathfrak{m}_{\mathbb{F}^n,0}$ of $\mathscr{O}_{\mathbb{F}^n,0}$ is generated by the coordinate functions $u_1, \ldots, u_n$ of $\mathbb{F}^n$ (this is easy to prove in the complex analytic case, although is not so trivial in the C^∞ real case, where it is a consequence of the Hadamard's lemma, see Appendix C for details). As a consequence, $\mathfrak{m}_{X,x_0}$ is generated in $\mathscr{O}_{X,x_0}$ by $x_i := \phi^*(u_i) = u_i \circ \phi$, for $i = 1, \ldots, n$. Thus, we have:

Lemma 2.2 *The maximal ideal $\mathfrak{m}_{X,x_0}$ is generated by the coordinate functions $x_1, \ldots, x_n$ of the chart $\phi \colon (X, x_0) \to (\mathbb{F}^n, 0)$.* □

A *derivation* of $\mathscr{O}_{X,x_0}$ is a linear map $D \colon \mathscr{O}_{X,x_0} \to \mathbb{F}$ such that

$$D(gh) = D(g)h(p) + g(p)D(h), \quad \forall g, h \in \mathscr{O}_{X,x_0}.$$

The following proposition ensures that tangent vectors have a double life: they can be seen at the same time as tangent vectors to curves (as we have already defined), or as derivations of functions.

Proposition 2.2 *The mapping $v \mapsto D_v$ gives a bijection between $T_{x_0}X$ and* $\mathrm{Der}(\mathscr{O}_{X,x_0})$*, the set of all derivations of $\mathscr{O}_{X,x_0}$.*

Proof Let $v = [\alpha] \in T_{x_0}X$. For all $g, h \in \mathscr{O}_{X,x_0}$ and $\lambda, \mu \in \mathbb{F}$, we have

$$\begin{aligned} D_v(\lambda g + \mu h) &= ((\lambda g + \mu h) \circ \alpha)'(0) = \lambda(g \circ \alpha)'(0) + \mu(h \circ \alpha)'(0) \\ &= \lambda D_v(g) + \mu D_v(h), \end{aligned}$$

and

$$\begin{aligned} D_v(gh) &= ((gh) \circ \alpha)'(0) = (g \circ \alpha)'(0)(h \circ \alpha)(0) + (g \circ \alpha)(0)(h \circ \alpha)'(0) \\ &= D_v(g)h(p) + g(p)D_v(h). \end{aligned}$$

Hence, $D_v \in \mathrm{Der}(\mathscr{O}_{X,x_0})$. Moreover, the mapping $v \mapsto D_v$ is one-to-one by construction. It only remains to show that it is surjective.

Take a chart $\phi\colon (X, x_0) \to (\mathbb{F}^n, 0)$ and let $x_1, \dots, x_n$ be its coordinate functions. Given $D \in \mathrm{Der}(\mathscr{O}_{X,x_0})$, we set $\lambda_i := D(x_i) \in \mathbb{F}$, for each $i = 1, \dots, n$. We define $v = [\alpha] \in T_{x_0}X$, where $\alpha(t) = \phi^{-1}(t(\lambda_1, \dots, \lambda_n))$. We have

$$D_v(x_i) = (x_i \circ \alpha)'(0) = (t\lambda_i)'(0) = \lambda_i = D(x_i), \quad i = 1, \dots, n.$$

By Lemma 2.2, $\mathfrak{m}^2_{X,x_0}$ is generated by the functions $x_i x_j$, with $1 \le i, j \le n$. Thus, every function $h \in \mathscr{O}_{X,x_0}$ can be written as

$$h = a_0 + \sum_{i=1}^{n} a_i x_i + \sum_{i,j=1}^{n} x_i x_j g_{ij},$$

for some $a_i \in \mathbb{F}$ and $g_{ij} \in \mathscr{O}_{X,x_0}$, with $i, j = 1, \dots, n$. Since D is a derivation, $D(a_0) = 0$ and $D(x_i x_j g_{ij}) = 0$, for all i, j and the same is true for D_v. Thus,

$$D(h) = \sum_{i=1}^{n} a_i D(x_i) = \sum_{i=1}^{n} a_i D_v(x_i) = D_v(h),$$

and hence, $D = D_v$. □

Since $\mathrm{Der}(\mathscr{O}_{X,x_0})$ has a natural structure of $\mathbb{F}$-vector space, we can use the above bijection to add tangent vectors or to multiply a tangent vector by an element of $\mathbb{F}$.

Definition 2.5 The tangent space $T_{x_0}X$ has a unique structure of $\mathbb{F}$-vector space such that the mapping $v \mapsto D_v$ is an isomorphism of $\mathbb{F}$-vector spaces between $T_{x_0}X$ and $\mathrm{Der}(\mathscr{O}_{X,x_0})$.

Note that if X is a smooth submanifold of some $\mathbb{F}^N$, as in most of the early examples one meets, then $T_{x_0}X$ is naturally a vector subspace of the vector space $T_{x_0}\mathbb{F}^N \simeq \mathbb{F}^N$. As we show below in Sect. 2.4, the vector space structure defined by Definition 2.5 coincides with this subspace structure.

Let $\phi\colon U \to A$ be a chart in X with $x_0 \in U$ and let $u_0 = \phi(x_0) \in A$. Let $\{e_1, \dots, e_n\}$ be the canonical basis of $\mathbb{F}^n$. For each $i = 1, \dots, n$ we define the *i-th coordinate tangent vector* as:

$$\left.\frac{\partial}{\partial x_i}\right|_{x_0} := [\gamma_i] \in T_{x_0}X,$$

where $\gamma_i \in C_{X,x_0}$ is the curve $\gamma_i(t) = \phi^{-1}(u_0 + te_i)$.

Proposition 2.3

1. *For each $h \in \mathscr{O}_{X,x_0}$, the derivative of h in the direction of $\partial/\partial x_i|_{x_0}$ is*

$$\frac{\partial h}{\partial x_i}(x_0) := \frac{\partial (h \circ \phi^{-1})}{\partial u_i}(u_0),$$

where $u_1, \ldots, u_n$ are the coordinates in $\mathbb{F}^n$.

2. *For every $v \in T_{x_0}X$ we have*

$$v = \sum_{i=1}^{n} v(x_i) \left. \frac{\partial}{\partial x_i} \right|_{x_0},$$

where x_i is the i-th coordinate function of the chart ϕ.

3. *The coordinate tangent vectors form a basis of $T_{x_0}X$, in particular,* $\dim T_{x_0}X = n$.

Proof

1. By definition,

$$\left. \frac{\partial}{\partial x_i} \right|_{x_0} (h) = (h \circ \gamma_i)'(0) = \left. \frac{d}{dt} h \circ \phi^{-1}(u_0 + te_i) \right|_{t=0}$$
$$= \lim_{t \to 0} \frac{h \circ \phi^{-1}(u_0 + te_i) - h \circ \phi^{-1}(u_0)}{t} = \frac{\partial (h \circ \phi^{-1})}{\partial u_i}(u_0).$$

2. Let $w_i := \sum_{i=1}^{n} v(x_i) \left. \frac{\partial}{\partial x_i} \right|_{x_0}$. As in the proof of Proposition 2.2, in order to show that $v = w$, it is enough to show that $v(x_j) = w(x_j)$, for all $j = 1, \ldots, n$. By item 1,

$$w(x_j) = \left(\sum_{i=1}^{n} v(x_i) \left. \frac{\partial}{\partial x_i} \right|_{x_0} \right) (x_j) = \sum_{i=1}^{n} v(x_i) \frac{\partial x_j}{\partial x_i}(x_0)$$
$$= \sum_{i=1}^{n} v(x_i) \frac{\partial u_j}{\partial u_i}(u_0) = \sum_{i=1}^{n} v(x_i) \delta_i^j = v(x_j).$$

3. By item 2, $T_{x_0}X$ is generated by the vectors $\partial/\partial x_i|_{x_0}$, $i = 1, \ldots, n$. Suppose that

$$\sum_{i=1}^{n} a_i \left. \frac{\partial}{\partial x_i} \right|_{x_0} = 0,$$

with $a_1, \ldots, a_n \in \mathbb{F}$. For each $j = 1, \ldots, n$,

$$0 = \left(\sum_{i=1}^{n} a_i \left. \frac{\partial}{\partial x_i} \right|_{x_0} \right)(x_j) = \sum_{i=1}^{n} a_i \frac{\partial x_j}{\partial x_i}(x_0) = a_j.$$

Thus, the vectors $\partial/\partial x_i|_{x_0}$, $i = 1, \ldots, n$ are also linearly independent. □

Remark 2.1 There is another equivalent definition of the tangent space $T_{x_0}X$ in terms of the maximal ideal $\mathfrak{m} := \mathfrak{m}_{X,x_0}$ of $\mathscr{O}_{X,x_0}$. In fact, there is a canonical isomorphism between $T_{x_0}X$ and $(\mathfrak{m}/\mathfrak{m}^2)^*$, the dual vector space of $\mathfrak{m}/\mathfrak{m}^2$, defined as follows: to each tangent vector $v \in T_{x_0}X$ we associate an element $\hat{v} \in (\mathfrak{m}/\mathfrak{m}^2)^*$ given by $\hat{v}(h + \mathfrak{m}^2) = v(h)$, for each $h \in \mathfrak{m}$. This alternative definition of the tangent space is very useful in algebraic geometry, since it allows us to extend the concept of tangent space to varieties over any field. ◊

Remark 2.2 When X is a complex manifold of dimension n, the tangent space is usually called the *holomorphic tangent space*, and we can denote it by $T'_{x_0}X$, if we want to distinguish between it and the (real) tangent space $T_{x_0}X$ as a (real) vector space of dimension $2n$. The relationship between the two spaces is obtained via the complexification of $T_{x_0}X$. This is the complex space $T_{x_0}X \otimes_{\mathbb{R}} \mathbb{C}$, which admits the following decomposition:

$$T_{x_0}X \otimes_{\mathbb{R}} \mathbb{C} = T'_{x_0}X \oplus \overline{T'_{x_0}X},$$

where $\overline{T'_{x_0}X}$ is the so-called *anti-holomorphic tangent space*. Given a chart $\phi\colon U \to A$ with $x_0 \in U$, we denote by $z_1, \ldots, z_n$ the complex coordinates of ϕ. The associated real coordinates are $x_1, y_1, \ldots, x_n, y_n$, where x_k and y_k are the real and imaginary parts of each z_k, respectively. A basis over $\mathbb{C}$ of $T_{x_0}X \otimes_{\mathbb{R}} \mathbb{C}$ is then given by the coordinate tangent vectors

$$\left\{ \frac{\partial}{\partial x_1}, \frac{\partial}{\partial y_1}, \ldots, \frac{\partial}{\partial x_n}, \frac{\partial}{\partial y_n} \right\}.$$

For each $k = 1, \ldots, n$, the tangent vectors

$$\frac{\partial}{\partial z_k} = \frac{1}{2}\left(\frac{\partial}{\partial x_k} - i \frac{\partial}{\partial y_k} \right), \quad \frac{\partial}{\partial \overline{z}_k} = \frac{1}{2}\left(\frac{\partial}{\partial x_k} + i \frac{\partial}{\partial y_k} \right),$$

are called *holomorphic* and *anti-holomorphic* coordinate tangent vectors, respectively. It follows that

$$\left\{ \frac{\partial}{\partial z_1}, \frac{\partial}{\partial \overline{z}_1}, \ldots, \frac{\partial}{\partial z_n}, \frac{\partial}{\partial \overline{z}_n} \right\}$$

is also a basis of $T_{x_0}X \otimes_{\mathbb{R}} \mathbb{C}$ and the holomorphic and anti-holomorphic tangent spaces in this new basis are precisely

$$T'_{x_0}X = \mathrm{Sp}_{\mathbb{C}}\left\{\frac{\partial}{\partial z_1}, \dots, \frac{\partial}{\partial z_n}\right\}, \quad \overline{T'_{x_0}X} = \mathrm{Sp}_{\mathbb{C}}\left\{\frac{\partial}{\partial \overline{z}_1}, \dots, \frac{\partial}{\partial \overline{z}_n}\right\}.$$

◊

Let $f : A \to \mathbb{F}^p$ be a smooth mapping where $A \subset \mathbb{F}^n$ is an open subset. We recall that the differential of f at a point $u_0 \in A$ is the mapping $d_{u_0}f : \mathbb{F}^n \to \mathbb{F}^p$ defined as

$$d_{u_0}f(v) = \lim_{t \to 0} \frac{f(u_0 + tv) - f(u_0)}{t}. \tag{2.1}$$

This is a linear mapping whose matrix in the canonical bases of $\mathbb{F}^n$ and $\mathbb{F}^p$ is the Jacobian matrix

$$\left(\frac{\partial f_i}{\partial u_j}(u_0)\right),$$

with $i = 1, \dots, p$ and $j = 1, \dots, n$. The tangent space we have introduced allows to extend the concept of differential to smooth mappings between manifolds.

Definition 2.6 Let $f : X \to Y$ be a smooth mapping between manifolds and let $x_0 \in X$. The *differential* of f at x_0 is the mapping $d_{x_0}f : T_{x_0}X \to T_{f(x_0)}Y$ given by

$$d_{x_0}f(v) = [f \circ \alpha],$$

for every tangent vector $v = [\alpha] \in T_{x_0}X$.

The main properties of the differential are given in the next proposition.

Proposition 2.4 *With the above notation, we have:*

1. $d_{x_0}f(v)(h) = v(h \circ f)$, *for every* $h \in \mathcal{O}_{Y,f(x_0)}$.
2. $d_{x_0}f(v)$ *is well defined, that is, it does not depend on the choice of* α *such that* $v = [\alpha]$.
3. *The differential is linear.*
4. *The chain rule: given another smooth mapping* $g : Y \to Z$*, then*

$$d_{x_0}(g \circ f) = d_{f(x_0)}g \circ d_{x_0}f.$$

5. *If* $\phi : U \to A$ *and* $\psi : V \to B$ *are charts on* X, Y*, respectively, such that* $x_0 \in U \subset f^{-1}(V)$*, then the matrix of* $d_{x_0}f$ *in the bases of coordinate tangent vectors*

of $T_{x_0}X$ and $T_{f(x_0)}Y$ coincides with the Jacobian matrix of $\psi \circ f \circ \phi^{-1}$ at $\phi(x_0)$ and is thus equal to:

$$\left(\frac{\partial f_i}{\partial x_j}(x_0) \right),$$

where $x_1, \ldots, x_n$ and $y_1, \ldots, y_p$ are the coordinate functions of the charts ϕ and ψ, respectively, and $f_i := y_i \circ f$ is the i-th coordinate function of f.

Proof

1. By construction,

$$d_{x_0} f(v)(h) = [f \circ \alpha](h) = (h \circ (f \circ \alpha))'(0) = ((h \circ f) \circ \alpha)'(0) = v(h \circ f).$$

2. This is an immediate consequence of item 1 and Proposition 2.2.
3. Given $v, w \in T_{x_0}X$, $\lambda, \mu \in \mathbb{F}$ and $h \in \mathscr{O}_{X,x_0}$,

$$\begin{aligned} d_{x_0} f(\lambda v + \mu w)(h) &= (\lambda v + \mu w)(h \circ f) = \lambda v(h \circ f) + \mu w(h \circ f) \\ &= \lambda d_{x_0} f(v)(h) + \mu d_{x_0} f(w)(h), \end{aligned}$$

hence $d_{x_0} f(\lambda v + \mu w) = \lambda d_{x_0} f(v) + \mu d_{x_0} f(w)$.
4. We use item 1: for every $h \in \mathscr{O}_{Z,g(f(x_0))}$,

$$\begin{aligned} d_{x_0}(g \circ f)(v)(h) &= v(h \circ (g \circ f)) = v((h \circ g) \circ f) \\ &= d_{x_0}(v)(h \circ g) = d_{f(x_0)} g(d_{x_0} f(v))(h), \end{aligned}$$

so $d_{x_0}(g \circ f)(v) = d_{f(x_0)} g(d_{x_0} f(v))$.
5. Let $w_i = d_{x_0} f \left(\frac{\partial}{\partial x_i}\Big|_{x_0} \right)$, then by item 2 of Proposition 2.3,

$$d_{x_0} f \left(\frac{\partial}{\partial x_i}\bigg|_{x_0} \right) = w_i = \sum_{j=1}^{p} w_i(y_j) \frac{\partial}{\partial y_j}\bigg|_{f(x_0)},$$

hence the components of the matrix of $d_{x_0} f$ are the derivatives $w_i(y_j)$. Now apply item 1 again:

$$w_i(y_j) = d_{x_0} f \left(\frac{\partial}{\partial x_i}\bigg|_{p} \right)(y_j) = \frac{\partial (y_j \circ f)}{\partial x_i}(x_0) = \frac{\partial f_j}{\partial x_i}(x_0).$$

Moreover, by item 1 of Proposition 2.3,

$$\frac{\partial f_j}{\partial x_i}(x_0) = \frac{\partial (f_j \circ \phi^{-1})}{\partial u_i}(\phi(x_0)) = \frac{\partial (u_j \circ (\psi \circ f \circ \phi^{-1}))}{\partial u_i}(\phi(x_0)),$$

which are the components of the Jacobian matrix of $\psi \circ f \circ \phi^{-1}$ at $\phi(x_0)$. □

Remark 2.3 Let $f : X \to Y$ be a holomorphic mapping between complex manifolds. As in Remark 2.2, we denote by $d'_{x_0} f : T'_{x_0} X \to T'_{f(x_0)} Y$ the *holomorphic* differential between the holomorphic tangent spaces and by $d_{x_0} f : T_{x_0} X \to T_{f(x_0)} Y$ the (real) differential between the (real) tangent spaces. The induced mapping between the complexifications is

$$d_{x_0} f \otimes \mathrm{id} : T_{x_0} X \otimes \mathbb{C} \longrightarrow T_{f(x_0)} Y \otimes \mathbb{C},$$

and it follows that $d'_{x_0} f$ is the restriction of $d_{x_0} f \otimes \mathrm{id}$ to the holomorphic tangent spaces. In fact, one can show more, namely, that $d_{x_0} f \otimes \mathrm{id}$ is the direct sum of $d_{x_0} f'$ with the *anti-holomorphic* differential:

$$\overline{d_{x_0} f'} : \overline{T'_{x_0} X} \longrightarrow \overline{T'_{f(x_0)} Y}.$$

If $d'_{x_0} f$ has matrix $A = (a_{ij})$ with respect to the holomorphic coordinate bases of $T'_{x_0} X$ and $T'_{f(x_0)} Y$, then the matrix of $\overline{d'_{x_0} f}$ with respect to the anti-holomorphic bases of $\overline{T'_{x_0} X}$ and $\overline{T'_{f(x_0)} Y}$ is the conjugate matrix $\overline{A} = (\overline{a_{ij}})$ and the matrix of $d_{x_0} f \otimes \mathrm{id}$ is

$$A \oplus \overline{A} = \begin{pmatrix} a_{ij} & 0 \\ 0 & \overline{a_{ij}} \end{pmatrix}.$$

One consequence of this fact is that $d'_{x_0} f$ has rank r over $\mathbb{C}$ if and only if $d_{x_0} f$ has rank $2r$ over $\mathbb{R}$. ◊

From now on we treat the real and complex cases together wherever possible, and use the notation $d_{x_0} f$ to refer both to the real differential in the real case, and the holomorphic differential hitherto $d'_{x_0} f$ in the complex case.

Exercises for Sect. 2.2

1 Show that an open U subset of a manifold X has an induced structure of manifold of the same dimension.

2 Show that a product $X \times Y$ of manifolds X, Y has an induced structure of manifold. What is its dimension?

3 Show that a topological space Y which is homeomorphic to a manifold X has an induced structure of manifold of the same dimension.

4 Show that every chart $\phi\colon U \to A$ of a manifold X is a diffeomorphism.

5 Conversely, show that every diffeomorphism $\phi\colon U \to A$, where $U \subset X$ and $A \subset \mathbb{F}^n$ are open subsets, can be considered as a chart of X.

6 Let U be an open subset of a manifold X. Show that the inclusion $U \hookrightarrow X$ is smooth.

7 Let $X \times Y$ be a product of manifolds X, Y. Show that the projections $X \times Y \to X$ and $X \times Y \to Y$ are smooth.

8 Show that when $X = \mathbb{F}^n$, for each $x_0 \in \mathbb{F}^n$ there exists a canonical identification between $T_{x_0}\mathbb{F}^n$ and $\mathbb{F}^n$ given by $[\alpha] \mapsto \alpha'(0)$.

9 Prove that for every $v \in T_{x_0}X$, $\hat{v}$ is a well-defined element of $\left(\mathfrak{m}/\mathfrak{m}^2\right)^*$ (see Remark 2.1) and that the mapping $v \mapsto \hat{v}$ gives an isomorphism between $T_{x_0}X$ and $\left(\mathfrak{m}/\mathfrak{m}^2\right)^*$.

10 Let U be an open subset of a manifold X. Show that there is a canonical isomorphism $T_{x_0}U \equiv T_{x_0}X$, for each $x_0 \in U$.

11 Let $X \times Y$ be a product of manifolds X, Y. Show that there is a canonical isomorphism $T_{(x_0,y_0)}(X \times Y) \equiv T_{x_0}X \times T_{y_0}Y$, for each $x_0 \in X$ and $y_0 \in Y$.

12 Let $f\colon A \to \mathbb{F}^p$ be a smooth mapping where $A \subset \mathbb{F}^n$ is an open subset. Show that via the canonical identifications $T_{x_0}A \equiv \mathbb{F}^n$ and $T_{f(x_0)}\mathbb{F}^p \equiv \mathbb{F}^p$, the definition of $d_{x_0}f$ in Definition 2.6 coincides with that of (2.1).

13 *Cauchy–Riemann equations in complex coordinates.* Let $f\colon A \to \mathbb{C}^p$ be any mapping, where $A \subset \mathbb{C}^n$ is an open subset. Show that f is holomorphic if and only if f is C^∞ and

$$\frac{\partial f_i}{\partial \overline{z}_j} \equiv 0,\ \forall i = 1, \ldots, p,\ \forall j = 1, \ldots, n.$$

14 *Coordinate-free version of Cauchy–Riemann equations.* Let $f\colon X \to Y$ be any mapping between complex manifolds. Show that f is holomorphic if and only if f is C^∞ and for all $x_0 \in X$,

$$(d_{x_0}f \otimes \mathrm{id})(T'_{x_0}X) \subset T'_{f(x_0)}Y.$$

2.3 Inverse Mapping Theorem and Consequences

Most of the results of this section are local, so it is more convenient to work with germs of mappings instead of mappings. A smooth germ $f\colon (X, x_0) \to (Y, y_0)$ is called a *diffeomorphism* if there exists a representative $f\colon U \to V$ which is a diffeomorphism, where U and V are open neighbourhoods of x_0 and y_0 in X and Y, respectively. This is equivalent to say that f is invertible as a smooth germ, that is, there exists another smooth germ $f^{-1}\colon (Y, y_0) \to (X, x_0)$ such that

$$f \circ f^{-1} = \mathrm{id}_{(Y,y_0)}, \quad f^{-1} \circ f = \mathrm{id}_{(X,x_0)}\,.$$

By the chain rule, we have

$$d_{x_0} f \circ d_{y_0} f^{-1} = \mathrm{id}_{T_{y_0}Y}, \quad d_{y_0} f^{-1} \circ d_{x_0} f = \mathrm{id}_{T_{x_0}X}\,.$$

In particular, $d_{x_0} f\colon T_{x_0}X \to T_{y_0}Y$ is an isomorphism whose inverse is $d_{y_0} f^{-1}$. The inverse mapping theorem gives the converse of this fact.

Theorem 2.1 (Inverse Mapping Theorem for Manifolds) *A smooth germ $f\colon (X, x_0) \to (Y, y_0)$ is a diffeomorphism if and only if $d_{x_0} f\colon T_{x_0}X \to T_{y_0}Y$ is an isomorphism.*

Proof We have already seen the "only if" part, so we need to only show "if". Consider first the case $\mathbb{F} = \mathbb{R}$. Here the statement is a simple consequence of the classical inverse mapping theorem for C^∞-mappings from $\mathbb{R}^n$ to $\mathbb{R}^n$. In fact, we take charts $\phi\colon U \to A$ in X and $\psi\colon V \to B$ in Y, with $x_0 \in U \subset f^{-1}(V)$ and consider their germs $\phi\colon (X, x_0) \to (\mathbb{R}^n, u_0)$ and $\psi\colon (Y, y_0) \to (\mathbb{R}^n, v_0)$, which are diffeomorphisms. By the chain rule, the differential of

$$g := \psi \circ f \circ \phi^{-1}\colon (\mathbb{R}^n, u_0) \to (\mathbb{R}^n, v_0)$$

is an isomorphism. By the classical inverse mapping theorem, g is a diffeomorphism and hence, so is f.

For the case $\mathbb{F} = \mathbb{C}$ we use the real version and Remark 2.3. By taking charts as above, we only need to prove that if $g\colon (\mathbb{C}^n, u_0) \to (\mathbb{C}^n, v_0)$ is a holomorphic germ whose holomorphic differential is a $\mathbb{C}$-isomorphism, then g is a biholomorphism (i.e., bijective with holomorphic inverse). Let $A = (a_{ij})$ be the Jacobian matrix of the holomorphic differential $d'_{u_0} g$ in complex coordinates. It follows that the (real) differential $d_{u_0} g$ is also an $\mathbb{R}$-isomorphism whose complexification has matrix

$$A \oplus \overline{A} = \begin{pmatrix} a_{ij} & 0 \\ 0 & \overline{a_{ij}} \end{pmatrix}.$$

By the classical inverse mapping theorem, g is a C^∞-diffeomorphism. Hence, g^{-1} is C^∞ and the complexification of $d_{v_0} g^{-1}$ has matrix of the form

$$A^{-1} \oplus \overline{A}^{-1} = \begin{pmatrix} b_{ij} & 0 \\ 0 & \overline{b_{ij}} \end{pmatrix}.$$

But this implies that $\partial g_i^{-1}/\partial \bar{z}_j \equiv 0$, for all $i, j = 1, \ldots, n$. Hence, g^{-1} is holomorphic, by the Cauchy–Riemann equations in complex coordinates (see Exercise 2.2.13). □

One of the consequences of the inverse mapping theorem is the constant rank theorem, which states that every smooth germ of constant rank between manifolds can be "linearised", that is, we can choose coordinates in the source and target such that the mapping in these coordinates is the restriction of a linear mapping. The *rank* of a smooth mapping at a point is the rank of the differential at that point. We say that a smooth germ $f\colon (X, x_0) \to (Y, y_0)$ has constant rank r if there exists a representative $f\colon U \to Y$ such that f has rank r at every point $x \in U$. The *corank* of a mapping at a point x_0 is the difference between the maximum possible and the rank at x_0. Thus if the dimension of the domain is less than or equal to the dimension of the target, the corank is the dimension of $\ker d_{x_0} f$; if greater than or equal, the corank is the dimension of $\operatorname{coker} d_{x_0} f$.

Lemma 2.3 *Let $f\colon (\mathbb{F}^n, 0) \to (\mathbb{F}^p, 0)$ be a smooth mapping of rank r at 0. There exist a diffeomorphism $\phi\colon (\mathbb{F}^n, 0) \to (\mathbb{F}^n, 0)$ and a linear isomorphism $\psi\colon \mathbb{F}^p \to \mathbb{F}^p$ such that*

$$\psi \circ f \circ \phi^{-1}(u) = (u_1, \ldots, u_r, g_{r+1}(u), \ldots, g_p(u)),$$

for some smooth functions $g_{r+1}, \ldots, g_p\colon (\mathbb{F}^n, 0) \to (\mathbb{F}, 0)$.

Proof After taking linear isomorphisms in $\mathbb{F}^n$ and $\mathbb{F}^p$ we can assume that the Jacobian matrix of f at 0 is

$$\begin{pmatrix} I_r & 0 \\ 0 & 0 \end{pmatrix}.$$

Let $\phi\colon (\mathbb{F}^n, 0) \to (\mathbb{F}^n, 0)$ be the smooth germ given by

$$\phi(u) = (f_1(u), \ldots, f_r(u), u_{r+1}, \ldots, u_n).$$

The Jacobian matrix of ϕ at 0 is the identity, so ϕ is a diffeomorphism by the inverse mapping theorem 2.1. Define $g := f \circ \phi^{-1}\colon (\mathbb{F}^n, 0) \to (\mathbb{F}^p, 0)$, which is written as

$$g(s) = (s_1 \ldots, s_r, g_{r+1}(s), \ldots, g_p(s)),$$

for some smooth functions $g_{r+1}, \ldots, g_p\colon (\mathbb{F}^n, 0) \to (\mathbb{F}, 0)$. □

Coordinates in which f takes the form specified in the lemma are called "linearly adapted coordinates".

Theorem 2.2 (Constant Rank Theorem) *Let $f\colon (X, x_0) \to (Y, y_0)$ be a smooth germ of constant rank r. There exist diffeomorphisms $\phi\colon (X, x_0) \to (\mathbb{F}^n, 0)$ and $\psi\colon (Y, y_0) \to (\mathbb{F}^p, 0)$ such that $\psi \circ f \circ \phi^{-1}\colon (\mathbb{F}^n, 0) \to (\mathbb{F}^p, 0)$ is the germ of the* *linear mapping*

$$(u_1, \ldots, u_n) \longmapsto (u_1, \ldots, u_r, 0, \ldots, 0).$$

Proof By taking charts in X and Y, it is enough to prove the statement for a smooth germ $f\colon (\mathbb{F}^n, 0) \to (\mathbb{F}^p, 0)$ of constant rank r. By Lemma 2.3, we can assume that f is written as

$$f(u) = (u_1 \ldots, u_r, g_{r+1}(u), \ldots, g_p(u)),$$

for some smooth functions $g_{r+1}, \ldots, g_p\colon (\mathbb{F}^n, 0) \to (\mathbb{F}, 0)$. The Jacobian matrix of f at u is equal to

$$\begin{pmatrix} I_r & 0 \\ \frac{\partial g_i}{\partial u_j}(u) & \frac{\partial g_i}{\partial u_k}(u) \end{pmatrix},$$

where $1 \leq j \leq r$ and $r + 1 \leq k \leq n$. Since f has constant rank r, $\frac{\partial g_i}{\partial u_k} \equiv 0$, for all $i = r + 1, \ldots, p$ and $k = 1 \ldots, n - r$. This implies that the functions $g_{r+1}, \ldots, g_p$ only depend on the first variables $u_1, \ldots, u_r$ and we can write $g_i(u) = g_i(u_1, \ldots, u_r)$, for all u and for each $i = r + 1, \ldots, p$.

Now consider the smooth germ $\psi\colon (\mathbb{F}^p, 0) \to (\mathbb{F}^p, 0)$ given by

$$\psi(v) = \big(v_1, \ldots, v_r, v_{r+1} - g_{r+1}(v_1, \ldots, v_r), \ldots, v_p - g_p(v_1, \ldots, v_r)\big).$$

It is obvious that ψ is a diffeomorphism, since its inverse mapping is

$$\psi^{-1}(w) = \big(w_1, \ldots, w_r, w_{r+1} + g_{r+1}(w_1, \ldots, w_r), \ldots, w_p + g_p(w_1, \ldots, w_r)\big),$$

which is also smooth. We have constructed ψ precisely so that

$$\psi \circ f(u) = (u_1, \ldots, u_r, 0, \ldots, 0),$$

which concludes the proof. □

Definition 2.7 A smooth mapping $f\colon X \to Y$ is called an *immersion* (resp. *submersion) at* $x_0 \in X$ if $d_{x_0} f$ is a monomorphism (resp. epimorphism). We say that f is an *immersion* (resp. *submersion*) if it is an immersion (resp. submersion) at every point $x_0 \in X$. A germ $f\colon (X, x_0) \to (Y, y_0)$ is called an *immersion* (resp. *submersion*) if $d_{x_0} f$ is a monomorphism (resp. epimorphism).

The rank of a smooth map $f\colon X \to Y$ at a point $x_0 \in X$ is an upper semi-continuous function. This means that if the rank of f at x_0 is r, then there exists an open neighbourhood U of x_0 in X such that the rank of f at every $x \in U$ is $\geq r$. In particular, if f is an immersion or submersion at x_0, then f has constant rank in a neighbourhood of x_0. The following corollary is a particular case of the constant rank theorem.

Corollary 2.1 (Local Form of the Immersion/Submersion) *Let $f\colon (X, x_0) \to (Y, y_0)$ be an immersion or submersion. There exist diffeomorphisms $\phi\colon (X, x_0) \to (\mathbb{F}^n, 0)$ and $\psi\colon (Y, y_0) \to (\mathbb{F}^p, 0)$ such that $\psi \circ f \circ \phi^{-1}\colon (\mathbb{F}^n, 0) \to (\mathbb{F}^p, 0)$ is the germ of the linear mapping*

$$(u_1, \ldots, u_n) \longmapsto \begin{cases} (u_1, \ldots, u_n, 0, \ldots, 0), & \text{if } f \text{ is an immersion,} \\ (u_1, \ldots, u_p), & \text{if } f \text{ is a submersion.} \end{cases}$$

Another consequence is that every immersion admits a left inverse which is a submersion and every submersion admits a right inverse which is an immersion.

Corollary 2.2 *Let X and Y be manifolds. Then:*

1. *If $f\colon (X, x_0) \to (Y, y_0)$ is an immersion, there exists a submersion $g\colon (Y, y_0) \to (X, x_0)$ such that $g \circ f = \mathrm{id}_{(X,x_0)}$.*
2. *If $f\colon (X, x_0) \to (Y, y_0)$ is a submersion, there exists an immersion $g\colon (Y, y_0) \to (X, x_0)$ such that $f \circ g = \mathrm{id}_{(Y,y_0)}$.* □

Exercises for Sect. 2.3

1 A smooth germ $f\colon (X, x_0) \to (Y, y_0)$ gives rise to a map $f^*\colon \mathcal{O}_{Y,y_0} \to \mathcal{O}_{X,x_0}$ given by $f^*(h) = h \circ f$, for each $h \in \mathcal{O}_{Y,y_0}$. Show:

(a) f^* is an $\mathbb{F}$-algebra homomorphism.
(b) For any other smooth germ $g\colon (Y, y_0) \to (Z, z_0)$, we have $(g \circ f)^* = f^* \circ g^*$.
(c) f is a diffeomorphism if and only if f^* is an isomorphism.
(d) f is an immersion if and only if f^* is an epimorphism.

2 Show that a submersion $f\colon X \to Y$ is an open map (that is, $f(U)$ is open in Y, for every open subset $U \subset X$).

3 Let $f\colon X \to Y$ be a submersion, where X is compact and Y is connected. Show that f is surjective.

4 A smooth mapping $f\colon X \to Y$ is called a local diffeomorphism if for each $x_0 \in X$, the germ $f\colon (X, x_0) \to (Y, f(x_0))$ is a diffeomorphism. Show that if f is a local diffeomorphism and is one-to-one, then $f(X)$ is open in Y and the restriction $f\colon X \to f(X)$ is a diffeomorphism.

5 Give an example of a mapping $f\colon \mathbb{F}^2 \to \mathbb{F}^2$ which is a local diffeomorphism but is not a diffeomorphism onto its image.

6 Show that up to right-left equivalence, the germ on the right-hand side in the equality of Lemma 2.3 can be assumed to be an unfolding of a germ of rank 0.

2.4 Submanifolds

A submanifold of a manifold Y is a subset $X \subset Y$ which is also a manifold whose smooth structure is compatible with that of Y. More precisely:

Definition 2.8 An immersion $f\colon X \to Y$ is called an embedding if the restriction $f\colon X \to f(X)$ is a homeomorphism. A manifold X is a submanifold of the manifold Y if $X \subset Y$ and the inclusion $X \to Y$ is an embedding.

In the definition of submanifold, we always assume that X has the topology induced as a subspace of Y, hence if X is a manifold, then the inclusion mapping $i\colon X \hookrightarrow Y$ is an embedding if and only if it is an immersion.

When X is a submanifold of Y, then at each point $x_0 \in X$ the inclusion mapping $i\colon X \hookrightarrow Y$ induces a canonical monomorphism $d_{x_0} i\colon T_{x_0}X \to T_{x_0}Y$. In fact, this monomorphism can be easily visualised: given a tangent vector $v = [\alpha] \in T_{x_0}X$, where $\alpha \in C_{X,x_0}$, the associated tangent vector in Y is $\tilde{v} = [i \circ \alpha] \in T_{x_0}Y$. Since α and $i \circ \alpha$ parameterise the same curve, but considered either in X or Y, we will identify the two tangent vectors $v \equiv \tilde{v}$ and hence, we consider $T_{x_0}X$ as a subspace of $T_{x_0}Y$.

Lemma 2.4 *Let $f\colon X \to Y$ and $g\colon Y \to Z$ be mappings between manifolds such that g is an embedding. If $g \circ f$ is smooth, then f is smooth.*

Proof The restriction $\bar{g}\colon Y \to g(Y)$ is a homeomorphism, hence it has a continuous inverse mapping $\bar{g}^{-1}\colon g(Y) \to Y$. We put $f = \bar{g}^{-1} \circ (g \circ f)$, so f is continuous. For each point $x_0 \in X$, the germ $g\colon (Y, f(x_0)) \to (Z, g(f(x_0)))$ is an immersion. By Corollary 2.2, there exists a submersion $h\colon (Z, g(f(x_0))) \to (Y, f(x_0))$ such that $h \circ g = \mathrm{id}_{(Y, f(x_0))}$. The germ $f\colon (X, x_0) \to (Y, f(x_0))$ can be written as $f = (h \circ g) \circ f = h \circ (g \circ f)$, so f is smooth at x_0. □

The following proposition is usually referred to as the uniqueness of the smooth structure of a submanifold.

Proposition 2.5 *Let X be any subset of a manifold Y. There exists at most one smooth structure on X such that X is a submanifold of Y.*

Proof Let $[\mathcal{A}]$ and $[\mathcal{A}']$ be two smooth structures in X such that $(X, [\mathcal{A}])$ and $(X, [\mathcal{A}])$ are both submanifolds of Y. We have two commutative diagrams:

$$
\begin{array}{ccc}
(X, [\mathcal{A}]) & \xrightarrow{\ i\ } & Y\,, \\
\Big\downarrow \mathrm{id} & \nearrow_{i'} & \\
(X, [\mathcal{A}']) & &
\end{array}
\qquad
\begin{array}{ccc}
(X, [\mathcal{A}']) & \xrightarrow{\ i'\ } & Y\,, \\
\Big\downarrow \mathrm{id}' & \nearrow_{i} & \\
(X, [\mathcal{A}]) & &
\end{array}
$$

where i, i' are the inclusion mappings and $\mathrm{id}, \mathrm{id}'$ the corresponding identity mappings. Since i' is an embedding and $i = i' \circ \mathrm{id}$ is smooth, id is smooth by Lemma 2.4. Analogously, id' is also smooth, which implies $[\mathcal{A}] = [\mathcal{A}']$. □

As a corollary, the property that X is a submanifold of Y is local. A set-germ (X, x_0) is called a *submanifold* of (Y, y_0) if there exists a representative X which is a submanifold of Y.

Corollary 2.3 *Let X be any subset of a manifold Y and assume that (X, x_0) is a submanifold of (Y, x_0), for each $x_0 \in Y$. Then X is a submanifold of Y.*

Proof Take an open covering $\{U_i\}_{i \in I}$ of X in Y such that $X \cap U_i$ is a submanifold of Y, for all $i \in I$. If $\mathcal{A}_i$ is an atlas in $X \cap U_i$, then by Proposition 2.5, $\mathcal{A} = \cup_{i \in I} \mathcal{A}_i$ is an atlas in X which provides it a structure of submanifold of Y. □

Definition 2.9 Let $f: X \to Y$ be a smooth mapping. We say that $x_0 \in X$ is a *critical point* if f is not a submersion at x_0 and the set C of all critical points of f is called the *critical set* of f. A *critical value* is the image of a critical point and the set of all critical values $\Delta = f(C)$ is called the *discriminant* of f. A *regular value* of f is a point $y_0 \in Y$ which is not a critical value (this includes also the case that y_0 is not in the image of f).

Theorem 2.3 (Regular Value Theorem) *Let $f: X \to Y$ be a smooth mapping and let $y_0 \in f(X) \setminus \Delta$. Then $Z = f^{-1}(y_0)$ is a submanifold of X of dimension* $\dim X - \dim Y$, *whose tangent space at each point $x_0 \in Z$ is*

$$T_{x_0} Z = \ker d_{x_0} f.$$

Proof Given $x_0 \in X$, f is a submersion at x_0 and by Corollary 2.1, there exist diffeomorphisms $\phi: (X, x_0) \to (\mathbb{F}^n, 0)$ and $\psi: (Y, y_0) \to (\mathbb{F}^p, 0)$ such that $\psi \circ f \circ \phi^{-1}: (\mathbb{F}^n, 0) \to (\mathbb{F}^p, 0)$ is the germ of the projection

$$(u_1, \ldots, u_n) \mapsto (u_1, \ldots, u_p).$$

In particular, $\phi(Z, x_0) = (\{0\} \times \mathbb{F}^{n-p}, 0)$. Now $\{0\} \times \mathbb{F}^{n-p}$ is obviously a submanifold of $\mathbb{F}^n$ of dimension $n - p$ and ϕ is a diffeomorphism, so (Z, x_0) is a submanifold of (X, x_0) of dimension $n - p$. By Corollary 2.3, Z is a submanifold of X dimension $n - p$.

To see the second part, consider the inclusion $i: Z \hookrightarrow X$, then the composition $f \circ i$ is the constant mapping y_0. By the chain rule, $d_{x_0} f \circ d_{x_0} i = 0$, for all $x_0 \in Z$. Thus, $T_{x_0} Z \subset \ker d_{x_0} f$ and since both spaces have the same dimension, we have the equality. □

The converse of Theorem 2.3 also holds locally:

Proposition 2.6 *If Z is a submanifold of X, of codimension k, then for each $x_0 \in Z$ there is a neighbourhood U of x_0 in X and a smooth map $g: U \to \mathbb{F}^k$ with 0 as regular value, such that $Z \cap U = g^{-1}(0)$.*

Proof The inclusion $Z \hookrightarrow Y$ is an immersion. It follows from the local normal form of immersions that there is a neighbourhood U of x_0 and a diffeomorphism $(U, Z \cap U) \xrightarrow{\phi} \left(V, V \cap (\{0\} \times \mathbb{F}^{n-k})\right)$ where V is open in $\mathbb{F}^n$. Let $y_1, \ldots, y_n$ be coordinates on V. Take, as g, the map $(y_1, \ldots, y_k) \circ \phi$. □

The component functions of the map g in the proposition are often referred to as *regular equations for* Z in the neighbourhood of x_0.

Exercises for Sect. 2.4

1 Suppose that $X^k \subset Y^n$, and that both are manifolds, with possibly unrelated smooth structures. Show that

(i) X is a submanifold of Y if the restriction to X of every smooth function on an open set in Y is smooth with respect to the smooth structure on X, and, conversely, every smooth function on X is locally the restriction of a smooth function on Y.
(ii) X is a submanifold of Y if for each $x \in X$ there is a neighbourhood U of x in Y and a chart $\varphi : U \to V \subset \mathbb{R}^n$ such that $\varphi|_{X \cap U}$ is a diffeomorphism of $X \cap U$ to $\mathbb{R}^k \cap V$.

2 Let $f : X \to Y$ be an embedding. Show that $f(X)$ is a submanifold of Y such that the restriction $f : X \to f(X)$ is a diffeomorphism.

3 Let Y be a manifold of dimension p and $X \subset Y$ a subset. A point $x_0 \in X$ is called a *regular* point of X of dimension n if (X, x_0) is a submanifold of (Y, x_0) of dimension n. If x_0 is not regular (of any dimension), then x_0 is called a *singular* point of X. Show that the following statements are equivalent:

(a) x_0 is a regular point of X of dimension n;
(b) there exists a chart $\phi : (Y, x_0) \to (\mathbb{F}^p, 0)$ such that $\phi(X, x_0) = (\mathbb{F}^n \times \{0\}, 0)$;
(c) there exists an immersion $f : (\mathbb{F}^n, 0) \to (Y, y_0)$ such that $f(\mathbb{F}^n, 0) = (X, x_0)$;
(d) there exists a submersion $g : (Y, x_0) \to (\mathbb{F}^{p-n}, 0)$ such that $g^{-1}(0) = (X, x_0)$.

4 Let X be the subset of $Y = \mathbb{F}^2$ of points (x, y) such that $x^2 = y^3$. Show that X has only a singular point at the origin. Hint: the hardest part is to prove that the origin is not regular. Assume that 0 is regular and take an immersion $f : (\mathbb{F}, 0) \to (\mathbb{F}^2, 0)$ such that $f(\mathbb{F}^m, 0) = (X, 0)$. Now take a Taylor expansion of f and substitute in the equation $x^2 = y^3$ to get a contradiction.

5 Find the singular points of the subset X of the manifold Y in the following cases:

(a) $Y = \mathbb{F}^2$ and X is the subset of points (x, y) such that $xy = 0$;

(b) $Y = \mathbb{F}^3$ and X is the subset of points (x, y, z) such that

$$x^2 + z^2 = y^3(1 - y)^3.$$

(c) $Y = \mathbb{R}^2$ and X is the boundary of the square $[0, 1] \times [0, 1]$.

6 Let $f : \mathbb{F}^3 \to \mathbb{F}$ be the function $f(x, y, z) = x^2 + y^2 - z^2$.

(a) Find the discriminant Δ of f.
(b) Show that if $a \in \Delta$, then $f^{-1}(a)$ is not a submanifold of $\mathbb{F}^3$.
(c) Show that if a and b are regular values in the same connected component of $\mathbb{F} \setminus \Delta$, then the submanifolds $f^{-1}(a)$ and $f^{-1}(b)$ are diffeomorphic.
(d) Is there any difference between the real C^∞ and the complex analytic cases in this example?
(e) Consider $\mathbb{F} = \mathbb{R}$ and make a picture of the transition between the submanifolds $f^{-1}(a)$ and $f^{-1}(b)$ when passing through a critical value.

2.5 Vector Fields and Flows

In order to classify functions and mappings up to right- or right-left equivalence, a central aim in singularity theory, one needs a good supply of diffeomorphisms. In almost all cases, these are obtained by integrating vector fields. A vector field on a manifold X is a rule ξ which assigns to each point $x \in X$ a tangent vector $\xi(x) \in T_xX$, in such a way that $\xi(x)$ depends smoothly on x. To give a precise meaning to the term "smooth" here, we need a smooth structure on the set of all tangent vectors to X.

Definition 2.10 Let X be a manifold. A *vector bundle* on X is a manifold E and a smooth mapping $\pi : E \to X$ such that:

1. For each $x \in X$, the fibre $E_x := \pi^{-1}(x)$ is a vector space over $\mathbb{F}$.
2. For each $x_0 \in X$, there exists an open neighbourhood $U \subset X$ of x_0 and a diffeomorphism $\varphi\colon U \times \mathbb{F}^k \to \pi^{-1}(U)$ such that

   a. $\pi \circ \varphi(x, v) = x$, for all $(x, v) \in U \times \mathbb{F}^k$,
   b. the map $\varphi_x : \mathbb{F}^k \to E_x$ given by $\varphi_x(v) = \varphi(x, v)$ is a linear isomorphism, for all $x \in U$.

The number k in item 2 is equal to the difference $\dim E - \dim X$ and is called the *rank* of the vector bundle. A *section* of the vector bundle is a smooth mapping $\xi\colon X \to E$ such that $\pi \circ \xi = \mathrm{id}_X$. This means that $\xi(x) \in E_x$ for all $x \in X$. It is common to use the notation $\xi_x := \xi(x)$.

Given two vector bundles $\pi : E \to X$ and $\pi' : F \to Y$, a *morphism* between them is a pair of smooth mappings (g, f) where $g : E \to F$ and $f : X \to Y$ such that

1. $f \circ \pi = \pi' \circ g$, that is, we have a commutative diagram

$$\begin{array}{ccc} E & \xrightarrow{g} & F \\ \downarrow{\scriptstyle \pi} & & \downarrow{\scriptstyle \pi'} \\ X & \xrightarrow{f} & Y. \end{array}$$

2. For each $p \in X$, the restriction $g_x : E_x \to F_{f(x)}$ is linear.

Definition 2.11 Let X be a manifold of dimension n. We define the *tangent bundle* of X as follows. As a set, it is the disjoint union of all tangent spaces, that is,

$$TX := \coprod_{x \in X} T_x X.$$

The projection $\pi : TX \to X$ is the mapping defined as $v \mapsto x$ such that $v \in T_x X$.

We will show that TX has a natural topology and a structure of smooth manifold induced by those of X. Let $\mathcal{A} = \{\phi_i : U_i \to A_i\}_{i \in I}$ be an atlas in X. For each $i \in I$ we define the subset $\tilde{U}_i := \pi^{-1}(U_i) \subset TX$ and the mapping $\tilde{\phi}_i : \tilde{U}_i \to A_i \times \mathbb{F}^n$ given by

$$\tilde{\phi}_i(v) = (\pi(v), v(x_1), \ldots, v(x_n)),$$

where $x_1, \ldots, x_n$ are the coordinate functions of ϕ_i on U_i. The mapping $\tilde{\phi}_i$ is bijective and its inverse is

$$\tilde{\phi}_i^{-1}(u, w) = \sum_{i=1}^{x} w_i \left. \frac{\partial}{\partial x_i} \right|_{\phi_i^{-1}(u)}.$$

We consider in TX the topology whose basis is the family of subsets of the form $\tilde{\phi}_i^{-1}(A)$, where $i \in I$ and $A \subset A_i \times \mathbb{F}^n$ is any open subset. It follows that each $\tilde{U}_i$ is open in TX and that each mapping $\tilde{\phi}_i$ is a homeomorphism.

Proposition 2.7 *The family* $\tilde{\mathcal{A}} = \{\tilde{\phi}_i : \tilde{U}_i \to A_i \times \mathbb{F}^n\}_{i \in I}$ *defines a smooth structure in* TX *such that the projection* $\pi : TX \to X$ *is a vector bundle of rank* n.

Proof For each $i, j \in I$, the mapping

$$\tilde{\phi}_j \circ \tilde{\phi}_i^{-1}(u, w) = (\phi_j \circ \phi_i^{-1}(u), d_u(\phi_j \circ \phi_i^{-1})(w)),$$

is smooth and thus, $\tilde{\mathscr{A}}$ defines a smooth structure in TX. Moreover, the projection $\pi : TX \to X$ is smooth since for each $i \in I$, we have

$$\phi_i \circ \pi \circ \tilde{\phi}_i^{-1}(u, w) = u.$$

The fact that $\pi : TX \to X$ is a vector bundle of rank n follows directly from the definitions. □

Given a smooth mapping $f : X \to Y$, the *differential* of f is the mapping $df : TX \to TY$ such that $df(v) = d_x f(v)$, for each $v \in T_x X$. By construction, we have a commutative diagram

$$\begin{array}{ccc} TX & \xrightarrow{df} & TY \\ \downarrow{\scriptstyle \pi} & & \downarrow{\scriptstyle \pi} \\ X & \xrightarrow{f} & Y \end{array}$$

Proposition 2.8 *The pair (df, f) is a morphism between the tangent bundles of X and Y.*

Proof We only need to prove that $df : TX \to TY$ is smooth. Given any $v \in TX$, let $x = \pi(v) \in X$. There exist charts $\phi : U \to A$ in X and $\psi : V \to B$ in Y with $x \in U \subset f^{-1}(V)$ such that $\bar{f} = \psi \circ f \circ \phi^{-1} : A \to B$ is smooth. Consider the induced charts $\tilde{\phi} : \tilde{U} \to A \times \mathbb{F}^n$ in TX and $\tilde{\psi} : \tilde{V} \to B \times \mathbb{F}^p$ in TY. We have $v \in \tilde{U} \subset (df)^{-1}(\tilde{V})$ and moreover,

$$\tilde{\psi} \circ df \circ \tilde{\phi}^{-1}(u, w) = (\bar{f}(u), d_u \bar{f}(w)),$$

which is smooth. □

Definition 2.12 A *vector field* on X is a section $\xi : X \to TX$ of the tangent bundle.

The set of vector fields on X is denoted by θ_X and has a structure of $\mathscr{O}_X$-module. Given any chart $\phi : U \to A$ in X, for each $i = 1, \dots, n$, the *ith-coordinate vector field* is the vector field $\frac{\partial}{\partial x_i} \in \theta_U$ given by the mapping $x \mapsto \left.\frac{\partial}{\partial x_i}\right|_x$. Since the coordinate tangent vectors at a point form a basis of the tangent space, it follows that θ_U is a free $\mathscr{O}_U$-module of rank n generated by the coordinate vector fields. In particular, every vector field $\xi \in \theta_X$ can be written locally in a unique way as

$$\xi|_U = \sum_{i=1}^{n} \xi_i \frac{\partial}{\partial x_i},$$

for some functions $\xi_1, \ldots, \xi_n \in \mathcal{O}_U$, which are called *components* of ξ in the chart $\phi\colon U \to A$. Given a subset $S \subset X$, we can also consider the set of germs of vector fields on X at S. This set is denoted by $\theta_{X,S}$ and has a structure of $\mathcal{O}_{X,S}$-module. Again, by taking charts, it follows that $\theta_{X,S}$ is a free $\mathcal{O}_{X,S}$-module of rank n generated by the germs of the coordinate vector fields $\frac{\partial}{\partial x_i} \in \theta_{X,S}$, with $i = 1, \ldots, n$.

Definition 2.13 Given a smooth mapping $f\colon X \to Y$, a *vector field along f* is a smooth mapping $\xi\colon X \to TY$ such that $\pi \circ \xi = f$, where $\pi\colon TY \to Y$ is the canonical projection.

The set of vector fields along f is denoted by $\theta(f)$ and again has the structure of $\mathcal{O}_X$-module. For instance, if we consider $\xi \in \theta_X$ and $\eta \in \theta_Y$, then both $df \circ \xi$ and $\eta \circ f$ are vector fields along f. In general not all vector fields along f are of this type. The quotient of the set of all vector fields along f by the set obtained in this way plays a major rôle in the subject, as we will see in Chap. 3.

Definition 2.14 Given a map $f : X \to Y$, we say that vector fields $\xi \in \theta_X$ and $\eta \in \theta_Y$ are *f-related* if $df \circ \xi = \eta \circ f$.

Given any chart $\psi\colon V \to B$ in Y with coordinate functions $y_1, \ldots, y_n$, the *coordinate vector fields* along f are the vector fields $\frac{\partial}{\partial y_i} \circ f \in \theta(f|_U)$, where $U = f^{-1}(V)$. As in the case of vector fields on a manifold, $\theta(f|_U)$ is a free $\mathcal{O}_U$-module of rank p generated by the coordinate vector fields. Any vector field $\xi \in \theta(f)$ can be written locally in a unique way as

$$\xi|_U = \sum_{i=1}^{p} \xi_i \left(\frac{\partial}{\partial y_i} \circ f \right),$$

for some functions $\xi_1, \ldots, \xi_p \in \mathcal{O}_U$. When we consider a smooth map-germ $f\colon (X, S) \to Y$ instead of a mapping, then we have the set of germs of vector fields along f which is also denoted by $\theta(f)$. It is a free $\mathcal{O}_{X,S}$-module of rank p generated by the germs of the coordinate vector fields.

Let X be a manifold and $\gamma\colon D \to X$ a smooth curve, where $D \subset \mathbb{F}$ is an open subset. The *tangent vector field* of γ is defined as

$$\dot{\gamma} := d\gamma \circ \frac{d}{dt} \in \theta(\gamma).$$

Given a vector field $\xi \in \theta_X$, we say that γ is an *integral curve* of ξ if $\xi \circ \gamma = \dot{\gamma}$. If $t_0 \in D$ and $\gamma(t_0) = x_0$, then we say that the integral curve γ *passes* through x_0 at $t = t_0$. All the above definitions can be adapted easily to germs instead of mappings or vector fields. The following theorem is better stated in terms of germs, due to its local nature.

Theorem 2.4 *Let $\xi \in \theta_{X,S}$ be a germ of vector field. There exists a unique smooth map-germ $\tilde{\phi}\colon (X \times \mathbb{F}, S \times \{0\}) \to (X, S)$ with the following property:*

- *for each x in a neighbourhood of S, the curve γ_x given by $\gamma_x(t) = \tilde{\phi}(x, t)$ is an integral curve of X passing through x at $t = 0$.*

Proof Take a chart $\psi : (X, S) \to (\mathbb{F}^n, S')$, then we can write

$$\xi = \sum_{i=1}^{n} \xi_i \frac{\partial}{\partial x_i},$$

for some functions $\xi_1, \ldots, \xi_n \in \mathscr{O}_{X,S}$. We put $A_i = \xi_i \circ \psi^{-1} \in \mathscr{O}_{\mathbb{F}^n,S'}$, for each $i = 1, \ldots, n$. On the other hand, any curve γ is expressed in the chart by its coordinate functions $\gamma_1, \ldots, \gamma_n$. The condition that γ is an integral curve of ξ passing through x at $t = 0$ is equivalent to its being a solution of the following system of ordinary differential equations:

$$\dot{\gamma}_i(t) = A_i(\gamma_1(t), \ldots, \gamma_n(t)),\ i = 1, \ldots, n,$$

with initial conditions $\gamma_i(0) = x_i$, $i = 1, \ldots, n$ (here $x_1, \ldots, x_m$ are the coordinates of x in the chart). The existence and uniqueness of solutions of system of ordinary differential equations with initial conditions imply the existence and uniqueness of integral curves γ_x passing through a point x at $t = 0$. Moreover, the fact that the solutions depend smoothly on the initial condition implies that the mapping $\tilde{\phi}(x, t) = \gamma_x(t)$ is also smooth. (These results on systems of differential equations are well known for the real C^∞ case; a holomorphic version of these results can be found for instance in [Car95].) □

Definition 2.15 The map-germ $\tilde{\phi}\colon (X \times \mathbb{F}, S \times \{0\}) \to (X, S)$ obtained in Theorem 2.4 is called the *local flow* of $\xi \in \theta_{X,S}$.

Let us denote by $\Phi\colon (X \times \mathbb{F}, S \times \{0\}) \to (X \times \mathbb{F}, S \times \{0\})$ the map-germ

$$\Phi(x, t) = (\tilde{\phi}(x, t), t). \tag{2.2}$$

This map-germ is determined uniquely by the two following properties:

1. $\Phi(x, 0) = (x, 0)$, for all x,
2. $\Phi \circ \frac{\partial}{\partial t} = \xi \circ \Phi$.

It follows from item 1 that the differential of Φ at each $(x, 0)$ is the identity, hence Φ is a diffeomorphism by the Inverse Mapping Theorem 2.1. Moreover, Φ is an unfolding of the identity in (X, S), that is, it makes commutative the following diagram:

$$\begin{array}{ccc} (X, S) & \xrightarrow{\quad \mathrm{id} \quad} & (X, S) \\ \big\downarrow i & & \big\downarrow i \\ (X \times \mathbb{F}, S \times \{0\}) & \xrightarrow{\quad \Phi \quad} & (X \times \mathbb{F}, S \times \{0\}) \\ & \searrow^{\pi} \qquad {}^{\pi}\swarrow & \\ & (\mathbb{F}, 0) & \end{array} \tag{2.3}$$

where $i(x) = (x, 0)$ and $\pi(x, t) = t$.

Definition 2.16 The diffeomorphism $\Phi\colon (X \times \mathbb{F}, S \times \{0\}) \to (X \times \mathbb{F}, S \times \{0\})$ obtained from the local flow as in (2.2) is called the *local unfolding* of $\xi \in \theta_{X,S}$.

Fix a representative $\Phi\colon U \times D \to X \times \mathbb{F}$ which is a diffeomorphism onto its image, where $U \subset X$ and $D \subset \mathbb{F}$ are open neighbourhoods of S and 0, respectively. For each $t \in D$, we denote by $\phi_t : U \to X$ the mapping $\phi_t(x) = \tilde{\phi}(x, t)$.

Lemma 2.5 *For all $t \in D$, $\phi_t(U) \subset X$ is open and $\phi_t : U \to \phi_t(U)$ is a diffeomorphism.*

Proof Let $i_t\colon X \to X \times \mathbb{F}$ be the mapping $i_t(y) = (y, t)$. Since $\Phi(U \times D)$ is open in $X \times \mathbb{F}$, it follows that $\phi_t(U) = i_t^{-1}(\Phi(U \times D))$ is open in X. Moreover, since the restriction $\Phi\colon U \times D \to \Phi(U \times D)$ is a diffeomorphism, it has an inverse $\Phi^{-1}\colon \Phi(U \times D) \to U \times D$. Necessarily, Φ^{-1} will be of the form $\Phi^{-1}(y, t) = (\tilde{\psi}(y, t), t)$ and it follows that $\psi_t : \phi_t(U) \to U$ given by $\psi_t(y) = \tilde{\psi}(y, t)$ is the inverse of ϕ_t. □

The 1-parameter family of diffeomorphisms $\phi_t : U \to \phi_t(U)$, with $t \in D$, is also sometimes referred to as the local flow of the vector field ξ.

The following property of flows is known as the *Thom–Levine lemma*. In [Lee13] it is called the property of *naturality of flows*. Let $f : X \to Y$ be a smooth mapping and consider $\xi \in \theta_X$ and $\eta \in \theta_Y$. For each $x \in X$, we take local flows of ξ, η at $x, f(x)$, respectively. By shrinking the neighbourhoods if necessary, we can assume that the local flows are, respectively, of the form $\phi_t : U \to \phi_t(U)$ and $\psi_t\colon V \to \psi_t(V)$, with $|t| < \epsilon$ and such that $f(U) \subset V$.

Lemma 2.6 *Assume ξ, η are f-related. Then the following diagram is commutative for all t, with $|t| < \epsilon$:*

$$\begin{array}{ccc} U & \xrightarrow{\ f\ } & V \\ \big\downarrow \phi_t & & \big\downarrow \psi_t \\ \phi_t(U) & \xrightarrow{\ f\ } & \psi_t(V) \end{array}$$

Proof Given $x \in U$ and $t \in D$, $\gamma_x(t) = \phi_t(x)$ is the integral curve of ξ passing through x at $t = 0$. That is, $\gamma_x(0) = 0$ and $\xi(\gamma_x(t)) = \dot{\gamma}_x(t)$. We claim that $(f \circ \gamma_x)(t)$ is the integral curve of η passing through $f(x)$ at $t = 0$. In fact, $f(\gamma_x(0)) = f(x)$ and also

$$\begin{aligned}\eta((f \circ \gamma_x)(t)) &= (\eta \circ f)(\gamma_x(t)) = (df \circ \xi)(\gamma_x(t)) \\ &= df(\xi(\gamma_x(t))) = df(\dot{\gamma}_x(t)) = (f \dot{\circ} \gamma_x)(t).\end{aligned}$$

But this implies that $\psi_t(f(x)) = (f \dot{\circ} \gamma_x)(t) = f(\gamma_x(t)) = f(\phi_t(x))$. □

When f is a diffeomorphism, then for each $\xi \in \theta_X$ there exists a unique $\eta \in \theta_N$ which is f-related to ξ and which is given by $\eta = df \circ \xi \circ f^{-1}$. This vector field is called the *push-forward* of ξ by f and is denoted by $f_* \xi$.

Unfortunately, the construction of local flow and local unfolding of a vector field is not enough for our purposes. We need these notions for a more general class, namely the time-dependent vector fields. We explain how the constructions of the local flow and the local unfolding can be adapted easily for this type of vector fields just by adding an extra parameter.

By definition, a *time-dependent vector field* on X is a smooth mapping $\tilde{\xi} : X \times D \to TX$, where $D \subset \mathbb{F}$ is open, such that for each $t \in D$, the mapping $\xi_t : X \to TX$ given by $\xi_t(p) = \tilde{\xi}(p, t)$ is a vector field on X. Given a time-dependent vector field $\tilde{\xi}$, we can associate another vector field $\xi \in \theta_{X \times D}$ just by setting

$$\xi := \tilde{\xi} + \frac{\partial}{\partial t},$$

where we identify $T(X \times D)$ with $TX \oplus TD$, and $\partial/\partial t$ is the coordinate vector field on D. Conversely, if $\xi \in \theta_{X \times D}$ is such that $dt(\xi) = \xi(t) = 1$, then we can associate a unique time-dependent vector field $\tilde{\xi}$ just by subtracting $\partial/\partial t$ to ξ. Hence, the set of time-dependent vector fields on X can be identified canonically with the subset of vector fields $\xi \in \theta_{X \times D}$ such that $\xi(t) = 1$.

Now, given a point $x \in X$, consider $(x, 0) \in X \times D$. The local flow of ξ at $(x, 0)$ is a map $\tilde{\psi} : U \times B_\epsilon(0) \times B_\delta(0) \to X \times D$, where U is an open neighbourhood of x in X and $\epsilon, \delta > 0$. Of course, we can assume without loss of generality that $\epsilon = \delta$ and since $X(t) = 1$, we have $\tilde{\psi}(x, t, s) = (\tilde{\psi}_1(x, t, s), s)$, where $\tilde{\psi}_1$ is the first component of $\tilde{\psi}$. We define $\tilde{\phi} : U \times B_\epsilon(0) \to X$ as

$$\tilde{\phi}(x, t) = \tilde{\psi}_1(x, t, t),$$

and we call it the *local flow* of the time-dependent vector field $\tilde{X}$ at p. Now, the construction of the 1-parameter family of diffeomorphisms $\phi_t : U \to \phi_t(U)$ and of the *local unfolding* $\Phi(x, t) = (\phi_t(x), t)$ follows the same steps as in the case of a vector field.

Exercises for Sect. 2.5

1 Let X be a smooth manifold. We have defined a vector field v (a map assigning a tangent vector $v(x) \in T_x X$ to each point $x \in X$) to be *smooth* if it is smooth as a section of the tangent bundle. Show that this is equivalent to the following characterisation: v is smooth at x if for all $f \in \mathcal{O}_{X,x}$, the germ $v \cdot f$ is smooth.

2 Let $\pi : E \to X$ be a vector bundle on a manifold X. The *dual* vector bundle is defined as $\pi' : E^* \to X$, where

$$E^* := \coprod_{x \in X} E_x^*,$$

where E_x^* is the dual vector space of E_x and π' is defined in the obvious way. Show that $\pi' : E^* \to X$ is indeed a vector bundle on X.

3 The *cotangent bundle* of X is defined as the dual of the tangent bundle, that is, $T^*X := (TX)^*$. Let $f : X \to \mathbb{F}$ be a smooth function. For each $x \in X$, we denote by $d_x f \in T_x^* X$ the linear form given by $d_x f(v) = v(f)$, for all $v \in T_x X$. Show that the mapping $df : X \to T^*X$ given by $x \mapsto d_x f$ is a section of the cotangent bundle (this is usually called the *differential 1-form* of f).

4 Let $f : (X, x_0) \to (Y, y_0)$ be an immersion. Show that for each $\xi \in \theta_{X,x_0}$ there exists $\eta \in \theta_{Y,y_0}$ such that ξ and η are f-related.

5 Let $f : (X, x_0) \to (Y, y_0)$ be a submersion. Show that for each $\eta \in \theta_{Y,y_0}$ there exists $\xi \in \theta_{X,x_0}$ such that ξ and η are f-related.

6 Find the local flow of the following vector fields on $\mathbb{F}^2$:

(a) $x\frac{\partial}{\partial x} + y\frac{\partial}{\partial y}$,
(b) $-y\frac{\partial}{\partial x} + x\frac{\partial}{\partial y}$.

Draw also a picture of these vector fields and of their local flows in the case $\mathbb{F} = \mathbb{R}$.

7 Let $\xi \in \theta_X$ have local flow $\phi_t : U \to \phi_t(U)$, $t \in D$. Show that:

(a) $\phi_s \circ \phi_t = \phi_{s+t}$ in the subset of U where the composition $\phi_s \circ \phi_t$ is defined;
(b) the inverse of ϕ_t is ϕ_{-t}.

8 Let $\xi \in \theta_X$ have local flow $\phi_t : U \to \phi_t(U)$, $t \in D$. Let $x \in U$ be a point such that $X(x) = 0$. Show that $\phi_t(x) = x$, for all $t \in D$.

9 Let $\xi \in \theta_{X \times D}$ be a vector field such that $\xi(t) = 1$. Show that the local unfolding of the associated time-dependent vector field at any point $x \in X$ is characterised uniquely by the following two conditions:

(a) Φ is an unfolding of the identity (see (2.3)),
(b) $\xi = \Phi_*(\partial/\partial t)$.

10 Let $\Phi\colon (X \times \mathbb{F}, S \times \{0\}) \to (X \times \mathbb{F}, S \times \{0\})$ be a diffeomorphism which is an unfolding of the identity (see (2.3)). Write $\Phi(x, t) = (\phi_t(x), t)$ and assume that $\phi_s \circ \phi_t = \phi_{s+t}$, whenever the composition is defined. Show that there exists $\xi \in \theta_{X,S}$ whose local unfolding is Φ.

2.6 Transversality

The Regular Value Theorem 2.3 says that if $y \in Y$ is a regular value of a smooth mapping $f\colon X \to Y$, then the inverse image $f^{-1}(y)$ is a submanifold of X. One can ask when the inverse image $f^{-1}(Z)$ of a submanifold $Z \subset Y$ is also a submanifold of X. It turns out that the answer is provided by the notion of *transversality*.

Definition 2.17 Let $f\colon X \to Y$ be a smooth mapping between manifolds and $Z \subset Y$ a submanifold. Given a point $x \in X$ such that $f(x) \in Z$, we say that f is *transverse* to Z at x if

$$d_x f(T_x X) + T_{f(x)} Z = T_{f(x)} Y. \tag{2.4}$$

We say that f is *transverse* to Z it is transverse at every point $x \in X$ such that $f(x) \in Z$. Note that in particular, if $f^{-1}(Z) = \emptyset$, then f is transverse to Z automatically.

We use the symbol $f \pitchfork Z$ to indicate that f is transverse to Z. The following properties are immediate.

1. If $Z \subset Y$ is an open submanifold, then $f \pitchfork Z$ for every smooth mapping $f\colon X \to Y$.
2. If $f\colon X \to Y$ is a submersion, then $f \pitchfork Z$ for every submanifold $Z \subset Y$.
3. For a single point $y \in Y$, $f : X \to Y$ is transverse to $\{y\}$ if and only if y is a regular value of f.
4. When $\dim X < \operatorname{codim} Z$ then $f \pitchfork Z$ if and only if $f(X) \cap Z = \emptyset$.

The next theorem extends the Regular Value Theorem 2.3 to transverse preimages.

Theorem 2.5 *Let $f\colon X \to Y$ be a smooth mapping between manifolds and $Z \subset Y$ a submanifold such that $f \pitchfork Z$. Then, provided it is not empty, $f^{-1}(Z)$ is a submanifold of X with* $\operatorname{codim} f^{-1}(Z) = \operatorname{codim} Z$ *and moreover, for each $x \in f^{-1}(X)$,*

$$T_x f^{-1}(Z) = (d_x f)^{-1}(T_{f(x)} Z).$$

Proof By Corollary 2.3, we only have to prove that $(f^{-1}(Z), x)$ is a submanifold of (X, x) for each $x \in f^{-1}(Z)$. Let $y = f(x)$ and let $g_1, \ldots, g_k$ be regular equations for Z in some neighbourhood U of y. From the hypothesis of transversality we have

$$T_0\mathbb{F}^k = d_y g(T_y Y) = d_y g(d_x f(T_x X) + T_y Z).$$

Since $T_y Z = \ker d_y g$, the right-hand side reduces to $d_y g(d_x f(T_x X))$, showing that $g \circ f$ is a submersion at x. Thus 0 is a regular value of $g \circ f$, and so $(g \circ f)^{-1}(0)$ is a smooth submanifold of $f^{-1}(U)$, of codimension k, by the Regular Value Theorem 2.3. But $(g \circ f)^{-1}(0) = f^{-1}(Z) \cap f^{-1}(U)$.

The statement on tangent spaces holds also by the regular value theorem:

$$T_x(f^{-1}(Z)) = \ker d_x(g \circ f) = (d_x f)^{-1}(\ker d_y g) = (d_x f)^{-1}(T_y Z).$$

□

Definition 2.18 Let $X, Z \subset Y$ be submanifolds and let $x \in X \cap Z$. We say that X and Z are *transverse* at x if the inclusion $i\colon X \hookrightarrow Y$ is transverse to Z at x. We say that X and Z are *transverse* if they are transverse at every point of $X \cap Z$.

Since the differential of the inclusion is the inclusion of tangent spaces, the transversality of X and Z at x means that

$$T_x X + T_x Z = T_x Y.$$

In particular, the condition is the same if we interchange the order of the submanifolds X, Z. We will represent the transversality between X and Z by the symbol $X \pitchfork Z$. The following corollary is an immediate consequence of Theorem 2.5.

Corollary 2.4 *Let $X, Z \subset Y$ be transverse submanifolds. Then $X \cap Z$ is a submanifold,* $\operatorname{codim}(X \cap Z) = \operatorname{codim} X + \operatorname{codim} Z$ *and for each $x \in X \cap Z$, we have*

$$T_x(X \cap Z) = T_x X \cap T_x Z.$$

Transversality is a crucial notion. A transversal intersection of submanifolds is "stable"—move either one slightly, and the transversality persists, and the intersection does not change significantly. Transversality is also a "dense" property: a non-transverse intersection can be perturbed an arbitrarily small amount to become transverse (see Fig. 2.1). These simple facts are at the root of the notions of stability and stable perturbation that are central in singularity theory. They are made more precise in the elementary transversality theorem, Exercise A.3 in Appendix A, and in Exercise B.1.1 in Appendix B.

We recall that $N \subset \mathbb{R}^k$ is a null subset if for each $\epsilon > 0$ there exists a countable family of cuboids $\{C_i\}_{i=1}^{\infty}$ such that $N \subset \bigcup_{i=1}^{\infty} C_i$ and moreover

$$\sum_{i=1}^{\infty} \operatorname{vol}(C_i) < \epsilon.$$

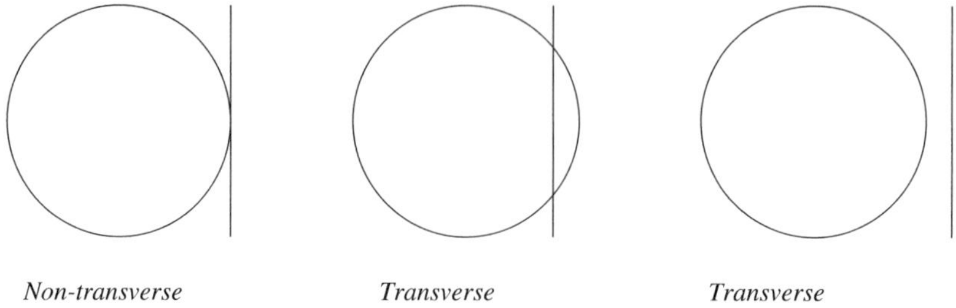

Fig. 2.1 Transverse perturbations of a non-transverse intersection

By a cuboid we mean any product of closed intervals $C = [a_1, b_1] \times \cdots \times [a_k, b_k]$ with $a_i < b_i$ for $i = 1, \ldots, k$. Its volume is $\mathrm{vol}(C) = (b_1 - a_1) \ldots (b_k - a_k)$. In the complex case, we say that $N \subset \mathbb{C}^k$ is a null subset if it is a null subset when considered as a subset of $\mathbb{R}^{2k}$.

Definition 2.19 We say that $N \subset X$ is a *null* subset of a manifold X if for each $x \in N$, there exists a chart $\phi \colon U \to A$ on X, with $x \in U$, such that $\phi(U \cap N)$ is a null subset of $\mathbb{F}^n$.

Theorem 2.6 (Sard's Theorem) *The discriminant Δ of a smooth mapping $f \colon X \to Y$ is a null subset of Y.*

A proof for the C^∞ real version of Sard's theorem can be found, for instance, in [GG73, Theorem 1.12]. The complex holomorphic case follows from the real one, since the discriminant of a holomorphic mapping coincides with the discriminant when considered as a C^∞ real mapping. It is common to say that a certain property holds "for almost every" point in a manifold if the property holds outside a null subset. Thus, Sard's theorem can be also stated by saying that almost every $y \in Y$ is a regular value of f.

The following result, known as the basic transversality lemma, is an easy consequence of Sard's theorem; it is a key ingredient in the proof of the Thom Transversality Theorem A.1, and, thus, in the proof of all genericity theorems , such as the density of stable mappings in the nice dimensions discussed in Sect. 5.2. Given a mapping $F \colon X \times S \to Y$, for each $s \in S$ we denote by $f_s \colon X \to Y$ the mapping $f_s(p) = F(p, s)$.

Theorem 2.7 (Basic Transversality Lemma) *Let $F \colon X \times S \to Y$ be a smooth mapping and $Z \subset Y$ a submanifold. If $F \pitchfork Z$, then $f_s \pitchfork Z$ for almost every $s \in S$.*

Proof By Theorem 2.5, $W = F^{-1}(Z)$ is a submanifold of $X \times S$. Let $\pi : W \to S$ be the restriction of the natural projection from $X \times S$ into S. We will show that $f_s \pitchfork Z$ whenever s is a regular value of π. Thus, the result follows from Sard's Theorem 2.6.

Suppose first that s is a regular value of π and let $x \in X$ such that $f_s(x) = y \in Z$. Since $F \pitchfork Z$, we have

$$d_{(x,s)}F(T_{(x,s)}(X \times S)) + T_y Z = T_y Y.$$

That is, given any $a \in T_y Y$ there exists $b \in T_{(x,s)}(X \times S)$ such that

$$d_{(x,s)}F(b) - a \in T_y Z.$$

We want to find a vector $v \in T_x X$ such that $d_x(f_s)(v) - a \in T_y Z$. Since

$$T_{(x,s)}(X \times S) = T_x X \times T_s S,$$

we can write $b = (w, e)$ with $w \in T_x X$ and $e \in T_s S$. Moreover, π is the restriction of the projection onto the second component, hence its differential

$$d_{(x,s)}\pi : T_{(x,s)}W \to T_s S$$

is also the restriction of the projection of $T_x X \times T_s S$ into $T_s S$. Since s is a regular value of π, $d_{(x,s)}\pi$ is an epimorphism and hence, there exists $u \in T_x X$ such that $(u, e) \in T_{(x,s)}W$. On the other hand, since $W = F^{-1}(Z)$, its tangent space is $T_{(x,s)}W = (d_{(x,s)}F)^{-1}(T_y Z)$, so $d_{(x,s)}F(u, e) \in T_y P$.

Let $v = w - u \in T_x X$, then:

$$\begin{aligned} d_x(f_s)(v) - a &= d_{(x,s)}F(v, 0) - a \\ &= d_{(x,s)}F[(w, e) - (u, e)] - a \\ &= d_{(x,s)}F(b) - a - d_{(x,s)}F(u, e) \in T_y Z. \end{aligned}$$

□

Exercises for Sect. 2.6

1 Suppose that $f : X \to Y$ is a smooth map of manifolds, $Z \subset Y$ is a submanifold, and $x_0 \in X$ such that $f(x_0) \in Z$. Suppose also that Z has regular defining equation $g : U \to \mathbb{F}^c$, where U is a neighbourhood of $f(x_0)$. Show that

$$f \pitchfork Z \text{ at } x_0 \quad \Longleftrightarrow \quad g \circ f \text{ is a submersion at } x_0.$$

This was already used in the proof of Theorem 2.5. The condition on the right is often easier to check than the definition of transversality.

2 Let X be a manifold of dimension n and let $N \subset X$ be a null subset. Show that for every $x \in N$ and for every chart $\phi : U \to A$ with $x \in U$, $\phi(N \cap U)$ is a null subset of $\mathbb{F}^n$.

3 Let $f: X \to \mathbb{F}^p$ be a smooth mapping and $Z \subset \mathbb{F}^p$ a submanifold. Show that for almost every $a \in \mathbb{F}^p$, the mapping $f + a: X \to \mathbb{F}^p$ is transverse to Z. Hint: consider the family $F : X \times \mathbb{R}^p \to \mathbb{R}^p$ defined by $F(x, a) = f(x) + a$.

4 Let $f: X \to \mathbb{F}$ be a smooth function. A critical point $x_0 \in X$ of f is called *non-degenerate* if given a chart $\phi: (X, x_0) \to (\mathbb{F}^n, u_0)$, the Hessian determinant of $f \circ \phi^{-1}$ is not zero at u_0, that is,

$$\det\left(\frac{\partial^2(f \circ \phi^{-1})}{\partial u_i \partial u_j}(u_0)\right) \neq 0.$$

Show that the definition of non-degenerate critical point does not depend on the choice of the chart ϕ.

5 Let x_0 be a critical point of $f: X \to \mathbb{F}$. Show that x_0 is non-degenerate if and only if the differential 1-form $df: X \to T^*X$ is transverse to the zero section $Z \subset T^*X$ at x_0 (see Exercise 2.5.3).

6 A *Morse function* is a smooth function $f: X \to \mathbb{F}$ whose critical points are all non-degenerate and have pairwise distinct critical values. Suppose that $X \subset \mathbb{F}^N$. Show

(i) that for almost every $a \in \mathbb{F}^N$, the function $f_a: X \to \mathbb{F}$ given by

$$f_a(x) = f(x) + a_1x_1 + \cdots + a_Nx_N,$$

has only non-degenerate critical points, and

(ii) for almost all $a \in \mathbb{F}^n$, f_a is a Morse function.

2.7 Local Conical Structure

A further justification for the use of the notion of germ in singularity theory comes from the fact that analytic spaces are "locally conical". That is, for every point x_0 in a closed analytic subset X of some open set in $\mathbb{C}^n$ (or, indeed, in a closed semialgebraic subset of an open set of $\mathbb{R}^n$), the intersection of X with a sufficiently small ball $B_\varepsilon(x_0)$ centred at x_0 is homeomorphic to the cone on its boundary, $X \cap \partial B_\varepsilon(x_0)$—see Theorem B.7 in Sect. B.3. The locally conical structure is particularly important in the definition of the vanishing homology, so we discuss it in some detail in Sect. B.3.

Chapter 3
Left-Right Equivalence and Stability

As in Chap. 2, by a smooth map between manifolds we mean either a C^∞-map between C^∞-manifolds when $\mathbb{F} = \mathbb{R}$ or a holomorphic map between complex manifolds when $\mathbb{F} = \mathbb{C}$. In the complex case a smooth map with smooth inverse is usually called "bianalytic"; we prefer the term "diffeomorphism" for both the real and the complex case.

Definition 3.1 Let $f_i\colon (X_i, S_i) \to (Y_i, y_i)$, $i = 1, 2$ be germs of smooth maps between manifolds. They are

1. *right-equivalent*, if there exists a germ of diffeomorphism $\varphi\colon (X_1, S_1) \to (X_2, S_2)$ such that $f_2 = f_1 \circ \varphi^{-1}$;
2. *left-equivalent*, if there exists a germ of diffeomorphism $\psi\colon (Y_1, y_1) \to (Y_2, y_2)$ such that $f_2 = \psi \circ f_1$;
3. *left-right-equivalent*, if there exist germs of diffeomorphism $\varphi\colon (X_1, S_1) \to (X_2, S_2)$ and $\psi\colon (Y_1, y_1) \to (Y_2, y_2)$ such that $f_2 = \psi \circ f_1 \circ \varphi^{-1}$.

By taking charts in X and Y, any germ from X to Y is right-left equivalent to a germ $f\colon (\mathbb{F}^n, S) \to (\mathbb{F}^p, 0)$, so for our purposes it is enough to consider this class of germs. The advantage when the source and target are fixed is that the equivalences can be seen as group actions. Let $\mathscr{A} = \mathrm{Diff}(\mathbb{F}^n, S) \times \mathrm{Diff}(\mathbb{F}^p, 0)$ be the group of pairs of diffeomorphisms. We have an action of $\mathscr{A}$ on the set of germs $f\colon (\mathbb{F}^n, S) \to (\mathbb{F}^p, 0)$, given by

$$(\varphi, \psi) \cdot f = \psi \circ f \circ \varphi^{-1}.$$

Analogously, we can consider the groups $\mathscr{R} = \mathrm{Diff}(\mathbb{F}^n, S)$ and $\mathscr{L} = \mathrm{Diff}(\mathbb{F}^p, 0)$ and the corresponding actions. If $\mathscr{G} = \mathscr{R}$, $\mathscr{L}$ or $\mathscr{A}$, we say that f_1, f_2 are $\mathscr{G}$-*equivalent* if they are in the same $\mathscr{G}$-orbit. In this situation, we will use the terms $\mathscr{R}$, $\mathscr{L}$ or $\mathscr{A}$-equivalences instead of right, left or right-left equivalences, respectively. For a very good survey of these groups and their actions, see [Wal81].

D. Mond, J. J. Nuño-Ballesteros, *Singularities of Mappings*,
Grundlehren der mathematischen Wissenschaften 357,
https://doi.org/10.1007/978-3-030-34440-5_3

A big part of singularity theory has always been concerned with the problem of classification. Generally one classifies germs of smooth maps $(\mathbb{F}^n, S) \to (\mathbb{F}^p, 0)$ up to $\mathscr{A}$-equivalence, and up to $\mathscr{R}$-equivalence if $p = 1$.

A key ingredient in classification is the notion of *finite determinacy*. For a smooth germ $f: (\mathbb{F}^n, S) \to (\mathbb{F}^p, 0)$ where $S = \{x_1, \ldots, x_r\}$, and for each $k \geq 0$, we denote by $j^k f$ the k-jet of f at S, that is, $j^k f = (j^k f(x_1), \ldots, j^k f(x_r))$, where $j^k f(x_i)$ is the k-jet of f at x_i (see Appendix A for the definition of $j^k f(x_i)$).

Definition 3.2 Let $f: (\mathbb{F}^n, S) \to (\mathbb{F}^p, 0)$ be a smooth germ, and let $\mathscr{G}$ be one of the groups listed above, or a subgroup. We say f is *k-determined for $\mathscr{G}$-equivalence* if whenever the k-jet at S of another germ g coincides with that of f, then f, g are $\mathscr{G}$-equivalent, and *finitely determined* if it is k-determined for some $k \in \mathbb{N}$.

In [Mat68b], John Mather showed that for all of the groups listed above, in the complex analytic case, finite determinacy is equivalent to *isolated instability*. We will prove this for $\mathscr{R}$ and $\mathscr{A}$. The key is to understand how to construct diffeomorphisms. In all of singularity theory (except where the inverse and implicit function theorems apply), the diffeomorphisms we use arise as the local flow of vector fields, as described in Sect. 2.5.

3.1 Classification of Functions by Right Equivalence

The Thom–Levine Lemma 2.6 shows how an infinitesimal condition gives rise to a family of diffeomorphisms. The equality $df \circ \xi = \eta \circ f$ of Definition 2.14 is linear in ξ and η, and these vector fields can often be constructed by the methods of commutative algebra. This is the entry-point of commutative algebra, which, through it, has a huge input into singularity theory.

As an example of what is involved, let us prove the simplest of the determinacy theorems of John Mather [Mat68b], also proved independently by Jean-Claude Tougeron in [Tou68]. We will return to the theme of finite determinacy, this time for $\mathscr{A}$-equivalence, in Chap. 6.

Let $\mathscr{O}_n := \mathscr{O}_{\mathbb{F}^n,0}$ be the local ring of germs of smooth functions $f: (\mathbb{F}^n, 0) \to \mathbb{F}$, with maximal ideal $\mathfrak{m}_n := \mathfrak{m}_{\mathbb{F}^n,0}$. In the real C^∞ case $\mathscr{O}_n$ is often denoted by $\mathscr{E}_n$, but in order not to have to distinguish between real and complex except when necessary, we prefer to use $\mathscr{O}_n$ in both cases. Remark 3.2 and Proposition 3.8 below will justify treating the two cases together in this way. If $f \in \mathscr{O}_n$, then the ideal in $\mathscr{O}_n$ generated by the first order partial derivatives $\partial f/\partial x_1, \ldots, \partial f/\partial x_n$ is called the *Jacobian ideal* and denoted J_f.

Theorem 3.1 *Let $f \in \mathscr{O}_n$.*

(i) Suppose that f is k-determined for $\mathscr{R}$-equivalence. Then

$$\mathfrak{m}_n J_f \supset \mathfrak{m}^{k+1}.$$

(ii) Conversely, suppose that

$$\mathfrak{m}_n\, J_f \supset \mathfrak{m}^k\,. \tag{3.1}$$

Then f is k-determined for $\mathscr{R}$-equivalence.

Notice that different powers of $\mathfrak{m}_n$ appear in (i) and in (ii). It is *not* true that

$$\mathfrak{m}_n\, J_f \supset \mathfrak{m}_n^{k+1} \quad \Longrightarrow \quad f \text{ is } k\text{-determined for } \mathscr{R}.$$

Example 3.1 The Morse Lemma, which gave rise to Morse theory in topology (see e.g. [Mil63]), states that if a C^∞ function $f : M^n \to \mathbb{R}$ has a non-degenerate critical point at x_0 (i.e. the Hessian determinant of f at x_0 is non-zero) then in suitable coordinates around x_0, f takes the form

$$f(x_1, \ldots, x_n) = f(x_0) - \sum_{i=0}^{k} x_i^2 + \sum_{i=k+1}^{n} x_i^2. \tag{3.2}$$

This follows easily from Theorem 3.1. For as Exercise 3.1.2 shows, non-degeneracy is equivalent to having $J_f = \mathfrak{m}_n$, so that $\mathfrak{m}_n\, J_f = \mathfrak{m}_n^2$ and, by the theorem, f is 2-determined for right equivalence, and is thus right-equivalent to $f(x_0)$ + non-degenerate quadratic form. The form given in (3.2) then follows from the classification of quadratic forms. ◊

Proof of Theorem 3.1

(i) Let $h \in \mathfrak{m}_n^{k+1}$. Then since f is k-determined for $\mathscr{R}$, for all t there exists $\varphi_t \in \text{Diff}(\mathbb{F}^n, 0)$ such that $f + th = f \circ \varphi_t$. If we could assume the existence of a smoothly parameterised family of diffeomorphisms φ_t with $\varphi_0 = \text{id}$ such that $f \circ \varphi_t = f + th$ then we could reason as follows:

$$h = \frac{d(f + th)}{dt} = \frac{d(f \circ \varphi_t)}{dt} = \sum_i \left(\frac{\partial f}{\partial x_i} \circ \varphi_t\right) \frac{\partial \varphi_{t,i}}{\partial t}. \tag{3.3}$$

Note that since $\varphi_t(0) = 0$ for all t it follows that $\partial\varphi_{t,i}/\partial t \in \mathfrak{m}_n$. When $t = 0$, since $\varphi_0 = \text{id}$, this gives

$$h = \sum_i \frac{\partial f}{\partial x_i} \frac{\partial \varphi_{t,i}}{\partial t} \in \mathfrak{m}_n\, J_f \tag{3.4}$$

so that $\mathfrak{m}_n^{k+1} \subset \mathfrak{m}_n\, J_f$ as required.

However, our hypothesis does not allow us immediately to assert that the diffeomorphisms φ_t fit together to give a smooth family. So instead we look in jet space $J^{k+1}(n,1) = \mathfrak{m}_n / \mathfrak{m}_n^{k+2}$. As f is k-determined, the set

$$L := \{j^{k+1}(f+h) : h \in \mathfrak{m}_n^{k+1}\} \subset J^{k+1}(n,1)$$

lies entirely in the $\mathscr{R}^{(k+1)}$-orbit of f, where $\mathscr{R}^{(k+1)}$ is the finite-dimensional quotient of $\mathrm{Diff}(\mathbb{F}^n, 0)$ acting on jet space. Now $\mathscr{R}^{(k+1)}$ can be identified with the set

$$\{j^{k+1}\varphi(0) : \varphi \in \mathrm{Diff}(\mathbb{F}^n, 0)\}$$

and has a natural structure of algebraic group: the composite of two polynomial mappings depends polynomially on their coefficients, and in $\mathscr{R}^{(k+1)}$ one composes and then truncates at degree $k+1$. This group acts algebraically on $J^{k+1}(n,1)$. Thus, as the set L lies in the orbit of $j^{k+1}f(0)$, writing $z = j^{k+1}f(0)$, and $\mathscr{R}^{(k+1)}z$ for the $\mathscr{R}^{(k+1)}$-orbit of z, one has

$$\frac{\mathfrak{m}_n^{k+1}}{\mathfrak{m}_n^{k+2}} = T_z L \subset T_z(\mathscr{R}^{(k+1)}z) = \frac{\mathfrak{m}_n\, J_f + \mathfrak{m}_n^{k+2}}{\mathfrak{m}_n^{k+2}}, \tag{3.5}$$

and thus

$$\mathfrak{m}_n^{k+1} \subset \mathfrak{m}_n\, J_f + \mathfrak{m}_n^{k+2}.$$

The conclusion we want follows by Nakayama's Lemma C.1, which we apply taking $(\mathscr{O}_n, \mathfrak{m}_n)$ as $(R, \mathfrak{m})$, $\mathfrak{m}_n^k$ as M and $\mathfrak{m}_n\, J_f$ as N.

The second equality in (3.5) is important and not completely obvious. It can be obtained along the lines of the argument leading up to (3.4), but using the crucial fact that if the Lie group G acts on the manifold M and for $x \in M$ we denote by α_x the orbit map $g \in G \mapsto gx$, then for each $x \in M$ with smooth orbit Gx,

$$T_x Gx = d_e\alpha_x(T_eG).$$

Now

$$d_e\alpha_x(T_eG) = \left\{ \frac{d}{dt}(\gamma(t)\cdot x)\Big|_{t=0} : \gamma\colon (\mathbb{F},0) \to (G,e) \text{ is a curve germ} \right\};$$

every curve in $(\mathscr{R}^{(k+1)}, \mathrm{id})$ is of the form $j^{k+1}\varphi_t$ for a 1-parameter family of diffeomorphisms φ_t, so now it really is true that $T_z\mathscr{R}^{(k+1)}z$ is equal to

$$\left\{ \frac{d}{dt} j^{k+1}(f\circ\varphi_t)\Big|_{t=0} : \varphi_t \text{ is a 1-parameter family in } \mathrm{Diff}(\mathbb{F}^n,0) \text{ with } \varphi_0 = \mathrm{id} \right\},$$

which is equal to

$$\left\{ j^{k+1}\left(\frac{d}{dt}(f \circ \varphi_t)\bigg|_{t=0}\right) : \varphi_t \text{ is a 1-parameter family in } \mathrm{Diff}(\mathbb{F}^n, 0) \text{ with } \varphi_0 = \mathrm{id} \right\}.$$

This shows that (3.4) holds even without the assumption that there is a smoothly parameterised family of diffeomorphisms φ_t such that $f + th = f \circ \varphi_t$.

(ii) Suppose that g has the same degree k Taylor polynomial as f. Then $g - f \in \mathfrak{m}_n^{k+1}$. Let $F(x,t) = f(x) + t(g(x) - f(x))$, and write $f_t(x) = F(t,x)$. The idea of the proof is to show that for each value t_0 of t, there is a neighbourhood $U(t_0)$ of t_0 in $\mathbb{F}$ such that f_t and f_{t_0} are $\mathscr{R}$-equivalent for all $t \in U(t_0)$. Thus, the set of t such that f_t is $\mathscr{R}$-equivalent to f is open, and so is its complement. Since $\mathbb{F}$ is connected, $f = f_0$ and $g = f_1$ are $\mathscr{R}$-equivalent.

We construct the set U_{t_0} first for $t_0 = 0$. As F is a function of the $n+1$ variables $x_1, \ldots, x_n, t$, we consider the germ $F \in \mathscr{O}_{n+1}$. Notice that $\partial F/\partial t = g - f \in M_n^{k+1}$, where by M_n we mean the ideal in $\mathscr{O}_{n+1}$ generated by $(x_1, \ldots, x_n)$. Because $\mathfrak{m}_n J_f \supset \mathfrak{m}_n^k$,

$$\frac{\partial F}{\partial t} \in M_n\left(\frac{\partial f}{\partial x_1}, \ldots, \frac{\partial f}{\partial x_n}\right). \tag{3.6}$$

We would like to show

$$\frac{\partial F}{\partial t} \in M_n\left(\frac{\partial F}{\partial x_1}, \ldots, \frac{\partial F}{\partial x_n}\right). \tag{3.7}$$

For if we have

$$\frac{\partial F}{\partial t} = \xi_1 \frac{\partial F}{\partial x_1} + \cdots + \xi_n \frac{\partial F}{\partial x_n} \tag{3.8}$$

for some functions $\xi_i \in M_n$, then defining a germ of vector field ξ on $(\mathbb{F}^{n+1}, 0)$ by

$$\xi = \frac{\partial}{\partial t} - \sum_i \xi_i \frac{\partial}{\partial x_i},$$

(3.8) becomes

$$dF(\xi) = 0.$$

This is exactly the condition of Definition 2.14 with $\eta = 0$. Let $\Phi(x,t) = (\varphi_t(x), t)$ be the integral flow of the vector field ξ. Since $\xi_i \in M_n$ for $i = 1, \ldots, n$, we have $\Phi(0,t) = (0,t)$ for all t. The integral flow of the zero vector field is the identity

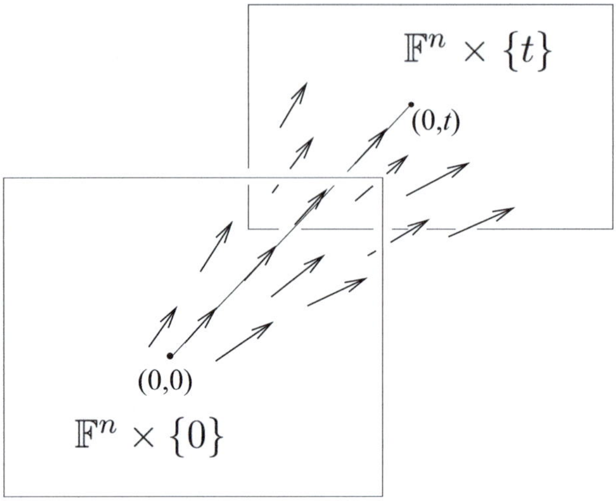

Fig. 3.1 The arrows show the vector field ξ of the proof. At all points of the t-axis, the vector field is tangent to the axis, so any trajectory beginning at a point on the axis remains on the axis. Thus $\varphi_t(0) = 0$

map, and therefore by the Thom–Levine lemma we have

$$F \circ \tilde{\Phi} = F. \tag{3.9}$$

Since the component of ξ in the t-direction is 1, it follows that φ_t maps $\mathbb{F}^n \times \{0\}$ to $\mathbb{F}^n \times \{t\}$. Restricting both sides of (3.9) to $\mathbb{F}^n \times \{0\}$ we therefore get

$$f_t \circ \varphi_t = f,$$

with $\varphi_t(0) = 0$ (see Fig. 3.1). Now take representatives of the germs involved in (3.9). Since the germs on the left- and right-hand sides are equal, it follows that the representatives coincide on some neighbourhood of $(0, 0)$ in $\mathbb{F}^n \times \mathbb{F}$. We deduce that there exists $\eta > 0$ such that for $|t| < \eta$, the germ of f_t is $\mathscr{R}$-equivalent to the germ of f.

Now we set about deducing (3.7) from (3.6). Since $\partial F/\partial t = g - f \in M_n^{k+1}$, to show (3.7), it will be enough to show

$$M_n^{k+1} \subset M_n \left(\frac{\partial F}{\partial x_1}, \ldots, \frac{\partial F}{\partial x_n} \right). \tag{3.10}$$

Because

$$\frac{\partial f}{\partial x_i} = \frac{\partial F}{\partial x_i} - t \frac{\partial (g - f)}{\partial x_i} \tag{3.11}$$

it follows that

$$\frac{\partial f}{\partial x_i} \equiv \frac{\partial F}{\partial x_i} \mod \mathfrak{m}_{n+1} M_n^k$$

and therefore

$$M_n^{k+1} \subset M_n \left(\frac{\partial f}{\partial x_1}, \ldots, \frac{\partial f}{\partial x_n} \right) \subset M_n \left(\frac{\partial F}{\partial x_1}, \ldots, \frac{\partial F}{\partial x_n} \right) + \mathfrak{m}_{n+1} M_n^{k+1}. \tag{3.12}$$

Again, Nakayama's Lemma C.1 comes to our aid. We apply it taking $(\mathcal{O}_{n+1}, \mathfrak{m}_{n+1})$ as $(R, \mathfrak{m})$, M_n^k as M and $M_n \left(\frac{\partial F}{\partial x_1}, \ldots, \frac{\partial F}{\partial x_n} \right)$ as N to conclude that

$$M_n^{k+1} \subset \left(\frac{\partial F}{\partial x_1}, \ldots, \frac{\partial F}{\partial x_n} \right). \tag{3.13}$$

This completes the proof that the deformation $f + t(g - f)$ is trivial for t in some neighbourhood of 0. The remainder of the proof involves showing that the same procedure can be employed for every value of t: we want to show that for any t_0 the deformation $f + t(g - f)$ is trivial in a neighbourhood of t_0. This deformation can be written in the form $(f + t_0(g - f)) + (t - t_0)(g - f)$, and taking as new parameter $s = t - t_0$, the problem reduces to what we have already discussed, except that instead of our original f we now have a new function, $f_{t_0} := f + t_0(g - f)$. In order that our earlier argument should apply, we have to show that f_{t_0} also satisfies the hypothesis of this argument: that

$$\mathfrak{m}_n^k \subset \mathfrak{m} \, J_{f_{t_0}}. \tag{3.14}$$

Once again this is done by a simple argument involving Nakayama's Lemma: we have

$$\frac{\partial f_{t_0}}{\partial x_i} \equiv \frac{\partial f}{\partial x_i} \mod \mathfrak{m}_n^k$$

and so

$$\mathfrak{m}_n^k \subset \mathfrak{m}_n J_f \subset \mathfrak{m}_n J_{f_{t_0}} + \mathfrak{m}_n \mathfrak{m}_n^k,$$

from which (3.14) follows by Nakayama's Lemma. □

The space

$$\left\{ \left. \frac{\partial f_t}{\partial t} \right|_{t=0} : f_0 = f \right\}$$

of derivatives of one-parameter deformations of f is evidently equal to $\mathcal{O}_n$, since each $g \in \mathcal{O}_n$ gives a one-parameter deformation $f_t = f + tg$. If we insist that $f_t(0) = 0$ for all t then clearly $\left.\frac{\partial f_t}{\partial t}\right|_{t=0} \in \mathfrak{m}_n$.

In the first part of the proof of Theorem 3.1 we saw that if $f_t = f \circ \varphi_t$ is a deformation of f obtained by composing it with a one-parameter family of diffeomorphisms φ_t with $\varphi_0 = \text{id}$, then

$$\left.\frac{df_t}{dt}\right|_{t=0} = \sum_{i=1}^{n} \left.\frac{d\varphi_{i,t}}{dt}\right|_{t=0} \frac{\partial f}{\partial x_i}.$$

If $\varphi_t(0) = 0$ for all t, then we can think of φ_t as a path in the group $\mathcal{R} = \text{Diff}(\mathbb{F}^n, 0)$, and of $f \circ \varphi_t$ as a path in the $\mathcal{R}$-orbit of f, passing through f itself when $t = 0$. Thus $\left.\frac{d(f\circ\varphi_t)}{dt}\right|_{t=0}$ is (heuristically only—the $\mathcal{R}$-orbit of f in $\mathfrak{m}_n$ is not really a manifold) a tangent vector to that orbit at f.

The derivative $\left.\frac{d\varphi_t}{dt}\right|_{t=0}$ defines a vector field germ on $(\mathbb{F}^n, 0)$; to each point x it associates the tangent vector at x to the trajectory $\varphi_t(x)$. The condition that $\varphi_t(0) = 0$ for all t means that in particular $\left.\frac{d\varphi_{i,t}}{dt}\right|_{t=0} \in \mathfrak{m}_n$ for $i = 1, \ldots, n$, and thus $\left.\frac{df_t}{dt}\right|_{t=0} \in \mathfrak{m}_n J_f$. Conversely, if $\xi_1, \ldots, \xi_n \in \mathfrak{m}_n$ then the germ at 0 of

$$\varphi_t(x) = x + t(\xi_1, \ldots, \xi_n)$$

is a diffeomorphism for t near 0, with $\left.\frac{df_t}{dt}\right|_{t=0} = \sum_i \xi_i \partial f/\partial x_i$. Thus we have shown

Proposition 3.1

$$\left\{ \left.\frac{df_t}{dt}\right|_{t=0} : f_0 = f \right\} = \mathcal{O}_n$$

$$\left\{ \left.\frac{df_t}{dt}\right|_{t=0} : f_0 = f,\ f_t(0) = 0 \textit{ for all } t \right\} = \mathfrak{m}_n$$

$$\left\{ \left.\frac{d(f \circ \varphi_t)}{dt}\right|_{t=0} : \varphi_t \textit{ is a path in } \mathcal{R}, \varphi_0 = id \right\} = \mathfrak{m}_n J_f.$$

This provides the heuristic for the following definition.

Definition 3.3 Given $f \in \mathcal{O}_n$ we define the *extended $\mathcal{R}$-tangent space $T\mathcal{R}_e f$* and the *$\mathcal{R}$-tangent space $T\mathcal{R} f$*, respectively, by

$$T\mathcal{R}_e f = J_f, \quad T\mathcal{R} f = \mathfrak{m}_n J_f,$$

and we define quotients

$$T^1_{\mathscr{R}_e} f = \frac{\mathscr{O}_n}{J_f}, \quad T^1_{\mathscr{R}} f = \frac{\mathfrak{m}_n}{\mathfrak{m}_n J_f},$$

and the $\mathscr{R}_e$-*codimension* and $\mathscr{R}_e$-*codimension* of f, respectively, as

$$\operatorname{codim}_{\mathscr{R}_e} f = \dim_{\mathbb{F}} T^1_{\mathscr{R}_e} f, \quad \operatorname{codim}_{\mathscr{R}} f = \dim_{\mathbb{F}} T^1_{\mathscr{R}} f.$$

Proposition 3.2 *Let $f \in \mathscr{O}_n$. The following statements are equivalent:*

1. *$codim_{\mathscr{R}} f < \infty$,*
2. *$codim_{\mathscr{R}_e} f < \infty$,*
3. *$\mathfrak{m}_n^k \subset J_f$, for some $k \in \mathbb{N}$,*
4. *f is finitely determined for $\mathscr{R}$-equivalence.*

Proof We first claim that if $I \subset \mathscr{O}_n$ is any ideal, then $\dim_{\mathbb{F}} \mathscr{O}_n/I < \infty$ if and only if $\mathfrak{m}_n^k \subset I$, for some $k \in \mathbb{N}$. In fact, if $\mathfrak{m}_n^k \subset I$, then $\dim_{\mathbb{F}} \mathscr{O}_n/I \leq \dim_{\mathbb{F}} \mathscr{O}_n/\mathfrak{m}_n^k < \infty$, since $\mathscr{O}_n/\mathfrak{m}_n^k$ is generated over $\mathbb{F}$ by all the monomials of degree $< k$. For the converse, we consider the sequence of ideals

$$I + \mathfrak{m}_n \supset I + \mathfrak{m}_n^2 \supset \cdots \supset I.$$

If $\dim_{\mathbb{F}} \mathscr{O}_n/I < \infty$ the sequence must stop at some point, so $I + \mathfrak{m}_n^k = I + \mathfrak{m}_n^{k+1}$, for some $k \geq 1$. In particular, $\mathfrak{m}_n^k \subset I + \mathfrak{m}_n^{k+1}$ and hence $\mathfrak{m}_n^k \subset I$ by Nakayama's Lemma C.1.

Now we see the equivalences:

(2) $\Longleftrightarrow$ (3) This follows directly from the claim when $I = J_f$.

(1) $\Longleftrightarrow$ (3) If $\operatorname{codim}_{\mathscr{R}} f < \infty$, then $\dim_{\mathbb{F}} \mathscr{O}_n/\mathfrak{m}_n J_f = \dim_{\mathbb{F}} \mathfrak{m}_n/\mathfrak{m}_n J_f + 1 < \infty$ and hence, $\mathfrak{m}_n^k \subset \mathfrak{m}_n J_f \subset J_f$ for some k. Conversely, if $\mathfrak{m}_n^k \subset J_f$, then $\mathfrak{m}_n^{k+1} \subset \mathfrak{m}_n J_f$ and thus, $\dim_{\mathbb{F}} \mathfrak{m}_n/\mathfrak{m}_n J_f = \dim_{\mathbb{F}} \mathscr{O}_n/\mathfrak{m}_n J_f - 1 < \infty$.

(4) $\Longleftrightarrow$ (3) If f is finitely determined, then $\mathfrak{m}_n^{k+1} \subset \mathfrak{m}_n J_f \subset J_f$ for some k, by Theorem 3.1. Moreover, if $\mathfrak{m}_n^k \subset J_f$, then $\mathfrak{m}_n^{k+1} \subset \mathfrak{m}_n J_f$ and hence, f is $(k+1)$-determined, again by Theorem 3.1. □

The extended codimension $\operatorname{codim}_{\mathscr{R}_e} f$ is usually called the *Milnor number* of f and it is denoted by $\mu(f)$. We say that f is $\mathscr{R}$-*finite* if $\mu(f) < \infty$. The $\mathscr{R}$-finiteness has now the following geometrical interpretation:

Corollary 3.1 *Let $f \in \mathscr{O}_n$. If $0 < \mu(f) < \infty$ then f has an isolated singularity at 0. When $\mathbb{F} = \mathbb{C}$ we also have the converse, that is, if f has an isolated singularity at 0 then $0 < \mu(f) < \infty$.*

Proof If $0 < \mu(f) < \infty$ then $\mathfrak{m}_n^k \subset J_f \subset \mathfrak{m}_n$ for some k, by Proposition 3.2. The inclusion $J_f \subset \mathfrak{m}_n$ implies that 0 is a singular point of f. The other inclusion

$\mathfrak{m}_n^k \subset J_f$ gives

$$x_i^k = \sum_{j=1}^{n} a_{ij} \frac{\partial f}{\partial x_j},$$

for some $a_{ij} \in \mathcal{O}_n$. If $\frac{\partial f}{\partial x_j} = 0$ for all j, then $x_i^k = 0$ for all i and hence, $x = 0$.

Assume now that $\mathbb{F} = \mathbb{C}$. If f has an isolated singularity at 0, then $V(J_f) = \{0\}$, where $V(J_f)$ is the zero locus of the Jacobian ideal. By the Rückert Nullstellensatz (see Theorem D.4), we have $\sqrt{J_f} = \mathfrak{m}_n$, where $\sqrt{J_f}$ is the radical of J_f. But this implies $\mathfrak{m}_n^k \subset J_f \subset \mathfrak{m}_n$ for some k and hence, $0 < \mu(f) < \infty$. □

In the real case, it is not true in general that if $f : (\mathbb{R}^n, 0) \to \mathbb{R}$ has isolated singularity, then it is $\mathcal{R}$-finite, even if f is real analytic. When f is real analytic, we can consider its complexification $f_{\mathbb{C}} : (\mathbb{C}^n, 0) \to \mathbb{C}$.

Proposition 3.3 *Let $f : (\mathbb{R}^n, 0) \to \mathbb{R}$ be real analytic. Then $\mu(f) < \infty$ if and only if $\mu(f_{\mathbb{C}}) < \infty$. Moreover, if $\mu(f) < \infty$ then $T^1_{\mathcal{R}_e} f_{\mathbb{C}} \cong T^1_{\mathcal{R}_e} f \otimes_{\mathbb{R}} \mathbb{C}$ and in particular $\mu(f) = \mu(f_{\mathbb{C}})$.*

Proof In order to distinguish between the real and complex cases, we use here the notations $(\mathcal{E}_n, \mathfrak{m}_n)$ and $(\mathcal{O}_n, \mathfrak{m}'_n)$ for the local rings of C^∞-germs $(\mathbb{R}^n, 0) \to \mathbb{R}$ and complex analytic germs $(\mathbb{C}^n, 0) \to \mathbb{C}$, respectively.

If $\mu(f) < \infty$, then $\mathfrak{m}_n^k \subset J_f$ for some k. This implies that for any multi-index α, with $|\alpha| = k$,

$$x^\alpha = \sum_{i=1}^{n} a_i \frac{\partial f}{\partial x_i},$$

for some $a_i \in \mathcal{E}_n$. Replacing each a_i by its k-jet, we get a polynomial version of these equalities modulo $\mathfrak{m}_n^{k+1}$. These equalities are also valid if we take the corresponding complexifications, giving the inclusion ${\mathfrak{m}'_n}^k \subset J_{f_{\mathbb{C}}} + {\mathfrak{m}'_n}^{k+1}$. By Nakayama's Lemma C.1, ${\mathfrak{m}'_n}^k \subset J_{f_{\mathbb{C}}}$ and hence, $\mu(f_{\mathbb{C}}) < \infty$.

Conversely, if $\mu(f_{\mathbb{C}}) < \infty$, then ${\mathfrak{m}'_n}^k \subset J_{f_{\mathbb{C}}}$. As above, for each α, with $|\alpha| = k$,

$$x^\alpha = \sum_{i=1}^{n} a_i \frac{\partial f_{\mathbb{C}}}{\partial x_i},$$

for some $a_{ij} \in \mathcal{O}_n$. Since the partial derivatives have real coefficients, the same equalities hold with a_i^*, the function obtained from a_i by complex conjugating the coefficients in its Taylor expansion, in place of a_i. It follows that

$$x^\alpha = \frac{1}{2} \sum_{i=1}^{n} (a_i + a_i^*) \frac{\partial f_{\mathbb{C}}}{\partial x_i},$$

where now the functions $a_i + a_i^*$ have real coefficients. We deduce that $\mathfrak{m}_n^k \subset J_f$ and thus, $\mu(f) < \infty$.

For the second part, take k such that $\mathfrak{m}_n^k \subset J_f$ and $\mathfrak{m}_n'^k \subset J_{f_{\mathbb{C}}}$. The polynomial complexification gives an isomorphism

$$\frac{\mathscr{E}_n}{\mathfrak{m}_n^k} \otimes_{\mathbb{R}} \mathbb{C} \cong \frac{\mathscr{O}_n}{\mathfrak{m}_n'^k},$$

which maps $(J_f/\mathfrak{m}_n^k) \otimes_{\mathbb{R}} \mathbb{C}$ onto $J_{f_{\mathbb{C}}}/\mathfrak{m}_n'^k$. Therefore, $\mu(f) = \mu(f_{\mathbb{C}})$. □

Example 3.2 Let $f : (\mathbb{R}^2, 0) \to \mathbb{R}$ be the function $f(x_1, x_2) = (x_1^2 + x_2^2)^2$. The singular locus of f is given by the equation $x_1^2 + x_2^2 = 0$. This gives only the origin in $\mathbb{R}^2$, so f has isolated singularity. However, the same equation in $\mathbb{C}^2$ gives the union of two lines passing through the origin and thus, $f_{\mathbb{C}}$ has not isolated singularity. By Corollary 3.1 and Proposition 3.3, f is not $\mathscr{R}$-finite. ♢

The heuristic justification of the extended tangent space is less immediately clear than that of $T\mathscr{R}f$; it can be obtained by the same argument as $T\mathscr{R}f$ *if we remove the requirement that* $\varphi_t(0) = 0$ *for all* t. This loosening of requirements is exactly what is needed "in the field", where there is no distinguished base-point. For both versions, something precise is true, nevertheless. We have already seen, in the first part of the proof of Theorem 3.1, that modulo $\mathfrak{m}_n^{k+1}$, $\mathfrak{m}_n J_f$ is the tangent space at $j^k f$ to the $\mathscr{R}^{(k)}$-orbit of f in $J^k(n, 1)$, and moreover, if f is k-determined for $\mathscr{R}$-equivalence then by Theorem 3.1(i), the codimension of this orbit in $J^k(n, 1)$ is equal to $\operatorname{codim}_{\mathscr{R}} f$. For the second, we need a preparatory lemma.

Lemma 3.1 *Suppose* $f \in \mathscr{O}_n$ *has an isolated singularity at* 0, *then*

(i) $\dim_{\mathbb{F}} J_f / \mathfrak{m}_n J_f = n$, *and*
(ii) $\mathit{codim}_{\mathscr{R}} f = \mathit{codim}_{\mathscr{R}_e} f + n - 1$.

Proof

(i) As every function $c \in \mathscr{O}_n$ can be written in the form $c = c(0) + \tilde{c}$, where $\tilde{c} \in \mathfrak{m}_n$, it follows that

$$\frac{J_f}{\mathfrak{m}_n J_f} = \mathrm{Sp}_{\mathbb{F}}\{\overline{\frac{\partial f}{\partial x_1}}, \ldots, \overline{\frac{\partial f}{\partial x_n}}\},$$

where $\overline{\partial f/\partial x_i}$ means the class of $\partial f/\partial x_i$ in the quotient. If the dimension of the quotient is less than n, the $\overline{\partial f/\partial x_i}$ are not linearly independent, so that there is some non-trivial relation

$$\sum_i c_i \overline{\frac{\partial f}{\partial x_i}} = 0,$$

or in other words

$$\sum_i c_i \frac{\partial f}{\partial x_i} = \sum_i \alpha_i \frac{\partial f}{\partial x_i}$$

for some functions $\alpha_1, \ldots, \alpha_n \in \mathfrak{m}_n$, where $c_1, \ldots, c_n \in \mathbb{F}$ are not all zero. Subtracting the right-hand side from the left, we get

$$\sum_i (c_i - \alpha_i) \frac{\partial f}{\partial x_i} = 0. \tag{3.15}$$

The vector field

$$\vartheta := \sum_i (c_i - \alpha_i) \frac{\partial}{\partial x_i}$$

has a smooth 1-dimensional orbit though 0, since not all of its coefficients vanish at 0. Since $df(\vartheta) = 0$, f is invariant under the integral flow of ϑ. This means that at all points of the orbit through 0, f has a critical point, since it has a critical point at 0. This contradicts the assumption that f has isolated singularity at 0. Thus, the $\overline{\partial f/\partial x_i}$ must be linearly independent in $J_f / \mathfrak{m}_n J_f$, which therefore has dimension n.

(ii) The exactness of

$$0 \to \frac{J_f}{\mathfrak{m}_n J_f} \to \frac{\mathfrak{m}_n}{\mathfrak{m}_n J_f} \to \frac{\mathfrak{m}_n}{J_f} \to 0 \quad \text{and} \quad 0 \to \frac{\mathfrak{m}_n}{J_f} \to \frac{\mathcal{O}_n}{J_f} \to \frac{\mathcal{O}_n}{\mathfrak{m}_n} \to 0$$

now shows that

$$\dim_{\mathbb{F}} \frac{\mathfrak{m}_n}{\mathfrak{m}_n J_f} = n + \dim_{\mathbb{F}} \frac{\mathfrak{m}_n}{J_f} \quad \text{and} \quad \dim_{\mathbb{F}} \frac{\mathcal{O}_n}{J_f} = \dim_{\mathbb{F}} \frac{\mathfrak{m}_n}{J_f} + 1,$$

and the conclusion follows. □

Remark 3.1 A theorem from commutative algebra proves a slightly stronger result in the complex analytic case. If the zero locus of the n partials $\partial f/\partial x_1, \ldots, \partial f/\partial x_n$ has codimension n, then these partials form a *regular sequence* (see Definition C.5). A *relation* (or *syzygy*) between the n partials in $\mathcal{O}_n$ is an n-tuple $(b_1, \ldots, b_n)$ such that $\sum_i b_i \partial f/\partial x_i = 0$. The fact that the $\partial f/\partial x_i$ form a regular sequence is equivalent to the property that the $\mathcal{O}_n$-module of relations between them is generated by the so-called trivial relations

$$(0, \ldots, 0, -\frac{\partial f}{\partial x_j}, 0, \ldots, 0, \frac{\partial f}{\partial x_i}, 0, \ldots, 0)$$

(where $-\partial f/\partial x_j$ is in the i'th place and $\partial f/\partial x_i$ is in the j'th place). In particular, all the b_j must lie in the ideal J_f, and thus in $\mathfrak{m}_n$, so that a relation like (3.15), where not all of the coefficients $c_i - \alpha_i$ are in $\mathfrak{m}_n$, is impossible.

Now we justify the label "$\operatorname{codim}_{\mathscr{R}} f$" for $\dim_{\mathbb{F}} \mathscr{O}_n / J_f$. Suppose that

$$F : X \times U \to \mathbb{F}, (x, u) \mapsto f_u(x)$$

is a family of functions, where X and U are open in $\mathbb{F}^n$ and $\mathbb{F}^d$, respectively. Let $\mathscr{R}_e^{(k)} f \subset J^k(X, \mathbb{F})$ be the set of k-jets of germs $g\colon (X, x_0) \to (\mathbb{F}, 0)$ which are right-equivalent to f in the sense that there is a germ of diffeomorphism $\varphi\colon (X, x_0) \to (\mathbb{F}^n, 0)$ such that $g = f \circ \varphi$. Similarly, let $\overline{\mathscr{R}_e}^{(k)} f$ be the set of jets with the same property except that now $g = f \circ \varphi + c$ for some constant c. Each fibre of the projection $J^k(X, \mathbb{F}) \to X \times \mathbb{F}$ is diffeomorphic to $J^k(n, 1)$ and contains a copy of $\mathscr{R}^{(k)} j^k f$, and $\overline{\mathscr{R}_e}^{(k)} f$ is the disjoint union of these copies. Thus

$$\begin{aligned} \text{codimension of } \overline{\mathscr{R}_e}^{(k)} f \text{ in } J^k(X, \mathbb{F}) &= \text{codimension of } \mathscr{R}^{(k)} f \text{ in } J^k(n, 1) \\ \text{codimension of } \mathscr{R}_e^{(k)} f \text{ in } J^k(X, \mathbb{F}) &= \text{codimension of } \mathscr{R}^{(k)} f \text{ in } J^k(n, 1) + 1. \end{aligned} \tag{3.16}$$

If f is k-determined for $\mathscr{R}$-equivalence then these codimensions are, respectively, $\operatorname{codim}_{\mathscr{R}} f$ and $\operatorname{codim}_{\mathscr{R}} f + 1$, by Theorem 3.1(i).

Let $j^k F/U : X \times U \to J^k(X, \mathbb{F})$ be the relative jet extension map, defined by $j^k F/U(x, u) = j^k f_u(x)$. Let $\pi : X \times U \to U$ be the projection.

Proposition 3.4 *Suppose that f is k-determined for $\mathscr{R}$-equivalence. Then*

(i) if $j^k F/U \pitchfork \mathscr{R}_e^{(k)} f$ and $(j^k F/U)^{-1}(\mathscr{R}_e^{(k)} f) \neq \emptyset$, then the codimension in U of $\pi((j^k F/U)^{-1}(\mathscr{R}_e^{(k)} f))$ is equal to $\operatorname{codim}_{\mathscr{R}_e} f$, *and*

(ii) if $j^k F/U \pitchfork \overline{\mathscr{R}_e}^{(k)} f$ and $(j^k F/U)^{-1}(\overline{\mathscr{R}_e}^{(k)} f) \neq \emptyset$, the codimension in U of $\pi((j^k F/U)^{-1}(\overline{\mathscr{R}_e}^{(k)} f)$ is equal to $\operatorname{codim}_{\mathscr{R}_e} f - 1$.

Thus, in a "generic family" f_u, the set of parameter values u such that f_u has a singularity right-equivalent to f has codimension $\operatorname{codim}_{\mathscr{R}_e}(f)$, *and the set of parameter values such that f_u has a singularity right-equivalent to $f + c$, for some constant c, has codimension* $\operatorname{codim}_{\mathscr{R}_e} f - 1$.

Proof As f has an isolated singularity, so does any germ equivalent to f. Thus in particular the projections to U of $(j^k F/U)^{-1}(\mathscr{R}_e^{(k)} f)$ and $(j^k F/U)^{-1}(\overline{\mathscr{R}_e}^{(k)} f)$ are both finite-to-one. It follows that projection to U does not reduce the dimension of either set, and thus reduces the codimension of each by $\dim X = n$. By the assumption of transversality, and by (3.16), their codimensions in $X \times U$ are $\operatorname{codim}_{\mathscr{R}} f + 1$ and $\operatorname{codim}_{\mathscr{R}} f$, respectively. The conclusion follows by Lemma 3.1. □

Exercises for Sect. 3.1

1 Show

(i) If $f(x_1, \ldots, x_n) = x_1^2 + \cdots + x_n^2$ then J_f is the maximal ideal $\mathfrak{m}_n$ and $\dim_{\mathbb{F}} \mathscr{O}_n / J_f = 1$.
(ii) If $f(x_1, \ldots, x_n) = x_1^2 + \cdots + x_n^2 +$ higher order terms, then $J_f = \mathfrak{m}_n$ (use Nakayama's Lemma C.1).
(iii) If $\dim_{\mathbb{F}} \mathscr{O}_n / J_f = 1$ then $J_f = \mathfrak{m}_n$.
(iv) $J_f = \mathfrak{m}_n$ if and only if the Hessian determinant $\det(\partial^2 f/\partial x_i \partial x_j)_{1 \le i,j \le n}$ does not vanish at 0. To prove this, use Nakayama's Lemma and the fact that modulo $\mathfrak{m}_n^2$,

$$\begin{pmatrix} \partial f/\partial x_1 \\ \vdots \\ \partial f/\partial x_n \end{pmatrix} = \begin{pmatrix} \partial^2 f/\partial x_1^2 & \cdots & \partial^2 f/\partial x_n \partial x_1 \\ \vdots & \vdots & \vdots \\ \partial^2 f/\partial x_n \partial x_1 & \cdots & \partial^2 f/\partial x_n^2 \end{pmatrix} \begin{pmatrix} x_1 \\ \vdots \\ x_n \end{pmatrix}.$$

2 Show

(i) If $f(x_1, x_2) = x_1^2 + x_2^{k+1}$ then $J_f = (x_1, x_2^k)$ and $\dim_{\mathbb{F}} \mathscr{O}_n / J_f = k$.
(ii) If $f(x_1, x_2) = x_1^2 x_2 + x_2^{k-1}$ then $J_f = (x_1 x_2, x_1^2 + (k-1)x_2^{k-2})$ and $\dim_{\mathbb{F}} \mathscr{O}_n / J_f = k$.
(iii) If $f(x_1, x_2) = x_1^2 x_2$ then $J_f = (x_1 x_2, x_1^2)$ and $\dim_{\mathbb{F}} \mathscr{O}_n / J_f = \infty$.

3 Find the lowest value of k for which (3.1) holds for

(i) $f_1(x_1, x_2) = x_1^2 + x_2^{k+1}$
(ii) $f_2(x_1, x_2) = x_1^2 x_2 + x_2^{k-1}$
(iii) $f_3(x_1, x_2) = x_1^k + x_2^\ell$.

4 Show by an example that it is *not* true that if $\mathfrak{m}_n J_f \supset \mathfrak{m}_n^{k+1}$ then f is k-determined for $\mathscr{R}$. You do not have to look far.

5 At what point does the proof of part (ii) of Theorem 3.1 fail if we assume only the weaker hypothesis that $\mathfrak{m}_n J_f \supset \mathfrak{m}_n^{k+1}$?

3.2 Left-Right Equivalence and Stability

In this book we are interested in $\mathscr{A}$-equivalence more than $\mathscr{R}$-equivalence. But Theorem 3.1 is a good indication of what is true and how, in principle, one goes about proving it. In this section we define a tangent space for $\mathscr{A}$-equivalence, which will play a rôle in the theory similar to that of $T\mathscr{R}f = \mathfrak{m}_n J_f$ for $\mathscr{R}$-equivalence. In Chap. 6 we then develop criteria for finite determinacy in terms of it. For $\mathscr{A}$ the process is more complicated than for $\mathscr{R}$, since one has simultaneously to produce families of diffeomorphisms of source and target. But the overall strategy is the same.

Mather and Thom, in their work in the 1960s on smooth maps, thought in global terms: a C^∞ map of manifolds $f : N \to P$ is *stable* if its orbit under the natural action of Diff(N) $\times$ Diff(P) is open in $C^\infty(N, P)$, with respect to a suitable topology. Here, we are interested in local geometry, and so we give a local version of this definition: a map-germ $f : (\mathbb{F}^n, S) \to (\mathbb{F}^p, 0)$ is stable if every deformation is trivial: roughly speaking, if f_t is a deformation of f then there should exist deformations of the identity maps of $(\mathbb{F}^n, S)$ and $(\mathbb{F}^p, 0)$, φ_t and ψ_t, such that

$$f_t = \psi_t \circ f \circ \varphi_t^{-1}. \tag{3.17}$$

A substantial part of Mather's six papers on the stability of C^∞ mappings [Mat68a, Mat68b, Mat69a, Mat69b, Mat70, Mat71] is devoted to showing that if all the germs of a proper mapping f are stable in this local sense then f is stable in the global sense. We will discuss global stability later in Chap. 5, but do not prove the statement of Mather just referred to.

Definition 3.4

(1) A *d-parameter unfolding* of f is a map-germ

$$F : (\mathbb{F}^n \times \mathbb{F}^d, S \times \{0\}) \to (\mathbb{F}^p \times \mathbb{F}^d, 0)$$

of the form

$$F(x, u) = (\tilde{f}(x, u), u)$$

such that $\tilde{f}(x, 0) = f(x)$. If we denote the map $x \mapsto \tilde{f}(x, u)$ by f_u, then the above condition becomes simply $f_0 = f$.

Retaining the parameters u in the second component of the map makes the following definition easier to write down:

(2) Two unfoldings F, G of f are *equivalent* if there exist germs of diffeomorphisms

$$\Phi : (\mathbb{F}^n \times \mathbb{F}^d, S \times \{0\}) \to (\mathbb{F}^n \times \mathbb{F}^d, S \times \{0\})$$

and

$$\Psi : (\mathbb{F}^p \times \mathbb{F}^d, 0) \to (\mathbb{F}^p \times \mathbb{F}^d, 0),$$

which are themselves unfoldings of the identity in $\mathbb{F}^n$ and $\mathbb{F}^p$, respectively, such that

$$G = \Psi \circ F \circ \Phi^{-1}. \tag{3.18}$$

(3) The unfolding F is *trivial* if it is equivalent to $f \times \mathrm{id}$ (the "constant" unfolding $(x, u) \mapsto (f(x), u)$).
(4) The map-germ $f : (\mathbb{F}^n, S) \to (\mathbb{F}^p, 0)$ is *stable* if every unfolding of f is trivial.

By writing $\Phi(x, u) = (\varphi_u(x), x)$ and $\Psi(y, u) = (\psi_u(y), u)$, from (3.18) we recover the heuristic definition (3.17) in the particular case $d = 1$. We do not insist that the mappings φ_u and ψ_u preserve the base-points of the germ f. After all, if the interesting behaviour merely changes its location, we should not regard the unfolding as non-trivial.

Example 3.3

(i) Consider the map-germ $f : (\mathbb{F}, 0) \to (\mathbb{F}, 0)$, $f(x) = x^2$, and its unfolding $F(x, u) = (x^2 + ux, u)$. This is trivialised by the families of diffeomorphisms $\Phi(x, u) = (x + u/2, u)$, $\Psi(y, u) = (y - u^2/4, u)$. Both Φ and Ψ are just families of translations.
(ii) The unfolding $F(x, u) = (x^3 + ux, u)$ of $f(x) = x^3$ is not trivial. Two arguments are available to prove this. The first is that if we write $f_u(x) = x^3 + ux$, then for $u \neq 0$, f_u has two distinct critical points, whereas f has only one. If F were equivalent to $f \times 1$ then this could not happen. We clarify this: if $\Psi \circ F = (f \times 1) \circ \Phi$, where Φ and Ψ are diffeomorphisms unfolding the identity, then Φ restricts to a diffeomorphism from the critical set Σ_F of F to the critical set $\Sigma_{f \times 1}$ of $f \times 1$, and moreover since Φ is an unfolding, the diagram

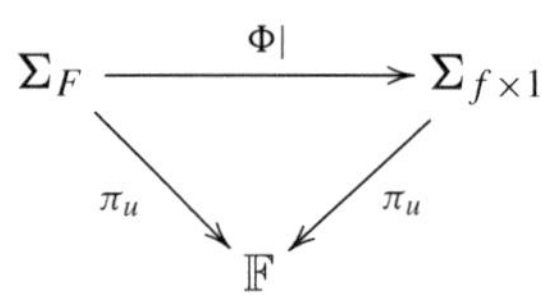

commutes, where π_u is projection to the parameter space. The projection on the left is two-to-one on the complement of 0, whereas the projection on the right is one-to-one.

The second argument involves determining precisely what is the space

$$\left\{ \left.\frac{df_t}{dt}\right|_{t=0} : F(x, t) = (f_t(x), t) \text{ is a trivial unfolding of } f \right\}. \tag{3.19}$$

and showing that the derivative we obtain from the unfolding F above, x, does not belong to it. We cannot yet make this calculation, but turn to it shortly. ◊

The space (3.19) can also be described as

$$\left\{ \left.\frac{df_t}{dt}\right|_{t=0} : f_t = \psi_t \circ f \circ \varphi_t^{-1}, \varphi_0 = \mathrm{id},\ \psi_0 = \mathrm{id} \right\}. \tag{3.20}$$

The families of diffeomorphisms ψ_t and φ_t here do not necessarily preserve the base-points of the germ f. The germs Φ and Ψ are unfoldings of the identity, so that $\varphi_0(S) = S$ and $\psi_0(0) = 0$ but are not required to satisfy $\varphi_t(S) = S$ or $\psi_t(0) = 0$ for $t \neq 0$.

Determining an algebraic formula for the space (3.19) will also provide us with a simple and computable criterion for stability.

We consider (3.20) as a subspace of the space of all *infinitesimal deformations* of f, which we define as

$$ID(f) := \left\{ \left. \frac{df_t}{dt} \right|_{t=0} : F(x,t) = (f_t(x), t) \text{ is any unfolding of } f \right\}. \tag{3.21}$$

As we will see later, $ID(f)$ has a richer algebraic structure, but for the moment, we only consider in $ID(f)$ the structure of vector space over $\mathbb{F}$. A special case of infinitesimal deformation is when we require the unfolding F to preserve the base-points of the germ f, that is, when $f_t(S) = \{0\}$, for all t. We will denote by $ID_0(f)$ the subspace of $ID(f)$ of infinitesimal deformations which preserve the base-points.

In the following definition, $\mathbb{F}$ is either $\mathbb{R}$ or $\mathbb{C}$. When $\mathbb{F} = \mathbb{R}$ then f and the deformations f_t are assumed to be C^∞, and when $\mathbb{F} = \mathbb{C}$ then f and $\mathbb{F} = \mathbb{C}$ then f and the f_t are assumed to be complex analytic.

Definition 3.5 Given a map-germ $f\colon (\mathbb{F}^n, S) \to (\mathbb{F}^p, 0)$, we define

$$T\mathscr{A}_e f = \left\{ \left. \frac{d}{dt}(\psi_t \circ f \circ \varphi_t^{-1}) \right|_{t=0} : \varphi_0 = \mathrm{id},\ \psi_0 = \mathrm{id} \right\},$$

$$T\mathscr{A} f = \left\{ \left. \frac{d}{dt}(\psi_t \circ f \circ \varphi_t^{-1}) \right|_{t=0} : \varphi_0 = \mathrm{id},\ \psi_0 = \mathrm{id},\ \varphi_t(S) = S,\ \psi_t(0) = 0,\ \forall t \right\},$$

and the quotients

$$T^1_{\mathscr{A}_e} f = \frac{ID(f)}{T\mathscr{A}_e f}, \quad T^1_{\mathscr{A}} f = \frac{ID_0(f)}{T\mathscr{A} f}.$$

The space $T\mathscr{A}_e f$ plays a central rôle in the subject, and is called the *extended tangent space of f*. Its subspace $T\mathscr{A} f$ is the (heuristic) *tangent space to the $\mathscr{A}$-orbit of f*. For just as in the discussion preceding Definition 3.3, if both φ_t and ψ_t fix the base-points, then the families φ_t and ψ_t can be regarded as curves in the groups $\mathscr{R}$ and $\mathscr{L}$, so that $\psi_t \circ f \circ \varphi_t$ is a curve in the $\mathscr{A}$-orbit of f.

We use the term"$T^1_{\mathscr{A}_e} f$" here because, as we shall see in Chap. 5 and in particular in Theorem 5.1 , it is the tangent space at 0 to the base space of a miniversal (i.e. semi-universal) unfolding of f; the term "T^1" is the standard notation for the corresponding tangent space, in the deformation theory of analytic spaces. The upper index 1 is slightly misleading because it suggests that we might later meet a T^2 or a T^3. The deformation theory of map-germs that we study here, developed by Thom, Mather, Martinet and others, is, by construction, unobstructed—we place no requirement that any of the spaces involved should be flat over the base of the deformation—and there is no need for these higher cotangent functors.

Definition 3.6 (Continued) We define the *$\mathscr{A}_e$-codimension* and the *$\mathscr{A}$-codimension* of f, respectively, as

$$\operatorname{codim}_{\mathscr{A}_e}(f) = \dim_{\mathbb{F}} T^1_{\mathscr{A}_e} f, \quad \operatorname{codim}_{\mathscr{A}}(f) = \dim_{\mathbb{F}} T^1_{\mathscr{A}} f.$$

Definition 3.7 A germ $f: (\mathbb{F}^n, S) \to (\mathbb{F}^p, 0)$ is *$\mathscr{A}$-finite* if $\dim_{\mathbb{F}} T^1_{\mathscr{A}_e} f < \infty$.

Proposition 3.5 *If $f: (\mathbb{F}^n, S) \to (\mathbb{F}^p, 0)$ is stable, then $T^1_{\mathscr{A}_e} f = 0$.*

Proof Let $F(x, t) = (f_t(x), t)$ be a 1-parameter unfolding of f. By hypothesis it is trivial, so $F = \Psi \circ (f \times 1) \circ \Phi$ for some Φ and Ψ, unfoldings of the identity. Writing $\Phi(x, t) = (\varphi_t(x), t)$ and $\Psi(y, t) = (\psi_t(y), t)$, we have $f_t = \psi_t \circ f \circ \varphi_t$. □

If $T^1_{\mathscr{A}_e} f = 0$ we say that f is *infinitesimally stable*. Mather [Mat69a] proved the converse of Proposition 3.5, giving

Theorem 3.2 Infinitesimal stability is equivalent to stability: *f is stable if and only if $T^1_{\mathscr{A}_e} f = 0$.*

We will prove this later in the chapter. For an unstable germ f, the $\mathscr{A}_e$-codimension of f is an important invariant, measuring the failure of stability of f. In this chapter we develop techniques for calculating $T\mathscr{A}_e f$, and apply them in some examples.

Lemma 3.2 *If φ_t and ψ_t are parameterised families of diffeomorphisms, we have*

$$\frac{d}{dt}\left(\psi_t \circ f \circ \varphi_t^{-1}\right)\bigg|_{t=0} = df \circ \left(\frac{d\varphi_t^{-1}}{dt}\bigg|_{t=0}\right) + \left(\frac{d\psi_t}{dt}\bigg|_{t=0}\right) \circ f.$$

Proof Although this statement uses nothing but the chain rule, there is some subtlety in its application, so we give the proof. Define Φ and $\tilde{\phi}$, Ψ and $\tilde{\psi}$, and G and $\tilde{g}$, by the equalities $\Phi(x, t) = (\tilde{\varphi}(x, t), t) = (\varphi_t(x), t)$, $\Psi(y, t) = (\tilde{\psi}(y, t), t) = (\psi_t(y), t)$, and $G(x, t) = (\tilde{g}(x, t), t) = (f(\varphi_t^{-1}(x)), t)$. Pick representatives of Φ, f and Ψ. Let $(x, 0)$ be a point in the domain of $\Psi \circ (f \times 1) \circ \Phi^{-1}$. Then

$$\frac{d}{dt}\left(\psi_t \circ f \circ \varphi_t^{-1}\right)\bigg|_{t=0}(x) = \frac{\partial}{\partial t}(\tilde{\psi} \circ G)(x, 0).$$

By the chain rule this is equal to

$$\sum_{i=1}^{p} \frac{\partial \tilde{\psi}}{\partial y_i}(G(x, 0))\frac{\partial \tilde{g}_i}{\partial t}(x, 0) + \frac{\partial \tilde{\psi}}{\partial t}(G(x, 0)).$$

Because $\tilde{\psi}(y, 0) = y$, and because $\tilde{g}(x, 0) = f(x)$, this sum is just

$$\frac{\partial \tilde{g}}{\partial t}(x, 0) + \left.\frac{d\psi_t}{dt}\right|_{t=0} (f(x)).$$

Again by the chain rule, and because $\varphi_0(x) = x$, the first summand here is

$$\sum_{i=1}^{n} \frac{\partial f}{\partial x_i}(x) \left.\frac{d(\varphi_t^{-1})_i}{dt}\right|_{t=0} (x),$$

and this is equal to $d_x f\left(\left.\frac{d\varphi_t^{-1}}{dt}\right|_{t=0}\right)$. □

As discussed before Proposition 3.1, both $(d\varphi_t/dt)\,|_{t=0}$ and $(d\psi_t/dt)|_{t=0}$ determine germs of vector fields, on $(\mathbb{F}^n, S)$ and $(\mathbb{F}^p, 0)$, respectively: $(d\varphi_t(x)/dt)|_{t=0}$ is the tangent vector at x to the trajectory $\varphi_t(x)$. The set of all germs of vector fields on $(\mathbb{F}^n, S)$ is denoted by $\theta_n := \theta_{\mathbb{F}^n,S}$. It is a module over $\mathscr{O}_n := \mathscr{O}_{\mathbb{F}^n,S}$, the ring of smooth functions $(\mathbb{F}^n, S) \to \mathbb{F}$. The subset of those $(d\varphi_t(x)/dt)|_{t=0}$ which preserve the base-points is equal to the submodule $\mathfrak{m}_n\theta_n$, where $\mathfrak{m}_n := \mathfrak{m}_{\mathbb{F}^n,S}$ is the ideal of $\mathscr{O}_n$ of functions vanishing on S.

In the same way, the infinitesimal deformations of f should be thought of as vector fields along f (*cf.* Definition 2.13); $(df_t/dt)\,|_{t=0}$ is the tangent vector at $f(x)$ to the trajectory $x \mapsto f_t(x)$. By associating to $(df_t/dt)|_{t=0}$ the map

$$\hat{f} : x \mapsto \left(f(x), \left.\frac{df_t(x)}{dt}\right|_{t=0}\right) \in T\mathbb{F}^p,$$

we obtain a commutative diagram:

$$\begin{array}{ccc} T\mathbb{F}^n & \xrightarrow{df} & T\mathbb{F}^p \\ \downarrow & \overset{\hat{f}}{\nearrow} & \downarrow \\ \mathbb{F}^n & \xrightarrow{f} & \mathbb{F}^p \end{array} \tag{3.22}$$

in which the vertical maps are the bundle projections.

Conversely, any vector field ξ along f is a smooth map $\xi : (\mathbb{F}^n, S) \to T\mathbb{F}^p$ of the form $\xi(x) = (f(x), \tilde{\xi}(x))$, for some $\tilde{\xi} : (\mathbb{F}^n, S) \to \mathbb{F}^p$. It follows that $\tilde{\xi}$ is the infinitesimal deformation $(df_t/dt)|_{t=0}$ associated to the deformation $f_t = f + t\tilde{\xi}$. Thus, we have a canonical identification $ID(f) \equiv \theta(f)$, the space of vector fields along f. The space $\theta(f)$ is also a module over $\mathscr{O}_n$ and the infinitesimal deformations which preserve the base-points correspond to the submodule $\mathfrak{m}_n\theta(f)$.

From now on, we will refer to vector fields along f rather than infinitesimal deformations of f, and abandon the notation $ID(f)$ in favour of $\theta(f)$.

As an $\mathcal{O}_n$-module, θ_n is freely generated by the germs of the coordinate vector fields $\partial/\partial x_1, \ldots, \partial/\partial x_n$. Thus an element ξ of θ_n can be written in various ways: as a sum

$$\xi(x) = \sum_{j=1}^{n} \xi_j(x)\partial/\partial x_j,$$

where each ξ_j is a germ of function at S, or, suppressing mention of the coordinate system, simply as an n-tuple,

$$\xi = (\xi_1(x), \ldots, \xi_n(x))$$

(sometimes as a column rather than a row). In calculations below we will usually write the elements of $\theta(f)$ and of θ_n as columns. We discuss the modifications to the notation needed in dealing with multi-germs in Sect. 3.4 below.

Important Piece of Notation We denote by

$$tf : \theta_n \to \theta(f)$$

the map $\xi \mapsto df \circ \xi$ and by

$$\omega f : \theta_p \to \theta(f)$$

the map $\eta \mapsto \eta \circ f$.

The notation "tf" is slightly fussy. We use it instead of df here because we think of df as the bundle map between tangent bundles induced by f, as in the diagram (3.22), whereas tf is the map "left composition with df" from θ_n to $\theta(f)$. Some authors use "df" for both.

Now using this notation, from Lemma 3.2 we deduce

Corollary 3.2 *For any map-germ* $f\colon (\mathbb{F}^n, S) \to (\mathbb{F}^p, 0)$, *we have*

$$T\mathscr{A}_e f = tf(\theta_n) + \omega f(\theta_p), \quad T\mathscr{A} f = tf(\mathfrak{m}_n\,\theta_n) + \omega f(\mathfrak{m}_p\,\theta_p)$$

and also

$$T^1_{\mathscr{A}_e} f = \frac{\theta(f)}{T\mathscr{A}_e f}, \quad T^1_{\mathscr{A}} f = \frac{\mathfrak{m}_n\theta(f)}{T\mathscr{A} f}.$$

Remark 3.2 (Real or Complex?) The singularity theory of real maps is usually concerned with smooth maps and map-germs, whereas complex singularity theory is concerned with complex analytic germs. Our notation, and indeed our definitions, do not distinguish between the two cases, but at first sight they are rather different. Suppose that $f\colon (\mathbb{F}^n, 0) \to (\mathbb{F}^p, 0)$ is a smooth (real) germ or a complex analytic germ. In the real smooth case, $\theta(f)$ is a free module of rank p over the ring $\mathscr{E}_n$ of germs of smooth functions, whereas in the complex analytic case, $\theta(f)$ is a free module of rank p over the ring $\mathscr{O}_n$ of germs of complex analytic functions. There is no simple relation between $\mathscr{O}_n$ and $\mathscr{E}_n$. In particular, $\mathscr{E}_n$ contains non-zero germs whose Taylor series about 0 is identically zero, which have no counterpart in $\mathscr{O}_n$. Equally, for every formal power series, convergent or not, there is a smooth germ with this as its Taylor series [Brö75, 4.9]. Thus there is no reason to expect a straightforward relation between the $\mathscr{R}$ or $\mathscr{A}$ tangent spaces and T^1's of a smooth germ $(\mathbb{R}^n, 0) \to (\mathbb{R}^p, 0)$ defined by real power series, and the corresponding tangent spaces of its complexification $f_{\mathbb{C}}\colon (\mathbb{C}^n, 0) \to (\mathbb{C}^p, 0)$ defined by the same power series, now with complex variables. Nevertheless, for $\mathscr{A}$-finite germs the relation is transparent: Proposition 3.8 below shows that a real analytic germ is $\mathscr{A}$-finite in the C^∞ category if and only if its complexification is $\mathscr{A}$-finite in the complex analytic category, and in this case, $T^1_{\mathscr{A}_e} f_{\mathbb{C}} = T^1_{\mathscr{A}_e} f \otimes_{\mathbb{R}} \mathbb{C}$—so a basis for the T^1 of a real analytic germ is also a basis for the T^1 of its complexification. In particular, when we discuss $\mathscr{A}$-finite germs, it is not necessary to distinguish in our notation between the real smooth and the complex analytic cases. $\diamondsuit$

Remark 3.3 Since $tf : \theta_n \to \theta(f)$ is $\mathscr{O}_n$-linear, $\theta(f)/tf(\theta_n)$ is naturally an $\mathscr{O}_n$-module. By means of the ring homomorphism $f^* : \mathscr{O}_p \to \mathscr{O}_n$ (right composition with f), every $\mathscr{O}_n$-module becomes an $\mathscr{O}_p$-module, and every $\mathscr{O}_n$-linear map becomes $\mathscr{O}_p$-linear. Thus $\theta(f)$ and $\theta(f)/tf(\theta_n)$ become $\mathscr{O}_p$-modules. As $\theta(f)$ is a finitely generated $\mathscr{O}_n$-module, it is finitely generated over $\mathscr{O}_p$ when $\mathscr{O}_n$ is. By the Preparation Theorem D.2, this holds if and only if $\mathscr{O}_n / f^*(\mathfrak{m}_p)\, \mathscr{O}_n$ is a finite-dimensional $\mathbb{F}$-vector space. In the complex case, by the Nullstellensatz, this holds if and only if the germ of $f^{-1}(0)$ consists just of S. This cannot happen when $n > p$. However, even when $n > p$ then provided f is $\mathscr{A}$-finite, $\theta(f)/tf(\theta_n)$ *is* finitely generated over $\mathscr{O}_p$ (Exercise 7 below). The property of being finitely generated is important in many of the algebraic arguments in singularity theory; for example, it is required in Nakayama's Lemma (Proposition C.1 in the Appendix on Commutative Algebra). $\diamondsuit$

Proposition 3.6 *Suppose* $f\colon (\mathbb{F}^n, S) \to (\mathbb{F}^p, 0)$, *with* $S = \{s_1, \ldots, s_r\} \subset C(f)$. *Then:*

(i) $\operatorname{codim}_{\mathscr{A}_e} f < \infty$ *if and only if* $\operatorname{codim}_{\mathscr{A}} f < \infty$.
(ii) If $0 < \operatorname{codim}_{\mathscr{A}_e} f < \infty$ *then* $\operatorname{codim}_{\mathscr{A}} f = \operatorname{codim}_{\mathscr{A}_e} f + p - r(n-p)$.

Proof Denote the coordinates near the i'th base-point s_i by $x_1^{(i)}, \ldots, x_n^{(i)}$, for $i = 1, \ldots, r$. The quotient $T\mathscr{A}_e f / T\mathscr{A} f$ is generated as vector space over $\mathbb{F}$ by $rn + p$ elements:

$$tf\left(\frac{\partial}{\partial x_k^{(i)}}\right), i = 1, \ldots, r,\ k = 1, \ldots, n, \qquad \text{and} \qquad \omega f\left(\frac{\partial}{\partial y_j}\right), j = 1, \ldots, p.$$

If they are not linearly independent, then there are constants a_{ik} and b_j, not all zero, such that

$$tf\left(\sum_{ik} a_{ik}\frac{\partial}{\partial x_k^{(i)}}\right) + \omega f\left(\sum_j b_j \frac{\partial}{\partial y_j}\right) \in T\mathscr{A} f,$$

and hence vector fields $\xi_0 \in \mathfrak{m}_n\,\theta_n$ and $\eta_0 \in \mathfrak{m}_p\,\theta_p$ such that

$$tf\left(\sum_{ik} a_{ik}\frac{\partial}{\partial x_k^{(i)}}\right) + \omega f\left(\sum_j b_j \frac{\partial}{\partial y_j}\right) = tf(\xi_0) + \omega f(\eta_0).$$

Rearranging, we get

$$tf\left(\sum_{ik} a_{ik}\frac{\partial}{\partial x_k^{(i)}} - \xi_0\right) = \omega f\left(\eta_0 - \sum_j b_j \frac{\partial}{\partial y_j}\right). \tag{3.23}$$

Write $\xi := \sum_{ik} a_{ik}\partial/\partial x_k^{(i)} - \xi_0$ and $\eta := \eta_0 - \sum_j b_j\partial/\partial y_j$. Integrating the vector fields ξ and η we get flows φ_t in the source and η_t in the target such that $f \circ \varphi_t = \psi_t \circ f$ (here we are using the Thom–Levine Lemma, 2.6). It follows that for each value of t, the germ $f\colon (\mathbb{F}^n, \varphi_t(S)) \to (\mathbb{F}^p, \psi_t(0))$ is left-right equivalent to the germ of f at S. Because not all of the a_{ik} and b_j are equal to 0, either ξ is not zero at some point of S, or η is not zero at 0. If the former, say $\xi(s_i) \neq 0$, so that the trajectory $\varphi_t(s_i)$ is a smooth curve germ, which lies in the germ of the critical set of f at s_i. The germ of f *at* s_i is $\mathscr{A}$-finite (Exercise 3.4.3.4 below) and therefore finite-to-one on its critical set. It follows that we cannot have $f(\varphi_t(s_i)) \equiv 0$ for all t. Hence *the trajectory of* η *through* 0 *is a curve and not just a point* (and thus in fact $\eta(0) \neq 0$). But this is impossible: because $0 < \operatorname{codim}_{\mathscr{A}_e} f < \infty$, the germ at $0 \in \mathbb{F}^p$ of the set of points

$$\{y \in (\mathbb{F}^p, 0) : \text{the germ of} f \text{ at } (\mathbb{F}^n, (f^{-1}(y) \cap C(f)) \text{ is not stable }\}$$

consists only of 0, by the Mather–Gaffney criterion for $\mathscr{A}$-finiteness, Theorem 4.5. This contradiction shows that all of the a_{ik} and b_j must be zero, so that $\dim_{\mathbb{F}} T\mathscr{A}_e f/T\mathscr{A} f = rn + p$. As in the proof of Lemma 3.1(ii), the conclusion follows from the exactness of the sequences

$$0 \to \frac{T\mathscr{A}_e f}{T\mathscr{A} f} \to \frac{\theta(f)}{T\mathscr{A} f} \to \frac{\theta(f)}{T\mathscr{A}_e f} \to 0$$

and

$$0 \to \frac{\mathfrak{m}_n\,\theta(f)}{T\mathscr{A}f} \to \frac{\theta(f)}{T\mathscr{A}f} \to \frac{\theta(f)}{\mathfrak{m}_n\,\theta(f)} \to 0$$

and the fact that $\dim_{\mathbb{F}} \theta(f)/\mathfrak{m}_n\theta(f) = rp$. □

Remark 3.4 When f is a mono-germ then $r = 1$, and $\operatorname{codim}_{\mathscr{A}} f = \operatorname{codim}_{\mathscr{A}_e} f + n$. The number $p - r(p - n)$ which gives the difference between the $\mathscr{A}$ and $\mathscr{A}_e$-codimensions in the multi-germ case will appear later in Chap. 9, where it is denoted by $d_r(n, p)$. It is the dimension of the set of r-multiple points in X^r of an immersion with normal crossings $f: X \to Y$ with $\dim X = n$ and $\dim Y = p$.

◊

The proof of the Proposition uses the Mather–Gaffney criterion for $\mathscr{A}$-finiteness, which we discuss in detail in Sect. 4.5. For now, we highlight a consequence, and introduce an important piece of terminology.

Proposition 3.7 *Let $f: (\mathbb{F}^n, S) \to (\mathbb{F}^p, 0)$ be a smooth germ and suppose that $S \subset C(f)$ (where $C(f)$ is the critical set of f). If $0 < \operatorname{codim}_{\mathscr{A}_e} f < \infty$ then the germ at 0 of the set of points*

$$\left\{y \in \mathbb{F}^p : \textit{the germ of } f \textit{ at } f^{-1}(y) \cap C(f) \textit{ is left-right equiv. to the germ of } f \textit{ at } S\right\}$$

consists only of 0. □

The set-germ defined in the proposition is called the *isosingular locus* of f. The following Lemma is proved in Chap. 5, but may be proved using a sequence of exercises at the end of this section.

Lemma 3.3 *In both the real C^∞ and complex analytic cases, a germ $f: (\mathbb{F}^n, S) \to (\mathbb{F}^p, 0)$ is $\mathscr{A}$-finite if and only if $T\mathscr{A}_e f \supset \mathfrak{m}_n^k\,\theta(f)$ for some finite k.*

Proof "If" is trivial in both cases, since $\theta(f)/\,\mathfrak{m}_n^k\,\theta(f)$ is generated by the finite collection of monomials $x^\alpha \partial/\partial y_i$, $i = 1, \ldots, p$, $|\alpha| < k$. The opposite implication is proved in Exercises 5–8 below. □

In Chap. 6 we will prove the general version of Mather's Finite Determinacy Theorem 6.2. For $\mathscr{G} = \mathscr{R}$ and $p = 1$, we already proved this theorem in Sect. 3.1. For $\mathscr{G} = \mathscr{A}$, we get the following equivalent statements:

1. f is $\mathscr{A}$-finite,
2. f is finite determined for $\mathscr{A}$-equivalence,
3. $\mathfrak{m}_n^k\theta(f) \subset T\mathscr{A}_e f$, for some $k \in \mathbb{N}$.

We now use the equivalence of (i) and (iii) in order to prove the version for $\mathscr{A}$-equivalence of Proposition 3.3.

Proposition 3.8 *Suppose that* $f\colon (\mathbb{R}^n, S) \to (\mathbb{R}^p, 0)$ *is a real analytic germ and let* $f_{\mathbb{C}}$ *be its complexification. Then*

$$f \text{ is } \mathscr{A}\text{-finite} \iff f_{\mathbb{C}} \text{ is } \mathscr{A}\text{-finite},$$

and when they are $\mathscr{A}$*-finite,*

$$T^1_{\mathscr{A}_e} f_{\mathbb{C}} \cong \left(T^1_{\mathscr{A}_e} f\right) \otimes_{\mathbb{R}} \mathbb{C} \qquad \text{and} \qquad T^1_{\mathscr{A}} f_{\mathbb{C}} \cong \left(T^1_{\mathscr{A}} f\right) \otimes_{\mathbb{R}} \mathbb{C}. \tag{3.24}$$

In particular

$$\operatorname{codim}_{\mathscr{A}_e} f = \operatorname{codim}_{\mathscr{A}_e} f_{\mathbb{C}} \quad \text{and} \quad \operatorname{codim}_{\mathscr{A}} f = \operatorname{codim}_{\mathscr{A}} f_{\mathbb{C}}.$$

Proof In the proof, all of the objects associated with f are in the C^∞ category, while those associated with $f_{\mathbb{C}}$ are complex analytic.

Suppose that f is $\mathscr{A}$-finite. Then by Exercises 8 and 9 below,

$$tf(\theta_n) + f^* \mathfrak{m}_p \theta(f) \supset \mathfrak{m}_n^\ell \theta(f) \quad \text{for some finite } \ell \tag{3.25}$$

and

$$tf(\theta_n) + \omega f(\theta_p) \supset \mathfrak{m}_n^k \theta(f) \quad \text{for some finite } k. \tag{3.26}$$

We claim that the same statements hold with $f_{\mathbb{C}}$ in place of f. For the first, it is enough, by Nakayama's Lemma, to show that $tf_{\mathbb{C}}(\theta_n) + f_{\mathbb{C}}^* \mathfrak{m}_p \theta(f_{\mathbb{C}}) + \mathfrak{m}_n^{\ell+1} \theta(f_{\mathbb{C}}) \supset \mathfrak{m}_n^\ell \theta(f_{\mathbb{C}})$. This is easy: we know that for each α with $|\alpha| = \ell$ and for $i = 1, \ldots, p$, there exist smooth $\xi \in \theta_n$ and $\beta_j \in \theta(f)$ such that

$$x^\alpha \frac{\partial}{\partial y_i} = tf(\xi) + \sum f_j \beta_j. \tag{3.27}$$

Replacing ξ and the β_j by their Taylor polynomials of degree ℓ, we deduce that a polynomial version of (3.25) holds modulo $\mathfrak{m}_n^{\ell+1} \theta(f)$. This can equally be regarded as an equality for $f_{\mathbb{C}}$. Then Nakayama's Lemma proves the required inclusion.

For the second, we can use the same argument, for it is enough, by Lemma 6.3, to show that

$$tf_{\mathbb{C}}(\theta_n) + \omega f_{\mathbb{C}}(\theta_p) + \mathfrak{m}_n^{k+\ell} \theta(f_{\mathbb{C}}) \supset \mathfrak{m}_n^k \theta(f_{\mathbb{C}}). \tag{3.28}$$

We use Lemma 6.3 with $C = \theta(f_{\mathbb{C}})/tf_{\mathbb{C}}(\theta_n)$, $A = T\mathscr{A}_e f_{\mathbb{C}}/tf_{\mathbb{C}}(\theta_n)$, $I = \mathfrak{m}_n^k$ and $\mathscr{M} = \mathfrak{m}_m^\ell$. Thus, smooth vector fields ξ and η giving exact solutions to

$$x^\alpha \frac{\partial}{\partial y_i} = tf(\xi) + \omega f(\eta)$$

can be replaced by polynomial vector fields giving solutions modulo $\mathfrak{m}_n^{k+\ell}\,\theta(f)$, which can be viewed as solutions modulo $\mathfrak{m}_n^{k+\ell}\,\theta(f_{\mathbb{C}})$.

The same arguments work in reverse. It is necessary to add one extra step: if ξ and η are complex analytic vector fields such that $tf_{\mathbb{C}}(\xi) + \omega f_{\mathbb{C}}(\eta) = x^\alpha \partial/\partial y_i$, then we can assume that ξ and η have real analytic coefficients. For if ξ^* and η^* are obtained from ξ and η by complex conjugating the coefficients in their Taylor series, then also $tf_{\mathbb{C}}(\xi^*) + \omega f_{\mathbb{C}}(\eta^*) = x^\alpha \partial/\partial y_i$, since the coefficient on the right-hand side is real. Thus, we can replace ξ and η by $\frac{1}{2}(\xi + \xi^*)$ and $\frac{1}{2}(\eta + \eta^*)$.

So (3.25) and (3.26) hold for f if and only if they hold with $f_{\mathbb{C}}$ in place of f. In particular, f is $\mathscr{A}$-finite if and only if $f_{\mathbb{C}}$ is.

The last part of the statement is now straightforward. We have

$$\frac{\theta(f_{\mathbb{C}})}{\mathfrak{m}_n^{k+\ell}\,\theta(f_{\mathbb{C}})} = \frac{\theta(f)}{\mathfrak{m}_n^{k+\ell}\,\theta(f)} \otimes_{\mathbb{R}} \mathbb{C} \tag{3.29}$$

and

$$\frac{tf_{\mathbb{C}}(\theta_n) + \omega f_{\mathbb{C}}(\theta_p) + \mathfrak{m}_n^{k+\ell}\,\theta(f_{\mathbb{C}})}{\mathfrak{m}_n^{k+\ell}\,\theta(f_{\mathbb{C}})} = \frac{tf(\theta_n) + \omega f(\theta_p) + \mathfrak{m}_n^{k+\ell}\,\theta(f)}{\mathfrak{m}_n^{k+\ell}\,\theta(f)} \otimes_{\mathbb{R}} \mathbb{C}, \tag{3.30}$$

so taking quotients, $T^1_{\mathscr{A}_e} f_{\mathbb{C}} = T^1_{\mathscr{A}_e} f \otimes_{\mathbb{R}} \mathbb{C}$. The proof for $T^1_{\mathscr{A}} f$ is analogous. □

The finite determinacy theorem (6.2 below) says that $\mathscr{A}$-finiteness is equivalent to finite $\mathscr{A}$-determinacy. Thus, by Proposition 3.8, a real analytic germ f is finitely determined for $\mathscr{A}$-equivalence if and only if its complexification $f_{\mathbb{C}}$ is. It is not evident that the determinacy degrees are the same, but in fact this is the case—we briefly discuss a theorem of [BdPW87] which shows this in 6.11 below.

The next proposition is the analogue for $\mathscr{A}$ of Proposition 3.4, and highlights the importance of $\mathscr{A}_e$–codimension. To state it we need a definition.

Definition 3.8 Given a germ $f\colon (\mathbb{F}^n, S) \to (\mathbb{F}^p, 0)$ and manifolds N and P of dimensions n and p, for each $k \in \mathbb{N}$ we define $\mathscr{A}_e^{(k)} f$ to be the set of k-jets in ${}_rJ^k(N, P)$ of germs $g\colon (N, S') \to (P, y_0)$ which are left-right equivalent to f.

As mentioned above, the finite determinacy theorem says that if f is $\mathscr{A}$-finite then it is finitely determined for $\mathscr{A}$-equivalence. Suppose now that f is k-determined. An argument like the one in the proof of Theorem 3.1(i) shows that $T\mathscr{A} f \supseteq \mathfrak{m}_n^{k+1}\,\mathscr{O}_n^{\times p}$ (see Theorem 6.2(ii) below). It follows that the codimension in ${}_rJ^k(n, p)$ of $\mathscr{A}^{(k)} f$ is equal to $\operatorname{codim}_{\mathscr{A}} f$ (see Corollary 6.1). Each fibre of the natural projection ${}_rJ^k(N, P) \to N^{(r)} \times P^r$ is isomorphic to ${}_rJ^k(n, p)$ and is contained in the submanifold W of multi-jets with the same target in P, which has codimension $(r-1)p$. Moreover, because each fibre meets $\mathscr{A}_e^{(k)} f$ in an orbit isomorphic to $\mathscr{A}^{(k)} f$, it follows that the codimension of $\mathscr{A}_e^{(k)} f$ in ${}_rJ^k(N, P)$ is equal to $\operatorname{codim}_{\mathscr{A}} f + (r-1)p$.

Proposition 3.9 *Suppose that* $0 < \operatorname{codim}_{\mathscr{A}_e} f < \infty$. *Let* $F : N \times U \to P$ *be a family of mappings, and suppose that* ${}_r j^k(F/U) : N^{(r)} \times U \to {}_r J^k(N, P)$ *is transverse to* $\mathscr{A}_e^{(k)} f$. *Then if* f *is k-determined for* $\mathscr{A}$, *the codimension in* U *of the projection of* $({}_r j^k(F/U))^{-1}(\mathscr{A}_e^{(k)} f)$ *is equal to* $\operatorname{codim}_{\mathscr{A}_e} f$.

That is, for a generic family $F(x, u) = (f_u(x), u)$, *the set of parameter values in* U *for which* f_u *has a singularity in* P *left-right equivalent to* f, *has codimension equal to* $\operatorname{codim}_{\mathscr{A}_e} f$.

Proof By the preceding remarks and by transversality, the codimension of the submanifold $({}_r j^k(F/U))^{-1}(\mathscr{A}_e^{(k)} f)$ in $N^{(r)} \times U$ is equal to $\operatorname{codim}_{\mathscr{A}} f + (r-1)p$. Because the isosingular locus of f at 0 consists of the single point 0, the isosingular locus of f_u at any point $y \in P$ where the germ of f_u is left-right equivalent to the germ of f consists of the single point y. It follows that the projection $N^{(r)} \times U \to U$ is locally finite on $({}_r j^k(F/U))^{-1}(\mathscr{A}_e^{(k)} f)$, so that projection does not reduce its dimension. Hence

$$\begin{aligned}\operatorname{codim} \pi({}_r j^k(F/U))^{-1}(\mathscr{A}_e^{(k)} f) &= \operatorname{codim}({}_r j^k(F/U))^{-1}(\mathscr{A}_e^{(k)} f) - rn \\ &= \operatorname{codim}_{\mathscr{A}} f - rn + (r-1)p \\ &= \operatorname{codim}_{\mathscr{A}_e} f,\end{aligned}$$

by Proposition 3.6. □

Example 3.4 (Looking at Bent Wires II) In the Introduction we discussed the different local views one can see by looking at a knot in $\mathbb{R}^3$ from different positions with one eye closed, and encouraged the reader to experiment, and make a list of "essentially distinct" classes of local views. For a generic knot, one observes three classes of local views whose viewing set (the set of points in $\mathbb{R}^3$ from which one can see a local view of the given class) has codimension 1, corresponding to the three Reidemeister moves in knot theory. With the help of Proposition 3.9 we can recover this observation as a consequence of the theory we are developing, by finding parameterisations of the three local views shown in the second diagram in Example 1.1, and showing that their $\mathscr{A}_e$-codimension is indeed equal to 1. The three local views at the centre of the Reidemeister moves are

R1: First order cusp: a plane curve germ $f : (\mathbb{R}, 0) \to (\mathbb{R}^2, 0)$ having a non-immersive point at 0, but doing so in the least degenerate way possible. Clearly $j^2 f$ must take the form $x \mapsto (ax^2, bx^2)$ and the assumption of minimal degeneracy means that at least one of a, b is different from 0. Suppose $a \neq 0$. Then a scale change, and a target coordinate change subtracting a multiple of y_1 from y_2, brings $j^2 f$ to the form $x \mapsto (x^2, 0)$. Thus f takes the form $f(x) = (x^2(1 + \alpha(x)), x^3(\beta(x)))$ where $\alpha \in \mathfrak{m}_1$. The coordinate change in the domain $\bar{x} = x(1 + \alpha)^{1/2})$, together with the assumption of minimal degeneracy and scale change on the y_2 coordinate, brings f to the form $f(x) = (x^2, x^3 + \text{h.o.t.})$. By subtracting monomials in y_1 and y_2 from y_2, one can increase the order of the higher order terms as far as one likes, so it is at least plausible that we can suppose $f(x) = (x^2, x^3)$. Later we will see that this germ has $\mathscr{A}_e$-codimension 1 and is 3-determined for $\mathscr{A}$-equivalence.

Interruption to Discuss Notation The remaining Reidemeister moves involve multi-germs, germs $f: (\mathbb{F}^n, S) \to (\mathbb{F}^p, 0)$ with $|S| > 1$. Notation can become a problem here. Suppose $S = \{s_1, \ldots, s_r\}$. For each point $s_i \in S$, we will denote the germ of f at s_i by $f^{(i)}$, and describe $f^{(i)}$ using local coordinates centred on s_i (so that s_i itself becomes $(0, \ldots, 0)$). Thus, to represent f, we use $|S|$ distinct local coordinate systems in the source. It would be possible to denote the coordinates around s_i by $x_1^{(i)}, \ldots, x_n^{(i)}$, but the profusion of upper indices quickly becomes tiresome, especially when we want to raise $x_j^{(i)}$ to the power k. So we will denote each set of coordinates in the source simply by $x_1, \ldots, x_n$. This may seem like a recipe for confusion, but seems to work. Our notation uses "place value" (see below) to distinguish between the different coordinate systems.

R2: a tacnode, two immersive plane branches having a first order tangency at 0. A target coordinate change turns one of the two branches into the y_1-axis, and we take $y_1 \circ f_1$ as new coordinate in the domain of the first branch. With respect to the new coordinates the second branch has equation $y_2 = g(y_1)$, with $g(0) = g'(0) = 0$, and the minimality of the tangency implies $g''(0) \neq 0$. Because the two branches are tangent we can take $y_1 \circ f_2$ as coordinate on the domain of the second branch. After a linear coordinate change, f_2 has the form $x \mapsto (x, x^2(1 + \alpha))$. Replacing y_1 by $\bar{y}_1 = y_1(1 + \alpha)^{1/2}$ brings f to the form

$$\begin{cases} x \mapsto (x, 0) \\ x \mapsto (x, x^2) \end{cases}.$$

R3: a triple point, three immersive plane branches meeting two-by-two transversely at 0. Here a change of coordinates in the target turns two of the branches into the coordinate axes. The third, being transverse to each of the other two, has equation $y_2 = g(y_1)$ with respect to the new coordinates, for some function g with $g(0) = 0$, $g'(0) \neq 0$. The target coordinate change $\bar{y}_2 = y_2 - g(y_1)$, followed by a suitable scale change, brings f to the form

$$\begin{cases} x \mapsto (x, 0) \\ x \mapsto (0, x) \\ x \mapsto (x, x) \end{cases}.$$

◊

3.2.1 *Right Equivalence and Left Equivalence*

In this book we are mainly interested in right-left equivalence of map-germs, but for some specific applications we may ask ourselves questions about right or left equivalences. In fact, in Sect. 3.1 we already considered right equivalence, but only for functions. At this point of the book, it will not be difficult for the reader to predict how the definitions and properties of this section can be adapted for the groups $\mathscr{R}$ and $\mathscr{L}$. Given a map-germ $f\colon (\mathbb{F}^n, S) \to (\mathbb{F}^p, 0)$, the extended tangent spaces and the tangent spaces for $\mathscr{R}$ and $\mathscr{L}$ are defined as follows:

$$T\mathscr{R}_e f = tf(\theta_n), \quad T\mathscr{R} f = tf(\mathfrak{m}_n\theta_n),$$
$$T\mathscr{L}_e f = \omega f(\theta_p), \quad T\mathscr{L} f = \omega f(\mathfrak{m}_p\theta_p).$$

Observe that $T\mathscr{A}_e f = T\mathscr{R}_e f + T\mathscr{L}_e f$ and $T\mathscr{A} f = T\mathscr{R} f + T\mathscr{L} f$, which is compatible with the fact that $\mathscr{A} = \mathscr{R} \times \mathscr{L}$. The corresponding spaces T^1 are

$$T^1_{\mathscr{G}_e} f = \frac{\theta(f)}{T\mathscr{G}_e f}, \quad T^1_{\mathscr{G}} f = \frac{\mathfrak{m}_n\theta(f)}{T\mathscr{G} f},$$

as well as the codimensions

$$\operatorname{codim}_{\mathscr{G}_e}(f) = \dim_{\mathbb{F}} T^1_{\mathscr{G}_e} f, \quad \operatorname{codim}_{\mathscr{G}}(f) = \dim_{\mathbb{F}} T^1_{\mathscr{G}} f,$$

where $\mathscr{G} = \mathscr{R}$ or $\mathscr{L}$. We will say that f is $\mathscr{G}$-finite if $\operatorname{codim}_{\mathscr{G}_e}(f) < \infty$. Details of how to adapt the properties of this section for $\mathscr{R}$ and $\mathscr{L}$ are left for the reader.

Exercises for Sect. 3.2

1 Check that in part (i) of Example 3.3, $F = \Psi \circ (f \times \mathrm{id}) \circ \Phi^{-1}$.

2 Check that the diagram of part (ii) of Example 3.3 does indeed commute.

3 Show that every unfolding of the identity is a germ of diffeomorphism.

4 Germs of submersions and immersions are infinitesimally stable and therefore stable. This is an easy calculation using the normal forms of Theorems 1.1 and 1.2. For submersions, show that $tf : \theta_n \to \theta(f)$ is surjective, and for immersions, show that $\omega f : \theta_p \to \theta(f)$ is surjective.

The following exercises are straightforward but useful consequences of the definitions. They are used only very occasionally in this chapter, but do give an introduction to the kind of algebra which plays an increasing role as the book goes on.

5

(i) If M is a finitely generated $\mathcal{O}_n$-module, then $\mathfrak{m}_n^k M$ is also finitely generated.
(ii) If f is $\mathcal{A}$-finite then for every $k, \ell < \infty$,

$$\dim_{\mathbb{F}} \frac{\theta(f)}{tf(\mathfrak{m}_n^k \theta_n) + \omega f(\mathfrak{m}_p^\ell \theta_p)} < \infty.$$

Hint: $\dim_{\mathbb{F}} \theta_n / \mathfrak{m}_n^k \theta_n < \infty$, $\dim_{\mathbb{F}} \theta_p / \mathfrak{m}_p^\ell \theta_p < \infty$
(iii) If f is $\mathcal{A}$-finite then for every $k, \ell < \infty$,

$$\dim_{\mathbb{F}} \frac{\theta(f)}{tf(\mathfrak{m}_n^k \theta_n) + f^* \mathfrak{m}_p^\ell \theta(f)} < \infty.$$

Hint: compare $f^* \mathfrak{m}_p^\ell \theta(f)$ with $\omega f(\mathfrak{m}_p^\ell \theta_p)$.

6

(i) If M is a finitely generated $\mathcal{O}_n$-module, and $\dim_{\mathbb{F}} M = k < \infty$, then $\mathfrak{m}^k M = 0$. Hint: Consider the chain

$$M \supset \mathfrak{m} M \supset \mathfrak{m}_n^2 M \supset \cdots \supset \mathfrak{m}^{k+1} M.$$

Not all the inclusions can be strict. Use Nakayama's Lemma.
(ii) If $N \subset M$ are modules over $\mathcal{O}_n$, with M finitely generated, and $\dim_{\mathbb{F}} M/N = k < \infty$, then $N \supset \mathfrak{m}^k M$.

7

(i) If $f\colon (\mathbb{F}^n, 0) \to (\mathbb{F}^p, 0)$ is $\mathcal{A}$-finite, then $\theta(f)/tf(\theta_n)$ is a finitely generated module over $\mathcal{O}_p$. Hint: if the classes of $g_1, \ldots, g_d$ generate $T^1_{\mathcal{A}_e} f$ then

$$\frac{\theta(f)}{tf(\theta_n)} = \mathcal{O}_p \cdot \left\langle \omega f(\frac{\partial}{\partial y_1}), \ldots, \omega f(\frac{\partial}{\partial y_p}) \right\rangle + \mathrm{Sp}_{\mathbb{F}} \langle g_1, \ldots, g_d \rangle.$$

(ii) If $\dim_{\mathbb{F}} T^1_{\mathcal{A}_e} f = k < \infty$ then $T\mathcal{A}_e f \supset f^* \mathfrak{m}_p^k \theta(f)$. Hint: consider $T^1_{\mathcal{A}_e} f$ as $\mathcal{O}_p$-module and use Exercise 6(ii) (with $\mathcal{O}_p$ in place of $\mathcal{O}_n$).

8

(i) If $\dim_{\mathbb{F}} T^1_{\mathcal{A}_e} f = k < \infty$ then $tf(\theta_n) + f^* \mathfrak{m}_p^k \theta(f) \supset \mathfrak{m}_n^\ell \theta(f)$ for some finite ℓ. Hint: Use Exercises 5, 6 and 7.
(ii) If $\dim_{\mathbb{F}} T^1_{\mathcal{A}_e} f = k < \infty$ then $T\mathcal{A}_e f \supset \mathfrak{m}_n^\ell \theta(f)$ for some $\ell < \infty$. Note: ℓ may be considerably greater than k.

9 Let $f : (\mathbb{F}^n, S) \to (\mathbb{F}^p, 0)$ be $\mathscr{A}$-finite. Show that

(i) $tf(\theta_n) + f^* \mathfrak{m}_p \theta(f) \supset \mathfrak{m}_n^k \theta(f)$ for some finite k.
(ii) Suppose that $tf(\theta_n) + f^* \mathfrak{m}_p \theta(f) \supset \mathfrak{m}_n^k \theta(f)$. Let J_f be the ideal in $\mathscr{O}_n$ generated by the maximal minors of the Jacobian matrix of f. Show that $J_f + f^* \mathfrak{m}_p \mathscr{O}_n \supset \mathfrak{m}_n^{kp}$.
(iii) Suppose that $J_f + f^* \mathfrak{m}_p \mathscr{O}_n \supset \mathfrak{m}_n^{kp}$. Show that f is finite-to-one on its critical set. Hint: $V(J_f + f^* \mathfrak{m}_p \mathscr{O}_n) = f^{-1}(0) \cap C(f)$. Now use the Nullstellensatz, Theorem D.4.
(iv) Let $f : (\mathbb{F}^n, 0) \to (\mathbb{F}^p, 0)$ be $\mathscr{A}$-finite. Show that f is finite-to-one on its critical set.

3.3 First Calculations

We determine an $\mathbb{F}$-basis for $T^1_{\mathscr{A}_e} f = \theta(f)/T\mathscr{A}_e f$ for a series of simple examples. In each case, we may assume that $\theta(f)/T\mathscr{A}_e f$ is generated by monomials—terms of the form $x_1^{\alpha_1} \cdots x_n^{\alpha_n} \partial/\partial y_j$. Although it is not true that $\theta(f)$ itself is generated by monomials (an infinite power series is not in the linear span of monomials), in each example it will be clear "by inspection" that f is stable outside 0, and from this it follows by the Mather–Gaffney criterion (Theorem 4.5 below) that $T\mathscr{A}_e f \supseteq \mathfrak{m}_n^k \theta(f)$ for some finite k. From this it follows that the quotient $\theta(f)/T\mathscr{A}_e f$ has a basis consisting of monomials, of degree $\leq k-1$. Later in the chapter we develop techniques which remove the need for this kind of geometrical insight.

Our calculations here are naive, and hard to implement in a computer algebra system. In later chapters we develop more effective techniques for calculation. In particular in Chap. 8 we give an algorithm, Algorithm 8.1, for the $\mathscr{A}_e$-codimension of a polynomial map-germ $(\mathbb{F}^n, 0) \to (\mathbb{F}^{n+1}, 0)$ with rational coefficients, which one can implement in the algebraic geometry software packages *Macaulay 2* and *Singular*.

Example 3.5

(1) Consider the germ of the first Reidemeister move, $f : (\mathbb{R}, 0) \to (\mathbb{R}^2, 0)$, defined by $f(x) = (x^2, x^3)$. Note that this is an injective immersion outside 0. We write elements of θ_2 and $\theta(f)$ as column vectors. Every monomial x^k except x itself can be written as a product of powers of x^2 and x^3 and hence as a composite $a \circ f$. It follows that

$$\omega f(\theta_2) + \mathrm{Sp}_{\mathbb{R}} \left\{ \begin{pmatrix} x \\ 0 \end{pmatrix}, \begin{pmatrix} 0 \\ x \end{pmatrix} \right\} = \theta(f),$$

where we use $\mathrm{Sp}_{\mathbb{R}}$ to denote the $\mathbb{R}$-vector space spanned by the vectors which follow. Since

$$tf\left(\frac{\partial}{\partial x}\right) = \begin{pmatrix} 2x \\ 3x^2 \end{pmatrix},$$

it follows that

$$tf(\theta_1) + \omega f(\theta_2) + \mathrm{Sp}_{\mathbb{R}}\left\{\begin{pmatrix} 0 \\ x \end{pmatrix}\right\} = \theta(f).$$

Finally, it is easy to see that the missing term $\begin{pmatrix} 0 \\ x \end{pmatrix}$ does not lie in $T\mathscr{A}_e f$, since the order of the second component of any element in $T\mathscr{A}_e f$ is at least 2. Hence $T^1_{\mathscr{A}_e} f$ has as basis the class of $\begin{pmatrix} 0 \\ x \end{pmatrix}$. It is also easy to see that the unfolding $F(x, t) = (x^2, x^3 + tx, t) =: (f_t(x), t)$ is not trivial. The reader should sketch the image curves for f_t for different values of t and compare the sequence, as t varies, with the sequence of views of a space curve as one's eye crosses the tangent developable surface.

(2) The map-germ

$$f(x_1, x_2) = (x_1, x_2^2, x_1 x_2)$$

parameterises the Whitney umbrella (also known as pinch-point). We will show that it is stable by showing that $T^1_{\mathscr{A}_e} f = 0$. From now on we will refer to this germ as a Whitney umbrella. We use coordinates (x_1, x_2) on the source and (y_1, y_2, y_3) on the target.

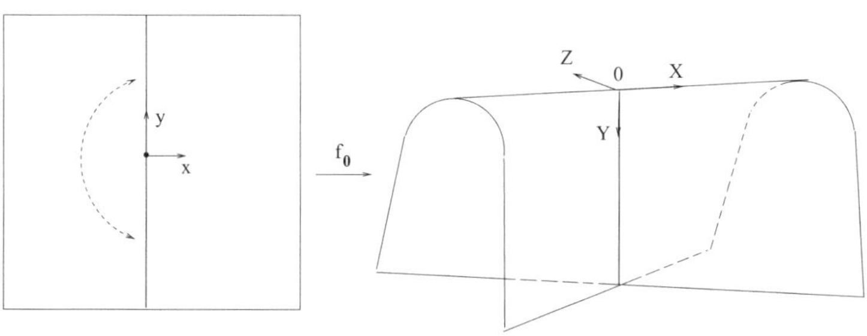

Elements of $\theta_{\mathbb{F}^2,0}, \theta_{\mathbb{F}^3,0}$ and $\theta(f)$ will be written as column vectors. We divide $\mathscr{O}_{\mathbb{F}^2,0}$ into *even* and *odd* parts with respect to the x_2 variable, and denote them by $\mathscr{O}^e$ and $\mathscr{O}^o$. Every element of $\mathscr{O}^e$ can be written in the form $a(x_1, x_2^2)$, and every element of $\mathscr{O}^o$ in the form $x_2 a(1, x_2^2)$. This is obvious in the analytic case. It is also true for C^∞ germs (see Exercise 3.3.3 below). Then (we hope the notation is self-explanatory)

$$\theta(f) = \begin{pmatrix} \mathcal{O}^e \oplus \mathcal{O}^o \\ \mathcal{O}^e \oplus \mathcal{O}^o \\ \mathcal{O}^e \oplus \mathcal{O}^o \end{pmatrix}$$

and since

$$\omega f \begin{pmatrix} a(y_1, y_2) \\ b(y_1, y_2) \\ c(y_1, y_2) \end{pmatrix} = \begin{pmatrix} a(x_1, x_2^2) \\ b(x_1, x_2^2) \\ c(x_1, x_2^2) \end{pmatrix} \tag{3.31}$$

we see that the even part of $\theta(f)$ is indeed contained in $T\mathscr{A}_e f$, and we need look only for the odd part. Since

$$tf\left(a(x_1, x_2^2)\frac{\partial}{\partial x}\right) = \begin{pmatrix} 1 & 0 \\ 0 & 2x_2 \\ x_2 & x_1 \end{pmatrix} \begin{pmatrix} a(x_1, x_2^2) \\ 0 \end{pmatrix} = \begin{pmatrix} a(x_1, x_2^2) \\ 0 \\ x_2 a(x_1, x_2^2) \end{pmatrix}, \tag{3.32}$$

$T\mathscr{A}_e f$ contains all of the odd part of the third row. Since

$$tf\left(a(x_1, x_2^2)\frac{\partial}{\partial x_2}\right) = \begin{pmatrix} 1 & 0 \\ 0 & 2x_2 \\ x_2 & x_1 \end{pmatrix} \begin{pmatrix} 0 \\ a(x_1, x_2^2) \end{pmatrix} = \begin{pmatrix} 0 \\ 2x_2 a(x_1, x_2^2) \\ x_1 a(x_1, x_2^2) \end{pmatrix}, \tag{3.33}$$

$T\mathscr{A}_e f$ contains the odd part of the second row. Since

$$tf\left(x_2 a(x_1, x_2^2)\frac{\partial}{\partial x_1}\right) = \begin{pmatrix} 1 & 0 \\ 0 & 2x_2 \\ x_2 & x_1 \end{pmatrix} \begin{pmatrix} x_2 a(x_1, x_2^2) \\ 0 \end{pmatrix} = \begin{pmatrix} x_2 a(x_1, x_2^2) \\ 0 \\ x_2^2 a(x_1, x_2^2) \end{pmatrix}, \tag{3.34}$$

$T\mathscr{A}_e f$ contains all of the odd part of the first row. So $T\mathscr{A}_e f = \theta(f)$, $T^1_{\mathscr{A}_e} f = 0$ and f is infinitesimally stable.

(3) The map-germ $f(x_1, x_2) = (x_1, x_2^2, x_2^3 + x_1^2 x_2)$ is not stable. It has a non-immersive point at (0, 0), which splits into two in the unfolding $F(u, x_1, x_2) = (u, x_1, x_2^2, x_2^3 + x_1^2 x_2 + yx_2) =: (u, f_u(x_1, x_2))$, for $u \neq 0$.

In determining $T^1_{\mathscr{A}_e} f$, the calculation of (3.31), (3.33) and (3.34) still apply, with insignificant modifications. The only change from (3.33) is that (3.32) now shows that

$$T\mathscr{A}_e f \supset \left(x\,\mathscr{O}^o\right)\partial/\partial y_3 \tag{3.35}$$

and we need an extra calculation

$$tf\left(x_2 a(x_1, x_2^2)\frac{\partial}{\partial y}\right) = \begin{pmatrix} 1 & 0 \\ 0 & 2x_2 \\ 2x_1x_2 & x_1^2 + 3x_2^2 \end{pmatrix}\begin{pmatrix} 0 \\ x_2 a(x_1, x_2^2) \end{pmatrix}$$
$$= \begin{pmatrix} 0 \\ 2x_2^2 a(x_1, x_2^2) \\ x_1^2 x_2 a(x_1, x_2^2) + 3x_2^3 a(x_1, x_2^2) \end{pmatrix}.$$

In view of (3.35) and what we know about the even terms, this completes the proof that

$$T^1_{\mathscr{A}_e} f = \begin{pmatrix} \mathscr{O}^e + \mathscr{O}^o \\ \mathscr{O}^e + \mathscr{O}^o \\ \mathscr{O}^e + x_1\,\mathscr{O}^o + x_2^2\,\mathscr{O}^o \end{pmatrix}. \tag{3.36}$$

It follows that $T^1_{\mathscr{A}_e} f$ is generated, as a vector space over $\mathbb{F}$, by $x_2\partial/\partial y_3$. ◊

Notation We will often use the following notation to display the results of our calculations of tangent spaces. Let $\alpha \in \mathbb{N}^n$ be a multi-index, and $x^\alpha = x_1^{\alpha_1}\cdots x_n^{\alpha_n}$ the corresponding monomial. For a collection $\alpha^{(1)}, \dots, \alpha^{(r)}$ of such multi-indices, we denote by

$$\mathscr{O}_n \setminus \{x^{\alpha^{(1)}}, \dots, x^{\alpha^{(r)}}\}$$

the set of convergent power series in n complex variables in which the coefficients of $x^{\alpha^{(1)}}, \dots, x^{\alpha^{(r)}}$ are all equal to 0. Thus for example $\mathscr{O}_n \setminus \{1\}$ is just $\mathfrak{m}_n$. The expression

$$\begin{pmatrix} \mathscr{O}_n \setminus \{x^{\alpha^{(1)}}, \dots, x^{\alpha^{(r)}}\} \\ \mathscr{O}_n \setminus \{x^{\beta^{(1)}}, \dots, x^{\beta^{(s)}}\} \\ \vdots \end{pmatrix},$$

where there are p rows, will denote the set of p-tuples of elements of $\mathscr{O}_n$ in which the first component has coefficients of $x^{\alpha^{(1)}}, \dots, x^{\alpha^{(r)}}$ all equal to zero, the second has coefficients of $x^{\beta^{(1)}}, \dots, x^{\beta^{(s)}}$ all equal to zero, etc. Then for $f\colon (\mathbb{F}^n, 0) \to$

$(\mathbb{F}^p, 0)$, $T\mathscr{A}_e f$ can be written in the form

$$\begin{pmatrix} \mathscr{O}_n \setminus \{x^{\alpha^{(1)}}, \ldots, x^{\alpha^{(r)}}\} \\ \mathscr{O}_n \setminus \{x^{\beta^{(1)}}, \ldots, x^{\beta^{(s)}}\} \\ \vdots \end{pmatrix} + \mathrm{Sp}_{\mathbb{F}}\{v_1, \ldots, v_k, \ldots\}, \tag{3.37}$$

where each v_i is a column-vector in $(\mathscr{O}_n)^p$. The first summand here simply lists the monomials $x^\alpha \partial u_i$ which do not lie in $T\mathscr{A}_e f$, and the second summand lists the linear combinations of these monomials which *do* lie in $T\mathscr{A}_e f$. For example, if the expression (3.37) is irredundant then the $\mathscr{A}_e$-codimension of f is the number of $\alpha^{(i)}$ plus the number of $\beta^{(j)}$ plus $\cdots$ (we are listing the missing monomials row by row here) minus the number of v_k.

Example 3.6

1. If $f(x_1, x_2) = (x_1, x_2^3 + x_1 x_2)$ then $\mathscr{O}_2 / f^* \mathfrak{m}_2 \mathscr{O}_2$ is generated over $\mathbb{F}$ by $1, x_2$ and x_2^2. It follows from the preparation theorem that $\mathscr{O}_2$ (source) is generated over $\mathscr{O}_2$ (target) by these germs of functions, and that $\theta(f) = \mathscr{O}_2 \oplus \mathscr{O}_2$ is generated over $\mathscr{O}_2$ (target) by

$$\begin{pmatrix} 1 \\ 0 \end{pmatrix}, \begin{pmatrix} x_2 \\ 0 \end{pmatrix}, \begin{pmatrix} x_2^2 \\ 0 \end{pmatrix}, \begin{pmatrix} 0 \\ 1 \end{pmatrix}, \begin{pmatrix} 0 \\ x_2 \end{pmatrix}, \begin{pmatrix} 0 \\ x_2^2 \end{pmatrix}. \tag{3.38}$$

Since $T\mathscr{A}_e f$ is a module over $\mathscr{O}_2$ (target), to prove that $\theta(f) = T\mathscr{A}_e f$ it is enough to find each of these six generators in $T\mathscr{A}_e f$. This is the objective of Exercise 3.3.7 below.

2. A similar calculation shows that if $f(x_1, x_2) = (x_1, x_2^3 + x_1^2 x_2)$ then the same six elements generate $\theta(f)$ over $\mathscr{O}_2$ (target). In Exercise 3.3.8 you are asked to show that all but $\begin{pmatrix} 0 \\ x_2 \end{pmatrix}$ are in $T\mathscr{A}_e f$. This shows that $T^1_{\mathscr{A}_e} f$ is generated over $\mathscr{O}_2$ (target) by $\begin{pmatrix} 0 \\ x_2 \end{pmatrix}$. However, the calculations in Exercise 3.3.8 show that $\mathfrak{m}_2$ (target) times $\begin{pmatrix} 0 \\ x_2 \end{pmatrix}$ is contained in $T\mathscr{A}_e f$, so that now $\begin{pmatrix} 0 \\ x_2 \end{pmatrix}$ generates $T^1_{\mathscr{A}_e} f$ over $\mathscr{O}_2 / \mathfrak{m}_2$ (target), i.e. over $\mathbb{F}$. ◊

Exercises for Sect. 3.3

1 Using the methods of Example 3.5(2) and (3), calculate the $\mathscr{A}_e$-codimension, and an $\mathbb{F}$-basis for $T^1_{\mathscr{A}_e} f$, when

(i) $f(x) = (x^2, x^5)$
(ii) $f(x) = (x^2, x^{2k+1})$

(iii) $f(x) = (x^3, x^4)$
(iv) $f(x_1, x_2) = (x_1, x_2^2, x_2^3 + x_1^{k+1}x)$
(v) $f(x_1, x_2) = (x_1, x_2^2, x_1^2x_2 + x_2^5)$
(vi) $f(x_1, x_2) = (x_1, x_2^2, x_1^2x_2 + x_2^{2k+1})$.

2 Generalise the method of Example 3.5(2) and (3) to show that if $f\colon (\mathbb{F}^n, 0) \to (\mathbb{F}^{n+1}, 0)$ has the form

$$f(\mathbf{x}, y) = (\mathbf{x}, y^2, x_n p(\mathbf{x}, y^2))$$

(where $\mathbf{x} \in \mathbb{F}^{n-1}$) with respect to target coordinates $(X_1, \ldots, X_{n-1}, Y, Z)$, then

$$T\mathscr{A}_e f = \left(\bigoplus_{i=1}^{n-1} \mathscr{O}_n \frac{\partial}{\partial X_i}\right) \oplus \mathscr{O}_n \frac{\partial}{\partial Y} \oplus \left(\mathscr{O}^e + y\,\mathscr{O}^e \left(y^2 \frac{\partial p}{\partial y} \circ h,\ \frac{\partial p}{\partial x} \circ h,\ p \circ h\right)\right) \frac{\partial}{\partial Z}.$$

3 Use the preparation theorem (Corollary D.1) to prove that if $f\colon (\mathbb{F}^n, 0) \to (\mathbb{F}^n, 0)$ is the "fold map"

$$f(x_1, \ldots, x_n) = (x_1, \ldots, x_{n-1}, x_n^2)$$

then every germ $h \in \mathscr{O}_n$ can be written uniquely in the form

$$h = h_1 \circ f + x_n h_2 \circ f$$

for some smooth germs h_1, h_2. Hint: find an $\mathbb{F}$-basis for $\mathscr{O}_2 / f^* \mathfrak{m}_2 \mathscr{O}_2$.

4 Let $f(x_1, \ldots, x_n) = x_1^2 + \cdots + x_n^2$; show f is stable. From the Morse Lemma and Exercise 3.3 it follows that the germ of any Morse function is stable.

5 Show that if $f\colon (\mathbb{F}^n, 0) \to (\mathbb{F}, 0)$ is a stable germ which is not a submersion then f has a non-degenerate critical point at 0. Hint: since $f \in \mathfrak{m}_n^2$, stability implies that $J_f = \mathfrak{m}_n$.

6 Show that the fold map $f(x_1, \ldots, x_n) = (x_1, \ldots, x_k, x_{k+1}^2 + \cdots + x_n^2)$ is stable.

7 The *Whitney cusp* map is the germ

$$f(x_1, x_2) = (x_1, x_1x_2 + x_2^3).$$

Show that the elements of $\theta(f)$ listed in (3.38) in fact lie in $T\mathscr{A}_e f$. In Example 3.6, we used the preparation theorem to deduce from this that this germ is therefore stable.

8 Let $f(x_1, x_2) = (x_1, x_2^3 \pm x_1^2x_2)$.

(i) Show that all of the elements of $\theta(f)$ listed in (3.38), except for the fifth (which we call g_5), lie in $T\mathscr{A}_e f$.

(ii) Show that x_1g_5 and $(x_2^3 + x_1x_2)g_5$ both lie in $T\mathscr{A}_e f$. In Example 3.6 we used the Preparation Theorem to conclude that $\mathscr{A}_e$-codimension of f is 1.

9 The calculation of a basis for $T^1_{\mathscr{A}_e} f$ in Example 3.5(1) above suggests that the unfolding $F(u, x) = (u, x^2, x^3 + ux)$ should be interesting. Make drawings of the images of f_u for $u < 0$, $u = 0$ and $u > 0$, and show that as u passes through 0, the family f_u parameterises the first Reidemeister move of knot theory.

10 Make an analogous sequence of drawings of the images of the maps in the family

$$f_u(x_1, x_2) = (x_1, x_2^2, x_2^3 + x_1^2 x_2 + ux_2)$$

suggested by Example 3.5(3). We will return to this example in Chap. 8.

11 Show

(i) that if $f: (\mathbb{F}, 0) \to (\mathbb{F}^2, 0)$ is a non-immersive germ such that the two vectors $f''(0)$ and $f'''(0)$ are linearly independent then f is left-right-equivalent to the cusp $t \mapsto (t^2, t^3)$.
(ii) if $f: (\mathbb{F}, 0) \to (\mathbb{F}^2, 0)$ is a non-immersive germ such that the two vectors $f''(0)$ and $f'''(0)$ are *not* linearly independent then $\operatorname{codim}_{\mathscr{A}_e} f > 1$.

12 Let f be the germ of Example 3.5(2).

(i) Check that $j^1 f: (\mathbb{R}^2, 0) \to L(\mathbb{R}^2, \mathbb{R}^3)$ meets, and is transverse to, the submanifold Σ^1 of linear maps of corank 1. Suggestion: rather than directly applying the definition of transversality, Definition 2.17, use the characterisation of transversality in Exercise 2.6.1.
(ii) In fact transversality to Σ^1 *characterises* the Whitney umbrella: any map-germ $g: (\mathbb{F}^2, 0) \to (\mathbb{F}^3, 0)$ with this property is $\mathscr{A}$-equivalent to f. To prove this, begin by choosing coordinates in which g takes the form $g(u, v) = (u, g_2(u, v), g_3(u, v))$ (*cf* Lemma 2.3). It is then not hard to find explicit coordinate changes which reduce g to the form $g(u, v) = f(u, v) +$ higher order terms.
(iii) Using the fact that f is 2- determined for $\mathscr{A}$-equivalence, (see Chap. 6 below), deduce that any stable non-immersive germ $(\mathbb{F}^2, 0) \to (\mathbb{F}^3, 0)$ is $\mathscr{A}$-equivalent to f. Hint: use Proposition A.1.

13 If $f: (\mathbb{F}^2, 0) \to (\mathbb{F}^3, 0)$ is not an immersion then the ideal f^*m_3 generated in $\mathscr{O}_2$ by the three component functions of f is strictly contained in m_2. It follows that $\dim_{\mathbb{F}} \mathscr{O}_2 / f^*m_3 \geq 2$. Show that every germ for which this dimension is exactly 2 (as in all the examples above) is $\mathscr{A}$-equivalent to one of the form $f(u, x) = (u, x^2, xp(u, x^2))$. Details can be found in [Mon85].

14 Suppose that $f: (\mathbb{F}^n, 0) \to (\mathbb{F}^p, 0)$, of the form $f(x) = (x_1, \ldots, x_{p-1}, f_p(x))$ has finite $\mathscr{A}_e$-codimension. Then the map-germ $F: (\mathbb{F}^{n+1}, 0) \to (\mathbb{F}^p, 0)$ defined by $F(x, y) = (x_1, \ldots, x_{p-1}, f_p(x) + y^2)$ has $T^1_{\mathscr{A}_e} F \simeq T^1_{\mathscr{A}_e} f$.

15 Given a germ $f\colon (\mathbb{F}^n,0) \to (\mathbb{F}^p,0)$ and diffeomorphisms $\varphi\colon (\mathbb{F}^n,0) \to (\mathbb{F}^n,0)$ and $\psi\colon (\mathbb{F}^p,0) \to (\mathbb{F}^p,0)$, find

(i) a natural isomorphism of $\mathscr{O}_p$-modules $K : \theta(f) \to \theta(f \circ \varphi)$, for which $K(T\mathscr{A}_e f) = T\mathscr{A}_e(f \circ \varphi)$
(ii) a natural isomorphism of $\mathscr{O}_p$-modules $L : \theta(f) \to \theta(\psi \circ f)$, for which $L(T\mathscr{A}_e f) = T\mathscr{A}_e(\psi \circ f)$.

For (i), the diagram

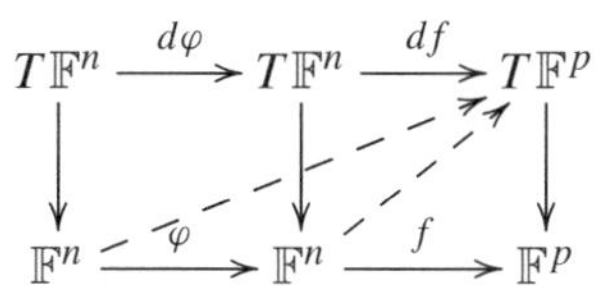

in which elements of $\theta(f)$ and $\theta(f \circ \varphi)$ are shown as dashed arrows, can help to guide the definition of K. A similar diagram will help with (ii).

Conclude that if f and g are $\mathscr{G}$-equivalent, for $\mathscr{G} = \mathscr{R}, \mathscr{L}$ or $\mathscr{A}$, then $\theta(f)/T\mathscr{A}_e f$ and $\theta(g)/T\mathscr{G}_e g$ are isomorphic as $\mathscr{O}_p$-modules.

16 Use the approach of Example 3.6 to find a $\mathbb{F}$-basis for $T^1_{\mathscr{A}_e} f$ when $f(x,y) = (x, y^3 + x^k y)$.

3.4 Multi-Germs

We now turn to multi-germs, beginning with the Reidemeister moves R2 and R3. The first requirement is an adequate notation.

3.4.1 Notation

For each positive integer r we define

$$(\mathbb{F}^n_r, \mathbf{0}_r) = \coprod_{i=1}^{r} (\mathbb{F}^n, 0).$$

We identify a multi-germ $f\colon (\mathbb{F}^n_r, \mathbf{0}_r) \to (\mathbb{F}^p, 0)$ with an r-tuple of map-germs $f^{(i)}\colon (\mathbb{F}^n,0) \to (\mathbb{F}^p,0)$, $i = 1,\dots,r$. When necessary, we will refer to the j'th base-point as 0_j. We will not distinguish in our notation between the local coordinates around 0_i and 0_j for $i \neq j$; each set of coordinates will be denoted $x_1,\dots,x_n$. The $\mathbb{F}$-algebra of germs of functions on $(\mathbb{F}^n_r, \mathbf{0}_r)$ (analytic or C^∞, depending on the case) will be denoted $\mathcal{O}_{n,r}$. Clearly

$$\mathcal{O}_{n,r} = \bigoplus_{i=1}^{r} \mathcal{O}_{\mathbb{F}^n,0_i};$$

we write its elements as row vectors. It is a semi-local ring, with maximal ideals

$$\mathfrak{m}_{n,i} = \{g \in \mathcal{O}_{n,r} : g_i(0_i) = 0\}, i = 1, \ldots, r$$

and Jacobson radical (the intersection of its maximal ideals) $\mathfrak{m}_J$.

Let $f : (\mathbb{F}^n_r, \mathbf{0}_r) \to (\mathbb{F}^p, 0)$ be a multi-germ. Evidently

$$\theta(f) = \bigoplus_{i=1}^{r} \theta(f^{(i)}).$$

In keeping with our notation up to now, we write elements of $\theta(f^{(i)})$ as column vectors, and therefore elements of $\theta(f)$ as $p \times r$ matrices, in which the j'th column represents an element of $\theta(f^{(j)})$, and has entries in $\mathcal{O}_{\mathbb{F}^n,0_i}$. The space of germs of vector fields on $(\mathbb{F}^n_r, \mathbf{0}_r)$ is denoted $\theta_{n,r}$, and once again is a direct sum. The homomorphism $tf : \theta_{n,r} \to \theta(f)$ is the direct sum of the homomorphisms $tf^{(i)} : \theta_{\mathbb{F}^n,0_i} \to \theta(f^{(i)})$.

The homomorphism $\omega f : \theta_p \to \theta(f)$ has the form

$$\omega f(\eta) = (\eta \circ f^{(1)}, \ldots, \eta \circ f^{(r)}) = \begin{pmatrix} \eta_1 \circ f^{(1)} & \cdots & \eta_1 \circ f^{(r)} \\ \vdots & \vdots & \vdots \\ \eta_p \circ f^{(1)} & \cdots & \eta_p \circ f^{(r)} \end{pmatrix}.$$

Warning In this notation, the action of $\mathcal{O}_{n,r}$ on $\theta(f)$ takes a slightly off-putting form in which the r-tuple $(a_1, \ldots, a_r) \in \mathcal{O}_{\mathbb{C}^n,\mathbf{0}_r}$ acts on a $p \times r$ matrix representing an element of $\theta(f)$ by multiplying the j'th column in the matrix by a_j for $j = 1, \ldots, r$.

Example 3.7

(i) According to Example 3.4, the bi-germ of Reidemeister move R2 is equivalent to the bi-germ

$$\begin{cases} x \mapsto (x, 0) \\ x \mapsto (x, h(x)) \end{cases} \tag{3.39}$$

for $h(x) = x^2$. Let us calculate $T\mathcal{A}_e f$. We have

$$tf\left(a(x)\frac{\partial}{\partial x}, 0\right) = \begin{pmatrix} a(x) & 0 \\ 0 & 0 \end{pmatrix}$$

so

$$T\mathscr{A}_e f \supset \begin{pmatrix} \mathscr{O}_1 & 0 \\ 0 & 0 \end{pmatrix}; \tag{3.40}$$

now since

$$\omega f\left(a(y_1)\frac{\partial}{\partial y_1}\right) = \begin{pmatrix} a(x) & a(x) \\ 0 & 0 \end{pmatrix}$$

it follows from (3.40) that

$$T\mathscr{A}_e f \supseteq \begin{pmatrix} 0 & \mathscr{O}_1 \\ 0 & 0 \end{pmatrix} \tag{3.41}$$

also.

Since

$$tf\left(0, b(x)\frac{\partial}{\partial x}\right) = \begin{pmatrix} 0 & b(x) \\ 0 & h'(x)b(x) \end{pmatrix}, \tag{3.42}$$

it follows from (3.41) that

$$T\mathscr{A}_e f \supset \begin{pmatrix} 0 & 0 \\ 0 & J_h \end{pmatrix}, \tag{3.43}$$

where J_h is the Jacobian ideal of h. Contributions to the bottom left-hand entry in $T\mathscr{A}_e f$ come only from ωf:

$$\omega f\left(\eta_2\frac{\partial}{\partial y_2}\right) = \begin{pmatrix} 0 & 0 \\ \eta_2(x,0) & \eta_2(x,h(x)) \end{pmatrix}. \tag{3.44}$$

We have $\eta_2(x, 0) = 0$ if and only if $\eta_2(y_1, y_2)$ is divisible by y_2, in which case $\eta_2(x, h(x)) \in (h) \subset J_h$ (recall that for $h \in \mathscr{O}_1$, $h \in J_h$ always). Thus

$$T\mathscr{A}_e f \bigcap \begin{pmatrix} 0 & 0 \\ 0 & \mathscr{O}_1 \end{pmatrix} = \begin{pmatrix} 0 & 0 \\ 0 & J_h \end{pmatrix}. \tag{3.45}$$

A map

$$\ell : \frac{\mathscr{O}_1}{J_h} \to \frac{\theta(f)}{T\mathscr{A}_e f}$$

may now be defined as follows. For $a \in \mathcal{O}_1$ denote the class of a modulo J_h by $\bar{a}$. Then

$$\ell(\bar{a}) = \begin{pmatrix} 0 & 0 \\ 0 & a \end{pmatrix} + T\mathscr{A}_e f.$$

It is well defined and injective by (3.45). It is surjective, by (3.40), (3.41) and (3.44).

Note that ℓ is $\mathcal{O}_2$-linear, where $\mathcal{O}_1$ is an $\mathcal{O}_2$-module via $f^{(2)}$. We have proved

Proposition 3.10 *$T^1_{\mathscr{A}_e} f$ is isomorphic as $\mathcal{O}_2$-module to $\mathcal{O}_1 \,/\, J_h$.* □

Recall that for R2, $h(x) = x^2$, and $\dim_{\mathbb{C}} \mathcal{O}_1 \,/\, J_h = 1$. Thus we have shown that R2 has A_e-codimension 1, and that the node (the transverse crossing of two immersed branches), where $h(x) = x$, is infinitesimally stable.

Since the germ h in (3.39) can be perturbed to have $\nu = \text{order}\, h$ non-degenerate zeros, the germ f of this example can be perturbed to a bi-germ with ν nodes.

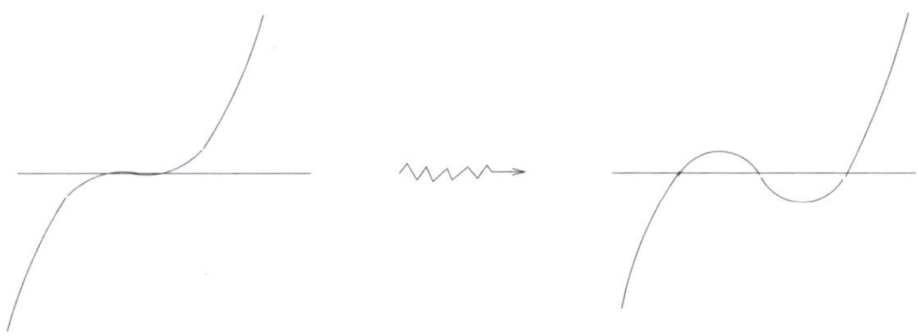

So the number of nodes is one more than the codimension. In fact the image of a ν-nodal perturbation is homotopy-equivalent to a wedge of $\nu - 1$ circles, and so we conclude

Corollary 3.3 *The bi-germ f of Example 3.7 can be perturbed to a germ whose image has the homotopy-type of a wedge of circles, with the number of circles in the image equal to the $\mathscr{A}_e$-codimension of f.*

We will return to this theme many times! The relation between the $\mathcal{A}_e$-codimension of a map-germ and the geometry and topology of a stable perturbation is one of the most interesting aspects of the subject. It is clearest in the complex analytic case, but we can see it in the real C^∞ case if we are observant—in particular in the three Reidemeister moves.

Example 3.7 (continued)

(ii) The third Reidemeister move, R3, is left-right equivalent to

$$\begin{cases} x \mapsto (x, 0) \\ x \mapsto (0, x) \\ x \mapsto (x, x). \end{cases}$$

Let $a \in \mathscr{O}_1$. We have

$$tf\,(a\partial_x, 0, 0) = \begin{pmatrix} a & 0 & 0 \\ 0 & 0 & 0 \end{pmatrix} \qquad tf\,(0, a\partial_x, 0) = \begin{pmatrix} 0 & 0 & 0 \\ 0 & a & 0 \end{pmatrix}$$

so

$$T\mathscr{A}_e f \supset \begin{pmatrix} \mathscr{O}_1 & 0 & 0 \\ 0 & \mathscr{O}_1 & 0 \end{pmatrix}. \tag{3.46}$$

Let $a, b \in \mathscr{O}_2$ be functions of y_1 and y_2, respectively. Then

$$\omega f\,(a\partial_{y_1}) = \begin{pmatrix} a & a(0) & a \\ 0 & 0 & 0 \end{pmatrix} \qquad \omega f\,(b\partial_{y_2}) = \begin{pmatrix} 0 & 0 & 0 \\ b(0) & b & b \end{pmatrix}$$

so taking into account (3.46),

$$T\mathscr{A}_e f \supset \begin{pmatrix} 0 & 0 & \mathfrak{m}_1 \\ 0 & 0 & \mathfrak{m}_1 \end{pmatrix}. \tag{3.47}$$

Similarly,

$$\omega f\left(a\frac{\partial}{\partial y_2}\right) = \begin{pmatrix} 0 & 0 & 0 \\ a & a(0) & a \end{pmatrix} \qquad \omega f\left(b\frac{\partial}{\partial y_1}\right) = \begin{pmatrix} b(0) & b & b \\ 0 & 0 & 0 \end{pmatrix}$$

so taking into account (3.46) and (3.47),

$$T\mathscr{A}_e f \supset \begin{pmatrix} 0 & \mathfrak{m}_1 & 0 \\ \mathfrak{m}_1 & 0 & 0 \end{pmatrix} \quad \text{and thus} \quad T\mathscr{A}_e f \supset \mathfrak{m}_J\,\theta(f). \tag{3.48}$$

Let L be the 6-dimensional complement to $\mathfrak{m}_J\,\theta(f)$ in $\theta(f)$,

$$L = \mathrm{Sp}_{\mathbb{F}}\left\{\begin{pmatrix} 1 & 0 & 0 \\ 0 & 0 & 0 \end{pmatrix}, \ldots, \begin{pmatrix} 0 & 0 & 0 \\ 0 & 0 & 1 \end{pmatrix}\right\}.$$

The intersection of $T\mathscr{A}_e f$ with L is generated by the five elements

$$tf(\partial/\partial x, 0, 0), \;\; tf(0, \partial/\partial x, 0), \;\; tf(0, 0, \partial/\partial x), \;\; \omega f(\partial y_1), \;\; \omega f(\partial y_2).$$

These are linearly independent. So $\dim T^1_{\mathscr{A}_e} f = 1$. ◊

Exercises for Sect. 3.4

1 We say that the multi-germ $g\colon (\mathbb{F}^n_r, \mathbf{0}_r) \to (\mathbb{F}^p, 0)$ *is contained in* (*is strictly contained in*) the multi-germ $g\colon (\mathbb{F}^n_s, \mathbf{0}_s) \to (\mathbb{F}^p, 0)$, and write $g \subseteq f$ ($g \subset f$) if the mono-germs making up g form a subset (form a proper subset) of the mono-germs making up f.

(i) Show that if $g \subseteq f$ then $\operatorname{codim}_{\mathscr{A}_e} g \leq \operatorname{codim}_{\mathscr{A}_e} f$.
(ii) Show that if $n < p$, $g \subset f$, and g is not stable then $\operatorname{codim}_{\mathscr{A}_e} g < \operatorname{codim}_{\mathscr{A}_e} f$.

2 Show that if $f\colon (\mathbb{F}_r, \mathbf{0}_r) \to (\mathbb{F}^2, 0)$ has $r > 3$ then $\operatorname{codim}_{\mathscr{A}_e} f > 1$. Suggestion: consider

$$\dim_{\mathbb{F}} \frac{\theta(f)}{\mathfrak{m}_J\,\theta(f) + T\mathscr{A}_e f}.$$

3 Show that R1, R2 and R3 are the only map-germs $(\mathbb{F}_r, \mathbf{0}_r) \to (\mathbb{F}^2, 0)$ of $\mathscr{A}_e$-codimension 1. Most of the proof can be assembled from earlier exercises.

4 Show that if $f\colon (\mathbb{F}^n_2, \mathbf{0}_2) \to (\mathbb{F}, 0)$ is a bi-germ with both branches non-submersive then f cannot be stable. What is the smallest possible $\mathscr{A}_e$-codimension for such a germ?

5 *Classifying bi-germs of immersions*: we measure the contact between the two branches of a bi-germ $(\mathbb{F}_2, \mathbf{0}_2) \to (\mathbb{F}^2, 0)$

$$g : \begin{cases} x \mapsto g_1(x) \\ x \mapsto g_2(x) \end{cases}$$

as follows: pick an equation h_1 for the image of g_1, and define the order of contact $\nu(g_1, g_2)$ to be the order (lowest non-zero derivative at 0) of $g_2^*(h_1)$.

(i) Find ν for the bi-germs

$$\text{(a)} \begin{cases} x \mapsto (x, 0) \\ x \mapsto (x, r(x)) \end{cases} \quad \text{(b)} \begin{cases} x \mapsto (x^2, x^3) \\ x \mapsto (x, r(x)) \end{cases} \quad \text{(c)} \begin{cases} x \mapsto (x, r(x)) \\ x \mapsto (x^2, x^3) \end{cases}$$

(ii) Show

(a) The definition of $\nu(g_1, g_2)$ is independent of choice of h_1;
(b) $\nu(g_2, g_1) = \nu(g_1, g_2)$;
(c) If g_1 and g_2 are both immersions then g is $\mathscr{A}$-equivalent to the germ

$$g^{(\nu)} : \begin{cases} x \mapsto (x, 0) \\ x \mapsto (x, x^\nu) \end{cases}.$$

6 Suppose that $f\colon (\mathbb{F}^n_2, \mathbf{0}_2) \to (\mathbb{F}^{n+1}, 0)$ is a bi-germ of immersions whose two branches meet transversely. Show that f is $\mathscr{A}$-equivalent to

$$\begin{cases} (x_1, \ldots, x_n) \mapsto (x_1, \ldots, x_{n-1}, x_n, 0) \\ (x_1, \ldots, x_n) \mapsto (x_1, \ldots, x_{n-1}, 0, x_n) \end{cases}.$$

Hint: let g_1 and g_2 be equations for the images of the two branches. Show that the transversality of the two branches is equivalent to the linear independence of $d_0 g_1$ and $d_0 g_2$. So transversality implies that g_1 and g_2 can be taken as the last two members of a system of coordinates on $\mathbb{F}^{n+1}$. In this system, the image of f has equation $x_n x_{n+1} = 0$. Now compose the new coordinates on $\mathbb{F}^{n+1}$ with $f^{(i)}$ to get new coordinates on $(\mathbb{F}^n, 0_i)$.

7 Generalise the approach of the last question to the cases

(a) where $f\colon (\mathbb{F}^n_r, \mathbf{0}_r) \to \mathbb{F}^{n+1}, 0)$, $r \geq 3$, is a multi-germ of immersion with the images of the r branches meeting in general position, and
(b) where $f\colon (\mathbb{F}^n_r, \mathbf{0}_r) \to (\mathbb{F}^{n+k}, 0)$, $k \geq 2$, is a multi-germ of immersion with the images of the r branches meeting in general position.

8 Show, by a direct calculation, that a multi-germ of immersions whose images meet in general position is infinitesimally stable. That this is true follows also from the much more general Theorem 3.3, below.

9 Let $f\colon (\mathbb{F}^2_3, 0_3) \to (\mathbb{F}^4, 0)$ be the tri-germ

$$\begin{cases} (x_1, x_2) \mapsto (x_1, x_2, 0, 0) \\ (x_1, x_2) \mapsto (x_1, x_2, x_1, x_2) \\ (x_1, x_2) \mapsto (0, 0, x_1, x_2) \end{cases}.$$

(a) Show that every tri-germ in these dimensions, consisting of three immersions meeting two-by-two transversely, is left-right equivalent to f.
(b) Deduce that the $\mathscr{A}_e$-codimension of f is 2.

10 Let $f\colon (\mathbb{F}^n_2, \mathbf{0}_2) \to (\mathbb{F}^{n+1}, 0)$ be a bi-germ of immersions whose images meet tangentially at 0.

(a) Show that in suitable coordinate f can be written in the form

$$\begin{cases} x \mapsto (x, 0)) \\ x \mapsto (x, h(x)) \end{cases}, \tag{3.49}$$

where $h \in \mathfrak{m}^2$.

(b) Show that

$$T\mathscr{A}_e f \bigcap \begin{pmatrix} 0 & 0 \\ 0 & \mathscr{O}_n \end{pmatrix} = \begin{pmatrix} 0 & 0 \\ 0 & (h) + J_h \end{pmatrix} \tag{3.50}$$

and deduce that

$$T^1_{\mathscr{A}_e} f \simeq \frac{\mathscr{O}_n}{(h) + J_h}.$$

Hint: the calculations of Example 3.7 go through unchanged until the line corresponding to (3.45). We call the function h the *separation function* of the bi-germ f.

11 Here we assume familiarity with the Milnor fibre of an isolated hypersurface singularity, and introduce the notion of *stable perturbation* without a formal definition. It is sufficient to consider a perturbation of the bi-germ in which the two immersions become transverse.

(a) Show that when $\mathbb{F} = \mathbb{C}$, the image X_t of a stable perturbation of a bi-germ of the form (3.49) has reduced homology satisfying

$$\widetilde{H}_q(X_t) = \begin{cases} \mathbb{Z}^{\mu(h)} & \text{if } q = n \\ 0 & \text{otherwise} \end{cases}. \tag{3.51}$$

(b) Show further that this image is homotopy-equivalent to a wedge of $\mu(h)$ $(n-1)$-spheres.

3.5 Infinitesimal Stability Implies Stability

Next, we prove some lemmas which are needed in order to prove Theorem 3.2. The first is a Thom–Levine type result similar to Lemma 2.6, namely, an infinitesimal condition for the triviality of a 1-parameter unfolding. Given any k, we consider germs of vector fields ζ on $\mathbb{F}^k \times \mathbb{F}$ such that $\zeta(t) = 1$, that is, of the form

$$\zeta = \sum_{i=1}^{k} \zeta_i(x,t)\frac{\partial}{\partial x_i} + \frac{\partial}{\partial t}. \tag{3.52}$$

Lemma 3.4 *Let $f : (\mathbb{F}^n, S) \to (\mathbb{F}^p, 0)$ be a smooth germ and F a 1-parameter unfolding of f. Then F is trivial if and only if there exist germs of vector fields ξ on $(\mathbb{F}^n \times \mathbb{F}, S \times \{0\})$ and η on $\big(\mathbb{F}^p \times \mathbb{F}, (0,0)\big)$ such that $\xi(t) = 1$, $\eta(t) = 1$ and*

$$dF \circ \xi = \eta \circ F.$$

Proof Suppose first that F is trivial. Then there exist diffeomorphisms Φ, Ψ which are unfoldings of the identity in $\mathbb{F}^n$, $\mathbb{F}^p$, respectively, such that $F = \Psi \circ G \circ \Phi^{-1}$, where $G = f \times \mathrm{id}$. We define ξ, η as the vector fields given by

$$\xi = d\Phi \circ \frac{\partial}{\partial t} \circ \Phi^{-1}, \quad \eta = d\Psi \circ \frac{\partial}{\partial t} \circ \Psi^{-1}. \tag{3.53}$$

Since Φ, Ψ are unfoldings of the identity, we have $\xi(t) = 1$, $\eta(t) = 1$. We use the fact that

$$dG \circ \frac{\partial}{\partial t} = \frac{\partial}{\partial t} \circ G,$$

and the chain rule:

$$\begin{aligned} dF \circ \xi &= dF \circ d\Phi \circ \frac{\partial}{\partial t} \circ \Phi^{-1} = d(F \circ \Phi) \circ \frac{\partial}{\partial t} \circ \Phi^{-1} = d(\Psi \circ G) \circ \frac{\partial}{\partial t} \circ \Phi^{-1} \\ &= d\Psi \circ dG \circ \frac{\partial}{\partial t} \circ \Phi^{-1} = d\Psi \circ \frac{\partial}{\partial t} \circ G \circ \Phi^{-1} = \eta \circ \Psi \circ G \circ \Phi^{-1} = \eta \circ F. \end{aligned}$$

Conversely, assume there exist vector fields ξ, η such that $\xi(t) = 1$, $\eta(t) = 1$ and $dF \circ \xi = \eta \circ F$. By taking the integral flows of ξ, η we define Φ, Ψ as the unique diffeomorphisms which satisfy Eqs. (3.53). Since $\xi(t) = 1$ and $\eta(t) = 1$, it follows that Φ, Ψ are unfoldings of the identity on $\mathbb{F}^n$, $\mathbb{F}^p$, respectively.

Now let $G = \Psi^{-1} \circ F \circ \Phi$. Again, by the chain rule we have

$$\begin{aligned} dG \circ \frac{\partial}{\partial t} &= d(\Psi^{-1} \circ F \circ \Phi) \circ \frac{\partial}{\partial t} = d\Psi^{-1} \circ dF \circ d\Phi \circ \frac{\partial}{\partial t} = d\Psi^{-1} \circ dF \circ \xi \circ \Phi \\ &= d\Psi^{-1} \circ \eta \circ F \circ \Phi = d\Psi^{-1} \circ \eta \circ \Psi \circ G = \frac{\partial}{\partial t} \circ G. \end{aligned}$$

If the unfolding G is written as $G(x,t) = (\tilde{g}(x,t), t)$, then the above condition means that $\partial \tilde{g}/\partial t = 0$. Thus, G is the constant unfolding. □

If ζ is a germ of vector field in $\mathbb{F}^k \times \mathbb{F}$ as in (3.52), then we write

$$\tilde{\zeta} = \zeta - \frac{\partial}{\partial t} = \sum_{i=1}^{k} \zeta_i(x,t) \frac{\partial}{\partial x_i}.$$

We can view $\tilde{\zeta}$ as a vector field along the projection $\pi_k : \mathbb{F}^k \times \mathbb{F} \to \mathbb{F}^k$ onto the first factor—that is, as an element of $\theta(\pi_k)$. We can also view $\tilde{\zeta}$ as a time-dependent vector field on $\mathbb{F}^k$: given a representative of ζ in an open neighbourhood $U \times D \subset \mathbb{F}^k \times \mathbb{F}$, for each $t \in D$, ζ_t is the vector field on U given by $(\zeta_t)_x = \tilde{\zeta}_{(x,t)}$.

Let $F\colon (\mathbb{F}^n \times \mathbb{F}, S \times \{0\}) \to (\mathbb{F}^p \times \mathbb{F}, 0)$ be a 1-parameter unfolding given by $F(x,t) = (\tilde{f}(x,t), t)$. The infinitesimal condition $dF \circ \xi = \eta \circ F$ can be written now in matrix notation as

$$\begin{pmatrix} df_t & \frac{\partial \tilde{f}}{\partial t} \\ 0 & 1 \end{pmatrix} \begin{pmatrix} \xi_t \\ 1 \end{pmatrix} = \begin{pmatrix} \eta_t \circ f_t \\ 1 \end{pmatrix}$$

which turns out to be equivalent to

$$\frac{\partial \tilde{f}}{\partial t} + df_t \circ \xi_t = \eta_t \circ f_t. \tag{3.54}$$

In the next definition we introduce relative versions of the maps tf and ωf and of the module $T^1_{\mathscr{A}_e} f$ for unfoldings. We define them for unfoldings with several parameters, since we will need them later.

Definition 3.9 We denote by $\theta_{n+d/d}$ the set of germs of relative vector fields: vector fields on $\mathbb{F}^n \times \mathbb{F}^d$ "in the $\mathbb{F}^n$ direction", i.e. of the form

$$\sum_{i=1}^{n} \xi_i(x, u)\partial/\partial x_i.$$

We will use the same notation for germs at $S \times \{0\}$. Likewise, $\theta(F/d)$ will denote the space of relative infinitesimal deformations of F: infinitesimal deformations in the $\mathbb{F}^p$ direction, i.e. vector fields along F (i.e.in $\theta(F)$) of the form

$$\sum_j \eta_j(x, u)\partial/\partial y_j.$$

We will sometimes denote the domain and target of a map-germ $f\colon (\mathbb{F}^n, S) \to (\mathbb{F}^p, 0)$ by X and Y, and the parameter space of its unfolding F by U; in this case $\theta_{n+d/d}$ becomes $\theta_{X\times U/U}$, $\theta_{p+d/d}$ becomes $\theta_{Y\times U/U}$ and $\theta(F/d)$ becomes $\theta(F/U)$.

Let F be any d-parameter unfolding of f given by $F(x, u) = (\tilde{f}(x, u), u) = (f_u(x), u)$. Then tF and ωF restrict to morphisms $t_{\mathrm{rel}}F : \theta_{n+d/d} \to \theta(F/d)$ and $\omega_{\mathrm{rel}}F : \theta_{p+d/d} \to \theta(F/d)$. We set

$$T^1_{\mathscr{A}_e} F/d = \frac{\theta(F/d)}{t_{\mathrm{rel}}F(\theta_{n+d/d}) + \omega_{\mathrm{rel}}F(\theta_{p+d/d})}.$$

This is an $\mathscr{O}_{p+d}$-module via F, in the same way that $T^1_{\mathscr{A}_e} f$ is an $\mathscr{O}_p$-module via f. It is called the *relative* T^1 *of the unfolding* F.

In effect, $T^1_{\mathscr{A}_e}F/d$ is just a version of $T^1_{\mathscr{A}_e}f$ with parameters $u_1, \ldots, u_d$, and moreover

$$\frac{T^1_{\mathscr{A}_e}F/d}{\{u_1, \ldots, u_d\} \cdot T^1_{\mathscr{A}_e}F/d} = T^1_{\mathscr{A}_e}f. \tag{3.55}$$

Lemma 3.5 *Let* $F\colon (\mathbb{F}^n \times \mathbb{F}^d, S \times \{0\}) \to (\mathbb{F}^p \times \mathbb{F}^d, (0,0))$ *be an unfolding of the* $\mathscr{A}$*-finite germ* $f\colon (\mathbb{F}^n, S) \to (\mathbb{F}^p, 0)$*, let* $G_1, \ldots, G_k \in \theta(F/d)$*, and for* $i = 1, \ldots, k$ *let* $g_i(x) = G_i(x, 0)$*. The following statements are equivalent:*

(a) *the classes of* $G_1, \ldots, G_k$ *generate* $T^1_{\mathscr{A}_e}F/d$ *over* $\mathscr{O}_d$*,*
(b) *the classes of* $g_1, \ldots, g_k$ *generate* $T^1_{\mathscr{A}_e}f$ *over* $\mathbb{F}$*.*

Proof The implication (a) ⇒ (b) is immediate from (3.55).

To see the converse, we define

$$M := \frac{\theta(F/d)}{t_{\mathrm{rel}}F(\theta_{n+d/d})}, \qquad M_0 := \frac{M}{\{u_1, \ldots, u_d\} \cdot M} \cong \frac{\theta(f)}{tf(\theta_n)}.$$

We apply the multi-germ version of the preparation theorem (Corollary D.2) simultaneously to M and M_0. These are finitely generated, over $\mathscr{O}_{n+d}$ and $\mathscr{O}_n$, respectively, (recall that $\theta(F/d) \cong \mathscr{O}^p_{n+d}$ and $\theta(f) \cong \mathscr{O}^p_n$).

Assuming condition (b),

$$\begin{aligned}\theta(f) &= tf(\theta_n) + \omega f(\theta_p) + \mathrm{Sp}_{\mathbb{F}}\{g_1, \ldots, g_k\}\\ &= tf(\theta_n) + \mathscr{O}_p \cdot \{\partial/\partial y_1, \ldots, \partial/\partial y_p\} + \mathrm{Sp}_{\mathbb{F}}\{g_1, \ldots, g_k\},\end{aligned}$$

which shows that M_0 is finitely generated over $\mathscr{O}_p$. Hence $M_0/(f^*\mathfrak{m}_{\mathbb{F}^p,0}M_0)$ is finitely generated over $\mathbb{F}$. However,

$$\frac{M}{F^*\mathfrak{m}_{p+d}M} \cong \frac{M_0}{f^*\mathfrak{m}_p M_0},$$

since F is an unfolding of f, and it now follows by the preparation theorem that M is finitely generated over $\mathscr{O}_{p+d}$.

Next, we use the preparation theorem again, but this time with the module $T^1_{\mathscr{A}_e}F/U$ and the germ of the projection $\pi\colon (\mathbb{F}^p \times \mathbb{F}^d, 0) \to (\mathbb{F}^d, 0)$. Since $T^1_{\mathscr{A}_e}F/d$ is a quotient of M, it also is finitely generated over $\mathscr{O}_{p+d}$. Moreover,

$$\frac{T^1_{\mathscr{A}_e}F/d}{\pi^*\mathfrak{m}_d \cdot T^1_{\mathscr{A}_e}F/d} = \frac{T^1_{\mathscr{A}_e}F/d}{\{u_1, \ldots, u_d\} \cdot T^1_{\mathscr{A}_e}F/d} \cong T^1_{\mathscr{A}_e}f,$$

which is generated over $\mathbb{F}$ by the classes of $g_1, \ldots, g_k$. Since g_i is the image of the class of G_i under the isomorphism of the previous line, it follows by the preparation theorem that $T^1_{\mathscr{A}_e} F/d$ is generated over $\mathscr{O}_d$ by the classes of $G_1, \ldots, G_k$. □

Proof *(of Theorem 3.2)* Assume f is stable. For each $\zeta \in \theta(f)$, we consider the 1-parameter unfolding $F(x,t) = (f(x,t), t)$ given by $f(x,t) = f + t\zeta$. Because f is stable, F is trivial and by Lemma 3.4 there exist vector fields ξ, η such that $\xi(t) = 1$, $\eta(t) = 1$ and $dF \circ \xi = \eta \circ F$. By (3.54) this means that

$$\frac{\partial f(x,t)}{\partial t} + d_x f(x,t) \circ \xi = \eta \circ F,$$

and evaluating at $t = 0$,

$$\zeta = \left.\frac{\partial f_t}{\partial t}\right|_{t=0} = -df \circ \xi_0 + \eta_0 \circ f \in T\mathscr{A}_e f.$$

Conversely, assume that $T^1_{\mathscr{A}_e} f = 0$. We first prove that any 1-parameter unfolding $F(x,t) = (f_t(x), t)$ is trivial. In fact, we know by Lemma 3.5 that $T^1_{\mathscr{A}_e} F/d = 0$. Hence, there exist vector fields $\tilde{\xi}$ and $\tilde{\eta}$ such that

$$\frac{\partial f_t}{\partial t} = df_t \circ \xi_t + \eta_t \circ f_t.$$

Again by displayed equation (3.54) we have $dF \circ \xi = \eta \circ F$, where $\xi = -\tilde{\xi} + \partial/\partial t$ and $\eta = \tilde{\eta} + \partial/\partial t$, so F is trivial by Lemma 3.4.

We show now that any r-parameter unfolding $F(x,u) = (f_u(x), u)$ is trivial, by induction on r. We have already proved the case $r = 1$. Assume now the result is true for $r - 1$. Consider the $(r-1)$-parameter unfolding F_1 obtained from F by taking $u_r = 0$. By the induction hypothesis, F_1 is trivial and hence, equivalent to $f \times \mathrm{id}$. But this implies that F_1 is also $\mathscr{A}$-equivalent to $f \times \mathrm{id}$ as a map-germ and hence $T^1_{\mathscr{A}_e} F_1 = 0$. Since F is a 1-parameter unfolding of F_1 we deduce that F is a trivial unfolding of F_1 and hence, a trivial unfolding of f. □

Exercises for Sect. 3.5

1 Carefully justify Eq. (3.55).

3.6 Stability of Multi-Germs

Suppose $S = \{s_1, \ldots, s_r\}$. A natural question is how the stability of $f\colon (\mathbb{F}^n, S) \to (\mathbb{F}^p, 0)$ is related to the stability of each branch $f_i := f|_{(\mathbb{F}^n, x_i)}\colon (\mathbb{F}^n, s_i) \to (\mathbb{F}^p, 0)$, $i = 1, \ldots, r$. To answer this question we need to introduce a new concept.

Definition 3.10 For each multi-germ $f\colon (\mathbb{F}^n, S) \to (\mathbb{F}^p, 0)$, we define

$$\tau(f) = \mathrm{ev}\left((\omega f)^{-1}((f^*\mathfrak{m}_{\mathbb{F}^p,0})\theta(f) + tf(\theta_{\mathbb{F}^n,S}))\right),$$

where $\mathrm{ev} : \theta_{\mathbb{F}^p,0} \to T_0\mathbb{F}^p$ is evaluation at 0, given by $\mathrm{ev}(\eta) = \eta_0$.

Clearly, $\tau(f)$ is an $\mathbb{F}$-vector subspace of $T_0\mathbb{F}^p$.

In the case that f is stable, $\tau(f)$ has a very nice geometrical interpretation as follows. Let us fix a representative $f : U \to V$ of the germ, where U, V are open neighbourhoods of S and 0 in $\mathbb{F}^n$ and $\mathbb{F}^p$, respectively, and let C be the critical set of f. Denote by $\mathrm{Iso}(f)$ the subset of points $y \in V$ such that the multi-germ of f at $f^{-1}(y) \cap C$ is $\mathscr{A}$-equivalent to the multi-germ of f at S. This subset is called the *isosingular locus* of f. Then $\mathrm{Iso}(f)$ is in fact a submanifold of V, whose tangent space at the origin is precisely $\tau(f)$. We will prove this in Sect. 7.3, but for now it may help to understand the arguments.

Example 3.8 In the case of the fold germ $(\mathbb{F}^n, 0) \to (\mathbb{F}^{k+1}, 0)$ given by

$$(x_1, \ldots, x_n) \mapsto (x_1, \ldots, x_k, x_{k+1}^2 + \cdots + x_n^2),$$

the isosingular locus $\mathrm{Iso}(f)$ is evidently the hyperplane $\{(y_1, \ldots, y_{k+1}) : y_{k+1} = 0\}$. We compare this with the space $\tau(f)$. We have

$$\begin{aligned}
\tau(f) &= \mathrm{ev}\left((\omega f)^{-1}((f^*\mathfrak{m}_{\mathbb{F}^{k+1},0})\theta(f) + tf(\theta_{\mathbb{F}^n,S}))\right) \\
&= \mathrm{ev}\left((\omega f)^{-1}\left((x_1, \ldots, x_k, x_{k+1}^2 + \cdots + x_n^2)\theta(f)\right.\right. \\
&\quad + \mathscr{O}_n\{\partial y_1, \ldots, \partial y_k, x_{k+1}\partial y_{k+1}, \ldots, x_n\partial y_{k+1}\})) \\
&= \mathrm{ev}\left((\omega f)^{-1}(\mathscr{O}_n \oplus \cdots \oplus \mathscr{O}_n \oplus \mathfrak{m}_n)\right) \\
&= \mathrm{ev}\,(\mathscr{O}_{k+1} \oplus \cdots \oplus \mathscr{O}_{k+1} \oplus \mathfrak{m}_{k+1}) \\
&= \mathbb{F}^k \times 0,
\end{aligned}$$

This is indeed the tangent space at 0 to $\mathrm{Iso}(f)$. ◊

We introduce now an important mapping due to Mather, which can be seen as an analogue of the Kodaira–Spencer map in the context of complex manifolds. Note that $\omega f(\mathfrak{m}_{\mathbb{F}^p,0}\theta_{\mathbb{F}^p,0}) \subset (f^*\mathfrak{m}_{\mathbb{F}^p,0})\theta(f)$, so ωf induces the homomorphism

$$\overline{\omega} f : \frac{\theta_{\mathbb{F}^p,0}}{\mathfrak{m}_{\mathbb{F}^p,0}\theta_{\mathbb{F}^p,0}} \longrightarrow \frac{\theta(f)}{tf(\theta_{\mathbb{F}^n,S}) + (f^*\mathfrak{m}_{\mathbb{F}^p,0})\theta(f)}.$$

Moreover, by means of the evaluation map, we can identify $\theta_{\mathbb{F}^p,0}/\mathfrak{m}_{\mathbb{F}^p,0}\theta_{\mathbb{F}^p,0}$ with $T_0\mathbb{F}^p$, and thus consider $\overline{\omega} f$ as a map

$$T_0\mathbb{F}^p \longrightarrow \frac{\theta(f)}{tf(\theta_{\mathbb{F}^n,S}) + (f^*\mathfrak{m}_{\mathbb{F}^p,0})\theta(f)},$$

whose kernel is precisely $\tau(f)$. The following lemma now characterises stability in terms of $\overline{\omega} f$.

Lemma 3.6 *A multi-germ* $f\colon (\mathbb{F}^n, S) \to (\mathbb{F}^p, 0)$ *is stable if and only if* $\overline{\omega} f$ *is an epimorphism.*

Proof In order to simplify the notation, we set:

$$R = \mathscr{O}_{\mathbb{F}^p,0},\ \mathfrak{m} = \mathfrak{m}_{\mathbb{F}^p,0},\ S = \mathscr{O}_{\mathbb{F}^n,S},\ N = \theta_{\mathbb{F}^p,0},\ M = \theta_{\mathbb{F}^n,S},\ L = \theta(f).$$

If f is stable then $T^1_{\mathscr{A}_e} f = 0$, that is, $\omega f(N) + tf(M) = L$. For any $\zeta \in L$, there exist $\xi \in M$ and $\eta \in N$ such that $\zeta = tf(\xi) + \omega f(\eta)$. This gives $\overline{\omega} f([\eta]) = [\zeta]$ and thus, $\overline{\omega} f$ is surjective.

Conversely, suppose now that $\overline{\omega} f$ is surjective. This implies that

$$\omega f(N) + tf(M) + (f^*\mathfrak{m})L = L. \tag{3.56}$$

We define $L' = L/tf(M)$ and denote by $\pi : L \to L'$ the canonical projection. Note that L' is a finitely generated S-module (since in fact $L \cong S^p$). Then (3.56) may be rewritten as

$$\pi \circ \omega f(N) + (f^*\mathfrak{m})L' = L'. \tag{3.57}$$

Considering L' as an R-module via f, (3.57) becomes

$$\pi \circ \omega f(N) + \mathfrak{m}L' = L'. \tag{3.58}$$

Since $\pi \circ \omega f(N)$ is finitely generated over R, it follows that $L'/\mathfrak{m}L'$ is also finitely generated over R, and hence finitely generated over $\mathbb{F} = R/\mathfrak{m}$ (since $\mathfrak{m}$ annihilates it). Therefore by the Preparation Theorem D.2, L' is finitely generated over R. Now from (3.58) it follows by Nakayama's Lemma that $\pi \circ \omega f(N) = L'$ and hence, $\omega f(N) + tf(M) = L$. □

We recall the definition of regular intersection of subspaces $E_1, \ldots, E_r$ of a vector space F of finite dimension.

Definition 3.11 We say that $E_1, \ldots, E_r$ have *regular intersection* (or *meet in general position*) if

$$\operatorname{codim}(E_1 \cap \cdots \cap E_r) = \operatorname{codim}(E_1) + \cdots + \operatorname{codim}(E_r).$$

An equivalent definition is that the canonical mapping

$$F \longrightarrow (F/E_1) \oplus \cdots \oplus (F/E_r),$$

is surjective. This follows easily from the fact that the kernel is precisely $E_1 \cap \cdots \cap E_r$.

Theorem 3.3 *A multi-germ $f : (\mathbb{F}^n, S) \to (\mathbb{F}^p, 0)$ is stable if and only if each branch $f_i : (\mathbb{F}^n, s_i) \to (\mathbb{F}^p, 0)$ is stable and $\tau(f_1), \ldots, \tau(f_r)$ meet in general position.*

Proof We continue with the notation introduced in the proof of Lemma 3.6. By the lemma, we have

$$f \text{ is stable if and only if } \overline{\omega} f : T_0\mathbb{F}^p \longrightarrow \frac{L}{(f^*\mathfrak{m})L + tf(M)} \text{ is surjective,} \tag{3.59}$$

and for each $i = 1, \ldots, r$

$$f_i \text{ is stable if and only if } \overline{\omega} f_i : T_0\mathbb{F}^p \longrightarrow \frac{L_i}{(f_i^*\mathfrak{m})L_i + tf_i(M_i)} \text{ is surjective,} \tag{3.60}$$

where now $M_i = \theta_{\mathbb{F}^n, s_i}$ and $L_i = \theta(f_i)$. Note that the kernel of $\overline{\omega} f_i$ is $\tau(f_i)$. Thus, if f_i is stable we have

$$\frac{T_0\mathbb{F}^p}{\tau(f_i)} \cong \frac{L_i}{(f_i^*\mathfrak{m})L_i + tf_i(M_i)}. \tag{3.61}$$

On the other hand, we also have an isomorphism

$$\frac{L}{(f^*\mathfrak{m})L + tf(M)} \cong \bigoplus_{i=1}^{r} \frac{L_i}{(f_i^*\mathfrak{m})L_i + tf_i(M_i)}, \tag{3.62}$$

from which it follows, by (3.61), that we can write $\overline{\omega} f$ in the form

$$\overline{\omega} f : T_0\mathbb{F}^p \longrightarrow (T_0\mathbb{F}^p/\tau(f_1)) \oplus \cdots \oplus (T_0\mathbb{F}^p/\tau(f_r)). \tag{3.63}$$

It is now immediate that $\overline{\omega} f$ is surjective if and only if each $\overline{\omega} f_i$ is surjective and the $\tau(f_i)$ have regular intersection. By (3.59) and (3.60), this proves the theorem. □

Exercises for Sect. 3.6

1 Show that if $f : (\mathbb{F}^n, S) \to (\mathbb{F}^{n+1}, 0)$ is stable then $|S| \leq n + 1$.

2 If $f : (\mathbb{F}^n, S) \to (\mathbb{F}^{n+k}, 0)$ is stable, how big can $|S|$ be?

3 If $f\colon (\mathbb{F}^n, S) \to (\mathbb{F}^p, 0)$ is stable, with $n \geq p$, and each point of S is a critical point (i.e., f is not a submersion), how big can $|S|$ be?

4 Find the complete list of stable singularities of multi-germs $f\colon (\mathbb{F}^n, S) \to (\mathbb{F}^p, 0)$, where all the points $s_i \in S$ are critical points of f, in the following cases:

(i) $p = 1$ (scalar functions),
(ii) $n = 1$, $p = 2$ (plane curves),
(iii) $n = 2$, $p = 2$ (plane-to-plane singularities),
(iv) $n = 2$, $p = 3$ (surfaces to 3-space).

5

(i) Suppose that $f : (\mathbb{F}^n, S) \to (\mathbb{F}^p, 0)$ is a stable multi-germ with branches $f^{(i)} : (\mathbb{F}^n, s_i) \to (\mathbb{F}^p, 0)$. Show that

$$\tau(f) = \bigcap_{i=1}^{|S|} \tau(f^{(i)}).$$

(ii) Suppose that $f_1 : (\mathbb{F}^n, S_1) \to (\mathbb{F}^p, 0)$ and $f_2 : (\mathbb{F}^n, S_2) \to (\mathbb{F}^p, 0)$ are stable multi-germs, with $S_1 \cap S_2 = \emptyset$. Show that the multi-germ $\{f_1, f_2\} : (\mathbb{F}^n, S_1 \cup S_2) \to (\mathbb{F}^p, 0)$ is stable if and only if $\tau(f_1) \pitchfork \tau(f_2)$.

Chapter 4
Contact Equivalence

4.1 The Contact Tangent Space

Our calculations in Chap. 3 are somewhat atypical. Calculating $T\mathscr{A}_e f$ is generally rather complicated. Checking that a given map-germ is *stable*, however, is made much easier by a theorem of John Mather, which makes use of an auxiliary module known as the *contact tangent space*, denoted $T\mathscr{K}_e f$ and defined by

$$T\mathscr{K}_e f = tf(\theta_{\mathbb{F}^n,S}) + f^*\mathfrak{m}_p\theta(f). \tag{4.1}$$

Note that f^*m_p is simply the ideal in $\mathscr{O}_{\mathbb{C}^n,0}$ generated by the component functions of f. Of course, we have already met $T\mathscr{K}_e f$, though not by that name, in Theorem 3.3, in the proof of Proposition 3.8, and, implicitly, in the definition of $\tau(f)$ in Definition 3.10.

When $p = 1$, $T\mathscr{K}_e f$ is just the ideal $(f, \partial f/\partial x_1, \ldots, \partial f/\partial x_n)$ of $\mathscr{O}_{\mathbb{F}^n,S}$. In any case it is always an $\mathscr{O}_{\mathbb{F}^n,S}$-module, which makes calculating with it very much easier than calculating $T\mathscr{A}_e f$. Like $T\mathscr{A}_e f$, $T\mathscr{K}_e f$ is the "extended" tangent space to the orbit of f under the action of a group, in this case the contact group $\mathscr{K}$, which we will discuss below. The role of $T\mathscr{K}_e f$ here does not involve this geometrical interpretation, so it can safely be left until we have seen the significance of $T\mathscr{K}_e f$.

Mather's theorem is

Theorem 4.1 *For any multi-germ* $f\colon (\mathbb{F}^n, S) \to (\mathbb{F}^p, 0)$*, the following are equivalent:*

1. $T\mathscr{A}_e f = \theta(f)$ *(i.e.* $T^1_{\mathscr{A}_e} f = 0$*, so* f *is stable).*
2. $T\mathscr{K}_e f + \mathrm{Sp}_{\mathbb{F}}\{\partial/\partial y_1, \ldots, \partial/\partial y_p\} = \theta(f)$
3. $T\mathscr{K}_e f + \mathrm{Sp}_{\mathbb{F}}\{\partial/\partial y_1, \ldots, \partial/\partial y_p\} + \mathfrak{m}^{p+1}_{\mathbb{F}^n,S}\theta(f) = \theta(f)$.

D. Mond, J. J. Nuño-Ballesteros, *Singularities of Mappings*,
Grundlehren der mathematischen Wissenschaften 357,
https://doi.org/10.1007/978-3-030-34440-5_4

Proof (1) $\Longrightarrow$ (2) and (2) $\Longrightarrow$ (3) are trivial, since the left-hand sides of the equalities increase from each statement to the next.

The statement that (2) $\Longrightarrow$ (1) is just Lemma 3.6.

To see that (3) $\Longrightarrow$ (2), first we simplify the notation. We will write as follows:

$$\mathscr{O}_n = \mathscr{O}_{\mathbb{F}^n,S}, \quad \mathfrak{m}_n = \mathfrak{m}_{\mathbb{F}^n,S}, \quad \theta_n = \theta_{\mathbb{F}^n,S},$$
$$\mathscr{O}_p = \mathscr{O}_{\mathbb{F}^p,0}, \quad \mathfrak{m}_p = \mathfrak{m}_{\mathbb{F}^p,0}, \quad \theta_p = \theta_{\mathbb{F}^p,0}.$$

Now suppose that (3) holds and let $\alpha_1, \ldots, \alpha_p \in \mathfrak{m}_n$. We will show that $\alpha_1 \cdots \alpha_p \partial/\partial y_i \in T\mathscr{K}_e f + \mathfrak{m}_n^{p+1}\,\theta(f)$. Because every member of $\mathfrak{m}_n^p\,\theta(f)$ is a sum of such elements, it will follow that

$$\mathfrak{m}_n^p\,\theta(f) \subset T\mathscr{K}_e f + \mathfrak{m}_n\,\mathfrak{m}_n^p\,\theta(f),$$

and therefore, by Nakayama's Lemma, that

$$\mathfrak{m}_n^p\,\theta(f) \subset T\mathscr{K}_e f.$$

To see that $\alpha_1 \cdots \alpha_p \partial/\partial y_i \in T\mathscr{K}_e f + \mathfrak{m}_n^{p+1}\,\theta(f)$, observe that because, by (3),

$$\dim_{\mathbb{C}} \frac{\theta(f)}{T\mathscr{K}_e f + \mathfrak{m}_n^{p+1}\,\theta(f)} \leq p,$$

the $p+1$ elements

$$\partial/\partial y_i, \alpha_1 \partial/\partial y_i, \ldots, \alpha_1 \cdots \alpha_p \partial/\partial y_i$$

cannot be linearly independent. Thus there exist $c_0, \ldots, c_p \in \mathbb{F}$, not all zero, such that

$$c_0 \partial/\partial y_i + c_1 \alpha_1 \partial/\partial y_i + \cdots + c_p \alpha_1 \cdots \alpha_p \partial/\partial y_i = 0 \tag{4.2}$$

in $\theta(f)/T\mathscr{K}_e f + \mathfrak{m}_n^{p+1}\,\theta(f)$. Let c_j be the first of the c_i to be non-zero. Then (4.2) can be rewritten as

$$(c_j \alpha_1 \cdots \alpha_j + \cdots + c_p \alpha_1 \cdots \alpha_p) \partial/\partial y_i \in T\mathscr{K}_e f + \mathfrak{m}_n^{p+1}\,\theta(f).$$

The left-hand side here is an $\mathscr{O}_n$-unit times $\alpha_1 \cdots \alpha_j \partial/\partial y_i$, and thus $\alpha_1 \cdots \alpha_j \partial/\partial y_i$, and hence also $\alpha_1 \cdots \alpha_p \partial/\partial y_i$, are members of $T\mathscr{K}_e f + \mathfrak{m}_n^{p+1}\,\theta(f)$. □

Corollary 4.1 *Whether or not* $f\colon (\mathbb{F}^n, S) \to (\mathbb{F}^p, 0)$ *is stable is determined by its* $(p+1)$*-jet.*

Proof If $j^{p+1}f = j^{p+1}g$ then

$$T\mathscr{K}_e f + \mathfrak{m}_n^{p+1}\theta(f) = T\mathscr{K}_e g + \mathfrak{m}_n^{p+1}\theta(g).$$

So (3) holds for f if and only if it holds for g. □

Example 4.1 In the following calculations, we use the letter L to denote the subspace $\mathrm{Sp}_{\mathbb{F}}\{\frac{\partial}{\partial y_1}, \ldots, \frac{\partial}{\partial y_p}\}$.

(1) We apply Theorem 4.1 to the map-germ $f(u, x) = (u, x^2, ux)$ of Example 3.5(2). We have

$$\begin{aligned} T\mathscr{K}_e f &= tf(\theta_2) + f^* m_3\theta(f) \\ &= \mathscr{O}_2 \cdot \{\partial f/\partial a, \partial f/\partial x\} + (u, x^2)\theta(f) \\ &= \mathscr{O}_2 \cdot \left\{ \begin{pmatrix} 1 \\ 0 \\ x \end{pmatrix}, \begin{pmatrix} 0 \\ 2x \\ u \end{pmatrix} \right\} + (u, x^2)\theta(f) \end{aligned}$$

It is easy to show that $T\mathscr{K}_e f + L = \theta(f)$; in particular, since $(x, y^2)\theta(f) \subset T\mathscr{K}_e f$, it is necessary only to check for terms of the form $x\partial/\partial y_i$. We have

$$\begin{pmatrix} x \\ 0 \\ 0 \end{pmatrix} = x \begin{pmatrix} 1 \\ 0 \\ x \end{pmatrix} \quad \text{mod } (u, x^2)\theta(f) \quad \text{and so lies in } T\mathscr{K}_e f$$

$$\begin{pmatrix} 0 \\ x \\ 0 \end{pmatrix} = \frac{1}{2} \begin{pmatrix} 0 \\ 2x \\ u \end{pmatrix} \quad \text{mod } (u, x^2)\theta(f) \quad \text{and so lies in } T\mathscr{K}_e f$$

$$\begin{pmatrix} 0 \\ 0 \\ x \end{pmatrix} = \begin{pmatrix} 1 \\ 0 \\ x \end{pmatrix} - \begin{pmatrix} 1 \\ 0 \\ 0 \end{pmatrix} \in T\mathscr{K}_e f + L$$

(2) The same theorem can easily be used to show that the map-germs

1. $f\colon (\mathbb{F}^3, 0) \to (\mathbb{F}^3, 0)$, $f(a, b, x) = (a, b, x^4 + ax^2 + bx)$
2. $f\colon (\mathbb{F}^4, 0) \to (\mathbb{F}^5, 0)$ $f(a, b, c, x) = (a, b, c, x^3 + ax, bx^2 + cx)$
3. $f\colon (\mathbb{F}^6, 0) \to (\mathbb{F}^7, 0)$ $f(x, y, a, b, c, d) = (x^2 + ay, xy + bx + cy, y^2 + dx, a, b, c, d)$
4. $f\colon (\mathbb{F}^5, 0) \to (\mathbb{F}^4, 0)$ $f(a, b, c, x, y) = (a, b, c, x^3 + y^3 + axy + bx + cy, a, b, c)$

are stable. We do the calculation for the first of these and leave the rest as exercises. We have

$$T\mathscr{K}_e f = \mathscr{O}_3 \cdot \left\{ \begin{pmatrix} 1 \\ 0 \\ x^2 \end{pmatrix}, \begin{pmatrix} 0 \\ 1 \\ x \end{pmatrix}, \begin{pmatrix} 0 \\ 0 \\ 4x^3 + 2ax + b \end{pmatrix} \right\} + f^* \mathfrak{m}_3 \, \theta(f)$$

$$= \mathscr{O}_3 \cdot \left\{ \begin{pmatrix} 1 \\ 0 \\ x^2 \end{pmatrix}, \begin{pmatrix} 0 \\ 1 \\ x \end{pmatrix}, \begin{pmatrix} 0 \\ 0 \\ 4x^3 \end{pmatrix} \right\} + f^* \mathfrak{m}_3 \, \theta(f).$$

Since L generates $\theta(f)$ over $\mathscr{O}_3$, to show that $T\mathscr{K}_e f + L$ is equal to $\theta(f)$, it is enough to show that it is an $\mathscr{O}_3$-module. Because $f^* \mathfrak{m}_3$ contains the variables a and b, we need to check only that $x^k \partial/\partial y_i$ lies in $T\mathscr{K}_e f + L$ for each k and i. We have

$$x \begin{pmatrix} 1 \\ 0 \\ 0 \end{pmatrix} = x \cdot \begin{pmatrix} 1 \\ 0 \\ x^2 \end{pmatrix} - \frac{1}{4} \begin{pmatrix} 0 \\ 0 \\ 4x^3 \end{pmatrix}$$

$$x \begin{pmatrix} 0 \\ 1 \\ 0 \end{pmatrix} = x \begin{pmatrix} 0 \\ 1 \\ x \end{pmatrix} - \begin{pmatrix} 1 \\ 0 \\ x^2 \end{pmatrix} + \begin{pmatrix} 1 \\ 0 \\ 0 \end{pmatrix}$$

$$x \begin{pmatrix} 0 \\ 0 \\ 1 \end{pmatrix} = \begin{pmatrix} 0 \\ 1 \\ x \end{pmatrix} - \begin{pmatrix} 0 \\ 1 \\ 0 \end{pmatrix}$$

The first of these shows that $x\partial/\partial y_1 \in T\mathscr{K}_e f$, and since $T\mathscr{K}_e f$ is an $\mathscr{O}_3$-module, all multiples of this by powers of x are therefore contained in $T\mathscr{K}_e f + L$. The second and third terms, on the other hand, each have a contribution from L, so it is not immediate that multiples by powers of x are in $T\mathscr{K}_e f + L$, and further checks are required. We have

$$x^2 \begin{pmatrix} 0 \\ 1 \\ 0 \end{pmatrix} = x^2 \begin{pmatrix} 0 \\ 1 \\ x \end{pmatrix} - \frac{1}{4} \begin{pmatrix} 0 \\ 0 \\ 4x^3 \end{pmatrix} \in T\mathscr{K}_e f$$

$$x^2 \begin{pmatrix} 0 \\ 0 \\ 1 \end{pmatrix} = \begin{pmatrix} 1 \\ 0 \\ x^2 \end{pmatrix} - \begin{pmatrix} 1 \\ 0 \\ 0 \end{pmatrix}.$$

Again, the second once again uses a contribution from L, so a further check is required. But it is obvious that $x^3\partial/\partial y_3 \in T\mathscr{K}_e f$, so the proof is complete.

(3) Let $F\colon (\mathbb{F}^4_2, \mathbf{0}_2) \to (\mathbb{F}^3, 0)$ be defined by

$$\begin{cases} f^{(1)}(x_1, x_2, x_3, x_4) = (x_1, x_2, x_3^2 + x_4^2) \\ f^{(2)}(x_1, x_2, x_3, x_4) = (x_1, x_3^2 + x_4^2, x_2). \end{cases}$$

Each branch of F is a fold and hence stable, and the two 2-dimensional isosingular loci, $\mathbb{F}^2 \times \{0\}$ and $\mathbb{F} \times \{0\} \times \mathbb{F}$, are in general position, so by Theorem 3.3 F is stable. Here we calculate directly that condition 2 of Theorem 4.1 holds, partly as a means of familiarising the reader with the notation. We have

$$T\mathscr{K}_e F + L = T\mathscr{K}_e f^{(1)} \oplus T\mathscr{K}_e f^{(2)} + L, \tag{4.3}$$

where

$$T\mathscr{K}_e f^{(1)} = \mathscr{O}_{\mathbb{F}^4,0_1}\{\partial y_1, \partial y_2, 2x_3\partial y_3, 2x_4\partial y_3\} + (x_1, x_2, x_3^2 + x_4^2)\theta(f^{(1)})$$
$$T\mathscr{K}_e f^{(2)} = \mathscr{O}_{\mathbb{F}^4,0_2}\{\partial y_1, \partial y_3, 2x_3\partial y_2, 2x_4\partial y_2\} + (x_1, x_2, x_3^2 + x_4^2)\theta(f^{(2)}).$$

Recall that we write elements of $\theta(F)$ as 3×2 matrices, whose left column is in $\theta(f^{(1)})$ and right column in $\theta(f^{(2)})$. In this notation $\partial y_1, \partial y_2, \partial y_3$ become

$$\begin{pmatrix} 1 & 1 \\ 0 & 0 \\ 0 & 0 \end{pmatrix}, \quad \begin{pmatrix} 0 & 0 \\ 1 & 1 \\ 0 & 0 \end{pmatrix}, \quad \begin{pmatrix} 0 & 0 \\ 0 & 0 \\ 1 & 1 \end{pmatrix}$$

respectively. Recall also that $\mathscr{O}_{\mathbb{F}^4_2,\mathbf{0}_2} = \mathscr{O}_{\mathbb{F}^4,0_1} \oplus \mathscr{O}_{\mathbb{F}^4,0_2}$ and that we write its elements as ordered pairs in the usual way.[1]

As in the previous example, it is necessary to check only that the left-hand side of (4.3) is an $\mathscr{O}_{\mathbb{F}^4_2,\mathbf{0}_2}$-module. It is easy to see that for any h in $\mathfrak{m}_{0_1} \oplus \mathfrak{m}_{0_2}$, the Jacobson radical of $\mathscr{O}_{\mathbb{F}^4_2,\mathbf{0}_2}$, $h\partial y_i \in T\mathscr{K}_e F$. The elements $(1, 0)$ and $(0, 1)$ are not in $\mathfrak{m}_{0_1} \oplus \mathfrak{m}_{0_2}$, so we check:

$$(1,0)\begin{pmatrix} 1 & 1 \\ 0 & 0 \\ 0 & 0 \end{pmatrix} = \begin{pmatrix} 1 & 0 \\ 0 & 0 \\ 0 & 0 \end{pmatrix} \in T\mathscr{K}_e f^{(1)} \oplus 0, \quad (0,1)\begin{pmatrix} 1 & 1 \\ 0 & 0 \\ 0 & 0 \end{pmatrix} = \begin{pmatrix} 0 & 1 \\ 0 & 0 \\ 0 & 0 \end{pmatrix} \in 0 \oplus T\mathscr{K}_e f^{(2)},$$

[1] However, recall the warning about this notation, immediately before Example 3.7.

and similarly $(1,0)\partial y_2 \in T\mathscr{K} f^{(1)} \oplus 0$ and $(0,1)\partial y_3 \in 0 \oplus T\mathscr{K} f^{(2)}$. Finally

$$(0,1)\begin{pmatrix}0&0\\1&1\\0&0\end{pmatrix} = \begin{pmatrix}0&0\\1&1\\0&0\end{pmatrix} - \begin{pmatrix}0&0\\1&0\\0&0\end{pmatrix} \in L + (T\mathscr{K}_e f^{(1)} \oplus 0),$$

$$(1,0)\begin{pmatrix}0&0\\0&0\\1&1\end{pmatrix} = \begin{pmatrix}0&0\\0&0\\1&1\end{pmatrix} - \begin{pmatrix}0&0\\0&0\\0&1\end{pmatrix} \in L + (0 \oplus T\mathscr{K}_e f^{(2)}).$$

◊

4.2 Using $T\mathscr{K}_e f$ to Calculate $T\mathscr{A}_e f$

The contact tangent space turns out to be useful in calculating the $\mathscr{A}_e$ tangent space of unstable germs as well as in proving stability. In Chap. 3 we used only direct calculation. The following result gives a handle on more complex calculations. Recall that given a map-germ $f : (\mathbb{F}^n, S) \to (\mathbb{F}^p, 0)$, composition with f defines a ring homomorphism $f^* : \mathscr{O}_p \to \mathscr{O}_n$, making $\mathscr{O}_n$, and therefore any $\mathscr{O}_n$-module, into a module over $\mathscr{O}_p$.

Theorem 4.2 *Given a map-germ* $f : (\mathbb{F}^n, S) \to (\mathbb{F}^p, 0)$, *suppose that*

$$T\mathscr{K}_e f \supseteq \mathfrak{m}_n^\ell \theta(f) \tag{4.4}$$

for some finite ℓ *and that* C *is an* $\mathscr{O}_p$*-submodule of* $\theta(f)$ *such that*

$$C \supseteq \mathfrak{m}_n^k \theta(f), \tag{4.5}$$

Then $C \subseteq T\mathscr{A}_e f$ *if and only if*

$$C \subseteq T\mathscr{A}_e f + \mathfrak{m}_n^{k+\ell} \theta(f) + f^* \mathfrak{m}_p C. \tag{4.6}$$

and this equivalence holds also with equality in place of inclusion.

Proof We need to only prove "if". We have

$$\begin{aligned} f^* \mathfrak{m}_p C + tf(\theta_n) &\supseteq f^* \mathfrak{m}_p \left(\mathfrak{m}_n^k \theta(f)\right) + tf(\theta_n) \\ &\supseteq \mathfrak{m}_n^k \big(tf(\theta_n) + f^* \mathfrak{m}_p \theta(f)\big) + tf(\theta_n) \\ &\supseteq \mathfrak{m}_n^k \mathfrak{m}^\ell \theta(f) + tf(\theta_n) \quad = \quad \mathfrak{m}_n^{k+\ell} \theta(f) + tf(\theta_n). \end{aligned}$$

Therefore if (4.6) holds, it follows that

$$C \subseteq T\mathcal{A}_e f + f^* \mathfrak{m}_p C. \tag{4.7}$$

If C were a finitely generated $\mathcal{O}_p$-module, we would be able to apply Nakayama's Lemma. Since we cannot assume this, we argue as follows: let $M = \theta(f)/tf(\theta_n)$. Then

$$\frac{M}{f^* \mathfrak{m}_p M} = \frac{\theta(f)}{tf(\theta_n) + f^* \mathfrak{m}_p \theta(f)},$$

and by (4.4), this is finite-dimensional over $\mathbb{F}$. It follows by the preparation theorem that M is finitely generated over $\mathcal{O}_p$. Now $(C + tf(\theta_n))/tf(\theta_n)$ is a submodule of M. It is itself finitely generated over $\mathcal{O}_p$, for $\left(\mathfrak{m}_n^k \theta(f) + tf(\theta_n)\right)/tf(\theta_n) = \mathfrak{m}_n^k M$ is finitely generated over $\mathcal{O}_p$ (Exercise 3.2.5(i)) and the quotient of $(C + tf(\theta_n))/tf(\theta_n)$ by $\left(\mathfrak{m}_n^k \theta(f) + tf(\theta_n)\right)/tf(\theta_n)$ is a finite-dimensional vector space.

So we can apply Nakayama's Lemma to the inclusion

$$\frac{C + tf(\theta_n)}{tf(\theta_n)} \subseteq \frac{T\mathcal{A}_e f}{tf(\theta_n)} + f^* \mathfrak{m}_p \left(\frac{C + tf(\theta_n)}{tf(\theta_n)} \right)$$

which we obtain by quotienting (4.7) by $tf(\theta_n)$. This proves that $C \subseteq T\mathcal{A}_e f$.

The proof with equality in place of inclusion follows exactly the same lines. □

Example 4.2 We calculate $T\mathcal{A}_e f$ for the germ $f : (\mathbb{F}^2, 0) \to (\mathbb{F}^4, 0)$ defined by

$$f(x_1, x_2) = (x_1, x_1x_2 + x_2^3, x_1x_2^2 + x_2^5, x_2^4),$$

from the list of $\mathcal{A}_e$-simple singularities of maps from surfaces to 4-space in [KPR07]. Take coordinates $y_1, \ldots, y_4$ on $\mathbb{F}^4$. To simplify notation, we write ∂_k in place of $\frac{\partial}{\partial y_k}$. It is easy to see that $T\mathcal{K}_e f \supset \mathfrak{m}_2^3 \theta(f)$.

Step 1 It seems reasonable to guess that $T\mathcal{A}_e f \supset \mathfrak{m}_2^5 \theta(f)$. To prove that this is so, it is enough, by the theorem, to show that

$$T\mathcal{A}_e f + \mathfrak{m}_2^8 \theta(f) + f^* \mathfrak{m}_4 \cdot \mathfrak{m}_2^5 \theta(f) \supset \mathfrak{m}_2^5 \theta(f). \tag{4.8}$$

To check this, one works downwards in degree, checking first that modulo $\mathfrak{m}_2^8 \theta(f)$, the left-hand side in (4.8) contains $\mathfrak{m}_2^7 \theta(f)$—i.e. that we get all monomials $x_1^i x_2^j \partial_k$ with $i + j = 7$. Once this is checked, we can work modulo $\mathfrak{m}_2^7 \theta(f)$ to check that we get all monomials of degree 6, and so on down.

Since $x_1 \in f^* \mathfrak{m}_4 \mathcal{O}_2$, it follows that $f^* \mathfrak{m}_4 \cdot \mathfrak{m}_2^5 \theta(f)$ contains all monomials $x_1^i x_2^j \partial_k$ with $i + j \geq 6$ and $i \geq 1$. So the only monomials of degree 7 that we need to find are $x_2^7 \partial_k$ for $k = 1, \ldots, 4$. Working modulo $\mathfrak{m}_2^8 \theta(f)$ and those terms we have already found, we have

$$\omega f(y_2 y_4 \partial_k) = x_2^7 \partial_k.$$

Now we repeat the procedure, working modulo $\mathfrak{m}_2^7\theta(f)$. Once again, we need to find only $x_2^6\partial_k$ for $1 \leq k \leq 4$. This is easy: modulo terms we have already found,

$$\omega f((y_2^2 - 2y_1y_5 - y_1y_3)\partial_k = x_2^6\partial_k.$$

Now we work modulo $\mathfrak{m}_2^6\theta(f)$ to find the monomials of degree 5. Since, modulo $\mathfrak{m}_2^6\theta(f)$,

$$f^*(y_1^5) = x_1^5, \quad f^*(y_1^3 y_2) = x_1^4 x_2, \quad f^*(y_1^2 y_3) = x_1^3 x_2^2,$$
$$f^*(y_2 y_3) = x_1^2 x_2^3, \quad f^*(y_1 y_4) = x_1 x_2^4,$$

we get all degree 5 monomials except $x_2^5\partial_k$ from $\omega f(\theta_4)$. Then working cumulatively and modulo obvious terms such as x_2^4 which we can obtain using ωf,

from $tf\left(x_2^5\frac{\partial}{\partial x_1}\right)$ we get $x_2^5\frac{\partial}{\partial y_1}$

from $tf\left(x_1x_2^2\frac{\partial}{\partial x_1}\right)$, $\omega f\left(y_3\frac{\partial}{\partial y_1}\right)$, $tf\left(x_2^3\frac{\partial}{\partial x_2}\right)$ we get $x_2^5\frac{\partial}{\partial y_2}$

from $tf\left(x_1\frac{\partial}{\partial x_1}\right)$, $tf\left(x_2\frac{\partial}{\partial x_2}\right)$, $\omega f\left(y_2\frac{\partial}{\partial y_2}\right)$ and $\omega f\left(y_3\frac{\partial}{\partial y_3}\right)$ we get $x_2^5\frac{\partial}{\partial y_3}$

from $tf\left(x_2^2\frac{\partial}{\partial x_2}\right)$, $tf\left(x_2^3\frac{\partial}{\partial x_1}\right)$, $tf\left(x_1x_2\frac{\partial}{\partial x_1}\right)$, $\omega f\left(y_3\frac{\partial}{\partial y_2}\right)$ and $\omega f\left(y_2\frac{\partial}{\partial y_1}\right)$

we get $x_2^5\frac{\partial}{\partial y_4}$.

These calculations can be quite entertaining. They are best done using a score sheet to cross off all the monomials already obtained, which can then be ignored in further calculations. Now that we have shown that $T\mathscr{A}_e f \supset \mathfrak{m}_2^5\theta(f)$, by such means we can quickly proceed to calculate that $T\mathscr{A}_e f$ is equal to

$$\begin{pmatrix} \mathscr{O}_2 - \{x_2\} \\ \mathscr{O}_2 - \{x_2, x_2^2\} \\ \mathscr{O}_2 - \{x_2, x_1x_2, x_2^2, x_2^3\} \\ \mathscr{O}_2 - \{x_2, x_1x_2, x_2^2, x_2^3\} \end{pmatrix} + \mathrm{Sp}_{\mathbb{F}}\left\{\begin{pmatrix} 0 \\ x_2 \\ x_2^2 \\ 0 \end{pmatrix}, \begin{pmatrix} x_2 \\ x_2^2 \\ x_2^3 \\ 0 \end{pmatrix} \begin{pmatrix} 0 \\ 0 \\ x_1x_2 + x_2^3 \\ 0 \end{pmatrix} \begin{pmatrix} 0 \\ 3x_2^2 \\ 2x_1x_2 \\ 4x_2^3 \end{pmatrix} \begin{pmatrix} 0 \\ 0 \\ 0 \\ x_1x_2 + x_2^3 \end{pmatrix}\right\} \tag{4.9}$$

◊

It is important to note that the first step in the calculation was to guess that $T\mathscr{A}_e f \supset \mathfrak{m}_2^5\theta(f)$. This "strategy of guessing," which we learned of from Terry Gaffney, works for many calculations: make a sensible guess and then check it using Theorem 4.2. The more ambitious the guess, the easier it may be to check it, since a lower k in (4.5) means a lower $k+\ell$ in (4.6). And one can learn as much from an incorrect guess as from a correct one. The theorem can also be used to check a conclusion such as (4.9) (Exercise 1 below).

Exercises for Sect. 4.2

1 Use Theorem 4.2 to check equality in (4.9).

2 Use Theorem 4.2 to calculate $T\mathscr{A}_e f$ for the following germs.

(i) $f(x) = (x^3, x^4)$.
(ii) $f : (\mathbb{F}, 0) \to (\mathbb{F}^2, 0)$, $f(x) = (x^2, x^{2k+1})$.
(iii) $f : (\mathbb{F}^2, 0) \to (\mathbb{F}^3, 0)$, $f(x_1, x_2) = (x_1, x_2^3, x_1x_2+x_2^5)$ (H_2 in the classification of [Mon85]).
(iv) $f : (\mathbb{F}^2, 0) \to (\mathbb{F}^3, 0)$, $f(x, y) = (x, y^2, y^3 + x^{k+1}y)$ (S_k in the classification of [Mon85]).
(v) $f(x_1, x_2) = (x_1^2 - x_2^2 + x_1^5, x_1x_2)$ ($II_{2,2}^2$ in the classification of [RR91])
(vi) $f(x_1, x_2, x_3) = (x_1, x_2, x_1x_3+x_2^2x_3^2+x_3^5, x_3^2+x_2^2x_3)$ ($S_{1,2}$ in the classification of [HK99]).
(vii) $f : (\mathbb{F}^5, 0) \to (\mathbb{F}^4, 0)$, $f(x_1, x_2, u_1, u_2, u_3) = (x_1^2x_2 + \frac{1}{4}x_2^4 + 2u_1x_1 + u_2x_2 + u_3x_2^2, u_1, u_2, u_3)$ (unstable unfolding of D_5 from [Fuk99, Section 5])

3 Show that the statement of Theorem 4.2 continues to hold if we replace $T\mathscr{A}_e f$ by any $\mathscr{O}_p$ submodule N of $\theta(f)$, providing $N \supset tf(\theta_n)$.

4 Ditto, but now with the assumption that $N \supseteq tf(\mathfrak{m}_n^s\theta_n)$ for some finite s, and that $tf(\mathfrak{m}_n^s\theta_n) + f^*\mathfrak{m}_p\theta(f) \supseteq \mathfrak{m}_n^\ell\theta(f)$.

At this point, the most important application of Theorem 4.2 is where $N = T\mathscr{A}_e f$. Later we will make use of it with $N = tf(\mathfrak{m}_n^s\theta_n) + \omega f(\mathfrak{m}_p^s\theta_p)$ for some $s \in \mathbb{N}$.

4.3 Construction of Stable Germs as Unfoldings

The reader will note that each of the germs listed in Example 4.1(2) is itself an unfolding of a germ of rank 0 (i.e. whose derivative at 0 vanishes). Of course, thanks to the inverse function theorem *any* germ can be put in this form, in suitable coordinates (see Lemma 2.3 and Exercise 2.3.6). In fact there is a general procedure for finding *all* stable map-germs as unfoldings of lower dimensional germs of rank zero, due to Mather in [Mat69b]. The procedure is as follows:

1. Given $f\colon (\mathbb{F}^n, 0) \to (\mathbb{F}^p, 0)$ of rank 0, calculate $T\mathcal{K}_e f$. Because f_0 has rank 0 at 0, $T\mathcal{K}_e f_0 \subset \mathfrak{m}_n\,\theta(f_0)$.
2. Find a basis for the quotient $\mathfrak{m}_n\,\theta(f)/T\mathcal{K}_e f$.
3. If $g_1, \ldots, g_d \in \theta(f)$ project to a basis for the quotient $\mathfrak{m}_n\,\theta(f)/T\mathcal{K}_e f$, then the unfolding $F\colon (\mathbb{F}^n \times \mathbb{F}^d, (0,0)) \to (\mathbb{F}^p \times \mathbb{F}^d, (0,0))$ defined by

$$F(x, u_1, \ldots, u_d) = (f(x) + \sum_j u_j g_j(x), u_1, \ldots, u_d) \tag{4.10}$$

   is stable.

In Chap. 7 we explain this construction (see Theorem 7.2).

Example 4.3 We believe the following example is self-explanatory.

Let $f_0\colon (\mathbb{C}^2, 0) \to (\mathbb{C}^3, 0)$ be given by $f(x_1, x_2) = (x_1^2, x_2^3, x_1x_2)$. We show how to use *Macaulay 2* to find a stable unfolding following the recipe explained. Our script is deliberately pedestrian—it is of course possible to combine many of the steps into one.

Step 1 We find a basis for $\theta(f)/T\mathcal{K}_e f$.

```
S0=QQ[x1,x2]
T0=QQ[X,Y,Z]
f=map(S0,T0,matrix{{x1∧ 2,x2∧ 3,x1*x2}})
J=transpose jacobian matrix f
I=ideal matrix f
V=(module S0)∧3
W=mingens I*V
TT=J|W
basis cokernel TT
```

The last step returns the matrix

$$\begin{pmatrix} 1 & x_2 & x_2^2 & 0 & 0 & 0 & 0 & 0 & 0 \\ 0 & 0 & 0 & 1 & x_1 & x_2 & 0 & 0 & 0 \\ 0 & 0 & 0 & 0 & 0 & 0 & 1 & x_1 & x_2 \end{pmatrix}$$

whose columns form a basis for the quotient $\theta(f)/T\mathcal{K}_e f$.

Step 2 It follows that

$$F(x, u) = (x_1^2 + u_1x_2 + u_2x_2^2, x_2^3 + u_3x_1 + u_4x_2, x_1x_2 + u_5x_1 + u_6x_2, u_1, \ldots, u_6)$$

is a minimal stable unfolding of f. ◊

Mather showed that given n and p, a stable map-germ $(\mathbb{F}^n, S) \to (\mathbb{F}^p, 0)$ is determined, up to $\mathscr{A}$-equivalence, by its local algebra, so the procedure described here gives the unique stable map-germ for each algebra type and dimension-pair. We prove this theorem in Chap. 7.

Exercises for Sect. 4.3

1 Find an expression for a stable map-germ of corank 1, with local algebra of dimension k, from $\mathbb{F}^n$ to $\mathbb{F}^n$. These germs are called *Morin singularities*. Suggestion: the Whitney cusp map $(x, u) \mapsto (x^3 + ux, u)$ is the case $k = 3$. Try $k = 4$ next.

2 Find an expression for a stable map-germ of corank 1, with local algebra of dimension k, from $\mathbb{F}^n$ to $\mathbb{F}^{n+1}$. The Whitney umbrella map $(x, u) \mapsto (x^2, ux, u)$ is the case $k = 2$.

3 Use the procedure of this subsection to find stable unfoldings of

(i) $f(x_1, x_2) = (x_1^2, x_1x_2, x_2^2)$,
(ii) $f(x_1, x_2) = (x_1^2, x_2^4, x_1x_2)$,
(iii) $f(x_1, x_2) = (x_1^2, x_2^3, x_1x_2^2)$,
(iv) $f(x_1, x_2, x_3) = (x_1^2 + x_2x_3, x_2^2 + x_1x_3, x_3^2 + x_1x_2, x_1x_2x_3)$.

4

(i) Show that the singularities of stable maps $M^n \to N^{n+1}$ are all of corank 1 when $n \le 5$.
(ii) For which n are all the singularities of stable maps $M^n \to N^n$ of corank 1?
(iii) For which n are all the singularities of stable maps $M^n \to N^n$ of corank ≤ 2?
(iv) For which n are all the stable maps $M^n \to N^{n+2}$ immersions?
(v) For which n are all stable map-germs $(\mathbb{F}^{n+1}, 0) \to (\mathbb{F}^n, 0)$ unfoldings of function germs $(\mathbb{F}^2, 0) \to (\mathbb{F}, 0)$?

5

(i) What is the lowest possible $\mathscr{K}_e$-codimension of a corank 3 map-germ $(\mathbb{F}^3, 0) \to (\mathbb{F}^3, 0)$? Give an example of such a germ, and use *Macaulay 2* to find a stable unfolding.
(ii) Ditto, but now with germs $(\mathbb{F}^3, 0) \to (\mathbb{F}^4, 0)$.

6 What is the smallest value of n for which there can be a stable map-germ $M^n \to N^{n+1}$ consisting of two non-immersive branches? Give a formula for such a germ.

4.4 Contact Equivalence

Contact equivalence of map-germs $(\mathbb{F}^n, S) \to (\mathbb{R}^p, 0)$ is induced by the action of the group $\mathscr{K}$ of diffeomorphisms of $(\mathbb{F}^n \times \mathbb{F}^p, S \times \{0\})$ of the form

$$\Phi(x, y) = (\varphi(x), \psi(x, y))$$

where $\psi(x, 0) = 0$ for all x. It is obvious that this is a subgroup of $\mathrm{Diff}(\mathbb{F}^n \times \mathbb{F}^p, S \times \{0\})$. It acts on the set of multi-germs of maps $(\mathbb{F}^n, S) \to (\mathbb{F}^p, 0)$ via its action on their graphs: if $f : (\mathbb{F}^n, S) \to (\mathbb{F}^p, 0)$ and $\Phi \in \mathscr{K}$, then $\Phi \cdot f$ is the multi-germ $(\mathbb{F}^n, S) \to (\mathbb{F}^p, 0)$ whose graph is $\Phi(\mathrm{graph}(f))$. Since

$$\mathrm{graph}(f) = \{(x, f(x)) : x \in S\},$$

this means that

$$\mathrm{graph}\big((\Phi \cdot f)\big) = \{(\varphi(x), \psi(x, f(x)) : x \in S\}$$

and thus

$$(\Phi \cdot f)(\varphi(x)) = \psi(x, f(x)),$$

so that

$$(\Phi \cdot f)(x) = \psi\big(\varphi^{-1}(x), f(\varphi^{-1}(x))\big). \tag{4.11}$$

For any map $h : N \to P$, let $\mathrm{gr}(h) : N \to N \times P$ be the graph embedding of N in $N \times P$:

$$\mathrm{gr}(h)(x) = (x, h(x)).$$

Remark 4.1 By (4.11), the following diagram commutes:

$$\begin{array}{ccc} (\mathbb{F}^n, S) & \xrightarrow{\quad\varphi\quad} & (\mathbb{F}^n, S) \\ \Big\downarrow \mathrm{gr}(f) & & \Big\downarrow \mathrm{gr}(\Phi \cdot f) \\ (\mathbb{F}^n \times \mathbb{F}^p, S \times \{0\}) & \xrightarrow{\quad\Phi\quad} & (\mathbb{F}^n \times \mathbb{F}^p, S \times \{0\}) \end{array} \tag{4.12}$$

Thus two maps $(\mathbb{F}^n, S) \to (\mathbb{F}^p, 0)$ are $\mathscr{K}$-equivalent if and only if their graph embeddings are left-right equivalent via the pair $(\varphi, \Phi) \subset \mathrm{Diff}(\mathbb{F}^n, S) \times \mathrm{Diff}(\mathbb{F}^n \times \mathbb{F}^p, S \times \{0\})$. This will be used in the proof of finite determinacy for $\mathscr{K}$-equivalence. ◊

We will see shortly that multi-germs are contact-equivalent if and only if their fibres are isomorphic, and so contact equivalence has a clear geometric significance. Nevertheless its significance for the theory of singularities of mappings goes much further than this. Theorem 4.1 has already given a glimpse of this.

One simplifying feature of $\mathscr{K}$-equivalence is that for multi-germs we can work independently on each branch: multi-germs $(\mathbb{F}^n, S) \to (\mathbb{F}^p, 0)$ are contact-equivalent if and only if their mono-germ branches are contact-equivalent. Because of this, from now on in this section, we will consider only the case of mono-germs $f\colon (\mathbb{F}^n, 0) \to (\mathbb{F}^p, 0)$, unless otherwise stated.

We define the subgroup $\mathscr{C}$ of $\mathscr{K}$ as set of all those $\Phi = (\varphi, \psi) \in \mathscr{K}$ such that φ is the identity. Thus by (4.11), $\Phi = (\mathrm{id}, \psi) \in \mathscr{C}$ acts by

$$(\Phi \cdot f)(x) = \psi(x, f(x)).$$

Observe that $\mathscr{R}$ and $\mathscr{L}$ (and therefore $\mathscr{R} \times \mathscr{L} = \mathscr{A}$) are also naturally embedded in $\mathscr{K}$: given $\varphi \in \mathscr{R}$ and $\eta \in \mathscr{L}$, define Φ_φ and Φ_η by $\Phi_\varphi(x, y) = (\varphi(x), y)$, $\quad \Phi_\eta(x, y) = (x, \eta(y)$; then by (4.11)

$$(\Phi_\varphi \cdot f)(x) = f(\varphi^{-1}(x)), \quad (\Phi_\eta \cdot f)(x) = \eta \circ f(x).$$

Proposition 4.1 *$\mathscr{K}$ is the semi-direct product of $\mathscr{R}$ and $\mathscr{C}$.*

Proof First we show that $\mathscr{K} = \mathscr{C}\mathscr{R}$. Given $\Phi = (\varphi, \psi) \in \mathscr{K}$, define $\Phi_\varphi \in \mathscr{R} \subset \mathscr{K}$ by $\Phi_\varphi(x, y) = (\varphi(x), y)$, and $\Phi_1 \in \mathscr{C} \subset \mathscr{K}$ by $\Phi_1(x, y) = (x, \psi(\varphi^{-1}(x), y))$. Then $\Phi = \Phi_1 \circ \Phi_\varphi$.

In view of this, to show that $\mathscr{C}$ is normal, we need to only show that if $\mathscr{G} \in \mathscr{C}$ and $\Phi_\varphi \in \mathscr{R} \subset \mathscr{K}$, then

$$\Phi_{\varphi^{-1}} \mathscr{G} \Phi_\varphi \in \mathscr{C}.$$

This is straightforward:

$$\begin{aligned}\big(\Phi_{\varphi^{-1}} \mathscr{G} \Phi_\varphi\big)(x, y) &= \big(\Phi_{\varphi^{-1}} \mathscr{G}\big)(\phi(x), y) \\ &= \Phi_{\varphi^{-1}}(\phi(x), \psi(\varphi(x), y)) = (x, \psi(\varphi(x), y)).\end{aligned}$$ □

Let $\mathrm{Gl}_p(\mathscr{O}_n)$ be the group of invertible $p \times p$ matrices over $\mathscr{O}_n = \mathscr{O}_{\mathbb{F}^n,0}$. It acts on the space of germs of maps $(\mathbb{F}^n, 0) \to (\mathbb{F}^p, 0)$: if $A \in \mathrm{Gl}_p(\mathscr{O}_n)$ and $f\colon (\mathbb{F}^n, 0) \to (\mathbb{F}^p, 0)$, then $(A \cdot f)(x) = A(x) f(x)$. In fact the map $(x, y) \mapsto (x, A(x)y)$ is a diffeomorphism of $(\mathbb{F}^n \times \mathbb{F}^p, 0)$ and maps $\mathbb{F}^n \times \{0\}$ to itself, and as such lies in the group $\mathscr{C}$. We will denote by $\mathscr{C}_L$ the subgroup of $\mathscr{C}$ consisting of all such maps.

Proposition 4.2 *Map-germs $f, g\colon (\mathbb{F}^n, 0) \to (\mathbb{F}^p, 0)$ are $\mathscr{C}$-equivalent only if they are $\mathscr{C}_L$-equivalent.*

Proof Let $\Phi \in \mathscr{C}$ with $\Phi(x, y) = (x, \psi(x, y))$, and let ψ have components $\psi_1, \ldots, \psi_p$. Because $\psi(x, 0) = 0$, for each $i = 1, \ldots, p$ we have

$$\psi_i(x, y) = \sum_{j=1}^{p} y_j \psi_{ij}(x, y)$$

for some functions ψ_{ij}. It follows that for any $f\colon (\mathbb{F}^n, 0) \to (\mathbb{F}^p, 0)$ with components $f_1, \ldots, f_p$,

$$\begin{aligned}(\Phi \cdot f)(x) &= \psi(x, f(x)) \\ &= \big(\psi_1(x, f(x)), \ldots, \psi_p(x, f(x))\big) \\ &= \left(\sum_{j=1}^{p} \psi_{1j}(x, f(x)) f_j(x), \ldots, \sum_{j=1}^{p} \psi_{pj}(x, f(x)) f_j(x)\right). \end{aligned} \tag{4.13}$$

Let $a_{ij}(x) = \psi_{ij}(x, f(x))$ and define a $p \times p$ matrix A by $A = (a_{ij})$. Then $A \in \mathrm{Gl}_p(\mathscr{O}_n)$ since the matrix of the linear isomorphism $d_0\Phi$ is equal to

$$\begin{pmatrix} I_n & 0 \\ 0 & A(0) \end{pmatrix}.$$

Now let Φ_A be the corresponding element of $\mathscr{C}_L$. Then by (4.13), we have $\Phi_A \cdot f = \Phi \cdot f$. □

It is an odd feature of this proof that the element $\Phi_A \in \mathscr{C}_L$ that we construct depends on the map-germ f; we have not defined a retraction $\mathscr{C} \to \mathscr{C}_L$.

Recall that a parameterised family $\Psi_t \in \mathscr{G}$, where $\mathscr{G}$ is a group of diffeomorphisms of $(\mathbb{F}^m, x_0)$, is *smooth* if the resulting map $(\mathbb{F} \times \mathbb{F}^m, \{0\} \times \{x_0\}) \to (\mathbb{F}^m, x_0)$ defined by $(t, x) \mapsto \Phi_t(x)$ is smooth.

Lemma 4.1 *Let $f\colon (\mathbb{F}^n, 0) \to (\mathbb{F}^p, 0)$ be a smooth germ. Then*

$$\left\{\frac{d\Phi_t \cdot f}{dt}|_{t=0} : \Phi_t \in \mathscr{K} \text{ is smooth }, \Phi_0 = id\right\} = tf(\mathfrak{m}_n \theta_n) + f^* \mathfrak{m}_p \theta(f). \tag{4.14}$$

Proof Let Φ_t be a smooth path in $\mathscr{K}$. By Proposition 4.2, we may suppose that $\Psi_t \cdot f = A_t f \circ \varphi_t$, where $A_t \in \mathrm{Gl}_p(\mathscr{O}_n)$, $A_0 = I_{p\times p}$ and $\varphi_0 = \mathrm{id}$. Then

$$\frac{d\Phi_t \cdot f}{dt}|_{t=0} = \frac{dA_t}{dt} f + tf\left(\frac{d\varphi_t}{dt}|_{t=0}\right). \tag{4.15}$$

The first term on the right lies in $f^* \mathfrak{m}_n \theta(f)$ and the second in $tf(\mathfrak{m}_n \theta_n)$, since, as we have seen, $\frac{d\varphi_t}{dt}|_{t=0}$ is a vector field on $(\mathbb{F}^n, 0)$ which vanishes at 0 as $\Phi_t(0) = 0$

for all t. This shows that the left-hand side of (4.14) is contained in the right. For the opposite inclusion, observe that for any $p \times p$ matrix B over $\mathscr{O}_n$, the matrix $I_{p\times p} + tB$ is invertible for small t, and thus Bf is in the left-hand side. Similarly, for any vector field $\xi \in \mathfrak{m}_n \theta_n$, the germ $\varphi_t(x) = \mathrm{id} + t\xi(x)$ is the germ of a diffeomorphism for small enough t, and thus the left-hand side contains $tf(\mathfrak{m}_n \theta_n)$. □

The space $T\mathscr{K}_e f = tf(\mathfrak{m}_n \theta_n) + f^* \mathfrak{m}_p \theta(f)$ is thus the (heuristic) tangent space to the $\mathscr{K}$-orbit of f, and $T\mathscr{K}_e f = tf(\theta_n) + f^* \mathfrak{m}_p \theta(f)$ is the "extended $\mathscr{K}$ tangent space."

Definition 4.1 Let $f\colon (\mathbb{F}^n, 0) \to (\mathbb{F}^p, 0)$. Then

$$\operatorname{codim}_{\mathscr{K}} f = \dim_{\mathbb{F}} \frac{\mathfrak{m}_n \theta(f)}{T\mathscr{K} f}$$

$$\operatorname{codim}_{\mathscr{K}_e} f = \dim_{\mathbb{F}} \frac{\theta(f)}{T\mathscr{K}_e f}$$

$$\tau'(f) = \dim_{\mathbb{F}} \frac{\mathscr{O}_n}{J_f + f^* \mathfrak{m}_p \mathscr{O}_n}$$

Here J_f is the ideal generated by the $p \times p$ minors of the matrix of df.

The $\mathscr{K}_e$-codimension of f is also known as its Tjurina number, and denoted by $\tau(f)$.

Proposition 4.3 *With the above notation, we have:*

$$\tau(f) < \infty \quad \Longleftrightarrow \quad \tau'(f) < \infty.$$

Proof In the analytic case, this follows from the Hilbert–Rückert Nullstellensatz (see Theorem E.3 below), since both modules have the same support, namely the intersection of $f^{-1}(0)$ with the critical space $C(f)$. In fact an explicit algebraic argument proves the result also for the C^∞ case. If $\tau < \infty$, then $T\mathscr{K}_e f \supset \mathfrak{m}_n^k \theta(f)$ for some finite k. Now work modulo $f^* \mathfrak{m}_p \mathscr{O}_n$. For every monomial x^α with $|\alpha| \geq k$ and for $i = 1, \ldots, p$, $x^\alpha \partial/\partial_{y_i}$ is in the image of the Jacobian matrix $[df]$. It follows that $\operatorname{diag}(x^{\alpha_1}, \ldots, x^{\alpha_p}) = [df][B]$ for some $n \times p$ matrix $[B]$. The determinant of the left- hand side is a product of $p \times p$ minors of the two matrices on the right-hand side. Hence, $J_f + f^* \mathfrak{m}_p \mathscr{O}_n \supset \mathfrak{m}^{kp}$, so $\tau'(f) < \infty$.

Conversely, it is easy to show that $J_f \theta(f) \subset tf(\theta_n)$, so that if $\tau'(f) < \infty$, so is $\tau(f)$. □

Let $\mathscr{K}_e^{(k)} f$ be the set of k-jets in $J^k(N, P)$ of germs $(N, x_0) \to (P, y_0)$ which are $\mathscr{K}$-equivalent to f. A discussion like that preceding Proposition 3.9 shows

Proposition 4.4 *Let $F : N \times U \to P$ be a family of mappings, and suppose that $j^k F/U : N \times U \to J^k(N, P)$ is transverse to $\mathscr{K}_e^{(k)} f$. Then if f is k-determined*

for $\mathcal{K}$, the codimension in U of the projection of $(j^k F/U)^{-1}(\mathcal{K}_e^{(k)} f)$ is equal to $\operatorname{codim}_{\mathcal{K}_e} f$.

That is, for a generic family $F(x,u) = (f_u(x), u)$ on a parameter space U, the set of parameter values for which f_u has a singularity in N which is $\mathcal{K}$ equivalent to f has codimension in U equal to $\operatorname{codim}_{\mathcal{K}_e} f$.

Proposition 4.5 *Let $f, g: (\mathbb{F}^n, 0) \to (\mathbb{F}^p, 0)$ be map-germs and suppose that the ideals $(f_1, \ldots, f_p)$ and $(g_1, \ldots, g_p)$ of $\mathcal{O}_n$ are equal. Then the map-germs f and g are $\mathcal{C}$-equivalent.*

Proof Because the two ideals are equal, there exist $a_{ij} \in \mathcal{O}_n$ and $b_{ij} \in \mathcal{O}_n$, for $1 \le i, j \le p$, such that

$$f_i = \sum_j a_{ij} g_j \text{ and } g_i = \sum_j b_{ij} f_j \quad \text{for } 1 \le i \le p. \tag{4.16}$$

Defining matrices $A = (a_{ij})$ and $B = (b_{ij})$ and writing $\mathbf{f}$ and $\mathbf{g}$ for the column vectors $(f_1, \ldots, f_p)^t$ and $(g_1, \ldots, g_p)^t$, (4.16) becomes

$$A\mathbf{f} = \mathbf{g} \text{ and } B\mathbf{g} = \mathbf{f},$$

so that $BA\mathbf{f} = \mathbf{f}$ and $AB\mathbf{g} = \mathbf{g}$. Unfortunately, despite this, A and B need not be invertible (consider for example the case where $f_1 = f_2$ and $g_1 = g_2$; it is easy to find non-invertible A and B such that (4.16) holds); to find a suitable element of $\mathcal{C}$ transforming $\mathbf{f}$ to $\mathbf{g}$ we modify A to ensure its invertibility.

Let I_p be the $p \times p$ identity matrix. □

Lemma 4.2 *Let $A_0, B_0 : \mathbb{F}^p \to \mathbb{F}^p$ be linear maps. There exists a linear map $Q_0 : \mathbb{F}^p \to \mathbb{F}^p$ such that $A_0 + Q_0(I_p - B_0 A_0)$ is invertible.* □

Proof of Lemma Let W be a complement to the image of A_0 in $\mathbb{F}^p$, and choose $Q_0 : \mathbb{F}^p \to W$ such that $Q_0| : \ker A_0 \to W$ is an isomorphism. Define $C_0 = A_0 + Q_0(I_p - B_0 A_0)$. Then C_0 is injective and therefore an isomorphism. □

We apply the lemma by taking A_0 and B_0 to be $A(0)$ and $B(0)$, respectively. Define the $p \times p$ matrix C by $C = A + Q_0(I_p - BA)$. Then $C(0)$ is the matrix C_0 of the lemma, so C is invertible. Clearly $(I_p - BA)$ annihilates $\mathbf{f}$, so $C \cdot \mathbf{f} = \mathbf{g}$, and f and g are $\mathcal{C}$-equivalent (indeed, $\mathcal{C}_L$-equivalent), as required.

An immediate consequence of Proposition 4.5 is that the $\mathcal{C}$-class of a germ f uniquely determines the $\mathcal{C}$-class of all its unfoldings.

Corollary 4.2 *Let F be a d-parameter unfolding of a germ $f : (\mathbb{F}^n, 0) \to (\mathbb{F}^p, 0)$. Then*

(1) F is $\mathcal{C}$-equivalent to the constant unfolding $f \times \mathrm{id}$.
(2) $\operatorname{codim}_{\mathcal{K}_e}(F) = \operatorname{codim}_{\mathcal{K}_e} f$.

Proof (2): it is an easy calculation that $\operatorname{codim}_{\mathscr{K}_e}(f \times 1) = \operatorname{codim}_{\mathscr{K}_e}(f)$. Because F is $\mathscr{C}$-equivalent, and therefore $\mathscr{K}$-equivalent, to $f \times 1$, $\operatorname{codim}_{\mathscr{K}_e}(F) = \operatorname{codim}_{\mathscr{K}_e}(f \times 1)$. □

Definition 4.2 An *analytic algebra* is an $\mathbb{F}$-algebra of the form $A = \mathscr{O}_n/I$ for some ideal $I \subset \mathscr{O}_n$. We say that two analytic algebras $A = \mathscr{O}_n/I$ and $B = \mathscr{O}_n/J$ are *induced isomorphic* if there exists a diffeomorphism $\varphi\colon (\mathbb{F}^n, 0) \to (\mathbb{F}^n, 0)$ such that $\varphi^*(I) = J$ (in particular, φ^* induces an isomorphism of local algebras between A and B).

Definition 4.3 For a map-germ $f\colon (\mathbb{F}^n, 0) \to (\mathbb{F}^p, 0)$, the analytic algebra

$$Q(f) = \frac{\mathscr{O}_n}{f^*\mathfrak{m}_p} = \frac{\mathscr{O}_n}{(f_1, \ldots, f_p)}$$

is called the *local algebra* of f.

Proposition 4.6 *Two map-germs $f, g\colon (\mathbb{F}^n, 0) \to (\mathbb{F}^p, 0)$ are $\mathscr{K}$-equivalent if and only if their local algebras $Q(f)$, $Q(g)$ are induced isomorphic.*

Proof Suppose that f, g are $\mathscr{K}$-equivalent. There exists $\Phi = (\varphi, \psi) \in \mathscr{K}$ which transforms the graph of f to that of g. Then $g \circ \varphi$ and f are $\mathscr{C}$-equivalent and hence $\mathscr{C}_L$-equivalent. It follows immediately that the ideals $\big((g \circ \varphi)_1, \ldots, (g \circ \varphi)_p\big)$ and $\big(f_1, \ldots, f_p\big)$ are equal. In other words, $\varphi^*(g^*\mathfrak{m}_p) = f^*\mathfrak{m}_p$, and thus φ^* induces an isomorphism between $Q(f)$, $Q(g)$.

Conversely, if $Q(f)$, $Q(g)$ are induced isomorphic, there exists a diffeomorphism $\varphi : (\mathbb{F}^n, 0) \to (\mathbb{F}^n, 0)$ such that $\varphi^*(f^*\mathfrak{m}_p) = g^*\mathfrak{m}_p$. This implies $(f \circ \varphi)^*\mathfrak{m}_p = g^*\mathfrak{m}_p$ and by Proposition 4.5, $f \circ \varphi$ and f are $\mathscr{C}$-equivalent and hence $\mathscr{K}$-equivalent. Since f is also $\mathscr{R}$-equivalent to $f \circ \varphi$, it follows that f and g are $\mathscr{K}$-equivalent. □

When $\mathbb{F} = \mathbb{C}$, $Q(f)$ is nothing but the algebra of germs on the space-germ $(f^{-1}(0), 0)$, possibly with non-reduced structure. In fact, Proposition 4.6 can be improved as follows:

Theorem 4.3 *Two map-germs $f, g\colon (\mathbb{C}^n, 0) \to (\mathbb{C}^p, 0)$ are $\mathscr{K}$-equivalent if and only if their local algebras $Q(f)$, $Q(g)$ are isomorphic.*

Proof In view of Proposition 4.6 we only need to prove that if $Q(f)$ and $Q(g)$ are isomorphic, then they are induced isomorphic. We first show that it is enough to prove it in the case that both f and g have rank 0. In fact, if $Q(f)$, $Q(g)$ are isomorphic, then f and g have the same rank d by Exercise 4.4.1. Hence, we can choose coordinates in which both f and g are written as d-parameter unfoldings of germs f_0, g_0 of rank 0. Then $Q(f_0) \simeq Q(f) \simeq Q(g) \simeq Q(g_0)$. By Corollary 4.2, it will be enough to show that f_0 and g_0 are $\mathscr{K}$-equivalent.

Assume now that f, g both have rank 0 and let $\alpha : Q(f) \to Q(g)$ be an isomorphism of local algebras. Then, $\alpha(\mathfrak{m}) = \mathfrak{n}$, where $\mathfrak{m}, \mathfrak{n}$ are the maximal ideals of $Q(f)$, $Q(g)$, respectively. Both $\mathfrak{m}$, $\mathfrak{n}$ are generated by the classes of the coordinate functions $x_1, \dots, x_n$ in their corresponding quotients. Suppose that for each $i = 1, \dots, n$, $\alpha([x_i]) = [\varphi_i]$, for some $\varphi_i \in \mathcal{O}_n$ and let $\varphi \colon (\mathbb{C}^n, 0) \to (\mathbb{C}^n, 0)$ be the map-germ $\varphi = (\varphi_1, \dots, \varphi_n)$. As $\alpha(\mathfrak{m}) = \mathfrak{n}$, in the algebra $\mathcal{O}_n$ we have:

$$\mathfrak{m}_n \subset \varphi^*\mathfrak{m}_n + g^*\mathfrak{m}_p \subset \varphi^*\mathfrak{m}_n + \mathfrak{m}_n^2,$$

where the second inclusion follows from the fact that g has rank 0. By Nakayama's Lemma C.1, we get $\mathfrak{m}_n \subset \varphi^*\mathfrak{m}_n$, hence

$$x_i = \sum_{i=1}^{n} a_{ij}\varphi_j, \quad i = 1, \dots, n,$$

for some $a_{ij} \in \mathcal{O}_n$. By taking derivatives with respect to x_k and evaluating at $x = 0$, we get the following matrix equality

$$I_n = (a_{ij}(0)) \cdot \left(\frac{\partial \varphi_j}{\partial x_k}(0)\right),$$

so φ is a diffeomorphism.

Consider the induced isomorphism $\varphi^* : \mathcal{O}_n \to \mathcal{O}_n$ and the diagram:

$$\begin{array}{ccc} \mathcal{O}_n & \xrightarrow{\varphi^*} & \mathcal{O}_n \\ \downarrow & & \downarrow \\ Q(f) & \xrightarrow{\alpha} & Q(g), \end{array}$$

where the columns are the quotient maps. Let us see that this diagram is commutative. In fact, for the coordinate functions, we have $\alpha([x_i]) = [\varphi_i] = [\varphi^*(x_i)]$, for all $i = 1, \dots, n$. From this, we deduce that $\alpha([q]) = [\varphi^*(q)]$, for any polynomial function $q \in \mathcal{O}_n$. For any function $h \in \mathcal{O}_n$, let $j^k h$ denote its k'th Taylor polynomial. For each $k \in \mathbb{N}$, $h - j^k h \in \mathfrak{m}_n^{k+1}$, hence

$$\alpha([h]) - [\varphi^*(h)] \in \mathfrak{n}^{k+1}.$$

Since $Q(g)$ is a Noetherian local ring, $\cap_k \mathfrak{n}^{k+1} = 0$ (by Krull's Intersection Theorem) and thus, $\alpha([h]) = [\varphi^*(h)]$.

From the commutativity of the diagram, we deduce that $\varphi^*(f^*\mathfrak{m}_p) = g^*\mathfrak{m}_p$, so $Q(f)$ and $Q(g)$ are induced isomorphic. By Proposition 4.6, f and g are $\mathcal{K}$-equivalent. □

The proof here relies on the fact that $\bigcap_{k=1}^{\infty} \mathfrak{n}^k = 0$, and thus does not work in the real C^∞ case. Nevertheless, the result is also true with the additional hypothesis of finite $\mathcal{K}$-determinacy.

Theorem 4.4 *Let $f, g\colon (\mathbb{R}^n, 0) \to (\mathbb{R}^p, 0)$ be two C^∞ map-germs and assume that f is finitely $\mathscr{K}$-determined. Then f, g are $\mathscr{K}$-equivalent if and only if their local algebras $Q(f)$, $Q(g)$ are isomorphic.*

Proof Let $\alpha : Q(f) \to Q(g)$ be an isomorphism of local algebras. We proceed as in the proof of Theorem 4.3. We construct a C^∞-diffeomorphism $\varphi\colon (\mathbb{R}^n, 0) \to (\mathbb{R}^n, 0)$ such that $\alpha([x_i]) = [\varphi_i] = [\varphi^*(x_i)]$, for all $i = 1, \ldots, n$ and hence, $\alpha([q]) = [\varphi^*(q)]$, for any polynomial function $q \in \mathscr{O}_n$.

Assume that f is k-$\mathscr{K}$-determined. We have

$$\varphi^*(f^*\mathfrak{m}_p) + \mathfrak{m}_n^{k+1} = g^*\mathfrak{m}_p + \mathfrak{m}_n^{k+1}.$$

Let $f' = f \circ \varphi$, which is $\mathscr{K}$-equivalent to f. The above equality gives

$$(f')^*\mathfrak{m}_p + \mathfrak{m}_n^{k+1} = g^*\mathfrak{m}_p + \mathfrak{m}_n^{k+1}.$$

For each $i = 1, \ldots, p$, we can write

$$f_i' + h_i = \sum_{j=1}^{p} a_{ij} g_j,$$

for some $h_i \in \mathfrak{m}_n^{k+1}$ and $a_{ij} \in \mathscr{O}_n$. Let $f_i'' = f_i' + h_i$ and $f'' = (f_1'', \ldots, f_p'')$. Since $j^k f'' = j^k f'$ and f' is k-$\mathscr{K}$-determined, f'' and f' are $\mathscr{K}$-equivalent. But now $(f'')^*\mathfrak{m}_p = g^*\mathfrak{m}_p$ and thus f'' and g are $\mathscr{C}$-equivalent by Proposition 4.5. □

Remark 4.2 For a multi-germ $f\colon (\mathbb{F}^n, S) \to (\mathbb{F}^p, 0)$, the algebra $Q(f)$ can be defined analogously as follows:

$$Q(f) = \frac{\mathscr{O}_{\mathbb{F}^n,S}}{f^*\mathfrak{m}_p} = \frac{\mathscr{O}_{\mathbb{F}^n,S}}{(f_1, \ldots, f_p)}.$$

The main difference is that $Q(f)$ is not local, but semi-local. In fact, we have an isomorphism

$$Q(f) \cong Q(f_1) \oplus \cdots \oplus Q(f_r),$$

where $f_1, \ldots, f_r$ are the branches of f. Proposition 4.6 holds also for multi-germs: two multi-germs $f, g\colon (\mathbb{F}^n, S) \to (\mathbb{F}^p, 0)$ are $\mathscr{K}$-equivalent if and only if their algebras $Q(f)$, $Q(g)$ are induced isomorphic. The obvious versions of Theorems 4.3 and 4.4 also hold. ◊

Exercises for Sect. 4.4

1 Since the isomorphism type of the local algebra $Q(f)$ determines f up to contact equivalence, algebraic properties of $Q(f)$ must reflect contact-invariant properties of f, and one should be able to determine invariants of f from $Q(f)$ alone.

(i) Let $f\colon (\mathbb{F}^n, 0) \to (\mathbb{F}^p, 0)$ be a map-germ. Show that

$$\operatorname{rank} d_0 f = n - \dim_{\mathbb{F}} \frac{\mathfrak{m}}{\mathfrak{m}^2},$$

where $\mathfrak{m}$ is the maximal ideal of $Q(f)$. The number $\dim_{\mathbb{F}} \mathfrak{m}/\mathfrak{m}^2$ is the *embedding dimension* of $(f^{-1}(0), 0)$): it is the dimension of the smallest affine space over $\mathbb{F}$ containing a germ isomorphic to $f^{-1}(0)$.

(ii) Characterise the local algebra $Q(f)$ when f is an immersion, and when f is a submersion.

2 Given $\mathscr{K}$-equivalent germs f and g, find a natural isomorphism $\theta(f) \to \theta(g)$ which passes to the quotient to define an isomorphism $\theta(f)/T\mathscr{K}_e(f) \to \theta(g)/T\mathscr{K}_e(g)$. See Exercise 3.3.15 for hints.

3 Show that, as stated above, multi-germs $(\mathbb{F}^n, S) \to (\mathbb{F}^p, 0)$ are contact-equivalent if and only if their mono-germ branches are contact equivalent.

4.5 Geometric Criterion for Finite $\mathscr{A}_e$-Codimension

We give here a very simple geometric criterion due to Mather and Gaffney for a multi-germ $f\colon (\mathbb{F}^n, S) \to (\mathbb{F}^p, 0)$ to have finite $\mathscr{A}_e$-codimension. Here the real C^∞ and complex analytic theories diverge: the criterion is necessary in the first case, and necessary and sufficient in the second. Roughly speaking, in the complex analytic case, a multi-germ has finite $\mathscr{A}_e$-codimension if and only if it has an isolated instability. To make this precise, we recall first the notion of locally stable mapping.

Definition 4.4 Let $f : N \to P$ be a smooth map between manifolds, let $C(f)$ be its critical set and let $\Delta(f) = f(C(f))$. We say that f is *locally stable* if:

1. the restriction $f : C(f) \to P$ is finite (i.e. closed and finite-to-one see Definition 4.5 below),
2. for any $y \in \Delta(f)$, the multi-germ $f\colon (N, S) \to (P, y)$ is stable, where $S = f^{-1}(y) \cap C(f)$.

Note that this is not the same as the definition of stability of *germs*, in Definition 3.4(4). Below we prove a Proposition 4.8, linking the two notions.

Notation Given a map $f : N \to P$ as in the definition, and $y \in \Delta(f)$, we denote by $(f)_y$ the multi-germ $f : (N, S) \to (P, y)$, where $S = f^{-1}(y) \cap C(f)$.

We recall that the critical set $C(f)$ is by definition the set of points $x \in N$ such that $d_x f$ is not surjective, so that $C(f) = N$ when $\dim N < \dim P$, but $C(f)$ coincides with the singular set $\Sigma(f)$ otherwise. The fact that we only need to check the stability of the mapping at the critical points is based on the following fact: A germ $f : (\mathbb{F}^n, S) \to (\mathbb{F}^p, 0)$ is stable if and only if the germ at the smaller set, $S \cap C(f)$, namely $f : (\mathbb{F}^n, S \cap C(f)) \to (\mathbb{F}^p, 0)$, is stable (Exercise 4.5.1 below). For this reason, from now on *we consider only multi-germs* $f : (\mathbb{F}^n, S) \to (\mathbb{F}^p, 0)$ *for which* $S \subset C(f)$.

4.5.1 Sheafification

Let $f_0 : (\mathbb{C}^n, S) \to (\mathbb{C}^p, 0)$ be a germ, and $f : X \to Y$ a representative. In order to describe what happens in the neighbourhood of S in the source and 0 in the target, we extend the module $T^1_{\mathscr{A}_e} f_0$ to a coherent sheaf, $\mathscr{T}^1_{\mathscr{A}_e} f$, on a possibly smaller neighbourhood Y' of 0 with $Y' \subseteq Y$, and then use the properties of coherent sheaves to derive the information we want. The proof of the geometric criterion uses the finite coherence theorem of Grauert, (Theorem E.2 in Appendix E) which says that the push-forward of a coherent sheaf by a finite map is once again coherent, and a corollary of the Hilbert–Ruckert Nullstellensatz (Theorem E.3 in Appendix E), which says that that if $\mathscr{S}$ is a coherent sheaf on an analytic space X, then a point $x \in X$ is isolated in the support of $\mathscr{S}$ if and only if the $\mathbb{C}$-vector space dimension of its stalk at x is finite.

Definition 4.5 A continuous map $f : X \to Y$ is *finite* if it is closed ($f(C)$ is closed in Y whenever C is closed in X) and has finite fibres.

The following property will be also discussed in Theorem D.5.

Proposition 4.7 ([GR04, I.3.2]) *Let* $f : X \to Y$ *be a holomorphic map of complex spaces, such that* x_0 *is an isolated point in* $f^{-1}(f(x_0))$. *Then there exist neighbourhoods* X' *and* Y' *of* x_0 *in* X *and* $f(x_0)$ *in* Y *such that* $f(X') \subset Y'$ *and* $f : X' \to Y'$ *is finite.* □

We fix some notation which slightly extends what we have used previously. Let $f : (\mathbb{C}^n, S) \to (\mathbb{C}^p, 0)$ be a holomorphic multi-germ and fix a representative $f : X \to Y$, where X, Y are open sets in $\mathbb{C}^n, \mathbb{C}^p$, respectively. We consider the following sheaves and morphisms of holomorphic vector fields:

- $\theta_X, \theta(f)$ are the sheaves of $\mathscr{O}_X$-modules of germs of vector fields on X and of vector fields along $f|_X$, respectively, and $tf : \theta_X \to \theta(f)$ is the morphism defined by composition with df.
- θ_Y is the sheaf of $\mathscr{O}_Y$-modules of germs of vector fields on Y and $\omega f : \theta_Y \to f_*\theta(f)$ is the induced morphism defined by composition with f.

Proposition 4.8 *If the germ $f_0\colon (\mathbb{F}^n, S) \to (\mathbb{F}^p, 0)$ is stable, in the sense of Definition 3.4(4), then for every representative $f : X \to Y$, there is a sub-representative which is locally stable in the sense of Definition 4.4.*

Proof Suppose first that $\mathbb{F} = \mathbb{C}$. By Theorem 4.1,

$$\dim_{\mathbb{C}} \frac{\theta(f)}{tf(\theta_n) + f^* \mathfrak{m}_p \theta(f)} \leq p.$$

The support of this quotient is $C(f) \cap f^{-1}(0)$, and it follows, by Theorem E.3, that S is isolated in this support. By Proposition 4.7, after shrinking X and Y, we may assume that the restriction of f to its critical set is finite, and that $f^{-1}(0) \cap C(f) \subseteq S$ (see Lemma 4.3 below).

We define a sheaf $\mathscr{T}^1_{\mathscr{R}_e} f$ on X by

$$\mathscr{T}^1_{\mathscr{R}_e} f = \frac{\theta(f)}{tf(\theta_X)}.$$

As cokernel of the morphism $tf : \theta_X \to \theta(f)$, this is a coherent sheaf. It is supported on $C(f)$. By the finite mapping Theorem E.2, $f_*(\mathscr{T}^1_{\mathscr{R}_e} f)$ is a coherent sheaf of $\mathscr{O}_Y$-modules. The morphism $\omega f : \theta_Y \to f_*(\theta(f))$ induces a morphism of coherent sheaves $\theta_Y \to f_*\left(\mathscr{T}^1_{\mathscr{R}_e}(f)\right)$. We define

$$\mathscr{T}^1_{\mathscr{A}_e} f := \frac{f_*\left(\mathscr{T}^1_{\mathscr{R}_e}(f)\right)}{\omega f(\theta_Y)}.$$

As the cokernel of a morphism of coherent sheaves, this too is coherent. Since $f^{-1}(0) \cap X = S$, the stalk of $\mathscr{T}^1_{\mathscr{A}_e} f$ at $0 \in Y$ is naturally identified with $T^1_{\mathscr{A}_e} f_0$, and is thus equal to 0. Since the support of a coherent sheaf is closed (in fact this is a special case of Theorem E.3), there is a neighbourhood Y' of 0 in Y which does not meet the support of $\mathscr{T}^1_{\mathscr{A}_e} f$. Let $X' = f^{-1}(Y')$. Then the representative $f : X' \to Y'$ is locally stable.

Now suppose that $\mathbb{F} = \mathbb{R}$, and that f is a stable C^∞ germ. By the finite determinacy Theorem 6.2 below, f is right-left equivalent to a polynomial map-germ. Since the assertion of the proposition is invariant under right-left equivalence, we may assume that f is this polynomial germ. It therefore has a complexification, $f_{\mathbb{C}}\colon (\mathbb{C}^n, S) \to (\mathbb{C}^p, 0)$, defined by the same real formula, now with complex variables. This formula also defines a complexification of any real analytic representative of f. By what we have shown, there are open neighbourhoods $X'_{\mathbb{C}}$, $Y'_{\mathbb{C}}$ of S and $0 \in \mathbb{C}^p$ such that $f_{\mathbb{C}} : X'_{\mathbb{C}} \to Y'_{\mathbb{C}}$ is locally stable. Let $x_1, \ldots, x_r \in X'_{\mathbb{C}} \cap \mathbb{R}^n$ share the same image. The stability of the complex analytic multi-germ $f_{\mathbb{C}}\colon (X'_{\mathbb{C}}, \{x_1, \ldots, x_r\}) \to (Y'_{\mathbb{C}}, y)$ implies the stability of the real C^∞ multi-germ $f : (X'_{\mathbb{C}} \cap \mathbb{R}^n, \{x_1, \ldots, x_r\}) \to (Y'_{\mathbb{C}} \cap \mathbb{R}^p, y)$, by Proposition 3.8. Thus the representative $f : X'_{\mathbb{C}} \cap \mathbb{R}^n \to Y'_{\mathbb{C}} \cap \mathbb{R}^p$ has the required property. □

Lemma 4.3 *If $f\colon (\mathbb{C}^n, S) \to (\mathbb{C}^p, 0)$ has finite $\mathscr{A}_e$-codimension, then for a sufficiently small representative $f : U \to V$, $f^{-1}(0) \cap C(f) \subset S$.*

Proof By the hypothesis that $\dim_{\mathbb{C}} T^1_{\mathscr{A}_e} f < \infty$, there exist $h_1, \ldots, h_r \in \mathscr{O}_{\mathbb{C}^n,S}$ such that

$$T\mathscr{A}_e(f) + \mathrm{Sp}_{\mathbb{C}}\{h_1, \ldots, h_r\} = \theta(f).$$

Since $\omega f(\theta_{\mathbb{C}^p,0}) \subset (f^*\mathfrak{m}_{\mathbb{C}^p,0})\theta(f) + \{\partial/\partial y_1, \ldots, \partial/\partial y_p\}$, $\dim_{\mathbb{C}} \theta(f)/T\mathscr{K}_e(f) \leq r + p$. Therefore for each $x \in S$ we also have $\dim_{\mathbb{C}} \theta(f_x)/T\mathscr{K}_e(f_x) < \infty$, where f_x denotes the mono-germ of f at x.

Let $f : U \to V$ be any representative of the multi-germ and consider the sheaf of $\mathscr{O}_U$-modules

$$\mathscr{T}^1_{\mathscr{K}_e} f = \frac{\theta(f)}{tf(\theta_U) + (f^*\mathfrak{m}_{V,0})\theta(f)},$$

where $\mathfrak{m}_{V,0}$ is the ideal sheaf in $\mathscr{O}_V$ of functions vanishing at 0. Since all sheaves appearing in the definition of $\mathscr{T}^1_{\mathscr{K}_e} f$ are coherent (in fact, $\theta_U, \theta(f)$ are locally free of rank n, p, respectively), $\mathscr{T}^1_{\mathscr{K}_e} f$ is also coherent. The stalk of $\mathscr{T}^1_{\mathscr{K}_e} f$ at $x \in S$ is

$$\frac{\theta(f_x)}{tf_x(\theta_{U,x}) + (f_x^*\mathfrak{m}_{V,0})\theta(f_x)} = \frac{\theta(f_x)}{T\mathscr{K}_e(f_x)}.$$

Hence, for each $x \in S$, $\dim_{\mathbb{C}} \left(\mathscr{T}^1_{\mathscr{K}_e} f\right)_x < \infty$, and therefore by the Nullstellensatz Theorem E.3, x is isolated in the support of $\mathscr{T}^1_{\mathscr{K}_e} f$. Thus we can find an open neighbourhood U' of S in U such that

$$\mathrm{Supp}\, \mathscr{S} \cap U' \subset S.$$

To finish the proof, it only remains to show that $\mathrm{Supp}\, \mathscr{T}^1_{\mathscr{K}_e} f = C(f) \cap f^{-1}(0)$. In fact, if $x \in C(f) \cap f^{-1}(0)$, then $f(x) = 0$ and $d_x f$ is not surjective. This implies that $f_x^*\mathfrak{m}_{V,0} \subset \mathfrak{m}_{U,x}$ and that $\partial/\partial y_i \notin tf_x(\theta_{U,x})$ for some $i = 1, \ldots, p$. So $\left(\mathscr{T}^1_{\mathscr{K}_e} f\right)_x \neq 0$.

Conversely, if $x \notin C(f) \cap f^{-1}(0)$, then either $f(x) \neq 0$ or $d_x f$ is surjective. It follows that either $f_x^*\mathfrak{m}_{V,0} = \mathscr{O}_{U,x}$ or $\partial/\partial y_i \in tf_x(\theta_{U,x})$ for all $i = 1, \ldots, p$ and hence, $\left(\mathscr{T}^1_{\mathscr{K}_e} f\right)_x = 0$ in both cases. □

Theorem 4.5 (Mather–Gaffney Criterion) *A holomorphic multi-germ $f\colon (\mathbb{C}^n, S) \to (\mathbb{C}^p, 0)$ (with $S \subset C(f)$) has finite $\mathscr{A}_e$-codimension if and only if there is a small enough representative $f : X \to Y$ such that:*

1. *$f^{-1}(0) \cap C(f) = S$;*
2. *the restriction $f : X \setminus f^{-1}(0) \to Y \setminus \{0\}$ is locally stable.*

Proof Suppose first that f_0 has finite $\mathscr{A}_e$-codimension. By Lemma 4.3, we can choose a representative f satisfying condition (1). By Proposition 4.7, after shrinking the neighbourhoods X, Y if necessary, we can assume that the restriction to the critical set $f : C(f) \to Y$ is finite. The coherent sheaf $\mathscr{T}^1_{\mathscr{A}_e} f$ introduced in Proposition 4.8 has stalk at $y \in Y$ equal to

$$\frac{(f)_*(\mathscr{T}^1_{\mathscr{R}_e} f)_y}{\omega f(\theta_{Y,y})} = \frac{\left(\bigoplus_{x\in S'} \frac{\theta(f_x)}{tf_x(\theta_{X,x})}\right)}{\omega f(\theta_{Y,y})} \cong \frac{\theta(f_{S'})}{tf_{S'}(\theta_{X,S'}) + \omega f_{S'}(\theta_{Y,y})} = T^1_{\mathscr{A}_e}(f_{S'}),$$

where $S' = f^{-1}(y) \cap C(f)$ and $f_{S'}$ denotes the multi-germ of f at S'.

Now we prove the equivalence of finite $\mathscr{A}_e$-codimension with condition (2). The stalk at $0 \in Y$ of $\mathscr{T}_{\mathscr{A}_e} f$ is equal to $T^1_{\mathscr{A}_e} f_0$. If this has finite dimension over $\mathbb{C}$, then by the corollary to the Nullstellensatz, Theorem E.3, there is an open set Y', with $0 \in Y' \subset Y$, such that $\operatorname{Supp} \mathscr{T}^1_{\mathscr{A}_e} f \cap Y' \subset \{0\}$. But this means that $\left(\mathscr{T}^1_{\mathscr{A}_e} f\right)_y = 0$ for every $y \in Y' \setminus \{0\}$ and hence the restriction of f to $X' \setminus S$ is locally stable, where $X' = f^{-1}(Y')$.

Conversely, assume there is an open set X', $S \subset X' \subset X$, such that the restriction of f to $X' \setminus S$ is locally stable. Since $f : C(f) \to Y$ is finite, there is an open neighbourhood Y' of 0, contained in Y, such that $f(X' \cap C(f)) = Y' \cap f(C(f))$. Then for each $y \in Y' \setminus \{0\}$ we have $\left(\mathscr{T}^1_{\mathscr{A}_e} f\right)_y = 0$. Thus, $\operatorname{Supp} \mathscr{T}^1_{\mathscr{A}_e} f \cap Y' \subset \{0\}$ and again by Theorem E.3 we get

$$\dim_{\mathbb{C}} \left(\mathscr{T}^1_{\mathscr{A}_e} f\right)_0 = \dim_{\mathbb{C}} T^1_{\mathscr{A}_e}(f_S) < \infty.$$

□

An argument using the complexification and the finite determinacy theorem, as in the last part of the proof of Proposition 4.8, shows that if $f : (\mathbb{R}^n, S) \to (\mathbb{R}^p, 0)$ is $\mathscr{A}$-finite, then it has isolated instability. The converse does not hold: for example, the germ $(\mathbb{R}^2, 0) \to (\mathbb{R}, 0)$ defined by $(x, y) \mapsto (x^2 + y^2)^2$ is not $\mathscr{A}$-finite despite having isolated instability (in $\mathbb{R}^2$ though not in $\mathbb{C}^2$).

Corollary 4.3 *Let $f : (\mathbb{R}^n, S) \to (\mathbb{R}^p, 0)$ be $\mathscr{A}$-finite. Then there exists a representative $f : X \to Y$ such that:*

1. $f^{-1}(0) \cap C(f) = S$;
2. *the restriction $f : X \setminus f^{-1}(0) \to Y \setminus \{0\}$ is locally stable.*

Proof By the Finite Determinacy Theorem 6.2 we can assume that f is a polynomial and by Proposition 3.8, its complexification $f_{\mathbb{C}}$ is also $\mathscr{A}$-finite. Hence, there exists a representative $f_{\mathbb{C}} : \hat{X} \to \hat{Y}$ such that

1. $f_{\mathbb{C}}^{-1}(0) \cap C(f_{\mathbb{C}}) = S$;
2. the restriction $f_{\mathbb{C}} : \hat{X} \setminus f_{\mathbb{C}}^{-1}(0) \to \hat{Y} \setminus \{0\}$ is locally stable.

Let X and Y be the projections of $\hat{X}$ and $\hat{Y}$ on $\mathbb{R}^n$ and $\mathbb{R}^p$, respectively, and consider the representative $f : X \to Y$. For each $y \in Y$, the set $S' = f^{-1}(y) \cap C(f)$ is contained in $\hat{S}' = f_{\mathbb{C}}^{-1}(y) \cap C(f_{\mathbb{C}})$. For $y = 0$, we get $f^{-1}(0) \cap C(f) = S$. For $y \neq 0$, the complexification of the germ $f\colon (\mathbb{R}^n, S') \to (\mathbb{R}^p, y)$ is obtained as the restriction of $f_{\mathbb{C}}\colon (\mathbb{C}^n, \hat{S}') \to (\mathbb{C}^p, y)$ to the branches with real base-point. Since the restriction of a stable germ is also stable, $f\colon (\mathbb{R}^n, S') \to (\mathbb{R}^p, y)$ is stable, again by Proposition 3.8. □

Example 4.4 Consider the case of a holomorphic function $f\colon (\mathbb{C}^n, 0) \to (\mathbb{C}, 0)$. If f has finite $\mathscr{A}_e$-codimension, then by the Mather–Gaffney criterion Theorem 4.5, there is a small enough representative $f : X \to Y$ such that

1. $f^{-1}(0) \cap C(f) \subset \{0\}$;
2. the restriction $f : X \setminus f^{-1}(0) \to Y \setminus \{0\}$ is locally stable.

By Exercise 5, all the critical points of f on $X \setminus \{0\}$ are non-degenerate, and so are isolated. Since $C(f)$ is a closed analytic subset of X, it follows that $C(f)$ cannot accumulate at the origin. Thus, after shrinking the neighbourhood X if necessary, we can assume that $C(f) = \{0\}$. The converse is obviously true, so we have shown that f has finite $\mathscr{A}_e$-codimension if and only if it has an isolated critical point. ◊

Exercises for Sect. 4.5

1 Let $f\colon (\mathbb{C}^n, S) \to (\mathbb{C}^p, 0)$ be a holomorphic multi-germ, with S a finite set. Show that f is stable if and only if the germ at the smaller set, $S \cap C(f)$, namely $f\colon (\mathbb{F}^n, S \cap C(f)) \to (\mathbb{F}^p, 0)$, is stable.

2 Use an analogous argument to that of Example 4.4 to show that a germ $f\colon (\mathbb{C}, 0) \to (\mathbb{C}^2, 0)$ is $\mathscr{A}$-finite if and only if it is one-to-one.

3 Suppose that $f\colon (\mathbb{F}^n, 0) \to (\mathbb{F}^{n+1}, 0)$ is $\mathscr{A}$-finite. Show that

(i) the set of points where f is not an immersion has codimension ≥ 2. Hint: $J^1(n, n+1)$ can be identified with the space $L(n, n+1)$ of linear maps $\mathbb{F}^n \to \mathbb{F}^{n+1}$. Apply Proposition A.1(2) to each of the strata $W_i = \{\sigma \in J^1(n, n+1) : \dim \ker \sigma = i\}$.
(ii) if $\mathbb{F} = \mathbb{C}$, then the codimension of this set is exactly 2, unless it is empty (see Sect. C.5).

4 Suppose that $f\colon (\mathbb{F}^n, 0) \to (\mathbb{F}^p, 0)$, with $n \geq p$, is $\mathscr{A}$-finite. Show that

(i) the set $C(f)$ of points in the domain where f is not a submersion has codimension $\geq n - p + 1$.
(ii) if $\mathbb{F} = \mathbb{C}$, then the codimension of this set is exactly $n - p + 1$, unless it is empty.

5 Use Theorem 4.5 to give a necessary and sufficient geometrical criterion for a germ $(\mathbb{C}^2, 0) \to (\mathbb{C}^2, 0)$ to be $\mathscr{A}$-finite.

6 Use an analogous argument to that of Example 4.4 to show that a germ $f : (\mathbb{C}^2, 0) \to (\mathbb{C}^3, 0)$ has finite $\mathscr{A}_e$-codimension if and only if f is an immersion with only transverse double points away from the origin. Equivalently, $\operatorname{codim}_{\mathscr{A}_e}(f) = \infty$ if and only if either:

(i) there is a curve of non-immersive points in the domain;
(ii) there is a curve of non-transverse self-intersection in the target;
(iii) there is a curve of triple points in the target.

7 Show that the germ $f : (\mathbb{C}^2, 0) \to (\mathbb{C}^3, 0)$ given by $f(x, y) = (x, y^2, y(x^k+y^2))$ has finite $\mathscr{A}_e$-codimension. (Hint: Use the fact that the mapping $(x, y) \mapsto (x, y^2)$ has no triple points to deduce that the same is true for f).

8 Let $f : (\mathbb{C}^2, 0) \to (\mathbb{C}^3, 0)$ be the germ $f_c(x, y) = (x, y^3+xy, y^4+cxy^2)$, where $c \in \mathbb{C}$ is a constant. Show that f_c has infinite $\mathscr{A}_e$-codimension when $c = 2/3$, 1 or 2 by showing explicitly that the image contains a curve of one of the types listed in Exercise 6. Hint: in fact each of the three degeneracies (a), (b) and (c) of Exercise 2 occurs for one of these three values of c. When $c = 1$ the first coordinate axis in the target is triple-covered.

9 Prove the following geometric criterion for $\mathscr{K}$-finiteness: the germ $f : (\mathbb{C}^n, 0) \to (\mathbb{C}^p, 0)$ is $\mathscr{K}$-finite if and only if there is a representative which is finite-to-one on its critical set.

4.6 Transversality

The aim of this section is to prove that a map-germ is stable if and only if its multijet extension map is transverse to its $\mathscr{A}$-orbit in the jet space, if and only if it is transverse to its $\mathscr{K}$-orbit.

Let us denote by ${}_rJ^k(n, p)$ the space of k-jets of smooth map germs $f : (\mathbb{F}^n, S) \to (\mathbb{F}^p, 0)$, where $r = |S|$. Evidently ${}_rJ^k(n, p)$ is isomorphic to $({}_1J^k(n, p))^r$. In what follows we generally omit the initial subscript, since the value of r will be clear from the context. Since $J^k(n, p)$ is a finite-dimensional vector space, it has a natural structure of manifold. There is an epimorphism

$$j^k : \mathfrak{m}_n\theta(f) \to J^k(n, p),$$

whose kernel is $\mathfrak{m}_n^{k+1}\theta(f)$, which allows us to identify $J^k(n, p)$ with the quotient $\mathfrak{m}_n\theta(f)/\mathfrak{m}_n^{k+1}\theta(f)$. In fact, this quotient can be thought as the tangent space of $J^k(n, p)$ at $\sigma = j^k f$.

We can also describe the tangent vectors to $J^k(n, p)$ at σ by means of k-jets of 1-parameter unfoldings of f. Let $F : (\mathbb{F}^n \times \mathbb{F}, S \times \{0\}) \to (\mathbb{F}^p \times \mathbb{F}, 0)$ be any origin-preserving unfolding of f, given by $F(x,t) = (f_t(x), t)$. This gives rise to a smooth curve $\gamma\colon (\mathbb{F}, 0) \to J^k(n, p)$ defined by $\gamma(t) = j^k f_t$, such that $\gamma(0) = \sigma$. The associated tangent vector is:

$$\gamma'(0) = \frac{d}{dt}\{j^k f_t\}\bigg|_{t=0} = j^k\left(\frac{df_t}{dt}\bigg|_{t=0}\right),$$

where the last equality comes by changing the order of differentiation. In fact, it is easy to see that any tangent vector to $J^k(n, p)$ at σ can be obtained in this way.

For each $\ell \geq k$, there is a natural projection

$$\pi^k_\ell : J^\ell(n, p) \to J^k(n, p),$$

defined by $\pi^k_\ell(j^\ell f) = j^k f$. This map is obviously an epimorphism of vector spaces and hence a submersion when we consider the manifold structures.

For each group $\mathscr{G} = \mathscr{R}, \mathscr{L}, \mathscr{C}, \mathscr{A}$ or $\mathscr{K}$, denote by ${}_r\mathscr{G}^{(k)}$ the space of k-multi-jets of germs in $\mathscr{G}$. As above, we will in general omit the subindex r. In each case, $\mathscr{G}^{(k)}$ is an open set of some k-jet space, so it has again a natural structure of manifold. Moreover, composition induces a group structure on $\mathscr{G}^{(k)}$ so that it becomes a Lie group and the action of $\mathscr{G}$ on the set of map germs also induces a Lie group action of $\mathscr{G}^{(k)}$ on $J^k(n, p)$. For $\mathscr{G} = \mathscr{R}, \mathscr{C}$ and $\mathscr{K}$, the group ${}_r\mathscr{G}^{(k)}$ acting on ${}_rJ^k(n, p)$ is just $({}_1G^{(k)})^r$, with the product action, but note that ${}_r\mathscr{L}^{(k)}$ is just a single copy of ${}_1\mathscr{L}^{(k)}$ with the diagonal action, and ${}_r\mathscr{A}^{(k)} = {}_r\,\mathscr{R}^{(k)} \times_r \mathscr{L}^{(k)}$.

If G is a Lie group acting smoothly on a manifold M, the orbit of a point $x \in M$ is denoted by Gx, and is the image of the smooth map $\alpha_x : G \to M$ given by $\alpha_x(g) = gx$. We consider the differential $d_e\alpha_x : T_eG \to T_xM$ at the identity e of G. The tangent space T_eG is called the Lie algebra of G and is denoted by $\mathfrak{g}$. There is an induced map $\mathfrak{g} \times M \to TM$ given by $\eta \cdot x = d_e\alpha_x(\eta)$, for each $\eta \in \mathfrak{g}$ and $x \in M$, known as the *infinitesimal action* of $\mathfrak{g}$ on M.

It is well known that orbits are injectively immersed submanifolds, so that they always have a well-defined tangent space at each point. We include here a proof of this result.

Lemma 4.4 *Let G be a Lie group with Lie algebra $\mathfrak{g}$ acting smoothly on a manifold M. For each $x \in M$, the orbit Gx is an injectively immersed submanifold whose tangent space at x is $\mathfrak{g}x$.*

Proof Let G_x be the isotropy group of x, i.e. $G_x = \{g \in G : \; gx = x\}$. This is a closed subgroup of G, hence G/G_x has a unique smooth manifold structure and moreover G acts transitively on G/G_x by left multiplication. We factor the map α_x as a composition

$$\begin{array}{ccc} G & \xrightarrow{\;\pi\;} & G/G_x \\ & \searrow^{\alpha_x} & \downarrow i \\ & & M \end{array}$$

where π is surjective and i is injective. Since π and i are both equivariant with respect to the G action, they have constant rank. By the constant rank theorem, π must be a submersion and i an immersion. Thus, Gx is the image of i which is an injective immersion and its tangent space at x is the image of the differential of i, that is,

$$T_x(Gx) = d_{[e]}i(T_{[e]}(G/G_x)) = d_{[e]}i(d_e\pi(T_eG)) = d_e\alpha_x(T_eG) = \mathfrak{g}x. \qquad \square$$

We remark that in general the orbits may not be (embedded) submanifolds, unless they are closed subsets of M. In our case, the manifold $M = J^k(n, p)$ is a vector space, the Lie group $G = \mathscr{G}^{(k)}$ is a semialgebraic subset of a vector space and the action is given by a semialgebraic map (see Appendix B for the definitions of semialgebraic sets and maps). The next proposition shows that in that case, the orbits are in fact submanifolds.

Proposition 4.9 *Let G be a Lie group acting smoothly on a manifold M. If the action $G \times M \to M$ is semialgebraic, then the orbits are submanifolds of M.*

Proof For each $x \in M$, the map $\alpha_x : G \to M$ is semialgebraic and hence, its image Gx is also semialgebraic by the Tarski–Seidenberg Theorem B.4. By Theorem B.5, Gx has at least one regular point $y \in Gx$ of dimension $d = \dim Gx$. Let $z \in Gx$ be any other point and assume that $y = gx$ and $z = hx$, for some $g, h \in G$. The map $\phi : M \to M$ given by $u \to hg^{-1}u$ is a diffeomorphism such that $\phi(Gx) = Gx$ and $\phi(y) = z$. Thus, z is also a regular point of Gx of dimension d. This proves that Gx is a submanifold of M. $\square$

Lemma 4.5 *For each $\sigma = j^k f \in J^k(n, p)$, the orbit $\mathscr{G}^{(k)}\sigma$ is a submanifold of $J^k(n, p)$ whose tangent space at σ is*

$$T_\sigma(\mathscr{G}^{(k)}\sigma) = j^k(T\mathscr{G}(f)).$$

Proof By Lemma 4.4 and Proposition 4.9, $\mathscr{G}^{(k)}\sigma$ is a submanifold of $J^k(n, p)$ whose tangent space at σ is

$$T_\sigma(\mathscr{G}^{(k)}\sigma) = d_e\alpha_\sigma(T_e\mathscr{G}^{(k)}),$$

where $\alpha_\sigma : \mathscr{G}^{(k)} \to J^k(n, p)$ is the map $\alpha_\sigma(\varphi) = \varphi\sigma$. Thus, all tangent vectors in $T_\sigma(\mathscr{G}^{(k)}\sigma)$ can be obtained as follows:

$$\begin{aligned} d_e\alpha_\sigma\left(j^k\left(\left.\frac{d\varphi_t}{dt}\right|_{t=0}\right)\right) &= d_e\alpha_\sigma\left(\left.\frac{d}{dt}\{j^k\varphi_t\}\right|_{t=0}\right) = \left.\frac{d}{dt}\{\alpha_\sigma(j^k\varphi_t)\}\right|_{t=0} \\ &= \left.\frac{d}{dt}\{j^k(\varphi_t f)\}\right|_{t=0} = j^k\left(\left.\frac{d(\varphi_t f)}{dt}\right|_{t=0}\right), \end{aligned}$$

where φ_t is any origin-preserving unfolding of the identity in $\mathscr{G}$. To complete the proof, it only remains to show that $T\mathscr{G}(f)$ coincides with the set of all vector fields obtained as $d(\varphi_t f)/dt|_{t=0}$, where φ_t is any origin-preserving unfolding of the identity in $\mathscr{G}$.

Assume $\mathscr{G} = \mathscr{A}$. Let ϕ_t and ψ_t be origin-preserving unfoldings of the identity in $(\mathbb{F}^n, S)$ and $(\mathbb{F}^p, 0)$, respectively. As we saw in Sect. 3.2, we have

$$\frac{d}{dt}\{\psi_t \circ f \circ \phi_t^{-1}\}\bigg|_{t=0} = tf\left(\frac{d\phi_t^{-1}}{dt}\bigg|_{t=0}\right) + \omega f\left(\frac{d\psi_t}{dt}\bigg|_{t=0}\right) \in T\mathscr{A}(f).$$

Conversely, if $\vartheta \in T\mathscr{A}(f)$, there exist $\xi \in \mathfrak{m}_n\theta_n$ and $\eta \in \mathfrak{m}_p\theta_p$ such that $\vartheta = tf(\xi) + \omega f(\eta)$. Let ϕ_t and ψ_t be origin-preserving unfoldings of the identity in $(\mathbb{F}^n, S)$ and $(\mathbb{F}^p, 0)$, respectively, such that $\xi = d\phi_t^{-1}/dt|_{t=0}$ and $\eta = d\psi_t/dt|_{t=0}$, Then,

$$\vartheta = tf\left(\frac{d\phi_t^{-1}}{dt}\bigg|_{t=0}\right) + \omega f\left(\frac{d\psi_t}{dt}\bigg|_{t=0}\right) = \frac{d}{dt}\{\psi_t \circ f \circ \phi_t^{-1}\}\bigg|_{t=0}.$$

The proof for the other groups is analogous and is left as an exercise. □

Let $f\colon (\mathbb{F}^n, S) \to (\mathbb{F}^p, 0)$ be a smooth map-germ and let $\sigma = j^k f$ be its k-jet. Take a representative $f : X \to Y$, where X, Y are open neighbourhoods of S and 0 in $\mathbb{F}^n, \mathbb{F}^p$, respectively, and denote by ${}_rj^k f : X^{(r)} \to {}_rJ^k(X, Y)$ the multi-jet extension map of f. Recall that ${}_rJ^k(X, Y)$ is a smooth fibre bundle over $X^{(r)} \times Y^r$ with fibre $(J^k(n, p))^r$, so that its tangent space at σ can be described as

$$T_\sigma({}_rJ^k(X, Y)) \cong \mathbb{F}^{rn} \oplus \mathbb{F}^{rp} \oplus \frac{\mathfrak{m}_n\theta(f)}{\mathfrak{m}_n^{k+1}\theta(f)} \cong \mathbb{F}^{rn} \oplus \frac{\theta(f)}{\mathfrak{m}_n^{k+1}\theta(f)}. \tag{4.17}$$

For any $\mathscr{A}^{(k)}$-invariant subset $W \subset J^k(n, p)$, there is an associated subbundle $W(X, Y) \subset {}_rJ^k(X, Y)$ with fibre W. In particular, this applies when W is an orbit $\mathscr{A}^{(k)}$ or $\mathscr{K}^{(k)}$ orbit, in which case $W(X, Y)$ is a submanifold of ${}_rJ^k(X, Y)$.

Lemma 4.6 *Assume $S = \{s_1, \ldots, s_r\}$ and set $z = (s_1, \ldots, s_r) \in X^{(r)}$. Then:*

1. *For $W = \mathscr{A}^{(k)}\sigma$, ${}_rj^k f$ is transverse to $W(X, Y)$ at z if and only if*

$$T\mathscr{A}_e(f) + \mathfrak{m}_n^{k+1}\theta(f) = \theta(f).$$

2. *For $W = \mathscr{K}^{(k)}\sigma$, ${}_rj^k f$ is transverse to $W(X, Y)$ at z if and only if*

$$T\mathscr{K}_e(f) + T\mathscr{A}_e(f) + \mathfrak{m}_n^{k+1}\theta(f) = \theta(f).$$

Proof Consider first the case $W = \mathscr{A}^{(k)}\sigma$. By Lemma 4.5 and via the identification (4.17), the tangent space to $W(X, Y)$ at σ is

$$\mathbb{F}^{rn} \oplus \mathbb{F}^{rp} \oplus \frac{T\mathscr{A}(f) + \mathfrak{m}_n^{k+1}\theta(f)}{\mathfrak{m}_n^{k+1}\theta(f)} \cong \mathbb{F}^{rn} \oplus \frac{tf(\mathfrak{m}_n\theta_n) + \omega f(\theta_p) + \mathfrak{m}_n^{k+1}\theta(f)}{\mathfrak{m}_n^{k+1}\theta(f)}. \tag{4.18}$$

Since ${}_r j^k f$ is the r-fold product of $j^k f$, the image of $d_z({}_r j^k f)$ is the sum of the images of the differentials $d_{s_i}(j^k f)$, for $i = 1, \dots, r$. Take coordinates $x = (x_1, \dots, x_n)$ in X. Then, for each $j = 1, \dots, n$

$$\frac{\partial j^k f}{\partial x_j}(s_i) = \frac{d}{dt}\{j^k f_t(s_i)\}\bigg|_{t=0},$$

where $f_t(x) = f(x + te_j)$. By linearity of the k-jet, we get

$$\frac{\partial j^k f}{\partial x_j}(s_i) = j^k\left(\frac{df_t}{dt}\bigg|_{t=0}\right)(s_i) = j^k\left(\frac{\partial f}{\partial x_j}\right)(s_i).$$

By means of the identification (4.17), we deduce that the image of $d_z({}_r j^k f)$ is equal to

$$\mathbb{F}^{rn} \oplus \frac{tf(\mathrm{Sp}_{\mathbb{F}}\{e_1, \dots, e_n\}) + \mathfrak{m}_n^{k+1}\theta(f)}{\mathfrak{m}_n^{k+1}\theta(f)}. \tag{4.19}$$

The sum of (4.18) plus (4.19) gives

$$\mathbb{F}^{rn} \oplus \frac{T\mathscr{A}_e(f) + \mathfrak{m}_n^{k+1}\theta(f)}{\mathfrak{m}_n^{k+1}\theta(f)},$$

which proves part 1.

In the case $W = \mathscr{K}^{(k)}\sigma$, the tangent space to $W(U, V)$ at σ is

$$\mathbb{F}^{rn} \oplus \mathbb{F}^{rp} \oplus \frac{tf(\mathfrak{m}_n\theta_n) + f^*\mathfrak{m}_p\theta(f) + \mathfrak{m}_n^{k+1}\theta(f)}{\mathfrak{m}_n^{k+1}\theta(f)}. \tag{4.20}$$

Since $\mathbb{F}^{rp} \oplus f^*\mathfrak{m}_p\theta(f) = \omega f(\theta_p) + f^*\mathfrak{m}_p\theta(f)$, the sum of (4.20) plus (4.19) now gives

$$\mathbb{F}^{rn} \oplus \frac{T\mathscr{A}_e(f) + T\mathscr{K}_e(f) + \mathfrak{m}_n^{k+1}\theta(f)}{\mathfrak{m}_n^{k+1}\theta(f)},$$

so we have part 2. □

Theorem 4.6 *Let* $f\colon (\mathbb{F}^n, S) \to (\mathbb{F}^p, 0)$ *be a smooth map-germ, where* $S = \{s_1, \ldots, s_r\}$, *and let* $f : X \to Y$ *be a representative. Write* $z = (s_1, \ldots, s_r) \in X^{(r)}$, *and* $\sigma = j^{p+1} f$. *The following statements are equivalent:*

1. f *is stable;*
2. ${}_r j^{p+1} f$ *is transverse to* $W(X, Y)$ *at* z, *for* $W = \mathscr{A}^{(p+1)}\sigma$;
3. ${}_r j^{p+1} f$ *is transverse to* $W(X, Y)$ *at* z, *for* $W = \mathscr{K}^{(p+1)}\sigma$.

Proof By Theorem 3.2 and Lemma 4.6, we have to prove that the following statements are equivalent:

(a) $T\mathscr{A}_e(f) = \theta(f)$,
(b) $T\mathscr{A}_e(f) + \mathfrak{m}_n^{p+2}\theta(f) = \theta(f)$,
(c) $T\mathscr{K}_e(f) + T\mathscr{A}_e(f) + \mathfrak{m}_n^{p+2}\theta(f) = \theta(f)$.

The implications (a) $\Rightarrow$ (b) $\Rightarrow$ (c) are obvious. Assume (c). Then since $T\mathscr{A}_e(f) \subset T\mathscr{K}_e(f) + \mathrm{Sp}_{\mathbb{F}}\{e_1, \ldots, e_p\}$, the classes of $\{e_1, \ldots, e_p\}$ generate $\theta(f)/(T\mathscr{K}_e(f) + \mathfrak{m}_n^{p+2}\theta(f))$ and hence,

$$\dim_{\mathbb{F}} \frac{\theta(f)}{T\mathscr{K}_e(f) + \mathfrak{m}_n^{p+2}\theta(f)} \le p.$$

The submodules $B_k = T\mathscr{K}_e(f) + \mathfrak{m}_n^k\theta(f)$, with $k = 0, \ldots, p+2$, form a decreasing chain. Since $\dim_{\mathbb{F}} B_0/B_{p+2} \le p$, they cannot be all distinct. Thus, $\mathfrak{m}_n^k\theta(f) \subset T\mathscr{K}_e(f) + \mathfrak{m}_n^{k+1}\theta(f)$, for some $k = 0, \ldots, p+1$. By Nakayama's Lemma C.1, $\mathfrak{m}_n^k\theta(f) \subset T\mathscr{K}_e(f)$, and hence $\mathfrak{m}_n^{p+2}\theta(f) \subset T\mathscr{K}_e(f)$. Thus $T\mathscr{K}_e(f) + T\mathscr{A}_e(f) = \theta(f)$, or equivalently,

$$T\mathscr{A}_e(f) + f^*\mathfrak{m}_p\theta(f) = \theta(f). \tag{4.21}$$

Because $\theta(f)/\big(tf(\theta_n) + f^*\mathfrak{m}_p\theta(f)\big)$ is finite-dimensional as $\mathbb{F}$-vector space, it follows by the preparation theorem that $\theta(f)/tf(\theta_n)$ is finitely generated over $\mathscr{O}_p$. Hence so is $\theta(f)/T\mathscr{A}_e(f)$. But now from (4.21) it follows by Nakayama's Lemma that $T\mathscr{A}_e(f) = \theta(f)$. □

Corollary 4.4 *Let* $f : N \to P$ *be a smooth map between manifolds, with* $\dim P = p$ *and such that the restriction* $f : C(f) \to P$ *is finite. The following statements are equivalent:*

1. f *is locally stable;*
2. *for all* $r \ge 1$, ${}_r j^{p+1} f$ *is transverse to every submanifold* $W(N, P)$ *in* ${}_r J^{p+1}(N, P)$, *where* W *ranges over all of the* $\mathscr{A}^{(p+1)}$*-orbits;*
3. *for all* $r \ge 1$, ${}_r j^{p+1} f$ *is transverse to every submanifold* $W(N, P)$ *in* ${}_r J^{p+1}(N, P)$, *where* W *ranges over all the* $\mathscr{K}^{(p+1)}$*-orbits.*

4.7 Thom–Boardman Singularities

The Thom–Boardman singularities were introduced by Thom in [Tho56] for generic mappings $f: X \to Y$ as a way to find a partition of X which is finer than the rank partition and to take into account higher order derivatives. Recall that given an integer $i_1 \geq 0$, $\Sigma^{i_1}(f)$ is the set of points $x \in X$ where $d_x f$ has kernel rank i_1. Clearly, $\Sigma^{i_1}(f) = (j^1 f)^{-1}(\Sigma^{i_1})$, where $\Sigma^{i_1} \subset J^1(X, Y)$ is the submanifold $X \times Y \times$ {kernel rank i_1 matrices} and $j^1 f : X \to J^1(X, Y)$ is the jet extension map. If f is generic, then $j^1 f$ is transverse to Σ^{i_1}, so $\Sigma^{i_1}(f)$ is a submanifold of X. For each $i_2 \geq 0$ we can then define $\Sigma^{i_1,i_2}(f)$ as $\Sigma^{i_2}(f|_{\Sigma^{i_1}})$. In general, if we have constructed $\Sigma^{i_1,\dots,i_{k-1}}(f)$ and assume that it is a submanifold of X, then we can define $\Sigma^{i_1,\dots,i_k}(f)$ as $\Sigma^{i_k}\left(f|_{\Sigma^{i_1,\dots,i_{k-1}}}\right)$.

The unsatisfactory point of this construction is that we must ensure that f is sufficiently generic that at each step, $\Sigma^{i_1,\dots,i_k}(f)$ is a submanifold of X, in order to proceed to the next step. Boardman solved this problem in [Boa67] by constructing subsets $\Sigma^{i_1,\dots,i_k} \subset J^k(X, Y)$ which allow to define $\Sigma^{i_1,\dots,i_k}(f)$ as $(j^k)^{-1}(\Sigma^{i_1,\dots,i_k})$, for any smooth mapping $f: X \to Y$ (not necessarily generic). The main results of [Boa67] are that the $\Sigma^{i_1,\dots,i_k}$ are in fact submanifolds of $J^k(X, Y)$ and if $j^k f$ is transverse to $\Sigma^{i_1,\dots,i_k}$, then f is sufficiently generic for the earlier construction to work, and $\Sigma^{i_1,\dots,i_k}(f) = \Sigma^{i_k}\left(f|_{\Sigma^{i_1,\dots,i_{k-1}}}\right)$.

Example 4.5 Let $f_u: \mathbb{R}^2 \to \mathbb{R}^2$ be the mapping $f_u(x, y) = (x, y^3+x^2y+uy)$, with $u \in \mathbb{R}$. When $u = 0$, f_0 has an isolated instability at the origin known as the "lips" singularity. For all u, f_u has only kernel rank 1 singularities and the singular set $\Sigma^1(f_u)$ is given by $3y^2 + x^2 = -u$. If $u > 0$, this is the empty set so f_u is regular and hence, locally stable. Otherwise, if $u < 0$, $\Sigma^1(f_u)$ is an ellipse. To compute the singularities of the restriction of f_u to $\Sigma^1(f_u)$, we take the parameterisation $x(\theta) = \sqrt{u}\cos\theta$ and $y(\theta) = \sqrt{u/3}\sin\theta$, which gives

$$\eta(\theta) = f_u(x(\theta), y(\theta)) = \sqrt{-u}\left(\cos\theta, \frac{2}{3\sqrt{3}}u\cos\theta\sin^2\theta\right).$$

The Σ^1-points of the restriction correspond to $\eta'(\theta) = 0$. We obtain $(x, y) \in \Sigma^{1,1}(f_u)$ when $\theta = 0, \pi$ and $(x, y) \in \Sigma^{1,0}(f_u)$ otherwise.

If $(x, y) \in \Sigma^{1,0}(f_u)$, η is regular at θ and hence, the germ of f_u at (x, y) is $\mathscr{A}$-equivalent to the Whitney fold (Exercise 3.3.6). When $(x, y) \in \Sigma^{1,1}(f_u)$, we have $\eta'(\theta) = 0$ but also $\eta''(\theta) \neq 0$ and the germ of f_u at (x, y) is $\mathscr{A}$-equivalent to the Whitney cusp (Exercise 3.3.7). Furthermore, η is one-to-one, so f_u is locally stable. Notice that the partition of the source $\mathbb{R}^2$ given by the types Σ^0, $\Sigma^{1,0}$ and $\Sigma^{1,1}$ distinguishes between all the $\mathscr{A}$-types which appear in f_u. Figures 4.1 and 4.2 show the picture of $\Sigma^1(f_u)$ and its image, which explains the name "lips" for the singularity of f_0 at the origin. ◊

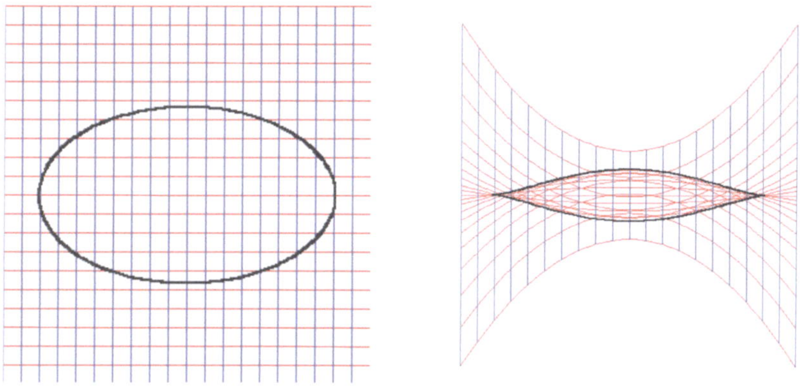

Fig. 4.1 Stable perturbation of the "lips" singularity. Points inside the lips on the right have three preimages, one inside the ellipse on the left and two outside

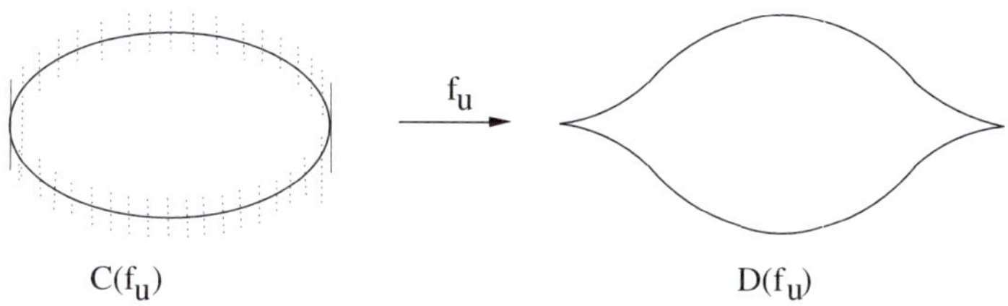

Fig. 4.2 The short vertical lines on the left show the kernel of df_u. At the right and left extremes of the critical set, and only there, the kernel of df_u is tangent to $C(f_u)$, and there the germ of f_u has Boardman symbol $\Sigma^{1,1}$. At the other points of $C(f_u)$, it has Boardman symbol $\Sigma^{1,0}$

Given any $q \times p$ matrix λ with entries in a ring R, we adopt the following conventions:

$$\min_{\ell}(\lambda) = \begin{cases} R, & \text{if } \ell \leq 0; \\ \text{ideal in } R \text{ generated by the } \ell \times \ell \text{minors of } \lambda, & \text{if } 1 \leq \ell \leq \min\{p, q\}; \\ 0, & \text{if } \ell > \min\{p, q\}. \end{cases}$$

Definition 4.6 Let $I \subset \mathscr{O}_n$ be a proper ideal generated by $f_1, \ldots, f_p \in \mathfrak{m}_n$. The *rank* of I is defined as the rank at 0 of J_f, the Jacobian matrix of $f = (f_1, \ldots, f_p)$. For each integer s, the *sth-Jacobian extension* of I is the ideal

$$\Delta^s I = I + \min_{n-s+1}(J_f).$$

This gives a chain of inclusions

$$I = \Delta^0 I \subset \Delta^1 I \subset \cdots \subset \Delta^n I \subset \Delta^{n+1} I = \mathscr{O}_n,$$

and $n - \text{rank}(I)$ is the last s such that $\Delta^s I$ is a proper ideal. The above definitions depend neither on the set of generators of I nor the choice of coordinates in $\mathbb{F}^n$. The proof is easy and is left as an exercise (see Exercises 4.7.1 and 4.7.2).

Definition 4.7 Let $I \subset \mathcal{O}_n$ be a proper ideal generated by $f_1, \ldots, f_p \in \mathfrak{m}_n$ and let $k \geq 1$. The *k'th Boardman symbol* of I is the sequence $\mathbf{i} = (i_1, \ldots, i_k)$ such that

1. I has rank $n - i_1$;
2. for each $\ell = 2, \ldots, k$, $\Delta^{i_{\ell-1}} \ldots \Delta^{i_1} I$ has rank $n - i_\ell$.

Given a smooth germ $f : (\mathbb{F}^n, 0) \to (\mathbb{F}^p, 0)$, the *k'th Boardman symbol* of f is defined as the k'th Boardman symbol of the ideal I_f generated by the components of f.

The same definition can also be given in terms of *critical Jacobian extensions*. By definition, the ideal $\Delta^s I = I + \min_{n-s+1} J_f$ is the *critical* Jacobian extension of I if $\Delta^s I \neq \mathcal{O}_n$ and $\Delta^{s+1} I = \mathcal{O}_n$. The k'th Boardman symbol of an ideal I is $\mathbf{i} = (i_1, \ldots, i_k)$ if $\Delta^{i_1} I$ is the critical jacobian extension of I, $\Delta^{i_2}(\Delta^{i_1} I)$ is the critical Jacobian extension of $\Delta^{i_1} I$ and so on.

Proposition 4.10 *The k'th Boardman symbol of a smooth germ* $f : (\mathbb{F}^n, 0) \to (\mathbb{F}^p, 0)$ *is determined by* $j^k f$*, for each* $k \geq 0$.

Proof Let $I = I_f$. For $\ell = 1$, the first term i_1 of the Boardman symbol is determined by $j^1 f$ and $\Delta^{i_1} I$ is generated by polynomials in the partial derivatives of order ≤ 1 of the components f_j, $1 \leq j \leq p$. Assume that for $\ell - 1$, the terms $i_1, \ldots, i_{\ell-1}$ are determined by $j^{\ell-1} f$ and that $\Delta^{i_{\ell-1}} \ldots \Delta^{i_1} I$ is generated by combinations of partial derivatives of order $\leq \ell - 1$ of the f_j. The term i_ℓ only depends on the rank of the generators and hence is determined by $j^\ell f$. Moreover, $\Delta^{i_\ell} \ldots \Delta^{i_1} I$ is generated by polynomials in the partial derivatives of order $\leq \ell$ of the f_j. □

Definition 4.8 A *Boardman symbol* of length k is any sequence of integers $\mathbf{i} = (i_1, \ldots, i_k)$. For each Boardman symbol $\mathbf{i}$, we define the *Thom–Boardman set* $\Sigma^{\mathbf{i}}$ as the subset of $J^k(n, p)$ of jets $\sigma = j^k f$ such that f has Boardman symbol $\mathbf{i}$.

Example 4.6 Consider the case $k = 1$. Then f has Boardman symbol i_1 if and only if it has kernel rank i_1. Thus Σ^{i_1} coincides with the definition given in Appendix A. In this case, $J^1(n, p)$ can be identified with $\text{Mat}_{p \times n}(\mathbb{F})$, the space of $p \times n$ matrices with coefficients in $\mathbb{F}$, and Σ^{i_1} is the subset of matrices of rank $n - i_1$. This is in fact a locally closed semialgebraic submanifold of codimension of $J^1(n, p)$ of codimension $(p - n + i_1)i_1$ (see Exercise 1). ◊

It is clear from the construction that all the Thom–Boardman sets $\Sigma^{\mathbf{i}}$ are locally closed and semialgebraic in $J^k(n, p)$. The next proposition gives a necessary and sufficient condition on $\mathbf{i}$ for $\Sigma^{\mathbf{i}}$ to be non-empty.

Proposition 4.11 *For each Boardman symbol* $\mathbf{i}$*,* $\Sigma^{\mathbf{i}}$ *is non-empty if and only if*

1. $n \geq i_1 \geq \cdots \geq i_k \geq 0$;
2. $i_1 \geq n - p$;
3. *if* $i_1 = n - p$ *then* $i_1 = \cdots = i_k$.

In particular, the sets $\Sigma^{\mathbf{i}}$ *provide a finite partition of* $J^k(n, p)$.

Proof Assume that $\Sigma^{\mathbf{i}} \neq \emptyset$ and let $\sigma = j^k f \in \Sigma^{\mathbf{i}}$. Since i_1 is the kernel rank of f, we have $n \geq i_1 \geq n - p$. Moreover, $\operatorname{rank} I \leq \operatorname{rank} \Delta^s I$ for any s and for any ideal I. This shows that $i_{\ell-1} \geq i_\ell$ for all $\ell = 2, \ldots, k$. Finally, if $s = n - p$ then $\Delta^s I = I$ and hence $i_1 = n - p$ implies that $i_1 = \cdots = i_k$.

Conversely, let $\mathbf{i}$ be a Boardman symbol satisfying conditions 1, 2 and 3. It is enough to find a germ f with Boardman symbol $\mathbf{i}$. If $i_1 = n - p$, we take $f_1 = x_1, \ldots, f_p = x_{n-p}$. Otherwise, if $i_1 > n - p$, we take

$$f_i = \begin{cases} x_i, & 1 \leq i \leq n - i_1, \\ \displaystyle\sum_{j=n-i_1+1}^{n-i_2} x_j^2 + \sum_{j=n-i_2+1}^{n-i_3} x_j^3 + \ldots, & i = n - i_1 + 1, \\ 0, & n - i_1 + 2 \leq i \leq p. \end{cases}$$

It is an easy computation to check that in both cases f has Boardman symbol $\mathbf{i}$. □

For each Boardman symbol $\mathbf{i}$, we define the number

$$\nu(\mathbf{i}) = (p - n + i_1)\mu(i_1, \ldots, i_k) - (i_1 - i_2)\mu(i_2, \ldots, i_k) + \cdots + (i_{k-1} - i_k)\mu(i_k),$$

where $\mu(i_1, \ldots, i_k)$ is the number of sequences $(j_1, \ldots, j_k)$ such that $j_1 \geq \cdots \geq j_k \geq 0$, $i_\ell \geq j_\ell$ for all $\ell = 1, \ldots, k$ and $j_1 > 0$.

Example 4.7 When $k = 1$, we have $\mu(i_1) = i_1$ and hence $\nu(i_1) = (p - n + i_1)i_1$. For $k = 2$, we have $\mu(i_1, i_2) = i_1(i_2 + 1) - j_2(i_2 - 1)/2$, so

$$\nu(i_1, i_2) = (p - n + i_1)i_1 + \frac{i_2}{2}\big((p - n + i_1)(2i_1 - i_2 + 1) - 2i_1 + 2i_2\big).$$

Another interesting particular case is when $i_1 = \cdots = i_k = 1$. We have $\mu(1, \ldots, 1) = k$ and hence $\nu(1, \ldots, 1) = (p - n + 1)k$. ◊

Now we can state the main result of Boardman in [Boa67]:

Theorem 4.7 *For every Boardman symbol* $\mathbf{i}$*,* $\Sigma^{\mathbf{i}}$ *is a submanifold of* $J^k(n, p)$ *of codimension* $\nu(\mathbf{i})$.

We refer to the paper of Mather [Mat73] for a proof simpler than Boardman's original proof. A consequence of the theorem is that the submanifolds $\Sigma^{\mathbf{i}}$ provide a stratification of $J^k(n, p)$. It is known as the *Thom–Boardman stratification*.

Lemma 4.7 *If $f, g\colon (\mathbb{F}^n, 0) \to (\mathbb{F}^p, 0)$ are $\mathscr{K}$-equivalent, then they have the same k'th Boardman symbol, for all $k \geq 1$.*

Proof By Theorem 4.3, there exists a diffeomorphism $\phi\colon (\mathbb{F}^n, 0) \to (\mathbb{F}^n, 0)$ such that $\phi^* I_f = I_g$. The result is now a consequence of Exercise 4.7.2. □

This lemma implies that $\Sigma^{\mathbf{i}}$ is $\mathscr{K}$-invariant (and hence $\mathscr{A}$-invariant) in $J^k(n, p)$. Thus, for any pair of smooth manifolds X, Y, there exists a unique subbundle $\Sigma^{\mathbf{i}}(X, Y) \subset J^k(X, Y)$ whose fibre is $\Sigma^{\mathbf{i}}$. If $f\colon X \to Y$ is any smooth mapping, we have the k-jet extension map $j^k f : X \to J^k(X, Y)$ and we define

$$\Sigma^{\mathbf{i}}(f) = (j^k)^{-1}(\Sigma^{\mathbf{i}}(X, Y)).$$

By construction, $\Sigma^{\mathbf{i}}(f)$ is the subset of X of points x such that the germ of f at x has k'th Boardman symbol $\mathbf{i}$. Analogously, if $f\colon (\mathbb{F}^n, S) \to (\mathbb{F}^p, 0)$ is any smooth germ, then we define

$$\Sigma^{\mathbf{i}}(f) = (j^k)^{-1}(\Sigma^{\mathbf{i}}).$$

where $j^k f : (\mathbb{F}^n, S) \to J^k(n, p)$ is the k-jet extension. Again, $\Sigma^{\mathbf{i}}(f)$ is the set-germ in $(\mathbb{F}^n, S)$ of points x such that the germ of f at x has k'th Boardman symbol $\mathbf{i}$.

Definition 4.9 A smooth mapping $f\colon X \to Y$ is called *Thom–Boardman generic* if $j^k f \pitchfork \Sigma^{\mathbf{i}}$, for all Boardman symbols $\mathbf{i}$ and for all $k \geq 1$.

A germ $f\colon (\mathbb{F}^n, S) \to (\mathbb{F}^p, 0)$ is called *Thom–Boardman generic* if there exists a representative $f\colon X \to Y$ which is Thom–Boardman generic.

The following are direct consequences of Theorem 4.7 and of Corollary A.1:

Corollary 4.5 *Let $f\colon X \to Y$ be a smooth mapping.*

1. *If f is Thom–Boardman generic, then $\Sigma^{\mathbf{i}}(f)$ is a submanifold of X of codimension $\nu(\mathbf{i})$.*
2. *If f is locally stable it is Thom–Boardman generic.*

We state without proof another important result of Boardman [Boa67] which says that his definition of Thom–Boardman singularities coincides with Thom's original construction when f is generic.

Theorem 4.8 *Let $f\colon X \to Y$ be Thom–Boardman generic. For all $\mathbf{i}$,*

$$\Sigma^{\mathbf{i}}(f) = \Sigma^{i_k}(f|_{\Sigma^{i_1,\dots,i_{k-1}}}).$$ □

Example 4.8 Let $f : \mathbb{R}^2 \to \mathbb{R}^2$ be the mapping $f(x, y) = (x^2, y^2)$. The Jacobian determinant is $4xy$, so we have

- $\Sigma^0(f) = \{(x, y) \in \mathbb{R}^2 : xy \neq 0\}$,
- $\Sigma^1(f) = \{(x, y) \in \mathbb{R}^2 : xy = 0\} \setminus \{0\}$ and
- $\Sigma^2(f) = \{0\}$.

The Jacobian extension $\Delta^1 I_f$ is generated by x^2, y^2, xy, which has rank 2 at every point $(x, y) \in \Sigma^1(f)$. Hence $\Sigma^1(f) = \Sigma^{1,0}(f)$. Observe that the germ of f at $(x, y) \in \Sigma^1(f)$ is $\mathscr{A}$-equivalent to the Whitney fold, which is stable and hence Thom–Boardman generic. Analogously, $\Delta^2 I_f$ is generated by x, y and has also rank 2. So, $\Sigma^2(f) = \Sigma^{2,0}(f)$. However, f is not Thom–Boardman generic at the origin, since if it was, $\Sigma^2(f)$ should have codimension 4.

Now let $F : \mathbb{R}^4 \to \mathbb{R}^4$ be the unfolding $F(x, y, u, v) = (x^2+uy, y^2+vx, u, v)$. It is a locally stable mapping, so it is Thom–Boardman generic. In fact, the Jacobian determinant is $4xy - uv$, so

- $\Sigma^0(F) = \{(x, y, u, v) \in \mathbb{R}^4 : 4xy - uv \neq 0\}$,
- $\Sigma^1(F) = \{(x, y, u, v) \in \mathbb{R}^4 : 4xy - uv = 0\} \setminus \{0\}$,
- $\Sigma^2(F) = \{0\}$ and
- $\Sigma^3(F) = \Sigma^4(F) = \emptyset$.

Since F is an unfolding of f we deduce that $\Sigma^2(F) = \Sigma^{2,0}(F)$. But $\Delta^1 I_F$ is generated by $x^2 + uy, y^2 + vx, u, v, 4xy - uv$. Taking the Jacobian matrix of this, we get

- $\Sigma^{1,1}(F) = \{(x, y, u, v) \in \Sigma^1(F) : 8x^2 - 4uy = 8y^2 - 4vx = 0\}$ and
- $\Sigma^{1,0}(F) = \Sigma^1(F) \setminus \Sigma^{1,1}(F)$.

We can proceed one step more: $\Delta^{1,1} I_F$ is generated by $\Delta^1 F$ and the two new equations $8x^2 - 4uy, 8y^2 - 4vx$. Again by taking the Jacobian of this, we arrive at

- $\Sigma^{1,1,1}(F) = \{(x, y, u, v) \in \Sigma^{1,1}(F) : x = y = uv = 0\}$ and
- $\Sigma^{1,1,0}(F) = \Sigma^{1,1}(F) \setminus \Sigma^{1,1,1}(F)$.

In the last step, we just check that $\Sigma^{1,1,1,0}(F) = \Sigma^{1,1,1}(F)$, which gives the complete Thom–Boardman stratification of F.

When $uv \neq 0$, $f_{u,v} : \mathbb{R}^2 \to \mathbb{R}^2$ is Thom–Boardman generic, as one can check easily. We have $\Sigma^2(f_{u,v}) = \Sigma^{1,1,1}(f_{u,v}) = \emptyset$ and the equations for $\Sigma^{1,0}(f_{u,v})$ and $\Sigma^{1,1}(f_{u,v})$ are the same as those of F. At a point $(x, y) \in \Sigma^{1,0}(f_{u,v})$ the germ of f is $\mathscr{A}$-equivalent to the Whitney fold. The equations of $\Sigma^{1,1}(f_{u,v})$ give exactly three solutions (one is real and the other two are complex) and at these points the germ of f is $\mathscr{A}$-equivalent to the Whitney cusp. Figure 4.3 shows a picture of the image this mapping. ♢

As the above example shows, the Thom–Boardman singular sets of a mapping can be obtained by means of the iterated Jacobian extensions. The procedure can be systematised easily by using the ideals $J_{\mathbf{i}}(f)$ introduced by Bernard Morin in [Mor75]. These ideals are more convenient if we want to compute not only the Boardman symbol at a specific point but also the sets $\Sigma^{\mathbf{i}}(f)$.

Definition 4.10 Let $f : (\mathbb{F}^n, 0) \to (\mathbb{F}^p, 0)$ be a smooth map-germ and let $I \subset \mathscr{O}_n$ be an ideal generated by $g_1, \ldots, g_r$. For each integer s, the *sth-Jacobian extension*

Fig. 4.3 The image of $f_{u,v}$, with $uv \neq 0$

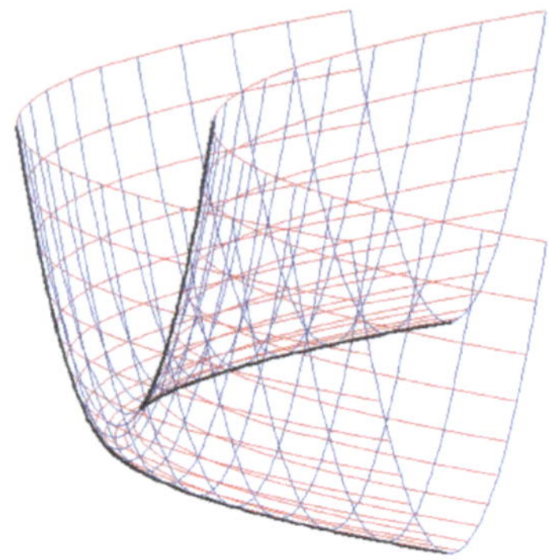

of the pair (f, I) is the ideal

$$\Delta^s(f, I) = I + \min_{n-s+1}(J_{(f,g)}).$$

For each Boardman symbol $\mathbf{i} = (i_1, \ldots, i_k)$, the *iterated Jacobian extension* of f is the ideal $J_{\mathbf{i}}(f)$ defined recursively as follows:

$$J_{\mathbf{i}}(f) = \begin{cases} \Delta^{i_1}(f, \{0\}), & \text{if } k = 1, \\ \Delta^{i_k}(f, J_{i_1,\ldots,i_{k-1}}(f)), & \text{if } k > 1. \end{cases}$$

The following lemma gives the relationship between $J_{\mathbf{i}}(f)$ and the ideals $\Delta^{i_k} \ldots \Delta^{i_1} I_f$ used to define the Boardman symbol of f.

Lemma 4.8 *For any germ* $f : (\mathbb{F}^n, 0) \to (\mathbb{F}^p, 0)$ *and for any Boardman symbol* $\mathbf{i} = (i_1, \ldots, i_k)$,

$$\Delta^{i_k} \ldots \Delta^{i_1} I_f = I_f + J_{\mathbf{i}}(f). \tag{4.22}$$

In particular, $\Delta^{i_k} \ldots \Delta^{i_1} I_f$ *is a proper ideal if and only if* $J_{\mathbf{i}}(f)$ *is.*

Proof The equality (4.22) is clear for $k = 1$. Assume that (4.22) is true for $k - 1$. If $J_{\mathbf{i}}(f) = (g_1, \ldots, g_r)$, then $\Delta^{i_{k-1}} \ldots \Delta^{i_1} I_f = (f_1, \ldots, f_p, g_1, \ldots, g_r)$. By construction and by the induction hypothesis,

$$\begin{aligned} \Delta^{i_k} \ldots \Delta^{i_1} I_f &= \Delta^{i_{k-1}} \ldots \Delta^{i_1} I_f + \min_{n-i_k+1}(J_{(f,g)}) \\ &= I_f + J_{i_1,\ldots,i_{k-1}}(f) + \min_{n-i_k+1}(J_{(f,g)}) \\ &= I_f + J_{\mathbf{i}}(f). \end{aligned}$$

□

We consider the lexicographical order in the set of all Boardman symbols: if $\mathbf{i} = (i_1, \dots, i_k)$ and $\mathbf{j} = (j_1, \dots, j_\ell)$, we write $\mathbf{i} \preceq \mathbf{j}$ if either $\mathbf{i} = \mathbf{j}$ or $i_{r_0} < j_{r_0}$, where r_0 is the least r such that $i_r \neq j_r$.

Lemma 4.9 *Let $f\colon (\mathbb{F}^n, 0) \to (\mathbb{F}^p, 0)$ be a germ and let $\mathbf{i}, \mathbf{j}$ be Boardman symbols such that $\mathbf{i} \preceq \mathbf{j}$. Then $J_{\mathbf{i}}(f) \subset J_{\mathbf{j}}(f)$.*

Proof We first claim that for any s, t and for any ideal $I \subset \mathcal{O}_n$, we have

$$\Delta^t(f, \Delta^s(f, I)) \subset \Delta^{s+1}(f, I). \tag{4.23}$$

In fact, let $I = (g_1, \dots, g_r)$. We denote by $h_1, \dots, h_m$ the $n - s + 1$-minors of $J_{(f,g)}$. Then $\Delta^t(f, \Delta^s(f, I))$ is generated by $g_1, \dots, g_r, h_1, \dots, h_m$ and by the $n - t + 1$-minors of $J_{(f,g,h)}$. Each h_i can be written as a linear combination of elements in $\Delta^{s+1}(f, I)$. It follows from the rule of derivation of the determinants, that any $n - t + 1$-minor of $J_{(f,g,h)}$ also belongs to $\Delta^{s+1}(f, I)$. This proves the claim.

Assume now that $\mathbf{i} \preceq \mathbf{j}$, with $\mathbf{i} = (i_1, \dots, i_k)$ and $\mathbf{j} = (j_1, \dots, j_\ell)$ and $\mathbf{i} \neq \mathbf{j}$. We have $i_{r_0} < j_{r_0}$, where r_0 is the least r such that $i_r \neq j_r$. By (4.23),

$$\begin{aligned} J_{\mathbf{i}}(f) &= J_{i_1,\dots,i_{r_0-1},i_{r_0},\dots,i_k}(f) \\ &\subset J_{i_1,\dots,i_{r_0-1},i_{r_0},\dots,i_{k-1}+1}(f) \\ &\subset J_{i_1,\dots,i_{r_0-1},i_{r_0}+1}(f) \\ &\subset J_{i_1,\dots,i_{r_0-1},j_{r_0}}(f) \\ &\subset J_{i_1,\dots,i_{r_0-1},j_{r_0},\dots,j_\ell}(f) = J_{\mathbf{j}}(f). \end{aligned}$$

□

Definition 4.11 Given a Boardman symbol $\mathbf{i}$ of length k, the *successor* $\mathbf{i}'$ is the least Boardman symbol $\mathbf{j}$ of length $\leq k$ such that $\mathbf{i} \preceq \mathbf{j}$ and $\mathbf{i} \neq \mathbf{j}$.

In the non-trivial cases, we only consider decreasing Boardman symbols, that is, Boardman symbols $\mathbf{i} = (i_1, \dots, i_k)$ such that $i_1 \geq \cdots \geq i_k$. Let r such that $i_r > i_{r+1} = \cdots = i_k$. Then the successor of $\mathbf{i}$ is $\mathbf{i}' = (i_1, \dots, i_r, i_{r+1} + 1)$.

Proposition 4.12 *Let $f\colon (\mathbb{F}^n, 0) \to (\mathbb{F}^p, 0)$ be a smooth map-germ. Then, for each Boardman symbol* $\mathbf{i}$,

$$\Sigma^{\mathbf{i}}(f) = V(J_{\mathbf{i}}(f)) \setminus V(J_{\mathbf{i}'}(f)).$$

Proof Given a point x in a neighbourhood of 0 in $\mathbb{F}^n$, we must show that f has k'th Boardman symbol $\mathbf{i}$ at x if and only if $x \in V(J_{\mathbf{i}}(f)) \setminus V(J_{\mathbf{i}'}(f))$. We can assume without loss of generality that $x = 0$. So, we must prove that the germ f has k'th Boardman symbol $\mathbf{i}$ if and only if $J_{\mathbf{i}}(f)) \neq \mathcal{O}_n$ and $J_{\mathbf{i}'}(f) = \mathcal{O}_n$.

Again we proceed by induction on k. The equivalence is clear for $k = 1$. Assume that the equivalence is true for $k - 1$. Suppose that f has k'th Boardman symbol

$\mathbf{i}$. In particular, f has $(k-1)$th Boardman symbol $\mathbf{j} := (i_1, \ldots, i_{k-1})$ and by the hypothesis induction, $J_{\mathbf{j}}(f) \neq \mathscr{O}_n$ and $J_{\mathbf{j}'}(f) = \mathscr{O}_n$. Moreover, $\Delta^{i_{k-1}} \ldots \Delta^{i_1} I_f$ has rank $n - i_k$ or equivalently,

$$\Delta^{i_k} \ldots \Delta^{i_1} I_f \neq \mathscr{O}_n, \quad \Delta^{i_k+1} \Delta^{i_{k-1}} \ldots \Delta^{i_1} I_f = \mathscr{O}_n.$$

By (4.22), this implies that

$$J_{\mathbf{i}}(f) \neq \mathscr{O}_n, \quad J_{i_1,\ldots,i_{k-1},i_k+1}(f) = \mathscr{O}_n.$$

If $i_{k-1} = i_k$, then $\mathbf{i}' = \mathbf{j}'$ and otherwise, if $i_{k-1} > i_k$, then $\mathbf{i}' = (i_1, \ldots, i_{k-1}, i_k + 1)$. In both cases, we get $J_{\mathbf{i}'}(f) = \mathscr{O}_n$.

Conversely, suppose that $J_{\mathbf{i}}(f) \neq \mathscr{O}_n$ and $J_{\mathbf{i}'}(f) = \mathscr{O}_n$ and let $\mathbf{j} := (i_1, \ldots, i_{k-1})$. We have $J_{\mathbf{j}}(f) \subset J_{\mathbf{i}}(f) \neq \mathscr{O}_n$. Assume first that $i_{k-1} = i_k$. Then $\mathbf{i}' = \mathbf{j}'$ and hence $J_{\mathbf{j}'}(f) = \mathscr{O}_n$. By the induction hypothesis, $\mathbf{j}$ is the $(k-1)$'th Boardman symbol of f. In particular, $\Delta^{i_{k-2}} \ldots \Delta^{i_1} I_f$ has rank $n - i_{k-1}$ and hence, $\Delta^{i_{k-1}} \ldots \Delta^{i_1} I_f$ has rank $\geq n - i_{k-1} = n - i_k$. On the other hand, from (4.22) we also see that $\Delta^{i_k} \ldots \Delta^{i_1} I_f \neq \mathscr{O}_n$ and thus, $\Delta^{i_{k-1}} \ldots \Delta^{i_1} I_f$ has rank $\leq n - i_k$. We deduce that $\Delta^{i_{k-1}} \ldots \Delta^{i_1} I_f$ has rank $n - i_k$, so f has k'th Boardman symbol $\mathbf{i}$.

Otherwise, if $i_{k-1} > i_k$, then $\mathbf{i}' = (i_1, \ldots, i_{k-1}, i_k + 1)$. Let r be such that $i_r > i_{r+1} = \cdots = i_{k-1}$, then $\mathbf{j}' = (i_1, \ldots, i_r, i_{r+1} + 1)$. We have $\mathbf{i}' \preceq \mathbf{j}'$ and hence, $\mathscr{O}_n = J_{\mathbf{i}'}(f) \subset J_{\mathbf{j}'}(f)$ by Lemma 4.9. By the induction hypothesis, $\mathbf{j}$ is the $(k-1)$'th Boardman symbol of f. Moreover, by (4.22),

$$\Delta^{i_k} \ldots \Delta^{i_1} I_f \neq \mathscr{O}_n, \quad \Delta^{i_k+1} \Delta^{i_{k-1}} \ldots \Delta^{i_1} I_f = \mathscr{O}_n.$$

This is equivalent to $\Delta^{i_{k-1}} \ldots \Delta^{i_1} I_f$ having rank $n - i_k$. Therefore, f has k'th Boardman symbol $\mathbf{i}$. □

Example 4.9 Consider the stable map $F : \mathbb{R}^4 \to \mathbb{R}^4$ given by $F(x, y, u, v) = (x^2 + uy, y^2 + vx, u, v)$ of Example 4.8. The first order Jacobian extensions give

$$J_1(F) = (4xy - uv), \quad J_2(F) = (x, y, u, v), \quad J_3(F) = J_4(f) = (1),$$

and the iterated Jacobian extensions are

$$J_{1,1}(F) = (4xy - uv, 8x^2 - 4uy, 8y^2 - 4vx), \quad J_{1,1,1}(F) = (x, y, uv),$$
$$J_{1,1,1,1}(F) = (1), \quad J_{2,1}(F) = (1).$$

This allows us to compute the Thom–Boardman stratification easily by Proposition 4.12:

- $\Sigma^1(F) = V(J_1(F)) \setminus V(J_2(F))$,
- $\Sigma^2(F) = V(J_2(F)), \quad \Sigma^3(F) = \Sigma^4(F) = \emptyset$,
- $\Sigma^{1,0}(F) = V(J_1(F)) \setminus V(J_{1,1}(F))$,

- $\Sigma^{1,1}(F) = V(J_{1,1}(F)) \setminus V(J_2(F))$,
- $\Sigma^{1,1,0}(F) = V(J_{1,1}(F)) \setminus V(J_{1,1,1}(F))$,
- $\Sigma^{1,1,1}(F) = V(J_{1,1,1}(F)) \setminus V(J_2(F))$,
- $\Sigma^{1,1,1,0}(F) = V(J_{1,1,1}(F)), \quad \Sigma^{1,1,1,1}(F) = \emptyset$,
- $\Sigma^{2,0}(F) = V(J_2(F)), \quad \Sigma^{2,1}(F) = \Sigma^{2,2}(F) = \emptyset$.

In the last part of this section we restrict ourselves to Thom–Boardman generic germs $f: (\mathbb{F}^n, 0) \to (\mathbb{F}^p, 0)$, with $n \le p$ and of type Σ^1. These are called Morin singularities, after Bernard Morin [Mor65]. We will see that in this class of singularities, the Thom–Boardman stratification is enough to characterise stability and the $\mathscr{A}$-class.

For each $k \ge 1$ we denote by $\mathbf{1}_k$ the Boardman symbol $(1, \ldots, 1, 0)$ with 1 repeated k-times. If f has $(k+1)$'st Boardman symbol $\mathbf{1}_k$, then its ℓ'th Boardman symbol is $(1, \ldots, 1, 0, \ldots, 0)$, for all $\ell \ge k+1$. In such a case we will say that f has *Boardman type* $\mathbf{1}_k$.

Lemma 4.10 *Let $f: (\mathbb{F}^n, 0) \to (\mathbb{F}^p, 0)$ be $\mathscr{K}$-finite, with $n \le p$ and type Σ^1. Then there exists $k \ge 1$ such that has Boardman type $\mathbf{1}_k$. Moreover, any other germ with Boardman type $\mathbf{1}_k$ is $\mathscr{K}$-equivalent to f.*

Proof Since f has type Σ^1, we can choose coordinates (x, u) in $\mathbb{F}^n = \mathbb{F} \times \mathbb{F}^{n-1}$ such that

$$f(x, u_1, \ldots, u_{n-1}) = (f_1(x, u), \ldots, f_{p-n+1}(x, u), u),$$

with $f_i \in \mathfrak{m}_n^2$, for all $i = 1, \ldots, p-n+1$. Thus, $Q(f) \cong \mathscr{O}_1/I$, where I is the ideal generated by the functions $h_i(x) = f_i(x, 0)$, $i = 1, \ldots, p-n+1$. But $Q(f)$ has finite $\mathbb{F}$-dimension, for f is $\mathscr{K}$-finite. Hence, $Q(f) \cong \mathscr{O}_1/(x^{k+1})$, for some $k \ge 1$. By Theorem 4.3 (see also Exercise 4.4.1), $h: (\mathbb{F}, 0) \to (\mathbb{F}^{p-n+1}, 0)$ is $\mathscr{K}$-equivalent to the germ $(x^{k+1}, 0, \ldots, 0)$, which has Boardman type $\mathbf{1}_k$ (see Exercise 4.7.5). But since f is an unfolding of h, f also has Boardman type $\mathbf{1}_k$, by Exercise 4.7.3 .

Let $g: (\mathbb{F}^n, 0) \to (\mathbb{F}^p, 0)$ be any germ with Boardman type $\mathbf{1}_k$. Then g has type Σ^1 and as above, $Q(g) \cong \mathscr{O}_1/I$, for some ideal I. If $I = 0$, then g should be $\mathscr{K}$-equivalent to an unfolding of the constant germ $(\mathbb{F}, 0) \to (\mathbb{F}^{p-n+1}, 0)$ which has ℓ'th Boardman symbol $(1, \ldots, 1)$ for all ℓ. Hence, $I \ne 0$ and necessarily $I = (x^{m+1})$, for some $m \ge 1$. Again by the above argument, this implies that g has Boardman type $\mathbf{1}_m$, so $m = k$ and hence g is $\mathscr{K}$-equivalent to f. □

An immediate consequence of the lemma is the following:

Corollary 4.6 *Let $\sigma \in J^{k+1}(n, p)$, with $n \le p$ and assume that $\sigma \in \Sigma^{\mathbf{1}_k}$. Then the $\mathscr{K}^{(k+1)}$-orbit of σ is $\Sigma^{\mathbf{1}_k}$.*

Definition 4.12 A Thom–Boardman generic germ $f: (\mathbb{F}^n, 0) \to (\mathbb{F}^p, 0)$ with $n \le p$ and type Σ^1 is called a *Morin singularity*. A *Morin map* is a Thom–Boardman generic map $f: X \to Y$ with $\dim X \le \dim Y$ and with only Σ^1-singularities.

Corollary 4.7 *Any Morin singularity* $f: (\mathbb{F}^n, 0) \to (\mathbb{F}^p, 0)$ *is stable. Moreover, two Morin singularities* $f, g: (\mathbb{F}^n, 0) \to (\mathbb{F}^p, 0)$ *with the same Boardman type are* $\mathscr{A}$*-equivalent.*

The first part of the corollary follows from Corollary 4.6 and Theorem 4.6. For the second part, we also need the fact that if two stable germs are $\mathscr{K}$-equivalent, then they are also $\mathscr{A}$-equivalent. This will be proved in Chap. 7 (see Theorem 7.1).

The above corollary is not true for multi-germs, since we need that the images of the Thom–Boardman submanifolds meet in general position in the target. This motivates the next definition.

Definition 4.13 Let $f: X \to Y$ be a Morin map. We say that f satisfies the *normal crossing* condition if given distinct points $x_1, \ldots, x_r \in X$ such that $f(x_1) = \cdots = f(x_r) = y$ and f has Boardman type $\mathbf{1}_{k_i}$ at each x_i, then the subspaces $d_{x_i} f\left(T_{x_i} \Sigma^{\mathbf{1}_{k_i}}(f)\right)$ meet in general position in $T_y Y$.

Now we have the global version of the above corollary, which follows easily from Corollary 4.4.

Corollary 4.8 *Let* $f: X \to Y$ *be a Morin map satisfying the normal crossing condition and such that the restriction* $f: C(f) \to Y$ *is finite. Then* f *is locally stable.*

Exercises for Sect. 4.7

1 Suppose that $f_1, \ldots, f_p \in \mathfrak{m}_n$ and $g_1, \ldots, g_q \in \mathfrak{m}_n$ generate the same ideal I in $\mathscr{O}_n$. Show that rank $J_f =$ rank J_g and that $I + \min_\ell(J_f) = I + \min_\ell(J_g)$, for all ℓ.

2 Let I be a proper finitely generated ideal in $\mathscr{O}_n$ and $\phi: (\mathbb{F}^n, 0) \to (\mathbb{F}^n, 0)$ a diffeomorphism. Show that rank $\phi^* I =$ rank I and that $\phi^*(\Delta^s I) = \Delta^s(\phi^* I)$, where $\phi^*: \mathscr{O}_n \to \mathscr{O}_n$ is the induced isomorphism.

3 Let $f: (\mathbb{F}^n, 0) \to (\mathbb{F}^p, 0)$ be smooth and let F be an unfolding of f. Show that F and f have the same k'th Boardman symbol, for all $k \geq 1$. Show also that f and the germ $(f, 0)$ obtained from f by adding $p+1$'st component 0 have the same k'th Boardman symbol for each k.

4 Let $f: (\mathbb{F}^n, 0) \to (\mathbb{F}, 0)$ be smooth. Show that f has Boardman symbol $(n, 0)$ if and only if it has a non-degenerate critical point.

5 Let $f: (\mathbb{F}, 0) \to (\mathbb{F}^p, 0)$ be the germ $f(x) = (x^{k+1}, 0, \ldots, 0)$. Show that its $(k+1)$'st Boardman symbol is $(1, \ldots, 1, 0)$.

6 Show that the k'th Boardman symbol of f is $(i_1, \ldots, i_{k-1}, 0)$ if and only if the $k-1$'st symbol is $(i_1, \ldots, i_{k-1})$ and $\Delta^{i_{k-1}} \cdots \Delta^{i_1} I$ is the maximal ideal $\mathfrak{m}$.

7 Show that if $f : (\mathbb{F}^n, 0) \to (\mathbb{F}^p, 0)$ is $\mathscr{K}$-finite then for some finite k, the k'th Boardman symbol $(i_1, \ldots, i_k)$ has $i_k = 0$. We call the first such k the *length* of the Boardman symbol of f.

8 Place bounds on the length of the Boardman symbol of a stable map-germ $(\mathbb{F}^n, 0) \to (\mathbb{F}^p, 0)$.

9 Let $f : \mathbb{F}^6 \to \mathbb{F}^5$ be the mapping

$$f(x, y, u, v, w, t) = (x^2y + y^4 + ux + vy + wx^2 + ty^2, u, v, w, t).$$

This is a stable unfolding of the function $x^2y + y^4$. Find the iterated Jacobian ideals of f and deduce the Thom–Boardman stratification by means of Proposition 4.12 as in Example 4.9.

10 Let $f : (\mathbb{F}^4, 0) \to (\mathbb{F}^4, 0)$ be Thom–Boardman generic. Show that f is stable.

11 Let $f : X \to Y$ be a smooth mapping with $\dim X = \dim Y = 4$. State and prove a result analogous to Corollary 4.8 but without the assumption that f has only Σ^1-singularities.

Chapter 5
Versal Unfoldings

Consider the parameterised plane curve $f(x) = (x^3, x^4)$. The 1-parameter deformations

$$f_t^{(1)}(x) = (x^3 - tx, x^4 - 0.6tx^2), \qquad f_t^{(2)}(x) = (x^3 - tx, x^4 - 0.8tx^2),$$

$$f_t^{(3)}(x) = (x^3 - tx, x^4 - 1.2tx), \qquad f_t^{(4)}(x) = (x^3 - tx, x^4 - 0.8tx^2 + 0.4tx),$$

have the following images for different values of the parameter t (Fig. 5.1).

Are these the only possible pictures we can get by deforming this curve? Can adding judicious combinations of other monomials produce curves with different images and other singularities? Which other singularities can appear? The main theorem of this chapter, Theorem 5.1, tells us that if f has finite $\mathscr{A}_e$-codimension, then there is a family f_u, where u varies in a finite-dimensional parameter space, which contains all possible deformations, up to a suitable notion of parameterised left-right equivalence. This needs some clarification: we are interested only in phenomena which are "infinitely close to f", that is, which appear in *germs* of deformations of f. This will be made clear in what follows. This notion is captured by the definition of *germ of unfolding*.

As already introduced in Definition 3.4, an unfolding of a germ $f_0 : (\mathbb{F}^n, S) \to (\mathbb{F}^p, 0)$ is a germ $F : (\mathbb{F}^n \times \mathbb{F}^d, S \times \{0\}) \to (\mathbb{F}^p \times \mathbb{F}^d, 0)$ of the form

$$F(x, u) = (\tilde{f}(x, u), u)$$

where $\tilde{f}(\_, 0) = f_0$. Evidently, from any germ of deformation f_u one can obtain an unfolding $F(x, u) = (f_u(x), u)$, and vice versa. Unfoldings are often written in the form $F(x, u) = (f_u(x), u)$ to emphasise that the first term is a deformation of some f_0, but it is important to realise that for $u \neq 0$, f_u is not a germ, since it does not have a base-point. If f_0 has finite $\mathscr{A}_e$-codimension then (with some care) it will be possible to view each f_u as a mapping on a more-or-less well-defined domain and target. We explore this further in Sect. 5.5.

D. Mond, J. J. Nuño-Ballesteros, *Singularities of Mappings*,
Grundlehren der mathematischen Wissenschaften 357,
https://doi.org/10.1007/978-3-030-34440-5_5

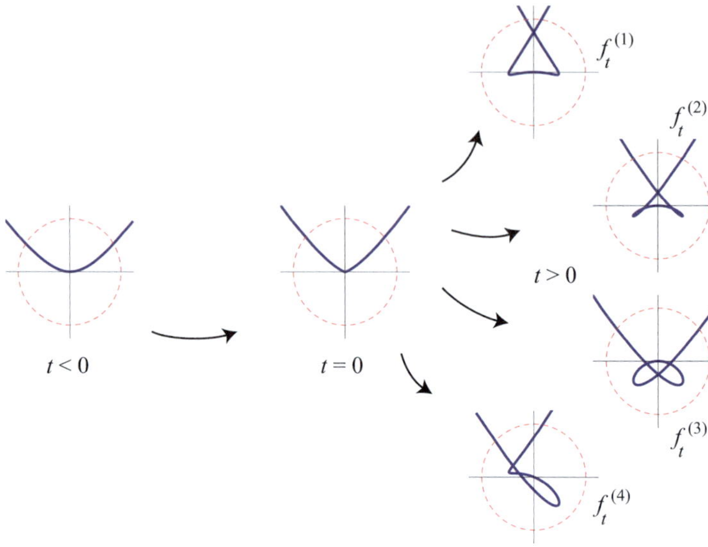

Fig. 5.1 Deformations of the curve $x \mapsto (x^3, x^4)$

A short remark at the end of the chapter compares what we do here, with the much more sophisticated deformation theory of singularities of varieties.

5.1 Versality

In this section we make precise sense of the notion of "unfolding of f_0 which contains all other unfoldings of f_0 up to parameterised equivalence," and study some examples. The first part of the following definition makes precise the notion of one unfolding "containing" another, and the second part gives the notion of parameterised left-right equivalence. Recall that we have defined equivalence of unfoldings in Definition 3.4.

Definition 5.1

(1) Let $F : (\mathbb{F}^n \times \mathbb{F}^d, S \times \{0\}) \to (\mathbb{F}^p \times \mathbb{F}^d, 0)$ be an unfolding of a multi-germ $f_0 : (\mathbb{F}^n, S) \to (\mathbb{F}^p, 0)$ and let $h : (\mathbb{F}^a, 0) \to (\mathbb{F}^d, 0)$ be a map-germ. The *pull-back* of F by h is the unfolding $h^*F : (\mathbb{F}^n \times \mathbb{F}^a, S \times \{0\}) \to (\mathbb{F}^p \times \mathbb{F}^a, 0)$ defined by

$$(x, v) \mapsto \big(\tilde{f}(x, h(v)), v\big).$$

The map-germ h in this context is often called the "base-change" map, and we say that h^*F is the unfolding *induced from* F by h.

(2) The unfolding F of f_0 is *versal* if for every unfolding G of f_0 on parameter space $(\mathbb{F}^a, 0)$, there is a base-change map $h : (\mathbb{F}^a, 0) \to (\mathbb{F}^d, 0)$ such that G is equivalent as unfolding to the unfolding h^*F (as defined in (1)). It is *miniversal* (or sometimes "semi-universal") if it is versal, and no unfolding with a lower-dimensional parameter space is versal.

The term "versal" is the intersection of the words "universal" and "transversal". Versal unfoldings were once upon a time called universal, but later it was decided that they did not deserve this term, because the base-change map h and the diffeomorphisms Φ and Ψ of part (2) of the definition are not in general unique. Uniqueness is an important ingredient in the "universal properties" which characterise many mathematical objects, and so universal unfoldings were stripped of their title. However the intersection with the word "transversal" is serendipitous, as we will see in Theorem 5.2.

Like stability, versality can be checked by means of an infinitesimal criterion. Let $F(x, u) = (f_u(x), u)$ be an unfolding of f_0. Write $\partial f_u/\partial u_j|_{u=0}$ as $\dot{F}_j$, for $j = 1, \dots, d$.

Theorem 5.1 ([Mar76]) *Versality is equivalent to infinitesimal versality: the unfolding F of f_0 is versal if and only if*

$$T\mathscr{A}_e(f_0) + Sp_{\mathbb{F}}\{\dot{F}_1, \dots, \dot{F}_d\} = \theta(f_0)$$

—in other words, if the images of $\dot{F}_1, \dots, \dot{F}_d$ in $T^1_{\mathscr{A}_e} f_0$ generate it as an $\mathbb{F}$-vector space.

Before proving the theorem, we look at two simple examples.

Example 5.1

1. If $f : (\mathbb{F}, 0) \to (\mathbb{F}, 0)$ is given by $f(x) = x^3$ then $T^1_{\mathscr{A}_e} f$ has $\mathbb{F}$-basis x. The theorem tells us that $F(x, u) = (x^3 + ux, u)$ is a miniversal unfolding. Consider the unfolding $G(x, v) = (g_v(x), v) = (x^3 + vx^2, v)$. We want to find coordinate changes in source and target which transform G into an unfolding induced from F by base change. To do this, we work to eliminate unfolding monomials other than those which appear in F. Write $x = \bar{x} - \frac{1}{3}v$. Then $x^3 + vx^2 = \bar{x}^3 - \frac{1}{3}v^2\bar{x} + \frac{2}{27}v^3$. By this substitution we have eliminated the x^2 term from g_v, but at the cost of introducing a term in x^0, which we would also like to eliminate. The target coordinate change $\bar{y} = y - \frac{2}{27}v^3$ does this. That is, setting $h(v) = -\frac{1}{3}v^2$, $\Phi(\bar{x}, v) = (\bar{x} - \frac{1}{3}v, v)$, $\Psi(y, v) = (y - \frac{2}{27}v^3, v)$, we have
$$\Psi \circ G \circ \Phi = h^*F.$$
2. Let $f(x) = (x^2, x^3)$. Then $T^1_{\mathscr{A}_e} f$ has basis $x\partial/\partial y_2$, and thus, by the theorem, $F(x, u) = (x^2, x^3 + ux, u)$ is a miniversal unfolding. Consider the unfolding $G(x, v) = (g_v(x), v) = (x^2 + vx, x^3, v)$. The following table shows a sequence

of coordinate changes, in the first two columns, and the effect of those changes on G. To describe each coordinate change we use $\bar{x}$ and $\bar{y}$ to denote the new coordinates in source and target, respectively. Since accumulating a series of bars would be typographically unpleasant, in the third column we always denote the source coordinates simply by x.

Source coord. change	Target coord. change	g_v in new coordinates
		$(x^2 + vx, x^3)$
$x = \bar{x} - \frac{1}{2}v$		$(x^2 - \frac{1}{4}v^2, x^3 - \frac{3}{2}vx^2 + \frac{3}{4}v^2x - \frac{1}{8}v^3)$
	$\bar{y}_1 = y_1 + \frac{1}{4}v^2$	$(x^2, x^3 - \frac{3}{2}vx^2 + \frac{3}{4}v^2x - \frac{1}{8}v^3)$
	$\bar{y}_2 = y_2 + \frac{3}{2}vy_1 + \frac{1}{8}v^3$	$(x^2, x^3 + \frac{3}{4}v^2x)$

(5.1)

Putting the coordinate changes together and thus defining

$$\Phi(x, v) = (x - \frac{1}{2}v, v), \quad \Psi(y_1, y_2) = (y_1 + \frac{1}{4}v^2, y_2 + \frac{3}{2}v(y_1 + \frac{1}{4}v^2) + \frac{1}{8}v^3),$$

we have

$$\Psi \circ G \circ \Phi(x, v) = (x^2, x^3 - \frac{3}{2}v^2x).$$

So we take $h(v) = -\frac{3}{2}v^2$ to get $\Psi \circ G \circ \Phi = h^*F$. ◊

These two examples are rather atypical. In general, given an unfolding G and a versal unfolding F, it is very hard to find explicit parameterised coordinate changes Φ and Ψ and a base-change map h for which $\Psi \circ G \circ \Phi = h^*F$. The reader who enjoys wrestling with a problem of this kind can try Exercise 5.1.1.

What the theorem offers, instead, is a way of finding how a germ f can deform. If $F(x, u) = (f_u(x), u)$ is a versal unfolding of f then by examining the geometry of the maps f_u for different u, one can understand the geometry of all the possible perturbations of f.

Now let us prove Theorem 5.1. One implication is straightforward:

Proof of 'only if' in Theorem 5.1 Let g be an arbitrary element of $\theta(f_0)$ and take, as G, the 1-parameter unfolding $G(x,t) = (f(x) + tg(x), t)$, so that $\dot{G} = g$. By definition, there exists a base change $h : (\mathbb{F}, 0) \to (\mathbb{F}^d, 0)$ such that G is equivalent to the pull-back h^*F. If $F(x,u) = (f_u(x), u)$, then $h^*F(x,t) = (f_{h(t)}(x), t)$ and by the chain rule

$$\dot{h^*F} = \left.\frac{d(f_{h(t)})}{dt}\right|_{t=0} = \sum_{i=1}^{d} h_i'(0)\left.\frac{\partial f_u}{\partial u_i}\right|_{u=0} = \sum_{i=1}^{d} h_i'(0)\dot{F}_i \in \mathrm{Sp}_{\mathbb{F}}\{\dot{F}_1, \ldots, \dot{F}_d\}.$$

Since G and h^*F are equivalent, there exist unfoldings of the identity Φ, Ψ, such that $G = \Psi \circ h^*F \circ \Phi^{-1}$. Write $\Phi(x,t) = (\varphi_t(x), t)$ and $\Psi(y,t) = (\psi_t(y), t)$, then $g_t = \psi_t \circ f_{h(t)} \circ \varphi_t^{-1}$. As in the proof of Lemma 3.2, by the chain rule, we have

$$\frac{d}{dt}\left(\psi_t \circ f_{h(t)} \circ \varphi_t^{-1}\right)\Big|_{t=0} = d(f_0) \circ \left(\left.\frac{d\varphi_t^{-1}}{dt}\right|_{t=0}\right) + \left(\left.\frac{d\psi_t}{dt}\right|_{t=0}\right) \circ f_0 + \left.\frac{d(f_{h(t)})}{dt}\right|_{t=0} \tag{5.2}$$

(here we use the fact that $f_{h(0)} = f_0$). From (5.2) it follows that $\dot{G} - \dot{h^*F} \in T\mathscr{A}_e(f_0)$, which concludes the proof. □

In order to prove the "if" part in Theorem 5.1, we need the following lemma, which is a generalisation of Lemma 3.4 and is in fact a Thom–Levine Lemma for unfoldings with several parameters.

Let us denote by $i_d : \mathbb{F}^{d-1} \to \mathbb{F}^d$ the map $i_d(u_1, \ldots, u_{d-1}) = (u_1, \ldots, u_{d-1}, 0)$. For each d-parameter unfolding F, we define $F_1 = (i_d)^*F$, that is, F_1 is the $(d-1)$-parameter unfolding obtained by restriction of F to $u_d = 0$.

Lemma 5.1 *The following statements are equivalent:*

1. *The unfolding F is equivalent to h^*F_1, where $h : (\mathbb{F}^d, 0) \to (\mathbb{F}^{d-1}, 0)$ is a submersion whose fibres are transverse to the hyperplane $u_d = 0$.*
2. *There exist vector fields ξ on $(\mathbb{F}^n \times \mathbb{F}^d, S \times \{0\})$ and η on $(\mathbb{F}^p \times \mathbb{F}^d, 0)$ of the form*

$$\xi = \sum_{i=1}^{n} \xi_i(x,u)\frac{\partial}{\partial x_i} + \sum_{k=1}^{d-1} \zeta_k(u)\frac{\partial}{\partial u_k} + \frac{\partial}{\partial u_d}, \tag{5.3}$$

$$\eta = \sum_{j=1}^{p} \eta_j(y,u)\frac{\partial}{\partial y_j} + \sum_{k=1}^{d-1} \zeta_k(u)\frac{\partial}{\partial u_k} + \frac{\partial}{\partial u_d}, \tag{5.4}$$

such that $dF \circ \xi = \eta \circ F$.

Proof The idea of the proof is to consider F as a 1-parameter unfolding of F_1 with respect to the u_d parameter and then proceed as in Lemma 3.4. The key point is that we have to preserve the product structure in $\mathbb{F}^n \times \mathbb{F}^{d-1}$ in order to obtain the required vector fields.

Assume F is equivalent to $h^* F_1$, where $h : (\mathbb{F}^d, 0) \to (\mathbb{F}^{d-1}, 0)$ is a submersion whose fibres are transverse to $u_d = 0$. There exist unfoldings of the identity Ψ, Φ such that $F = \Psi \circ (h^* F_1) \circ \Phi^{-1}$. Moreover, the hypothesis on h implies that the germ $H : (\mathbb{F}^d, 0) \to (\mathbb{F}^d, 0)$ defined as $H(u) = (h(u), u_d)$ is a diffeomorphism. We write $K = H^{-1}$. This defines a coordinate system in the parameter space in which the components of h are the new first $d-1$ coordinates. Thus, if we define $G := K^*(h^* F_1)$, then G is nothing but the 1-parameter constant unfolding of F_1, since $G = K^*(h^* F_1) = (h \circ K)^* F_1 = \pi_d^* F_1$, where $\pi_d(u) = (u_1, \ldots, u_{d-1})$.

Now we perform the same change of parameters in the diffeomorphisms. That is, we define new diffeomorphisms $A = \Phi \circ (\mathrm{id}_n \times K)$ and $B = \Psi \circ (\mathrm{id}_p \times K)$. Then A, B are 1-parameter unfoldings of the identity and

$$
\begin{aligned}
B \circ G \circ A^{-1} &= \Psi \circ (\mathrm{id}_p \times K) \circ G \circ (\mathrm{id}_n \times K)^{-1} \circ \Phi^{-1} \\
&= \Psi \circ (\mathrm{id}_p \times K) \circ K^*(h^* F_1) \circ (\mathrm{id}_n \times K)^{-1} \circ \Phi^{-1} \\
&= \Psi \circ h^* F_1 \circ \Phi^{-1} = F
\end{aligned}
$$

(in the last equality we are using the calculation of Exercise 5.1.3). Hence, A and B provide an equivalence of 1-parameter unfoldings between F and the constant unfolding G.

Next, we take the vector fields obtained by differentiating the diffeomorphisms with respect to the parameter u_d. We define

$$
\xi = dA \circ \frac{\partial}{\partial u_d} \circ A^{-1}, \quad \eta = dB \circ \frac{\partial}{\partial u_d} \circ B^{-1}, \quad \zeta = dK \circ \frac{\partial}{\partial u_d} \circ K^{-1},
$$

which are vector fields on $(\mathbb{F}^n \times \mathbb{F}^d, S \times \{0\})$, $(\mathbb{F}^p \times \mathbb{F}^d, 0)$, and $(\mathbb{F}^d, 0)$, respectively. By construction, ζ has the form

$$
\zeta = \sum_{k=1}^{d-1} \zeta_k(u) \frac{\partial}{\partial u_k} + \frac{\partial}{\partial u_d}.
$$

Moreover, ξ and η are liftings of ζ over the projections onto the parameter space. Hence, ξ and η will have the form given in (5.3) and (5.4).

Finally, we proceed as in Lemma 3.4. Since G is constant with respect to the u_d variable,

$$
dG \circ \frac{\partial}{\partial u_d} = \frac{\partial}{\partial u_d} \circ G.
$$

By the chain rule

$$
\begin{aligned}
dF \circ \xi &= dF \circ dA \circ \frac{\partial}{\partial u_d} \circ A^{-1} = d(F \circ A) \circ \frac{\partial}{\partial u_d} \circ A^{-1} \\
&= d(B \circ G) \circ \frac{\partial}{\partial u_d} \circ A^{-1} = dB \circ dG \circ \frac{\partial}{\partial u_d} \circ A^{-1} \\
&= dB \circ \frac{\partial}{\partial u_d} \circ G \circ A^{-1} = \eta \circ B \circ G \circ A^{-1} = \eta \circ F.
\end{aligned}
$$

This completes the proof that statement 1 implies statement 2.

Conversely, suppose there exist vector fields ξ and η on $(\mathbb{F}^n \times \mathbb{F}^d, S \times \{0\})$ and $(\mathbb{F}^p \times \mathbb{F}^d, 0)$, respectively, as in (5.3) and (5.4) and such that $dF \circ \xi = \eta \circ F$. We set ζ to be the vector field on $(\mathbb{F}^d, 0)$ given by

$$
\zeta = \sum_{k=1}^{d-1} \zeta_k(u) \frac{\partial}{\partial u_k} + \frac{\partial}{\partial u_d},
$$

so that ξ, η are liftings of ζ over the corresponding projections. By integrating ξ, η and ζ with respect to the u_d variable we obtain diffeomorphisms A, B and K, respectively, which are characterised by the following properties:

$$
\xi = dA \circ \frac{\partial}{\partial u_d} \circ A^{-1}, \quad \eta = dB \circ \frac{\partial}{\partial u_d} \circ B^{-1}, \quad \zeta = dK \circ \frac{\partial}{\partial u_d} \circ K^{-1},
$$

and furthermore, A, B and K are unfoldings of the identity when considered as 1-parameter unfoldings with respect to the variable u_d. Because ξ and η are liftings of ζ, the diffeomorphisms A, B and K will be of the form

$$
A(x, u) = (a_u(x), K(u)), \quad B(y, u) = (b_u(y), K(u)), \quad K(u) = (k(u), u_d),
$$

for some map-germ $k : (\mathbb{F}^d, 0) \to (\mathbb{F}^{d-1}, 0)$.

Let $G = B^{-1} \circ F \circ A$. Then G is an unfolding of f, and $G_1 = F_1$, where G_1, F_1 are the restrictions of G, F to $u_d = 0$, respectively. This unfolding G satisfies that

$$
\begin{aligned}
dG \circ \frac{\partial}{\partial u_d} &= d(B^{-1} \circ F \circ A) \circ \frac{\partial}{\partial u_d} = dB^{-1} \circ dF \circ dA \circ \frac{\partial}{\partial u_d} = dB^{-1} \circ dF \circ \xi \circ A \\
&= dB^{-1} \circ \eta \circ F \circ A = dB^{-1} \circ \eta \circ B \circ G = \frac{\partial}{\partial u_d} \circ G.
\end{aligned}
$$

Hence, G is constant as a 1-parameter unfolding with respect to the variable u_d, that is, $G = \pi_d^* G_1 = \pi_d^* F_1$.

The final step is to define new diffeomorphisms $\Phi = A \circ (\mathrm{id}_n \times K)^{-1}$ and $\Psi = B \circ (\mathrm{id}_p \times K)^{-1}$ and a map-germ $h : (\mathbb{F}^d, 0) \to (\mathbb{F}^{d-1}, 0)$ such that $H(u) = (h(u), u_d)$ is the inverse of K. Obviously, Φ and Ψ are unfoldings of the identity and

h is a submersion whose fibres are transverse to the hyperplane $u_d = 0$. Moreover, $h \circ K = \pi_d$, so that $G = \pi_d^* F_1 = (h \circ K)^* F_1 = K^*(h^* F_1)$. Finally, we see that

$$\begin{aligned}\Psi \circ h^* F_1 \circ \Phi^{-1} &= B \circ (\mathrm{id}_p \times K)^{-1} \circ h^* F_1 \circ (\mathrm{id}_n \times K) \circ A^{-1} \\ &= B \circ K^*(h^* F_1) \circ A^{-1} \\ &= B \circ G \circ A^{-1} = F,\end{aligned}$$

which concludes the proof. □

We can now complete the proof of Theorem 5.1.

Proof of "if" in Theorem 5.1 Assume that $T^1_{\mathscr{A}_e} f_0$ is generated over $\mathbb{F}$ by the classes of $\dot{F}_1, \ldots, \dot{F}_d$. Let G be an e-parameter unfolding of f and let L be the $(d+e)$-parameter unfolding defined as the "sum" of F and G, that is, $L(x, u, v) = (\ell(x, u, v), u, v)$, where

$$\ell(x, u, v) = f_u(x) + g_v(x) - f_0(x).$$

We also denote by $L_1, \ldots, L_e$ the unfoldings obtained by successive restrictions of L to $v_e = 0$, to $v_e = v_{e-1} = 0, \ldots,$ and to $v_e = \cdots = v_1 = 0$, so that $L_e = F$.

Since $\partial \ell / \partial u_i|_{u=v=0} = \dot{F}_i$, we deduce from Lemma 3.5 that $T^1_{\mathscr{A}_e} L/d$ is generated over $\mathscr{O}_{\mathbb{C}^d \times \mathbb{C}^e, 0}$ by the classes of $\partial \ell / \partial u_i$, $i = 1, \ldots, d$. In particular, we can write

$$\frac{\partial \ell}{\partial v_e} = t_{\mathrm{rel}} L(\tilde{\xi}) + \omega_{\mathrm{rel}} L(\tilde{\eta}) + \sum_{k=1}^{d-1} \zeta_k(u, v) \frac{\partial \ell}{\partial u_k},$$

for some vector fields $\tilde{\xi} \in \theta(n+d/d)$, $\tilde{\eta} \in \theta(p+d/d)$ and functions $\zeta_k \in \mathscr{O}_{\mathbb{C}^d \times \mathbb{C}^e, 0}$. Define ξ and η as the vector fields on $(\mathbb{F}^n \times \mathbb{F}^d, S \times \{0\})$ and $(\mathbb{F}^p \times \mathbb{F}^d, 0)$ which are liftings of $-\tilde{\xi}$ and $\tilde{\eta}$ and have the form given in (5.3) and (5.4), respectively. Then, $dL \circ \xi = \eta \circ L$ and by Lemma 5.1, L is equivalent to $(h_1)^* L_1$, for some submersion $h_1 : (\mathbb{F}^d \times \mathbb{F}^e, 0) \to (\mathbb{F}^d \times \mathbb{F}^{e-1}, 0)$.

Proceeding by induction, we deduce that L is equivalent to $(h_e)^* L_e = (h_e)^* F$, for some submersion $h_e : (\mathbb{F}^d \times \mathbb{F}^e, 0) \to (\mathbb{F}^d, 0)$. If $h : (\mathbb{F}^e, 0) \to (\mathbb{F}^d, 0)$ is the restriction of h_e to $u = 0$, then, G, which is the restriction of L to $u = 0$, is equivalent to $h^* F$. □

As a corollary of the theorem we have the following result about existence and unicity of versal unfoldings.

Definition 5.2 We say that two unfoldings F, G of f_0 are *isomorphic* if G is equivalent to $h^* F$, for some diffeomorphism $h : (\mathbb{F}^d, 0) \to (\mathbb{F}^d, 0)$.

Corollary 5.1 *Let f_0 be a multi-germ. Then*

1. *f_0 admits a versal unfolding if and only if* $\operatorname{codim}_{\mathscr{A}_e}(f_0) < \infty$*;*
2. $\operatorname{codim}_{\mathscr{A}_e}(f_0)$ *is the minimal number of parameters in a versal unfolding of f_0;*
3. *any two versal unfoldings with the same number of parameters are isomorphic.*

Proof If a d-parameter unfolding F is versal, then the classes of $\dot{F}_1, \ldots, \dot{F}_d$ in $T^1_{\mathscr{A}_e} f_0$ generate it over $\mathbb{F}$, hence $\operatorname{codim}_{\mathscr{A}_e}(f_0) \leq d$. Conversely, if $\operatorname{codim}_{\mathscr{A}_e}(f_0) = d$, then there exist $g_1, \ldots, g_d \in \theta(f)$ whose classes provide a basis of $T^1_{\mathscr{A}_e} f_0$ over $\mathbb{F}$. The d-parameter unfolding F defined as

$$F(x, u) = (f(x) + \sum_{i=1}^{d} u_i g_i(x), u)$$

is versal, since $\dot{F}_i = g_i$. This shows (1) and (2).

Now we consider (3). Let F, G be two d-parameter versal unfoldings. Suppose first that both have the minimal number of parameters. Since F is versal, G is equivalent to h^*F for some germ $h : (\mathbb{F}^d, 0) \to (\mathbb{F}^d, 0)$. But G is also versal, therefore so is h^*F. By the chain rule, we have

$$(\dot{h^*F})_i = \sum_{j=1}^{d} \frac{\partial h_j}{\partial u_i}(0)\dot{F}_j, \quad \text{for } i = 1, \ldots, d.$$

By Theorem 5.1, the classes of $\{\dot{F}_1, \ldots, \dot{F}_d\}$ and $\{(\dot{h^*F})_1, \ldots, (\dot{h^*F})_d\}$ are both bases of $T^1_{\mathscr{A}_e}(f_0)$ over $\mathbb{F}$, hence the matrix $(\frac{\partial h_j}{\partial u_i}(0))$ is invertible and thus, h is a diffeomorphism. It follows that G is isomorphic to F.

The general case, where F and G have more parameters, follows by Exercise 5 below. □

We end this section by showing that the $\mathscr{A}_e$-versality of an unfolding F of f is equivalent to the transversality of the family of jet extension maps $j^k f_u$, as u varies through the parameter space, to the $\mathscr{A}$-orbit of f in jet space.

Let $f : (\mathbb{F}^n, S) \to (\mathbb{F}^p, 0)$ be a smooth map-germ and let F be a d-parameter unfolding of f. Let $F : \mathscr{X} \to \mathbb{F}^p \times \mathbb{F}^d$ be any representative of the unfolding, where $\mathscr{X}$ is an open neighbourhood of $S \times \{0\}$ in $\mathbb{F}^n \times \mathbb{F}^d$. We take X and U open neighbourhoods of S and 0 in $\mathbb{F}^n$ and $\mathbb{F}^d$, respectively, such that $X \times U \subseteq \mathscr{X}$ and consider the restriction $F : X \times U \to \mathbb{F}^p \times \mathbb{F}^d$. Denote by ${}_r j^k F/U : X^{(r)} \times U \to {}_r J^k(X, \mathbb{F}^p)$ the map given by

$$(x'_1, \ldots, x'_r, u) \mapsto (j^k f_u(x'_1), \ldots, j^k f_u(x'_r)).$$

We call this the *relative multi-jet extension map* of F. Recall from Sect. 4.6 that given any $\mathscr{A}^{(k)}$-invariant subset $W \subseteq J^k(n, p)$, there is an associated subbundle

$W(X, \mathbb{F}^p) \subseteq {}_rJ^k(X, \mathbb{F}^p)$ over $X \times \mathbb{F}^p$, with fibre W. Here we apply this once again with W the $\mathscr{A}^{(k)}$-orbit of f.

Theorem 5.2 *Suppose that $T\mathscr{A}f \supseteq \mathfrak{m}_n^{k+1}\theta(f)$. The unfolding F of f is $\mathscr{A}_e$–versal if and only if ${}_rj^kF/U$ is transverse to the submanifold $W(X, \mathbb{F}^p)$, where W is the $\mathscr{A}^{(k)}$-orbit of $\sigma := j^kf$ in $J^k(n, p)$.*

Proof By following the steps of the proof of Lemma 4.6, we see that the image of the differential $d_z({}_rj^kF/U)$, where $z = (x_1, \dots, x_r, 0)$, is generated by the k-jets of the partial derivatives $tf(\partial/\partial x_1), \dots, tf(\partial/\partial x_n)$, and by $\dot{F}_1, \dots, \dot{F}_d$. If $W = \mathscr{A}^{(k)}\sigma$, we deduce that ${}_rj^kF/U$ is transverse to $W(X, \mathbb{F}^p)$ at z if and only if

$$T\mathscr{A}(f)+\mathrm{Sp}_{\mathbb{F}}\left\{tf\left(\frac{\partial}{\partial x_1}\right),\dots,tf\left(\frac{\partial}{\partial x_n}\right)\right\}+\mathrm{Sp}_{\mathbb{F}}\{\dot{F}_1,\dots,\dot{F}_d\}+\mathfrak{m}_n^{k+1}\theta(f)=\theta(f).$$

By our assumption on f, this is equivalent to

$$T\mathscr{A}_e(f) + \mathrm{Sp}_{\mathbb{F}}\{\dot{F}_1, \dots, \dot{F}_d\} = \theta(f),$$

which is equivalent to the versality of F, by Theorem 5.1. □

Example 5.2 (Looking at Bent Wires III) Consider a manifold $M \subseteq \mathbb{F}^N$. Radial projection P_q from a point q into a hyperplane H is defined in Fig. 5.2.

If the hyperplane H is replaced by another hyperplane H', then the corresponding projection $P'_q : M \to H'$ is left equivalent to P_q; composing P'_q with the restriction of P_q to H', we get P_q. On the other hand, if we vary the point q then we may well

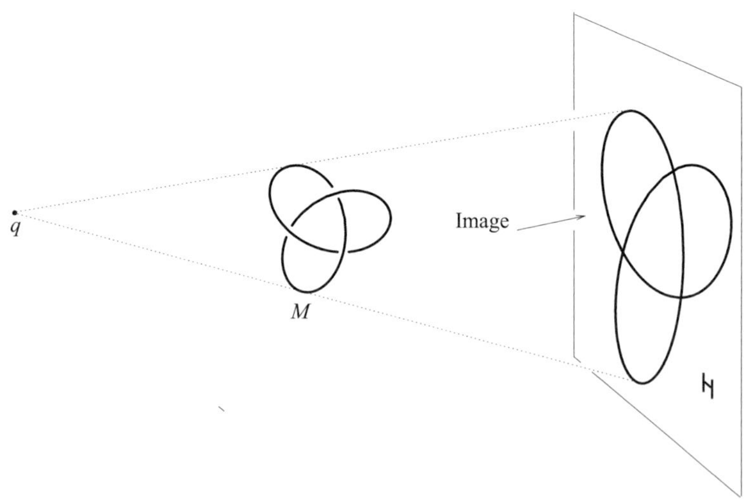

Fig. 5.2 Radial projection from a point into a hyperplane

deform the projection P_q non-trivially. So we consider the unfolding

$$P : M \times \mathbb{F}^N \to H \times \mathbb{F}^N.$$

It is instructive to look at this in the context of Examples 1.1 and 3.4, with the help of a visualiser or overhead projector, and to make drawings of what happens as one moves the point of projection. Are the unstable map-germs which appear from certain points of view versally unfolded in the family of all projections? In [Wal77] and [Dav83] it is shown that for a residual set of embeddings of S^1 into $\mathbb{R}^3$, the answer is Yes, except for the quadruple point. The proof uses the characterisation of versality in terms of transversality just proved, Theorem 5.2, together with a transversality theorem that guarantees that for any submanifold W of the jet bundle ${}_rJ^k(M, H)$, the set of embeddings $M \hookrightarrow \mathbb{R}^N$ for which the relative jet extension map ${}_rj^k(P/\mathbb{R}^N)$ is transverse to W is a residual subset of the set of embeddings with its Whitney topology (see Sect. 5.2 below). Since the intersection of finitely many residual subsets is residual, Theorem 5.2 together with the transversality theorem imply that for a "generic" embedding, any finite collection of singularity types will be versally unfolded whenever they appear.

We return to this topic in Example 11.4. ◊

Example 5.3 Let $f : (\mathbb{F}, 0) \to (\mathbb{F}^2, 0)$, $f(x) = (x^2, x^5)$. A straightforward calculation shows that

$$T\mathscr{A}_e f = \begin{pmatrix} \mathscr{O}_1 \\ \mathscr{O}_1 \smallsetminus \{x, x^3\} \end{pmatrix}.$$

By Theorem 5.1, the unfolding

$$F(x) = (x^2, x^5 + t_1x^3 + t_2x, t_1, t_2)$$

is miniversal. Which curves appear in the unfolding? We stratify the base space of the unfolding by the types of singularities which appear.

(1) *Non-immersive points:* Evidently f_t is a regular curve unless $t_2 = 0$, in which case it has a non-immersive point at $x = 0$.
(2) *Double points:* $f_t(x_1) = f_t(x_2)$ with $x_1 \neq x_2$ only if $x_1 = -x_2 \neq 0$. Since the second component is made up of odd powers of x, for it to take the same value at x_1 and $-x_1$ it must vanish there. Because we are assuming $x_1 \neq 0$ we may divide by x_1 to get

$$x_1^4 + t_1x_1^2 + t_2 = 0.$$

The formula for the roots of this quadratic in x_1^2 gives

$$x_1^2 = \frac{1}{2}\left(-t_1 \pm \sqrt{t_1^2 - 4t_2}\right).$$

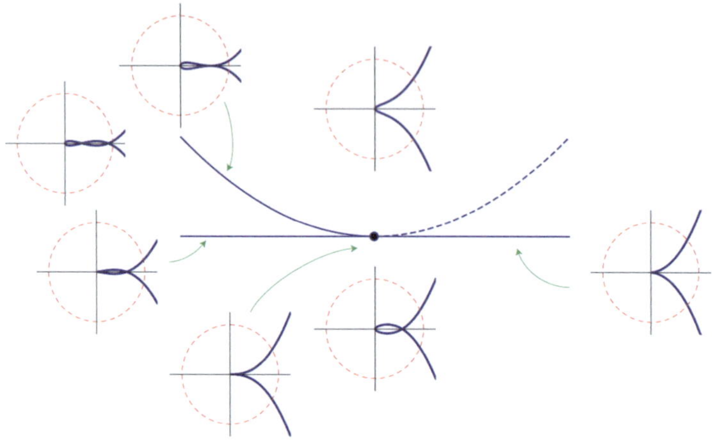

Fig. 5.3 Deformations of the curve $x \mapsto (x^2, x^5)$

For generic (t_1, t_2), there are two pairs of (complex) points sharing the same image:

$$x = \pm\sqrt{\frac{1}{2}\left(-t_1 + \sqrt{t_1^2 - 4t_2}\right)} \quad \text{and} \quad x = \pm\sqrt{\frac{1}{2}\left(-t_1 - \sqrt{t_1^2 - 4t_2}\right)}.$$

When the discriminant $t_1^2 - 4t_2$ vanishes, the two double points coincide, and the image curve has a self-tangency. Pictures of the images of the real maps $\mathbb{R} \to \mathbb{R}^2$ are shown in Fig. 5.3.

For parameter values on the dashed curve here, the complexification has a self-tangency, but over the reals this is not visible. ◊

Example 5.4 Let $f(x) = (x^3, x^4)$. An easy calculation shows that

$$T\mathscr{A}_e f = \begin{pmatrix} \mathscr{O}_1 \setminus \{x\} \\ \mathscr{O}_1 \setminus \{x, x^2\} \end{pmatrix}$$

and thus as versal unfolding we can take

$$F(x, a, b, c) = (x^3 + ax, x^4 + bx^2 + cx, a, b, c).$$

The analysis of the singularities of the deformations $f_{a,b,c}$, for all possible values of the parameters, is not easy. For this reason, we restrict ourselves to the section $c = 0$, which will allow us to proceed with some computations. To simplify the notation, we put

$$f_{a,b}(x) = f_{a,b,0}(x) = (x(a+x^2), x^2(b+x^2)).$$

(1) *Non-immersive points:* We have $f'_{a,b}(x) = (a+3x^2, 2x(b+2x^2))$. Thus, the non-immersive points appear at $x=0$ when $a=0$ and at $x = \pm\sqrt{-a/3}$ when $3b=2a$.

(2) *Double points:* As in Example 5.3, we look for pairs (x_1, x_2) such that $f_{a,b}(x_1) = f_{a,b}(x_2)$ with $x_1 \neq x_2$. To find them, we consider the "divided differences":

$$\frac{f_{a,b}(x_2) - f_{a,b}(x_1)}{x_2 - x_1} = (a + x_1^2 + x_1x_2 + x_2^2, (x_1+x_2)(b + x_1^2 + x_2^2)) = 0.$$

These two equations define a surface in x_1, x_2, a, b-space, with two components. The first is $x_1 + x_2 = a + x_1^2 = 0$, giving a double point

$$f_{a,b}(x_1) = (0, a(a-b)), \tag{5.5}$$

with $x_1 = \pm\sqrt{-a}$ and $a \neq 0$. The equations for the second component are

$$a + x_1^2 + x_1x_2 + x_2^2 = b + x_1^2 + x_2^2 = 0,$$

which after eliminating the variable x_2 give the quartic equation

$$(a-b)^2 + bx_1^2 + x_1^4 = 0,$$

with discriminant $\Delta = (a-b)^2(b^2 - 4(a-b)^2) = (a-b)^2(3b-2a)(2a-b)$. When $\Delta \neq 0$, there are four simple roots

$$x_1 = \pm\sqrt{\frac{-b \pm \sqrt{\Delta'}}{2}}, \quad \Delta' = (3b-2a)(2a-b),$$

which correspond to two more double points

$$f_{a,b}(x_1) = \left(\frac{\sqrt{-b \pm \sqrt{\Delta'}}}{2\sqrt{2}}(2a - b + \sqrt{\Delta'}), -(a-b)^2\right). \tag{5.6}$$

The fact that the roots are simple in the two components implies that there is no tangency between the branches meeting at the double points. Moreover, it is not difficult to see that the three double points we have found are distinct. Thus, when $a(a-b)(3b-2a)(2a-b) \neq 0$, the curve $f_{a,b}$ is stable with exactly three double points in the complex case or with up to three double points in the real case.

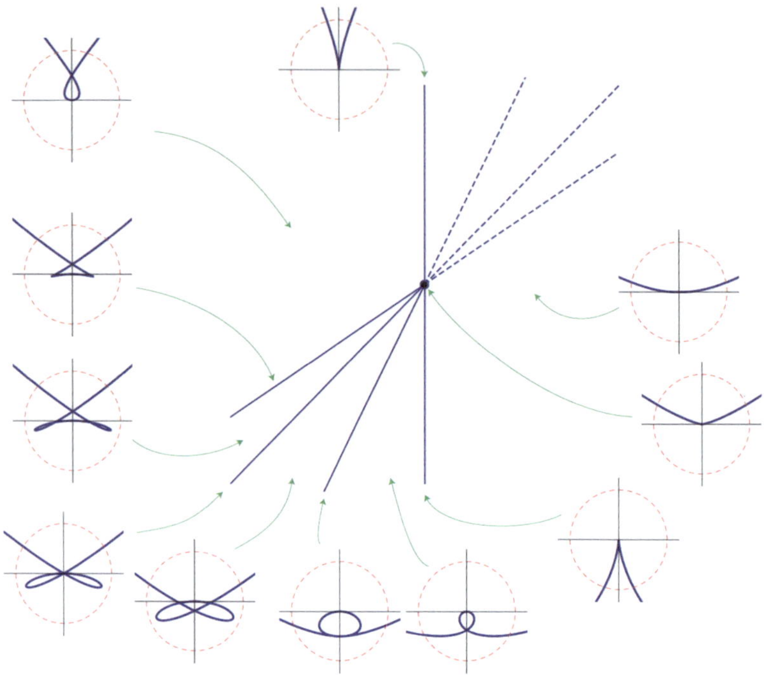

Fig. 5.4 Deformations of the curve $x \mapsto (x^3, x^4)$

We have already seen in (1) that the parameters on the lines $a = 0$ or $3b = 2a$ give non-immersive points. On the line $b = 2a$, the quartic has two double roots, which means that the curve has a self-tangency at each of the points in (5.6). Finally, when $a = b$, the quartic becomes $x^2(a + x^2) = 0$, so either $x = 0$ or $a + x^2 = 0$. In both cases, the image is $f_{a,a}(x) = (0, 0)$ and also coincides with the double point in (5.5). Thus, we have shown that the origin is in fact a triple point of $f_{a,a}$. Pictures of the images of the real maps $\mathbb{R} \to \mathbb{R}^2$ are shown in Fig. 5.4. As in Example 5.3, the dashed curves are the branches where the complexification has an instability which is not visible over the reals.

Note that because we have set $c = 0$, our image curve is always symmetric under reflection in the vertical axis. This means that some singularities appear in pairs, e.g. the two cusps on the second-from-top curve on the left in the diagram. In general the codimension of a stratum B_i of the bifurcation set is equal to the sum of the $\mathscr{A}_e$-codimensions of the singularities of f_u for $u \in B_i$ (see Propositions 5.1 and 5.2 below) so on a codimension 1 stratum one would expect to see only one unstable point. We leave the reader to ponder the nature of the intersection of the bifurcation set $\mathscr{B}$ (see Sect. 5.4 below) with $\{c = 0\}$ in this case. ◊

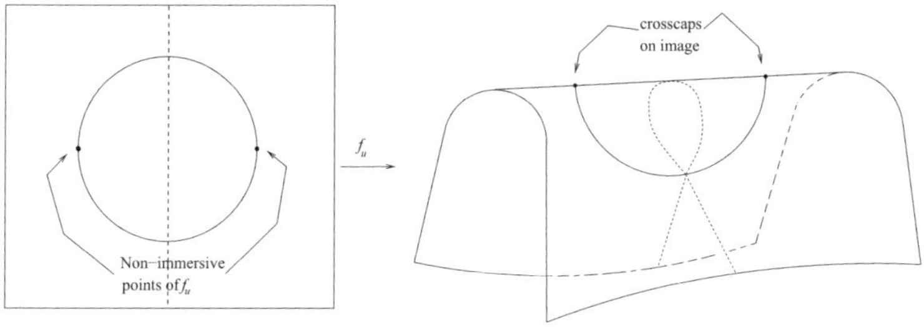

Fig. 5.5 Stable perturbation of $(x, y) \mapsto (x, y^2, y^3 + x^2 y)$

Example 5.5 Consider the map-germ $f_0(x, y) = (x, y^2, y^3 + x^2 y)$ of Example 3.5. We saw that $y\partial/\partial Z$ projects to a basis for $T^1_{\mathscr{A}_e} f_0$. So

$$F(x, y, u) = (x, y^2, y^3 + x^2 y + uy, u)$$

is a versal deformation. What is the geometry here? Think of F as a family of mappings,

$$f_u(x, y) = (x, y^2, y^3 + x^2 y + uy).$$

The *ramification ideal* $\mathscr{R}_{f_u} \subseteq \mathscr{O}_{\mathbb{F}^2}$ generated by the 2×2 minors of the matrix $[df_u]$ defines the set of points where f_u fails to be an immersion. Here $\mathscr{R}_{f_u} = (y, x^2 + u)$. So for $u \neq 0$, f_u has two non-immersive points. They are only visible over $\mathbb{R}$ when $u < 0$ (see Fig. 5.5). How does f_u behave in the neighbourhood of each of these points? At each, $\mathscr{R}_{f_u}$ is equal to the maximal ideal; it follows that df_u is transverse to the submanifold $\Sigma^1 \subseteq L(\mathbb{F}^2, \mathbb{F}^3)$ consisting of linear maps of rank 1. In fact this transversality *characterises* the map-germ f parameterising the Whitney umbrella (described in Example 3.5(2)) up to $\mathscr{A}$-equivalence, though here we are not yet able to show that. Using this characterisation, we see that in a neighbourhood of the image of each of the two points $(\pm\sqrt{-u}, 0)$, the image of f_u looks like the drawing in Example 3.5. The key to assembling the image of f_u from its constituent parts is the curve of self-intersection. The only points mapped 2–1 by f_u are the points of the curve $\{x^2 + y^2 + u = 0\}$; for $u < 0$ this is a circle when viewed over $\mathbb{R}$. Here points $(x, \pm y)$ share the same image. The two non-immersive points of f_u are the fixed points of the involution $(x, y) \mapsto (x, -y)$ which interchanges pairs of points sharing the same image.

The image contains a chamber; indeed it is homotopy-equivalent to a 2-sphere. This is no coincidence. Figure 8.2 at the start of Chap. 8 shows the images of stable perturbations of each of the remaining codimension 1 singularities of maps from surfaces into 3-space. Each is homotopy-equivalent to a 2-sphere. Some choices have been made regarding the real form: sometimes a change of sign which makes no difference to the right-left equivalence class over $\mathbb{C}$ does make a difference over $\mathbb{R}$. Nevertheless in all of these cases it is possible to choose a suitable real form whose perturbation is a homotopy 2-sphere. We study the topology of stable perturbations in Chap. 8. ◊

Exercises for Sect. 5.1

1 Let $f(x_1, x_2) = (x_1, x_2^3 + x_1^2 x_2)$.

(i) Find a miniversal unfolding of f.
(ii) (*Harder*) Is the unfolding $G(x_1, x_2, v) = (x_1 + vx_2^2, x_2^3 + x_1^2 x_2, v)$ trivial?

2 Let F be an unfolding with parameter space $(\mathbb{F}^d, 0)$ and let $K : (\mathbb{F}^d, 0) \to (\mathbb{F}^d, 0)$ be a diffeomorphism. Show that

$$(\mathrm{id}_p \times K)^{-1} \circ F \circ (\mathrm{id}_n \times K) = K^* F.$$

3 Show that if f_0 is a stable multi-germ, then any unfolding is versal. This is just a question of unravelling definitions.

4 Prove Equation (5.2) by using the chain rule as in the proof of Lemma 3.2.

5 Show that any versal unfolding is isomorphic to the constant unfolding of a miniversal unfolding.

6 Let F, G be two unfoldings of the same germ. In the proof of Theorem 5.1 we defined their "sum" as the unfolding L given by

$$L(x, u, v) = (f_u(x) + g_v(x) - f_0(x), u, v).$$

Show that if F is versal, then so is L.

7 Show that if F is versal, then it is stable (as a multi-germ).

8 Show that if the $\mathscr{A}_e$-codimension of f is 1, then any stable unfolding of f is versal.

5.2 Global Stability of C^∞ Mappings

In Examples 5.3 and 5.4, we see that it is possible, by suitably deforming the map-germs, to make their instability disappear—after an arbitrarily small deformation, all of the multi-germs of the deformed map are stable. In the dimension range we have looked at, every $\mathscr{A}$-finite germ has this property. These are "nice dimensions", to use a term due to John Mather: the dimension-pairs (n, p) such that for every $\mathscr{A}$-finite germ, the set of parameter values u in the base of a versal deformation, for which f_u has an unstable multi-germ (the "bifurcation set"), is a proper subset.

Mather's original definition of the nice dimensions was different, and referred to global stability of smooth (C^∞) mappings between smooth manifolds (in the real case). We now describe it and show that it is equivalent to the definition given here. We first need to define the Whitney C^∞ topology on the space $C^\infty(N, P)$ of smooth mappings from a smooth manifold N to a smooth manifold P.

Definition 5.3

(1) Let N and P be smooth manifolds, and let $W \subseteq {}_rJ^k(N, P)$ be an open subset. We define A^k_W as the set $\{f \in C^\infty(N, P) : {}_rj^k f(N) \subseteq W\}$.
(2) The Whitney C^k topology on $C^\ell(N, P)$, for $\ell \geq k$, is the topology generated by the sets A^k_W.
(3) The Whitney C^∞ topology on $C^\infty(N, P)$ is the topology generated by the sets A^k_W for $k \in \mathbb{N}$.

Remark 5.1 The following properties of the Whitney topology are easily proved. For details see the account in [GG73, Ch. 2 §3].

(1) If $W \subseteq {}_rJ^k(N, P)$ is open and $k \leq \ell$, let $W' = (\pi^\ell_k)^{-1}(W)$, where $\pi^\ell_k : {}_rJ^\ell(N, P) \to {}_rJ^k(N, P)$ is the usual projection. Then W' is open, and $A^k_W = A^\ell_{W'}$. It follows that the C^ℓ topology contains the C^k topology.
(2) The familiar *compact-open topology* on the space $C^0(N, P)$ of continuous maps from N to P has, as sub-base, the collection of all sets $A^0(K, U)$, where $K \subseteq N$ is compact and $U \subseteq P$ is open. Each of these sets is open in the Whitney C^0 topology: for each compact $K \subseteq N$, there is a family V_j, $j \in \mathbb{N}$ such that $K = \bigcap_j V_j$; let $W_j \subseteq J^0(N, P) = N \times P$ be defined by

$$W_j = (V_j \times U) \cup (N \setminus K) \times P;$$

then

$$A^0(K, U) = \bigcup_j A^0_{W_j}.$$

(3) It can be shown that if N is compact, then $C^\infty(N, P)$, with the Whitney C^∞ topology, is first countable (i.e. every member has a countable neighbourhood basis)—see [GG73, Ch. 2 §3]. If N is not compact, this no longer holds. Indeed, the C^k topology has the following rather rigid property: a sequence f_n of maps converges to f if and only if there is a compact subset $K \subseteq N$ such that $j^k f_n$ converges uniformly to $j^k f$ on K, and outside K, f_n is equal to f for all but a finite number of indices n. For this and other reasons, the Whitney topologies are most useful when N is compact.
(4) If M is compact, the composition mapping $C^\infty(M, N) \times C^\infty(N, P)$ sending (f, g) to $g \circ f$ is continuous with respect to the Whitney C^∞ topology [GG73, Ch. 2, Prop. 3.9].

(5) If $\varphi : N \to N$ is a diffeomorphism, and $\psi : P \to P$ is any smooth map, then the mapping $C^\infty(N, P) \to C^\infty(N, P)$ defined by $f \mapsto \psi \circ f \circ \varphi^{-1}$ is continuous. If also ψ is a diffeomorphism then this mapping is a homeomorphism. ◊

Definition 5.4 The smooth mapping $f : N \to P$ is *stable* if there is a neighbourhood A of f, in the Whitney C^∞ topology, such that every $g \in A$ is left-right equivalent to f.

It follows from Remark 5.1(5) that f is stable if and only if its orbit under the action of the group $\text{Diff}(N) \times \text{Diff}(P)$ is open in $C^\infty(N, P)$.

In his series of papers [Mat68a, Mat69b, Mat69a, Mat68b, Mat70], John Mather related this definition of stability to the stability of germs that we have been working with, in the case of proper maps. Let $C^\infty_{\text{pr}}(N, P)$ be the subspace of $C^\infty(N, P)$ consisting of proper maps.

Theorem 5.3 ([Mat70, Th. 4.1]) *Let $f : N \to P$ be a smooth proper mapping, let $r \geq p + 1$ and let $k \geq \dim P$. The following conditions are equivalent:*

1. *f is stable (in the sense of Definition 5.4).*
2. *${}_rj^k f$ is transverse to every contact orbit in ${}_rJ^k(N, P)$.*
3. *For each $y \in P$, and for every finite set $S \subseteq f^{-1}(y)$, the multi-germ $f : (N, S) \to (P, y)$ is stable, in the sense of Definition 3.4(4).*

The equivalence of (2) and (3) follows from Theorem 4.6 above. We will not discuss the proof that (1) and (2) are equivalent here. Suffice it to say that it involves a fourth equivalent condition: that f is *globally* infinitesimally stable, in the sense that the $C^\infty(N)$-module of *global* vector fields along f is equal to the global version of the right-left tangent space of f.

5.2.1 *Stable Maps Are Not Always Dense*

In 1958 H. Whitney conjectured that stable mappings are always dense in the space of proper mappings $C^\infty_{\text{pr}}(N, P)$. René Thom then found a counterexample, or rather a method for generating infinitely many counterexamples. The method is described in H. Levine's lectures in [Lev71, §11.2]. Consider germs of mappings $(\mathbb{F}^{n^2}, 0) \to (\mathbb{F}^{n^2}, 0)$. The codimension in $J^1(n^2, n^2)$ of the set Σ^n of 1-jets with corank n is n^2 (see Exercise A.1). It is easy to construct a germ f whose 1-jet $j^1 f$ meets Σ^n transversely at 0 (see Exercise A.12), though in fact a non-trivial theorem of Mather ([Mat70, Th. 6.1], together with remarks in the introduction to that paper) shows that for any N, P and k and submanifold W of jet space $J^k(N, P)$ of codimension $\leq N$, there exists a smooth map $f : N \to P$ for which the k-jet $j^k f$ meets W transversely. Thom showed that for any germ $(\mathbb{F}^{n^2}, 0) \to (\mathbb{F}^{n^2}, 0)$ of type Σ^n, the codimension in $J^2(n^2, n^2)$ of its $\mathscr{A}^{(2)}$-orbit is greater than n^2. This means that f cannot be stable, since by Theorem 4.6, stability implies transversality of the jet extension map $j^2 f$ to the $\mathscr{A}^{(2)}$ orbit of $j^2 f(0)$. This shows also that f has no stable perturbation. For since $j^1 f$ meets Σ^n transversely, any sufficiently small perturbation of f also meets Σ^n, and therefore cannot be stable.

Thanks to Theorem 4.1, we can slightly simplify Thom's argument to show that f cannot be stable. If f has type Σ^n then in linearly adapted coordinates it takes the form

$$f(x_1, \ldots, x_{n^2}) = (x_1, \ldots, x_{n^2-n}, f_1(x), \ldots, f_n(x)),$$

with the f_i of order at least 2. To simplify notation, we denote the first $n^2 - n$ coordinates on the source by $u_i, i = 1\ldots, n^2 - n$, and the last n coordinates by $v_j, j = 1, \ldots, n$. Thus, f is an $(n^2 - n)$-parameter unfolding of the germ $f_0 : (\mathbb{F}^n, 0) \to (\mathbb{F}^n, 0)$ defined by restricting the last n components of f to the zero-locus of the u_i. We show that for sufficiently big n it cannot be stable, by showing that a stable unfolding of f_0 requires more than $n^2 - n$ parameters. By Lemma 5.5 below, together with Theorem 4.1, the minimal number of parameters needed is

$$\dim_{\mathbb{F}} \frac{\mathfrak{m}_n\, \theta(f_0)}{tf_0 + f_0^*\, \mathfrak{m}_n\, \theta(f_0)}.$$

Denote the coordinates on the target of f_0 by $y_1, \ldots, y_n$. Now

$$\dim_{\mathbb{F}} \frac{\mathfrak{m}_n\, \theta(f_0)}{tf_0 + f_0^*\, \mathfrak{m}_n\, \theta(f_0)} \geq \dim_{\mathbb{F}} \frac{\mathfrak{m}_n\, \theta(f_0)}{tf_0 + f_0^*\, \mathfrak{m}_n\, \theta(f_0) + \mathfrak{m}_n^3\, \theta(f_0)}. \tag{5.7}$$

We have

$$\dim_{\mathbb{F}} \frac{\mathfrak{m}_n}{\mathfrak{m}_n^3} = \frac{1}{2} n(n+3)$$

and so

$$\dim_{\mathbb{F}} \frac{\mathfrak{m}_n\, \theta(f_0)}{\mathfrak{m}_n^3\, \theta(f_0)} = \frac{1}{2} n^2(n+3).$$

The dimension on the right of (5.7) is minimised by assuming that $f_1, \ldots, f_n$ are linearly independent quadrics—recall that they have zero linear part. Then

$$\frac{tf_0(\theta_n) + f_0^*\, \mathfrak{m}_n\, \theta(f_0) + \mathfrak{m}_n^3\, \theta(f_0)}{\mathfrak{m}_n^3\, \theta(f_0)}$$

is generated by n linear terms $tf_0(\partial/\partial v_i)$, $1 \leq i \leq n$, n^2 quadric terms $f_i\partial/\partial y_j$, $1 \leq i, j \leq n$ and n^2 quadric terms $tf_0(v_i\partial/\partial v_j)$, $1 \leq i, j \leq n$. Thus it has dimension $\leq 2n^2 + n$. So

$$\dim_{\mathbb{F}} \frac{\mathfrak{m}_n\,\theta(f_0)}{tf_0(\theta_n) + f_0^*\,\mathfrak{m}_n\,\theta(f_0) + \mathfrak{m}_n^3\,\theta(f_0)} \geq \frac{1}{2}n^2(n+3) - (2n^2+n) = \frac{1}{2}n(n^2 - n - 2). \tag{5.8}$$

If $n \geq 4$ then the right-hand side of (5.8) is greater than $n^2 - n$, and thus f cannot be stable.

Now suppose that N and P are manifolds of dimension n^2. By a simple argument using a partition of unity, one can glue a representative of the germ constructed in Exercise A.12 into a proper map $f_0 : N \to P$, to obtain a map f for which $j^1 f$ meets $\Sigma^n \subseteq J^1(N, P)$ transversely. By the argument of the last paragraph, f is not stable. The set U of maps f' for which $j^1 f'$ continues to meet Σ^n transversely contains an open neighbourhood of f in the Whitney C^∞ topology. And by what we have seen, U contains no stable map.

To summarise: in this example, the non-density of stable maps in $C^\infty(N, P)$ is shown by constructing a submanifold W of jet space $J^k(N, P)$ such that

1. codim $W \leq \dim N$, and
2. W is a union of $\mathscr{K}$-orbits all of which have codimension $> \dim N$,

and then showing that there exist mappings $N \to P$ such that $j^k f$ meets W transversely. Note that conditions (1) and (2) together mean that W must consist of uncountable many $\mathscr{K}$-orbits. Thus the non-density of stable mappings is related to the presence of "moduli" in the $\mathscr{K}$ classification.

5.2.2 *Mather's Nice Dimensions*

In [Mat70], Mather defined the "nice dimensions" as the range of dimension-pairs (n, p) such that for all smooth manifolds N and P of dimensions n and p, stable maps are dense in $C^\infty_{\mathrm{pr}}(N, P)$, and then used the strategy followed in the last subsection to determine which are the nice dimensions. He introduced the following definitions.

Definition 5.5

1. For $k, n, p, r \in \mathbb{N}$, let $W_r^k(n, p)$ be the set of k-jets $z \in J^k(n, p)$ such that the $\mathscr{K}^{(k)}$-codimension of z is $\geq r$. Then $W_r^k(n, p)$ is a closed algebraic subset of $J^k(n, p)$.
2. Let $W_r^k(n, p)^*$ denote the union of all irreducible components of $W_r^k(n, p)$ whose codimension is less than r, and let $\pi^k(n, p) = \bigcup_{r \geq 0} W_r^k(n, p)^*$. Then $\pi^k(n, p)$ is also a closed algebraic subset of $J^k(n, p)$.

3. Let $\sigma^k(n, p)$ be the codimension in $J^k(n, p)$ of $\pi^k(n, p)$. Clearly $\sigma^\ell(n, p) \geq \sigma^k(n, p)$ if $\ell \geq k$.
4. We define $\sigma(n, p) = \inf_k \sigma^k(n, p)$.

Example 5.6 Thom's example can be seen in this light. If we replace $tf_0(\theta_n)$ by $tf(\mathfrak{m}_n \theta_n)$ in the denominator of the left-hand side of (5.8), the dimension of the quotient is precisely the $\mathscr{K}^{(2)}$-codimension of the 2-jet $j^2 f(0)$. This replacement increases the dimension by n. Write $r = \frac{1}{2}n(n^2 - n)$. We have shown that for each 2-jet z of rank 0, the $\mathscr{K}^{(2)}$-codimension is $\geq r$. Thus the submanifold $M_{n^2} := (\pi_1^2)^{-1}(\Sigma^n)$ of $J^2(n^2, n^2)$ is contained in $W_r^2(n^2, n^2)$. The codimension of M_{n^2} is n^2, and therefore the codimension of the irreducible component of $W_r^2(n^2, n^2)$ containing it is $\leq n^2$. When $n \geq 4$, $r > n^2$, and thus M_{n^2} is contained in $W_r^2(n^2, n^2)^*$. Hence $\sigma(n^2, n^2) < \operatorname{codim} W_r^2(n^2, n^2)^* \leq \operatorname{codim} M_{n^2} = n^2$. ♢

Mather proved the following slightly surprising theorem.

Theorem 5.4 ([Mat70, Th. 8.1]) *Let N and P be smooth manifolds of dimensions n and p. The set of stable mappings $N \to P$ is dense in $C^\infty_{pr}(N, P)$ if and only if $n < \sigma(n, p)$.* □

In particular, the density of stable mappings has nothing to do with the topology or smooth structure of the manifolds concerned—it depends only on their dimensions. The following corollary is immediate.

Corollary 5.2 *(n, p) are nice dimensions if and only if $n < \sigma(n, p)$.* □

The idea behind the proof of Theorem 5.4 is the same as in Thom's example discussed in Sect. 5.2.1. In [Mat71], Mather computed the value of $\sigma(n, p)$ for each pair (n, p) and, as a consequence, deduced that (n, p) are nice dimensions if and only if they satisfy one of the following conditions:

1. $n < \frac{6}{7}p + \frac{8}{7}$ and $p - n \geq 4$,
2. $n < \frac{6}{7}p + \frac{9}{7}$ and $3 \geq p - n \geq 0$,
3. $p < 8$ and $p - n = -1$,
4. $p < 6$ and $p - n = -2$,
5. $p < 7$ and $p - n \leq -3$.

We go over some of his calculations in Sect. 7.4, in order to prove (or, rather, check, by going through lists of stable germs) that in Mather's nice dimensions, every stable germ is weighted homogeneous with respect to suitable coordinates.

In Fig. 5.6, the nice dimensions are contained in the region to the left and below the polygon.

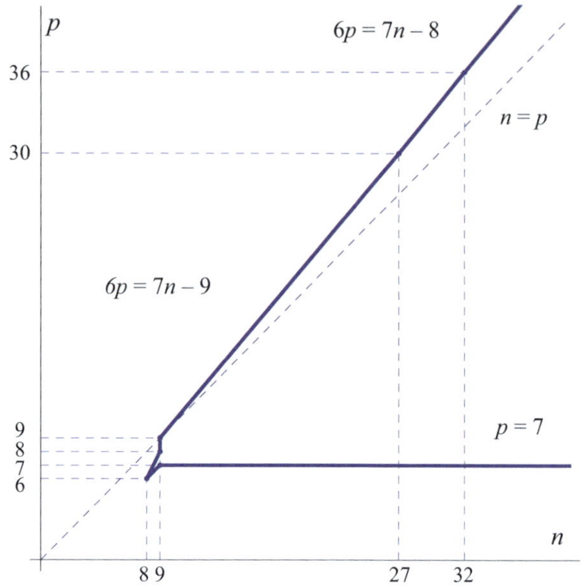

Fig. 5.6 The region of Mather's nice dimensions

5.3 Topological Stability

A map $f : N \to P$ is C^0-*stable*, or *topologically stable*, if it has a neighbourhood U in $C^\infty_{\rm pr}(N, P)$ (with the Whitney C^∞ topology) such that each $f' \in U$ is C^0-left-right equivalent to f—that is, there are homeomorphisms ψ and φ such that $f' = \psi \circ f \circ \varphi$. The notion of topological stability is very different from C^∞ stability because there is no group acting: the collection of C^∞ maps which are C^0-left-right equivalent to f is not the orbit of f under the action of a group of C^0-equivalences defined independently of f. Nevertheless, in the 1970s Mather outlined a proof that topologically stable mappings are dense in $C^\infty_{\rm pr}(N, P)$. His argument was not so different from the arguments used to prove C^∞ stability in the nice dimensions. He proved the following theorem, which we quote in the form given by Wall and du Plessis in [dPW95]:

Theorem 5.5 *For each $k \in \mathbb{N}$ there is a canonical $\mathscr{K}$-invariant Whitney regular stratification $\mathscr{A}^k(n, p)$ of $J^k(n, p) \setminus W^k(n, p)$, where $W^k(n, p)$ is a certain $\mathscr{K}$-invariant closed algebraic subset of $J^k(n, p)$ whose codimension tends to ∞ as k goes to ∞, with the following property: if $f \in C^\infty_{pr}(N, P)$ and for all k $j^k f$ avoids the subbundle $W^k(N, P)$ corresponding to $W^k(n, p)$ and $j^k f$ is multitransverse to $\mathscr{A}^k(N, P)$, then f is topologically stable.*

From this, the density of topological stable maps follows by the Mather multi-jet transversality theorem in the same way as does the density of C^∞ stable maps in the nice dimensions.

Wall and du Plessis comment that although Mather proves the existence of a canonical stratification, it is very hard to determine which are its strata, beyond the $\mathscr{K}$-simple orbits which make up the stratification used in the nice dimensions to prove C^∞ stability. A second problem is that transversality to the canonical stratification, although sufficient, is not necessary for C^0 stability, unlike the situation for C^∞ stability. The book [GWdPL76] gives a detailed proof of the density of topological stability, following Mather's outline; the later book, [dPW95], focuses on improving the theory in the light of these two criticisms. Mather himself filled in the details of his outline in [Mat76], though ultimately using a somewhat different strategy to complete the proof.

5.4 Bifurcation Sets

Let F be a d-parameter unfolding of an $\mathscr{A}$-finite germ $f_0 : (\mathbb{F}^n, S) \to (\mathbb{F}^p, 0)$. By Lemma 4.3, f has finite singularity type, and hence so does F. Therefore by Propositions 4.3 and 4.7, we may choose a representative $F : \mathscr{X}_0 \to \mathscr{Y}_0$, where $\mathscr{X}_0$ and $\mathscr{Y}_0$ are open neighbourhoods of $S \times \{0\}$ and 0 in $\mathbb{F}^n \times \mathbb{F}^d$ and $\mathbb{F}^p \times \mathbb{F}^d$, respectively, and such that

(i) $F^{-1}(0) \cap C(F) = S \times \{0\}$,
(ii) the restriction $F| : C(F) \to \mathscr{Y}_0$ is finite.

It would be preferable to replace this representative by another one which takes into account the product structure with the parameter space, but we have to be careful if we want to keep the above properties. In fact, we can choose $Y$and U, open neighbourhoods of the origin in $\mathbb{F}^p$, $\mathbb{F}^d$, respectively, such that $Y \times U \subseteq \mathscr{Y}_0$ and define $\mathscr{X} = F^{-1}(Y \times U)$. Then we take the sub-representative $F : \mathscr{X} \to Y \times U$ which still satisfies conditions (1) and (2) above. Let $X_u = \{x \in \mathbb{F}^n : (x, u) \in \mathscr{X}\}$. Now for each $u \in U$, we consider the mapping $f_u : X_u \to Y$. It is easy to see that the restriction $f_u| : C(f_u) \to Y$ is also finite.

For each $y \in Y$, we denote by $(f_u)_y$ the germ $f_u : (X_u, f_u^{-1}(y) \cap C(f_u)) \to (Y, y)$, and define $F_{(y,u)}$ analogously. Given any subset $Y' \subseteq Y$, we will say that f_u is *locally stable on* Y' if the multi-germ $(f_u)_y$ is stable, for all $y \in Y'$. After shrinking Y if necessary, we may also assume that f_0 is locally stable on $Y \setminus \{0\}$ by the Mather–Gaffney criterion Theorem 4.5.

Lemma 5.2 *For each compact subset $K \subseteq Y$, the subset*

$$\{u \in U : f_u \text{ is not locally stable on } K\}$$

is closed in U.

Proof Take a parameter $u_0 \in U$ such that f_{u_0} is locally stable on K. For each $y \in K$, the multi-germ $(f_{u_0})_y$ is stable, so the multi-germ $F_{(y,u_0)}$ is a trivial unfolding of it. In terms of representatives, this means we have a commutative diagram

$$\begin{array}{ccc} F^{-1}(Y_y \times U_y) & \xrightarrow{F} & Y_y \times U_y \\ \downarrow & & \downarrow \\ f_u^{-1}(Y_y) \times U_y & \xrightarrow{f_{u_0} \times \mathrm{id}} & Y_y \times U_y \end{array}$$

where Y_y and U_y are open neighbourhoods of y and u_0 in Y and U, respectively, and the columns are diffeomorphisms which are unfoldings of the identity. In particular, $f_u : f_u^{-1}(Y_y) \to Y_y$ is $\mathscr{A}$-equivalent to $f_{u_0} : f_{u_0}^{-1}(Y_y) \to Y_y$ and hence, locally stable on $Y_y \cap K$, for all $u \in U_y$. Since K is compact, we can choose finitely many points $y_1, \dots, y_r \in B_\epsilon$ such that $K \subseteq Y_{y_1} \cup \dots \cup Y_{y_r}$. Now, take $U' := U_{y_1} \cap \dots \cap U_{y_r}$. This is an open neighbourhood of u_0 in U, and for all $u \in U'$, f_u is locally stable on K. □

Remark 5.2 In the complex analytic case this follows from the coherence of the sheaf $\mathscr{T}^1_{\mathscr{A}_e}(F/U)$. By coherence, its support is closed, and then provided the projection π from $\mathrm{supp}\mathscr{T}^1_{\mathscr{A}_e}(F/U)$ to U is finite, $\pi(\mathrm{supp}\mathscr{T}^1_{\mathscr{A}_e}(F/U))$ is closed in U. ◊

Let $\epsilon > 0$ be such that $B_\epsilon \subseteq Y$, where B_ϵ is the closed ball of radius ϵ centred at the origin in $\mathbb{F}^p$. Let D_η denote the ball of radius η centred at 0 in U. It follows from Lemma 5.2 that there exists $\eta > 0$ such that $\mathring{D}_\eta \subseteq U$ and for all $u \in \mathring{D}_\eta$, f_u is locally stable at all points of the sphere $S_\epsilon = \partial B_\epsilon$.

Definition 5.6 We define the *bifurcation set* of F as

$$\mathscr{B}(F) := \{u \in \mathring{D}_\eta : \ f_u \text{ is not locally stable on } B_\epsilon\}.$$

We also denote by $(\mathscr{B}(F), 0)$ the set-germ of $\mathscr{B}(F)$ in $(\mathbb{F}^d, 0)$.

Again by Lemma 5.2, $\mathscr{B}(F)$ is a closed subset of $\mathring{D}_\eta$ and it is easy to see that $(\mathscr{B}(F), 0)$ is well defined as a set-germ, that is, it is independent of the choice of ϵ, η. In the complex case, we can prove more, namely, that $(\mathscr{B}(F), 0)$ is an analytic germ in $(\mathbb{C}^d, 0)$.

Lemma 5.3 *If* $f_0 : (\mathbb{C}^n, S) \to (\mathbb{C}^p, 0)$ *is* $\mathscr{A}$*-finite, then* $(\mathscr{B}(F), 0)$ *is an analytic germ in* $(\mathbb{C}^d, 0)$ *for any unfolding* F.

Proof The proof is similar to the proof of Mather–Gaffney criterion Theorem 4.5. Let $F : \mathscr{X} \to Y \times U$ be a representative such that $F^{-1}(0) \cap C(F) \subseteq S \times \{0\}$ and the restriction $F : C(F) \to Y \times U$ is finite. Recall the coherent sheaf $\mathscr{T}^1_{\mathscr{A}_e} f$ constructed in the proof of Proposition 4.8. Here we construct a relative or parameterised version of $\mathscr{T}^1_{\mathscr{A}_e} f$.

Recall from Definition 3.9 the modules of relative vector fields $\theta_{\mathscr{X}/U}$, $\theta_{Y\times Y/U}$ and $\theta(F/U)$, and the morphisms $t_{\mathrm{rel}}F : \theta_{\mathscr{X}/U} \to \theta(F/U)$ and $\omega_{\mathrm{rel}}F : \theta_{Y\times U/U} \to \theta(F/U)$. Let

$$\mathscr{T}^1_{\mathscr{R}_e}(F/U) = \frac{\theta(F/U)}{t_{\mathrm{rel}}F(\theta_{\mathscr{X}/U})}.$$

It is supported on the relative critical set of F, $\{(x, u) \in \mathscr{X} : x \in C(f_u)\}$, which, thanks to the definition of unfolding, is the same as its critical set. Because f_0 is finite on its critical set, so is F, and thus by the Finite Coherence Theorem E.2, $F_*\left(\mathscr{T}^1_{\mathscr{R}_e}(F/U)\right)$ is a coherent sheaf of $\mathscr{O}_{Y\times U}$-modules. The morphism $\omega_{\mathrm{rel}}F$: $\theta_{Y\times U/U} \to F_*\theta(F/U)$ induces a morphism to $F_*\left(\mathscr{T}^1_{\mathscr{R}_e}(F/U)\right)$, which we continue to denote by $\omega_{\mathrm{rel}}F$. We define $\mathscr{T}^1_{\mathscr{A}_e}(F/U)$ as its cokernel i.e.

$$\mathscr{T}^1_{\mathscr{A}_e}(F/U) = \frac{F_*\left(T^1_{\mathscr{R}_e}(F/U)\right)}{\omega_{\mathrm{rel}}F(\theta(Y\times U/U)}.$$

As cokernel of a morphism of coherent sheaves, $\mathscr{T}^1_{\mathscr{A}_e}(F/U)$ is a coherent sheaf of $\mathscr{O}_{Y\times U}$-modules. Its stalk at a point $(y, u) \in Y \times U$ vanishes exactly when $T^1_{\mathscr{A}_e}(f_u)_y = 0$, by Lemma 3.5, where $(f_u)_y$ is the multi-germ $f_u : (X_u, C(f_u) \cap f_u^{-1}(y)) \to (Y, y)$. Thus,

$$\mathrm{supp}(\mathscr{T}^1_{\mathscr{A}_e}(F/U)) = \{(y, u) \in Y \times U : (f_u)_y \text{ is not stable}\},$$

which is a closed analytic subset of $Y \times U$, by the coherence of $\mathscr{T}^1_{\mathscr{A}_e}(F/U)$. Let π : $\mathrm{supp}(\mathscr{T}^1_{\mathscr{A}_e}(F/U)) \to U$ be projection to the parameter space. We have $\pi^{-1}(0) = \{0\}$, since f_0 is unstable only at 0, and hence after shrinking the neighbourhoods if necessary, we can assume that π is finite. Again by the Finite Mapping Theorem E.2, the image $\mathscr{B}(F) = \pi(\mathrm{supp}(T^1_{\mathscr{A}_e}(F/d))$ is a closed analytic subset of U. □

We will prove at the end of this section that in the real case, if f has discrete stable type (see Definition 5.8) and F is polynomial, then $(\mathscr{B}(F), 0)$ is semialgebraic.

Example 5.7 Consider again the germ of Example 5.3: $f(x) = (x^2, x^5)$, with versal unfolding $F(x, t_1, t_2) = (x^2, x^5+t_1x^3+t_2x)$. It becomes clear from the calculations there that the bifurcation set in the complex case is

$$\mathscr{B}_{\mathbb{C}}(F) = \{(t_1, t_2) \in \mathbb{C}^2 : t_2(t_1^2 - 4t_2) = 0\}.$$

It has two components, the line $t_2 = 0$ corresponding to maps f_t having a non-immersive point and the parabola $t_1^2 - 4t_2 = 0$, where f_t presents a self-tangency.

However, in the real case we have

$$\mathscr{B}_{\mathbb{R}}(F) = (\mathbb{R} \times \{0\}) \cup \{(t_1, t_2) \in \mathbb{R}^2 : t_1^2 - 4t_2 = 0,\ t_1 \leq 0\},$$

since for $t_1 > 0$, the self-tangency occurs at a pair of complex conjugate points (see Fig. 5.3). We observe that $\mathscr{B}_{\mathbb{R}}(F)$ is not real analytic (although it is in fact semialgebraic). ◊

Example 5.8 Consider again the germ of Example 5.4, $f(x) = (x^3, x^4)$. with versal unfolding

$$F(x, a, b, c) = (x^3 + ax, x^4 + bx^2 + cx, a, b, c).$$

The bifurcation set $\mathscr{B}(F)$ is not easy to describe by equations in this case and we would need more sophisticated techniques to proceed with a detailed computation. Nevertheless, we include here the equations and pictures of $\mathscr{B}(F)$; all details can be found in [MNnB14]. In the complex case, $\mathscr{B}_{\mathbb{C}}(F)$ is the hypersurface in $\mathbb{C}^3$ (Fig. 5.7c) defined by $g_1 g_2 g_3 = 0$, where

1. $g_1 = 4a(2a - 3b)^2 + 27c^2$; this defines the component corresponding to parameters where $f_{a,b,c}$ presents non-immersive points (Fig. 5.7a).

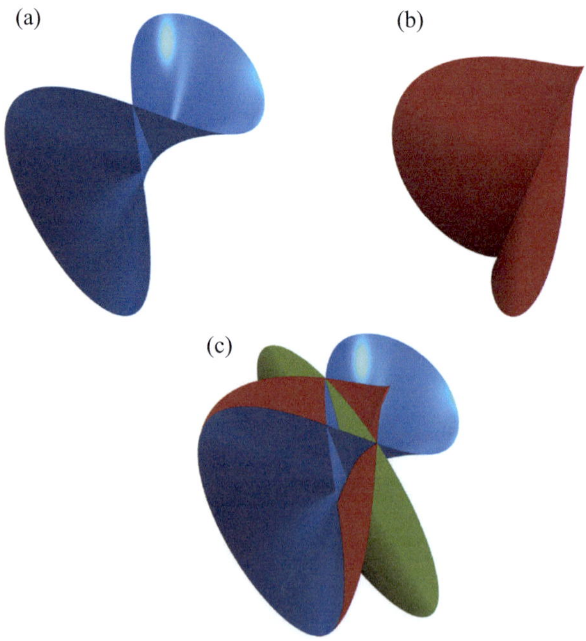

Fig. 5.7 Bifurcation set of $x \mapsto (x^3, x^4)$. (**a**) Component of non-immersive points. (**b**) Component of self-tangencies. (**c**) The whole bifurcation set

2. $g_2 = 4(2a-b)^3 + 27c^2$; this is the component corresponding to parameters where $f_{a,b,c}$ presents a self-tangency (Fig. 5.7b).
3. $g_3 = a - b$; this is the component of parameters where $f_{a,b,c}$ presents a triple point.

In the real case, the above three components together give the Zariski closure of $\mathscr{B}_{\mathbb{R}}(F)$ in $\mathbb{R}^3$. Observe also that the intersection of the bifurcation set with the plane $c = 0$ gives the four lines $a(a-b)(3b-2a)(2a-b) = 0$ which appear in Fig. 5.4. ◊

In Sects. 8.6 and 9.6 we return to this theme. In particular, in Sect. 8.6 we show that if $n \geq p$ then the bifurcation set is a hypersurface, when it is a proper subspace. In Sect. 9.6 we show that when $n < p$ the situation is more complicated.

Now we return to the theme of versality. A slight extension of the argument of the proof of Lemma 5.3 shows that versality is an open property.

Theorem 5.6 (Openness of Versality (C.T.C. Wall [Wal81, Theorem 3.7])) *Suppose that $F : \mathscr{X} \to Y \times U$ is a representative of a versal unfolding of a map-germ f. Then there is a sub-representative of the form $F : \mathscr{X}' \to Y' \times U'$ such that for every $(y, u) \in Y' \times U'$, the germ*

$$F : (\mathscr{X}', F^{-1}(y, u) \cap C(F)) \to (Y' \times U', (y, u))$$

is a versal unfolding of the germ

$$f_u : (X'_u, f_u^{-1}(y) \cap C(f_u)) \to (Y', y).$$

Proof Suppose first that $\mathbb{F} = \mathbb{C}$. We consider the Kodaira–Spencer map $\kappa : \theta_U \to \pi_*((\mathscr{T}^1_{\mathscr{A}_e}(F/U))$, defined by

$$\kappa(\vartheta) = \vartheta \cdot F = \text{class of } tF(\vartheta) \text{ in } \mathscr{T}^1_{\mathscr{A}_e}(F/U).$$

As versality is equivalent to infinitesimal versality (Theorem 5.1), versality of F is equivalent to the surjectivity of κ on the stalks at $0 \in U$. By coherence of the cokernel of κ, this implies that there is a neighbourhood U' of 0 in U on which κ is surjective.

Now suppose that $\mathbb{F} = \mathbb{R}$. Since f is $\mathscr{A}$-finite by Corollary 5.1, we can assume that it is polynomial (by the Finite Determinacy Theorem 6.2). Also by Corollary 5.1, we observe that it is enough to prove the statement for some other versal unfolding with the same number of parameters. Thus, we can also assume without loss of generality that F is polynomial (it follows from Theorem 5.1 that we can always construct a polynomial versal unfolding with the same number of parameters as F). Now the proof in this case follows easily by taking the complexification $\hat{F}$ of F, applying the complex case and then, taking the real projection of the obtained sub-representative. □

We look at some consequences of openness of versality. Suppose that $F : (\mathbb{F}^n \times \mathbb{F}^d, S \times \{0\}) \to (\mathbb{F}^p \times \mathbb{C}^d, (0, 0))$ is a versal unfolding, and choose a representative $F : \mathscr{X}' \to Y' \times U'$ as in Theorem 5.6—i.e. such that for each $(y, u) \in Y' \times U'$, the germ of F at $F^{-1}(y, u) \cap C(F)$ versally unfolds the germ of f_u at $f_u^{-1}(y) \cap C(f_u)$. Let $g : (\mathbb{F}^n, 0_r) \to (\mathbb{F}^p, 0)$ be a multi-germ. For $u \in U$, write $X_u = \{x \in \mathbb{F}^n : (x, u) \in \mathscr{X}\}$. We define $\widetilde{V}_g(F)$ as the subset of points $(x_1, \dots, x_r, u) \in (X_u)^{(r)} \times U'$ such that the germ of f_u at $\{x_1, \dots, x_r\}$ is right-left equivalent to g.

Note that any g for which $\widetilde{V}_g(F) \neq \emptyset$ has finite $\mathscr{A}_e$-codimension, since the corresponding germs of f_u are versally unfolded, by choice of the representative.

Proposition 5.1 *Under these circumstances, $\widetilde{V}_g(F)$ is smooth. If in addition g is not stable, then the restriction to $\widetilde{V}_g(F)$ of the projection $\mathbb{F}^n \times \mathbb{F}^d \to \mathbb{F}^d$ is an immersion, and the codimension of its image in $(\mathbb{F}^d, 0)$ is equal to the $\mathscr{A}_e$-codimension of g.*

Proof Let G be a miniversal unfolding of g, with base space $(\mathbb{F}^\ell, 0)$. Suppose $(x_1, \dots, x_r, u) \in \widetilde{V}(g)$. By Theorem 5.6, the germ of f_u at $\{x_1, \dots, x_r\}$ is versally unfolded by F. Since this germ is right-left equivalent to g, there is a germ of submersion $h : (\mathbb{F}^d, u) \to (\mathbb{F}^\ell, 0)$, such that the germ of F at $\{x_1, \dots, x_r\} \times \{u\}$ is equivalent to h^*G. It follows that there is a commutative diagram

$$\begin{array}{ccccc}
\big(\widetilde{V}_g(F), (x_1, \dots, x_r, u)\big) & \xrightarrow{\ \Phi\ } & \big(\widetilde{V}_g(h^*G), (\mathbf{0}_r, u)\big) & \xrightarrow{\ \mathrm{id}_{\mathbb{F}^n} \times h\ } & \big(\widetilde{V}_g(G), (\mathbf{0}_r, 0)\big) \\
\downarrow & & \downarrow & & \downarrow \\
(\mathbb{F}^d, u) & \xrightarrow{\ \mathrm{id}\ } & (\mathbb{F}^d, u) & \xrightarrow{\ h\ } & (\mathbb{F}^\ell, 0)
\end{array}$$

where Φ is a diffeomorphism and the vertical arrows are projections. Since G is miniversal, $\widetilde{V}_g(G)$ is a single point. The conclusion follows. □

Proposition 5.2 *In the circumstances of Proposition 5.1, suppose that f_u has a singularity of $\mathscr{A}$-type $g^{(i)}$ in $(U_0)_u$ at $x^{(i)} := (x_1^{(i)}, \dots, x_{r_i}^{(i)})$, for $i = 1, \dots, m$. Suppose also that $\{x_1^{(i)}, \dots, x_{r_i}^{(i)}\} \cap \{x_1^{(j)}, \dots, x_{r_j}^{(j)}\} = \emptyset$ if $i \neq j$. Then the germs of the immersions $\pi : (\widetilde{V}_F(g_i), (x^{i)}, u) \to (\mathbb{F}^d, u)$, for $i = 1, \dots, r$, meet in general position at u.*

Proof We have

$$\pi_*((\mathscr{T}^1_{\mathscr{A}_e}(F/U)))_u = \bigoplus_{(x,u)} (T^1_{\mathscr{A}_e}(F/U))_{(x,u)}.$$

The right-hand side has

$$\bigoplus_i (T^1_{\mathscr{A}_e}(F/U))_{(x^{(i)},u)}$$

as a natural quotient. The Kodaira–Spencer map $\theta_{\mathbb{F}^d,u} \to \pi_*((T^1_{\mathscr{A}_e}(F/U))$ is surjective, and therefore the composite

$$\theta_{\mathbb{F}^d,u} \to \bigoplus_i (T^1_{\mathscr{A}_e}(F/U))_{(x^{(i)},u)}$$

is surjective. The kernel of the further composite $\theta_{\mathbb{F}^d,u} \to (T^1_{\mathscr{A}_e}(F/U))_{(x^{(i)},u)}$ is the submodule of $\theta_{\mathbb{F}^d,u}$ consisting of germs of vector fields tangent to the image of the germ $\pi : \big(\widetilde{V}_F(g^{(i)}, (x^{(i)}, u)\big) \to (\mathbb{F}^d, u)$. It follows that these images meet in general position. □

Proposition 5.3 *Let $F : \mathscr{X} \times U \to Y \times U$ be a representative of a versal unfolding satisfying the conclusions of Theorem 5.6, and let W be any right-left invariant submanifold of ${}_rJ^k(\mathbb{F}^n, \mathbb{F}^p)$. Then ${}_rj^kF/U$ is transverse to W.*

Proof It follows, from the smoothness of each non-empty subspace $\widetilde{V}_F(g)$, that ${}_rj^kF/U$ is transverse to every $\mathscr{A}$-orbit. Since any right-left invariant submanifold of $W \subset_r J^k(\mathbb{F}^n, \mathbb{F}^p)$ is a union of orbits, transversality to W follows. □

Definition 5.7 A *stabilisation* of $f_0 : (\mathbb{F}^n, S) \to (\mathbb{F}^p, 0)$ is a 1-parameter unfolding F such that $(\mathscr{B}(F), 0) = (\{0\}, 0)$. This means that there exists a representative $F : U \to V \times D$, such that $f_t : U_t \to V$ is locally stable, for all $t \in D \setminus \{0\}$.

The following proposition gives a characterisation of the existence of a stabilisation in the complex analytic case. The real version will be seen later in Proposition 5.7.

Proposition 5.4 *Let $f_0 : (\mathbb{C}^n, S) \to (\mathbb{C}^p, 0)$ be a multi-germ of finite $\mathscr{A}_e$-codimension. The following statements are equivalent:*

1. *f_0 admits a stabilisation.*
2. *For some unfolding F, $(\mathscr{B}(F), 0)$ is a proper subset of the parameter space.*
3. *For any versal unfolding F, $(\mathscr{B}(F), 0)$ is a proper subset of the parameter space.*

Proof Let $F(x, u) = (f_u(x), u)$ be a d-parameter unfolding such that $(\mathscr{B}(F), 0) \subsetneq (\mathbb{C}^d, 0)$. Since $(\mathscr{B}(F), 0)$ is analytic by Lemma 5.3, there exists a line $L \subseteq \mathbb{C}^d$ passing through the origin such that $(L \cap \mathscr{B}(F), 0) = (\{0\}, 0)$. Take any point $a \in L$, with $a \neq 0$. Then, the 1-parameter unfolding defined by $(x, t) \mapsto (f_{ta}(x), t)$ is a stabilisation. This shows (2) $\Longrightarrow$ (1).

Assume now that $F(x, t) = (f_t(x), t)$ is a stabilisation and let $G(x, u) = (g_u(x), u)$ be any d-parameter versal unfolding. By definition of versality, there exists a map-germ $h : (\mathbb{C}, 0) \to (\mathbb{C}^d, 0)$ such that F is equivalent to h^*G. Obviously, h^*G is also a stabilisation, so $g_{h(t)}$ is locally stable for all $t \in \mathbb{C}^d \setminus \{0\}$.

But this implies that $(\mathscr{B}(G), 0) \subsetneq (\mathbb{F}^d, 0)$. Hence, we also have (1) $\Longrightarrow$ (3). The implication (3) $\Longrightarrow$ (2) is obvious. □

Let $f : (\mathbb{F}^n, S) \to (\mathbb{F}^p, 0)$ be a germ and F an unfolding of it. We use the same notation as in Theorem 5.2: take a representative of the form $F : X \times U \to \mathbb{F}^p \times \mathbb{F}^d$, where X, U are open neighbourhoods of S, 0 in $\mathbb{F}^n$, $\mathbb{F}^d$ respectively and denote by ${}_rj_x^kF : X^{(r)} \times U \to {}_rJ^k(\mathbb{F}^n, \mathbb{F}^p)$ the relative multi-jet extension map. We recall that using fixed coordinate systems on $\mathbb{F}^n$ and $\mathbb{F}^p$, we can identify ${}_rJ^k(X, \mathbb{R}^p)$ with $X^{(r)} \times (\mathbb{F}^p)^r \times J^k(n, p)^r$, and this allows us to define a projection $p\colon {}_rJ^k(X, \mathbb{R}^p) \to J^k(n, p)^r$.

Lemma 5.4 *For any germ $f : (\mathbb{F}^n, S) \to (\mathbb{F}^p, 0)$, there exists an unfolding F with parameter space U such that $p \circ j^k(F/U)$ is a submersion.*

Proof Take $U := J^k(n, p)$ as the parameter space of the unfolding and define $F(x, \sigma) = (f(x) + \sigma(x), \sigma)$. It follows that $p \circ j^k(F/U)(x, \sigma) = p(j^k f(x)) + \sigma$, clearly a submersion. □

Let $f : (\mathbb{F}^n, 0) \to (\mathbb{F}^p, 0)$ be a map-germ of finite $\mathscr{K}_e$-codimension and let F be an unfolding of f on parameters $u_1, \ldots, u_d$. Let $i : \mathbb{F}^n \to \mathbb{F}^{n+d}$ be the inclusion $x \mapsto (x, 0)$. We consider $\mathscr{O}_n$ as an $\mathscr{O}_{n+d}$-module via i^*.

Lemma 5.5 *There is an $\mathscr{O}_{n+d}$-linear isomorphism*

$$\beta : \frac{\theta(F)}{T\mathscr{K}_e(F)} \to \frac{\theta(f)}{T\mathscr{K}_e(f)},$$

sending the class of $\partial/\partial y_i$ to the class of $\partial/\partial y_i$ for $i = 1, \ldots, p$ and the class of $\partial/\partial u_j$ to the class of $-\dot{F}_j$ for $j = 1, \ldots, d$.

Proof Since $\theta(F)$ is a free $\mathscr{O}_{n+d}$-module with basis $\partial/\partial y_i, \partial/\partial u_j, i = 1, \ldots, n, j = 1, \ldots, d$, the $\mathscr{O}_{n+d}$-linear map $\beta_0 : \theta(F) \to \theta(f)$ sending $\partial/\partial y_i$ to $\partial/\partial y_i$ and $\partial/\partial u_j$ to $-\dot{F}_j$ is an epimorphism.

Let us see that $\beta_0(T\mathscr{K}_e(F)) \subset T\mathscr{K}_e(f)$. On one hand, $i^*(F^*\mathfrak{m}_{p+d}) = f^*\mathfrak{m}_p$, so $\beta_0(F^*\mathfrak{m}_{p+d}\theta(F)) = f^*\mathfrak{m}_p\theta(f)$. On the other hand, for $k = 1, \ldots, n$

$$\beta_0\left(\frac{\partial F}{\partial x_k}\right) = \beta_0\left(\sum_{i=1}^{p} \frac{\partial F_i}{\partial x_k}\frac{\partial}{\partial y_i}\right) = \sum_{i=1}^{p} \frac{\partial f_i}{\partial x_k}\frac{\partial}{\partial y_i} = \frac{\partial f}{\partial x_k}.$$

And for $j = 1, \ldots, d$

$$\beta_0\left(\frac{\partial F}{\partial u_j}\right) = \beta_0\left(\sum_{i=1}^{p} \frac{\partial F_i}{\partial u_j}\frac{\partial}{\partial y_i} + \frac{\partial}{\partial u_j}\right) = \sum_{i=1}^{p} \left.\frac{\partial F_i}{\partial u_j}\right|_{u=0} \frac{\partial}{\partial y_i} - \dot{F}_j = \dot{F}_j - \dot{F}_j = 0.$$

Thus, β_0 induces a well-defined epimorphism β between the quotients as desired. But β is in fact an isomorphism, because f has finite $\mathscr{K}_e$-codimension and by

Corollary 4.2,

$$\dim_{\mathbb{F}} \frac{\theta(F)}{T\mathscr{K}_e(F)} = \dim_{\mathbb{F}} \frac{\theta(f)}{T\mathscr{K}_e(f)} = \operatorname{codim}_{\mathscr{K}_e}(f) < \infty.$$

□

Lemma 5.6 *Suppose that F is a stable unfolding of the multi-germ $f : (\mathbb{F}^n, S) \to (\mathbb{F}^p, 0)$ and take a locally stable representative $F\colon X \times U \to \mathbb{F}^p \times \mathbb{F}^d$. Let $S = \{s_1, \ldots, s_r\}$ and write $f^{(i)}$ for the mono-germ of f at s_i. Then ${}_r j^k(F/U) : ((X)^{(r)} \times U \to {}_r J^k(X, \mathbb{F}^p)$ is transverse to the product of the $\mathscr{K}^k$-orbits of the $j^k(f^{(i)})$.*

Proof By Theorem 4.1, $\theta(F)/T\mathscr{K}_e(F)$ is generated over $\mathbb{F}$ by the tangent vectors to the target space $\mathbb{F}^p \times \mathbb{F}^d$, $\partial/\partial y_1, \ldots, \partial/\partial y_p, \partial/\partial u_1, \ldots, \partial/\partial u_r$. Hence, the isomorphism of Lemma 5.5 gives that $\theta(f_u)/T\mathscr{K}_e(f_u)$ is generated over $\mathbb{F}$ by $\partial/\partial y_1, \ldots, \partial/\partial y_p$ and $\dot{F}_1, \ldots, \dot{F}_d$.

Let $S = \{s_1, \ldots, s_r\}$, and $z = ((s_1, 0), \ldots, (s_r, 0))) \in (\mathbb{F}^n \times \mathbb{F}^d)^{(r)}$, and let $\sigma =_r j^k(F/U)(z)$ and $W = \mathscr{K}^{(k)}\sigma$. Then, via the usual identifications, the tangent space of $W(X, \mathbb{F}^p)$ at σ is

$$\mathbb{F}^{rn} \oplus \mathbb{F}^{rp} \oplus \frac{tf_u(\mathfrak{m}_n\theta_n) + f_u^*\mathfrak{m}_p\theta(f_u) + \mathfrak{m}_n^{k+1}\theta(f_u)}{\mathfrak{m}_n^{k+1}\theta(f_u)}$$

and the image of the differential of ${}_r j^k F/U$ at z is generated by the k-jets of all the partial derivatives $tf_0(\partial/\partial x_1), \ldots, tf_0(\partial/\partial x_n)$ and $\dot{F}_1, \ldots, \dot{F}_d$. The sum of these two spaces gives

$$\mathbb{F}^{rn} \oplus \mathbb{F}^{rp} \oplus \frac{T\mathscr{K}_e(f_u) + \mathrm{Sp}_{\mathbb{F}}\{\dot{F}_1, \ldots, \dot{F}_d\} + \mathfrak{m}_n^{k+1}\theta(f_u)}{\mathfrak{m}_n^{k+1}\theta(f_u)} = \mathbb{F}^{rn} \oplus \mathbb{F}^{rp} \oplus \frac{\theta(f_u)}{\mathfrak{m}_n^{k+1}\theta(f_u)},$$

so ${}_r j^k F/U$ is transverse to $W(X, \mathbb{F}^p)$. □

Corollary 5.3 *Let $f : (\mathbb{F}^n, S) \to (\mathbb{F}^p, 0)$ be a germ of finite singularity type and $F\colon X \times U \to \mathbb{F}^p \times \mathbb{F}^d$ a locally stable representative of a stable unfolding of f. Then ${}_r j^k(F/U)$ is transverse to every submanifold $W(X, \mathbb{F}^p)$ in ${}_r J^k(X, \mathbb{F}^p)$, where W ranges over all the products of $\mathscr{K}^{(k)}$-orbits.*

Proof The corollary follows from the Lemma by openness of stability, Proposition 4.8. □

The following lemma is due to Mather [Mat70]:

Lemma 5.7 *$J^k(n, p) \setminus \pi^k(n, p)$ is a finite union of $\mathscr{K}^{(k)}$-orbits.*

Proof That $J^k(n, p) \setminus \pi^k(n, p)$ is a union of $\mathscr{K}^{(k)}$-orbits is obvious. We set

$$V_r = W_r^k(n, p) \setminus (W_r^k(n, p)^* \cup W_{r+1}^k(n, p)).$$

Then V_r is a Zariski locally closed subset of $J^k(n, p)$ of codimension r. For each $\sigma \in V_r$, the orbit $\mathscr{K}^{(k)}\sigma$ is a submanifold of V_r whose codimension in $J^k(n, p)$ is r. Hence, $\mathscr{K}^{(k)}\sigma$ is open in V_r (with respect to the Euclidean topology). It follows that the $\mathscr{K}^{(k)}$-orbits form a partition of V_r by open subsets and hence, each connected component of V_r is contained in a single $\mathscr{K}^{(k)}$-orbit. By a theorem of Whitney (see Proposition B.2), any subset of $\mathbb{R}^N$ which is locally closed in the Zariski topology has only finitely many connected components with respect to the Euclidean topology. Thus, V_r contains only a finite number of $\mathscr{K}^{(k)}$-orbits. Finally, observe that

$$J^k(n, p) \setminus \pi^k(n, p) = V_0 \cup \cdots \cup V_d,$$

where $d = \dim J^k(n, p)$. Hence, there are only finitely many $\mathscr{K}^{(k)}$-orbits in $J^k(n, p) \setminus \pi^k(n, p)$. □

Proposition 5.5 *If $n < \sigma(n, p)$ then any germ $f : (\mathbb{F}^n, S) \to (\mathbb{F}^p, 0)$ of finite singularity type admits an unfolding F, such that $(\mathscr{B}(F), 0)$ is a null subset of the parameter space U.*

Proof Since f has finite singularity type, it admits an unfolding F which is stable (by (4.10) and Theorem 7.2) and also has the property that $p \circ j^k F/U$ is a submersion. In fact, we only need to take the sum of the two unfoldings of Lemmas 5.4 and 5.3. Let ${}_r\pi^k(X, \mathbb{F}^p)$ be the subbundle of ${}_rJ^k(X, \mathbb{F}^p)$ over $X \times \mathbb{F}^p$ having $\pi^k(n, p)$ as fibre. On one hand, ${}_rJ^k(X, \mathbb{F}^p) \setminus {}_r\pi^k(X, \mathbb{F}^p)$ is a finite union of $\mathscr{K}^{(k)}$-orbits by Lemma 5.7. On the other hand, for $r = 1$ we choose a stratification of $\pi^k(n, p)$ by strata of codimension $\geq \sigma(n, p) > n$ in $J^k(n, p)$.

By the Basic Transversality Lemma 2.7, for each $r \geq 1$, there exists a null set $N_r \subseteq U$ such that for all $u \in U \setminus N_r$, the map ${}_rj^k f_u : X^{(r)} \to_r J^k(X, \mathbb{F}^p)$ is transverse to all the (finitely many) $\mathscr{K}^{(k)}$-orbits of ${}_rJ^k(X, \mathbb{F}^p) \setminus {}_r\pi^k(X, \mathbb{F}^p)$, and also $p \circ j^k f_u : X \to J^k(n, p)$ is transverse to all the (finitely many) strata of $\pi^k(n, p)$. We put $N = \cup_{r\geq 1} N_r$ and prove that the multi-germ of $f_u : X \to \mathbb{F}^p$ at any point in the target is locally stable, for all $u \in U \setminus N$. In fact, the strata of $\pi^k(n, p)$ have codimension $> n$ and the transversality of $p \circ j^k f_u$ means that it does not meet such strata. But then, for any $r \geq 1$, ${}_rj^k f_u$ does not meet the $\mathscr{K}^{(k)}$-orbits contained in ${}_r\pi^k(X, \mathbb{F}^p)$. We deduce that ${}_rj^k f_u$ is transverse to all the $\mathscr{K}^{(k)}$-orbits of ${}_rJ^k(X, \mathbb{F}^p)$ and hence, the result follows from Corollary 4.4. □

We will see in later in Chap. 7 that for a fixed pair of dimensions (n, p) with $n \leq p$ there exist only finitely many $\mathscr{K}$-classes of stable germs from $\mathbb{F}^n$ to $\mathbb{F}^p$. Hence, we can reproduce the arguments of Proposition 5.5 to prove a similar statement for corank 1 germs.

Proposition 5.6 *Any corank 1 germ $f : (\mathbb{F}^n, S) \to (\mathbb{F}^p, 0)$ with $n \leq p$ of finite singularity type admits an unfolding F, such that $(\mathscr{B}(F), 0)$ is a null subset of the parameter space.*

Proof Let $F : (X \times U, (0,0)) \to (\mathbb{F}^p \times U, (0,0))$ be a representative of a stable unfolding of f. Let $r \geq 1$. By Lemma 5.7, on a possibly smaller representative, the multi-jet extension map ${}_r j^k F$ will meet only finitely many $\mathscr{K}^{(k)}$-orbits in ${}_r J^k(X \times U, \mathbb{F}^p \times \mathbb{F}^d)$. But this implies that the relative multi-jet extension map ${}_r j^k F/U$ will meet only finitely many $\mathscr{K}^{(k)}$-orbits in ${}_r J^k(X, \mathbb{F}^p)$. By the Basic Transversality Lemma 2.7, there exists a null set $N_r \subseteq U$ such that for all $u \in U \setminus N_r$, the map ${}_r j^k f_u : X^{(r)} \to_r J^k(X, \mathbb{F}^p)$ is transverse to all the $\mathscr{K}^{(k)}$-orbits of ${}_r J^k(X, \mathbb{F}^p)$. We put $N = \cup_{r\geq 1} N_r$ and thus, the multi-germ of $f_u : X \to \mathbb{F}^p$ at any point in the target is locally stable, for all $u \in U \setminus N$. □

We finish this section with a pair of definitions which will be useful later.

Definition 5.8 Let $f : (\mathbb{F}^n, S) \to (\mathbb{F}^p, 0)$ be an $\mathscr{A}$-finite germ and let F be its versal unfolding. We take a representative $F : \mathscr{X} \to Y \times U$, where $\mathscr{X}, Y, U$ are open neighbourhoods of $S \times \{0\}$ in $\mathbb{F}^n \times \mathbb{F}^d$ and of the origin in $\mathbb{F}^p, \mathbb{F}^d$, respectively. For each $(y, u) \in Y \times U$ we denote by $(f_u)_y$ the multi-germ of f_u at y. We say that

1. f is *simple*, if there is a sub-representative of F, and a finite collection of $\mathscr{A}$-classes, such that the germs $(f_u)_y$ all lie in these classes,
2. f has *discrete stable type*, if there is a sub-representative of F, and a finite collection of stable $\mathscr{A}$-classes, such that every stable germ $(f_u)_y$ lies in one of these classes.

Remark 5.3 Some comments on the above definitions:

1. When f is stable, the two concepts of simple and discrete stable type are equivalent. In general, any simple germ has discrete stable type but the converse is not true (see Example 5.9).
2. If f is simple and F is any unfolding of f, then F is equivalent to a pull-back of the versal unfolding and hence, $(f_u)_y$ only meets a finite number of $\mathscr{A}$-classes. The analogous statement is also true for discrete stable type.
3. If f has corank 1 and $n \leq p$, then f always has discrete stable type. This will be a consequence of the results of Sect. 7.
4. If (n, p) are nice dimensions, then any germ f always has discrete stable type. In fact, by Lemma 5.7, $J^k(n, p) \setminus \pi^k(n, p)$ only contains a finite number of $\mathscr{K}$-classes. If $n < \sigma(n, p)$ and $(f_u)_y$ is stable, then it lies in one of these $\mathscr{K}$-classes and hence, in a finite number of $\mathscr{A}$-classes. For, as we will see in Theorem 7.1 below, stable germs are $\mathscr{A}$-equivalent if and only if they are $\mathscr{K}$-equivalent.

◊

Example 5.9 Consider the function $f : (\mathbb{F}^2, 0) \to (\mathbb{F}, 0)$ given by $f(x, y) = xy(x - y)(x + y)$ and the 1-parameter unfolding F defined by

$$f_t(x, y) = (x + ty)y(x - y)(x + y).$$

If f_t is $\mathscr{A}$-equivalent to $f_{t'}$, then there exists a diffeomorphism $\varphi : (\mathbb{F}^2, 0) \to (\mathbb{F}^2, 0)$ such that $\varphi(f_t^{-1}(0)) = f_{t'}^{-1}(0)$. Since f_t and $f_{t'}$ are both homogeneous

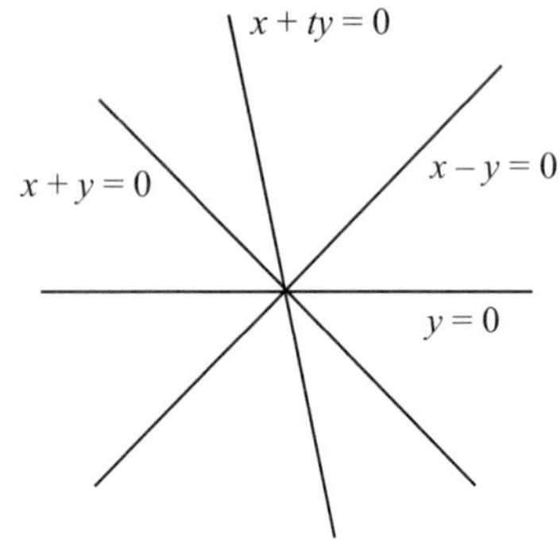

Fig. 5.8 The four lines defined by the zeros of f_t

of degree 4, this implies that $d_0\varphi(f_t^{-1}(0)) = f_{t'}^{-1}(0)$. Now $f_t^{-1}(0)$ and $f_{t'}^{-1}(0)$ are both a union of 4 lines in $\mathbb{F}^2$ passing through the origin and $d_0\varphi$ is a linear isomorphism which must preserve the crossratio between both sets (see Fig. 5.8). Hence $t = t'$, for $t, t' \in \mathbb{F}$ close to the origin. This shows that f is not simple. However, f has discrete stable type, since there are at most two stable types for functions from $\mathbb{F}^2$ to $\mathbb{F}$ (namely, the Morse singularities, equivalent to $x^2 \pm y^2$). ◊

We proved in Lemma 5.3 that in the complex case, the bifurcation set is analytic in the parameter space. We will see now that in the real case, the bifurcation set is semialgebraic, provided that f has discrete stable type and the unfolding is polynomial.

Lemma 5.8 *If* $f_0 : (\mathbb{R}^n, S) \to (\mathbb{R}^p, 0)$ *is* $\mathscr{A}$*-finite and has discrete stable type, then* $(\mathscr{B}(F), 0)$ *is semialgebraic in* $(\mathbb{R}^d, 0)$ *for any polynomial unfolding* F.

Proof By hypothesis, there exist a finite number of $\mathscr{A}$-classes $Q_1, \dots, Q_m$ such that if the multi-germ $(f_u)_y$ is stable, then it is $\mathscr{A}$-equivalent to one of these classes. We take a representative of F of the form $F : X \times U \to \mathbb{R}^p \times \mathbb{R}^d$ which allows us to consider the relative multi-jet extension maps ${}_r j^k F/U : X^{(r)} \times U \to {}_r J^k(X, \mathbb{R}^p)$, for each $r, k \geq 1$. Here, X and U are semialgebraic open neighbourhoods of S and 0 in $\mathbb{R}^n$ and $\mathbb{R}^d$, respectively.

Each $\mathscr{A}$-class Q_i is represented by a multi-germ of r points and thus, for each $k \geq p+1$, Q_i defines an orbit of ${}_r J^k(n, p)$ under the action of $\mathscr{A}^{(k)}$. But ${}_r J^k(n, p)$ is an affine space and $\mathscr{A}^{(k)}$ is a semialgebraic open subset of another affine space, acting algebraically on ${}_r J^k(n, p)$. This implies that the orbit Q_i is also semialgebraic, and hence, so is the subset $\tilde{Q}_i := X^{(r)} \times (\mathbb{R}^p)^r \times Q_i$ of ${}_r J^k(X, \mathbb{R}^p)$. The inverse image, $({}_r j^k F/U)^{-1}(\tilde{Q}_i)$ and its image S_i through F will be also semialgebraic in $X^{(r)}$ and $\mathbb{R}^p \times \mathbb{R}^d$, respectively.

Observe that S_i is the subset of $\Delta(F)$ of pairs (y, u) such that the multi-germ $(f_u)_y$ is $\mathscr{A}$-equivalent to Q_i and the semialgebraic set $T := \Delta(F) \setminus (S_1 \cup \cdots \cup S_m)$ is the subset of pairs (y, u) such that $(f_u)_y$ is not stable. In particular, its projection onto the parameter space, which is precisely $\mathscr{B}(F)$, is semialgebraic in $\mathbb{R}^d$. □

The following proposition is the real counterpart of Proposition 5.4.

Proposition 5.7 *Assume* $f_0 : (\mathbb{R}^n, S) \to (\mathbb{R}^p, 0)$ *is* $\mathscr{A}$*-finite and has discrete stable type. The following statements are equivalent:*

1. f_0 *admits a stabilisation.*
2. *For some unfolding* F, $(\mathscr{B}(F), 0)$ *is a null subset of the parameter space.*
3. *For any versal unfolding* F, $(\mathscr{B}(F), 0)$ *is a null subset of the parameter space.*

Proof The proof of (1) $\implies$ (3) is the same as those of Proposition 5.4 and (3) $\implies$ (2) is obvious, so we only have to prove (2) $\implies$ (1). It is clear that the two properties are preserved under $\mathscr{A}$-equivalence, hence we can assume that f_0 is polynomial by the Finite Determinacy Theorem 6.2. Moreover, we deduce from Theorem 5.1 that it is possible to construct a versal unfolding F of f_0 which is also polynomial (just take a monomial basis of a supplementary subspace of $T\mathscr{A}_e(f_0)$ in $\theta(f_0)$ and construct F by summing up these monomial vectors). By Lemma 5.8 and (3), $(\mathscr{B}(F), 0)$ is a closed, semialgebraic and null subset of the parameter space $(\mathbb{R}^d, 0)$. In particular, $\dim(\mathscr{B}(F), 0) \leq d - 1$ and hence, there exists a line $L \subseteq \mathbb{R}^d$ passing through the origin such that $(L \cap \mathscr{B}(F), 0) = (\{0\}, 0)$. Take any point $a \in L$, with $a \neq 0$. Then, the 1-parameter unfolding defined by $(x, t) \mapsto (f_{ta}(x), t)$ is a stabilisation. □

As a consequence, we can guarantee the existence of a stabilisation either when (n, p) are nice dimensions or the map-germ has corank 1 and $n \leq p$. The proof follows directly from Propositions 5.4, 5.5, 5.6 and 5.7.

Corollary 5.4 *Let* $f_0 : (\mathbb{F}^n, S) \to (\mathbb{F}^p, 0)$ *be an* $\mathscr{A}$*-finite germ and assume that either* (n, p) *are nice dimensions or* f_0 *has corank 1 and* $n \leq p$*. Then,* f_0 *admits a stabilisation.*

Exercises for Sect. 5.4

1 Give an example of an $\mathscr{A}$-finite germ f_0 and two stabilisations F and G which are not isomorphic.

2 Assume $\mathbb{F} = \mathbb{C}$ and show that any stabilisation F is also $\mathscr{A}$-finite (as a germ).

3 Let F be the miniversal unfolding of f_0. Up to isomorphism, any 1-parameter unfolding is of the form h^*F, for some curve in the parameter space $h : (\mathbb{F}, 0) \to (\mathbb{F}^d, 0)$. Show that h^*F is a stabilisation if and only if $C \cap \mathscr{B}(F) = \{0\}$, where $(C, 0)$ is the image of h.

4 Let $f_0 : (\mathbb{F}, 0) \to (\mathbb{F}, 0)$ be the germ $f_0(x) = x^{k+1}$.

(i) Show that $f_t(x) = x^{k+1} + tx$ defines a stabilisation of f_0.
(ii) How many real and complex critical points has f_t for $t \neq 0$?
(iii) Find a stabilisation f_t which exhibits the same number of real and complex critical points for $t \neq 0$.

5.5 The Notion of Stable Perturbation of a Map-Germ

In Example 1.1 we encouraged the reader to count into how many nodes (simple crossings) each singularity of parameterised plane curve could be split by deforming the curve. The simplicity of the question, and the apparent ease with which we can answer it visually, disguise the fact that implicit in it are some rather subtle notions. Consider the diagram in Fig. 5.9.

In (1), it shows a conical neighbourhood of a singularity of a curve C—the intersection of C with a suitable Milnor ball B_ϵ—see Theorem B.7 for an earlier discussion. In (2), (3) and (4) it shows perturbations of the curve. In (2), the cusp component of the singularity of C has been perturbed to give a node, as appears in the usual Reidemeister move, and the non-singular component of the singularity has been shifted up slightly, to allow three nodes in the Milnor ball. In (3) we see four nodes in the Milnor ball. But something is wrong: one of the nodes on C'' in (3) "has come from outside the ball". We should not count that node. The maximum number of nodes into which the singularity of C splits is three. In (4), something else has gone wrong: the node coming from the perturbation of the cusp has exited the Milnor ball. How do we distinguish between the correct perturbation C' and the misleading perturbations C'' and C'''? The answer we will adopt is the following: we insist that the perturbation we count should be part of a parameterised family, say C_t, with $C_0 = C$ and $C_1 = C'$, and that the new phenomena which we consider, and try, in this case, to count, should "emerge from within the Milnor ball". The meaning of this last phrase is a little difficult to pin down. We adopt the following

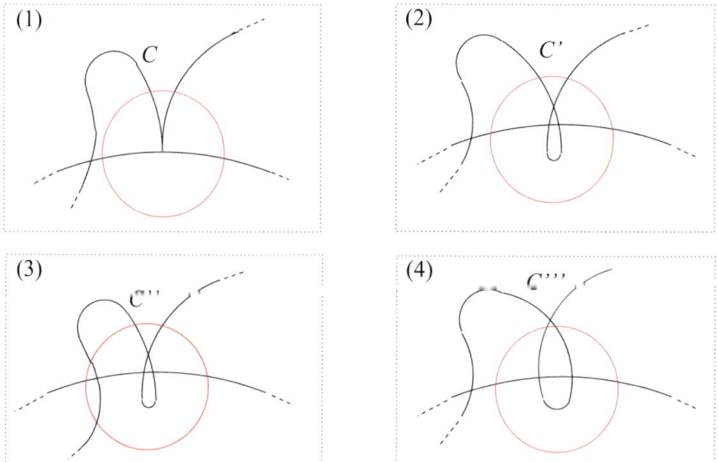

Fig. 5.9 Good and bad perturbations

requirement: in the family C_t, the intersection $C_t \cap \partial B_\epsilon$ does not change as t varies between 0 and 1. We can guarantee this by insisting that C_t should be transverse to ∂B_ϵ for $t \in [0, 1]$. In our example above, transversality of the image to ∂B_ϵ will fail for the parameter values t for which the node lies on ∂B_ϵ in (4), and for any parameter value t at which the curve is tangent to ∂B_ϵ, which must occur in the deformation from C to C''' in (4).

When we are dealing with parameterised curves, transversality is meant in the usual sense, but when we consider maps in higher dimensions, images and discriminants will often have non-isolated singularities, and thus we have to consider "stratified transversality" with respect to some convenient stratification. We will prove the following two theorems. Here the term "discriminant" means "image" when $n < p$, and the stratification implicitly referred to is the stratification by stable types introduced in Chap. 7.

Theorem 5.7 *Let $F : (\mathbb{F}^n \times \mathbb{F}^d, S \times \{0\}) \to (\mathbb{F}^p \times \mathbb{F}^d, 0)$ be an unfolding of an $\mathscr{A}$-finite germ $f_0 : (\mathbb{F}^n, S) \to (\mathbb{F}^p, 0)$, with (n, p) in the nice dimensions, and let $F : \mathscr{X} \to Y \times U$ be a representative.*

1. *There exists $\epsilon > 0$ such that the discriminant $\Delta(f)$ is stratified transverse to the sphere $S_{\epsilon'}$, for all ϵ' with $0 < \epsilon' \leq \epsilon$.*
2. *There exists $\delta > 0$ such that for $\|u\| < \delta$, the discriminant $\Delta(f_u)$ is stratified transverse to S_ϵ.*
3. *The stratified map $\pi : \Delta(F) \cap (B_\epsilon \times (B_\delta \smallsetminus \mathscr{B}(F))) \to B_\delta \smallsetminus \mathscr{B}(F)$ is a locally C^0 trivial fibre bundle.*

The theorem is a simple consequence of the Thom–Mather first isotopy lemma (see Theorem B.2). We prove it as Proposition 8.2. When $\mathbb{F} = \mathbb{C}$, $\mathscr{B}$ is a complex analytic subvariety of the parameter space, and this has topological codimension at least 2. It therefore does not separate B_δ, so that the base space of the locally trivial fibration in the theorem is path connected. It follows that all of its fibres are stratified homeomorphic to one another. The fibre is what we mean by "the discriminant of a stable perturbation". In the real case, the bifurcation set typically does separate the parameter space; to each path component of its complement there corresponds a different fibre, and so there is no topologically unique "discriminant of a stable perturbation".

But we have still not defined what we mean by "stable perturbation". To make this precise, we have to penetrate further into the theory of stratified mappings. We first consider the complex case. Given a germ of unfolding $F : (\mathbb{C}^n \times \mathbb{C}^d, S \times \{0\}) \to (\mathbb{C}^p \times \mathbb{C}^d, 0)$ of a germ $f : (\mathbb{C}^n, S) \to (\mathbb{C}^p, 0)$ of finite singularity type, the bifurcation set $\mathscr{B} \subseteq \mathbb{C}^d$ is an analytic subset. Now assume that $\mathscr{B}$ is a *proper* subset—for instance, if F is $\mathscr{A}_e$-versal and (n, p) are in Mather's nice dimensions. Now pick a representative of F, and ϵ, δ as in Theorem 5.7. Then

Theorem 5.8 *The diagram*

$$\begin{array}{ccc} \left(B_\rho \times B_\delta\right) \cap F^{-1}\left(B_\epsilon \times \left(B_\delta \smallsetminus \mathscr{B}(F)\right)\right) & \xrightarrow{F} & B_\epsilon \times \left(B_\delta \smallsetminus \mathscr{B}(F)\right) \\ \downarrow & \swarrow & \\ B_\delta \smallsetminus \mathscr{B}(F) & & \end{array} \tag{5.9}$$

is a C^0-left-right-locally trivial family of mappings.

As we are taking $\mathbb{F} = \mathbb{C}$, $\mathscr{B}$ does not separate B_δ, and so the base of the fibration is path connected. It follows that for all pairs $u_1, u_2 \in B_\delta \smallsetminus \mathscr{B}$, $f_{u_1} : f_{u_1}^{-1}(B_\epsilon) \to B_\epsilon$ and $f_{u_2} : f_{u_2}^{-1}(B_\epsilon) \to B_\epsilon$ are C^0-left-right equivalent. A member of this family, finally, is what we mean by a stable perturbation of f.

Theorem 5.8 follows from the Thom–Mather second Isotopy Lemma (Theorem B.3), but we omit the rather lengthy details.

Remark 5.4 The theory of unfoldings developed by Mather and Martinet, which we have explained in this chapter, is very much simpler than the general theory of deformations of singularities of varieties. If a variety is defined implicitly by equations, then careless modification of the equations can change the properties of the zero-locus in a way which does not really merit the term "deformation". For example, consider the ideal defining the union, X, of the three coordinate axes in 3-space,

$$I = (yz, xz, xy).$$

Even though it has three generators, I defines a set of codimension only two. It is not a "complete intersection ideal" (see Definition C.7 in Appendix C). This gives rise to some strange phenomena. The 3-parameter "deformation" of I given by fibring the variety of zeros of

$$\mathscr{I} = (yz - u, xz - v, xy - w)$$

over u, v, w-space has the following anomalies:

- the fibre of $V(\mathscr{I})$ over the point (t, t, t) in u, v, w-space with $t \neq 0$ consists just of the two points $\pm\sqrt{t}(1, 1, 1)$;
- the fibre of $V(\mathscr{I})$ over the point $(0, v, w)$ with $v \neq 0 \neq w$ is empty.

Thus, our curve X has been deformed to a pair of points, or deformed away altogether. To avoid such anomalies, a "deformation of a space-germ (X, x_0)" is taken to mean a family $\pi : (\mathscr{X}, x_0) \to (S, 0)$ with the properties that $\pi^{-1}(0) = X$ and π is a flat map, i.e. $\mathscr{O}_{\mathscr{X},x_0}$ is a flat module over $\mathscr{O}_{S,0}$ (see [Mat89, Chapter 8] for the definition of flatness). Flatness is automatic in the case of complete intersection ideals, such as in the case of hypersurfaces, but for other singularities it is not. And its introduction as an additional requirement complicates everything. Isolated singularities have versal deformations with finite-dimensional base spaces, but these need not be smooth or even irreducible. Indeed a famous theorem of Mnëv and Vakil (see [Vak06]) shows that every complex space-germ can arise as the base space of a versal deformation of some singularity.

For us, these complications never arise. All deformations are admissible, and all versal base spaces are smooth. ◊

Chapter 6
Finite Determinacy

Many of the applications of singularity theory, such as Thom's list of the seven elementary catastrophes, involve classification up to equivalence under the action of some group $\mathscr{G}$ of diffeomorphisms, usually $\mathscr{R}$, $\mathscr{K}$ or $\mathscr{A}$. Often, one wishes to find a list of germs whose orbits together contain all germs of $\mathscr{G}$-codimension less than or equal to some given bound, to give simple recognition criteria for a germ to lie in one of the listed orbits, and to determine the adjacencies among the orbits. In almost all cases, these involve *finite determinacy*.

Throughout this chapter we denote by $\mathscr{G}$ any one of Mather's groups $\mathscr{R}, \mathscr{L}, \mathscr{C}, \mathscr{A}$ or $\mathscr{K}$. Let $f\colon (\mathbb{F}^n, S) \to (\mathbb{F}^p, 0)$ be a smooth[1] multi-germ, with $S = \{s_1, \ldots, s_r\}$. For each $k \geq 0$, we denote by $j^k f$ the k-jet of f at S, that is, $j^k f = (j^k f(s_1), \ldots, j^k f(s_r))$, where $j^k f(s_i)$ is the k-jet of f at s_i. To avoid repetition, all germs will be assumed smooth except where otherwise indicated.

Definition 6.1 We say that f is *k-$\mathscr{G}$-determined* if any other multi-germ $g\colon (\mathbb{F}^n, S) \to (\mathbb{F}^p, 0)$ having the same k-jet at S is $\mathscr{G}$-equivalent to f. We say that f is *finitely $\mathscr{G}$-determined* if it is k-$\mathscr{G}$-determined for some $k < \infty$. The *$\mathscr{G}$-determinacy degree* of f is the lowest k such that f is k-$\mathscr{G}$-determined.

If f is $k - \mathscr{G}$-determined then up to $\mathscr{G}$-equivalence it can be replaced by its k'th-order Taylor polynomial. This can be studied by the methods of algebraic geometry, and, in particular, using computer algebra systems. Computation is not possible with analytic maps in general, since infinite power series require infinite amounts of data to specify them.

[1]Recall from Chap. 1 that "$\mathbb{F}$" refers to both $\mathbb{R}$ and $\mathbb{C}$, and that in the latter case "smooth" means complex analytic.

D. Mond, J. J. Nuño-Ballesteros, *Singularities of Mappings*,
Grundlehren der mathematischen Wissenschaften 357,
https://doi.org/10.1007/978-3-030-34440-5_6

The central result of this chapter is the following important theorem, known as Mather's infinitesimal criterion for finite determinacy [Mat68b]:

Theorem 6.1 *Finite $\mathscr{G}$-determinacy is equivalent to finite $\mathscr{G}$-codimension.*

Mather's original proof gave rather poor estimates for the determinacy degree when $\mathscr{G} = \mathscr{A}$. So instead, we follow the proof due to Wall [Wal81, Wal95], which builds on results of Gaffney [Gaf79] and du Plessis [dP80].

In the course of proving this, we give an upper bound for the determinacy degree for each of the standard groups, in Theorem 6.2. Later in the chapter we give methods for improving on these estimates.

First of all, we define the tangent and the extended tangent spaces for each one of the groups $\mathscr{G}$ (some of them were already introduced in Chap. 3.2). Later, we justify this definition by a calculation, in Lemma 4.5. In order to simplify the notation, we set

$$\mathscr{O}_n = \mathscr{O}_{\mathbb{F}^n,S}, \quad \mathfrak{m}_n = \mathfrak{m}_{\mathbb{F}^n,S}, \quad \theta_n = \theta_{\mathbb{F}^n,S},$$
$$\mathscr{O}_p = \mathscr{O}_{\mathbb{F}^p,0}, \quad \mathfrak{m}_p = \mathfrak{m}_{\mathbb{F}^p,0}, \quad \theta_p = \theta_{\mathbb{F}^p,0}.$$

The tangent and extended tangent spaces for each group are defined, respectively, as

$$\begin{aligned}
T\mathscr{R}(f) &= tf(\mathfrak{m}_n\theta_n), & T\mathscr{R}_e(f) &= tf(\theta_n),\\
T\mathscr{L}(f) &= \omega f(\mathfrak{m}_p\theta_p), & T\mathscr{L}_e(f) &= \omega f(\theta_p),\\
T\mathscr{C}(f) &= f^*\mathfrak{m}_n\theta(f), & T\mathscr{C}_e(f) &= T\mathscr{C}(f),\\
T\mathscr{A}(f) &= T\mathscr{R}(f) + T\mathscr{L}(f), & T\mathscr{A}_e(f) &= T\mathscr{R}_e(f) + T\mathscr{L}_e(f),\\
T\mathscr{K}(f) &= T\mathscr{R}(f) + T\mathscr{C}(f), & T\mathscr{K}_e(f) &= T\mathscr{R}_e(f) + T\mathscr{C}_e(f).
\end{aligned}$$

We also define the codimension and the extended codimension for each group $\mathscr{G}$, respectively, as

$$\operatorname{codim}_{\mathscr{G}}(f) = \dim_{\mathbb{F}} \frac{\mathfrak{m}_n\theta(f)}{T\mathscr{G}(f)}, \quad \operatorname{codim}_{\mathscr{G}_e}(f) = \dim_{\mathbb{F}} \frac{\theta(f)}{T\mathscr{G}_e(f)}.$$

We will often write $\mathscr{O}_n^{\times p}$ in place of $\theta(f)$, especially in discussions involving more than one germ.

Theorem 6.2 *The following statements are equivalent:*

(a) *f is finitely $\mathscr{G}$-determined,*
(b) *for some k, $\mathfrak{m}_n^k\theta(f) \subset T\mathscr{G}(f)$,*
(c) *for some k, $\mathfrak{m}_n^k\theta(f) \subset T\mathscr{G}_e(f)$,*
(d) $\operatorname{codim}_{\mathscr{G}}(f) < \infty$,
(e) $\operatorname{codim}_{\mathscr{G}_e}(f) < \infty$.

More precisely, if we set $\epsilon = 2$ *for* $\mathscr{G} = \mathscr{L}, \mathscr{A}$ *and* $\epsilon = 1$ *for* $\mathscr{G} = \mathscr{R}, \mathscr{C}, \mathscr{K}$*:*

(i) *If* $\mathfrak{m}_n^{k+1}\theta(f) \subset T\mathscr{G}(f)$*, then* f *is* $(\epsilon k + 1)$*-determined.*
(ii) *If* f *is* k*-determined, then* $\mathfrak{m}_n^{k+1}\theta(f) \subset T\mathscr{G}(f)$*.*
(iii) *If* $\operatorname{codim}_{\mathscr{G}}(f) = d < \infty$*, then* $\mathfrak{m}_n^{d+1}\theta(f) \subset T\mathscr{G}(f)$*, for* $\mathscr{G} = \mathscr{R}, \mathscr{C}$ *and* $\mathscr{K}$*, and* $\mathfrak{m}_n^{(rp+d)^2}\theta(f) \subset T\mathscr{G}(f)$ *for* $\mathscr{G} = \mathscr{A}$ *and* $\mathscr{L}$*.*

Observe that (i), (ii) and (iii), respectively, yield the implications (b) $\Rightarrow$ (a), (a) $\Rightarrow$ (b) and (d) $\Rightarrow$ (b), whereas (b) $\Rightarrow$ (d) and (c) $\Rightarrow$ (e) are trivial since $\mathfrak{m}_n^k\theta(f)$ has finite codimension in $\theta(f)$. The implication (b) $\Rightarrow$ (c) is also trivial because $T\mathscr{G}(f) \subset T\mathscr{G}_e(f)$. Hence, assuming (i),(ii) and (iii), to complete the proof of all the equivalences it is enough to show that (e) $\Rightarrow$ (d).

Assume $\operatorname{codim}_{\mathscr{G}_e}(f) < \infty$, then it suffices to prove that $\dim_{\mathbb{F}} T\mathscr{G}_e(f)/T\mathscr{G}(f) < \infty$, since this would imply

$$\begin{aligned}\operatorname{codim}_{\mathscr{G}}(f) = \dim_{\mathbb{F}} \frac{\mathfrak{m}_n\theta(f)}{T\mathscr{G}(f)} &\le \dim_{\mathbb{F}} \frac{\theta(f)}{T\mathscr{G}(f)} \\ &= \dim_{\mathbb{F}} \frac{\theta(f)}{T\mathscr{G}_e(f)} + \dim_{\mathbb{F}} \frac{T\mathscr{G}_e(f)}{T\mathscr{G}(f)} < \infty.\end{aligned}$$

For each $i = 1, \ldots, r$, we denote by $\mu_i \in \mathscr{O}_n$ the function germ which is constant equal to 1 in a neighbourhood of s_i and 0 elsewhere. Then,

$$\mathscr{O}_n = \mathfrak{m}_n \oplus \operatorname{Sp}_{\mathbb{F}}\{\mu_1, \ldots, \mu_r\}, \quad \mathscr{O}_p = \mathfrak{m}_p \oplus \operatorname{Sp}_{\mathbb{F}}\{1\}.$$

We also denote by $e_1, \ldots, e_p$ the standard basis $\partial/\partial y_1, \ldots, \partial/\partial y_p$ of θ_p. When $\mathscr{G} = \mathscr{A}$, we have

$$\begin{aligned}T\mathscr{A}_e(f) &= tf(\theta_n) + \omega f(\theta_p) \\ &= \mathscr{O}_n\left\{\frac{\partial f}{\partial x_i}\right\}_i + \mathscr{O}_p\{e_j\}_j \\ &= \mathfrak{m}_n\left\{\frac{\partial f}{\partial x_i}\right\}_i + \operatorname{Sp}_{\mathbb{F}}\left\{\mu_\ell \frac{\partial f}{\partial x_i}\right\}_{i,\ell} + \mathfrak{m}_p\{e_j\}_j + \operatorname{Sp}_{\mathbb{F}}\{e_j\}_j \\ &= tf(\mathfrak{m}_n\theta_n) + \operatorname{Sp}_{\mathbb{F}}\left\{\mu_\ell \frac{\partial f}{\partial x_i}\right\}_{i,\ell} + \omega f(\mathfrak{m}_p\theta_p) + \operatorname{Sp}_{\mathbb{F}}\{e_j\}_j \\ &= T\mathscr{A}(f) + \operatorname{Sp}_{\mathbb{F}}\left\{\mu_\ell \frac{\partial f}{\partial x_i}\right\}_{i,\ell} + \operatorname{Sp}_{\mathbb{F}}\{e_j\}_j.\end{aligned}$$

This implies that $\dim_{\mathbb{F}} T\mathscr{A}_e(f)/T\mathscr{A}(f) \le rn + p$. The proof in the other cases is analogous.

6.1 Proof of the Finite Determinacy Theorem

Let $f, g\colon (\mathbb{F}^n, S) \to (\mathbb{F}^p, 0)$ be two smooth multi-germs. In the following, by taking coordinates on $\mathbb{F}^p$, we identify $\theta(f)$ and $\theta(g)$ with $\mathscr{O}_n^{\times p}$.

Lemma 6.1 *If $j^\ell f = j^\ell g$, then*

$$T\mathscr{G}(f) + \mathfrak{m}_n^{\ell+1}\mathscr{O}_n^{\times p} = T\mathscr{G}(g) + \mathfrak{m}_n^{\ell+1}\mathscr{O}_n^{\times p}.$$

Proof It is easy to see that

1. if $\xi \in \theta_n$, then $tf(\xi) - tg(\xi) \in \mathfrak{m}_n^\ell \mathscr{O}_n^{\times p}$,
2. if $\phi \in \mathscr{O}_p$, then $f^*(\phi) - g^*(\phi) \in \mathfrak{m}_n^{\ell+1}$,
3. if $\eta \in \theta_p$, then $\omega f(\eta) - \omega g(\eta) \in \mathfrak{m}_n^{\ell+1}\mathscr{O}_n^{\times p}$.

So the claim follows directly from the definitions of $T\mathscr{G}(f)$. □

Let $F\colon (\mathbb{F}^n \times \mathbb{F}, S \times \{0\}) \to (\mathbb{F}^p \times \mathbb{F}, 0)$ be any 1-parameter unfolding of f given by $F(x,t) = (\tilde{f}(x,t), t)$. Associated with F, we introduced in Sect. 3.5 two morphisms, $t_{\rm rel}F : \theta_{n+1/1} \to \theta(F/1)$ and $\omega_{\rm rel}F : \theta_{p+1/1} \to \theta(F/1)$. In terms of the notation we are using here, these become $t_{\rm rel}F : \mathscr{O}_{n+1}^{\times n} \to \mathscr{O}_{n+1}^{\times p}$ and $\omega_{\rm rel}F : \mathscr{O}_{p+1}^{\times p} \to \mathscr{O}_{n+1}^{\times p}$. By using these two morphisms, we can now define adapted parameterised versions of the tangent spaces:

$$T\mathscr{R}(F/1) = t_{\rm rel}F(M_n\theta_{n+1/1}),$$

$$T\mathscr{L}(F/1) = \omega_{\rm rel}F(M_p\theta_{p+1/1}),$$

$$T\mathscr{C}(F/1) = F^*M_p\theta(F/1),$$

where $M_n = \mathfrak{m}_n\mathscr{O}_{n+1} \subset \mathscr{O}_{n+1}$ and $M_p = \mathfrak{m}_p\mathscr{O}_{p+1} \subset \mathscr{O}_{p+1}$. The tangent spaces $T\mathscr{A}(F/1)$ and $T\mathscr{K}(F/1)$ are defined as sums as usual.

We say that the unfolding F is *origin-preserving* if $f_t(S) = \{0\}$ for all $t \in \mathbb{F}$. If F is origin-preserving, then we can see F as a 1-parameter family of multi-germs $f_t\colon (\mathbb{F}^n, S) \to (\mathbb{F}^p, 0)$, $t \in \mathbb{F}$. By arguments similar to those of the proof of Lemma 6.1, we have the following:

Lemma 6.2 *Let F, G be two origin-preserving unfoldings of f and assume that $j^\ell f_t = j^\ell g_t$, for all $t \in \mathbb{F}$. Then,*

$$T\mathscr{G}(F/1) + M_n^{\ell+1}\mathscr{O}_{n+1}^{\times p} = T\mathscr{G}(G/1) + M_n^{\ell+1}\mathscr{O}_{n+1}^{\times p}.$$

Lemma 6.3 *Let $f\colon (\mathbb{F}^n, S) \to (\mathbb{F}^p, 0)$ be a smooth multi-germ, C a finitely generated $\mathscr{O}_n$-module, A an $\mathscr{O}_p$-submodule of $f^*\mathfrak{m}_pC$. Let $I \subset \mathfrak{m}_n$ and $\mathscr{M} \subset \mathfrak{m}_n$*

be ideals of $\mathscr{O}_n$ such that $\dim_{\mathbb{F}}(\mathscr{O}_n/\mathscr{M}) < \infty$ and I is finitely generated. Assume that

(a) $\mathscr{M}C \subset (f^*\mathfrak{m}_p + \mathfrak{m}_n\mathscr{M})C$ *and*
(b) $IC \subset A + \mathscr{M}IC$.

Then $IC \subset A$.

Proof We first set $E = (f^*\mathfrak{m}_p + \mathfrak{m}_n\mathscr{M})C/f^*\mathfrak{m}_pC$. Since C is finitely generated, E is also a finitely generated $\mathscr{O}_n$-module. Moreover, (a) implies that $\mathfrak{m}_nE = E$, so by Nakayama's lemma C.1, $E = 0$ and thus, $\mathscr{M}C \subset f^*\mathfrak{m}_pC$. It follows from this and (b) that

(c) $IC \subset A + f^*\mathfrak{m}_pIC$.

Next set $E' = (A + IC)/A$. It follows from (c) that $f^*\mathfrak{m}_pE' = E'$. The desired result $E' = 0$ will thus follow (by Nakayama's lemma) if we prove that E' is finitely generated over $\mathscr{O}_p$. Clearly, it is enough to show that IC is.

Now IC is finitely generated over $\mathscr{O}_n$ and since $I\mathscr{M}C \subset f^*\mathfrak{m}_pIC$,

$$\dim_{\mathbb{F}} \frac{IC}{f^*\mathfrak{m}_pIC} \leq \dim_{\mathbb{F}} \frac{IC}{\mathscr{M}IC}.$$

This is finite because $\dim_{\mathbb{F}}(\mathscr{O}_n/\mathscr{M}) < \infty$. By the preparation theorem D.2, IC is finitely generated over $\mathscr{O}_p$. □

Proof of Part (i) of Theorem 6.2 Set $\ell = \epsilon k + 1$ and let g be a smooth multi-germ such that $j^\ell f = j^\ell g$. Consider the 1-parameter family of multi-germs obtained by linear interpolation between f and g, that is, $f_t \colon (\mathbb{F}^n, S) \to (\mathbb{F}^p, 0)$, where $f_t = (1-t)f + tg$ and $t \in \mathbb{F}$. It is enough to show that this family f_t is locally $\mathscr{G}$-trivial, which means that for each $t_0 \in \mathbb{F}$, the multi-germs f_t are $\mathscr{G}$-equivalent to f_{t_0} for all t in a neighbourhood of t_0. Since $\mathbb{F}$ is connected, we deduce that the multi-germs f_t are $\mathscr{G}$-equivalent to each other for all $t \in \mathbb{F}$. In particular, $f_0 = f$ is $\mathscr{G}$-equivalent to $f_1 = g$.

To prove that the family f_t is locally $\mathscr{G}$-trivial, we use integration of vector fields, as we did in Sect. 3.5. We fix $t_0 \in \mathbb{F}$ and denote by $F \colon (\mathbb{F}^n \times \mathbb{F}, S \times \{t_0\}) \to (\mathbb{F}^p \times \mathbb{F}, (0, t_0))$ the unfolding of f_{t_0} associated with the family f_t, that is, $F(x,t) = (f_t(x), t)$. We also write $G = f \times \mathrm{id}$, for the constant unfolding. Since $j^\ell f_t = j^\ell f$, for all t, Lemma 6.2 gives

$$T\mathscr{G}(F/1) + M_n^{\ell+1}\mathscr{O}_{n+1}^{\times p} = T\mathscr{G}(G/1) + M_n^{\ell+1}\mathscr{O}_{n+1}^{\times p}. \tag{6.1}$$

Lemma 6.4 *If $\mathfrak{m}_n^{k+1}\mathscr{O}_n^{\times p} \subset T\mathscr{G}(f)$ then $M_n^{k+1}\mathscr{O}_{n+1}^{\times p} \subset T\mathscr{G}(G/1)$.*

Proof We give the proof for $\mathscr{A}$; the proofs for $\mathscr{R}$, $\mathscr{C}$ and $\mathscr{K}$ are easy, relying on nothing more than Nakayama's lemma. Indeed, the deduction of (3.7) from (3.6) back in Chap. 3 proves both this statement for $\mathscr{R}$, and the corresponding part of the approximation Lemma 6.5 below. The proof for $\mathscr{G} = \mathscr{L}$ is a simplification of the proof for $\mathscr{A}$.

Write

$$C = \frac{\theta(F/1)}{T\mathscr{R}(G/1)}, \qquad A = \frac{T\mathscr{A}(G/1)}{T\mathscr{R}(G/1)}.$$

Then it is enough to show that $M^{k+1}C \subset A$. Clearly C is finitely generated over $\mathscr{O}_{n+1}$. As

$$\frac{C}{G^*\mathfrak{m}_{p+1}C} = \frac{\theta(f)}{T\mathscr{R}(f) + f^*\mathfrak{m}_p\theta(f)} \tag{6.2}$$

is finite-dimensional over $\mathbb{F}$, C is therefore finitely generated over $\mathscr{O}_{p+1}$, by the preparation theorem. Moreover, because of the equality (6.2), we can choose $\xi_1, \ldots, \xi_s \in \theta(f)$ (i.e. independent of t) whose classes $\bar{\xi}_1, \ldots, \bar{\xi}_s$ in C generate it over $\mathscr{O}_{p+1}$. A typical element h of $M_n^{k+1}C$ is of the form $h = \sum_{|\alpha|=k+1} x^\alpha a_\alpha$, where $a_\alpha \in C$. We can write

$$a_\alpha = \sum_i b_{\alpha,i}\bar{\xi}_i$$

for some $b_{\alpha,i} \in \mathscr{O}_{p+1}$. Since $x^\alpha \xi_i \in \mathfrak{m}_n^{k+1}\theta(f) \subset T\mathscr{A}(f)$, and G is a constant unfolding, it follows that $h = \sum_{\alpha,i} b_{\alpha,i}x^\alpha\bar{\xi}_i \in A$. □

From (6.1) and the lemma, we get

$$M_n^{k+1}\mathscr{O}_{n+1}^{\times p} \subset T\mathscr{G}(F/1) + M_n^{\ell+1}\mathscr{O}_{n+1}^{\times p}. \tag{6.3}$$

Lemma 6.5 *The inclusion* (6.3) *implies that* $M_n^{k+1}\mathscr{O}_{n+1}^{\times p} \subset T\mathscr{G}(F/1)$.

Proof We consider first the simplest cases $\mathscr{G} = \mathscr{R}, \mathscr{C}, \mathscr{K}$ so that $\ell = k+1$. We set $A = T\mathscr{G}(F/1)$ and $B = M_n^{k+1}\mathscr{O}_{n+1}^{\times p}$. Then (6.3) implies that $B \subset A + \mathfrak{m}_{n+1}B$; then $B \subset A$ by Nakayama's lemma C.1. Note that here we are using the fact that for $\mathscr{G} = \mathscr{R}, \mathscr{C}$ and $\mathscr{K}$, $T\mathscr{G}(F/1)$ is an $\mathscr{O}_{n+1}$-module.

Now take $\mathscr{G} = \mathscr{A}$ or $\mathscr{L}$ and $\ell = 2k+1$. The situation is now more complicated, since $T\mathscr{G}(F)$ is no longer an $\mathscr{O}_{n+1}$-module. This and $\mathscr{O}_{n+1}^{\times p}$ are $\mathscr{O}_{p+1}$-modules, but now $\mathscr{O}_{n+1}^{\times p}$ is not necessarily finitely generated. We use Lemma 6.3. As in Lemma 6.4, we set

$$C = \mathscr{O}_{n+1}^{\times p}, \quad A = T\mathscr{L}(F/1), \quad I = M_n^{k+1}, \quad \mathscr{M} = \mathfrak{m}_{n+1}^{k+1},$$

when $\mathscr{G} = \mathscr{L}$, and

$$C = \frac{\mathscr{O}_{n+1}^{\times p}}{T\mathscr{R}(F/1)}, \quad A = \frac{T\mathscr{A}(F/1)}{T\mathscr{R}(F/1)}, \quad I = M_n^{k+1}, \quad \mathscr{M} = \mathfrak{m}_{n+1}^{k+1},$$

when $\mathscr{G} = \mathscr{A}$. In both cases, C is finitely generated over $\mathscr{O}_{p+1}$. To see this for $\mathscr{G} = \mathscr{A}$, it is enough, by the preparation theorem, to show that $T\mathscr{R}(F/1) + F^*\mathfrak{m}_{p+1}\mathscr{O}_{n+1}^{\times p}$ contains $\mathfrak{m}_{n+1}^k \mathscr{O}_{n+1}^{\times p}$. We have

$$T\mathscr{R}(F/1) + F^*\mathfrak{m}_{p+1}\mathscr{O}_{n+1}^{\times p} = T\mathscr{A}(F/1) + F^*\mathfrak{m}_{p+1}\mathscr{O}_{n+1}^{\times p}$$

so by (6.3),

$$T\mathscr{R}(F/1) + F^*\mathfrak{m}_{p+1}\mathscr{O}_{n+1}^{\times p} + M_n^{\ell+1}\mathscr{O}_{n+1}^{\times p} \supset (M_n^k + F^*\mathfrak{m}_{p+1})\mathscr{O}_{n+1}^{\times p} \supset \mathfrak{m}_{n+1}^k \mathscr{O}_{n+1}^{\times p}.$$

Since $\mathfrak{m}_{n+1}^{\ell+1} \supset M_n^{\ell+1}$, the required inclusion follows by Nakayama's lemma. The argument for $\mathscr{L}$ is similar but simpler.

In both cases (6.3) implies that $IC \subset A + I^2C \subset A + \mathscr{M}IC$ and thus we have condition (b) of Lemma 6.3. We also have condition (a). In fact, $\mathscr{M}$ is generated by all the monomials of degree $k+1$ in the variables $x_1, \dots, x_n, t - t_0$. If a monomial $h \in \mathscr{M}$ is not divisible by the variable $t - t_0$, then it belongs to I, hence

$$hC \subset IC \subset A + I^2C \subset F^*\mathfrak{m}_{p+1}C + \mathfrak{m}_{n+1}\mathscr{M}C = (F^*\mathfrak{m}_{p+1} + \mathfrak{m}_{n+1}\mathscr{M})C.$$

Otherwise, if h is divisible by $t - t_0$, then $h \in F^*\mathfrak{m}_{p+1}$, so that

$$hC \subset F^*\mathfrak{m}_{p+1}C \subset F^*\mathfrak{m}_{p+1}C + \mathfrak{m}_{n+1}\mathscr{M}C = (F^*\mathfrak{m}_{p+1} + \mathfrak{m}_{n+1}\mathscr{M})C.$$

By Lemma 6.3, we get $IC \subset A$ and thus $M_n^{k+1}\mathscr{O}_{n+1}^{\times p} \subset T\mathscr{G}(F/1)$. □

By the lemma, we see that

$$\frac{\partial \tilde{f}}{\partial t} = g - f \in M_n^{\ell+1}\theta(\tilde{f}) \subset M_n^{k+1}\theta(\tilde{f}) \subset T\mathscr{G}(F/1).$$

In the case $\mathscr{G} = \mathscr{A}$, this implies that there exist vector fields $\tilde{\xi} \in M_n\theta_{n+1/1}$ and $\tilde{\eta} \in M_p\theta_{p+1/1}$ such that

$$\frac{\partial \tilde{f}}{\partial t} + t_{\mathrm{rel}}F(\tilde{\xi}) = \omega_{\mathrm{rel}}F(\tilde{\eta}).$$

By (3.54), this is equivalent to $dF \circ \xi = \eta \circ F$, where $\xi = \tilde{\xi} + \partial/\partial t$ and $\eta = \tilde{\eta} + \partial/\partial t$, which implies that F is trivial by Lemma 3.4. In fact, if Φ, Ψ are the flows of ξ, η, respectively, we have that $\Psi \circ F \circ \Phi^{-1} = G$, or equivalently, that $\psi_t \circ f_t \circ \phi_t^{-1} =$

f, for all t in a neighbourhood of t_0. Moreover, since $\tilde{\xi} \in M_n\theta_{n+1/1}$ and $\tilde{\eta} \in M_p\theta_{p+1/1}$, the unfoldings Φ, Ψ are origin-preserving. In particular, we have germs of diffeomorphisms $\phi_t\colon (\mathbb{F}^n, S) \to (\mathbb{F}^n, S)$ and $\psi_t\colon (\mathbb{F}^p, 0) \to (\mathbb{F}^p, 0)$ and hence, f_t is $\mathscr{A}$-equivalent to f, for all t in a neighbourhood of t_0. This completes the proof that the unfolding F is locally trivial, and therefore the proof of the theorem, for the case $\mathscr{G} = \mathscr{A}$. The proof for $\mathscr{R}$ and $\mathscr{L}$ is a subset of the proof for $\mathscr{A}$; the proof for $\mathscr{C}$ and $\mathscr{K}$ is left as a guided exercise.

This completes the proof of part (i) of Theorem 6.2. □

Proof of Part (ii) of Theorem 6.2 Assume that f is k-$\mathscr{G}$-determined and let $\ell \geq k$. Consider the set $(\pi_\ell^k)^{-1}(\sigma)$ of all ℓ-jets in $J^\ell(n,p)$ whose k-jet is equal to $\sigma = j^k f$. This is a submanifold of $J^\ell(n,p)$. Since f is k-$\mathscr{G}$-determined, $(\pi_\ell^k)^{-1}(\sigma)$ is contained in the orbit $\mathscr{G}^{(\ell)}\sigma'$, where $\sigma' = j^\ell f$. In particular, by Lemma 4.5,

$$\frac{\mathfrak{m}_n^{k+1}\theta(f)}{\mathfrak{m}_n^{\ell+1}\theta(f)} = T_{\sigma'}\left((\pi_\ell^k)^{-1}(\sigma)\right) \subset T_{\sigma'}(\mathscr{G}^{(\ell)}\sigma') = \frac{T\mathscr{G}(f) + \mathfrak{m}_n^{\ell+1}\theta(f)}{\mathfrak{m}_n^{\ell+1}\theta(f)},$$

or equivalently,

$$\mathfrak{m}_n^{k+1}\theta(f) \subset T\mathscr{G}(f) + \mathfrak{m}_n^{\ell+1}\theta(f).$$

If $\mathscr{G} = \mathscr{R}, \mathscr{C}$ or $\mathscr{K}$, $T\mathscr{G}(f)$ is a finitely generated $\mathscr{O}_n$-module. By taking $\ell = k+1$, Nakayama's Lemma C.1 implies that $\mathfrak{m}_n^{k+1}\theta(f) \subset T\mathscr{G}(f)$ and we are done. If $\mathscr{G} = \mathscr{L}$ or $\mathscr{A}$, we set $\ell = 2k+1$ and use Lemma 6.3 with

$$C = \frac{\theta(f)}{T\mathscr{R}(f)}, \quad A = \frac{T\mathscr{A}(f)}{T\mathscr{R}(f)}, \quad I = M = \mathfrak{m}_n^{k+1},$$

when $\mathscr{G} = \mathscr{A}$ and

$$C = \theta(f), \quad A = T\mathscr{L}(f), \quad I = M = \mathfrak{m}_n^{k+1},$$

when $\mathscr{G} = \mathscr{L}$. In both cases, we have $MC \subset A + M^2C$, which implies conditions (a) and (b) of Lemma 6.3, so $MC \subset A$ and thus, $\mathfrak{m}_n^{k+1}\theta(f) \subset T\mathscr{G}(f)$. □

A consequence of part (ii) of Theorem 6.2 is the following geometrical interpretation of the $\mathscr{G}$-codimension as the codimension of the orbit in the jet space.

Corollary 6.1 *Assume f is k-$\mathscr{G}$-determined and let $\sigma = j^k f$. Then,* $\operatorname{codim}_{\mathscr{G}}(f)$ *is equal to the codimension of the orbit $\mathscr{G}^{(k)}\sigma$ in $J^k(n,p)$.*

Proof It follows from Lemma 4.5 and Theorem 6.2.(ii):

$$\begin{aligned}\operatorname{codim}\, \mathscr{G}^{(k)}\sigma &= \dim_{\mathbb{F}} T_\sigma(J^k(n,p)) - \dim_{\mathbb{F}} T_\sigma(\mathscr{G}^{(k)}\sigma)\\ &= \dim_{\mathbb{F}} \frac{\mathfrak{m}_n\theta(f)/\mathfrak{m}_n^{k+1}\theta(f)}{(T\mathscr{G}(f)+\mathfrak{m}_n^{k+1}\theta(f))/\mathfrak{m}_n^{k+1}\theta(f)}\\ &= \dim_{\mathbb{F}} \frac{\mathfrak{m}_n\theta(f)}{T\mathscr{G}(f)}\\ &= \operatorname{codim}_{\mathscr{G}}(f).\end{aligned}$$

Proof of Part (iii) of Theorem 6.2 We assume $\operatorname{codim}_{\mathscr{G}}(f) = \dim_{\mathbb{F}} \frac{\mathfrak{m}_n\theta(f)}{T\mathscr{G}(f)} = d$. We consider first the case $\mathscr{G} = \mathscr{R}, \mathscr{C}$ or $\mathscr{K}$, where $T\mathscr{G}(f)$ is a finitely generated $\mathscr{O}_n$-module. Set $C = \mathfrak{m}_n\theta(f)$ and $A = T\mathscr{G}(f)$, then

$$\dim_{\mathbb{F}} \frac{C}{A+\mathfrak{m}_n^{d+1}C} \le \dim_{\mathbb{F}} \frac{C}{A} = d.$$

The submodules $B_k = A + \mathfrak{m}_n^k C$, with $k = 0,\dots,d+1$, form a decreasing chain. Since $\dim_{\mathbb{F}} B_0/B_{d+1} \le d$, they cannot be all distinct. Hence, for some k between 0 and d, $A + \mathfrak{m}_n^k C = A + \mathfrak{m}_n^{k+1}C$. In particular, $\mathfrak{m}_n^k C \subset A + \mathfrak{m}_n^{k+1}C$ and by Nakayama's Lemma C.1, $\mathfrak{m}_n^k C \subset A$. Therefore, $\mathfrak{m}_n^{d+1}\theta(f) \subset T\mathscr{G}(f)$.

Consider now the case $\mathscr{G} = \mathscr{A}$ or $\mathscr{L}$. Here we set $d' = d + rp$ and

$$C = \frac{\theta(f)}{T\mathscr{R}(f)}, \quad A = \frac{T\mathscr{A}(f)}{T\mathscr{R}(f)},$$

when $\mathscr{G} = \mathscr{A}$ or

$$C = \theta(f), \quad A = T\mathscr{L}(f),$$

when $\mathscr{G} = \mathscr{L}$. We have again

$$\begin{aligned}\dim_{\mathbb{F}} \frac{C}{A+\mathfrak{m}_n^{d'+1}C} &\le \dim_{\mathbb{F}} \frac{C}{A}\\ &= \dim_{\mathbb{F}} \frac{\theta(f)}{T\mathscr{G}(f)}\\ &= \dim_{\mathbb{F}} \frac{\theta(f)}{\mathfrak{m}_n\theta(f)} + \dim_{\mathbb{F}} \frac{\mathfrak{m}_n\theta(f)}{T\mathscr{G}(f)}\\ &= rp + d\\ &= d'.\end{aligned}$$

As in the previous case, we consider the submodules $B_k = A + \mathfrak{m}_n^k C$, with $k = 0, \ldots, d'+1$. For some k between 0 and d', $\mathfrak{m}_n^k C \subset A + \mathfrak{m}_n^{k+1} C$. Set $M = \mathfrak{m}_n^k$, then $MC \subset A + \mathfrak{m}_n MC \subset (f^*\mathfrak{m}_p + \mathfrak{m}_n M)C$, so we have condition (a) of Lemma 6.3.

Finally, we apply the same argument again, but this time with the submodules $A+M^\ell C$, with $\ell = 0, \ldots, d'+1$. For some ℓ between 0 and d', $M^\ell C \subset A+M^{\ell+1}C$. Set $I = M^\ell$, then $IC \subset A + MIC$, which gives condition (b) of Lemma 6.3. Therefore, $IC \subset A$ and thus, $\mathfrak{m}_n^{(d')^2}\theta(f) \subset T\mathscr{G}(f)$. □

Exercises for Sect. 6.1

1 Complete the final step of the proof of Theorem 6.2(ii) for the group $\mathscr{K}$, following the lines of Mather's original proof in [Mat68b]. This uses the description of $\mathscr{K}$-equivalence of two germs f and g as a special kind of $\mathscr{A}$-equivalence applied to their graph embeddings, which we gave here as Remark 4.1, thus reducing the final step in the proof, where one integrates vector fields to get the required $\mathscr{K}$-triviality of the unfolding $F(t,x) = (t, f(x) + t(g-f)(x))$, to an application of the Thom–Levine Lemma 3.4.

Step 1 Suppose that $T\mathscr{K}f \supset \mathfrak{m}^k\mathscr{O}_n^{\times p}$ and that $j^k g = j^k f$, and let $\tilde{f}(t,x) = f(x) + t(g-f)(x)$ and $F(t,x) = (t, \tilde{f}(t,x))$. Show that $\frac{\partial \tilde{f}}{\partial t} \in T\mathscr{K}(F/1)$.

Step 2 Suppose that

$$\frac{\partial \tilde{f}}{\partial t} = t_{\text{rel}}F(\xi) + \sum_{i,j} \tilde{c}_{ij} f_j \frac{\partial}{\partial y_i},$$

where the $c_{ij} \in \mathscr{O}_{n+1}$. Let $G\colon (\mathbb{F} \times \mathbb{F}^n, (0,0)) \to (\mathbb{F} \times \mathbb{F}^n \times \mathbb{F}^p, (0,0,0))$ be the graph embedding of F:

$$G(t,x) = (t, x, \tilde{f}(t,x)).$$

Let $\eta \in \theta_{1+n+p}$ be the vector field

$$\eta = \sum_{ij} c_{ij} y_j \frac{\partial}{\partial y_i} + \frac{\partial}{\partial t} - \xi;$$

show that

$$tG\left(\frac{\partial}{\partial t} - \xi\right) = \omega G(\eta) \tag{6.4}$$

and deduce from (3.4) that G is a trivial unfolding of the graph embedding of f, with the trivialisation being of the special kind of $\mathscr{A}$-equivalence described in (4.1).

Step 3 Conclude that the germ at (0, 0) of the unfolding F is $\mathscr{K}$-trivial.
Step 4 Show that the same argument applies to the germ of F at $(t_0, 0)$ for any $t_0 \in \mathbb{F}$, and conclude that $f \sim_{\mathscr{K}} g$.

2 For $\mathscr{G} = \mathscr{C}$, the finite determinacy theorem has a simple direct proof which does not require the integration of vector fields. Use Proposition 4.5 to show that if $T\mathscr{C}f \supset \mathfrak{m}_n^k \mathscr{O}_n^{\times p}$ then for any g such that $j^k g = j^k f$, $g^*\mathfrak{m}_p\mathscr{O}_n = f^*\mathfrak{m}_p\mathscr{O}_n$. Deduce that f is k-determined for $\mathscr{C}$. Is f k-determined if $T\mathscr{C}f \supset \mathfrak{m}^{k+1}\mathscr{O}_n^{\times p}$?

3 We have already proved the determinacy theorem for $\mathscr{R}$ in the case where the target of f is 1-dimensional, in Theorem 3.1. Show that there is no need for a more general proof by showing that if $p > 1$, a map-germ $f: (\mathbb{F}^n, 0) \to (\mathbb{F}^p, 0)$ which is neither a submersion nor an immersion cannot be $\mathscr{R}$-finite.

4 Assume that f is k-$\mathscr{G}$-determined. Show that if f and g are $\mathscr{G}^{(k)}$-equivalent, then they are $\mathscr{G}$-equivalent.

6.2 Estimates for the Determinacy Degree

In practice, one should not expect to obtain the precise determinacy degree of a map-germ from a general theorem like Theorem 6.2. Instead, one can often significantly improve an estimate by using another result due to Mather (in [Mat68b, Lemma 3.1]) and known as "Mather's Lemma". Theorem 6.2 reduces the proof of finite determinacy to a verification of conditions of the form

$$\mathfrak{m}_n^k\theta(f) \subset T\mathscr{G}(f) + \mathfrak{m}_n^\ell\theta(f), \quad (k < \ell).$$

Since $\theta(f)/\mathfrak{m}_n^\ell\theta(f)$ is finite-dimensional, the classification problem can be stated in terms of some finite-dimensional Lie group acting on some finite-dimensional manifold.

Theorem 6.3 ("Mather's Lemma") *Suppose the Lie group G acts smoothly on the manifold M, and that $W \subset M$ is a smooth connected submanifold. Then a necessary and sufficient condition for W to be contained in a single orbit is that*

1. *for all $x \in W$, $T_xW \subset T_x(Gx)$ and*
2. *the dimension of $T_x(Gx)$ is the same for all $x \in W$.*

Proof The necessity is obvious, so we prove the sufficiency. By Lemma 4.4, Gx is an injectively immersed submanifold of M, whose tangent space at x is $T_xGx = d_e\alpha_x(T_eG)$, where $\alpha_x : G \to M$ is the map $\alpha_x(g) = gx$. For simplicity, we write $\mathfrak{g} = T_eG$ (this is the so-called Lie algebra of G). Hence, conditions (1) and (2) can be rewritten, respectively, as

(a) for all $x \in W$, $T_xW \subset d_e\alpha_x(\mathfrak{g})$ and
(b) the dimension of $d_e\alpha_x(\mathfrak{g})$ is the same for all $x \in W$.

By condition (b),

$$E = \bigcup_{x \in W} (\{x\} \times \ker d_e\alpha_x) \subset W \times \mathfrak{g}$$

is a smooth vector subbundle of the trivial vector bundle $W \times \mathfrak{g}$. We choose F any smooth vector subbundle such that

$$E \oplus F = W \times \mathfrak{g},$$

and denote by F_x the fibre of F on each $x \in W$. Now, by condition (a), we can define F_0 as the smooth vector subbundle of F given by

$$F_0 = \bigcup_{x \in W} \left(\{x\} \times (d_e\alpha_x)^{-1}(T_xW) \cap F_x\right).$$

The map $\omega : F_0 \to TW$ given by $\omega(x, \eta) = d\alpha_x(\eta)$ is obviously an isomorphism of smooth vector bundles. Let $\pi : W \times \mathfrak{g} \to \mathfrak{g}$ be the projection and consider the smooth map $\pi \circ \omega^{-1} : TW \to \mathfrak{g}$, which satisfies

$$d\alpha_x(\pi \circ \omega^{-1}(\eta)) = \eta, \quad \forall \eta \in T_xW, \quad \forall x \in W.$$

To prove that W is contained in a single orbit we need to see that any two points $x_1, x_2 \in W$ are contained in the same orbit. Since W is connected there exists a smooth curve $\gamma : [0, 1] \to W$ connecting x_1 and x_2. It is enough to show that for any $t_0 \in [0, 1]$ there exists $\epsilon > 0$ such that for any t with $t_0 - \epsilon < t < t_0 + \epsilon$, $\gamma(t)$ and $\gamma(t_0)$ are equivalent under the action of G.

Let $\gamma'(t) \in T_{\gamma(t)}W$ be the tangent vector of γ and let $X : [0, 1] \to \mathfrak{g}$ be the smooth map given by

$$X(t) = \pi \circ \omega^{-1}(\gamma'(t)),$$

which satisfies

$$d\alpha_x(X(t)) = \gamma'(t), \quad \forall t \in [0, 1].$$

We define $\tilde{X} : G \times [0, 1] \to TG$ as

$$\tilde{X}(g, t) = \tilde{X}_t(g) = d_eR_g(X(t)),$$

where $R_g : G \to G$ is the right translation in the group, $R_g(h) = hg$. Then $\tilde{X}_t$ is a time-dependent vector field on G (in fact, $\tilde{X}_t$ is the unique right invariant vector field on G which extends $X(t)$). For each t_0, let $\mu\colon (t_0-\epsilon, t_0+\epsilon) \to G$ be the integral curve of $\tilde{X}_t$, passing through e at $t = t_0$. This means that $\mu(t_0) = e$ and that

$$\mu'(t) = \tilde{X}_t(\mu(t)), \quad \forall t \in (t_0-\epsilon, t_0+\epsilon).$$

We will prove that $\mu(t)^{-1}\gamma(t) = \gamma(t_0)$, which implies that $\gamma(t)$ and $\gamma(t_0)$ are contained in the same orbit, for all $t \in (t_0-\epsilon, t_0+\epsilon)$. By the chain rule, we have

$$\frac{d}{dt}\{\mu(t)^{-1}\gamma(t)\} = d\alpha_{\gamma(t)}\left(\frac{d}{dt}\{\mu(t)^{-1}\}\right) + d\beta_{\mu(t)^{-1}}\left(\gamma'(t)\right), \tag{6.5}$$

where $\beta_g : M \to M$ is the map $\beta_g(x) = gx$. Note that $\mu(t)^{-1}\mu(t) = e$, by taking derivatives and using again the chain rule:

$$0 = \frac{d}{dt}\{\mu(t)^{-1}\mu(t)\} = dR_{\mu(t)}\left(\frac{d}{dt}\{\mu(t)^{-1}\}\right) + dL_{\mu(t)^{-1}}\left(\mu'(t)\right),$$

where now $L_g : G \to G$ is left translation in the group, $L_g(x) = gx$. Since R_g is a diffeomorphism whose inverse is $R_{g^{-1}}$, we obtain that the first term in the right-hand side of (6.5) is

$$\begin{aligned}
d\alpha_{\gamma(t)}\left(\frac{d}{dt}\{\mu(t)^{-1}\}\right) &= -d\alpha_{\gamma(t)} \circ dR_{\mu(t)^{-1}} \circ dL_{\mu(t)^{-1}}(\mu'(t)) \\
&= -d\beta_{\mu(t)^{-1}} \circ d\alpha_{\gamma(t)} \circ dR_{\mu(t)^{-1}}(\mu'(t)) \\
&= -d\beta_{\mu(t)^{-1}} \circ d\alpha_{\gamma(t)} \circ dR_{\mu(t)^{-1}}(\tilde{X}_t(\mu(t)) \\
&= -d\beta_{\mu(t)^{-1}} \circ d\alpha_{\gamma(t)} \circ dR_{\mu(t)^{-1}} \circ dR_{\mu(t)}(X(t)) \\
&= -d\beta_{\mu(t)^{-1}} \circ d\alpha_{\gamma(t)}(X(t)) \\
&= -d\beta_{\mu(t)^{-1}}(\gamma'(t)).
\end{aligned}$$

For the second equality above, we have used the fact that

$$\alpha_{\gamma(t)} \circ R_{\mu(t)^{-1}} \circ L_{\mu(t)^{-1}} = \beta_{\mu(t)^{-1}} \circ \alpha_{\gamma(t)} \circ R_{\mu(t)^{-1}}.$$

By substituting in (6.5) we get $\frac{d}{dt}\{\mu(t)^{-1}\gamma(t)\} = 0$ and hence, $\mu(t)^{-1}\gamma(t)$ must be constant. Thus, for all $t \in (t_0-\epsilon, t_0+\epsilon)$, we have

$$\mu(t)^{-1}\gamma(t) = \mu(t_0)^{-1}\gamma(t_0) = \gamma(t_0).$$

□

We remark that in general, condition 2 in Theorem 6.3 is not a consequence of condition 1 as we can see in the following example: Consider G the Lie group of matrices of the form

$$\begin{pmatrix} a & b \\ 0 & d \end{pmatrix}$$

with $a, b, d \in \mathbb{R}$ and $ad \neq 0$, acting on $M = \mathbb{R}^2$ in the standard representation, and let W be the parabola given by $y = (x-1)^2$. For any $p = (x, y) \in W$, the orbit Gp is either $\mathbb{R}^2 \setminus \mathbb{R} \times \{0\}$ if $y \neq 0$, or $\mathbb{R} \times \{0\}$ if $y = 0$. We see that $T_pW \subset T_p(Gp)$ for all $p \in W$, but the dimension of $T_p(Gp)$ depends on whether or not $p \in W$, and indeed W is clearly *not* contained in a single orbit.

However, in many cases in which we use Theorem 6.3, condition 2 follows from condition 1 and hence it is unnecessary to check it separately. This is the case in the following corollary.

Corollary 6.2 *Let G be a Lie group acting smoothly on two manifolds M, N and let $f : M \to N$ be a G-equivariant submersion (i.e. a submersion such that $f(gx) = gf(x)$ for all $x \in M$ and $g \in G$). Let $y \in N$ be such that $W = f^{-1}(y)$ is connected. Then a necessary and sufficient condition for W to be contained in a single orbit is that condition 1 of Theorem 6.3 is satisfied.*

Proof For each $x \in W$, $d_xf : T_xM \to T_yN$ is an epimorphism whose kernel is $\ker d_xf = T_xW$. Moreover, since $f(Gx) \subset Gy$, we can consider the restriction $d_xf : T_x(Gx) \to T_y(Gy)$. If $T_xW \subset T_x(Gx)$, then

$$\dim T_x(Gx) = \dim T_xW + \dim T_y(Gy).$$

Hence, the dimension of $T_x(Gx)$ is independent of $x \in W$. □

Example 6.1 The following example is due to Siersma [Sie73] . Let us consider the singularity of type E_7 , given by $f\colon (\mathbb{F}^2, 0) \to \mathbb{F}$ with $f(x, y) = x^3 + xy^3$. It is not difficult to check that $\mathfrak{m}_2^6 \subseteq \mathfrak{m}_2 J_f$, which implies that f is 5-$\mathscr{R}$-determined, by the Finite Determinacy Theorem 6.2. We use Corollary 6.2 to see that f is in fact 4-determined, which cannot be deduced directly from the estimates in Theorem 6.2.

Let g be another function with the same 4-jet as f. It is enough to show that g is $\mathscr{R}^{(5)}$-equivalent to f (Exercise 6.1.4). So, we take $M = J^5(2, 1)$, $G = \mathscr{R}^{(5)}$, $N = J^4(2, 1)$ and the submersion $\pi_4^5 : J^5(2, 1) \to J^4(2, 1)$. Observe that $\mathscr{R}^{(5)}$ also acts on $J^4(2, 1)$ by taking the 4-jet of the corresponding diffeomorphism in $\mathscr{R}$ and that π_4^5 is obviously equivariant. Let

$$W = (\pi_4^5)^{-1}(j^4 f) = \{\sigma = x^3 + xy^3 + a_0x^5 + a_1x^4y + \cdots + a_5y^5 \,:\, a_i \in \mathbb{F},\ i = 0, \ldots, 5\}.$$

If we can show that $T_\sigma W \subseteq T_\sigma(\mathscr{R}^{(5)} \cdot \sigma)$, for all $\sigma \in W$, then by Corollary 6.2, W is contained in a single orbit and we are done.

Since W is an affine subspace of $J^5(2, 1)$, its tangent space at any σ coincides with W itself (after translation to the origin), so $T_\sigma W$ is spanned by $\{x^5, \ldots, y^5\}$. On the other hand, $T_\sigma(\mathscr{R}^{(5)} \cdot \sigma)$ is the 5-jet image of the Jacobian ideal

$$J_\sigma = \langle 3x^2 + y^3 + O(4), 3xy^2 + O(4)\rangle,$$

where $O(4)$ means terms of order 4 or higher. We have

$$\begin{aligned}
x^5 &= \frac{1}{3}x^3(3x^2 + y^3 + O(4)) \mod \mathfrak{m}_2^5,\\
x^4y &= \frac{1}{3}x^2y(3x^2 + y^3 + O(4)) \mod \mathfrak{m}_2^5,\\
x^3y^2 &= \frac{1}{3}xy^2(3x^2 + y^3 + O(4)) \mod \mathfrak{m}_2^5,\\
x^2y^3 &= \frac{1}{3}y^2(3x^2 + y^3 + O(4)) \mod \mathfrak{m}_2^5,\\
xy^4 &= \frac{1}{3}y^2(3xy^2 + O(4)) \mod \mathfrak{m}_2^5,\\
x^2y^2 &= \frac{1}{3}x(3xy^2 + O(4)) \mod \mathfrak{m}_2^5,\\
y^5 &= y^2(3x^2 + y^3 + O(4)) - x^2y^2 \mod \mathfrak{m}_2^5,
\end{aligned}$$

which proves the inclusion of tangent spaces. ◊

Example 6.2 We show, using Mather's lemma, that every finitely determined germ $f\colon (\mathbb{F}^2, 0) \to (\mathbb{F}^3, 0)$ with $j^3 f = (x, y^3, xy)$ is equivalent to the germ $f_\ell(x, y) = (x, y^3, xy + y^{3\ell+2})$, for some finite $\ell \geq 2$. The method is iterative: we show

1. that if $j^k f = (x, y^3, xy)$ then $j^{k+1} f$ is $\mathscr{A}^{(k+1)}$-equivalent to (x, xy, y^3) if $k \equiv 0 \mod 3$ or $k \equiv 2 \mod 3$,
2. that if $k = 3\ell + 1$ then $j^{k+1} f$ is $\mathscr{A}^{(k+1)}$-equivalent to (x, y^3, xy) or to $(x, y^3, xy + y^{3\ell+2})$, and
3. if $j^k f = (x, y^3, xy + y^{3\ell+2})$ then $j^{k+1} f$ is $\mathscr{A}^{(k+1)}$-equivalent to $(x, y^3, xy + y^{3\ell+2})$—in other words, the germ f_ℓ is $3\ell + 2$-determined for $\mathscr{A}$–equivalence.

For each value of $k \geq 3$ we let σ_k denote the k-jet (x, y^3, xy), π_k^{k+1} denote the projection from $J^{k+1}(2, 3)$ to $J^k(2, 3)$, and W denote $(\pi_k^{k+1})^{-1}(\sigma_k)$.

1. Suppose $k = 3\ell$. We use Corollary 6.2 to show that W is contained in a single $\mathscr{A}^{(k+1)}$-orbit. Members of W are $k + 1$-jets of the form $(x, y^3, xy) + O(k + 1)$, where $O(k + 1)$ denotes a triple of homogenous polynomials of degree $k + 1$. As in the previous example, the group $\mathscr{A}^{(k+1)}$ acts via its quotient $\mathscr{A}^{(k)}$ on $J^k(2, 3)$, and π_k^{k+1} is equivariant with respect to the actions on $J^{k+1}(2, 3)$ and $J^k(2, 3)$. Let us take coordinates u_1, u_2, u_3 on $\mathbb{F}^3$. The tangent space to W at each point is

the linear span of the monomials $x^i y^j \partial u_s$ for $1 \le s \le 3$ and $i + j = 3\ell + 1$. We have to show that for each $\sigma \in W$, $T\mathscr{A}^{(3\ell+1)}\sigma$ contains $T_\sigma W$. Let $\sigma \in W$. In the following calculations we regard σ as a polynomial mapping. Every monomial $x^i y^j$ with $i + j = 3\ell + 1$ lies in $\sigma^{-1}\mathfrak{m}_3^2 \mod \mathfrak{m}_2^{3k+2}$, except for $y^{3\ell+1}$:

$$xy^{3\ell} \equiv (u_1 u_2^\ell) \circ \sigma \mod \mathfrak{m}_2^{3\ell+2},$$

and if $j = 3q + r$ with $0 \le r \le 2$ and $i > 1$ then

$$x^i y^j \equiv (u_1^{i-r} u_2^q u_3^r) \circ \sigma \mod \mathfrak{m}_2^{3\ell+2}.$$

For these equalities we use the fact that that for any monomial $m \in \mathfrak{m}_3^2$, $m \circ \sigma \equiv m \circ \sigma_k \mod \mathfrak{m}_2^{3k+2}$, since σ and σ_k differ only by terms of degree $3\ell + 1$. To find $y^{3\ell+1}\partial u_s$ in $T\mathscr{A}^{(3\ell+1)}\sigma$, we calculate that, modulo $\mathfrak{m}_2^{3\ell+2}\theta(\sigma)$,

$$\begin{aligned} y^{3\ell+1}\partial_{u_1} &\equiv t\sigma(y^{3\ell+1}\partial_x) \\ 3y^{3\ell+1}\partial_{u_2} &\equiv t\sigma(y^{3\ell-1}\partial_y - xy^{3k-2}\partial_x) + \omega\sigma(u_2^{\ell-1}u_3) \\ y^{3\ell+1}\partial_{u_3} &\equiv t\sigma(y^{3\ell}\partial_x) - \omega\sigma(u_2^\ell \partial_{u_1}), \end{aligned}$$

concluding the proof that $T_\sigma W \subset T_\sigma \mathscr{A}^{(3\ell+1)}\sigma$. Thus by Corollary 6.2, W lies in a single orbit, and in particular that of σ_{k+1}.

Now suppose $k = 3\ell+2$. We apply the same reasoning as in the previous case. Let $\sigma \in W = (\pi_k^{k+1})^{-1}(\sigma_k)$. An argument similar to the previous case shows that modulo $\mathfrak{m}_2^{3\ell+4}$, every monomial of degree $3\ell + 3$ is in $\sigma^{-1}(\mathfrak{m}_3^2)$, except for $xy^{3\ell+2}$. Working modulo $\mathfrak{m}_2^{3\ell+4}\theta(\sigma)$, we have

$$\begin{aligned} xy^{3\ell+2}\partial_{u_1} &\equiv t\sigma(xy^{3\ell+2}\partial_x) \\ 3xy^{3\ell+2}\partial_{u_2} &\equiv t\sigma(xy^{3\ell}\partial_y) - \omega\sigma(u_1^2 u_2^\ell \partial_{u_3}) \\ xy^{3\ell+2}\partial_{u_3} &\equiv t\sigma(xy^{3\ell+1}\partial_x) - \omega\sigma(u_2^\ell u_3 \partial_{u_1}). \end{aligned}$$

Once again, by Corollary 6.2, W lies in a single orbit.

2. If $k = 3\ell + 1$, the argument becomes a little more delicate. Elements of $(\pi_k^{k+1})^{-1}(\sigma_k)$ have the form $\sigma = \sigma_k + (a, b, c)$ where a, b, c are homogeneous polynomials in x and y of degree $k + 1 = 3\ell + 2$. Let $c_{0,3\ell+2}$ be the coefficient of $y^{3\ell+2}$ in the polynomial c. We divide $(\pi_k^{k+1})^{-1}$ into two parts,

$$W_0 = \sigma \in (\pi_k^{k+1})^{-1}(\sigma_k) \text{ such that } c_{0,3\ell+2} = 0$$

$$W_1 = \sigma \in (\pi_k^{k+1})^{-1}(\sigma_k) \text{ such that } c_{0,3\ell+2} \ne 0.$$

We show that each of W_0 and W_1 lies in a single orbit. Since neither consists of all of $(\pi_k^{k+1})^{-1}(\sigma_k)$, we cannot use Corollary 6.2 directly. However W_1 is open in $(\pi_k^{k+1})^{-1}(\sigma_k)$ and clearly connected, so by the argument of the proof of Corollary 6.2, to show that W_1 is contained in a single orbit, we need only show that $T_\sigma W_1 \subset T\mathscr{A}^{(3\ell+2)}\sigma$ for all $\sigma \in W_1$. On the other hand, to show that W_0 is contained in a single orbit we need to show that both conditions of Mather's lemma hold. As before, for all $\sigma \in W$, every monomial in x and y of degree $3\ell + 2$, in this case except for $y^{3\ell+2}$, is contained in $\sigma^{-1}(\mathscr{O}_3) + \mathfrak{m}_2^{3\ell+3}$. As before, $y^{3\ell+2}\partial_{u_1}$ and $y^{3\ell+2}\partial_{u_2}$ are contained in $T\mathscr{A}^{(3\ell+2)}\sigma$:

$$y^{3\ell+2}\partial_{u_1} \equiv t\sigma(y^{3\ell+2}\partial_x)$$
$$3y^{3\ell+2}\partial_{u_2} \equiv t\sigma(y^{3\ell}\partial_y) - \omega\sigma(u_1u_2^\ell\partial_{u_3}).$$

Now suppose $\sigma \in W_1$. Then modulo $\mathfrak{m}_2^{3\ell+2}\theta(\sigma)$ *and terms of degree* $3\ell+2$ *have already shown to lie in* $T\mathscr{A}^{(3\ell+2)}\sigma$,

$$t\sigma(y\partial_y) - \omega\sigma(3u_2\partial_{u_2}) = (xy + (3\ell+2)c_{0,3\ell_2}y^{3\ell+2})\partial_{u_3}$$
$$\omega\sigma(u_3\partial_{u_3}) = (xy + c_{0,3\ell_2}y^{3\ell+2})\partial_{u_3}$$

and thus because the two right-hand sides are linearly independent, $y^{3\ell+2}\partial_{u_3} \in T\mathscr{A}^{(3\ell+2)}\sigma$. Hence $T_\sigma W_1 \subset T\mathscr{A}^{(3\ell+2)}\sigma$. Thus the conditions for Mather's lemma apply to W_1, which is therefore contained in a single orbit.

From what we have already shown, it is clear that $T_\sigma W_0 \subset T\mathscr{A}^{(3\ell+2)}\sigma$ for each $\sigma \in W_0$. We now show that the dimension of $T\mathscr{A}^{(3\ell+2)}\sigma$ is the same for all $\sigma \in W_0$. The polynomials a, b and c contribute only to the degree $3\ell + 2$ part of certain generators of $T\mathscr{A}^{(3\ell+2)}\sigma$, namely the images under $t\sigma$ and $\omega\sigma$ of vector fields in θ_2 and θ_3 with linear coefficients. However as we have seen above, all terms of degree $3\ell + 2$, except for $y^{3\ell+2}\partial_{u_3}$, are contained in $\omega\sigma(\mathfrak{m}_3^2\theta_3)$, and therefore a, b and c do not contribute to $T\mathscr{A}^{(3\ell+2)}\sigma$, with one exception: the missing monomial $y^{3\ell+2}\partial_{u_3}$ appears with coefficient $c_{1,3\ell+1}$ in $t\sigma(y\partial_x)$. However, $t\sigma(y\partial_x)$ is the only generator of $T\mathscr{A}^{(3\ell+2)}\sigma$ in which $y\partial_{u_1}$ appears with non-zero coefficient, and so although the value of $c_{1,3\ell+1}$ affects the *space* $T\mathscr{A}^{(3\ell+2)}\sigma$, it does not affect its dimension.

3. We postpone the proof of $(3\ell+2)$-determinacy of f_ℓ until Example 6.4 in the next section, where we introduce the ideas and techniques of the paper [BdPW87]. ♢

The following proposition and theorem appear in Wall [Wal81], although Wall himself acknowledges that the results are implicit in Mather's work and have been used by other authors.

Proposition 6.1 *We set $\epsilon = 2$ for $\mathscr{G} = \mathscr{L}, \mathscr{A}$ and $\epsilon = 1$ for $\mathscr{G} = \mathscr{R}, \mathscr{C}, \mathscr{K}$. Then f is k-$\mathscr{G}$-determined if and only if for any g such that $j^k f = j^k g$, we have*

$$\mathfrak{m}_n^{k+1}\theta(g) \subset T\mathscr{G}(g) + \mathfrak{m}_n^{\epsilon k+2}\theta(g). \tag{6.6}$$

Proof Assume f is k-determined. For any g such that $j^k f = j^k g$, g is $\mathscr{G}$-equivalent to f. Hence g is also $\mathscr{G}$-determined. By Theorem 6.2 (ii) we have

$$\mathfrak{m}_n^{k+1}\theta(g) \subset T\mathscr{G}(g) \subset T\mathscr{G}(g) + \mathfrak{m}_n^{\epsilon k+2}\theta(g).$$

Let us see the converse. We first use (6.6) in the case $g = f$:

$$\mathfrak{m}_n^{k+1}\theta(f) \subset T\mathscr{G}(f) + \mathfrak{m}_n^{\epsilon k+2}\theta(f).$$

If $\mathscr{G} = \mathscr{R}, \mathscr{C}, \mathscr{K}$, we use Nakayama's lemma C.1 and get $\mathfrak{m}_n^{k+1}\theta(f) \subset T\mathscr{G}(f)$. Otherwise, if $\mathscr{G} = \mathscr{L}, \mathscr{A}$, we use Lemma 6.3 with

$$C = \frac{\theta(f)}{T\mathscr{R}(f)}, \quad A = \frac{T\mathscr{A}(f)}{T\mathscr{R}(f)}, \quad I = M = \mathfrak{m}_n^{k+1},$$

when $\mathscr{G} = \mathscr{A}$ and

$$C = \theta(f), \quad A = T\mathscr{L}(f), \quad I = M = \mathfrak{m}_n^{k+1},$$

when $\mathscr{G} = \mathscr{L}$. In both cases, we have $MC \subset A + M^2C$, which implies conditions (a) and (b) of Lemma 6.3, so $MC \subset A$ and thus, $\mathfrak{m}_n^{k+1}\theta(f) \subset T\mathscr{G}(f)$. By Theorem 6.2 (i), f is ℓ-determined, where $\ell = \epsilon k + 1$.

Since f is ℓ-determined, to show that it is k-determined it is enough to prove that for any g such that $j^k g = j^k f$, $j^\ell g$ is $\mathscr{G}^{(\ell)}$-equivalent to $j^\ell f$. In fact, this would imply that there exists $\varphi \in \mathscr{G}$ such that $j^\ell g = (j^\ell \varphi)(j^\ell f) = j^\ell(\varphi f)$. Since φf is $\mathscr{G}$-equivalent to f, φf is also ℓ-determined. Thus, g would be $\mathscr{G}$-equivalent to φf and also to f.

We prove this by means of Corollary 6.2. Consider the epimorphism $\pi_k^\ell : J^\ell(n, p) \to J^k(n, p)$ and $W = (\pi_k^\ell)^{-1}(\sigma)$, where $\sigma = j^k f$. Note that π_k^ℓ satisfies the hypothesis of Corollary 6.2 and that W is obviously connected, since it is a vector subspace. For any $\sigma' \in W$, $\sigma' = j^\ell g$ with $j^k g = j^k f$, so we can use the inclusion (6.6):

$$T_{\sigma'}W = \frac{\mathfrak{m}_n^{k+1}\theta(g)}{\mathfrak{m}_n^{\ell+1}\theta(g)} \subset \frac{T\mathscr{G}(g) + \mathfrak{m}_n^{\ell+1}\theta(g)}{\mathfrak{m}_n^{\ell+1}\theta(g)} = T_{\sigma'}(\mathscr{G}^{(\ell)}\sigma').$$

By Corollary 6.2, W is contained in a single orbit and we have the desired property. □

In the case $\mathscr{G} = \mathscr{A}$, we have the following useful theorem.

Theorem 6.4 *Suppose there is an $\mathscr{O}_n$-module D such that*

1. $D \subset T\mathscr{A}_e(f) + \mathfrak{m}_n^{k+1}\theta(f)$
2. $\mathfrak{m}_n^{k+1}\theta(f) \subset tf(\mathfrak{m}_n\theta_n) + f^*\mathfrak{m}_p D + \mathfrak{m}_n^{k+2}\theta(f)$.

Then, f is k-$\mathscr{A}$-determined.

Proof Let g such that $j^k g = j^k f$. By Lemma 6.1, we see that conditions 1 and 2 are also true if we put g instead of f. Applying Nakayama's lemma C.1 to condition 2,

$$\mathfrak{m}_n^{k+1}\theta(g) \subset tg(\mathfrak{m}_n\theta_n) + g^*\mathfrak{m}_p D,$$

which combined with condition 1 gives

$$\begin{aligned}\mathfrak{m}_n^{k+1}\theta(g) &\subset tg(\mathfrak{m}_n\theta_n) + \omega g(\mathfrak{m}_p\theta_p) + (g^*\mathfrak{m}_p)\mathfrak{m}_n^{k+1}\theta(g)\\ &= T\mathscr{A}(g) + (g^*\mathfrak{m}_p)\mathfrak{m}_n^{k+1}\theta(g).\end{aligned}$$

If we put

$$A = \frac{\mathfrak{m}_n^{k+1}\theta(g)}{\mathfrak{m}_n^{2k+2}\theta(g)}, \quad B = \frac{T\mathscr{A}(g) + \mathfrak{m}_n^{2k+2}\theta(g)}{\mathfrak{m}_n^{2k+2}\theta(g)},$$

then the above inclusion is written as $A \subset B + g^*\mathfrak{m}_p A$. Since A is finite-dimensional, we can use Nakayama's lemma again, so that $A \subset B$ and hence,

$$\mathfrak{m}_n^{k+1}\theta(g) \subset T\mathscr{A}(g) + \mathfrak{m}_n^{2k+2}\theta(g).$$

By Proposition 6.1, f is k-$\mathscr{A}$-determined. □

As a consequence of the theorem, we get the following estimate due to Gaffney [Gaf79].

Corollary 6.3 *If $\mathfrak{m}_n^k\theta(f) \subset T\mathscr{A}_e(f)$ and $\mathfrak{m}_n^\ell\theta(f) \subset T\mathscr{K}_e(f)$ with $k \geq 1$, then f is $(k+\ell)$-$\mathscr{A}$-determined.*

Proof Take $D = \mathfrak{m}_n^k\theta(f)$ and $s = k+\ell$, then $D \subset T\mathscr{A}_e(f)$ and

$$\mathfrak{m}_n^{k+\ell}\theta(f) \subset \mathfrak{m}_n^k T\mathscr{K}_e(f) \subset tf(\mathfrak{m}_n\theta_n) + f^*\mathfrak{m}_p D.$$

Another important consequence is the following estimate of Mather for stable map-germs.

Corollary 6.4 *Any stable map-germ is $(p+1)$-$\mathscr{A}$-determined.*

Proof If f is stable, then $\mathfrak{m}_n\theta(f) \subset T\mathscr{A}_e(f)$ and $\mathfrak{m}_n^p\theta(f) \subset T\mathscr{K}_e(f)$ (see the proof of Theorem 4.1). Gaffney's estimate Corollary 6.3 gives that f is $(p+1)$-$\mathscr{A}$-determined. □

6.3 Determinacy and Unipotency

The gap between the power of the maximal ideal in the *necessary* condition for k-determinacy for $\mathscr{A}$, that $T\mathscr{A}f \supset \mathfrak{m}_n^{k+1}\theta(f)$, and in the *sufficient* condition given by Theorem 6.2, that $T\mathscr{A}f \supset \mathfrak{m}_n^{\ell}\theta(f)$ for some $\ell \le \frac{1}{2}(k-1)+1$, is unbridgeable. Despite the improvements due to Gaffney and du Plessis, some gap is intrinsic to the group $\mathscr{A}$. However for the group $\mathscr{A}_1$, where the diffeomorphisms of source and target are required to have 1-jet equal to the 1-jet of the identity, it turns out that the corresponding necessary condition is also sufficient. What makes $\mathscr{A}_1$ different is the fact that its orbits in jet space are Zariski-closed, and this in turn follows from the property of unipotency, which we discuss below. In 1987 Bruce, du Plessis and Wall built on this insight and developed a very general theory in the paper [BdPW87], giving major improvement in determinacy estimates and vastly increasing the power of classification techniques in singularity theory. It allowed determinacy estimates of a new kind, in which $\mathscr{G}$-orbits (for $\mathscr{G} = \mathscr{R}, \mathscr{C}, \mathscr{K}, \mathscr{L}, \mathscr{A}$) and orbits of certain subgroups could be shown to contain spaces of germs of a wider range than just the usual $\mathfrak{m}_n^{k+1}\mathscr{O}_n^{\times p}$. Below we give the definition of unipotency, and briefly discuss the background in the theory of linear algebraic groups. Before doing so, we highlight the key ingredient in [BdPW87]:

Theorem 6.5 ([BdPW87, Proposition 3.3]) *Let U be a unipotent affine algebraic group over $\mathbb{F}$, acting algebraically on an affine variety V and let $\mathbb{F} = \mathbb{R}$ or $\mathbb{C}$. Then the orbits of $U(\mathbb{F})$ are Zariski-closed in $V(\mathbb{F})$.*

Zariski-closedness of the orbits of a unipotent group over an algebraically closed field had previously been proved by Kostant and Rosenlicht—see references in [BdPW87]. We give Springer's proof of this (from [Spr09]) in Theorem 6.8 below, but omit the proof for $\mathbb{F} = \mathbb{R}$, which requires a deeper study of algebraic groups. For this readers are urged to consult the original source [BdPW87, Theorem 3.3].

Examples of unipotent groups in singularity theory are the subgroups $\mathscr{G}_s^{(k)}$ of the standard jet groups $\mathscr{R}^{(k)}, \mathscr{C}^{(k)}, \mathscr{K}^{(k)}, \mathscr{L}^{(k)}, \mathscr{A}^{(k)}$ consisting of k-jets of diffeomorphisms whose s-jet coincides with the s-jet of the identity, for some $k \ge s \ge 1$ (see Proposition 6.4 below). By contrast, none of the groups $\mathscr{R}^{(k)}, \mathscr{C}^{(k)}, \mathscr{K}^{(k)}, \mathscr{L}^{(k)}, \mathscr{A}^{(k)}$ is unipotent, and indeed their orbits are never closed. For example, if $f\colon (\mathbb{F},0) \to (\mathbb{F},0)$ is given by $f(x) = x^\ell$, then $j^\ell f$ is $\mathscr{R}$-equivalent to $j^\ell f_t(x) = tf(x)$ for $t \in (0,1]$, but certainly not for $t = 0$. Thus the $\mathscr{R}^{(\ell)}$-orbit of $j^\ell f$ contains the curve $j^\ell(f_t), t \in (0,1]$, but not its limit point $j^\ell(0)$.

The significance of unipotency for determinacy estimates is not hard to see. In Proposition 6.2 we give the easy proof (based on the ideas of the proofs of Lemma 6.4 and Theorem 6.2(i)) that if $\mathscr{H}$ is a suitable subgroup of $\mathscr{A}$ and $T\mathscr{H}f \supset \mathfrak{m}_n^{k+1}\mathscr{O}_n^{\times p}$, then whenever $j^k g = j^k f$, $f_u := f + u(g-f)$ is $\mathscr{H}$-equivalent to f for *small enough* $u \in \mathbb{F}$. This of course is not enough to prove that g and f are $\mathscr{H}$-equivalent—for this we need the equivalence of $f + u(g-f)$ and f to hold all the way to $u = 1$. To highlight the difference between the two properties, we call the first property *local finite determinacy*, in contrast to true finite determinacy, which is *global* in u. Indeed, the difference between the (global) $\mathscr{G}$ determinacy degree, ℓ, and the local $\mathscr{G}$ determinacy degree (the lowest k such that $\mathfrak{m}_n^{k+1}\mathscr{O}_n^{\times p} \subset T\mathscr{G}f$, for the groups $\mathscr{G}$ we are interested in), is due to the fact that to prove k-determinacy, we need also that $\mathfrak{m}_n^{k+1}\mathscr{O}_n^{\times p} \subset T\mathscr{G}f_{u_0}$ for each $u_0 \in [0, 1]$, and this may not hold if f_{u_0} merely agrees with f up to degree k. What we can deduce from local determinacy if $\mathfrak{m}^{k+1}\mathscr{O}_n^{\times p} \subset T\mathscr{G}f$, by passing to representatives, is that in any jet space $J^\ell(n, p)$ with $\ell \geq k+1$, $\mathscr{H}^{(\ell)}j^\ell f$ contains the set $\{j^\ell(f + u(g - f)) : |u| < \eta\}$ for some positive η. This is a (Euclidean) open neighbourhood of $j^\ell f$ in the affine line $L := \{j^\ell(f + u(g - f)) : u \in \mathbb{F}\}$, and therefore Zariski-dense in L. If $\mathscr{H}^{(\ell)}j^\ell f$ is Zariski-closed, then so is $\mathscr{H}^{(\ell)}j^\ell f \cap L$, and it follows at once that $\mathscr{H}^{(\ell)}j^\ell f \supset L$. Thus, indeed $j^\ell f$ is $\mathscr{H}^{(\ell)}$-equivalent to $j^\ell g$. By taking ℓ greater than or equal to the $\mathscr{H}$-determinacy degree of f, we conclude that f and g are $\mathscr{H}$-equivalent.

From now on we restrict our discussion to subgroups of $\mathscr{A}$. We say that $\mathscr{H}$ is a *Lie subgroup* of $\mathscr{A}$ if it is a subgroup and for each $k \leq 1$, the k-jet projection $\mathscr{H}^{(k)}$ is a Lie subgroup of $\mathscr{A}^{(k)}$. When $\mathscr{H}$ is a Lie subgroup of $\mathscr{A}$, then for each germ f we can define the *tangent space* $T\mathscr{H}f$ in the obvious way: $T\mathscr{H}f$ is the sub-$\mathscr{O}_p$-module of $T\mathscr{A}f$ of vector fields of the form

$$\left.\frac{d(\varphi_t f)}{dt}\right|_{t=0},$$

where φ_t is any origin-preserving unfolding of the identity in $\mathscr{H}$. As in Lemma 4.5, we have that $j^k(T\mathscr{H}f) = T\mathscr{H}^{(k)}\sigma$, where $\sigma = j^k f$.

In order to state and prove Proposition 6.2 for subgroups of $\mathscr{A}$, we need to clarify and formalise the relation between the $\mathscr{H}$-tangent space of a germ $f\colon (\mathbb{F}^n, 0) \to (\mathbb{F}^p, 0)$ and the parameterised or "relative" tangent space $T\mathscr{H}(F/d)$ for an unfolding $F\colon (\mathbb{F}^n \times \mathbb{F}^d, 0) \to (\mathbb{F}^p \times \mathbb{F}^d, 0)$ of f. The latter was described for the classical groups $\mathscr{R}, \mathscr{L}, \mathscr{C}, \mathscr{K}, \mathscr{A}$ at the start of Sect. 6.1. Then we assume the following two properties: (G1) and (G2).

(G1) There exist a sub-$\mathscr{O}_n$-module L_s of $\mathfrak{m}_n\theta_n$, with $L_s \supset \mathfrak{m}_n^r\theta_n$ for some finite $r \geq 1$, and a sub $\mathscr{O}_p$-module L_t of $\mathfrak{m}_p\theta(p)$, such that for all f,

$$T\mathscr{H}f = tf(L_s) + \omega f(L_t).$$

This holds for all of the usual subgroups of $\mathscr{A}$, and in particular for the subgroups

$$\mathscr{A}_s = \{g \in \mathscr{A} : j^s g = j^s \mathrm{id}\},$$

for which $L_s = \mathfrak{m}_n^{s+1}\theta_n$ and $L_t = \mathfrak{m}_p^{s+1}\theta_p$.

Let $\pi_n : \mathbb{F}^n \times \mathbb{F}^d \to \mathbb{F}^d$ be the standard projection and let $\theta_{n+d/d}$ be the kernel of $t\pi : \theta_{n+d} \to \theta(\pi)$. Given the product structure on $\mathbb{F}^n \times \mathbb{F}^d$, each vector field

$\xi \in L_s = \sum_i \xi_i \partial/\partial x_i$ extends canonically to a vector field $\tilde{\xi} = \sum_i \pi_n^*(\xi_i)\partial/\partial x_i$ in $\theta_{n+d/d}$. We define

$$\tilde{L}_s = \{\sum_j \tilde{a}_j \tilde{\xi}_j : \xi_j \in L_s, \tilde{a}_j \in \mathscr{O}_{n+d}\}, \quad \tilde{L}_t = \{\sum_j \tilde{b}_j \tilde{\eta}_j : \eta_j \in L_t, \tilde{b}_j \in \mathscr{O}_{p+d}\}.$$

Morphisms $t_{\text{rel}}F : \tilde{L}_s \to \theta(F/d) = \mathscr{O}_{n+d}^{\times p}$ and $\omega_{\text{rel}}F : \tilde{L}_t \to \theta(F/d)$ are defined in the obvious way, as composition with dF and composition with F, respectively, as in Sect. 3.2 . Then we also assume:

(G2) If $\tilde{\xi} \in \tilde{L}_s$ and $\tilde{\eta} \in \tilde{L}_t$, and $\Phi(x,u) = (\varphi_u(x), u)$ and $\Psi(y,u) = (\psi_u(y), u)$ are their integral flows, then for all u the pair (φ_u, ψ_u) lies in $\mathscr{H}$.

Note that since $L_s \subset \mathfrak{m}_n\theta_n$ and $L_t \subset \mathfrak{m}_p\theta_p$, $\varphi_u(0) = 0$ and $\psi_u(0) = 0$ for all u.

Proposition 6.2 *Suppose that $\mathscr{H}$ is a subgroup of $\mathscr{A}$ with properties (G1) and (G2), and that $\mathfrak{m}_n^{k+1}\mathscr{O}_n^{\times p} \subset T\mathscr{H}f$. Then f is locally k-determined for $\mathscr{H}$. That is, if $g \in \mathfrak{m}_n^{k+1}\mathscr{O}_n^{\times p}$ and F is the germ of 1-parameter unfolding of f defined by $F(x,u) = (f(x) + ug(x), u)$, then F is $\mathscr{H}$-trivial.*

Proof We show first that

$$M_n^{k+1}\mathscr{O}_{n+1}^{\times p} \subset T\mathscr{H}(F/1). \tag{6.7}$$

Elements of $M_n^{k+1}\mathscr{O}_{n+1}^{\times p}$ are linear combinations of terms of the form $x^\alpha m(x,u)$, where $|\alpha| = k+1$ and $m \in \mathscr{O}_{n+1}^{\times p}$. By hypothesis, for each one there exist $\xi \in L_s$ and $\eta \in L_p$ such that

$$x^\alpha m(x,0) = tf(\xi) + \omega f(\eta).$$

As before, denote by $\tilde{\xi}$ and $\tilde{\eta}$ the extensions of ξ and η to $(\mathbb{F}^n \times \mathbb{F}, 0)$ and $(\mathbb{F}^p \times \mathbb{F}, 0)$. Then

$$t_{\text{rel}}F(\tilde{\xi}) = \sum_{i,j}\left(\xi_i \frac{\partial f_j}{\partial x_i}\frac{\partial}{\partial y_j} + u\xi_i \frac{\partial g_j}{\partial x_i}\frac{\partial}{\partial y_j}\right). \tag{6.8}$$

Because $g_j \in \mathfrak{m}_n^{k+1}$ for $j = 1, \ldots, p$, $\partial g_j/\partial x_i \in \mathfrak{m}_n^k$; since $\mathscr{H} \subset \mathscr{A}$, $\xi \in \mathfrak{m}_n\theta_n$ and $\tilde{\xi}_i \in M_n$ for $i = 1, \ldots, n$. Therefore the second summand on the right in (6.8) lies in $u \cdot M_n^{k+1}\mathscr{O}_{n+1}^{\times p} \subset \mathfrak{m}_{p+1}M_n^{k+1}\mathscr{O}_{n+1}^{\times p}$. Similarly

$$\omega_{\text{rel}}F(\tilde{\eta}) = \omega f(\eta) + \text{terms involving } ug_j$$

and once again the second summand on the right belongs to $\mathfrak{m}_{p+1}M_n^{k+1}\mathscr{O}_{n+1}^{\times p}$. Moreover

$$x^\alpha m(x,u) = x^\alpha m(x,0) + x^\alpha(m(x,u) - m(x,0)),$$

and here again the second summand on the right belongs to $\mathfrak{m}_{p+1}M_n^{k+1}\mathscr{O}_n^{\times p}$. We conclude that

$$x^\alpha m(x,u) \in T\mathscr{H}(F/1) + \mathfrak{m}_{p+1}M_n^{k+1}\mathscr{O}_{n+1}^{\times p}.$$

By linearity it follows that

$$M_n^{k+1}\mathscr{O}_{n+1}^{\times p} \subset T\mathscr{H}(F/1) + \mathfrak{m}_{p+1}M_n^{k+1}\mathscr{O}_{n+1}^{\times p}. \tag{6.9}$$

The proof concludes by using Nakayama's lemma, after the standard step to ensure finiteness: taking

$$N_1 = \frac{M_n^{k+1}\mathscr{O}_{n+1}^{\times p} + t_{\text{rel}}F(\tilde{L}_s)}{t_{\text{rel}}F(\tilde{L}_s)}, \quad N_2 = \frac{T\mathscr{H}(F/1)}{t_{\text{rel}}F(\tilde{L}_s)}$$

from (6.9) we see that

$$N_1 \subset N_2 + \mathfrak{m}_{p+1}N_1. \tag{6.10}$$

To conclude from Nakayama's lemma that $N_1 \subset N_2$, we have just to show that N_1 is a finitely generated $\mathscr{O}_{p+1}$-module. We apply the preparation theorem, using the fact that N_1 is a finitely generated $\mathscr{O}_{n+1}$-module. We have

$$\frac{N_1}{\mathfrak{m}_{p+1}N_1} = \frac{M_n^{k+1}\mathscr{O}_{n+1}^{\times p} + t_{\text{rel}}F(\tilde{L}_s)}{\mathfrak{m}_{p+1}M_n^{k+1}\mathscr{O}_{n+1}^{\times p} + t_{\text{rel}}F(\tilde{L}_s)} \simeq \frac{\mathfrak{m}_n^{k+1}\mathscr{O}_n^{\times p} + tf(L_s)}{\mathfrak{m}_p\mathfrak{m}_n^{k+1}\mathscr{O}_n^{\times p} + tf(L_s)}.$$

By (G1), the right-hand term is a quotient of

$$\frac{\mathfrak{m}_n^{k+1}\mathscr{O}_n^{\times p} + tf(\mathfrak{m}_n^r\theta_n)}{\mathfrak{m}_p\mathfrak{m}_n^{k+1}\mathscr{O}_n^{\times p} + tf(\mathfrak{m}_n^r\theta_n)}.$$

which is a subset of

$$\frac{\mathscr{O}_n^{\times p}}{\mathfrak{m}_p\mathfrak{m}_n^{k+1}\mathscr{O}_n^{\times p} + tf(\mathfrak{m}_n^r\theta_n)}$$

which is in turn a quotient of

$$\frac{\mathscr{O}_n^{\times p}}{\mathfrak{m}_n^{k+r+1}\left(\mathfrak{m}_p\mathscr{O}_n^{\times p} + tf(\theta_n)\right)}.$$

As $\mathscr{H}$-finiteness implies $\mathscr{A}$-finiteness implies $\mathscr{K}$-finiteness, this final quotient is finite-dimensional as $\mathbb{F}$-vector space, so by the preparation theorem, N_1 is finitely generated over $\mathscr{O}_{p+1}$, and therefore from (6.10) we conclude, by Nakayama's lemma, that $N_1 \subset N_2$. This, in turn, implies that $M_n^{k+1}\mathscr{O}_{n+1}^{\times p} \subset T\mathscr{H}(F/1)$, as required.

Since $g \in M_n^{k+1}\mathscr{O}_{n+1}^{\times p}$, by (6.7) there exist germs of vector fields $\tilde{\xi} \in M_n\theta(\pi_n)$ and $\tilde{\eta} \in M_p\theta(\pi_p)$ such that $h = t_{\text{rel}}F(\tilde{\xi}) + \omega_{\text{rel}}F(\tilde{\eta})$. By (3.54), this is equivalent to $dF \circ \xi = \eta \circ F$, where $\xi = \tilde{\xi} + \partial/\partial u$ and $\eta = \tilde{\eta} + \partial/\partial u$, which implies that F is $\mathscr{A}$—trivial by Lemma 3.4. By property (G2), F is in fact $\mathscr{H}$-trivial. □

We will also need the following generalisation of Theorem 6.2.

Theorem 6.6 *Let $\mathscr{H}$ be a subgroup of $\mathscr{A}$ satisfying (G1) and (G2), and suppose that $T\mathscr{H}f \supset \mathfrak{m}_n^{k+1}\theta(f)$. Then f is $2k+1$-determined for $\mathscr{H}$.*

Proof The proof is an easy modification of the proof of the corresponding result for $\mathscr{A}$, and follows the same steps. To prove the appropriate version of Lemma 6.5, with $\mathscr{H}$ in place of $\mathscr{A}$, it is necessary only to replace $\mathscr{R}$ by $\mathscr{R}_{r-1}$, where the r is the r of property (G1). □

The term "unipotent" refers to linear maps of finite-dimensional spaces and to affine algebraic groups. We will say that a subgroup $\mathscr{H}$ of one of the standard groups $\mathscr{G}$ of diffeomorphisms is unipotent if $j^k\mathscr{H}$ is a unipotent algebraic subgroup of $J^k\mathscr{G}$. The necessary and sufficient criterion for unipotency of $j^k\mathscr{H}$ (proved as [BdPW87, Proposition 1.8]—see Proposition 6.4 below) shows that this holds for all k if and only if it holds when $k = 1$.

Theorem 6.7 *Let $\mathscr{H}$ be a unipotent subgroup of $\mathscr{A}$ satisfying (G1) and (G2). Then f is k-determined for $\mathscr{H}$ if and only if $T\mathscr{H}f \supset \mathfrak{m}_n^{k+1}\theta(f)$.*

Proof "Only if" is proved as before. Suppose, conversely, that $T\mathscr{A}_s f \supset \mathfrak{m}_n^{k+1}\mathscr{O}_n^{\times p}$. Then f is (globally) $2k+1$-determined for $\mathscr{H}$, by Theorem 6.6, and locally k-determined for $\mathscr{A}_s$ by Proposition 6.2. Because $\mathscr{H}$ is unipotent, the difference between the two estimates can be removed by applying Theorem 6.5 in the jet space $J^{2k+1}(n, p)$: if $j^k g = j^k f$ then $j^{2k+1}g$ is $\mathscr{H}_s^{(2k+1)}$-equivalent to $j^{2k+1}f$ by the argument sketched in the paragraphs after Theorem 6.5, and g is therefore $\mathscr{H}$-equivalent to f by the $2k+1$-determinacy of f for $\mathscr{H}$. □

6.3.1 Unipotent Affine Algebraic Groups

Our account here is a summary of Springer's account in [Spr09, Chapter 2]. Each of the groups $\mathscr{G}^{(s)}$ of jets of diffeomorphisms acting on jet space is an affine variety for which the group operations of product $\mathscr{G}^{(s)} \times \mathscr{G}^{(s)} \to \mathscr{G}^{(s)}$ and inverse $\mathscr{G}^{(s)} \to \mathscr{G}^{(s)}$ are morphisms of varieties. Such objects are known as *affine algebraic groups* or as *linear algebraic groups*. The second name arises because every affine algebraic group G over $\mathbb{F}$ acts linearly and faithfully by right translation, $g \cdot f(g_1) = f(g_1 g)$, on its own coordinate ring $\mathbb{F}[G]$, the ring of polynomial functions on G. Indeed, if G

is embedded in affine space $\mathbb{A}^N_{\mathbb{F}}$, then the restriction of this action to the coordinate functions, $\mathbb{F}[G]_1$, is already faithful. Thus G is isomorphic to an algebraic subgroup of $\mathrm{Gl}(\mathbb{F}[G]_1)$.

Let V be a finite-dimensional $\mathbb{F}$-vector space. For the purposes of the next definitions it is convenient to embed $\mathrm{Gl}(V)$ into $\mathrm{End}(V)$. An element $A \in \mathrm{End}(V)$ is *semisimple* if it is diagonalisable, *nilpotent* if all of its eigenvalues are 0, or, equivalently, if $A^n = 0$ for some $n \geq 1$, and *unipotent* if all of its eigenvalues are equal to 1, or, equivalently, $(A - I)^n = 0$ for some n. The (additive) *Jordan decomposition* taught in every undergraduate mathematics curriculum expresses every $A \in \mathrm{End}(V)$ uniquely as a sum $A = A_s + A_n$ where A_s is semisimple, A_n is nilpotent and $A_n A_s = A_s A_n$. The less familiar *multiplicative Jordan decomposition* of an element of $\mathrm{Gl}(V)$ may be obtained by setting $A_u = 1 + A_s^{-1} A_n$; then $A = A_u A_s = A_s A_u$, A_u is unipotent and A_u and A_s are the unique semisimple and unipotent elements of $\mathrm{End}(V)$ with this property. Note that in the Jordan decomposition A_s has the same eigenvalues as A, hence A_s is invertible if A is.

Every upper uni-triangular matrix A is unipotent. Conversely, it is not hard to show (see e.g. [Spr09, 2.4.12] that every subgroup of $\mathrm{Gl}(V)$ consisting of unipotent elements is a subgroup of the group U_n of upper uni-triangular matrices with respect to a suitable basis. From this it follows that if $G \subset \mathrm{Gl}(V)$ is a group of unipotent matrices, then there is a non-zero element fixed by all of the members of G—in the case of U_n, the first basis vector. Surprisingly, as we will see below, it is this last property that gives Zariski-closedness of the orbits. Note that we can replace "upper unitriangular" by "lower unitriangular" in all of this discussion, since reversing the order of the vectors in a basis interchanges the two types of matrix.

To define unipotent *groups*, we need to penetrate a little further in the theory of algebraic groups. An endomorphism A of a (not necessarily finite-dimensional) vector space V is *locally finite* if V is a union of finite-dimensional A-stable subspaces. In this case, A is *semisimple* if its restriction to each finite-dimensional A-stable subspace is semisimple. If A is locally finite there is once again an additive Jordan decomposition: to define $A_s x$ and $A_n x$ for $x \in V$, pick a finite-dimensional subspace W containing x and set $A_s x = (A|_W)_s x$, $A_n x = (A|_W)_n x$. This is independent of the choice of W [Spr09, 2.4.4]. This additive Jordan decomposition gives rise to a multiplicative Jordan decomposition on each finite-dimensional A-stable subspace, and thus to a multiplicative Jordan decomposition $A = A_s A_u$ on V itself. Here A_u is locally unipotent—unipotent on finite-dimensional A-stable subspaces.

As remarked above, an affine algebraic group G acts linearly on its coordinate ring $\mathbb{F}[G]$. Each element $g \in G$ acts faithfully and locally finitely by right translation on $\mathbb{F}[G]$, and therefore *with respect to this action* g has a multiplicative Jordan decomposition $g = g_u g_s$. In fact this decomposition is the same for any other faithful representation:

Proposition 6.3 ([Spr09, Theorem 2.4.8]) *If $\varphi : G_1 \to G_2$ is a morphism of linear algebraic groups then $\varphi(g_s) = \varphi(g)_s$ and $\varphi(g_u) = \varphi(g)_u$.*

Definition 6.2 An element g of a linear algebraic group G is *unipotent* (respectively, *semisimple*) if $g = g_u$ (respectively, $g = g_s$) in the multiplicative Jordan decomposition just described. A linear algebraic group is *unipotent* (respectively, *semisimple*) if every element is unipotent (semisimple).

A particular case of Proposition 6.3 is the following:

Corollary 6.5 *Let G be a unipotent linear algebraic group and $\rho : G \to Gl(V)$ a rational representation. Then all members of $\rho(G)$ are unipotent.*

Theorem 6.8 (Kostant–Rosenlicht) *Let G be a unipotent linear algebraic group over an algebraically closed field and X an affine variety on which G acts algebraically. Then all orbits of G are Zariski-closed.*

Proof Let $O = G \cdot x$ be an orbit and let $\bar{O}$ be its Zariski closure. This too is an affine variety. Replace X by $\bar{O}$. The G-action on x determines an algebraic morphism $G \to \bar{O}$. The image, Z, of any morphism contains a non-empty subset open in the closure of Z [Spr09, Theorem 1.9.5]. So O contains a non-empty subset, say A, which is Zariski-open in $\bar{O}$. By the transitivity of G on O, every point in O lies in such a subset, and so O is open in $\bar{O}$. Evidently O is dense in $\bar{O}$. Let Y be the complement of O in $\bar{O}$, and I_Y the ideal in $\mathbb{F}[\bar{O}]$ of functions vanishing on Y. The action of G on $\bar{O}$ gives rise to a locally finite action on I_Y [Spr09, Proposition 2.3.6]. Because G is unipotent, there is a non-zero $f \in I_Y$ which is fixed by this action. Since the action is transitive on O, f is constant on O. As O is dense in $\bar{O}$, f is constant on $\bar{O}$. But this means that I_Y contains a non-zero constant function, and thus is equal to $\mathbb{F}[\bar{O}]$, so that $Y = \emptyset$ and $O = \bar{O}$. □

In [BdPW87, Proposition 3.3], Zariski-closedness is proved in the real case also. The proof involves a deeper study of algebraic groups, and we omit it here.

6.3.2 Unipotent Groups of k-Jets of Diffeomorphisms

Fortunately, unipotent groups of k-jets of diffeomorphisms are plentiful, and easy to recognise.

Proposition 6.4 ([BdPW87, Prop.1.8]) *Let $\mathscr{H}$ be a Zariski-closed subgroup of $\mathscr{K}^{(k)}$ for some $k \geq 1$. Then $\mathscr{H}$ is a unipotent affine algebraic group if and only if $\mathscr{H}^{(1)}$ is unipotent.*

Here $\mathscr{H}^{(1)}$ is the image of $\mathscr{H}$ under the standard projection of $\mathscr{K}^{(k)}$ to $\mathscr{K}^{(1)} = \mathrm{Gl}(n) \times \mathrm{Gl}(p)$. If we regard each element of $h \in \mathscr{H}$ as a germ of polynomial diffeomorphism of $\mathbb{F}^n \times \mathbb{F}^p$, then j^1h is its linear part.

Proof Necessity follows from Proposition 6.5. To prove sufficiency, note that $\mathscr{H}$ acts faithfully by right translation on the space $\mathbb{F}[\mathbb{F}^n \times \mathbb{F}^p]_k$ of polynomial functions on $\mathbb{F}^n \times \mathbb{F}^p$ of degree $\leq k$ (this is not completely obvious, since this is not the same as the usual action of $\mathscr{K}$), so by Proposition 6.3, it is enough to show that each element $h \in \mathscr{H}$ acts unipotently on $\mathbb{F}[\mathbb{F}^n \times \mathbb{F}^p]_k$. Choose a basis $x_1, \ldots, x_{n+p}$ of $(\mathbb{F}^n \times \mathbb{F}^p)^*$ with respect to which the action of j^1h is lower unitriangular. That is, for each j,

$$j^1h \cdot x_j = x_j + \text{scalar multiples of later variables.}$$

Now $h = j^1h +$ higher order terms, so for each monomial x^α,

$$h \cdot x^\alpha = (x_1 + \text{scalar multiples of later variables})^{\alpha_1} \cdots (x_{n+p})^{\alpha_{n+p}}$$
$$+ \text{higher-order monomials.}$$

The monomials of degree $\leq k$ form a basis for $\mathbb{F}[\mathbb{F}^n \times \mathbb{F}^p]_k$; order them by setting

$$x^\alpha \prec x^\beta \text{ if } \begin{cases} |\alpha| < |\beta|, \text{ or} \\ |\alpha| = |\beta| \text{ and } \alpha_1 = \beta_1, \cdots, \alpha_{i-1} = \beta_{i-1}, \alpha_i < \beta_i, \text{ for some } i. \end{cases} \tag{6.11}$$

Then with respect to this basis, the matrix of the action of h on $\mathbb{F}[\mathbb{F}^n \times \mathbb{F}^p]_k$ is also lower unitriangular: it follows from (6.11) that for each monomial x^α,

$$h \cdot x^\alpha = x^\alpha + \text{scalar multiples of later monomials.}$$

□

By Proposition 6.4, the following theorem of [BdPW87] now follows, for $\mathscr{G} = \mathscr{A}$, from Theorem 6.7.

Theorem 6.9 ([BdPW87]) *Let $f : (\mathbb{F}^n, S) \to (\mathbb{F}^p, 0)$. For each of the groups $\mathscr{G} = \mathscr{R}, \mathscr{K}, \mathscr{C}, \mathscr{A}, \mathscr{L}$, and $s \geq 1$, f is k-determined for $\mathscr{G}_s$ if and only if $T\mathscr{G}_s f \supset \mathfrak{m}_n^{k+1}\mathscr{O}_n^{\times p}$.*

Example 6.3 ([BdPW87, Example 2.4(i)]) The function germ $f_1(x, y) = x^3 + xy^3$ satisfies

$$T\mathscr{R}_1 f + \mathrm{Sp}_{\mathbb{F}}(x\partial f/\partial y) \supset \mathfrak{m}_2^4\mathscr{O}_2. \tag{6.12}$$

As remarked in Example 6.1, by Theorem 6.2, f is 5-determined for $\mathscr{R}$. The left-hand side of (6.12) is the tangent space $T\mathscr{U} f$ for the subgroup $\mathscr{U}$ of $\mathscr{R}$ obtained as the semi-direct product $\mathscr{R}_1 \ltimes U_2$, where U_2 is the subgroup of $\mathrm{Gl}_2 \subset \mathscr{R}$ consisting of 2×2 upper uni-triangular matrices. The group $\mathscr{U}$ has properties (G1) and (G2), and is unipotent, since $J^1\mathscr{U} = U_2$ is unipotent. It follows that f is 4-determined for $\mathscr{U}$, and therefore for $\mathscr{R}$—a result we obtained with considerably more effort in Example 6.1. ◊

Example 6.4 (Continuation of Example 6.2) We show that the germ $f_\ell(x, y) = (x, y^3, xy + y^{3\ell+2})$ is $3\ell+2$-determined for $\mathscr{A}$ equivalence, by proving the stronger result that it is $3\ell + 2$-determined for $\mathscr{A}_1$-equivalence. By Proposition 6.2 and Theorem 6.9, it is necessary to show only that

$$T\mathscr{A}_1 f \supset \mathfrak{m}_2^{3\ell+3}\theta(f). \tag{6.13}$$

We use part 3 of Theorem 4.2, taking $C = \mathfrak{m}^{3\ell+3}\theta(f)$, $s = 2$ and $N = T\mathscr{A}_1 f$. Clearly $f^*\mathfrak{m}_3\theta(f) \supset \mathfrak{m}_2^3\theta(f)$, so to apply Theorem 4.2 we need to show

$$T\mathscr{A}_1 f + f^*\mathfrak{m}_3\mathfrak{m}_2^{3\ell+3}\theta(f) + \mathfrak{m}_2^{3\ell+6}\theta(f) \supset \mathfrak{m}_2^{3\ell+3}\theta(f). \tag{6.14}$$

All monomials of the form $x^a y^b \partial_{u_i}$, with $i > 0$ and $i + j > 3\ell + 3$, lie in $f^*\mathfrak{m}_3\mathfrak{m}_2^{3\ell+3}\theta(f)$, and thus to show (6.14) we have to find only $y^{3\ell+5}\partial_{u_i}$, $y^{3\ell+4}\partial_{u_i}$, and $x^a y^b \partial_{u_i}$, for $i = 1, 2, 3$ and $a + b = 3\ell + 3$, in the left-hand side of (6.14). Working modulo $\mathfrak{m}_2^{3k+6}$, we have

$$\begin{aligned} y^{3k+5}\partial_{u_1} &= tf(y^{3k+5}\partial_x) \\ y^{3k+5}\partial_{u_2} &= tf(y^{3k+3}\partial_y) - \omega f(u_1 u_2^{k+1}\partial_{u_3}) \\ (3k+1)y^{3k+5}\partial_{u_3} &= tf(y^4\partial_y) - \omega f(3u_2^2\partial_{u_2} + u_2 u_3 \partial_{u_3}). \end{aligned}$$

Then working modulo $\mathfrak{m}_2^{3k+5}$,

$$\begin{aligned} y^{3k+4}\partial_{u_1} &= tf(y^{3k+4}\partial_x) \\ 3y^{3k+4}\partial_{u_2} &= tf(x^2\partial_x + y^{3k+2}\partial_y) + \omega f(-u_1^2\partial_{u_1} - u_1 u_3 \partial_{u_3}) \\ y^{3k+4}\partial_{u_3} &= tf(y^{3k+3}\partial_x) + \omega f(-u_2^{k+1}\partial_{u_1}). \end{aligned}$$

Finally, working modulo $\mathfrak{m}_2^{3k+4}$, all monomials of degree $3k + 3$ of the form $x^a y^b$ with $a \neq 1$ lie in $f^{-1}(\mathfrak{m}_3^2)$, so to complete the proof we need to find only the terms $xy^{3k+2}\partial_{u_i}$, for $i = 1, 2, 3$, in the left-hand side of (6.14). We have

$$\begin{aligned} xy^{3k+2}\partial_{u_1} &= tf(xy^{3k+2}\partial x) \\ 3xy^{3k+2}\partial_{u_2} &= tf(xy^{3k}\partial_y) + \omega f(-u_1^2 u_2^k) \\ xy^{3k+2}\partial_{u_3} &= tf(-x^2\partial_x) + \omega f(u_1^2\partial_{u_1} + u_1 u_3 \partial_{u_3}). \end{aligned}$$

This completes the proof. ◊

6.3.3 When Is a Closed Affine Space of Germs Contained in a $\mathscr{G}$-Orbit?

The paper [BdPW87] emphasises the ubiquity of unipotent groups. In particular, it shows that if, for suitable $\mathscr{G}$, a closed affine subspace V of jet space $J^r(n, p)$ is contained in the $\mathscr{G}^{(r)}$-orbit of $j^r f$, then there is a unipotent algebraic subgroup of $\mathscr{U} \subset \mathscr{G}^{(r)}$ such that $V \subset \mathscr{U} \cdot j^r f$. In particular,

Theorem 6.10 *Let $\mathscr{G}$ be a strongly Zariski-closed subgroup of one of $\mathscr{R}, \mathscr{C}, \mathscr{K}, \mathscr{L}$ or $\mathscr{A}$. Let $f\colon (\mathbb{F}^n, 0) \to (\mathbb{F}^p, 0)$ be a smooth map-germ. Then for any $r < \infty$, f is $r - \mathscr{G}$-determined if and only if there is a strongly closed unipotent subgroup $\mathscr{U} \subset \mathscr{G}$ for which*

$$\mathfrak{m}^{r+1}\mathscr{O}_n^{\times p} \subset T\mathscr{U} f.$$

Here $\mathscr{G}$ is said to be a *strongly* Z-closed subgroup of one of the classical groups (say $\mathscr{G}'$) if it is Zariski-closed and in addition $\mathscr{G}_s = \mathscr{G}'_s$ for some finite s. In Example 6.4, $\mathscr{G}$ is $\mathscr{A}$ itself and $\mathscr{U}$ is $\mathscr{A}_1$.

6.3.4 Complexification and Determinacy Degrees

One of the unexpected consequences of the results of [BdPW87] is the following.

Theorem 6.11 ([BdPW87, Theorem 5.8]) *Let $\mathscr{H}$ be a strongly Z-closed subgroup of one of the standard groups $\mathscr{R}, \mathscr{L}, \mathscr{K}, \mathscr{C}, \mathscr{A}$, and let f be a real analytic germ. Then f is k-determined for $\mathscr{H}$ if and only if $f_{\mathbb{C}}$ is k-determined for the complexification of $\mathscr{H}$.* □

6.3.5 Notes

1. Our statement of Proposition 6.4 differs slightly from [BdPW87, Proposition 1.8]. The version there assumes only that $\mathscr{H}$ is closed in $J^s\mathscr{K}$ in the Euclidean topology, and invokes an algebraisation theorem proved in [Hoc81, Theorem XVI 4.2], though it does this at the cost of the additional assumption that $\mathscr{H}$ is connected.
2. Jim Damon has extended many of the results of this and the preceding chapter to a very large class of subgroups of the classical groups $\mathscr{K}$ and $\mathscr{A}$, see [Dam84].
3. Since the paper [BdPW87] appeared, various authors have adapted its results to other contexts. We cite in particular [DGH11].

6.4 Complete Transversals

The method of complete transversals, used in conjunction with the determinacy results described in the previous sections, is probably the most efficient technique known for carrying out the classification of map-germs with respect to various subgroups of the Mather groups $\mathscr{R}, \mathscr{L}, \mathscr{C}, \mathscr{A}$ or $\mathscr{K}$. It was developed by Bruce, Kirk and du Plessis in [BKdP97], based on the PhD Thesis of Kirk [Kir93]. Kirk even wrote a computer package called `Transversal` [Kir96] to be used with the symbolic algebra system `Maple`.

The main idea is to consider a k-jet σ_k in $J^k(n, p)$ and then look at the $(k+1)$-jets $\sigma \in J^{k+1}(n, p)$ whose projection in $J^k(n, p)$ is equal to σ_k. A complete transversal is a subspace T of the space of homogeneous polynomials of degree $k+1$, transverse to the tangent space to the orbit of σ in $J^{k+1}(n, p)$. Any $(k + 1)$-jet whose k-jet is $\mathscr{A}^{(k)}$-equivalent to σ_k is itself $\mathscr{A}^{(k+1)}$-equivalent to $\sigma_k + h$, for some $h \in T$ (Lemma 6.6, below). We can iterate this process until we get an ℓ-jet σ_ℓ which is ℓ-determined, in which case σ_ℓ represents one of the orbits we are looking for. The following is the crucial result:

Proposition 6.5 *Let G be a Lie group with Lie algebra $\mathfrak{g}$ acting smoothly on an affine space A and let W be a subspace of V_A such that*

$$\mathfrak{g}(x + w) = \mathfrak{g}x, \tag{6.15}$$

for all $x \in A$ and $w \in W$. Then

1. *For any $x \in A$ we have*

$$x + \mathfrak{g}x \cap W \subset Gx \cap \{x + W\}.$$

2. *If $x_0 \in A$ and T is a vector subspace of W satisfying*

$$W \subset T + \mathfrak{g}x_0, \tag{6.16}$$

then for any $w \in W$ there exist $g \in G$ and $t \in T$ such that $g(x_0 + w) = x_0 + t$.

Proof The first part is a consequence of Mather's Lemma 6.3. The set $x + \mathfrak{g}x \cap W$ is an affine subspace of A, with tangent space $\mathfrak{g}x \cap W$. For each $x + w \in x + \mathfrak{g}x \cap W$,

$$T_{x+w}(G(x + w)) = \mathfrak{g}(x + w) = \mathfrak{g}x,$$

by (6.15). So, $T_{x+w}(G(x + w))$ has constant dimension along the points of $x + \mathfrak{g}x \cap W$ and also contains the tangent space to $x + \mathfrak{g}x \cap W$. By Mather's lemma,

$$x + \mathfrak{g}x \cap W \subset Gx,$$

which gives item 1. Item 2 now follows:

$$
\begin{aligned}
\bigcup_{t\in T} G(x_0+t) &\supset \bigcup_{t\in T}\{x_0+t+\mathfrak{g}(x_0+t)\cap W\} && \text{by item 1}\\
&= \bigcup_{t\in T}\{x_0+t+\mathfrak{g}x_0\cap W\} && \text{by (6.15)}\\
&= x_0+T+\mathfrak{g}x_0\cap W\\
&= x_0+(T+\mathfrak{g}x_0)\cap W && \text{since } T\subset W\\
&= x_0+W && \text{by (6.16).}
\end{aligned}
$$

□

For each k, let H^k be the vector subspace of $J^k(n,p)$ of homogeneous k-jets of degree k, that is,

$$H^k = \frac{\mathfrak{m}_n^k \mathscr{O}_n^{\times p}}{\mathfrak{m}_n^{k+1} \mathscr{O}_n^{\times p}}.$$

Let $\mathscr{G}$ be any one of Mather's groups $\mathscr{R}, \mathscr{L}, \mathscr{C}, \mathscr{A}$ or $\mathscr{K}$. A subgroup $\mathscr{H}$ of $\mathscr{G}$ is called a *Lie subgroup* of $\mathscr{G}_1$ if for each k the k-jet group $\mathscr{H}^{(k)}$ is a Lie subgroup of $\mathscr{G}_1^{(k)}$.

Proposition 6.6 *Let $\mathscr{H}$ be a Lie subgroup of $\mathscr{G}_1$ and let σ be a k-jet in $J^k(n,p)$. If $T \subset H^{k+1}$ is a vector subspace such that*

$$H^{k+1} \subset T + T\mathscr{H}^{(k+1)}\sigma, \tag{6.17}$$

then for each $(k+1)$-jet τ with $j^k\tau = \sigma$ there exists $t \in T$ such that τ is $\mathscr{H}^{(k+1)}$-equivalent to $\sigma + t$.

Proof We apply Proposition 6.5 with $A = J^{k+1}(n,p)$, $W = H^{k+1}$ and $G = \mathscr{H}^{(k+1)}$. Since $\mathscr{H}^{(k+1)} \subset \mathscr{G}_1^{(k+1)}$, we have $\varphi(h) = h$, for all $h \in H^{k+1}$ and $\varphi \in \mathscr{H}^{(k+1)}$. But this implies that $\eta(\sigma + h) = \eta\sigma$, for all $h \in H^{k+1}$ and $\eta \in \mathfrak{g}$, which gives condition (6.15). □

Definition 6.3 A subspace $T \subset H^{k+1}$ satisfying (6.17) is called a *complete transversal* for $\sigma \in J^k(n,p)$.

An example which illustrates very well the power and simplicity of the method of complete transversals is to apply it to the classification of functions $f\colon (\mathbb{F}^n,0) \to (\mathbb{F},0)$ under $\mathscr{R}$-equivalence. This is a classical topic which can be found in every textbook on catastrophe theory. We recall that the $\mathscr{R}_e$-codimension of f is

$$\operatorname{codim}_{\mathscr{R}_e} f = \dim_{\mathbb{F}} \frac{\mathscr{O}_n}{J_f}.$$

This number is also called the *Milnor number* of f and is usually denoted by $\mu(f)$. The first result is known as the splitting lemma.

Lemma 6.6 (Splitting Lemma) *Let $f\colon (\mathbb{F}^n, 0) \to (\mathbb{F}, 0)$ be such that $0 < \mu(f) < \infty$. If its Hessian matrix has corank c, then f is $\mathscr{R}$-equivalent to a function of the form*

$$g(x_1, \ldots, x_c) \pm x_{c+1}^2 \pm \cdots \pm x_n^2, \tag{6.18}$$

for some $g \in \mathfrak{m}_c^3$ such that $\mu(g) = \mu(f)$.

Proof By assumption, $j^2 f$ is equivalent, up to linear change of coordinates, to $\pm x_{c+1}^2 \pm \cdots \pm x_n^2$. We claim that for each $k \geq 2$, $j^k f$ is $\mathscr{R}^{(k)}$-equivalent to a k-jet of the form

$$\tau_k = \sigma_k \pm x_{c+1}^2 \pm \cdots \pm x_n^2,$$

for some $\sigma_k \in J^k(c, 1)$ such that $\sigma_k \in \mathfrak{m}_c^3$. We proceed by induction on k. The case $k = 2$ is clear, just take $\sigma_2 = 0$. We assume it is true for k and prove it for $k + 1$. We have that $T\mathscr{R}^{(k+1)}\tau_k$ is generated by the partial derivatives of σ_k and by the linear functions $x_{c+1}, \ldots, x_n$. Thus, a complete transversal for τ_k in $J^{k+1}(n, 1)$ is given by H_c^{k+1}, the subspace of $J^{k+1}(c, 1)$ of homogeneous polynomials of degree $k + 1$ in the variables $x_1, \ldots, x_c$. By Proposition 6.6, $j^{k+1} f$ will be $\mathscr{R}^{(k+1)}$-equivalent to

$$\tau_{k+1} = \sigma_k + h \pm x_{c+1}^2 \pm \cdots \pm x_n^2,$$

for some $h \in H_c^{k+1}$, which proves our claim. Since $\mu(f) < \infty$, f is finitely determined, hence $\mathscr{R}$-equivalent to $j^k f$, for some k big enough. We conclude that f is $\mathscr{R}$-equivalent to (6.18). The fact that $\mu(f) = \mu(g)$ follows easily:

$$\mu(f) = \dim_{\mathbb{F}} \frac{\mathscr{O}_n}{J_f} = \dim_{\mathbb{F}} \frac{\mathscr{O}_n}{(\frac{\partial g}{\partial x_1}, \ldots, \frac{\partial g}{\partial x_c}, x_{c+1}, \ldots, x_n)} = \dim_{\mathbb{F}} \frac{\mathscr{O}_c}{J_g} = \mu(g).$$

□

The following theorem includes the characterization of the A_k and D_k types and the classification of all singularities up to codimension 5. These were obtained by René Thom in his famous book [Tho76] and are known as the "seven elementary catastrophes".

Theorem 6.12 *Let $f\colon (\mathbb{F}^n, 0) \to (\mathbb{F}, 0)$ be such that $\mu(f) = k$ for some $2 \leq k < \infty$, and assume that its Hessian matrix has corank c.*

1. If $c = 1$, then f is $\mathscr{R}$-equivalent to the A_k-singularity:

$$\pm x_1^{k+1} \pm x_2^2 \pm \cdots \pm x_n^2.$$

2. *Suppose that $c = 2$ and f is given as in* (6.18). *If j^3g is a cubic form of elliptic, hyperbolic or parabolic type, then f is $\mathscr{R}$-equivalent to the $D_k^{\pm}$-singularity:*

$$x_1^2x_2 \pm x_2^{k-1} \pm x_3^2 \pm \cdots \pm x_n^2.$$

(In fact, D_k^+ and D_k^- are $\mathscr{R}$-equivalent if k is odd).

3. *If $\mu(f) \leq 5$, then f is $\mathscr{R}$-equivalent to one of the following seven types: A_2, A_3, A_4, A_5, D_4^+, D_4^- or D_5.*

Proof We assume from the beginning that f is given as in (6.18). Thus, we only need to classify g up to $\mathscr{R}$-equivalence. If $c = 1$ and $\mu(g) = k$, g is $\mathscr{R}$-equivalent to $\pm x_1^{k+1}$, which gives item 1.

To prove item 2, we assume $c = 2$ and look at the 3-jet j^3g. A non-zero homogeneous polynomial of degree 3 has three, two or one distinct complex roots, and those of each type form a single orbit under $\mathrm{Gl}_2(\mathbb{C})$. Over $\mathbb{R}$, the first of these breaks up into two (three real roots or one real and one complex conjugate pair of roots), so j^3g is equivalent, up to linear change of coordinates, to $x_1^2x_2 \pm x_2^3$, $x_1^2x_2$, x_1^3 or 0. That j^3g is elliptic/hyperbolic means that j^3g is equivalent to $x_1^2x_2 \pm x_2^3$. We get the $D_4^{\pm}$ singularity which is 3-determined and we are done.

When j^3g is parabolic, it is equivalent to $x_1^2x_2$. We consider $\sigma = x_1^2x_2$ as a k-jet in $J^k(2, 1)$, with $k \geq 3$. A complete transversal is $x_1^2x_2 + ax_2^{k+1}$, with $a \in \mathbb{F}$. If $a \neq 0$, we get $D_{k+2} = x^2y + y^{k+1}$ if k is odd or $D_{k+2}^{\pm} = x^2y \pm y^{k+1}$ if k is even. In both cases, the jet is $(k+1)$-determined, hence f is $\mathscr{R}$-equivalent to D_k or $D_k^{\pm}$, respectively. If $a = 0$, we get again $x_1^2x_2$ and proceed to the next jet space. Since f is $\mathscr{R}$-finite, this process must stop at some k and hence this gives all possible orbits when j^3g is parabolic.

Finally, we see item 3. If $c \geq 3$, then an easy computation shows that $\mu(g) \geq 8$. If $c = 2$ and $j^3g = 0$, we also have $\mu(g) \geq 9$. Assume that $c = 2$ and j^3g is equivalent to x_1^3. We have that J_g has colength k and is contained in $\mathfrak{m}_2^2$, so $\mathfrak{m}_2 J_g$ has colength $k+2$ and is contained in $\mathfrak{m}_2^3$. Then,

$$k+2 = \dim_{\mathbb{F}} \frac{\mathscr{O}_2}{\mathfrak{m}_2 J_g} = \dim_{\mathbb{F}} \frac{\mathscr{O}_2}{\mathfrak{m}_2^3} + \dim_{\mathbb{F}} \frac{\mathfrak{m}_2^3}{\mathfrak{m}_2 J_g} = 6 + \dim_{\mathbb{F}} \frac{\mathfrak{m}_2^3}{\mathfrak{m}_2 J_g}$$

$$\geq 6 + \dim_{\mathbb{F}} \frac{\mathfrak{m}_2^3}{\mathfrak{m}_2 J_g + \mathfrak{m}_2^4} = 6 + \dim_{\mathbb{F}} \frac{\mathfrak{m}_2^3}{(x_1^3, x_2^2x_2) + \mathfrak{m}_2^4} = 8,$$

which implies $\mu(g) = k \geq 6$. We have shown that if $\mu(g) \leq 5$ then either $c = 1$ or $c = 2$ and j^3g is equivalent to $x_1^2x \pm x_2^3$ or $x_1^2x_2$ and we can use items 1 and 2. □

Remark 6.1 In the complex case $\mathbb{F} = \mathbb{C}$, all the $\pm$ in the above normal forms can be converted into $+$. That is, the A_k-singularity is written as $x_1^{k+1} + x_2^2 + \cdots + x_n^2$ and the D_k-singularity (for any k) as $x_1^2x_2 + x_2^{k-1} + x_3^2 + \cdots + x_n^2$. ◊

Other easy applications of the method of complete transversals are left as exercises.

In [BKdP97], Proposition 6.6 is generalised by replacing $\mathscr{G}_1$ by any unipotent subgroup of $\mathscr{G}$, in which case one must replace the filtration $\cdots \subset J^k(n, p) \subset J^{k+1}(n, p) \subset \ldots$ by a filtration adapted to the group. Further generalisations allow one to use the filtration by weights associated to a weighting of the variables. In this latter case, the result substantially generalises Arnold's spectral sequence method.

The labels A_k and D_k of the singularity types of Theorem 6.12 were introduced by Arnold. In fact, Arnold extended in [Arn72] the list of Thom's seven elementary catastrophes to obtain a complete classification of all simple singularities of functions. This list is known as the ADE-classification. It says that any simple function germ $f: (\mathbb{F}^n, 0) \to (\mathbb{F}, 0)$ is $\mathscr{R}$-equivalent to A_k, D_k or one of the three exceptional germs E_6, E_7 or E_8. The ADE terminology is due to the connection with the classification of simple Lie algebras (see [Arn72] for details).

The classification of simple singularities of mappings $f: (\mathbb{F}^n, 0) \to (\mathbb{F}^p, 0)$ was an active field of research, especially in the 1980s and 1990s. This type of theorem is attractive since it provides a classification which is complete and is "solvable in finite time" (using the terminology of computer science). The lists obtained in these classifications typically include some series of infinite singularity types together with some exceptional germs, as happens in the Arnold's ADE-classification. Here we include a list of some references (in chronological order) where the reader can find complete classifications of simple germs under $\mathscr{A}$-equivalence for a given pair of dimensions (n, p):

1. Bruce and Gaffney [BG82], $(n, p) = (1, 2)$ (simple mono-germs);
2. Goryunov [Gor84], (simple singularities of mono-germs of projections);
3. Mond [Mon85], $(n, p) = (2, 3)$ (simple mono-germs);
4. Rieger [Rie87], $(n, p) = (2, 2)$ (simple mono-germs);
5. Rieger and Ruas [RR91], $p = 2$ and $n \geq 2$ (simple mono-germs);
6. Gibson and Hobbs [GH93], $(n, p) = (1, 3)$ (simple mono-germs);
7. Marar and Tari [MT96], $(n, p) = (3, 3)$ (simple mono-germs);
8. Houston and Kirk [HK99], $(n, p) = (3, 4)$ (simple mono-germs);
9. Wik-Atique [Ati00], $(n, p) = (2, 3)$ (simple multi-germs);
10. Zhitomirski [Zhi08], $n = 1$, $p = 2, 3$ (fully simple multi-germs).

Other notable classifications for $\mathscr{K}$-equivalence are

1. simple singularities of mappings $f: (\mathbb{F}^n, 0) \to (\mathbb{F}^p, 0)$ under $\mathscr{K}$-equivalence for all pairs (n, p) with $n \geq p$ due to Giusti [Giu83], which extends Arnold's ADE-classification;
2. simple singularities of Cohen–Macaulay codimension 2 varieties by Frühbis-Krüger [FK99] and Frühbis-Krüger and Neumer [FKN10]. Here an adapted version of $\mathscr{K}$-equivalence for matrix singularities is used.

Exercises for Sect. 6.4

1 Continue the classification of germs of functions given in Theorem 6.12 up to (i) $\mu \leq 6$; (ii) $\mu \leq 7$; (iii) $\mu \leq 8$. This is easier than you might guess!

2 Let $f\colon (\mathbb{F}^2, 0) \to (\mathbb{F}^2, 0)$ be $\mathscr{A}$-finite with $j^3 f$ equivalent to (x, y^3). Show that f is $\mathscr{A}$-equivalent to $(x, y^3 \pm x^k y)$ for some $k \geq 2$.

3 Let $f\colon (\mathbb{F}^2, 0) \to (\mathbb{F}^3, 0)$ be $\mathscr{A}$-finite with $j^2 f$ equivalent to $(x, y^2, 0)$. Show

(i) If $j^3 f \sim (x, y^2, y^3)$, then f is $\mathscr{A}$-equivalent to the $S_k^{\pm}$-singularity $(x, y^2, y(y^2 \pm x^k))$, for some $k \geq 2$.
(ii) If $j^3 f \sim (x, y^2, x^2 y)$, then f is $\mathscr{A}$-equivalent to the $B_k^{\pm}$-singularity $(x, y^2, y(x^2 \pm y^{2k}))$, for some $k \geq 2$.

(These two types appear in the classification paper [Mon85].)

6.5 Notes and Further Developments

The k-determinacy of a germ f is proved by showing the analytical triviality of the linear interpolation $f_t := f + tg$ between f and $f + g$, where g has order $k + 1$. A great deal of work has been and continues to be done, to give criteria for the triviality of these and other parameterised families f_t with respect to the weaker relations of topological triviality and Whitney equisingularity. We cite in particular [LêR76, LêT83, Tei82], and more recently the work of Terry Gaffney and his collaborators, usefully surveyed in [Gaf18].

We have not touched on the question of how general is finite determinacy. This is discussed in C.T.C.Wall's paper [Wal81]. A property of germs f which can be characterised by $j^k f$ lying in the complement of an algebraic subset A_k of the jet space $J^k(n, p)$ for all k is said to hold *in general* if the codimension of A_k in $J^k(n, p)$ tends to infinity with k. Whereas $\mathscr{K}$-finiteness holds in general in all dimensions, $\mathscr{A}$-finiteness does not hold in general even in Mather's nice dimensions. Instead, there is a smaller range of dimensions, which can be calculated by a procedure very like Mather's calculation of the nice dimensions, within which $\mathscr{A}$-finiteness holds in general. This theorem is attributed by Wall in [Wal81] to A. du Plessis. In [dP80] du Plessis gave a detailed version of a sketched proof of R. Thom that finite determinacy for topological left-right equivalence holds in general in all dimensions.

Another aspect of finite determinacy considers finite determinacy within subclasses of germs. An attempt to account for the phenomenon of series of singularities led in [Mon87] to the notion of *stem*: a non-finitely determined germ f is a stem if it is finitely determined within the class of non-finitely determined germs—see [Pel89]. For singularities of varieties, Möhring and van Straten showed that for a germ $X = V(J)$ of variety with arbitrary singularities, if the singular locus of X

contains the variety of zeros of an ideal I, then for sufficiently high k, $V(J + I^k)$ is isomorphic to X. We do not know of any extension of this result to determinacy for $\mathscr{A}$-equivalence of mappings.

Finally, in the monograph [Dam84], Jim Damon considers questions of finite determinacy and versality with respect to a very wide class of subgroups, the so-called *geometric subgroups*, of the standard groups $\mathscr{K}$ and $\mathscr{A}$; he continues this for topological equivalence in [Dam88].

Chapter 7
Classification of Stable Germs by Their Local Algebras

This chapter contains four significant results, all due to John Mather. The first (Theorem 7.1 below) is that *stable* map-germs are $\mathscr{A}$-equivalent to one another if and only if they are $\mathscr{K}$-equivalent. Together with Theorem 4.3, that germs are $\mathscr{K}$-equivalent if and only if their fibres are isomorphic, this gives an easy algebraic criterion for the $\mathscr{A}$-equivalence of stable germs. In the second section of the chapter, we justify the method given in Sect. 4.1 to construct stable map-germs, as unfoldings of rank 0 germs. The third section studies the isosingular loci of stable multi-germs, and shows that any stable germ is left-right equivalent to a trivial unfolding, with the isosingular locus as parameter space. The final section distils, from Mather's paper [Mat71], a proof of the "folklore" fact that in his nice dimensions, every stable germ is weighted homogeneous with respect to suitable coordinates. This fact plays a surprising role in the proof of the "Milnor–Tjurina relation" between the discriminant Milnor number and the $\mathscr{A}_e$-codimension of map-germs in the nice dimensions—see Theorems 8.1 and 8.2 below. In the process, we review some of the arguments from [Mat71].

7.1 Stable Germs Are Classified by Their Local Algebras

Given any smooth map-germ $f\colon (\mathbb{F}^n, S) \to (\mathbb{F}^p, 0)$ we always have the inclusion $T\mathscr{A}f \subset T\mathscr{K}f$. Our first result here is that this inclusion is in fact an equality when f is stable.

Proposition 7.1 *If* $f\colon (\mathbb{F}^n, S) \to (\mathbb{F}^p, 0)$ *is stable, then* $T\mathscr{A}f = T\mathscr{K}f$.

Proof To show that

$$tf(\mathfrak{m}_n\theta_n) + \omega f(\mathfrak{m}_p\theta_p) = tf(\mathfrak{m}_n\theta_n) + f^*\mathfrak{m}_p\theta(f),$$

D. Mond, J. J. Nuño-Ballesteros, *Singularities of Mappings*,
Grundlehren der mathematischen Wissenschaften 357,
https://doi.org/10.1007/978-3-030-34440-5_7

it is necessary only to show that $f^*\mathfrak{m}_p\theta(f)$ is contained in the left-hand side of this equality. This is easy: because f is stable,

$$\begin{aligned} f^*\mathfrak{m}_p\theta(f) = f^*\mathfrak{m}_p(tf(\theta_n) + \omega f(\theta_p)) &= tf(\mathfrak{m}_p\theta_n) + \omega f(\mathfrak{m}_p\theta_p) \\ &\subset tf(\mathfrak{m}_n\theta_n) + \omega f(\mathfrak{m}_p\theta_p). \end{aligned}$$

□

Theorem 7.1 ([Mat69b]) *Stable map-germs $(\mathbb{F}^n, S) \to (\mathbb{F}^p, 0)$ are $\mathscr{A}$-equivalent if and only if they are $\mathscr{K}$-equivalent, and thus are classified by the isomorphism classes of their algebras.*

Proof for $\mathbb{F} = \mathbb{C}$ The $(p+1)$-determinacy of stable germs (see Corollary 6.4) reduces questions about their classification under the group $\mathscr{A}$ to questions about the classification of jets in $J^{p+1}(n, p)$ under the action of $\mathscr{A}^{(p+1)}$. Let $\mathrm{St}^{(p+1)}$ be the subset of $J^{p+1}(n, p)$ of $(p+1)$-jets of stable germs $(\mathbb{F}^n, S) \to (\mathbb{F}^p, 0)$, which is well defined by Corollary 4.1. Then, it suffices to prove that for any $\sigma \in \mathrm{St}^{(p+1)}$,

$$\mathscr{A}^{(p+1)}\sigma = \mathscr{K}^{(p+1)}\sigma \cap \mathrm{St}^{(p+1)}. \tag{7.1}$$

The inclusion of $\mathscr{A}^{(p+1)}\sigma$ in $\mathscr{K}^{(p+1)}\sigma \cap \mathrm{St}^{(p+1)}$ is obvious. From Lemma 4.5 and Proposition 7.1 we know that $\mathscr{A}^{(p+1)}\sigma$ and $\mathscr{K}^{(p+1)}\sigma$ are both submanifolds of $J^{p+1}(n, p)$ such that

$$T_\sigma(\mathscr{A}^{(p+1)}\sigma) = T_\sigma(\mathscr{K}^{(p+1)}\sigma).$$

Thus, $\mathscr{A}^{(p+1)}\sigma$ is open in $\mathscr{K}^{(p+1)}\sigma$ and hence also open in $\mathscr{K}^{(p+1)}\sigma \cap \mathrm{St}^{(p+1)}$. But $\mathscr{K}^{(p+1)}\sigma \cap \mathrm{St}^{(p+1)}$ is the union of the $\mathscr{A}^{(p+1)}$-orbits of its members. Each is open in $\mathscr{K}^{(p+1)}\sigma \cap \mathrm{St}^{(p+1)}$, so $\mathscr{A}^{(p+1)}\sigma$, as the complement of the union of the other orbits, is also closed. To conclude the proof, we only need to show that $\mathscr{K}^{(p+1)}\sigma \cap \mathrm{St}^{(p+1)}$ is connected.

The complex algebraic group $\mathscr{K}^{(p+1)}$ itself is connected, and hence its continuous image $\mathscr{K}^{(p+1)}\sigma$ is also connected. The complement of $\mathrm{St}^{(p+1)}$ in $J^{p+1}(n, p)$ is a closed algebraic subset, and its intersection with the connected complex manifold $\mathscr{K}^{(p+1)}\sigma$ is therefore a closed complex algebraic subset of $\mathscr{K}^{(p+1)}\sigma$. It is a proper subset, since σ itself lies in $\mathrm{St}^{(p+1)}$. Hence it does not separate $\mathscr{K}^{(p+1)}\sigma$. We have shown that $\mathscr{K}^{(p+1)}\sigma \cap \mathrm{St}^{(p+1)}$ is connected. □

The prof just given uses the fact that a proper complex analytic subset does not separate a connected complex manifold. The corresponding statement for real analytic spaces is of course false. Nevertheless, Mather proved Theorem 7.1 also for real C^∞ germs. The proof involves extra complications, since $\mathscr{K}^{(p+1)}z$ may no longer be connected and the complement of $\mathrm{St}^{(p+1)}$ may separate it further. Instead of Mather's original proof of Theorem 7.1 in [Mat69b], we give here a different proof, simplifying the arguments used in Martinet's book [Mar82]. The following lemma gives a criterion for an unfolding of a stable germ to be trivialised by diffeomorphisms which are origin-preserving.

Lemma 7.1 *Let F be a 1-parameter unfolding of a stable germ $f\colon (\mathbb{F}^n, S) \to (\mathbb{F}^p, 0)$. Write $F(x,t) = (f_t(x), t)$ and assume that*

$$\frac{\partial f_t}{\partial t} \in T\mathscr{K}(F/1). \tag{7.2}$$

Then:

1. *F is origin-preserving, that is, $f_t(S) = 0$, for all t small enough;*
2. *the germ f_t is $\mathscr{A}$-equivalent to f, for all t small enough.*

Proof By definition,

$$T\mathscr{K}(F/1) := M_n \left\{ \frac{\partial f_t}{\partial x_1}, \ldots, \frac{\partial f_t}{\partial x_n} \right\} + F^* M_p \mathscr{O}_{n+1} \left\{ \frac{\partial}{\partial y_1}, \ldots, \frac{\partial}{\partial y_p} \right\},$$

where $M_n = \mathfrak{m}_n \mathscr{O}_{n+1}$ and $M_p = \mathfrak{m}_p \mathscr{O}_{p+1}$. By (7.2), $\frac{\partial f_t}{\partial t}(s_i) = 0$ and hence $f_t(s_i) = f(s_i) = 0$, for all t and $s_i \in S$. This shows item 1.

To see item 2, we proceed as in the proof of the Finite Determinacy Theorem 6.2.(i). It is enough to show that there exist vector fields $\xi \in \theta_{n+1}$ and $\eta \in \theta_{n+1}$ of the form

$$\xi = \sum_{i=1}^{n} \xi_i \frac{\partial}{\partial x_i} + \frac{\partial}{\partial t},$$

$$\eta = \sum_{j=1}^{p} \eta_j \frac{\partial}{\partial y_j} + \frac{\partial}{\partial t},$$

with $\xi_i \in M_n$ and $\eta_j \in M_p$ and such that $dF \circ \xi = \eta \circ F$. A simple computation shows such vector fields ξ, η can be obtained if

$$\frac{\partial f_t}{\partial t} \in M_n \left\{ \frac{\partial f_t}{\partial x_1}, \ldots, \frac{\partial f_t}{\partial x_n} \right\} + M_p \left\{ \frac{\partial}{\partial y_1}, \ldots, \frac{\partial}{\partial y_p} \right\} = T\mathscr{A}(F/1).$$

Thus, we only need to show that $T\mathscr{A}(F/1) = T\mathscr{K}(F/1)$. Observe that this is nothing but the relative version of Proposition 7.1. As in Proposition 7.1, the inclusion $T\mathscr{A}(F/1) \subset T\mathscr{K}(F/1)$ is obvious and it suffices to prove the opposite inclusion.

Consider the relative module of F with respect to $\mathscr{R}_e$:

$$T^1_{\mathscr{R}_e}(F/1) := \frac{\theta(F/1)}{\mathscr{O}_{n+1}\left\{\frac{\partial f_t}{\partial x_1}, \ldots, \frac{\partial f_t}{\partial x_n}\right\}}.$$

We have

$$\frac{T^1_{\mathscr{R}_e}(F/1)}{t \cdot T^1_{\mathscr{R}_e}(F/1)} \cong T^1_{\mathscr{R}_e}(f) := \frac{\theta(f)}{\mathscr{O}_n\left\{\frac{\partial f}{\partial x_1}, \ldots, \frac{\partial f}{\partial x_n}\right\}}.$$

Since f is stable, this module is generated over $\mathscr{O}_p$ by the classes of $\frac{\partial}{\partial y_1}, \ldots, \frac{\partial}{\partial y_p}$. It follows by Corollary D.3 that $T^1_{\mathscr{R}_e}(F/1)$ is generated over $\mathscr{O}_{p+1}$ by the classes of $\frac{\partial}{\partial y_1}, \ldots, \frac{\partial}{\partial y_p}$. This means that

$$T\mathscr{A}_e(F/1) := \mathscr{O}_{n+1}\left\{\frac{\partial f_t}{\partial x_1}, \ldots, \frac{\partial f_t}{\partial x_n}\right\} + \mathscr{O}_{p+1}\left\{\frac{\partial}{\partial y_1}, \ldots, \frac{\partial}{\partial y_p}\right\} = \theta(F/1).$$

Now we proceed as in the proof of Proposition 7.1. We have

$$\begin{aligned}
F^*M_p\theta(F/1) &= F^*M_p\left(\mathscr{O}_{n+1}\left\{\frac{\partial f_t}{\partial x_1}, \ldots, \frac{\partial f_t}{\partial x_n}\right\} + \mathscr{O}_{p+1}\left\{\frac{\partial}{\partial y_1}, \ldots, \frac{\partial}{\partial y_p}\right\}\right)\\
&= F^*M_p\left\{\frac{\partial f_t}{\partial x_1}, \ldots, \frac{\partial f_t}{\partial x_n}\right\} + M_p\left\{\frac{\partial}{\partial y_1}, \ldots, \frac{\partial}{\partial y_p}\right\}\\
&\subset M_n\left\{\frac{\partial f_t}{\partial x_1}, \ldots, \frac{\partial f_t}{\partial x_n}\right\} + M_p\left\{\frac{\partial}{\partial y_1}, \ldots, \frac{\partial}{\partial y_p}\right\}\\
&= T\mathscr{A}(F/1),
\end{aligned}$$

which gives $T\mathscr{A}(F/1) \supset T\mathscr{K}(F/1)$. □

Proof of Theorem 7.1 for $\mathbb{F} = \mathbb{R}$ ***and*** $\mathbb{C}$ If f and g are $\mathscr{K}$-equivalent, then by Proposition 4.2 there exists a diffeomorphism $\varphi\colon (\mathbb{F}^n, S) \to (\mathbb{F}^n, S)$ and a matrix $A \in \mathrm{Gl}_p(\mathscr{O}_n)$ such that $g \circ \varphi = A \cdot f$. After a linear change of coordinates in $\mathbb{F}^p$, we can assume that $A(0) = I_p$, the identity matrix. For each $t \in \mathbb{F}$, let $C_t := I_p + t(A - I_p) \in \mathrm{Gl}_p(\mathscr{O}_n)$ and $f_t := C_t \cdot f$. We have a 1-parameter family of germs $f_t\colon (\mathbb{F}^n, S) \to (\mathbb{F}^p, 0)$, with $t \in \mathbb{F}$ such that $f_0 = f$, $f_1 = g \circ \varphi$ and f_t is $\mathscr{C}$-equivalent to f, for all $t \in \mathbb{F}$.

Let F be the associated 1-parameter unfolding $F(x, t) = (f_t(x), t)$. We have $F^*M_{p+1} = f_t^*\mathfrak{m}_p\mathscr{O}_{n+1} = f^*\mathfrak{m}_p\mathscr{O}_{n+1}$. Thus,

$$\frac{\partial f_t}{\partial t} = \frac{\partial C_t}{\partial t} \cdot f \in F^*M_p\theta(F/1) \subset T\mathscr{K}(F/1).$$

Next, we show that for each $t \in \mathbb{F}$, f_t is stable. In fact, by Theorem 4.1, the stability of f_t is equivalent to

$$T\mathscr{K}_e(f_t) + \mathrm{Sp}_{\mathbb{F}}\left\{\frac{\partial}{\partial y_1}, \ldots, \frac{\partial}{\partial y_p}\right\} = \mathscr{O}_n^{\times p}. \tag{7.3}$$

But $C_t \in \mathrm{Gl}_p(\mathscr{O}_n)$ and hence, (7.3) is equivalent to

$$T\mathscr{K}_e f_t + C_t \cdot \left(\mathrm{Sp}_{\mathbb{F}}\left\{\frac{\partial}{\partial y_1}, \ldots, \frac{\partial}{\partial y_p}\right\}\right) = \mathscr{O}_n^{\times p}. \tag{7.4}$$

As $f_t = C_t \cdot f$ and $C_t \in \mathrm{Gl}_p(\mathscr{O}_n)$, a simple computation shows $T\mathscr{K}_e f_t = C_t \cdot T\mathscr{K}_e f$. Hence, (7.4) is equivalent to

$$T\mathscr{K}_e f + \mathrm{Sp}_{\mathbb{F}}\left\{\frac{\partial}{\partial y_1}, \ldots, \frac{\partial}{\partial y_p}\right\} = \mathscr{O}_n^{\times p}. \tag{7.5}$$

That is, we have shown that f_t is stable if and only if f is stable, which is true by hypothesis.

To conclude the proof, we only have to apply Lemma 7.1. For each $t_0 \in \mathbb{F}$, f_t is $\mathscr{A}$-equivalent to f_{t_0}, for all t in a neighbourhood of t_0. Since $\mathbb{F}$ is connected, all germs f_t with $t \in \mathbb{F}$ are $\mathscr{A}$-equivalent. In particular, $f_0 = f$ and $f_1 = g \circ \varphi$ are $\mathscr{A}$-equivalent. But $g \circ \varphi$ is also $\mathscr{A}$-equivalent to g. □

If $f\colon (\mathbb{F}^n, S) \to (\mathbb{F}^p, 0)$ and $n > p$ then $Q(f)$ cannot be finite-dimensional, by the Hauptidealsatz. It is therefore of interest that Theorem 7.1 can be strengthened as follows. Let $\mathfrak{m}$ be the maximal ideal in $Q(f)$, and for each $k \in \mathbb{N}$, let

$$Q_k(f) = Q(f)/\mathfrak{m}^{k+1} = \mathscr{O}_n/(f^*\mathfrak{m}_p + \mathfrak{m}_n^{k+1})\mathscr{O}_n.$$

Corollary 7.1 ([Mat69b, Theorem A]) *Stable map-germs* $f, g\colon (\mathbb{F}^n, S) \to (\mathbb{F}^p, 0)$ *are* $\mathscr{A}$*-equivalent if and only* $Q_{p+1}(f) \cong Q_{p+1}(g)$.

The proof of Corollary 7.1 from Theorem 7.1 is very different from the proof (3) $\Longrightarrow$ (2) in 4.1. The stability of f does *not* imply that $f^*\mathfrak{m}_p\mathscr{O}_n \supset \mathfrak{m}_n^{p+1}$. What is obvious is that $Q_{p+1}(f)$ depends only on the $(p+1)$-jet of f; for if f and g agree up to degree $p+1$ then

$$(f^*\mathfrak{m}_p + \mathfrak{m}_n^{p+2})\mathscr{O}_n = (g^*\mathfrak{m}_p + \mathfrak{m}_n^{p+2})\mathscr{O}_n.$$

One can deduce Corollary 7.1 from Theorem 7.1 as follows:

1. If $Q_{p+1}(f) \cong Q_{p+1}(g)$ then there exists a diffeomorphism $\varphi\colon (\mathbb{F}^n, S) \to (\mathbb{F}^n, S)$ such that $\varphi^*(f^*\mathfrak{m}_p\mathscr{O}_n)) + \mathfrak{m}_n^{p+1} = g^*\mathfrak{m}_p\mathscr{O}_n + \mathfrak{m}_n^{p+1}$. By the argument of Proposition 4.5, there exists a matrix $C \in \mathrm{Gl}_p(\mathscr{O}_n)$ such that $C \cdot f \circ \varphi = g$

mod $\mathfrak{m}_n^{p+1}$. Thus f is $\mathscr{K}$-equivalent to a germ g_1 which agrees with g up to degree $p+1$. Because $j^{p+1}g_1 = j^{p+1}g$, g_1 is stable, by Corollary 4.1.
2. Stable germs $(\mathbb{F}^n, S) \to (\mathbb{F}^p, 0)$ are $(p+1)$-determined for $\mathscr{A}$-equivalence (see Corollary 6.4). From this, it follows that g and g_1 are $\mathscr{A}$-equivalent. Now by Theorem 7.1, g_1 and f are $\mathscr{A}$-equivalent, and the $\mathscr{A}$-equivalence of f and g follows.

Example 7.1 Suppose that $f\colon (\mathbb{F}^2, 0) \to (\mathbb{F}^3, 0)$ has corank 1. In linearly adapted coordinates, f takes the form

$$f(x, y) = (x, a(x, y), b(x, y)),$$

where $\partial a/\partial y$ and $\partial b/\partial y$ both vanish at $(0, 0)$. By a coordinate change in the target we can remove from a and b any pure power of x. So we may assume that the 2-jet of f takes the form

$$j^2 f = (x, a_{11}xy + a_{02}y^2, b_{11}xy + b_{02}y^2).$$

The non-immersive locus of f is defined by the two functions $\partial a/\partial y$ and $\partial b/\partial y$. We denote by $\mathscr{R}_f$ the ideal they generate - the *ramification ideal*. By an easy application of Nakayama's Lemma,

$$\begin{aligned} \mathscr{R}_f = \mathfrak{m} \quad &\Longleftrightarrow \quad \mathscr{R}_f + \mathfrak{m}^2 = \mathfrak{m} \\ &\Longleftrightarrow \quad (a_{11}x + 2a_{02}y, b_{11}x + 2b_{02}y) + \mathfrak{m}^2 = \mathfrak{m} \\ &\Longleftrightarrow \quad \begin{vmatrix} a_{11} & a_{02} \\ b_{11} & b_{02} \end{vmatrix} \neq 0. \end{aligned}$$

It follows that if $\mathscr{R}_f = \mathfrak{m}$ then after a linear change of coordinates in the target, involving only the second and third coordinates, we have

$$j^2 f = (x, y^2, xy). \tag{7.6}$$

For the polynomial germ defined by the right-hand side of this equality it is obvious that

$$f^*\mathfrak{m}_2\mathscr{O}_2 = (x, y^2). \tag{7.7}$$

In fact, by Nakayama's Lemma, this holds for any germ with this 2-jet. It is now easy to show that condition (2) of Theorem 4.1 holds, so that f is stable. Because of (7.7), the local algebra of f is isomorphic to that of the cross cap. Hence by Theorem 7.1, f is equivalent to the cross cap. ◊

Exercises for Sect. 7.1

1 Let $n \leq p$ and let $f\colon (\mathbb{F}^n, 0) \to (\mathbb{F}^p, 0)$ given by

$$f_i(x) = \begin{cases} x_i, & \text{if } 1 \leq i \leq n-1, \\ x_n^{k+1} + \sum_{j=1}^{k-1} x_j x_n^j, & \text{if } i = n, \\ \sum_{j=1}^{k} x_{k(i-n)+j-1} x_n^j, & \text{if } n+1 \leq i \leq p, \end{cases}$$

with $1 \leq k \leq n/(p-n+1)$.

(a) Show that f is stable and has Boardman type $\mathbf{1}_k = (1, \ldots, 1, 0)$, with 1 repeated k times.
(b) Show that any other stable germ $g\colon (\mathbb{F}^n, 0) \to (\mathbb{F}^p, 0)$ with Boardman type $\mathbf{1}_k$ must be $\mathscr{A}$-equivalent to f.

2 Let $n > p$ and let $f\colon (\mathbb{F}^n, 0) \to (\mathbb{F}^p, 0)$ given by

$$f_i(x) = \begin{cases} x_i, & \text{if } 1 \leq i \leq p-1, \\ \pm x_p^2 \pm \cdots \pm x_{n-1}^2 \pm x_n^{k+1} + \sum_{j=1}^{k-1} x_j x_n^j, & \text{if } i = p, \end{cases}$$

with $1 \leq k \leq p$.

(a) Show that f is stable and has Boardman type $\mathbf{i} = (n-p+1, \ldots, n-p+1, 0)$, with $n-p+1$ repeated k times.
(b) Assume that $\mathbb{F} = \mathbb{C}$ and show that any other stable germ $g\colon (\mathbb{F}^n, 0) \to (\mathbb{F}^p, 0)$ with Boardman type $\mathbf{i}$ must be $\mathscr{A}$-equivalent to f.

3 Use the characterisation of the stability of a multi-germ in Theorem 3.3 to deduce the statement of Theorem 7.1 for multi-germs from the statement for mono-germs.

7.2 Construction of Stable Germs as Unfoldings

In this section we explain a procedure due to Mather in [Mat69b] for finding all stable map-germs as unfoldings of lower-dimensional germs of rank zero. This procedure was already presented in Sect. 4.1; here we prove that the map-germs it constructs are stable.

Any map-germ $f\colon (\mathbb{F}^n, 0) \to (\mathbb{F}^p, 0)$ can be seen, after taking suitable coordinates in the source and target, as an unfolding of another map-germ of rank 0. So, we assume first that f itself has rank 0, which implies that $T\mathscr{K}_e(f) \subset \mathfrak{m}_n\theta(f)$.

Theorem 7.2 *Let $f\colon (\mathbb{F}^n, 0) \to (\mathbb{F}^p, 0)$ be a map-germ of rank 0 and finite $\mathscr{K}_e$-codimension. Consider $g_1, \ldots, g_d \in \theta(f)$ whose classes generate over $\mathbb{F}$ the* *quotient* $\mathfrak{m}_n\theta(f)/T\mathscr{K}_e f$. *Then, the d-parameter unfolding F given by*

$$F(x, u_1, \ldots, u_d) = \left(f(x) + \sum_{j=1}^{d} u_j g_j(x), u_1, \ldots, u_d \right) \tag{7.8}$$

is stable. If $g_1, \ldots, g_d$ *form a basis for the quotient, then the unfolding F is a minimal stable unfolding (none exists with fewer parameters).*

Proof By hypothesis, $\theta(f)/T\mathscr{K}_e(f)$ is generated over $\mathbb{F}$ by the classes of $g_1, \ldots, g_d$ and $\partial/\partial y_1, \ldots, \partial/\partial y_p$. Since $g_j = \dot{F}_j$, by means of the isomorphism β of Lemma 5.5, $\theta(F)/T\mathscr{K}_e(F)$ is generated over $\mathbb{F}$ by the classes of $\partial/\partial y_1, \ldots, \partial/\partial y_p$ and $\partial/\partial u_1, \ldots, \partial/\partial u_d$. Therefore, F is stable by Theorem 4.1. □

As a consequence, we have the following theorem of existence and uniqueness of stable unfoldings for map-germs of finite $\mathscr{K}_e$-codimension (also called usually, map-germs of *finite singularity type*).

Proposition 7.2 *Any map-germ* $f\colon (\mathbb{F}^n, S) \to (\mathbb{F}^p, 0)$ *of finite singularity type admits a stable unfolding. Moreover, if we fix the number of parameters, the stable unfolding is unique up to* $\mathscr{A}$*-equivalence.*

Proof We first prove the existence of a stable unfolding when f is a mono-germ, that is, when $S = \{0\}$. After choosing suitable coordinates in the source and the target, we can assume that f is an unfolding of another map-germ g of rank 0. By Theorem 7.2, there exists a stable unfolding G of g. We take F as the sum of f and G, which is also stable by Exercise 5.1.6.

Now consider the general case $S = \{s_1, \ldots, s_r\}$. By the mono-germ case, we can assume that each branch $f_i\colon (\mathbb{F}^n, s_i) \to (\mathbb{F}^p, 0)$ is already stable. Let $\tau(f_i)$ be the $\mathbb{F}$-subspace of $T_0\mathbb{F}^p$ of Definition 3.10 and consider the induced morphism:

$$T_0\mathbb{F}^p \to \bigoplus_{i=1}^{r} \frac{T_0\mathbb{F}^p}{\tau(f_i)}, \qquad v \mapsto (v + \tau(f_1), \ldots, v + \tau(f_r)). \tag{7.9}$$

Let $(v_{11}, \ldots, v_{1r}), (v_{21}, \ldots, v_{2r}), \ldots, (v_{k1}, \ldots, v_{kr})$ be members of $(T_0\mathbb{F}^p)^r$ whose classes in

$$\bigoplus_{i=1}^{r} \frac{T_0\mathbb{F}^p}{\tau(f_i)}$$

project to a basis of the cokernel of (7.9). We claim that the k-parameter unfolding F with branches $F_i\colon (\mathbb{F}^n \times \mathbb{F}^k, (s_i, 0)) \to (\mathbb{F}^p \times \mathbb{F}^k, 0)$, $i = 1, \ldots, r$, given by

$$F_i(x, u) = \left(f_i(x) + \sum_{j=1}^{k} u_j v_{ji}, u \right)$$

is stable. In fact, for each $i = 1, \dots, r$, $\tau(F_i)$ is the subspace of $T_0\mathbb{F}^p \times T_0\mathbb{F}^k$ given by

$$\tau(F_i) = \tau(f_i) \times \{0\} + \mathrm{Sp}_{\mathbb{F}}\left\{\left(v_{1i}, \frac{\partial}{\partial u_1}\right), \dots, \left(v_{ki}, \frac{\partial}{\partial u_k}\right)\right\}.$$

A vector in $\tau(F_1) \cap \cdots \cap \tau(F_r)$ must be of the form $(\hat{v}, \hat{u})$, with

$$\hat{u} = \sum_{j=1}^{k} \lambda_j \frac{\partial}{\partial u_j},$$

$$\hat{v} = w_1 + \sum_{j=1}^{k} \lambda_j v_{j1} = \cdots = w_r + \sum_{j=1}^{k} \lambda_j v_{jr},$$

where $w_i \in \tau(f_i)$ for all $i = 1, \dots, r$. By the hypothesis that

$$(v_{11}, \dots, v_{1r}), (v_{21}, \dots, v_{2r}), \dots, (v_{k1}, \dots, v_{kr})$$

form a basis of the cokernel of (7.9), this implies that $\hat{v} = \hat{u} = 0$. Hence $\tau(F_1), \dots, \tau(F_r)$ are now in general position. Hence, F is stable by Theorem 3.3.

The uniqueness follows from Theorem 7.1, since stable unfoldings of the same germ with the same number of parameters have isomorphic local algebras, so they are $\mathscr{A}$-equivalent. □

Warning! With the number of parameters fixed, the stable unfolding is unique up to $\mathscr{A}$-equivalence *as a mapping*, but not as an unfolding. That is, two different stable unfoldings on the same number of parameters may not be equivalent to one another as unfoldings in the sense of Definition 3.4.

Example 7.2 Consider the bi-germ $f\colon (\mathbb{F}^2_2, \mathbf{0}_2) \to (\mathbb{F}^3, 0)$ with branches $f_1(x_1, x_2) = (x_1, x_2^2, x_1x_2)$ and $f_2(x_1, x_2) = (x_1, x_2, 0)$. Both branches f_1 and f_2 are stable as mono-germs, with $\tau(f_1) = \{0\}$ and $\tau(f_2) = \mathbb{F}^2 \times \{0\}$, but $\tau(f_1)$ and $\tau(f_2)$ do not meet transversely in $T_0\mathbb{F}^3$, so f is not stable. To find a stable unfolding F of f, we have to find unfoldings F_1 and F_2 of f_1 and f_2, with $\tau(F_1) \pitchfork \tau(F_2)$. Taking

$$F_1(x_1, x_2, t) = (x_1, x_2^2, x_1x_2 + t, t), \quad F_2(x_1, x_2, t) = (x_1, x_2, 0, t),$$

$\tau(F_1)$ is the line generated by $(0, 0, 1, 1)$ and $\tau(F_2) = \mathbb{F}^2 \times \{0\} \times \mathbb{F}$. As these are transverse to one another, F is stable. ◊

We recall that an analytic algebra is an $\mathbb{F}$-algebra of the form $A = \mathscr{O}_n/I$ for some ideal $I \subset \mathscr{O}_n$ (see Definition 4.3).

Definition 7.1 Let $A = \mathcal{O}_n/I$ be an analytic algebra, where $I \subset \mathcal{O}_n$ is an ideal. Let $f_1, \ldots, f_p \in \mathcal{O}_n$ be generators of I and write $f = (f_1, \ldots, f_p)$. We define

$$\tau_f(A) := \dim_{\mathbb{F}} \frac{A^p}{J(f)},$$

where $J(f)$ is the submodule of A^p generated by the projection of $\left\{\frac{\partial f}{\partial x_1}, \ldots, \frac{\partial f}{\partial x_n}\right\}$.

When $p = 1$ and $\mathbb{F} = \mathbb{C}$, A is the local algebra of a hypersurface $(X, 0)$ and $\tau_f(A)$ is known as the *Tjurina number* of $(X, 0)$.

In general, if $A = \mathcal{O}_n/I$ and I is generated by $f_1, \ldots, f_p \in \mathcal{O}_n$, it follows from the definition that $A = Q(f)$, where $f = (f_1, \ldots, f_p)\colon (\mathbb{F}^n, 0) \to (\mathbb{F}^p, 0)$ and that

$$\tau_f(Q(f)) = \operatorname{codim}_{\mathcal{K}_e}(f). \tag{7.10}$$

If we consider an r-tuple of analytic algebras $(A_1, \ldots, A_r)$, then we can assume that $A_i = \mathcal{O}_n/I_i$, for some ideal $I_i \subset \mathcal{O}_n$ (with the same n for all of them). Moreover, we can also assume that all the ideals I_i have a system of generators $f^i = (f^i_1, \ldots, f^i_p)$ (with the same p for all of them). It follows that we can construct a multi-germ $f\colon (\mathbb{F}^n, S) \to (\mathbb{F}^p, 0)$ such that $Q(f) \cong A_1 \oplus \cdots \oplus A_r$. Thus, we have the following immediate consequence of the above results.

Corollary 7.2 *Let $(A_1, \ldots, A_r)$ be an r-tuple of analytic algebras, where $A_i = \mathcal{O}_n/I_i$ and such that $\tau_{f^i}(A_i) < \infty$ for some system of generators $f^i = (f^1_1, \ldots, f^i_p)$ of I_i. Then there exists a stable multi-germ $F\colon (\mathbb{F}^{n+d}, S) \to (\mathbb{F}^{p+d}, 0)$ such that $Q(F) \cong A_1 \oplus \cdots \oplus A_r$.*

Remark 7.1 We have to allow for the possibility that the rank 0 germ that we are unfolding has one or more redundant components, which do not contribute to the ideal $f_0^*\mathfrak{m}_p\mathcal{O}_n$. Up to contact equivalence such components may be taken equal to 0. A germ $(f_0, 0)$ is $\mathcal{K}$-finite only if f_0 is itself a finite map, for the critical set of $(f_0, 0)$ is all of its domain. Thus $(f_0, 0)$ can be $\mathcal{K}$-finite only when $n \leq p$. Suppose that f_0 has stable unfolding F. Then the germ $(f_0, 0)$ has stable unfolding

$$F_1(x, u, v_1, \ldots, v_t) = (F(x, u), \sum_{j=1}^{t} v_j g_j, v_1, \ldots, v_t),$$

where $g_1, \ldots, g_t$ form a basis for $\mathfrak{m}_n/f_0^*\mathfrak{m}_p\mathcal{O}_n$ and the v_j are new unfolding parameters. ◊

Exercises for Sect. 7.2

1 For each algebra A in the following list, find a stable germ $f\colon (\mathbb{F}^n, S) \to (\mathbb{F}^p, 0)$ such that $Q(f)$ is isomorphic to A:

(a) $\mathscr{O}_1/(x_1^{k+1})$, with $k \geq 1$, with n as small as possible and $p = n$. Do the same with $p = n + 1$ and n as small as possible.
(b) $\mathscr{O}_2/(x_1^4 + \lambda x_1^2 x_2^2 + x_2^4)$, with $\lambda \in \mathbb{F}$, $\lambda \neq \pm 2$.
(c) $\mathscr{O}_1/(x_1^2) \oplus \mathscr{O}_2/(x_1^2, x_2^2, x_1 x_2)$.

2 Show that there exists a one-to-one correspondence between the set of all $\mathscr{K}$-orbits of germs $f\colon (\mathbb{F}^n, 0) \to (\mathbb{F}^p, 0)$ of rank 0 and $\mathscr{K}_e$-codimension $\leq d + p$ and the set of all $\mathscr{A}$-orbits of stable germs $f\colon (\mathbb{F}^{n+d}, 0) \to (\mathbb{F}^{p+d}, 0)$ of rank d.

3 Let $f\colon (\mathbb{F}^n, S) \to (\mathbb{F}^{n+1}, 0)$ be stable, and such that all its branches have corank ≤ 1. Show that

$$Q(f) \cong \frac{\mathscr{O}_1}{(x_1^{k_1+1})} \oplus \cdots \oplus \frac{\mathscr{O}_1}{(x_1^{k_r+1})},$$

for some $k_1, \ldots, k_r \in \mathbb{N}$ such that $k_1 + \cdots + k_r \leq n + 1$.

4

(i) Find a stable unfolding of the germ $(\mathbb{F}, 0) \to (\mathbb{F}, 0)$ given by $x \mapsto x^2$, and of the germ $(\mathbb{F}, 0) \to (\mathbb{F}^{r+1}, 0)$ given by $x \mapsto (x^2, 0, \ldots, 0)$.
(ii) Find an explicit expression for a stable germ $(\mathbb{F}^n, 0) \to (\mathbb{F}^{2n-1}, 0)$ with local algebra $\mathscr{O}_1/(x^2)$, for each $n \geq 2$.
(iii) Find a stable unfolding of the germ $(\mathbb{F}, 0) \to (\mathbb{F}, 0)$ given by $x \mapsto x^3$, and of the germ $(\mathbb{F}, 0) \to (\mathbb{F}^{r+1}, 0)$ given by $x \mapsto (x^3, 0, \ldots, 0)$.
(iv) In which dimensions is there a stable germ with local algebra $\mathscr{O}_1/(x^3)$?
(v) Find all stable mono germs $(\mathbb{F}^n, 0) \to (\mathbb{F}^{2n-1}, 0)$.

7.3 The Isosingular Locus

In this section, we show that the isosingular locus (also called the *analytic stratum* by Mather) of a stable multi-germ $f\colon (\mathbb{F}^n, S) \to (\mathbb{F}^p, 0)$ is a submanifold of $(\mathbb{F}^p, 0)$ whose tangent space at the origin is $\tau(f)$. This was stated in Sect. 3.6. Here we give the proof by using the results about finite determinacy. We also prove that any stable multi-germ is $\mathscr{A}$-equivalent to a Cartesian product of another stable multi-germ times the identity map on the isosingular locus. These are analogous to the results about isosingular loci of complex analytic spaces due to R. Ephraim [Eph78].

Given a multi-germ $f : (\mathbb{F}^n, S) \to (\mathbb{F}^p, 0)$, we introduced in Sect. 3.6 the mapping

$$\overline{\omega} f : T_0\mathbb{F}^p \to \frac{\theta(f)}{T\mathscr{K}_e(f)},$$

whose kernel is $\tau(f)$ and with the property that f is stable if and only if $\overline{\omega} f$ is an epimorphism (see Lemma 3.6).

Theorem 7.3 *Let* $f : (\mathbb{F}^n, S) \to (\mathbb{F}^p, 0)$ *be a stable multi-germ with* $\dim_{\mathbb{F}} \tau(f) = d$. *Then* f *is* $\mathscr{A}$*-equivalent to a product* $g_0 \times \mathrm{id}$, *where* $g_0 : (\mathbb{F}^{n-d}, S') \to (\mathbb{F}^{p-d}, 0)$ *is also a stable multi-germ and* $\mathrm{id} : (\mathbb{F}^d, 0) \to (\mathbb{F}^d, 0)$ *is the identity map.*

Proof We first choose linear coordinates in $\mathbb{F}^p$ such that $\tau(f) = \{0\} \times \mathbb{F}^d$. Assume $S = \{s_1, \ldots, s_r\}$ and let $f_i := f|_{(\mathbb{F}^n, s_i)} : (\mathbb{F}^n, s_i) \to (\mathbb{F}^p, 0)$, $i = 1, \ldots, r$. For each i, we have

$$\{0\} \times \mathbb{F}^d = \tau(f) \subset \tau(f_i) \subset \operatorname{im} d_{x_i} f_i.$$

We can choose coordinates in $\mathbb{F}^n$ near each point x_i such that f is written as an unfolding of some multi-germ $g_0 : (\mathbb{F}^{n-d}, S') \to (\mathbb{F}^{p-d}, 0)$, that is,

$$f(z, u) = (g_u(z), u), \quad \forall z \in \mathbb{F}^{n-d}, \ \forall u \in \mathbb{F}^d.$$

Then the local algebras of f and g_0 are isomorphic. So by Theorem 7.1 it will be enough to show that g_0 is stable. Consider the following commutative diagram:

$$\begin{array}{ccc}
\dfrac{T_0\mathbb{F}^p}{\tau(f)} & \longrightarrow & \dfrac{\theta(f)}{T\mathscr{K}_e(f)} \\
\Big\downarrow & & \Big\downarrow \beta \\
T_0\mathbb{F}^{p-d} & \xrightarrow{\overline{\omega} g_0} & \dfrac{\theta(g_0)}{T\mathscr{K}_e(g_0)}
\end{array}$$

The top arrow is an isomorphism induced by $\overline{\omega} f$ and the columns are also isomorphisms: the left arrow is induced by the projection and is an isomorphism because $\tau(f) = \{0\} \times \mathbb{F}^d$; the right arrow is the isomorphism of Lemma 5.5. Therefore $\overline{\omega} g_0$ is also an isomorphism, so g_0 is stable by Lemma 3.6. □

The following lemma gives the relationship between the codimension and the extended codimension for the contact group.

Lemma 7.2 *Assume* $f : (\mathbb{F}^n, S) \to (\mathbb{F}^p, 0)$ *is* $\mathscr{K}$*-finite, with* $S = \{s_1, \ldots, s_r\} \subset C(f)$. *Then,*

$$\operatorname{codim}_{\mathscr{K}}(f) = \operatorname{codim}_{\mathscr{K}_e}(f) + r(n - p).$$

Proof For each $j = 1, \dots, r$, we denote by $\mu_j \in \mathscr{O}_n$ the germ whose germ at s_j is constant equal to 1 and 0 elsewhere. We have $T\mathscr{K}_e(f) = T\mathscr{K}(f) + \mathrm{Sp}_{\mathbb{F}}\{\mu_i \frac{\partial f}{\partial x_i}\}$, hence,

$$\begin{aligned}\mathrm{codim}_{\mathscr{K}}(f) &= \dim_{\mathbb{F}} \frac{\mathfrak{m}_n\theta(f)}{T\mathscr{K}(f)} = \dim_{\mathbb{F}} \frac{\theta(f)}{T\mathscr{K}(f)} - rp \\ &= \dim_{\mathbb{F}} \frac{\theta(f)}{T\mathscr{K}_e(f)} + \dim_{\mathbb{F}} \frac{T\mathscr{K}_e(f)}{T\mathscr{K}(f)} - rp \\ &\leq \mathrm{codim}_{\mathscr{K}_e}(f) + r(n-p).\end{aligned}$$

Assume that the inequality is strict, there is a vector field $\xi \in \theta_n$, not zero at some point s_i and such that $tf(\xi) \in f^*\mathfrak{m}_p\theta(f)$. After a change of coordinates in $\mathbb{F}^n$ near s_i, we can take $\xi = \partial/\partial u_1$, where $u_1, \dots, u_n$ are the coordinates in $\mathbb{F}^n$ near s_i. Working only in $(\mathbb{F}^n, s_i)$, we have

$$\frac{\partial f_k}{\partial u_1} = \sum_{i=1}^{p} \alpha_{kj} f_j, \ \forall k = 1, \dots, p,$$

for some functions $\alpha_{kj} \in \mathscr{O}_{\mathbb{F}^n, s_i}$. But this implies that $f^{-1}(0) \subset C(f)$, which is in contradiction with the $\mathscr{K}$-finiteness of f at s_i, by the Nullstellensatz Theorem E.3. □

Let $f : N \to P$ be a smooth map between manifolds such that the restriction $f : C(f) \to P$ is finite (i.e., closed and finite-to-one) and let $\Delta(f) = f(C(f))$. For each $y \in \Delta(f)$, the multi-germ of f at y will mean the multi-germ $f : (N, S) \to (P, y)$, where $S = f^{-1}(y) \cap C(f)$. When $\dim N \geq \dim P$, we will refer to $\Delta(f)$ as the *discriminant* of f. When $\dim N < \dim P$ then $\Delta(f)$ is just the image of f.

Recall from Sect. 3.6 that if $f : N \to P$ is a smooth map between manifolds such that the restriction $f : C(f) \to P$ is finite, then for each $y_0 \in \Delta(f)$, the *isosingular locus* of f at y_0 is defined as the set $\mathrm{Iso}(f, y_0)$ of points $y \in \Delta(f)$ such that the multi-germ of f at y is $\mathscr{A}$-equivalent to the multi-germ of f at y_0.

In case $f : (\mathbb{F}^n, S) \to (\mathbb{F}^p, 0)$ is a $\mathscr{K}$-finite multi-germ with $S \subset C(f)$, we denote by $\mathrm{Iso}(f)$ the germ of the isosingular locus at 0 of a small enough representative of f.

Theorem 7.4 *Let $f : (\mathbb{F}^n, S) \to (\mathbb{F}^p, 0)$ be a stable multi-germ with $S \subset C(f)$. Then* $\mathrm{Iso}(f)$ *is a submanifold of* $(\mathbb{F}^p, 0)$ *whose tangent space at the origin is* $\tau(f)$.

Proof Suppose that $\dim_{\mathbb{F}} \tau(f) = d$. By Theorem 7.3, we can assume that f is a product $g_0 \times \mathrm{id}$, where $g_0 : (\mathbb{F}^{n-d}, S') \to (\mathbb{F}^{p-d}, 0)$ is also a stable multi-germ and $\mathrm{id} : (\mathbb{F}^d, 0) \to (\mathbb{F}^d, 0)$. Moreover, if we follow the arguments of the proof of the theorem, we see that $\tau(f) = \{0\} \times \mathbb{F}^d$ and $\tau(g_0) = \{0\}$. As $\mathrm{Iso}(f) = \mathrm{Iso}(g_0) \times (\mathbb{F}^d, 0)$, it is enough to show that $\mathrm{Iso}(g_0) = (\{0\}, 0)$.

Since $\tau(g_0) = \{0\}$, $\overline{\omega} g_0$ is an isomorphism, so $\operatorname{codim}_{\mathscr{K}_e}(g_0) = p - d$. By Lemma 7.2 and since g_0 is stable,

$$\operatorname{codim}_{\mathscr{A}}(g_0) = \operatorname{codim}_{\mathscr{K}}(g_0) = p - d + r(n - p),$$

where $r = \#S'$. We know from Corollary 6.1 that this is equal to the codimension of the orbit $W = \mathscr{A}^{(k)}\sigma$ in ${}_rJ^k(n-d, p-d)$, where $k = p - d + 1$ and $\sigma = j^k g_0$. Fix a representative $g_0 : U \to V$, where U, V are open neighbourhoods of $S', 0$ in $\mathbb{F}^{n-d}, \mathbb{F}^{p-d}$, respectively. Since in ${}_rJ^k(n-d, p-d)$ we force all the multi-jets to preserve the origin, the codimension of $W(U, V)$ in ${}_rJ^k(U, V)$ is

$$\operatorname{codim} W(U, V) = \operatorname{codim} W + (r-1)(p-d) = p-d+r(n-p)+(r-1)(p-d) = r(n-d).$$

By Theorem 4.6, ${}_rj^k g_0$ is transverse to $W(U, V)$, hence $C := ({}_rj^k g_0)^{-1}(W(U, V))$ is a 0-dimensional submanifold of $U^{(r)}$. We have $\operatorname{Iso}(g_0, 0) = g_0 \circ \pi_1(C)$, where π_1 is the projection onto the first factor, so $\operatorname{Iso}(g_0, 0)$ is a discrete subset of V. □

Remark 7.2 As a byproduct of the proof of Theorem 7.4, it follows that if $f : (\mathbb{F}^n, S) \to (\mathbb{F}^p, 0)$ is stable with $S \subset C(f)$ and $\dim \operatorname{Iso}(f) = d$, then:

$$\operatorname{codim}_{\mathscr{K}_e}(f) = p - d, \quad \operatorname{codim}_{\mathscr{K}}(f) = \operatorname{codim}_{\mathscr{A}}(f) = p - d + r(n - p).$$

◊

The following corollary follows immediately from theorems 7.3 and 7.4.

Corollary 7.3 *Let $f : N \to P$ be a locally stable map between smooth manifolds. The isosingular loci $\{\operatorname{Iso}(f, y)\}_{y \in \Delta(f)}$ provide a partition of $\Delta(f)$ into submanifolds. Moreover, for each $y \in \Delta(f)$, the multi-germ of f at y is $\mathscr{A}$-equivalent to a product multi-germ $g \times \mathrm{id}$, where $g : N' \to P'$ is also stable and* $\mathrm{id} : \operatorname{Iso}(f, y) \to \operatorname{Iso}(f, y)$.

Corollary 7.4 *Let $f : (\mathbb{F}^n, S) \to (\mathbb{F}^p, 0)$ be a multi-germ of finite $\mathscr{A}$-codimension with $S \subset C(f)$. There exists a representative $f : U \to V$ such that the isosingular loci $\{\operatorname{Iso}(f, y)\}_{y \in \Delta(f)}$ provide a partition of $\Delta(f)$ into submanifolds. Moreover, for each $y \in \Delta(f) \setminus \{0\}$, the multi-germ of f at y is $\mathscr{A}$-equivalent to a product multi-germ $g \times \mathrm{id}$, where $g : (\mathbb{F}^{n'}, S') \to (\mathbb{F}^{p'}, 0)$ is also stable and* $\mathrm{id} : \operatorname{Iso}(f, y) \to \operatorname{Iso}(f, y)$.

Proof If f is stable, then the result is obvious. Otherwise, by the Mather–Gaffney criterion Theorem 4.5, there exists a representative $f : U \to V$ such that

1. $f^{-1}(0) \cap C(f) \subset S$;
2. the restriction $f|_{U \setminus f^{-1}(0)} : U \setminus f^{-1}(0) \to V \setminus \{0\}$ is locally stable.

We apply Corollary 7.3 to $f|_{U \setminus f^{-1}(0)}$. For each $y \in \Delta(f) \setminus \{0\}$, $\operatorname{Iso}(f, y) = \operatorname{Iso}(f|_{U \setminus f^{-1}(0)}, y)$ and $\operatorname{Iso}(f, 0) = \{0\}$ so we have the result. □

In the case that f has discrete stable type (either when f is a stable map or an $\mathscr{A}$-finite germ), then we have only finitely many submanifolds in the isosingular partition, so we have in fact a stratification of $\Delta(f)$. We recall that the condition of discrete stable type always holds when f has corank 1 and $n \leq p$ or when we are in the range of the nice dimensions.

Definition 7.2 Assume that f has discrete stable type, then the partition of $\Delta(f)$ by the isosingular loci $\{\mathrm{Iso}(f, y)\}_{y\in\Delta(f)}$ is called the *stratification by stable types*.

For each $y \in \Delta(f)$ (or $y \in \Delta(f) \setminus \{0\}$ in the case of a germ), the multi-germ of f at y is $\mathscr{A}$-equivalent to a product. It follows that the stratification is locally analytically trivial, and it is easy to see that it satisfies Whitney's conditions (see Definition B.3).

Corollary 7.5 *If f has discrete stable type, then the stratification by stable types is a Whitney stratification of $\Delta(f)$.*

7.3.1 Weighted Homogeneity and Local Quasihomogeneity

Definition 7.3 A germ of function $f: (\mathbb{F}^n, 0) \to (\mathbb{F}, 0)$ is *quasihomogeneous*, or *weighted homogeneous*, if there are positive integers $w_1, \dots, w_n$ (the *weights*) and d (the *degree*) such that for $\lambda \in \mathbb{F}^*$,

$$f(\lambda^{w_1} x_1, \dots, \lambda^{w_n} x_n) = \lambda^d f(x_1, \dots, x_n).$$

If in addition f is smooth, then it follows that f is a polynomial of the form

$$f(x_1, \dots, x_n) = \sum_{w_1 i_1 + \cdots + w_n i_n = d} c_{i_1,\dots,i_n} x_1^{i_1} \cdots x_n^{i_n}.$$

A germ of mapping $f: (\mathbb{F}^n, 0) \to (\mathbb{F}^p, 0)$, $f(x) = (f_1(x), \dots, f_p(x))$, is quasihomogeneous on weights $w_1, \dots, w_n$ and with degrees $d_1, \dots, d_p$, if f_i is quasihomogeneous of degree d_i, for $i = 1, \dots, p$, all with respect to the same set of weights $w_1, \dots, w_n$.

If f is quasihomogeneous then it is equivariant with respect to $\mathbb{F}^*$ actions on $\mathbb{F}^n$ and $\mathbb{F}^p$ defined by

$$\lambda \cdot (x_1, \dots, x_n) = (\lambda^{w_1} x_1, \dots, \lambda^{w_n} x_n)$$
$$\lambda \cdot (y_1, \dots, y_p) = (\lambda^{d_1} y_1, \dots, \lambda^{d_p} y_p),$$

quasihomogeneity means simply that $f(\lambda \cdot x) = \lambda \cdot f(x)$. Conversely, if f is equivariant with respect to good $\mathbb{F}^*$ actions in source and target (actions for which 0 is in the closure of every orbit) then in suitable coordinates it is quasihomogeneous.

Definition 7.4 Let $X \subset \mathbb{F}^p$ be an analytic subspace, in the complex analytic case, or a subspace defined by smooth equations, in the real C^∞ case. We say X is *locally quasihomogeneous* if for each point $x_0 \in X$ there are smooth local coordinates around x_0, and a set of positive weights for those coordinates, with respect to which the ideal defining X is generated by quasihomogeneous functions.

Remark 7.3 In [Mat71], Mather used the results described in this chapter to classify all stable map-germs in the nice dimensions, and on their boundary. Inspection of his classification reveals the following remarkable fact: in the nice dimensions, and only in the nice dimensions, every stable map-germ is weighted homogeneous, with strictly positive weights, with respect to suitable coordinates. One may speculate as to the reason for this, which surely deserves an explanation. In the next section we check the part of this statement concerning the nice dimensions. For now, we note that it implies an interesting property of stable discriminants and images in the nice dimensions. ♢

Theorem 7.5 *Let $f\colon (\mathbb{F}^n, S) \to (\mathbb{F}^p, 0)$ be a stable map-germ, with (n, p) nice dimensions. Then the discriminant of f (if $n \geq p$), and the image of f (if $n < p$) are locally quasihomogeneous.* □

Proof This follows at once from Theorem 7.6, below, together with the "openness of stability": that in any representative $f : U \to V$, the set of points $y \in V$ such that the germ $f\colon (U, f^{-1}(y) \cap S(f)) \to (V, y)$ is stable, is an open set containing 0. □

7.4 Quasihomogeneity and the Nice Dimensions

The aim of this section is to prove the following theorem.

Theorem 7.6 *If $F\colon (\mathbb{F}^n, S) \to (\mathbb{F}^p, 0)$ is stable, and (n, p) are nice dimensions in the sense of Mather, then with respect to suitable coordinates, F is quasihomogeneous.*

Unfortunately we have no theoretical argument to prove this. It is possible to show, without too much difficulty, that a locally quasihomogeneous analytic subset X of $\mathbb{C}^n$ has a locally analytically trivial stratification—in other words, the partition determined by the maximal integral submanifolds of the logarithmic distribution $\mathrm{Der}(-\log X)$ is locally finite. This strongly suggests that the range of dimension-pairs (N, P) in which all stable germs are quasihomogeneous, should be contained in the nice dimensions—in other words, for every dimension-pair (N, P) outside Mather's nice dimensions, there should exist a non-quasihomogeneous stable germ—but it does not prove it. In any case, this statement is the reciprocal of the result we want.

Instead, we prove Theorem 7.6 simply by checking that each of the stable germs in the nice dimensions is quasihomogeneous. Mather's paper [Mat71] contains a more or less explicit list of all rank zero germs f_0 for which a minimal stable unfolding F takes place in the nice dimensions, so in principle the proof is a straightforward verification. The unfolding F can be constructed by the procedure of Sect. 4.3. We choose a basis for $\mathfrak{m}_n\theta(f_0)/T\mathscr{K}_e f_0$ consisting of monomials—that is, elements of the form $x^\alpha\partial/\partial y_j$, where x^α is a monomial in the variables $x_1, \ldots, x_n$ in the usual sense, and y_j is one of the target coordinates. The verification consists of computing, for each f_0, a monomial basis for $\mathfrak{m}_n\theta(f_0)/T\mathscr{K}_e f_0$, and then checking that it is possible to assign weights to the variables in the domain of f_0 with the following two properties:

1. f_0 is weighted homogeneous with respect to these weights, and
2. for each unfolding monomial $x^\alpha\partial/\partial y_j$, the weighted degree of x^α is less than the weighted degree of the corresponding component $(f_0)_j$ of f_0.

If this can be done, then by assigning the unfolding parameter $u_{\alpha,j}$ the weight weight$(f_0)_j$ − weight(x^α), the stable unfolding

$$F(x,u) = (f_0(x) + \sum_{\alpha,j} u_{\alpha,j}x^\alpha e_j, u) \tag{7.11}$$

becomes weighted homogeneous.

To find the list of rank zero germs, recall from Sect. 5.2 that $W_r^k(n,p)$ is the union of the all the $\mathscr{K}^{(k)}$-orbits in $J^k(n,p)$ of codimension $\geq r$, and $W_r^k(n,p)^*$ is the union of its irreducible components of codimension $< r$. Then the set $\Pi^k(n,p)$, defined by

$$\Pi^k(n,p) = \bigcup_{r\geq 0} W_r^k(n,p)^*,$$

is the smallest subset of $J^k(n,p)$ whose complement contains only a finite number of $\mathscr{K}^{(k)}$-orbits. In [Mat71], Mather defines $\sigma^k(n,p) = \operatorname{codim}\Pi^k(n,p)$, and $\sigma(n,p) = \inf_k \sigma^k(n,p)$. It is clear that $\sigma^k(n,p) \geq \sigma^{k+1}(n,p)$, so the infimum is the limit as $k \to \infty$.

Mather determines each set $\Pi^k(n,p)$ in two steps. After picking a likely choice, say Z, he has to show that every one of its open subsets meets uncountably many contact classes ($\mathscr{K}^{(k)}$-orbits). Any such set Z must be contained in $\Pi^k(n,p)$. The second step is to pick such a Z with the additional property that its complement in $J^k(n,p)$ consists of finitely many contact classes. Then $Z = \Pi^k(n,p)$. The second step involves listing the contact classes in $J^k(n,p) \setminus Z$—hence Mather's (almost complete) classification of all contact classes of stable germs in the nice dimensions. The algebra types whose corresponding $\mathscr{K}^{(k)}$ orbits are contained in $\Pi^k(n,p)$ can be ignored. For by Mather's theorem (our Theorem 5.4, above), the nice dimensions are those dimensions (n,p) for which $n < \sigma(n,p)$. If $n < \sigma(n,p)$ then the jet extension map $j^k f : N^n \to J^k(N, P)$ can only be transverse to a $\mathscr{K}^{(k)}$ orbit contained in $\Pi^k(n, p)$ by not meeting it. And a germ f is stable if and only if its jet extension map $j^k f$ is transverse to the $\mathscr{K}^{(k)}$-orbits it meets, for all k (Theorem 4.6).

Mather's paper [Mat71] is broken into two sections, one dealing with the case $n \le p$ and the other with the case $n > p$. The organising principle is the use of Boardman classes, or Thom–Boardman sets, in the nomenclature of our Sect. 4.7.

Every stable mono-germ F is an unfolding of a rank germ 0 germ f_0; we adopt the convention that $f_0 : (\mathbb{F}^n, 0) \to (\mathbb{F}^p, 0)$ and $F: (\mathbb{F}^N, 0) \to (\mathbb{F}^P, 0)$, so by implication $N - P = n - p$. We write $N(f_0)$ and $P(f_0)$ for the dimension of the source and target of a minimal stable unfolding of f_0, and $(N, P)(f_0)$ in place of $(N(f_0), P(f_0))$. Similarly, $n(f_0)$ and $p(f_0)$ are the dimensions of the source and target of f_0 itself. We have to show that if a minimal stable unfolding of f_0 is not quasihomogeneous then $(N, P)(f_0)$ is not in the nice dimensions. The *stable codimension* of a germ f_0, $\text{sc}(f_0)$, is the number of unfolding parameters needed for a minimal stable unfolding. If f_0 has rank 0 then $\text{sc}(f_0) = \mathscr{K}_e\text{-codim}(f_0) - p$, so a minimal stable unfolding takes place in dimensions

$$(N, P)(f_0) = (n, p)(f_0) + \text{sc}(f_0)(1, 1).$$

Remark 7.4 Suppose that f_0 has stable unfolding F. Then as observed in Remark 7.1, the germ $x \mapsto (f_0(x), 0)$ has stable unfolding

$$F_1(x, u, v_1, \ldots, v_t) = (F(x, u), \sum_{j=1}^{t} v_j g_j, v_1, \ldots, v_t),$$

where $g_1, \ldots, g_t$ form a basis for $\mathfrak{m}_n / f_0^* \mathfrak{m}_p \mathscr{O}_n$ and the v_j are new unfolding parameters. Provided that the stable unfolding F is weighted homogeneous, then we can choose the g_j, and weights w_j for the extra parameters v_j, which make F_1 weighted homogeneous, since in particular f_0 itself is weighted homogeneous. Take, as the g_j, a monomial basis for $\mathfrak{m}_n / f_0^* \mathfrak{m}_p \mathscr{O}_n$; if the highest weight of any of the g_j is $w_{\max}$ then to the unfolding parameter v_j we can assign weight $w_{\max} - \text{weight}(g_j) + 1$. Thus, if the minimal stable unfolding of f_0 is weighted homogeneous, so is a minimal stable unfolding of $(f_0, 0)$. It is therefore enough to prove quasihomogeneity for stable unfoldings of rank 0 germs without redundant components. We call such stable unfoldings "irredundant".

However, there is a catch. As explained above, let $(N, P)(f_0)$ denote the dimensions of a minimal stable unfolding of f_0. Adding the redundant component to f_0, and following this procedure to construct a stable unfolding of the new germ, (N, P) increases by $(t, t+1)$. Even if $(N, P)(f_0)$ lies outside the nice dimensions, it might be possible that $(N, P) + m(t, t+1)$ lies *inside* the nice dimensions for some positive integer m. We have to show that no non-quasihomogeneous stable unfolding outside the nice dimensions moves into the nice dimensions by this means. In fact this is not hard. An easy calculation using the inequalities which describe the boundary of the nice dimensions (Exercise 3 below) shows that such a movement is only possible if $t < 6$. Mather's list of algebra types shows that for every algebra with $t < 6$, a minimal irredundant stable unfolding is weighted homogeneous. $\Diamond$

7.4.1 Multi-Germs

A multi-germ $f\colon (\mathbb{F}^n, S) \to (\mathbb{F}^p, 0)$ with branches $f^{(i)}$ is quasihomogeneous if there are strictly positive integer weights $w_1, \ldots, w_p$ (for the target) and $w_1^{(i)}, \ldots, w_n^{(i)}$, for $i = 1, \ldots, |S|$, for the source, such that each $f^{(i)}$ is equivariant with respect to the resulting $\mathbb{F}^*$-actions on $(\mathbb{F}^n, s_i)$ and $(\mathbb{F}^p, 0)$.

Proposition 7.3 *Suppose that the stable multi-germ $F\colon (\mathbb{F}^N, S) \to (\mathbb{F}^P, 0)$ has mono-germ branches $F^{(i)}\colon (\mathbb{F}^n, s_i) \to (\mathbb{F}^p, 0)$. If all of the $F^{(i)}$ are quasi-homogeneous, then it is possible to choose coordinates and weights making F quasihomogeneous.*

Proof Suppose that $F\colon (\mathbb{F}^n, S) \to (\mathbb{F}^p, 0)$ and $G\colon (\mathbb{F}^n, T) \to (\mathbb{F}^p, 0)$ are stable multi-germs, each quasihomogeneous in the sense just defined, and with $S \cap T = \emptyset$. Suppose also that the multi-germ $(F, G)\colon (\mathbb{F}^n, S \cup T) \to (\mathbb{F}^p, 0)$ is stable. Let I_F and I_G be their isosingular loci in $\mathbb{F}^p$. By the characterisation of the stability of multi-germs in Theorem 3.3, $I_F \pitchfork I_G$. As F and G are trivial unfoldings with parameter spaces I_F and I_G, (F, G) is a trivial unfolding, with parameter space $I_F \cap I_G$. Discarding this trivial factor in source and target, we may suppose that $I_F \cap I_G = \{0\}$. Since the two are transverse, near 0 in any representative, $\mathbb{F}^p$ is isomorphic to a product $I_F \times I_G$. Then up to right-left equivalence, F and G are trivial unfoldings, of the form

$$F_0 \times \mathrm{id} : F^{-1}(I_G) \times I_F, (S \times \{0\}) \to (I_G \times I_F, (0, 0))$$

and

$$\mathrm{id} \times G_0\colon (I_G \times G^{-1}(I_F), \{0\} \times T) \to (I_G \times I_F, (0, 0).$$

Since F and G are quasihomogeneous, so are F_0 and G_0. It follows that we can equip $\mathbb{F}^p$ with the target weights of F_0 on I_G and the target weights of G_0 on I_F. On $(\mathbb{F}^n, s_i) \simeq (F^{-1}(I_G) \times I_F, (s_i, 0))$ we take the source weights of $F_0^{(i)}$ on $F^{-1}(I_G)$ and the target weights of G_0 on I_F. We do the same, *mutatis mutandis*, on $(\mathbb{F}^p, t_j)$. Now (F, G) is weighted homogeneous.

The proposition now follows, by induction on $|S|$. $\square$

The proposition shows that to prove Theorem 7.6, we need just to carry out the verification described above for mono-germs.

7.4.2 *The Case* $n \geq p$

We need only show that every stable germ $F: (\mathbb{F}^N, 0) \to (\mathbb{F}^P, 0)$ with $N \geq P \leq 6$ is weighted homogeneous in some local coordinate system. Suppose that f_0 is a germ of rank 0. Then $P(f_0) = \operatorname{codim}_{\mathscr{K}_e}(f_0)$.

If $p(f_0) = 1$ and $P(f_0) \leq 6$ then it follows that f_0 is a function of type A_k for $1 \leq k \leq 6$, D_k for $k = 4, 5, 6$ or E_6 (see Theorem 6.12 and Exercise 6.4.1). All of these have weighted homogeneous stable unfoldings (see Exercise 7.4.1 below).

The rank 0 germs $f_0: (\mathbb{F}^n, 0) \to (\mathbb{F}^p, 0)$ with $p = 2$ and $p = 3$ and $\operatorname{codim}_{\mathscr{K}_e}(f_0) < \infty$ are classified in Section 7 of [Mat71]. The following table shows Mather's list, together with their unfolding monomials and the dimensions $(N, P)(f_0)$. As can easily be seen, all are weighted homogeneous. Among them, only $I_{a,b}$ and $II_{a,b}$ with $a + b \leq 6$, and IV_a with $a \leq 3$, have $N = P \leq 6$. Of course, the germs of type III and V do not arise when $n \geq p$ (Table 7.1).

If $(n, p)(f_0) = (n, 2)$ with $n \geq 4$ then a calculation shows

$$\operatorname{codim}_{\mathscr{K}_e^{(2)}}(f_0) \geq 7,$$

so $(N, P)(f_0)$ is not in the nice dimensions. When $n = 4$ the calculation is as follows:

$$\dim \theta(f_0)/\mathfrak{m}_4^3\theta(f_0) = 2 \dim_{\mathbb{F}} \mathscr{O}_4/\mathfrak{m}_4^3 = 30,$$

while $(T\mathscr{K}_e f_0 + \mathfrak{m}_4^3\theta(f_0))/\mathfrak{m}_4^3\theta(f_0)$ is generated over $\mathbb{F}$ by

$$\partial f_0/\partial x_j,\ 1 \leq j \leq 4,\quad x_i\partial f_0/\partial x_j,\ 1 \leq i, j \leq 4,\quad (f_0)_i\partial/\partial y_j, 1 \leq i, j \leq 2,$$

Table 7.1 Stable unfoldings of germs of type $\Sigma^{2,0}$ [Mat71, §7]

Label	f_0	Unfolding monomials	QH? (weights) (N, P)
$I_{a,b}, II_{a,b}$ $b \geq a \geq 2$	$(xy, x^a \pm y^b)$	$\langle x, y\rangle\partial_1$, $\langle x, \ldots, x^{a-2}, y, \ldots, y^{b-2}\rangle\partial_2$	Yes (b, a) $(a+b, a+b)$
$III_{a,b}$ $b \geq a \geq 2$	(xy, x^a, y^b)	$\langle x, y\rangle\partial_1$, $\langle x, \ldots, x^{a-2}, y, \ldots, y^{b-1}\rangle\partial_2$, $\langle x, \ldots, x^{a-1}, y \ldots, y^{b-2}\rangle\partial_3$	Yes $(1, 1)$ $(2a+2b-1, 2a+2b)$
$IV_a, a \geq 3$	(x^2+y^2, x^a)	$\langle x\rangle\partial_1$, $\langle x, xy, \ldots, xy^{a-2}$, $y, \cdots, y^{a-2}\rangle\partial_2$	Yes $(1, 1)$ $(2a, 2a)$
$V_a, a \geq 3$	(x^2+y^2, x^a, yx^{a-1})	$\langle x, y\rangle\partial_1$, $\langle xy^0, \ldots, xy^{a-2}, y, \ldots, y^{a-1}\rangle\partial_2$ $\langle xy^0, \ldots, xy^{a-3}, y, \ldots, y^{a-2}\rangle\partial_3$	Yes $(1, 1)$ $(4a-1, 4a)$

subject (since $j^2 f_0$ is homogeneous) to the relation

$$\sum x_i \partial f/\partial x_i \equiv 2\Big((f_0)_1\partial/\partial y_1 + (f_0)_2\partial/\partial y_2\Big) \mod \mathfrak{m}_4^3\theta(f_0).$$

Thus the quotient $\theta(f_0)/(\mathfrak{m}_4^3\theta(f_0)+T\mathscr{K}_e f_0)$ has codimension $\geq 30-23=7$. The corresponding dimension when $n > 4$ is greater than this.

If $(n,p)(f_0) = (3,3)$, a similar calculation shows that $\operatorname{codim}_{\mathscr{K}_e^{(2)}}(f_0) \geq 10$. If $n > p = 3$ or $n \geq p > 3$ then the corresponding codimension is greater. Thus $(N,P)(f_0)$ is not in the nice dimensions.

We are left only with the case $(n,p)(f_0) = (3,2)$. Rank zero germs in these dimensions are partially classified in Section 12 of [Mat71]. The 2-jets with $\mathscr{K}_e^{(2)}$–codimension less than 7 are in the $\mathscr{K}^{(2)}$-orbits of

$$(x^2-y^2, x^2 \pm z^2)),\ (2xy+z^2, z(x-y)),\ (x(y-z), yz) \text{ and } (x(y-z), y^2+z^2),$$

with $\mathscr{K}_e^{(2)}$-codimension 5, 5, 6 and 6, respectively. Of these, $(x^2-y^2, x^2 \pm z^2))$ and $(2xy+z^2, z(x-y))$, are in fact 2-determined for $\mathscr{K}$-equivalence. Germs with 2-jet in the remaining classes are $\mathscr{K}$-equivalent to $(x(y-z), yz \pm x^k)$ for $k \geq 3$, and to $(x(y-z), y^2+z^2+x^k)$ for $k \geq 3$. The following table shows their unfolding monomials. As can be seen, all are weighted homogeneous with respect to suitable strictly positive weights (Table 7.2).

Table 7.2 Stable unfoldings of rank zero germs $(\mathbb{F}^3, 0) \to (\mathbb{F}^2, 0)$ [Mat71, §12]

f_0	Unfolding monomials	QH? (weights) (N, P)
$(x^2-y^2, x^2 \pm z^2)$	$\langle x, z\rangle\partial y_1$, $\langle y\rangle\partial y_2$	Yes (1, 1) (6, 5)
$(2xy+z^2, z(x-y))$	$\langle z\rangle\partial y_1$, $\langle y, z\rangle\partial y_2$	Yes (1, 1) (6, 5)
$(x(y-z), yz \pm x^k)$, $k \geq 3$	$\langle y, z\rangle\partial y_1$, $\langle x, \ldots, x^{k-2}, z\rangle\partial y_2$	Yes $(2, k, k)$ $(k+2, k+1)$
$(x(y-z), y^2+z^2+x^k)$, $k \geq 3$	$\langle y, z\rangle\partial y_1$, $\langle x, \ldots, x^{k-2}, z\rangle\partial y_2$	Yes $(2, k, k)$ $(k+2, k+1)$

7.4.3 *The Case* n < p

To determine the collection of stable germs whose quasihomogeneity we have to prove, we first remove all those types whose k-jet lies in Mather's set $\Pi^k(n, p)$, in the process listing the isomorphism classes of stable germs whose k-jet lies in the complement of $\Pi^k(n, p)$. At the end of this subsection, we check that all of these stable germs are weighted homogeneous.

We describe $\Pi^k(n, p)$ first for $k = 2$. In Section 3 of [Mat71], Mather shows

(A) If $r = 1$ or 2, then $\Pi^2_r(n, p) = \emptyset$.
(B) $\Pi^2_3(n, p)$ is the set of 2-jets of corank 3 for which the kernel of the linearised intrinsic second derivative $S^2K \to C$ has dimension at least 3.
(C) if $p - n \leq \frac{r(r-1)}{2} - 2$, then $\Pi^2_r(n, p)$ is the set of 2-jets of corank $r \geq 4$.
(D) if $p - n \geq \frac{r(r-1)}{2} - 2$ then $\Pi^2_r(n, p)$ is the set of 2-jets of corank $r \geq 4$ for which the kernel of the linearised intrinsic second derivative has dimension at least 2.

Note that in case (C), the dimension of this kernel is always at least 2.

The Set Described Is Contained in $\Pi^2_r(n, p)$

The argument is the same in each case: the dimension of the set of jets is greater than the dimension of the quotient of the contact group acting on it. Thus, it consists of uncountably many orbits. In more detail: for any germ $f : (\mathbb{F}^n, 0) \to (\mathbb{F}^p, 0)$, the *intrinsic second derivative* is a bilinear map $d^2 f : K \times K \to C$, where K and C are the kernel and cokernel of $d_0 f$, obtained by restricting the second derivative $D^2_0 f : \mathbb{F}^n \times \mathbb{F}^n \to \mathbb{F}^p$ to $K \times K$ and composing with the projection $\mathbb{F}^p \to C$. It is "intrinsic" because it transforms properly under the derivatives of coordinate changes in source and target, and thus its isomorphism class under the action of $\mathrm{Gl}(K) \times \mathrm{Gl}(C)$ is independent of the choice of coordinates. The *linearised* intrinsic second derivative is the linear map $S^2K \to C$ it induces, where S^2K is the second symmetric product of K. It is easily seen that the intrinsic second derivative of a germ is determined by its 2-jet, and thus we can refer to the intrinsic second derivative of a 2-jet z. Mather shows that the 2-jets z and z' of the same corank, r, are contact-equivalent if and only if their linearised intrinsic second derivatives are equivalent under the action of $\mathrm{Gl}(K) \times \mathrm{Gl}(C)$. Let $S(s)$ denote the set of 2-jets for which the kernel of the linearised second intrinsic derivative has dimension s. The 2-jets z and z' in $S(s)$ are equivalent if and only if the kernels of their linearised intrinsic second derivatives are in the same orbit of the action of $\mathrm{Gl}(K)$ on the Grassmannian $\mathrm{Gr}(s, S^2K)$ of s-planes in S^2K. The dimension of $\mathrm{Gr}(s, S^2K)$ is $s(\frac{1}{2}r(r+1) - s)$, while the dimension of the group acting, $\mathbb{P}\mathrm{Gl}(K)$, is $r^2 - 1$. Thus, if

$$s(\frac{1}{2}r(r+1) - s) > r^2 - 1 \tag{7.12}$$

then there are uncountably many orbits of $\mathbb{P}\mathrm{Gl}(K)$ in $\mathrm{Gr}(s, S^2K)$, and so $S(s)$ is contained in $\Pi_r^2(n, p)$. Since $\Pi_r^k(n, p)$ is closed, by definition, it also contains $S(s')$ for $s' > s$.

The Set Described Is All of $\Pi_r^2(n, p)$

Mather shows in each case that for each $k \geq 2$ there are only finitely many $\mathscr{K}^{(k)}$ orbits outside the candidate for $\Pi_r^2(n, p)$.

(A) When $r = 1$, the local algebra of a $\mathscr{K}$-finite germ of corank r is isomorphic to $\mathbb{F}[x]/(x^k)$ for some $k < \infty$. Thus for each value of n, p, k there is only one contact class, and hence, up to $\mathscr{A}$-equivalence, only one stable germ of this type.
When $r = 2$, $\dim S^2K = 3$. Denote by $S(s)$ the set of 2-jets for which the dimension of the kernel of the linearised intrinsic second derivative $d^2z : S^2K \to C$ is s. Then $S(0)$ and $S(3)$ consist of just one orbit, since in these cases $\mathrm{Gr}(s, S^2K)$ consists of a single point. Both $S(1)$ and $S(2)$ consist of finitely many orbits, for the classification of line and planes in $S^2K = \mathbb{F}^3$ under the action of $\mathrm{Gl}(K)$ is the same as the classification of non-zero quadratic forms in 2 variables.

(B) The case $r = 3$ is the most complicated. The classification of $S(1)$ is the classification of quadratic forms in three variables, and therefore finite. The classification of $S(2)$ is the classification of pencils (1-dimensional spaces) of quadratic forms in three variables. Mather shows that over $\mathbb{C}$ there are eight classes, which break up into 13 classes over $\mathbb{R}$. We describe the corresponding local algebra types below.

(C) When $r \geq 4$ and $p - n \leq \frac{1}{2}r(r-1) - 2$, then $p - n + r \leq \frac{1}{2}r(r+1) - 2$ so the kernel of the linearised intrinsic second derivative has dimension ≥ 2. It follows that $\dim S(s) > \dim \mathbb{P}\mathrm{Gl}(K)$ and so all 2-jets of this type lie in $\Pi_r^2(n, p)$.

(D) When $r \geq 4$ and $p-n \geq \frac{1}{2}r(r-1)-2$, we have to consider also $S(0)$ and $S(1)$, which may be non-empty. A line in S^2K corresponds to a quadratic form in r variables, and the elements of $S(1)/\mathrm{Gl}(K)$ are in bijection with equivalence classes of quadratic forms under the action of $\mathrm{Gl}(K)$. Hence there are finitely many classes. When $s = 0$, the Grassmannian consists only of one point. Again, we describe the corresponding algebra types below.

We now consider in turn the classification of the local algebras corresponding to the dimension ranges specified in cases (A)-(D), whose 2-jet does not lie in the subset we have identified as $\Pi^2(n, p)$. We will use Mather's notation, in which $V_k(I)$, for an ideal $I \subset \mathfrak{m}_n/\mathfrak{m}_n^{k+1}$, denotes the set of all k-jets contact-equivalent to a jet whose component functions generate the ideal I, and in which, for an ideal $I \subset \mathbb{F}[[x_1, \ldots, x_n]]$, $V(I)$ denotes the set of infinity-jets whose component formal power series generate an ideal formally contact-equivalent to I. In this notation, the dimension of the target, p, of the jet is not determined. Thus, for example, the 2-jets of $f(x, y) = (x^2, xy, y^2)$ and $g(x, y) = (x^2, xy, y^2, 0)$ both belong to $V_2(x^2, xy, y^2)$. The value of n is implicit, but is not specified by the notation. Thus, $V_2(x^2)$, in the context of ideals in $\mathfrak{m}_2/\mathfrak{m}_2^3$, means something different from $V_2(x^2)$ in the context of ideals in $\mathfrak{m}_1/\mathfrak{m}_1^3$ or $\mathfrak{m}_3/\mathfrak{m}_3^3$. Provided attention is paid to the context, this need not cause confusion.

A1: Classification of Jets of Type Σ^1

In the case of germs of corank 1, the notation just described is not needed. The local algebra of a germ of corank 1 is isomorphic to $\mathbb{F}[x]/(x^k)$ for some finite k. Thus, an irredundant germ of corank 1 and rank 0 is $\mathscr{K}$-equivalent to $x \mapsto x^k$. An irredundant minimal stable germ of this type is therefore $\mathscr{A}$-equivalent to

$$(x, u) \mapsto (x^k + u_{k-2}x^{k-2} + \cdots u_1 x, u).$$

A2: Classification of Jets of Type Σ^2

Mather shows first that the set of 2-jets of rank 0 is equal to the union of

$$V_2(x^2, xy, y^2),\ V_2(xy, x^2 \pm y^2),\ V_2(xy, x^2),\ V_2(xy),\ V_2(x^2 + y^2)$$
$$V_2(x^2),\ V_2(0). \tag{7.13}$$

Those on the first row here are of type $\Sigma^{2,0}$, while 2-jets in $V_2(x^2)$ or $V_2(0)$ are of type $\Sigma^{2,1}$ and $\Sigma^{2,2}$, respectively.

Theorem 7.7 ([Mat71, Section 7]) *If $F: (\mathbb{F}^N, 0) \to (\mathbb{F}^P, 0)$, $N < P$, is a minimal stable germ of type $\Sigma^{2,0}$ then it is one of the unfoldings described in Table 7.1.*

Theorem 7.8 ([Mat71, Section 8]) *If $F: (\mathbb{F}^N, 0) \to (\mathbb{F}^P, 0)$, $N < P$, is a minimal stable germ of type $\Sigma^{2,1}$ and (N, P) lie in Mather's nice dimensions then it is one of the unfoldings described in Tables 7.3 and 7.4.*

In [Mat71, §8], Mather lists some more $\mathscr{K}^{(4)}$ classes of germs of type $\Sigma^{2,1}$. These are shown in Table 7.5. None of them give rise to stable germs in the nice dimensions, since in each case, unfolding f_0 by even the monomials making up a basis for $\mathfrak{m}_n\theta(f_0)/T\mathscr{K} f_0 + \mathfrak{m}^5\theta(f_0)$ gives an unfolding for which (source dimension, target dimension) already lies outside the nice dimensions.

Section 9 of [Mat71] considers the classification of jets of type $\Sigma^{2,2}$. Let

$$W_i = \{z \in V_2(0) : \dim \mathfrak{m}(z)^3/\mathfrak{m}(z)^4 = i\}.$$

Then $V_2(0) = W_1 \cup W_2 \cup W_3 \cup W_4$.

Table 7.3 Stable unfoldings of germs of type $\Sigma^{2,1}$ [Mat71, §8]

f_0	Unfolding monomials	QH? (weights) (N, P)
(x^2, xy^2, y^3)	$\langle x, y, y^2\rangle\partial_1, \langle x, xy, y, y^2\rangle\partial_2, \langle x, y\rangle\partial_3$	Yes (1, 1) (11, 12)
(x^2, y^3)	$\langle y, y^2\rangle\partial_1, \langle x, xy, y\rangle\partial_2$	Yes (2, 1) (7, 7)
$(x^2 + y^3, xy^2, y^4)$	$\langle x, xy, y, y^2\rangle\partial_1, \langle x, xy, y, y^2\rangle\partial_2,$ $\langle x, y, y^2\rangle\partial_3$	Yes (3, 2) (13, 14)
$(x^2 + y^3, xy^2)$	$\langle x, xy, y, y^2\rangle\partial_1, \quad \langle x, y\rangle\partial_2$	Yes (3, 2) (8, 8)
(x^2, xy^2, y^k)	$\langle x, xy, y, \ldots, y^{k-1}\rangle\partial_1, \langle x, xy, y\rangle\partial_2,$ $\langle x, xy, y, \ldots, y^{k-2}\rangle\partial_3$	Yes $(k, 2)$ $(2k+6, 2k+7)$
$(x^2 \pm y^k, xy^2, y^{k+1})$	$\langle x, xy, y, y^2\rangle\partial_1,$ $\langle x, x^2, y, \ldots, y^{k-1}\rangle\partial_2,$ $\langle x, xy, y, \ldots, y^{k-2}\rangle\partial_3$	Yes if $k < 6$ $(k, 2)$ $(2k+7, 2k+8)$ not in nice dims. if $k \geq 4$
$(x^2 \pm y^k, xy)$	$\langle x, xy, y\rangle\partial_1, \langle x, xy, y, \cdots, y^{k-2}\rangle\partial_2$	Yes $(k, 2)$ $(k+5, k+5)$
$(x^2 + y^3, y^k, xy^{k-1})$	$\langle x, y, y^2\rangle\partial_1,$ $\langle$monomials of wt $< 2k\rangle\partial_2,$ $\langle$monomials of wt $< 2k+1\rangle\partial_3$	Yes (3, 2) $(4k-1, 4k)$

Clearly W_0 is a single $\mathscr{K}^{(3)}$-orbit; it is represented by the 3-jet

$$(x^3, x^2y, xy^2, y^3),$$

which is 3-determined for contact equivalence.

The $\mathscr{K}^{(3)}$-classification of W_1 is the classification of hyperplanes in the space of binary cubics under the contragredient action of Gl(2). This is isomorphic to the classification of binary cubics themselves; over $\mathbb{R}$ there are four classes of hyperplanes, represented by $H_1^{\pm} = \{a \pm c = 0\}$, $H_2 = \{c = 0\}$ and $H_3 = \{a = 0\}$. Thus

$$W_1 = V_3(x^3 \pm xy^2, x^2y, y^3) \bigcup V_3(x^3, xy^2, y^3) \bigcup V_3(x^3, x^2y, xy^2).$$

All but the last of these is 3-determined for contact equivalence.

Table 7.4 Stable unfoldings of germs of type $\Sigma^{2,1}$ [Mat71, §8] (cont.)

f_0	Unfolding monomials	QH? (weights) (N, P)
$(x^2 + y^3, y^k)$ $k \geq 3$	$\langle y, y^2 \rangle \partial_1$, $\langle y, y^2, x \ldots, x^\alpha, xy, \ldots, x^\beta y, xy^2, \ldots, x^\gamma y^2 \rangle \partial_2$ $\max\{3\alpha, 3\beta + 2, 3\gamma + 4\} < 2k$	Yes $(3, 2)$ $(2k + 1, 2k + 1)$
$(x^2 + y^3, xy^{k-1}, y^{k+1})$ $k \geq 4$	$\langle x, xy, y, y^2 \rangle \partial_1$, $\langle y, y^2, x, \ldots, x^{\alpha_1}, xy, \ldots, x^{\beta_1} y, xy^2, \ldots x^{\gamma_1} y^2 \rangle \partial_2$, $\langle y, y^2, x, \ldots, x^{\alpha_2}, xy, \ldots, x^{\beta_2} y, xy^2, \cdots, x^{\gamma_2} y^2 \rangle \partial_3$ $\max\{3\alpha_1, 3\beta_1 + 2, 3\gamma_1 + 4\} < 2k + 1$ $\max\{3\alpha_2, 3\beta_2 + 2, 3\gamma_2 + 4\} < 2k + 2$	Yes $(3, 2)$ $(4k - 3, 4k - 2)$
$(x^2 + y^3, xy^k + y^{k+2})$ $k \geq 3$		No: f_0 not QH $(N, P) \geq (10, 10)$ not in nice dims.
$(x^2 + y^3, xy^k)$ $k \geq 3$	$\langle x, xy, y, y^2 \rangle \partial_1$ $\langle x, \ldots, x^\alpha, y, xy, \ldots, x^\beta y, y^2, xy^2, \ldots, x^\gamma y^2 \rangle \partial_2$ $\max\{3\alpha, 3\beta + 2, 3\gamma + 4\} < 2k + 3$	Yes $(3, 2)$ $(2k + 4, 2k + 4)$

Here we have not attempted to write down a minimal set of unfolding monomials, since the inequalities on degrees are too lengthy to include in the table

Table 7.5 Stable unfoldings of germs of type $\Sigma^{2,1}$ outside the nice dimensions [Mat71, §8]

f_0	Unfolding monomials	QH? (weights) (N, P)
(x^2, xy^3, y^4)	$\langle x, y, y^2, y^3 \rangle \partial_1$, $\langle x, xy, xy^2, y, y^2 \rangle \partial_2$ $\langle x, xy, xy^2, y, y^2, y^3 \rangle \partial_3$	Yes $(5, 3)$ $(16, 17)$
(x^2, y^4)	$\langle y, y^2, y^3 \rangle \partial_1$, $\langle x, xy, xy^2, y, y^2 \rangle \partial_2$	Yes $(5, 3)$ $(10, 10)$
$(x^2 \pm y^4, xy^3 + \alpha y^5)$	$\langle x, y, xy, y^2, y^3 \rangle \partial_1$ $\langle x, xy, y, y^2, y^5 \rangle \partial_2$	No $(10, 10)$
$(x^2 \pm y^4, 0)$	$\mathscr{K}_4^{(4)}$ – codim $= 10$	$(N, P) \geq (10, 10)$ not in nice dims.
(x^2, xy^3)	$\mathscr{K}_4^{(4)}$ – codim $= 10$	$(N, P) \geq (12, 12)$ not in nice dims.
$(x^2, 0)$	$\mathscr{K}_4^{(4)}$ – codim $= 12$	$(N, P) \geq (14, 14)$ not in nice dims.

Table 7.6 Stable unfoldings of germs of type $\Sigma^{2,2}$ [Mat71, §9]

f_0	Unfolding monomials	QH? (weights) (N, P)
(x^3, x^2y, xy^2, y^3)	$\langle x, xy, y, y^2\rangle\partial_1$ $\langle x, xy, y, y^2\rangle\partial_2$ $\langle x, x^2, xy, y, y^2\rangle\partial_3$ $\langle x, x^2, xy, y, y^2\rangle\partial_3$	Yes (1, 1) (20, 22)
$(x^3 \pm xy^2, x^2y, y^3)$	$\langle x, xy, y, y^2\rangle\partial_1$ $\langle x, xy, y, y^2\rangle\partial_2$ $\langle x, x^2, xy, y, y^2\rangle\partial_3$	Yes (1, 1) (15, 16)
(x^3, x^2y, y^3)	$\langle x, xy, y, y^2, xy^2\rangle\partial_1$ $\langle x, xy, y, y^2\rangle\partial_2$ $\langle x, x^2, xy, y, y^2\rangle\partial_3$	Yes (3, 2) (16, 17)
(x^3, x^2y, xy^2, y^k) $k \geq 4$	$\langle x, xy, y, \ldots, y^{k-1}\rangle\partial_1$ $\langle x, x^2, xy, y, \ldots, y^{k-1}\rangle\partial_2$ $\langle x, x^2, xy, y, y^2\rangle\partial_3$ $\langle x, x^2, xy, y, \ldots, y^{k-2}\rangle\partial_4$	Yes $(k, 2)$ $(3k + 11, 3k + 13)$
$(x^3 + y^k, x^2y, xy^2)$ $k \geq 4$	$\langle x, xy, y, \ldots, y^{k-2}\rangle\partial_1$ $\langle x, x^2, xy, y, \ldots, y^{k-1}\rangle\partial_2$ $\langle x, x^2, xy, y, y^2\rangle\partial_3$	Yes if $k < 6$ $(k, 3)$ $(2k + 9, 2k + 10)$
$(x^3, x^2y + y^k, xy^2)$ $k \geq 4$	$\langle x, xy, y, \ldots, y^{k-2}\rangle\partial_1$ $\langle x, x^2, xy, y, \ldots, y^{k-1}\rangle\partial_2$ $\langle x, x^2, xy, y, y^2\rangle\partial_3$	Yes $(k - 1, 2)$ $(2k + 10, 2k + 11)$

A jet in W_2 corresponds to a 2-plane in the space C of binary cubics. The dimension of the Grassmannian of 2-planes in C is 4; the quotient of $\mathcal{K}^{(3)}$ acting is just $\mathbb{P}\mathrm{Gl}(3)$, of dimension 3. Thus there are uncountably many orbits, so that $W_2 \cup W_3 \cup W_4 = \overline{W_2}$ is contained in $\Pi^3_2(n, p)$ and thus contributes no stable germs in the nice dimensions.

Section 9 of [Mat71] ends with the conclusion that $\Pi_2(n, p) = \overline{W_2} \cup \overline{V_4}(x^2 + y^4, xy^3)$ (Table 7.6).

B: Classification of Jets of Type Σ^3

In §10, Mather identifies a set W of 3-jets not contained in $(\pi^3_2)^{-1}(\Pi^2_2(n, p))$ and shows that for $k \geq 3$,

$$\Pi^k_3(n, p) = (\pi^k_2)^{-1}(\Pi^2_3(n, p)) \bigcup (\pi^k_3)^{-1}(W);$$

Table 7.7 Stable unfoldings of germs of type $\Sigma^3$with $\dim_{\mathbb{F}} \mathfrak{m}_3^2/f_0^*\mathfrak{m}_p = 1$ [Mat71, §10]

f_0	Unfolding monomials	QH? (weights) (N, P)
$D_{0,0}$ $(x^2, y^2, z^2, xy, yz, zx)$	$\langle y, z\rangle\partial_1$, $\langle x, y, z\rangle\partial_2$, $\langle x, y, z\rangle\partial_3$, $\langle y, z\rangle\partial_4$ $\langle x, y, z\rangle\partial_5$, $\langle y, z\rangle\partial_6$	Yes $(1, 1, 1)$ $(18, 21)$
$D_{3,0}$ $(x^2 - y^2, y^2 - z^2, xy, yz, zx)$	$\langle y, z\rangle\partial_1$, $\langle x, y, z\rangle\partial_2$, $\langle y, z\rangle\partial_3$, $\langle x, y, z\rangle\partial_4$ $\langle y, z\rangle\partial_5$	Yes $(1, 1, 1)$ $(15, 17)$
$D_{2,0}$ $(x^2 - y^2, z^2, xy, yz, zx)$	$\langle y, z\rangle\partial_1$, $\langle x, y, y^2, z\rangle\partial_2$, $\langle y, z\rangle\partial_3$, $\langle x, y, z\rangle\partial_4$, $\langle y, z\rangle\partial_5$	Yes $(2, 2, 3)$ $(N, P) = (16, 18)$
$D_{1,0}(2, \ell)$ $(y^2 \pm x^\ell, z^2 \pm x^\ell, xy, xz, yz)$ for $\ell \geq 3$	$\langle y, z, x, \ldots, x^{\ell-1}\rangle\partial_1$, $\langle y, z, x, \ldots, x^{\ell-2}\rangle\partial_2$ $\langle y, z, x, \ldots, x^{\ell-2}\rangle\partial_3$, $\langle y, z\rangle\partial_4$, $\langle y, z\rangle\partial_5$	Yes $(2, \ell, \ell)$ $(3\ell + 9, 3\ell + 11)$
$D_{1,0}(1, \ell)$ $(y^2 \pm x^\ell, z^2, xy, xz, yz)$ for $\ell \geq 3$	$\langle y, z, x, \ldots, x^{\ell-2}\rangle\partial_1$, $\langle y, z, x, \ldots, x^{\ell-1}\rangle\partial_2$ $\langle y, z, y^2, x, \ldots, x^{\ell-1}\rangle\partial_3$, $\langle y, z\rangle\partial_4$ $\langle y, z\rangle\partial_5$	Yes $(2, \ell, \ell + 1)$ $(3\ell + 10, 3\ell + 12)$
$D_{1,0}(\ell)$ $(x^\ell, y^2, z^2, xy, xz, yz)$ for $\ell \geq 3$	$\langle y, z\rangle\partial_1$ $\langle y, z\rangle\partial_2$ $\langle y, z, x, \ldots, x^{\ell-1}\rangle\partial_3$, $\langle y, z, x, \ldots, x^{\ell-1}\rangle\partial_4$ $\langle y, z, x, \ldots, x^{\ell-1}\rangle\partial_5$ $\langle y, z, x, \ldots, x^{\ell-2}\rangle\partial_6$	Yes $(2, \ell, \ell)$ $(4\ell + 10, 4\ell + 13)$

that is, for every k there are only finitely many contact classes of corank 3 in the complement of $(\pi_2^k)^{-1}(\Pi_3^2(n, p)) \bigcup (\pi_3^k)^{-1}(W)$. The set W will become clear in the calculations which now follow.

We have to show that every stable germ $F\colon (\mathbb{F}^N, 0) \to (\mathbb{F}^P, 0)$ in one of these contact classes is quasihomogeneous, if (N, P) are nice dimensions.

We know that the complement in Σ^3 of $\Pi_3^2(n, p)$ is the set of 2-jets of corank 3 for which $\dim \mathfrak{m}(z)^2/\mathfrak{m}(z)^3 \leq 2$. If $\mathfrak{m}(z)^2/\mathfrak{m}(z)^3 = 0$ then for any germ f_0 with $j^2 f_0 = z$, $f_0^*\mathfrak{m}_p\mathscr{O}_3 + \mathfrak{m}_3^3 = \mathfrak{m}_3^2$ and so by Nakayama's Lemma, $f_0^*\mathfrak{m}_p\mathscr{O}_3 = \mathfrak{m}_3^2$. This type occupies the first row in Table 7.7. It does not appear in explicit coordinates in [Mat71]. We give it the label $D_{0,0}$, consistent with the labelling in the next paragraph.

Consider next the case $\dim \mathfrak{m}(z)^2/\mathfrak{m}(z)^3 = 1$. The following analysis applies in all dimensions r, though here in particular we are interested in the case $r = 3$.

Since we consider only jets of rank 0, the jet corresponds to an ideal I such that $I \subset \mathfrak{m}_r^2$, and $\dim \mathfrak{m}_r^2/I = 1$. A 2-jet z of this type can be represented as a tuple $(q_1, \ldots, q_c)$ of quadratic forms in r variables, whose linear span $\langle q_1, \ldots, q_c \rangle$ has codimension 1 in the space Q_r of quadratic forms in r variables. If $\langle q_1, \ldots, q_c \rangle = \langle q_1', \ldots, q_c' \rangle$ then the 2-jets z and z' are equivalent via a linear change of coordinates in the target, and thus the classification of these 2-jets under contact equivalence reduces to the classification of hyperplanes in Q_r under the action of Gl_r by linear coordinate changes in the r variables. Hyperplanes are determined by their equations, which lie in the dual projective space $\mathbb{P}Q_r^\vee$, and thus the classification of the 2-jets we are concerned with reduces to the classification of the projectivised space of symmetric matrices under the contragredient action of Gl_r. Over $\mathbb{R}$, each class is represented by a diagonal matrix with ± 1 or 0 in each diagonal entry, and at least one non-zero entry. Thus, the classes are $D_{ab} := \mathrm{diag}(1, \ldots, 1, -1, \ldots, -1, 0, \ldots, 0)$ with a 1's and b -1's, which corresponds to the hyperplane

$$\left\{ \sum_{i \le j} \alpha_{ij} x_i x_j \in S_r : \alpha_{11} + \cdots + \alpha_{aa} - \alpha_{a+1,a+1} - \cdots - \alpha_{a+b,a+b} = 0 \right\}.$$

There are only finitely many of these classes, and they fall into two subclasses:

(i) If $a + b > 1$ then the ideal $I := (q_1, \ldots, q_c) \subset \mathscr{O}_r$ contains $\mathfrak{m}_r^3$. In this case the jet $z = (q_1, \ldots, q_c)$ is sufficient for contact equivalence, in the sense that all germs $(\mathbb{F}^r, 0) \to (\mathbb{F}^c, 0)$ with this 2-jet are contact-equivalent to the germ $(q_1, \ldots, q_c)$. For by an easy application of Nakayama's Lemma, any ideal $(q_1 + h_1, \ldots, q_c + h_c) \subset \mathscr{O}_r$ with the $h_i \in \mathfrak{m}^3$ is equal to I. Two germs $(\mathbb{F}^r, 0) \to (\mathbb{F}^c, 0)$ with the same ideal are contact-equivalent, by Proposition 4.6. Over $\mathbb{C}$, $D_{a,b}$ collapses to $D_{a+b,0}$. The unfolding monomials over $\mathbb{R}$ and over $\mathbb{C}$ are the same, so in Table 7.7 we list only one representative of each class over $\mathbb{C}$. Since these classes are not described explicitly in [Mat71], they appear in Table 7.7 with the label $D_{a+b,0}$ we have given here.

(ii) If $a + b = 1$ then we may assume $a = 1, b = 0$, and then the ideal I contains all quadratic forms with x_1^2 coefficient equal to 0. In [Mat71, Proposition 4.1], Mather shows that the ideal in the ring of formal power series generated by the Taylor series of the component functions of any germ with a 2-jet of this type is formally contact-equivalent to one of

$$\left(x_1^\ell, \{x_i x_j : 2 \le i, j \le r\} \cup \left\{ x_1 x_i : i \ge 2 \right\} \right) \tag{7.14}$$

for some $\ell > 2$, or

$$\left(\left\{ x_i x_j + \alpha_{ij} x_1^\ell : 2 \le i, j \le r \right\} \cup \left\{ x_1 x_i : i \ge 2 \right\} \right) \tag{7.15}$$

for some $\ell > 2$, with $\alpha_{ij} \in \mathbb{F}$ and at least one $\alpha_{ij} \neq 0$, or to

$$\left(\left\{x_i x_j : 2 \leq i, j \leq r\right\} \cup \left\{x_1 x_i : i \geq 2\right\}\right). \tag{7.16}$$

We denote a germ of the type (7.14) by $D_{1,0}(\ell)$. Ideals of type (7.16) need not concern us, since a germ of this type is not $\mathscr{K}$-finite and therefore has no stable unfolding. Germs of type (7.14) and (7.15) are ℓ-determined for contact equivalence. In [Mat71, Proposition 5.1], Mather shows that two germs

$$\left(\left\{x_i x_j + \alpha_{ij} x_1^{\ell} : 2 \leq i, j \leq r\right\} \cup \left\{x_1 x_i : i \geq 2\right\}\right)$$

and

$$\left(\left\{x_i x_j + \alpha'_{ij} x_1^{\ell'} : 2 \leq i, j \leq r\right\} \cup \left\{x_1 x_i : i \geq 2\right\}\right)$$

of type (7.15) are contact-equivalent if and only if $\ell = \ell'$, and the symmetric matrices determined by the $\alpha_{i,j}$ and the $\alpha'_{i,j}$ are equivalent under the symmetric action of $\mathrm{Gl}_{n-1}(\mathbb{F})$ and multiplication by the ℓ'th power of a scalar in $\mathbb{F}$. Orbits under this action are represented by diagonal matrices $\mathrm{diag}(\pm 1, \ldots, \pm 1, 0, \ldots, 0)$ with ± 1 in k places and $r - k$ zeroes. We denote the corresponding germs by $D_{1,0}(k, \ell)$. Thus, when $r = 3$, we obtain the following collection of stable map-germs. All are weighted homogeneous with respect to suitable weights.

Now we consider the case where $\dim_{\mathbb{F}} \mathfrak{m}(z)^2/\mathfrak{m}(z)^3 = 2$. Here we must classify 2-jets consisting of c-tuples of quadrics in 3 variables whose linear span has codimension 2 in the 6-dimensional space $S^2\mathbb{F}^3$ of ternary quadrics. After discarding redundant components, $c = 4$. A 4-plane is determined by the 2-plane in $\mathbb{P}(S^2\mathbb{F}^3)^\vee$ of linear forms annihilating it, and thus the classification here reduces to the classification of pencils of ternary quadrics. Over $\mathbb{C}$ there are 8 classes, which split into 13 classes over $\mathbb{R}$. The ideals corresponding to them are listed in §10 of [Mat71] (Table 7.8).

Section 10 of [Mat71] continues with a list of further algebra types. None gives rise to any stable germ in the nice dimensions. The following table shows the list, in which members have one or two integer parameters; the second column shows the value of (N, P) for a stable germ of the type given when the parameters take their smallest value. The right-hand column shows $\dim_{\mathbb{F}} Q(f_0) - 1$.

In addition to the strata shown in Tables 7.9 and 7.10, there is one further stratum, of codimension $6(p - n) + 10$. Since $\sigma(n, p) = 6(p - n) + 9$, this means that no stable map in the nice dimensions can meet it.

Table 7.8 Stable unfoldings of germs of type Σ^3 with $\dim_{\mathbb{F}} \mathfrak{m}_3^2/f_0^*\mathfrak{m}_p = 2$, in the nice dimensions [Mat71, §10]

f_0	Unfolding monomials	QH? (weights) (N, P)
$(x^2+y^2\pm z^2, xy, xz, yz)$	$\langle y,z\rangle\partial_1,\quad \langle y,z\rangle\partial_2,$ $\langle x,y,z\rangle\partial_3,\quad \langle y,z\rangle\partial_4$	Yes $(1,1,1)$ $(12,13)$
$(x^2, y^2, z^2-xy, xz+yz)$	$\langle y,z\rangle\partial_1,\quad \langle x,y,z\rangle\partial_2,$ $\langle y,z\rangle\partial_3,\quad \langle y,z\rangle\partial_4$	Yes $(1,1,1)$ $(12,13)$
$(x^2, y^2, z^2, xy+xz)$	$\langle y,z,yz\rangle\partial_1,\quad \langle x,z\rangle\partial_2,$ $\langle x,y,z\rangle\partial_3,\quad \langle y,z\rangle\partial_4$	Yes $(3,2,2)$ $(13,14)$
$(x^2, y^2-z^2, xy+xz, yz)$	$\langle y,z,z^2\rangle\partial_1,\quad \langle x,y,z\rangle\partial_2,$ $\langle y,z\rangle\partial_3,\quad \langle x,z\rangle\partial_4$	Yes $(2,2,3)$ $(13,14)$
(x^2, y^2, xz, yz, z^k) $k\geq 3$	$\langle y,z,\ldots,z^{k-1}\rangle\partial_1,$ $\langle x,z,\ldots,z^{k-1}\rangle\partial_2,$ $\langle x,y,z,\rangle\partial_3,\quad \langle x,y,z\rangle\partial_4,$ $\langle x,y,xy,z,\ldots,z^{k-2}\rangle\partial_5$	Yes $(k,k,2)$ $(3k+10, 3k+12)$
$(x^2\pm z^k, y^2\pm z^k, xz, yz)$ $k\geq 3$	$\langle y,z.\ldots,z^{k-1}\rangle\partial_1,$ $\langle x,z,\ldots,z^{k-2}\rangle\partial_2,$ $\langle x,y,z\rangle\partial_3,\quad \langle x,y,z\rangle\partial_4$	Yes $(k,k,2)$ $(2k+8, 2k+9)$
$(x^2-y^2, xy, xz, yz, z^k)$ $k\geq 3$	$\langle y,z,\ldots,z^{k-1}\rangle\partial_1,$ $\langle y,z,\ldots,z^{k-1}\rangle\partial_2,$ $\langle x,y,z\rangle\partial_3,\quad \langle x,y,z\rangle\partial_4,$ $\langle x,y,y^2,z,\ldots,z^{k-2}\rangle\partial_5$	Yes $(k,k,2)$ $(3k+10, 3k+12)$
$(x^2\pm z^k, y^2, xz, yz)$ $k\geq 3$	$\langle y,z,\ldots,z^{k-2}\rangle\partial_1,$ $\langle x,x^2,z,\ldots,z^{k-1}\rangle\partial_2,$ $\langle x,y,z\rangle\partial_3,\quad \langle x,y,z\rangle\partial_4$	Yes $(k,k,2)$ $(2k+10, 2k+10)$
$(x^2-y^2+z^k, xy, xz, yz)$ $k\geq 3$	$\langle y,z,\ldots,z^{k-2},\rangle\partial_1,$ $\langle y,z,\ldots,z^{k-1}\rangle\partial_2,$ $\langle x,y,z\rangle\partial_3, \langle x,y,z\rangle\partial_4$	Yes $(k,k,2)$ $(2k+8, 2k+9)$
(x^2+yz, xz, y^2, z^2)	$\langle y,z\rangle\partial_1\quad \langle y,z\rangle\partial_2,$ $\langle x,z\rangle\partial_3, \langle x,y,z,xy,yz\rangle\partial_4$	Yes $(4,3,5)$ $(14,15)$
$(x^2+yz, xy, xz, z^2, y^k)$ $k\geq 3$	$\langle y,z\rangle\partial_1,\quad \langle x,y,z\rangle\partial_2,$ $\langle z,y,\ldots,y^{k-1}\rangle\partial_3,$ $\langle x,z,yz,y\ldots,y^{k-1}\rangle\partial_4,$ $\langle x,z,yz,y,\ldots,y^{k-2}\rangle\partial_5$	Yes $(4k+3, 8k-6, 12k)$ $(3k+11, 3k+13)$

Table 7.9 Contact classes of germs of type Σ^3 with $\dim_{\mathbb{F}} \mathfrak{m}_3^2/f_0^*\mathfrak{m}_p = 2$ outside the nice dimensions [Mat71, §10]

f_0	(N, P)	$\dim_{\mathbb{F}} Q(f_0)$
$(x^2+yz, xy, xz, z^2 \pm y^k)$ $k \geq 3$	$(2k+9, 2k+10)$	$k+3$
$(x^2+yz, xy, xz+y^k, z^2)$ $k \geq \ell \geq 3$	$(2k+10, 2k+11)$	$k+3$
$(xy, xz, z^2, x^k, y^\ell)$ $\ell \geq k \geq 3$	$(26, 28)$	$k+\ell+2$
$(xy, xz, yz, z^2 \pm y^\ell, x^k \pm y^\ell)$ $k, \ell \geq 3$	$(21, 23)$	$k+\ell$
$(xy, xz, yz, z^2, x^k \pm y^\ell)$ $k, \ell \geq 3$	$(22, 24)$	$k+\ell$
$(xy, xz, yz, z^2 \pm y^\ell, x^k)$ $k \geq 3$	$(22, 24)$	$k+\ell$
$(xy, xz+y^\ell, yz, z^2, x^k)$ $k, \ell \geq 3$	$(23, 25)$	$k+\ell$
$(xy, xz, yz, z^2 \pm y^\ell \pm x^k)$ $k, \ell \geq 3$	$(16, 17)$	$k+\ell+1$
$(xy, xz+y^\ell, yz, z^2 \pm x^k)$ $k, \ell \geq 3$	$(17, 18)$	7
$(xy, xz+y^\ell, yz+x^k, z^2)$ $k \geq 3$	(18,19)	7
$(x^2+y^2, xz, yz, z^2, x^k)$ $k \geq 3$	(22,24)	$2k$
$(x^2+y^2, xz, yz, z^2 \pm x^k)$ $k \geq 3$	$(16, 17)$	$2k+1$
$(x^2+y^2, xz+x^k, yz, z^2)$ $k \geq 3$	$(4k+6, 4k+7)$	$2k+1$
(x^2, y^2, yz, z^2) $k \geq 3$	(16,17)	5
$(y^2, xz, yz, z^2, x^3, x^2y)$ $k \geq 3$	(27,30)	5

C and D: Germs of Corank ≥ 4

If now $r \geq 4$ then taking $s = 2$, the inequality (7.12) holds. It follows that the only stable germs of corank ≥ 4 in the nice dimensions will be those for which $s = 0$ or $s = 1$.

By the analysis in case B, the germs are those listed in the following table.

Table 7.10 Stable unfoldings of corank ≥ 4 in the nice dimensions [Mat71, §5]

Label	f_0	Unfolding monomials	QH? (weights)
$D_{0,0}$	All monomials of degree 2	Monomials of degree 1	Yes $(1, \ldots, 1)$
$D_{r,0}$	$\begin{pmatrix} x_1^2 - x_i^2 : 1 < i, \\ x_i x_j : i < j \end{pmatrix}$	Monomials of degree 1	Yes $(1, \ldots, 1)$
$D_{r-1,0}$	$\begin{pmatrix} x_1^2, x_2^2 - x_i^2 : 2 < i \\ x_i x_j : i < j \end{pmatrix}$	Monomials of degree 1, $x_n^2 \partial_1$	Yes $(3, 2, \ldots, 2)$
$D_{r-2,0}$	$\begin{pmatrix} x_1^2, x_2^2, x_3^2 - x_i^2 : 3 < i \\ x_i x_j : i < j \end{pmatrix}$	Monomials of degree 1, $x_n^2 \partial_{1,1}, x_n^2 \partial_{2,2}, x_n^2 \partial_{1,2}$ (see explanation below)	Yes $(3, 3, 2, \ldots, 2)$
$\vdots$	$\vdots$	$\vdots$	$\vdots$
$D_{2,0}$	$\begin{pmatrix} x_1^2, \ldots, x_{n-2}^2, \\ x_{n-1}^2 - x_n^2 \\ x_i x_j : i < j \end{pmatrix}$	Monomials of degree 1, $x_n^2 \partial_{i,j} : 1 \leq i \leq j \leq n-2$ (see explanation below)	Yes $(3, \ldots, 3, 2, 2)$
$D_{1,0}(\ell)$ $\ell \geq 3$	$\begin{pmatrix} x_1^\ell, x_2^2, \ldots, x_n^2, \\ x_i x_j : i < j \end{pmatrix}$	Monomials of degree 1 $\langle x^2, \ldots, x^{\ell-2} \rangle \partial_1$, $\langle x^2, \ldots, x^{\ell-1} \rangle \partial_i, i \geq 2$	Yes $(2, \ell, \ldots, \ell)$
$D_{1,0}(\ell, k)$ $\ell \geq 3$, $n \geq k \geq 1$	$\begin{pmatrix} x_1^\ell + x_2^2, \ldots, x_1^\ell + x_k^2, \\ x_{k+1}^2, \ldots x_n^2, \\ x_i x_j : i < j \end{pmatrix}$	Monomials of degree 1 $\langle x_1^2, \ldots, x_1^{\ell-2} \rangle \partial_1$ $\langle x_1^2, \ldots, x_1^{\ell-1} \rangle \partial_i, i \geq 2$ $\langle x_2^2 \rangle \partial_i, k+1 \leq i \leq n$	Yes $(2, \underbrace{\ell, \ldots, \ell}_{k-1},$ $\underbrace{\ell+1, \ldots, \ell+1}_{n-k})$

1. The classification of germs of type $D_{i,0}(\ell)$ and $D_{i,0}(k, \ell)$ is described above in the text preceding displayed equations (7.14) and (7.15)
2. "Monomials of degree 1": not all are required for a minimal stable unfolding, since the $\partial f/\partial x_i$ provide n linear combinations of them. Listing a minimal basis is left to the reader
3. In the list of unfolding monomials for $D_{r-2,0}$ and $D_{2,0}$, "$\partial_{i,j}$" refers to the coordinate vector field for the component carrying the monomial $x_i x_j$ in f_0. Thus, in the case of $D_{r-2,0}$, $\partial_{1,1}$, $\partial_{2,2}$ and $\partial_{1,2}$ are the first, second and n'th coordinate vector fields

Exercises for Sect. 7.4

1 Show that all of the singularities simple singularities of functions (that is, functions of type A_k, D_k, E_6, E_7 and E_8) have weighted homogeneous stable unfoldings.

2 Suppose that f_1 and f_2 are $\mathscr{A}$-equivalent, with $f_1 = \psi \circ f_2 \circ \phi$. Express the intrinsic second derivative of f_1 in terms of the intrinsic second derivative of f_2, together with the derivatives of ψ and ϕ.

3 (*cf.* Remark 7.4 above) Suppose that

(i) $n \geq \frac{6}{7}p + \frac{8}{7}$ and
(ii) $n + mt < \frac{6}{7}(p + m(t+1)) + \frac{8}{7}$.

Show that $t < 6$.

4 Using the lists given in this section, or otherwise, find all local algebras of $\mathbb{F}$-vector-space dimension ≤ 5, and verify that a stable unfolding of an irredundant rank 0 germ with this local algebra is quasihomogeneous.

5 A locally closed analytic subspace $X \subset \mathbb{F}^N$ is *locally quasihomogeneous* if for all $x \in X$, there exist local coordinates centred at x, and a set of positive weights for those coordinates, with respect to which the defining ideal of X is generated by quasihomogeneous functions.

Show that if $f : N^n \to P^p$ is a stable mapping of real or complex manifolds, and (n, p) are in the nice dimensions, then the image of f (if $n < p$) or the discriminant of f (if $n \geq p$) is locally quasihomogeneous.

Part II
Images and Discriminants: The Topology of Stable Perturbations

Chapter 8
Stable Images and Discriminants

8.1 Introduction

In low-dimensional examples, it is possible to make drawings of the images and discriminants of stable perturbations of real $\mathscr{A}$-finite map germs. In some simple cases, by judicious choice of values for the unfolding parameters (several topologically inequivalent stable perturbations are available over $\mathbb{R}$) it is possible to arrange that the real image or discriminant is a deformation retract of the image or discriminant of a stable perturbation of the complexified germ. This is the case with the Reidemeister moves, shown in Fig. 1.4, where the good real picture is shown in each case on the right. It is the case also for germs $(\mathbb{R}^2, S) \to (\mathbb{R}^3, 0)$ of $\mathscr{A}_e$-codimension 1, whose stable images are shown in Figs. 8.1 and 8.2.

Example 8.1 Figure 8.3 shows the discriminant of a stable perturbation of the $\mathscr{A}_e$-codimension 1 bi-germ

$$\begin{cases}(x, y, z) \mapsto (x, y, z^3 - xz),\\ (x, y, z) \mapsto (x, y^3 + xy, z).\end{cases}$$

The picture shows, with a dotted line, two irreducible components, the discriminants of the two component mono-germs. Each is isomorphic to the product of a first-order cusp $\{(u, v) : u^2 = v^3\}$ and a line. Their intersection and their cuspidal edges are drawn with a continuous line. Their union carries a non-trivial 2-cycle, which appears in the drawing as the curvilinear tetrahedron whose edges are made up of the intersection of the two discriminants, together with their cuspidal edges.

As can be seen, despite their very different combinatorial structures, each image or discriminant has the homotopy type of a sphere of dimension $p - 1$.

D. Mond, J. J. Nuño-Ballesteros, *Singularities of Mappings*, Grundlehren der mathematischen Wissenschaften 357, https://doi.org/10.1007/978-3-030-34440-5_8

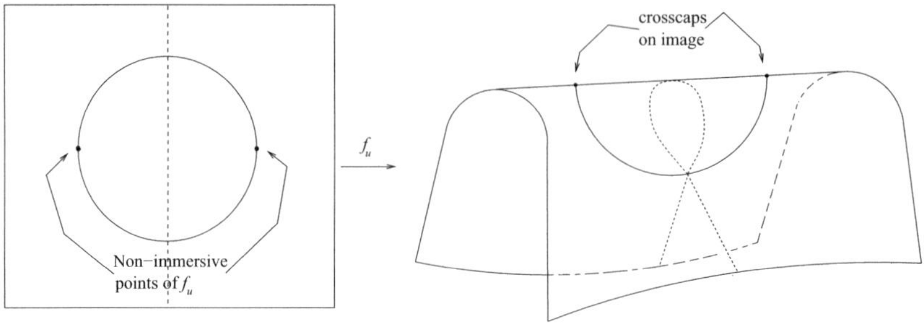

Fig. 8.1 Stable perturbation of $(x, y) \mapsto (x, y^2, y^3 + x^2 y)$

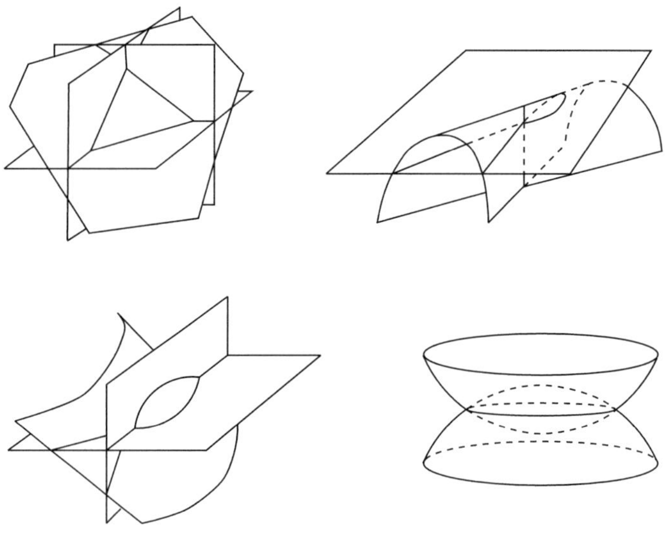

Fig. 8.2 Images of stable perturbations of codimension 1 multi-germs of maps from the plane to 3-space

Figure 8.4, from a drawing by Victor Goryunov in [Gor91], shows the image of a stable perturbation of a germ of type H_2, of $\mathscr{A}_e$-codimension 2. Here the image has the homotopy type of a wedge of two 2-spheres. Again, we see the coincidence of the number of spheres in the wedge and the $\mathscr{A}_e$-codimension.

In Sect. 8.3 we use Morse theory to prove that in the complex case, the discriminant of a stable perturbation of an $\mathscr{A}$-finite germ $f : (\mathbb{C}^n, S) \to (\mathbb{C}^p, 0)$ with $n \geq p$ always has the homotopy type of a wedge of spheres of dimension $p - 1$, and, similarly, that the image of a stable perturbation of an $\mathscr{A}$-finite germs $(\mathbb{C}^n, S) \to (\mathbb{C}^{n+1}, 0)$ has the homotopy type of a wedge of spheres of dimension n. The number of spheres in the wedge is called the *discriminant Milnor number* in the first case, and the *image Milnor number* in the second, and denoted by μ_Δ and μ_I, respectively. It is the key geometric invariant in each case.

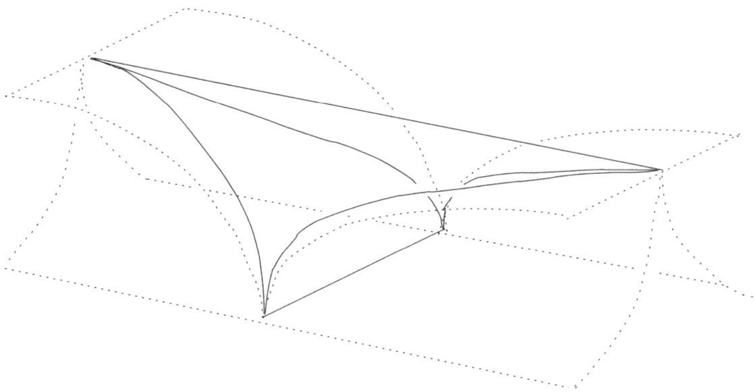

Fig. 8.3 Discriminant of the bi-germ of Example 8.1 (after [CMWA02])

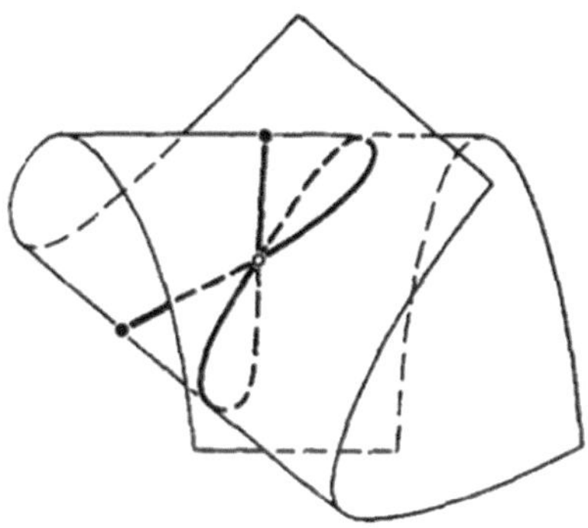

Fig. 8.4 Image of stable perturbation of the germ of type H_2, $(u, x) \mapsto (u, x^3, ux + x^5)$ (after [Gor91])

The resemblance to the classical theory of isolated hypersurface singularities is clear, and, as might be guessed at from these examples, extends to the "Milnor–Tjurina relation": in Sect. 8.5 we prove

Theorem 8.1 ([DM91]) *Let $f : (\mathbb{C}^n, S) \to (\mathbb{C}^p, 0)$ have finite $\mathcal{A}_e$-codimension, with $n \geq p$ and (n, p) nice dimensions. Then*

$$\mu_\Delta(f) \geq \operatorname{codim}_{\mathscr{A}_e}(f)$$

with equality if f is weighted homogeneous.

The relation between image Milnor number and $\mathscr{A}_e$-codimension for a germ $(\mathbb{C}^n, S) \to (\mathbb{C}^{n+1}, 0)$ is conjecturally the same, but the results here are less complete, and require more technical development, so that proofs are postponed to Chap. 11.

Theorem 8.2 ([dJvS91, Mon95]) *Let $f : (\mathbb{C}^n, S) \to (\mathbb{C}^{n+1}, 0)$ ($n = 1$ or 2) have finite $\mathcal{A}_e$-codimension. Then*

$$\mu_I(f) \geq \operatorname{codim}_{\mathscr{A}_e}(f) \tag{8.1}$$

with equality if f is weighted homogeneous.

Conjecture 8.1 The statement of Theorem 8.2 holds for all n such that $(n, n+1)$ are nice dimensions.

Example 8.2 (Looking at Bent Wires IV) Theorem 8.2 explains an empirical observation that the reader will perhaps have made following the suggestions in Examples 1.1 and 3.4. If C is a generic smooth simple curve in $\mathbb{R}^3$, then beside the three singularities at the centre of the Reidemeister moves (Fig. 1.4), the following singularities, and only these, may appear on a planar projection from some viewpoints. Each is characterised geometrically, and, except for the quadruple crossing, each forms a single $\mathscr{A}$-class. The quadruple crossing has a modulus (a continuous left-right invariant), namely the cross ratio of the four tangent lines at the crossing point, and thus this singularity type is made up of an uncountable number of $\mathscr{A}$-classes.

For each type X,

- let $\operatorname{View}(X)$ be the set of centres of projection $u \in \mathbb{R}^3$ for which a singularity of this type appears on the radial projection of C from u into a plane; that is, it is the set of points in $\mathbb{R}^3$ from which a singularity of type X can be seen when looking with one eye closed;
- let $n_{\max}(X)$ be the maximum number of nodes into which a singularity of this type splits on moving the centre of projection, and
- let r be the number of branches (local irreducible components) of a singularity of this type.

We encourage the reader to check that, with the exception of the quadruple crossing,

$$n_{\max} - r + 1 = \operatorname{codim} \operatorname{View}(X), \tag{8.2}$$

and when X is the quadruple crossing,

$$n_{\max} - r + 1 = \operatorname{codim} \operatorname{View}(X) + 1. \tag{8.3}$$

It is possible to check this using a (very thin) bent wire, but easier, and in some ways more instructive, to derive a suitable stable perturbation of each singularity from its geometrical description. This approach is justified by the fact that for a generic space-curve, the family of radial projections $\mathscr{A}_e$-versally unfolds each of its singularities, except for the quadruple point. In the case of the quadruple crossing, the cross ratio of the four tangent lines does not change as the centre of projection moves along the quadrisecant line. So, except for the quadruple crossing, any perturbation which can occur in theory will occur in the family of projections. Versality for the other singularities is proved by means of a transversality theorem proved by J.M.S. David in [Dav83]. Here we give a slightly sharpened version from [Mon86]:

Theorem 8.3 *Let M be a smooth manifold which can be embedded in $\mathbb{R}^p$, let H be an affine hyperplane in $\mathbb{R}^p$ and let W be a smooth submanifold of the jet bundle ${}_rJ^k(M, H)$. Then for a residual set of embeddings $i : M \to \mathbb{R}^p$, the r-jet extension map of the family of radial projections of $i(M)$ to H is transverse to W.*

Note that the domain of the family of projections excludes points $(q, x) \in \mathbb{R}^p \times M$ such that $q - i(x)$ is parallel to H.

Because $\mathscr{A}_e$ versality can be characterised in terms of the transversality of a suitable jet extension map to the relevant $\mathscr{A}$-orbit in jet space (Theorem 5.2), Theorem 8.3 implies that for any finite collection of $\mathscr{A}$-classes, every singularity of projection in those classes which occurs in the family of projections is versally unfolded. The singularity type "quadruple crossing" is made up of uncountably many $\mathscr{A}$-classes, because of the cross ratio. None can be versally unfolded in the family of projections, but equally, any finite number of them can be avoided, by the transversality theorem, for transversality holds vacuously if there is no intersection. Even though no individual orbit of a quadruple crossing can be met transversely by the family of projections, the submanifold of jet space consisting of the union of the $\mathscr{A}$-orbits can be.

The link between (8.1) and (8.2) is as follows. The image of a stable perturbation of an $\mathscr{A}$-finite singularity of projection of a space curve is a finite graph (1-dimensional CW complex) and therefore homotopy equivalent to a wedge of circles. A calculation with the Euler characteristic shows that for a graph with n vertices which can be assembled from r disjoint line segments, the number of circles in the wedge is equal to $n - r + 1$. Thus, the left-hand side of (8.2) is a real version of the image Milnor number μ_I which appears in Theorem 8.2. In fact the link is not just an analogy. Gusein-Zade [GZ74] and A'Campo [A'C75] proved that if $f : (\mathbb{C}, S) \to (\mathbb{C}^2, 0)$ is real (is defined by power series with real coefficients), then there exists a real stable perturbation all of whose nodes are located in $\mathbb{R}^2$. For any such perturbation, the inclusion of real image into its complexification is a homotopy-equivalence, and thus $n_{\max} - r + 1$ is indeed equal to the invariant μ_I of the complex parameterised curve singularity, which appears on the left-hand side of (8.1).

The right-hand side of (8.2), meanwhile, is equal to the $\mathscr{A}_e$ codimension of a germ of type X, by Proposition 3.9, and so (8.2) is just an instance of (8.1). The non-strict inequality in (8.1) is an equality here because all of the germs concerned are weighted homogeneous in suitable coordinates. ◊

Proof of Theorems 8.1 and 8.2 requires a deeper link between T^1_f and the geometry of the discriminant and the image. In both cases this is achieved by giving an alternative description of T^1_f in terms of this geometry. We develop this in

Sects. 8.5 and 8.4, after first studying the topology of images and discriminants in Sect. 8.3.

Discriminants, and the images of germs $(\mathbb{C}^n, S) \to (\mathbb{C}^{n+1}, 0)$, are hypersurfaces, and thus have a single defining equation. It is this that opens the way to the use of Morse theory. The images of germs $(\mathbb{C}^n, S) \to (\mathbb{C}^p, 0)$ with $p > n+1$ are not even complete intersections, and Morse theory cannot be used. The only calculation of the topology of the image of a stable perturbation available to us in this case uses the multiple point spaces of the germ, which we study in Chap. 9, and the Image Computing Spectral Sequence, developed in Chap. 10.

8.1.1 Complex Not Real

In this second part of the book we work almost entirely in the complex analytic case. There are several reasons. First, some key results, like conservation of multiplicity (Sect. E.3), hold only over $\mathbb{C}$. The second is that one of the principal techniques we use is the Morse theory of real-valued functions arising from complex analytic functions as real part, imaginary part or modulus. For all of these, the index of any non-degenerate critical point is always the same—the ambient complex dimension. These two crucial facts give a uniformity and simplicity to the complex case, which is absent from the real case. Why should complex singularity theory be simpler than the real? Perhaps the deepest reason is that in any complex analytic deformation with smooth complex parameter space, the bifurcation set (the set of parameters over which the object being deformed is not stable) is a complex analytic subvariety. As a consequence, it has real codimension at least two, and this means that it does not separate its complement into distinct connected components. Real bifurcation sets often do separate, and this is what allows the menagerie of different stable perturbations one observes near a real singularity. In the complex case there is only one stable perturbation, and the "nearby stable object" is unique up to homeomorphism. It is to this which we now turn our attention.

Exercises for Sect. 8.1

1 For each of the singularities shown in Fig. 8.5, find a weighted homogeneous parameterisation.

2 The following *Mathematica* script produces a numerically accurate picture of the surface drawn in Fig. 8.4.

```
ParametricPlot3D[{x, y^3 - 0.65y, xy + y^5 - 0.65y^3}, {x, -1, 1}, {y, -1, 1},
Mesh- > None, Boxed- > False, Axes- > False]}
```

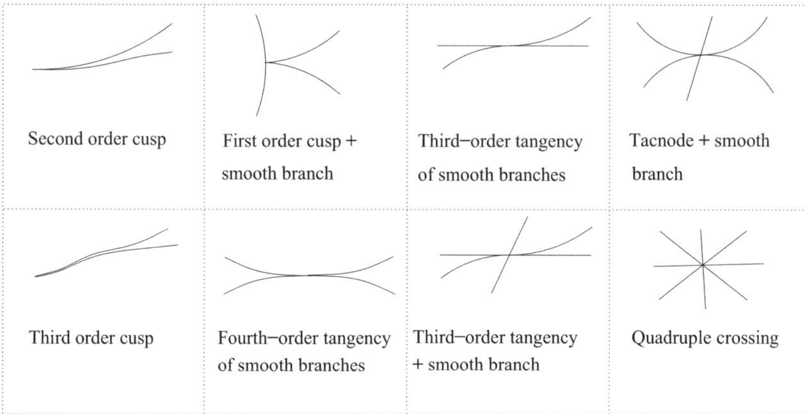

Fig. 8.5 Singularities of codimension > 1 of projection of a generic space-curve

Use it to explore the geometry of the surface. Is there a view which makes its topology as clear as the hand drawn picture?

3 For each of the germs $(\mathbb{R}^2, 0) \to (\mathbb{R}^2, 0)$ in the following list:

(i) Calculate $T\mathscr{A}_e f$, and then, using Theorem 8.1, find its discriminant Milnor number.

(ii) Draw pictures of its discriminant, and of the discriminant of a real stable perturbation. Over $\mathbb{R}$, there are usually several distinct stable perturbations, since the bifurcation set separates the parameter space of an $\mathscr{A}_e$-versal unfolding into several connected components. Is any of them a "good real picture", with discriminant in $\mathbb{R}^2$ having the same homology as the discriminant of the complexification? For information on the homology of a stable perturbation of the complexified map, see Theorem 8.1 above, and the discussion preceding it.

(i) $f(x_1, x_2) = (x_1, x_2^3 - x_1^2 x_2)$; (ii) $f(x_1, x_2) = (x_1, x_2^4 + x_1 x_2)$;
(iii) $f(x_1, x_2) = (x_1, x_2^3 + x_1^3 x_2)$.

8.2 Review of the Milnor Fibre

In the theory of isolated hypersurface singularities a key role is played by the Milnor fibre. Here is a very brief account. We suggest the book of Arnol'd, Gusein-Zade and Varchenko [AGZV12b] as a reference.

1. Let f be a complex analytic function defined on some neighbourhood of 0 in $\mathbb{C}^{n+1}$, and suppose it has isolated singularity at 0. Then by the curve selection lemma, there exists $\varepsilon > 0$ such that for ε' with $0 < \varepsilon' \leq \varepsilon$, the sphere of radius ε' centred at 0 is transverse to $f^{-1}(0)$. Let B_ε be the closed ball centred at 0

and with radius ε. Then from the transversality it follows that $f^{-1}(0) \cap B_\varepsilon$ is homeomorphic (indeed, diffeomorphic except at 0) to the cone on its boundary $f^{-1}(0) \cap S_\varepsilon$. The ball B_ε is a *Milnor ball* for the singularity.

2. By an argument involving properness, one can show that for suitably small $\eta > 0$, all fibres $f^{-1}(t)$ with $|t| < \eta$ are transverse to S_ε. Let D_η be the closed ball in $\mathbb{C}$ with radius η and centre 0, and let $D^*_\eta = D_\eta \setminus \{0\}$. By the Ehresmann fibration theorem,

$$f| : B_\varepsilon \cap f^{-1}(D^*_\eta) \to D^*_\eta$$

is a C^∞-locally trivial fibration. It is known as the Milnor fibration. Up to fibre-preserving homeomorphism, it is independent of the choice of ε.

3. Its fibre is called the *Milnor fibre* of f. It has the homotopy type of a wedge of n-spheres, whose number μ, the *Milnor number of* f, is equal to the dimension of the Jacobian algebra of f,

$$\mathcal{O}_n/J_f.$$

The argument for the last statement is based on the following two facts:

1. If $\dim_{\mathbb{C}} \mathcal{O}_n/J_f = 1$ (in which case f is said to have a "non-degenerate" critical point), then $J_f = \mathfrak{m}$ and by the holomorphic Morse lemma (or by Theorem 3.1(ii)), f is right-equivalent to $x \mapsto x_1^2 + \cdots + x_{n+1}^2$. An explicit calculation (see e.g. [Loo84, Chapter 2]) now shows that the Milnor fibre is diffeomorphic to the unit ball subbundle of the tangent bundle of S^n. This has S^n as a deformation retract, so in this case $\mu = 1$.
2. f can be perturbed so that the critical point at 0 splits into non- degenerate critical points. There are exactly μ of them, by conservation of multiplicity—see Corollary E.6—and each contributes one sphere to the wedge.

The dimension of the Jacobian algebra plays a second, completely different, role in the theory. The quotient by which we measure instability

$$\frac{\{\frac{d}{dt} f_t|_{t=0} : f_0 = f\}}{\{\frac{d}{dt} f \circ \varphi_t|_{t=0}\}}$$

is the self-same Jacobian algebra, and indeed the Jacobian ideal itself is the extended tangent space for right equivalence—see the discussion in the proof of Theorem 3.1 and following Proposition 3.1. The analogue of Theorem 5.1—that infinitesimal versality implies versality—shows that one can construct a versal deformation of f (versal for right equivalence, that is) by taking $g_1, \ldots, g_\mu \in \mathcal{O}_{\mathbb{C}^{n+1},0}$ whose images in the Jacobian algebra span it as vector space, and defining

$$F(x, u_1, \ldots, u_\mu) = f(x) + \sum_j u_j g_j.$$

The Milnor fibration $f| : B_\varepsilon \cap f^{-1}(D_\eta \setminus 0) \to D_\eta \setminus 0$ extends to a fibration

$$F \times \mathrm{id}_{\mathbb{C}^\mu}| : (B_\varepsilon \times B_\nu) \cap (F \times \mathrm{id}_{\mathbb{C}^\mu})^{-1} \left(D_\eta \times B_\nu \setminus \Delta\right) \to \left(D_\eta \times B_\nu \setminus \Delta\right), \tag{8.4}$$

where B_ν is a suitably small ball in the parameter space $\mathbb{C}^\mu$ and Δ is the set of critical values of $F \times \mathrm{id}_{\mathbb{C}^\mu}$. There is also a closely related fibration

$$\pi : B_\varepsilon \times B_\nu \cap F^{-1}(0) \to B_\nu \setminus \Delta_0,$$

where $\Delta_0 = \Delta \cap B_\nu \times \{0\}$, with the same fibre as the first fibration, obtained from the projection $F^{-1}(0) \to \mathbb{C}^\mu$.

By taking the cohomology bundle associated to either of these, we obtain a holomorphic vector bundle of rank μ. It is equipped with a canonical flat connection, the *Gauss–Manin connection*.

In the second part of this book we show that many of these same features can be found in the theory of singularities of mappings.

8.3 The Homotopy Type of the Discriminant of a Stable Perturbation: Discriminant and Image Milnor Numbers

Let $f : (\mathbb{F}^n, S) \to (\mathbb{F}^p, 0)$ be an $\mathscr{A}$-finite germ and assume that it has discrete stable type and that it admits a stabilisation (both conditions are satisfied if f has corank 1 or if (n, p) are nice dimensions, see Proposition 5.5 and Remark 5.3). Since f is $\mathscr{A}$-finite, it is finitely determined by Theorem 6.2 and we may assume that it is polynomial. We consider the stratification in the discriminant $\Delta(f)$ by stable types, which consists of finitely many semialgebraic strata.

Proposition 8.1 *There exists a real number $\varepsilon > 0$, such that*

1. *$\Delta(f)$ is transverse to $S_{\varepsilon'}$, for all ε' with $0 < \varepsilon' \leq \varepsilon$.*
2. *$\Delta(f) \cap B_\varepsilon$ is homeomorphic to the cone on its boundary $\Delta(f) \cap S_\varepsilon$.*

Proof

(1) Let $A_1, \dots, A_m$ be the strata of the stratification by stable types of $\Delta(f)$ and consider the polynomial map $\rho : \mathbb{F}^p \to \mathbb{R}$ given by $\rho(y) = \|y\|^2$. By the Curve Selection Lemma B.1, each restriction $\rho|_{A_i}$ has a finite number of critical points. Therefore, we can choose $\varepsilon > 0$ such that ε' is a regular value of $\rho|_{A_i}$, for all $i = 1, \dots, m$ and for all $0 < \varepsilon' \leq \varepsilon$. But this implies that A_i is transverse to $S_{\varepsilon'}$, for all $i = 1, \dots, m$ and for all ε' with $0 < \varepsilon' \leq \varepsilon$.

(2) This is the conic structure lemma of Burghelea and Verona. The transversality condition is equivalent to saying that the map

$$\rho : \Delta(f) \cap (B_\varepsilon \setminus \{0\}) \to (0, \varepsilon^2] \tag{8.5}$$

is a proper stratified submersion, and hence by the Thom–Mather First Isotopy Lemma B.2, it is a locally C^0-trivial fibration. As the base is contractible, it is globally trivial, so that

$$\Delta(f) \cap (B_\varepsilon \setminus \{0\}) \simeq (\Delta(f) \cap S_\varepsilon) \times (0, \varepsilon^2].$$

The homeomorphism extends to a homeomorphism between $\Delta(f) \cap B_\varepsilon$ and the cone on $\Delta(f) \cap S_\varepsilon$. □

Now let F be a mini-versal unfolding of f. Since f has discrete stable type, so does F and we can consider again the stratification by stable types of the discriminant $\Delta(F)$. We denote by $\pi : \Delta(F) \to (\mathbb{F}^d, 0)$ the restriction of the projection onto the parameter space. The assumption on the existence of the stabilisation of f implies that the bifurcation set $(\mathscr{B}(F), 0)$ is a proper closed subset of $(\mathbb{F}^d, 0)$ by Proposition 5.4.

Proposition 8.2 *Let $\varepsilon > 0$ be as in Proposition 8.1. There exists a real number $\eta > 0$ such that*

1. *The multi-germ $(f_u)_y$ is stable for all $y \in S_\varepsilon$ and for all u such that $\|u\| < \eta$.*
2. *$\Delta(f_u)$ is transverse to S_ε, for all $u \in \mathbb{F}^d$ such that $\|u\| < \eta$.*
3. *The restriction*

$$\pi| : \Delta(F) \cap \left(B_\varepsilon \times (\mathring{B}_\eta \setminus \mathscr{B}(F))\right) \to \mathring{B}_\eta \setminus \mathscr{B}(F)$$

is locally C^0-trivial as a stratified map (in the sense of Definition B.6).

Proof (1) Since $(f)_y$ is stable for all $y \in S_\varepsilon$, the same is true of $(f_u)_y$ for $\|u\|$ smaller than some positive η. Let $\{A_1, \dots, A_m\}$ be the stratification by stable types of $\Delta(F)$, then the stratification by stable types of $\Delta(f_u)$ is obtained by taking the sections of the strata A_i in $\mathbb{F}^p \times \{u\}$. Let $g : \mathbb{F}^p \times \mathbb{F}^d \to \mathbb{R}$ be the function $g(y, u) = \|y\|^2$. The fact that $\Delta(f)$ is transverse to S_ε means that for each $i = 1, \dots, m$, the restriction $g|_{A_i}$ is a submersion at any $(y, 0) \in S_\varepsilon \times \mathbb{F}^d$. By continuity of the derivatives, we can choose $\eta > 0$, such that for each $i = 1, \dots, m$, the restriction $g|_{A_i}$ is a submersion at any $(y, u) \in S_\varepsilon \times \mathbb{F}^d$, for all $u \in \mathbb{F}^d$ such that $\|u\| < \eta$. But this implies item (2).

To see item (3), we just need to see that the map π is a proper stratified submersion and then apply Thom's first isotopy lemma (Theorem B.2). Properness is obvious, since it is the restriction in the target of a proper map. To see that it is a stratified submersion, we first have to declare which is the stratification we are considering. In fact, we have two types of strata: strata of the form $A_i \cap (S_\varepsilon \times W)$ and of the form $A_i \cap (\mathring{B}_\varepsilon \times W)$, where $i = 1, \dots, m$ and $W = \mathring{B}_\eta \setminus \mathscr{B}(F)$.

For the strata of type $A_i \cap (S_\varepsilon \times W)$, the projection is submersive if and only if A_i is transverse to $S_\varepsilon \times W$, but this follows from item (1). For strata of type $A_i \cap (\mathring{B}_\varepsilon \times W)$, given a point $(y, u) \in W$, the multi-germ of (f_u) at y is stable, so the multi-germ of F at (y, u) is a trivial unfolding. In particular, there is a local diffeomorphism in $\mathbb{F}^p \times \mathbb{F}^d$ which preserves the parameters and that maps $\Delta(F)$ into $\Delta(f_u) \times \mathbb{F}^d$. Since this diffeomorphism must preserve the stratification, it follows that the projection π is a submersion on $A_i \cap (\mathring{B}_\varepsilon \times W)$. □

In the complex case, $\mathring{B}_\eta \setminus \mathscr{B}(F)$ is connected and hence, all the fibres $\Delta(f_u)$, with $u \in \mathring{B}_\eta \setminus \mathscr{B}(F)$ are homeomorphic, by a homeomorphism preserving the stratification. We recall that the map f_u was called in Chap. 5 a *stable perturbation* of f. In the case $n < p$, where Δ_{f_u} is the image of f_u, it is called the *disentanglement* of f. The topological invariants of the strata of $\Delta(f_u)$ provide interesting invariants of the germ f.

Example 8.3

1. If $f : (\mathbb{C}^n, 0) \to (\mathbb{C}, 0)$ is a function germ with isolated singularity, then the stable perturbation f_u is a Morse function and $\Delta(f_u)$ is a finite set which contains the critical points of f_u. The number of such critical points is an invariant of f.
2. Let $f : (\mathbb{C}, S) \to (\mathbb{C}^2, 0)$ be an $\mathscr{A}$-finite germ, whose image is a plane curve with isolated singularity. The stable perturbation f_u is an immersion with only transverse double points and its discriminant $\Delta(f_u)$ is the image of f_u, which is a plane curve with only nodal singularities. The number of such nodes of $\Delta(f_u)$ is now the well-known δ *invariant* of f.

We have already seen, in Example 5.5, that the *real* image of a suitable real form of each $\mathscr{A}_e$-codimension 1 singularity f of map from a surface to 3-space grows a 2-dimensional homotopy-sphere when f is suitably perturbed.

Proposition 8.3

1. *Suppose that* $f : (\mathbb{C}^n, S) \to (\mathbb{C}^{n+1}, 0)$ *is* $\mathscr{A}$*-finite. Then the disentanglement of* f *(the image of a stable perturbation of* f*) has the homotopy type of a wedge of* n*-spheres.*
2. *Suppose that* $f : (\mathbb{C}^n, S) \to (\mathbb{C}^p, 0)$ *is* $\mathscr{A}$*-finite, with* $n \geq p$*. Then the discriminant of a stable perturbation of* f *has the homotopy type of a wedge of* $(p-1)$*-spheres.*

In both cases, the number of spheres in the wedge is independent of the choice of stable perturbation.

Definition 8.1 The number of spheres in the wedge is called the *image Milnor number*, μ_I, in case (1), and the *discriminant Milnor number*, μ_Δ, in case (2).

We give two proofs of Proposition 8.3.

Proof 1 of Proposition 8.3 Both statements are consequences of a fibration theorem of Lê Dũng Tráng [Lê87, (4.4)], that says, in effect, that if (X, x_0) is a p-dimensional complete intersection singularity and $\pi : (X, x_0) \to (\mathbb{C}, 0)$ is a function with isolated singularity, then the analogue of the Milnor fibre of π (i.e. the intersection of a non-zero level set (in X) with a Milnor ball around x_0) has the homotopy type of a wedge of spheres of dimension $p - 1$. The notion of isolated singularity used here is simply that the restriction of π to each stratum of the canonical Whitney stratification of X is non-singular, except at the point x_0. To apply this theorem here, we take, as (X, x_0), the germ at $(0, 0)$ of the image in case (1), or discriminant, in case (2), of a 1-parameter stabilisation of f: that is, an unfolding $F : (\mathbb{C}^n \times \mathbb{C}, S \times \{0\}) \to (\mathbb{C}^p \times \mathbb{C}, (0, 0))$ with $F(x, u) = (\tilde{f}(x, u), u) = (f_u(x), u)$ such that f_u is stable for $u \neq 0$. Then in each case $(X, 0)$ is a hypersurface singularity, and thus a complete intersection. We take, as π, the projection to the parameter space. Thus $\pi^{-1}(u)$ is the image (or discriminant) of f_u. The fact that π has isolated singularity is a consequence of the following two facts:

1. f_u is stable everywhere, for $u \neq 0$. This implies that the unfolding is trivial away from $u = 0$, so that at each point (y, u) with $u \neq 0$, the vector field $\partial/\partial u$ in the target of π lifts to a vector field tangent to X.
2. f_0 is stable at each point $(y, 0)$ with $y \neq 0$, for f has isolated instability at 0, by the Mather–Gaffney criterion for $\mathscr{A}$-finiteness. Thus, once again the germ of unfolding is trivial, and the vector field $\partial/\partial u$ on the target of π lifts to a vector field tangent to the discriminant (or image) of f at (y, u).

In each case, the lift at (y, u) of the vector $\partial/\partial u$ is tangent to the isosingular locus of the unfolding F, and thus tangent to the stratum of the canonical Whitney stratification containing (y, u). It follows that the restriction of π to the stratum is a submersion at (y, u). This shows that π has isolated singularity. Independence of the choice of stable perturbation follows immediately from the fact that any stabilisation is equivalent to one induced from a versal unfolding, and thus the discriminant of a stable perturbation is stratified homeomorphic to the fibre of the fibration of Proposition 8.2(3). As the base of this fibration is path connected, all fibres are homeomorphic. □

Proof 2 of Proposition 8.3 The second proof, due to Dirk Siersma in [Sie91], gives additional information about the number of spheres in the wedge, and therefore we give it in greater detail. Siersma shows that the number of spheres is equal to the sum of the Milnor numbers of the isolated critical points of the defining equation g of X which move off X as f (and with it g) is deformed.

It is necessary to be quite careful about the level of generality in which we work here. Although the theorem is true in considerable generality, a general version requires some rather complicated hypotheses. We aim, more modestly, for a statement that holds for the discriminants and images of finitely determined germs (from now on we refer uniformly to "discriminants") and their stabilisations.

We need a preliminary definition.

Definition 8.2 If $W \subset \mathbb{C}^p$ is open and $g \in \mathscr{O}(W)$, then $\mathrm{Der}(-\log g)$ is the sheaf of vector fields on W which annihilate g.

The sheaf $\mathrm{Der}(-\log g)$ is in a natural way a (coherent) sheaf of $\mathscr{O}_W$-modules, and, indeed, a subsheaf of θ_W. The curious name is due to its close relation with the module $\mathrm{Der}(-\log X)$ of vector fields on W which are tangent to $X := \{g = 0\}$, which is dual to the module of differential forms with logarithmic poles along X.

Now we suppose that $F : A \to V \times U$ is a representative of a stabilisation of a finitely determined germ $f_0 : (\mathbb{C}^n, S) \to (\mathbb{C}^p, 0)$, with (n, p) in the nice dimensions and $n \geq p - 1$. Let $g : (V \times U, (0, 0)) \to (\mathbb{C}, 0)$ be a reduced equation for the discriminant X of F, and for each $u \in U$ let $g_u = g(\_, u)$ be the associated equation for the discriminant of f_u. We write X_u for the discriminant of f_u.

We begin by giving X the stratification by stable types of F outside $(0, 0)$ (together with the stratum $\{(0, 0)\}$ if F is not stable there). This is a finite locally analytically trivial stratification. We denote it by $\mathcal{A} = \{A_1, \ldots, A_m\}$. For each u, the stratification by stable types of X_u is denoted by $\mathcal{A}_u = \{(A_1)_u, \ldots, (A_m)_u\}$

1. By Proposition 8.1, there exists $\varepsilon > 0$ such that X_0 is stratified transverse to $S_{\varepsilon'}$ for all ε' with $0 < \varepsilon' \leq \varepsilon$ and $X_0 \cap B_\varepsilon$ is homeomorphic to the cone on its boundary $X_0 \cap S_\varepsilon$. At all points $y \in X_0 \cap S_\varepsilon$, f is stable at y and the stratum $(A_i)_0$ which contains it has dimension > 0. Thus, X_0 is locally analytically isomorphic to a non-trivial product, and there is a germ of vector field $\chi_y \in \mathrm{Der}(-\log g_0)_y$ such that $\chi_y \notin T_y S_\varepsilon$.
2. There exists $\eta > 0$ (depending on ε) such that for all t with $0 < |t| \leq \eta$, t is a regular value of g_0 and $g_0^{-1}(t) \pitchfork S_\varepsilon$. By the curve selection lemma, all critical values of g_0 are isolated, so there is an $\eta_0 > 0$ such that for all t with $0 < |t| \leq \eta_0$, t is a regular value of g_0. The second part of the assertion holds "by continuity", although a little more argument is required, for we are talking about stratified transversality rather than transversality. So in more detail: by item 1, for each $y \in X_0 \cap S_\varepsilon$, any representative of the germ χ_y gives a vector field $\tilde{\chi}_y \in \mathrm{Der}(-\log g)(W_y)$ for some neighbourhood W_y of y in $\mathbb{C}^p$, and we may suppose, by continuity, that $\tilde{\chi}_y(y') \notin T_{y'} S_\varepsilon$ for all $y' \in W_y$. Again by continuity, and by compactness of S_ε, there exist $\eta > 0$ with $0 < \eta \leq \eta_0$ such that for $|t| \leq \eta$,
$$g^{-1}(t) \cap S_\varepsilon \subset \bigcup W_y.$$
Then for $|t| \leq \eta$, $g^{-1}(t) \pitchfork S_\varepsilon$ as claimed.
3. There exists $\rho > 0$ (depending on ε and η) such that for all u with $|u| < \rho$, $X_u \pitchfork S_\varepsilon$ and for all t with $|t| = \eta$, t is a regular value of g_u and $g_u^{-1}(t) \pitchfork S_\varepsilon$. The fact that there exists $\rho > 0$ such that t is a regular value of g_u for all u, t with $|u| < \rho$ and $|t| = \eta$ is obvious by continuity of the partial derivatives of g_u and the compactness of ∂D_η, where D_η is the closed disc in $\mathbb{C}$ with centre 0 and radius η. The transversality assertion for X_u and $g_u^{-1}(t)$ follows by using analogous arguments to those of item 2.
4. For ε, η and ρ as in the previous steps, the projection $\pi : (B_\varepsilon \times \mathring{D}_\rho) \cap g^{-1}(D_\eta) \to \mathring{D}_\rho$ is a proper stratified submersion, and thus, by the Thom–Mather First Isotopy Lemma (B.2) a locally C^0-trivial fibre bundle.

   In fact, $M := (B_\varepsilon \times \mathring{D}_\rho) \cap g^{-1}(D_\eta)$ is a compact topological manifold with boundary of (real) dimension $2p + 2$ in $\mathbb{C}^p \times \mathbb{C}$, whose boundary is
$$\partial M = (S_\varepsilon \times \mathring{D}_\rho) \cap g^{-1}(D_\eta) \cup (B_\varepsilon \times \mathring{D}_\rho) \cap g^{-1}(\partial D_\eta).$$

One stratum of M is its interior and since it is an open subset of $\mathbb{C}^p \times \mathbb{C}$, it follows that π is a submersion on this stratum. The stratification of the boundary ∂M is induced from the stratification $\mathcal{A} = \{A_1, \ldots, A_m\}$ of X, with the following four types of strata:

a. $\tilde{A}_i := A_i \cap (S_\varepsilon \times \mathring{D}_\rho)$; given any point $(y, u) \in \tilde{A}_i$, we have $y \in X_u$; let $(A_i)_u$ be the stratum of X_u which contains y. Since f_u is stable at y, we have a commutative diagram

$$\begin{array}{ccc} U' \cap A_i & \xrightarrow{\phi} & (V' \cap (A_i)_u) \times W' \\ & \searrow^{\pi} & \downarrow \pi \\ & & W' \end{array}$$

where ϕ is a diffeomorphism and U',V' and W' are open neighbourhoods of (y, u), y and u, respectively. Moreover, by item 3, $(A_i)_u \pitchfork S_\varepsilon$ and the above diagram restricts to

$$\begin{array}{ccc} U' \cap \tilde{A}_i & \xrightarrow{\phi} & (V' \cap (A_i)_u \cap S_\varepsilon) \times W' \\ & \searrow^{\pi} & \downarrow \pi \\ & & W' \end{array}$$

Hence, π is a submersion at (y, u) on $\tilde{A}_i$.

b. $g^{-1}(\mathring{D}_\eta \setminus \{0\}) \cap (S_\varepsilon \times \mathring{D}_\rho)$; since this is open in $S_\varepsilon \times \mathring{D}_\rho$, π is obviously a submersion on this stratum.

c. $g^{-1}(\partial D_\eta) \cap (S_\varepsilon \times \mathring{D}_\rho)$; here π is a submersion because for all $t \in \partial D_\eta$ and for all $u \in \mathring{D}_\rho$, t is a regular value of g_u and $g_u^{-1}(t) \pitchfork S_\varepsilon$, by item 3.

d. $g^{-1}(\partial D_\eta) \cap (\mathring{B}_\varepsilon \times \mathring{D}_\rho)$; here π is once again a submersion because for all $t \in \partial D_\eta$ and for all $u \in \mathring{D}_\rho$, t is a regular value of g_u, again by item 3.

Moreover, the Whitney conditions are satisfied for this stratification. In fact, for any two strata of type (a), this follows from the fact that $\mathcal{A}$ is a Whitney stratification and for all other pairs of strata, the Whitney condition is easily checked.

5. Finally, we prove that $g_0^{-1}(D_\eta) \cap B_\varepsilon$, and thus all of the fibres of the bundle of item 4, are contractible (see Fig. 8.6). For $g_0^{-1}(D_\eta) \cap B_\varepsilon$ has $X_0 \cap B_\varepsilon$ as deformation retract, and we have chosen ε so that $X_0 \cap B_\varepsilon$ is contractible. We obtain the deformation retraction as follows: by item 2, $|g_0|$ has no critical value in the interval $(0, \eta]$; we let $\vartheta(x) = -x\partial/\partial x$ be the radial vector field on the complex disc D_η, whose flow retracts the disc to 0 in unit time. Then ϑ can be lifted to a controlled vector field $\tilde{\vartheta}$ on $|g_0|^{-1}([0, \eta])$, by [GWdPL76, Prop. 3.3]. As ϑ is globally integrable, so is $\tilde{\vartheta}$, by [GWdPL76, 4.8]. The integral flow of $\tilde{\vartheta}$ provides a deformation retraction of $g_0^{-1}(D_\eta) \cap B_\varepsilon$ to $X_0 \cap B_\varepsilon$.

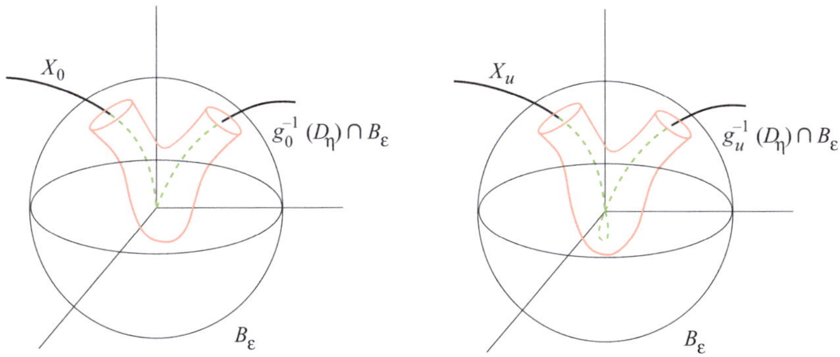

Fig. 8.6 The fibres $g_0^{-1}(D_\eta) \cap B_\varepsilon$ and $g_u^{-1}(D_\eta) \cap B_\varepsilon$

With $g, \varepsilon, \eta, \rho$ as just described, all fibres of g_u are smooth, except for $X_u = g_u^{-1}(0) \cap B_\varepsilon$, which in general has non-isolated singularities, and for some other fibres $g_u^{-1}(t) \cap B_\varepsilon$, with $t \neq 0$; these have isolated singularities, since the singular set of $g_u^{-1}(t)$ for $t \neq 0$ does not meet S_ε, and therefore must be zero-dimensional.

Theorem 8.4 ([Sie91]) *Under these circumstances, X_u has the homotopy type of a wedge of spheres of dimension $p-1$, and the number of these spheres is equal to the sum of the Milnor numbers of the singularities of g_u in the fibres different from X_u.*

Proof The proof uses Morse theory. Up to homotopy equivalence, $B_\varepsilon \cap g_u^{-1}(D_\eta)$ is obtained from $X_u = B_\varepsilon \cap g_u^{-1}(0)$ by progressively thickening it, i.e. by considering

$$|g_u|^{-1}([0, \eta'])$$

and increasing η'. For small enough η', $|g_u|^{-1}([0, \eta'])$ has X_u as deformation retract, by the argument of item 4 above. Now we consider the change in homotopy type of $|g_u|^{-1}([0, t]$ as t increases from η' to η. Except when t passes through a critical value of $|g_u|$, the thickening does not change the homotopy type. The critical points of $|g_u|$ off $g_u^{-1}(0)$ are the same as those of g_u (Exercise 1(i) below). Suppose first that each is non-degenerate. Each has Morse index equal to the ambient dimension, because of the complex structure (Exercise 1(ii)). Thus, the space $B_\varepsilon \cap g_u^{-1}(D_\eta)$ is obtained from $g_u^{-1}(0)$ by gluing in cells of dimension p.

Now suppose the critical point z_i off the zero level is degenerate, with Milnor number say μ_i. Since it is isolated, it can be morsified (split into μ_i non-degenerate critical points) by an arbitrarily small perturbation of g_u in the neighbourhood of z_i. This perturbation can be glued in, leaving g_u unchanged outside U_i, using a collared neighbourhood and a partition of unity. The argument then proceeds as before. Note that although holomorphicity in the collar is lost, the conclusion concerning the topology of $g_u^{-1}(0)$ stands.

Now we can easily calculate the homology of X_u by a standard argument. Let Z denote $B_\varepsilon \cap g_u^{-1}(D_\eta)$ and let Y denote the union of the p-cells which are glued in when we pass the critical values of $|g_u|$. Recall that Z is homeomorphic to $B_\varepsilon \cap g_0^{-1}(D_\eta)$, and is therefore contractible. Then $X_u \cup Y = Z$, and $X_u \cap Y$ is the disjoint union of the boundaries of the p-cells, each of which may be assumed to be a $p-1$-sphere.

Suppose first that $p \geq 3$. Since Z is simply connected, so too must be X_u, as Z is obtained from X_u by gluing in cells of dimension greater than 2. We may assume that these cells are disjoint. Because $\tilde{H}_*(Z) = 0$, the Mayer–Vietoris sequence of the pair X_u and Y consists of segments

$$0 \to \tilde{H}_k(X_u \cap Y) \to \tilde{H}_k(X_u) \oplus \tilde{H}_k(Y) \to 0;$$

from these we deduce that

$$\tilde{H}_k(X_u) = \begin{cases} \mathbb{Z}^\mu & \text{if } k = p-1 \\ 0 & \text{otherwise,} \end{cases}$$

where μ is the number of cells glued in. Now pick a point on each bounding $p-1$-sphere, and join all of these points to an arbitrary base-point $x_0 \in X_u$ by a collection of non-intersecting paths in X_u. Denote the union of these paths by Γ. Then $H_k(\Gamma \cup (X_u \cap Y)) = H_k(X_u \cap Y)$ except when $k = 0$: $\Gamma \cup (X_u \cap Y)$ is now path connected. Indeed $\Gamma \cup (X_u \cap Y)$ is simply connected, and now the inclusion

$$\Gamma \cup (X_u \cap Y) \hookrightarrow X_u$$

induces isomorphisms on all homology groups. Thus, by Whitehead's Theorem (e.g. [Hat02, Cor. 4.3.3]), this inclusion is a homotopy-equivalence. Evidently $\Gamma \cup (X_u \cap Y)$ is homotopy-equivalent to the wedge of the $(p-1)$-spheres.

In the case $p = 2$, X_u is 1-dimensional and therefore has the homotopy type of a finite connected 1-dimensional CW complex, and so of a wedge of 1-spheres. □

The counting procedure of Theorem 8.4 is essential for the proofs of the Theorems 8.1 and 8.2.

At the heart of the preceding proof is the fact that the Morse index of the modulus of a holomorphic function at a non-degenerate critical point is equal to the ambient complex dimension—see Exercise 1 below.

Exercises for Sect. 8.3

1 Suppose $g : (\mathbb{C}^p, y) \to \mathbb{C}$ is a complex analytic function with a non-degenerate critical point at P, and $g(y) \neq 0$. Show that

(i) y is a critical point of the real-valued function $|g|$ if and only if y is a critical point of g;

(ii) if $v \in T_y\mathbb{R}^{2p}$ is an eigenvector of $d^2|g|(y)$ with eigenvalue λ, then iv is an eigenvector with eigenvalue $-\lambda$,

(iii) the index of the Hessian of $|g|$ at y is equal to p, and
(iv) the negative eigenspace of $d^2|g|(y)$ spans $\mathbb{C}^p$ over $\mathbb{C}$.

8.4 Finding $T^1_{\mathscr{A}_e} f$ in the Geometry of f: Maps from n-Space to $n+1$-Space

Here we begin by giving the desired geometrical description of $T^1_{\mathscr{A}_e} f$. This leads on to a discussion of the conductor ideal, the annihilator of the quotient $\mathscr{O}_{n,r}/\mathscr{O}_D$, and to a simple algorithm for computing $\mathscr{A}_e$-codimension.

Let $f : (\mathbb{C}^n_r, 0_r) \to (\mathbb{C}^{n+1}, 0)$ be a finite map-germ which is an immersion outside of a codimension 2 subset, and let $g \in \mathscr{O}_{n+1}$ be a reduced defining equation for the image, X, of f. Define $\widetilde{dg} : \theta(f) \to J_g\mathscr{O}_{n,r}$ by

$$\widetilde{dg}\left(\sum_i \alpha_i \frac{\partial}{\partial y_i}\right) = \sum_i \alpha_i \left(\frac{\partial g}{\partial y_i} \circ f\right).$$

Lemma 8.1 *$\widetilde{dg}$ passes to the quotient to define an epimorphism of $\mathscr{O}_{n+1}$-modules*

$$\overline{dg} : \frac{\theta(f)}{T\mathscr{A}_e f} \to \frac{J_g\mathscr{O}_{n,r}}{J_g\mathscr{O}_{X,0}}. \tag{8.6}$$

If $n \geq 2$ then $\overline{dg}$ is an isomorphism.

Proof The proof is clearest when $|S| = 1$, and the reader may prefer to consider it in this special case at a first reading.

For the general case, let us clarify our notation. We denote the points of S by 0_i, $i = 1, \ldots, r$, and $\mathscr{O}_{n,r}$ denotes the direct sum $\bigoplus_{i=1}^r \mathscr{O}_{\mathbb{C}^n,0_i}$. The germ of f at 0_i will be denoted by $f^{(i)}$. The source, $\theta(f)$, of the morphism $\widetilde{dg}$ is the direct sum

$$\left(\mathscr{O}_{n,r}\right)^{\times(n+1)}.$$

We write its elements as r-tuples $(\alpha^{(1)}, \ldots, \alpha^{(r)})$. We regard $\mathscr{O}_{X,0}$ as a subring of $\mathscr{O}_{n,r}$, via composition with $f = (f^{(1)}, \ldots, f^{(r)})$. The partials $\partial g/\partial y_j$ generate an ideal in each summand $\mathscr{O}_{\mathbb{C}^n,0_i}$, and we denote their direct sum by $J_g\mathscr{O}_{n,r}$. A vector field $\beta \in \theta_{n+1}$ gives rise to an element of the Jacobian ideal $J_g\mathscr{O}_{n+1}$, and restriction to X gives an element of $J_g\mathscr{O}_{X,0}$. This is then composed with each $f^{(i)}$ to give an element of the subring $\mathscr{O}_{X,0}$ of $\mathscr{O}_{n,r}$. It is the quotient

$$\left(\bigoplus_{i=1}^r J_g\mathscr{O}_{\mathbb{C}^n,0_i}\right) / J_g\mathscr{O}_{X,0}$$

that is the target of our morphism $\overline{dg}$. This should not be confused with the direct sum of quotients

$$\bigoplus_{i=1}^{r}\left(J_g\mathscr{O}_{\mathbb{C}^n,0_i}/J_g\mathscr{O}_{X,0}\right)$$

in which the linear combinations of the partials of g in the denominators of the summands are independent of one another.

Now we proceed with the proof. First, it is obvious that $\widetilde{dg}:\theta(f)\to J_g\mathscr{O}_{n,r}$ is surjective.

If we apply $\widetilde{dg}$ to an element $tf(\xi)=\left(tf^{(1)}(\xi^{(1)}),\dots,tf^{(r)}(\xi^{(r)})\right)\in\theta(f)$, then by the chain rule

$$\widetilde{dg}(tf(\xi))=\left(d(g\circ f^{(1)})(\xi_1),\dots,d(g\circ f^{(r)})(\xi_r)\right)=(0,\dots,0)$$

since $g\circ f^{(i)}=0$ for $i=1,\dots,r$

If we apply $\widetilde{dg}$ to an element $\omega f(\eta)\in\theta(f)$, we get $\sum_i\eta_i\partial g/\partial y_i$, lying in $J_g\mathscr{O}_X$. Thus, $\widetilde{dg}(T\mathscr{A}_ef)\subset J_g\mathscr{O}_X$, and it follows that $\widetilde{dg}$ passes to the quotient to define an epimorphism $\overline{dg}$, as claimed. We have only to show that when $n\geq 2$, $\overline{dg}$ is injective.

Suppose that for some $\alpha\in\theta(f)$, $\widetilde{dg}(\alpha)\in J_g\mathscr{O}_X$. That is, there exists a vector field $\beta\in\theta_{\mathbb{C}^{n+1},0}$ such that $\widetilde{dg}(\alpha)=dg(\beta)$. We can write this as

$$\widetilde{dg}(\alpha-\beta\circ f)=0. \tag{8.7}$$

By assumption, the set Σ_f of points where f is not an immersion is of codimension ≥ 2. By the Hartogs' extension lemma [Kra01], it will be enough to construct an analytic vector field $\xi=(\xi^{(1)},\dots,\xi^{(r)})$ on the complement of Σ_f such that $\alpha=tf(\xi)-\omega f(\beta)$, i.e. such that for $i=1,\dots,r$,

$$\alpha^{(i)}=tf^{(i)}(\xi^{(i)})-\omega f^{(i)}(\beta). \tag{8.8}$$

From the extension lemma it follows that each $\xi^{(i)}$ extends to an analytic vector field on $(\mathbb{C}^n,0_i)$, which by continuity must also satisfy (8.8). That is, $\alpha\in T\mathscr{A}_ef$, showing that the map of quotients is indeed injective.

On $\mathbb{C}^n\setminus\Sigma_{f^{(i)}}$, we construct $\xi^{(i)}$ by completely elementary means, as follows: First, suppose that $f^{(i)}(x)\in X_{\text{reg}}$. Then $f^{(i)}$ is an immersion at x, and it follows that the vectors $\frac{\partial f^{(i)}}{\partial x_1}(x),\dots,\frac{\partial f^{(i)}}{\partial x_n}(x)$ form a basis for $\ker d_{f^{(i)}(x)}g$. Since $\alpha^{(i)}(x)-\beta(f^{(i)}(x))\in\ker d_{f^{(i)}(x)}g$, it follows that there exist unique $\xi_j^{(i)}(x)$, $j=1,\dots,n$, such that

$$\alpha^{(i)}(x)-\beta(f^{(i)}(x))=\sum_{j=1}^{n}\xi_j^{(i)}(x)\frac{\partial f^{(i)}}{\partial x_j}(x). \tag{8.9}$$

The $\xi_j^{(i)}$ vary analytically with x, and thus form the components of an analytic vector field $\xi^{(i)}$ on the preimage of X_{reg}. We wish to show that $\xi^{(i)}$ extends analytically to all of the complement of $\Sigma_{f^{(i)}}$. For all points $x \in \mathbb{C}^n \setminus \Sigma_{f^{(i)}}$, the non-zero $n+1$-tuple $\Delta := (\Delta_1, \ldots, \Delta_{n+1})$, with

$$\Delta_j = (-1)^{j-1} \frac{\partial(f_1^{(i)}, \ldots, \widehat{f_j^{(i)}}, \ldots, f_{n+1}^{(i)})}{\partial(x_1, \ldots, x_n)}$$

lies in, and therefore generates, the 1-dimensional subspace $(\text{Image}\, d_x f^{(i)})^\perp$ of $T^*_{f^{(i)}(x)} \mathbb{C}^{n+1}$. Since $d_{f^{(i)}(x)} g$ also lies in $(\text{Image}\, d_x f^{(i)})^\perp$, it follows that there exists some $\lambda(x)$ such that $d_{f^{(i)}(x)} g = \lambda(x)\Delta(x)$. On $(f^{(i)})^{-1}(X_{\text{reg}})$, $\lambda \neq 0$, so it follows from the fact that $dg(\alpha - \beta \circ f^{(i)}) = 0$ that $\Delta \cdot (\alpha - \beta \circ f^{(i)}) = 0$. By continuity, this equation must continue to hold on all of $\mathbb{C}^n, 0_i$. In particular, $\alpha - \beta \circ f^{(i)} \in \Delta^\perp$ on $\mathbb{C}^n \setminus \Sigma_{f^{(i)}}$. On $\mathbb{C}^n \setminus \Sigma_{f^{(i)}}$, $\Delta^\perp$ has, as basis, the vectors $\frac{\partial f^{(i)}}{\partial x_1}(x), \ldots, \frac{\partial f^{(i)}}{\partial x_n}(x)$. Hence the ξ_i extend to $\mathbb{C}^n \setminus \Sigma_{f^{(i)}}$. □

Note that provided $n \geq 2$, $\mathscr{A}$-finite germs $(\mathbb{C}^n, S) \to (\mathbb{C}^{n+1}, 0)$ satisfy all the hypotheses of the lemma.

When $n = 1$, $\overline{dg}$ is in general no longer injective. For future use, it is important to understand the kernel.

Example 8.4 Let $f : (\mathbb{C}, 0) \to (\mathbb{C}^2, 0)$ be defined by $f(x) = (x^2, x^3)$. The class of $tf(\partial/\partial x)$ in $\theta(f)$ is mapped to 0 by $\widetilde{dg}$, of course. Since it lies in $T\mathscr{A}_e f$, it does not give a non-trivial element of the kernel of $\overline{dg}$. However, $(1/x)tf(\partial/\partial x)$ also lies in $\theta(f)$, and is also in the kernel of $\widetilde{dg}$, but not in $T\mathscr{A}_e f$. It thus does contribute non-trivially to the kernel of $\overline{dg}$.

More generally, if, for some map-germ $f : (\mathbb{C}^n, 0) \to (\mathbb{C}^{n+1}, 0)$, there is an $h \in \mathfrak{m}_n$ dividing all components of the vector field $tf(\xi) \in \theta(f)$, then $(1/h)tf(\xi)$ lies in $\theta(f)$ and in $\ker \widetilde{dg}$. This vector field is not in $tf(\theta_n)$ unless the components of ξ are themselves divisible by h. However, if f is an immersion outside a codimension 2 set, then $tf(\xi)$ cannot vanish along a hypersurface without ξ itself vanishing along the hypersurface, and thus $(1/h)tf(\xi) = tf(\xi/h) \in T\mathscr{A}_e f$ and so does not contribute a non-trivial element of $\ker \overline{dg}$. ◊

Lemma 8.2

(i) Let $f = (f_1, f_2) : (\mathbb{C}, 0) \to (\mathbb{C}^2, 0)$ be $\mathscr{A}$-finite, and let $\overline{dg}$ be defined as in the preceding lemma. Then

$$\ker \overline{dg} \simeq \frac{\mathscr{O}_1}{(f_1', f_2')}.$$

(ii) If $f : (\mathbb{C}, S) \to (\mathbb{C}^2, 0)$ is a multi-germ, then the kernel of $\overline{dg}$ is the direct sum of the kernels of the individual branches of f.

Proof

(i) We may assume without loss of generality that the order of f_1 is less than or equal to the order of f_2. Thus $(f_1', f_2') = (f_1') = (x^p)$ for some $p \in \mathbb{N}$. Then the classes in $\theta(f)$ of the elements $(1/x^k)tf(\partial/\partial x)$ lie in $\ker \widetilde{dg}$ for $k = 1, \ldots, p$. They are linearly independent over $\mathbb{C}$ since all have different order, ranging from $\mathrm{ord}\big((1/x^p)tf(\partial/\partial x)\big) = 0$ to $\mathrm{ord}\big((1/x)tf(\partial/\partial x)\big) = p - 1$. Their classes in $\theta(f)/T\mathscr{A}_e f$ are still linearly independent over $\mathbb{C}$ since $T\mathscr{A}_e f \subset \mathfrak{m}^p\theta(f)$. We have just to show that they span $\ker \widetilde{dg}$. Suppose that $\alpha \in \ker \widetilde{dg}$. Then, as shown in the proof of the previous lemma, there is a vector field $\beta \in \theta_2$ such that $\widetilde{dg}(\alpha - \beta \circ f) = 0$. The restriction of β to the image of f is uniquely defined by this requirement, so the element $\alpha - \beta \circ f$ of $\theta(f)$ is determined by α. As in the proof of the lemma, on $\mathbb{C} \setminus \{0\}$ we have $\alpha - \beta \circ f = tf(\xi)$ for some uniquely defined ξ. Because we can write

$$\xi = \frac{\alpha_1 - \beta_1 \circ f}{f_1'},$$

ξ extends to a meromorphic vector field germ on $(\mathbb{C}, 0)$ with a pole at 0. If the order of this pole is ≤ 0, then $\alpha - \beta \circ f \in tf(\theta_1)$. If the order of the pole is greater than the order of f_1', then $\alpha - \beta \circ f \notin \theta(f)$. So the class of α in $\theta(f)/T\mathscr{A}_e f$ is in the linear span of the elements listed above.

(ii) Now suppose that $|S| = r > 1$. Denote the constituent mono-germ of f at the i'th point of S by $f^{(i)}$. Suppose the class in $\theta(f)/T\mathscr{A}_e f$ of $\underline{\alpha} = (\alpha^{(1)}, \ldots, \alpha^{(r)})$ is in the kernel of $\overline{dg}$. Then there exists $\beta \in \theta_2$ such that $\widetilde{dg}(\alpha - \beta \circ f) = 0$. This means that for $i = 1, \ldots, r$, $\widetilde{dg}(\alpha^{(i)} - \beta \circ f^{(i)}) = 0$. The argument of (i) shows that $\alpha^{(i)} - \beta \circ f^{(i)}$ is a linear combination of elements $(1/x^k)tf^{(i)}(\partial/\partial x)$ for $1 \leq k < \min\{\mathrm{ord}(f_1^{(i)}), \mathrm{ord}(f_2^{(i)})\}$, and, as before that these elements are linearly independent over $\mathbb{C}$. This proves (ii). □

Lemma 8.1 gives us a means of checking $\mathscr{A}$-finiteness, and leads to an algorithm for calculating the $\mathscr{A}_e$-codimension of a map-germ $f : (\mathbb{C}^n, 0) \to (\mathbb{C}^{n+1}, 0)$. We describe the algorithm, due to Ayse Altintas Sharland, and then illustrate its implementation in *Macaulay 2* with an example. In the following discussion we denote by f^* the ring homomorphism $\mathscr{O}_{n+1} \to \mathscr{O}_n$ induced by f. To avoid misunderstanding, given an ideal I in $\mathscr{O}_{n+1}$, we distinguish between the image $f^*(I) \subset \mathscr{O}_n$ and the ideal it generates in $\mathscr{O}_n$, by denoting the latter by $f^*(I)\mathscr{O}_n$.

Algorithm 8.1 Find

1. an equation, g, for the image of f.
2. the Jacobian ideal $J_g \subset \mathscr{O}_{n+1}$.
3. the ideal $f^*(J_g)\mathscr{O}_n$ generated by $f^*(J_g)$ in $\mathscr{O}_n$.
4. the preimage $(f^*)^{-1}(f^*(J_g)\mathscr{O}_n)$ in $\mathscr{O}_{n+1}$.

5. the dimension as $\mathbb{F}$-vector space of the quotient

$$\frac{(f^*)^{-1}(J_g\mathscr{O}_n)}{J_g+(g)} \tag{8.10}$$

The dimension of the quotient (8.10) is the $\mathscr{A}_e$-codimension of f.

Lemma 8.1 shows that the $\mathscr{A}_e$-codimension of f is the $\mathbb{F}$-vector-space dimension of the quotient $f^*(J_g)\mathscr{O}_n/f^*(J_g)$. Since (g) is the kernel of $f^* : \mathscr{O}_{n+1} \to \mathscr{O}_n$, this is equal to the dimension of $f^*(J_g)\mathscr{O}_n/f^*\big((J_g+(g))\mathscr{O}_{n+1}\big)$. The last step of the algorithm takes this quotient back to a subquotient of $\mathscr{O}_{n+1}$, using the fact, which we will prove shortly, that the ring homomorphism f^* induces an isomorphism

$$\frac{f^*(J_g)\mathscr{O}_n}{f^*(J_g)} \simeq \frac{(f^*)^{-1}\big(f^*(J_g)\mathscr{O}_n\big)}{(J_g+(g))\mathscr{O}_{n+1}} \tag{8.11}$$

We know that $f^* : \mathscr{O}_D \to \mathscr{O}_n$ is injective, so the proof requires simply showing that $f^* : \mathscr{O}_{n+1} \to \mathscr{O}_n$ maps surjectively onto $f^*(J_g)\mathscr{O}_n$. To show this, we show, in the next section, that $f^*(J_g)\mathscr{O}_n$ is contained in the *conductor ideal*: the ideal

$$\mathscr{C} = \{r \in \mathscr{O}_n : rh \in \mathscr{O}_D \text{ for all } h \in \mathscr{O}_n\}.$$

We note that everything is simplified when f is weighted homogeneous with respect to some set of positive weights, since then D also is weighted homogeneous and $g \in J_g$.

Implementation of the Algorithm We describe implementation in *Macaulay 2* for a polynomial map $F : \mathbb{F}^n \to \mathbb{F}^{n+1}$ defined over $\mathbb{Z}$. In the following, S and T will denote the rings $\mathbb{Q}[x_1,\ldots,x_n]$, and $\mathbb{Q}[y_1,\ldots,y_{n+1}]$, and "$f$" will be used to denote the ring homomorphism $T \to S$ induced by the map F.

When $f : T \to S$ is a ring homomorphism, and $Z \subset S$ is an ideal, the "preimage" command in *Macaulay 2* calculates the preimage $f^{-1}(Z)$. In particular, if f is the ring homomorphism $\mathbb{Q}[y_1,\ldots,y_p] \to \mathbb{Q}[x_1,\ldots,x_n]$ induced by a polynomial map $\mathbb{F}^n \to \mathbb{F}^p$, then $f^{-1}(0)$ is the ideal I of functions in $\mathbb{Q}[y_1,\ldots,y_p]$ vanishing on the image D. When $p = n+1$ then D is a hypersurface, so I is a principal ideal, and the preimage command returns a generator, i.e. an equation for the image D. In the following example the ideal is denoted by g.

Example 8.5 Consider the germ at 0 of the weighted homogeneous map $f : \mathbb{F}^2 \to \mathbb{F}^3$ defined by

$$f(x_1,x_2) = (x_1^2, x_2^2, x_1(x_1^2+x_2^2) + x_2(x_1^2-x_2^2))).$$

The sequence

```
i1 : S = QQ[x1, x2]
i2 : T = QQ[y1, y2, y3]
i3 : f = map(S, T, \matrix{{x1^2, x2^2, x1 * (x1^2 + x2^2)
        + x2 * (x1^2 - x2^2)}})
i4 : Z = ideal(promote(0, S))
i5 : g = preimage_f(Z)
```

returns an expression for a generator of the ideal defining the image of f,

$$y_1^6 + 2y_1^5y_2 + 7y_1^4y_2^2 + 4y_1^3y_2^2 + 7y_1^2y_2^4 - 6y_1y_2^5 + y_2^6 - 2y_1^3y_3^2 \\ - 6y_1^2y_2y_3^2 + 2y_1y_2^2y_3^2 - 2y_2^3y_3^2 + y_3^4.$$

(The definition of Z in input 4 is complicated by the fact that *Macaulay 2* regard 0 as an element the base ring $\mathbb{Z}$, and it is necessary to "promote" it to S).

The commands

```
i6 : J = ideal jacobian g
i7 : JS = f(J)
i8 : JJ = preimage_f(fJ)
i9 : b = basis(JJ/J)
```

(we have suppressed the output of i6–i8) then return the matrix

0	0	0	0	0	0	0	0	0	0	0	0
1	y2	0	0	0	0	0	0	0	0	0	0
0	0	1	0	0	0	0	0	0	0	0	0
0	0	0	0	0	0	0	0	0	0	0	0
0	0	0	0	0	0	0	0	0	0	0	0
0	0	0	1	y2	0	0	0	0	0	0	0
0	0	0	0	0	1	y2	0	0	0	0	0
0	0	0	0	0	0	0	1	0	0	0	0
0	0	0	0	0	0	0	0	1	y1	y2	y3

with 9 rows (corresponding to the 9 generators of JJ that Macaulay shows as the output to command `i8`) and 12 columns. The i'th column shows the coefficients of the linear combination of the 9 generators giving the i'th member of an $\mathbb{F}$-basis of JJ/J. The map is weighted homogeneous, so we do not need to add the ideal g to the denominator in Step `i9`. We conclude that the $\mathscr{A}_e$ codimension of f is 12. $\Diamond$

8.4.1 The Conductor Ideal

Let $f : (X, S) \to (\mathbb{C}^n, 0)$ be a germ with image D. For brevity we write $\mathscr{O}_D$ in place of $\mathscr{O}_{D,0}$. Once again we consider $\mathscr{O}_D$ as a subring of $\mathscr{O}_{X,S}$, by composing the elements of $\mathscr{O}_D$ with f.

Definition 8.3 The conductor ideal is the ideal $\mathscr{C} := \{h \in \mathscr{O}_{X,S} : h\mathscr{O}_{X,S} \subset \mathscr{O}_D\}$.

If $h \in \mathscr{C}$ then $h = h \cdot 1 \in h\mathscr{O}_{X,S} \subset \mathscr{O}_D$, so $\mathscr{C}$ is in fact an ideal of $\mathscr{O}_D$. It is the largest ideal of $\mathscr{O}_D$ which is also an ideal of $\mathscr{O}_{X,S}$.

Example 8.6

(1) Consider the germ $f : (\mathbb{C}, 0) \to (\mathbb{C}^2, 0)$ given by $f(x) = (x^2, x^3)$. Then

$$\mathscr{O}_D = \mathbb{C}\{x^2, x^3\} \subset \mathbb{C}\{x\} = \mathscr{O}_1.$$

By inspection, $\mathscr{C} = (x^2)$. If now $f(x) = (x^2, x^5)$ then $\mathscr{C} = (x^4)$. For $f(x) = (x^3, x^5)$, $\mathscr{C} = (x^8)$.

(2) Consider a bi-germ $f : (\mathbb{C}^n, S) \to (\mathbb{C}^{n+1}, 0)$, where $S = \{0^{(1)}, 0^{(2)}\}$, consisting of two germs of immersion, $f^{(1)}$ and $f^{(2)}$, with images $D^{(1)}$ and $D^{(2)}$ defined by the reduced equations $g^{(1)} = 0$ and $g^{(2)} = 0$. Then $D := D^{(1)} \cup D^{(2)}$ is the image of f. Each ring $\mathscr{O}_{\mathbb{C}^n,0^{(i)}}$ can be identified with $\mathscr{O}_{D^{(i)},0}$ and thus is a quotient of $\mathscr{O}_{n+1}$. Moreover

$$\mathscr{O}_{\mathbb{C}^n,0^{(1)}} = \mathscr{O}_{n+1} \cdot (1, 0), \quad \mathscr{O}_{\mathbb{C}^n,0^{(2)}} = \mathscr{O}_{n+1} \cdot (0, 1);$$

similarly viewed as a subring of $\mathscr{O}_{\mathbb{C}^n,S}$, $\mathscr{O}_D = \mathscr{O}_{n+1} \cdot (1, 1)$. Thus, as an ideal in $\mathscr{O}_{\mathbb{C}^n,S}$, $\mathscr{C}$ is the set of pairs $(c^{(1)}, c^{(2)})$ such that for all $(h^{(1)}, h^{(2)}) \in \mathscr{O}_{\mathbb{C}^n,S}$, $(c^{(1)}h^{(1)}, c^{(2)}h^{(2)}) = s(1, 1)$ for some $s \in \mathscr{O}_{n+1}$. Now

$$(c^{(1)}, c^{(2)}) \cdot (1, 1) \in \mathscr{O}_D \implies \text{there exists } r \in \mathscr{O}_{n+1} \text{ s.t. } c^{(1)} = r|_{D^{(1)}}, c^{(2)} = r|_{D^{(2)}}$$

$$(c^{(1)}, c^{(2)}) \cdot (1, 0) \in \mathscr{O}_D \implies \text{there exists } r_1 \in \mathscr{O}_{n+1} \text{ s.t. } c^{(1)} = r_1|_{D^{(1)}}, r_1|_{D^{(2)}} = 0$$

These two equations together show that $r = r_1$ on $D^{(1)}$, so that $r - r_1 \in (g^{(1)})$. The second equation implies that $r_1 \in (g^{(2)})$. Thus $r \in (g^{(1)}, g^{(2)})$, so

$$\mathscr{C} \subset (g^{(1)}, g^{(2)})\mathscr{O}_D.$$

Conversely, given any element $\alpha_1 g^{(1)} + \alpha_2 g^{(2)} \in (g^{(1)}, g^{(2)})$, and any $(h^{(1)}, h^{(2)}) \in \mathscr{O}_{\mathbb{C}^n,S}$, we have $(\alpha_1 g^{(1)} + \alpha_2 g^{(2)}) \cdot (h^{(1)}, h^{(2)}) = (\alpha_2 g^{(2)} h^{(1)}, \alpha_1 g^{(1)} h^{(2)})$. We may extend $h^{(1)}$ and $h^{(2)}$ to members of $\mathscr{O}_{n+1}$, which we denote by the same letters. The two components of the pair

$(\alpha_2 g^{(2)} h^{(1)}, \alpha_1 g^{(1)} h^{(2)})$ are the restrictions to D_1 and to D_2 of the function $\alpha_2 g^{(2)} h^{(1)} + \alpha_1 g^{(1)} h^{(2)} \in \mathcal{O}_{n+1}$. Thus $\alpha_1 g^{(1)} + \alpha_2 g^{(2)} \in \mathcal{C}$. So

$$\mathcal{C} = (g^{(1)}, g^{(2)}) \quad \text{as an ideal in } \mathcal{O}_D, \tag{8.12}$$

and

$$\mathcal{C} = (g^{(2)}) \oplus (g^{(1)}) \quad \text{as an ideal in } \mathcal{O}_{\mathbb{C}^n, S} = \mathcal{O}_{D^{(1)}} \oplus \mathcal{O}_{D^{(2)}}. \tag{8.13}$$

Note that although $\mathcal{O}_D \to \mathcal{O}_{D^{(1)}} \oplus \mathcal{O}_{D^{(2)}}$ is not an isomorphism overall, it does give an isomorphism from (8.12) to (8.13). Moreover, the image of $\mathcal{O}_D$ in $\mathcal{O}_{\mathbb{C},S}$ is not a direct sum, but the image of $\mathcal{C} \subset \mathcal{O}_D$ is.

(3) In the situation of (2), if $D^{(1)} \pitchfork D^{(2)}$ then in $\mathcal{O}_{D^{(1)}}$, $g^{(2)}$ is a reduced equation for $D^{(1)} \cap D^{(2)}$, and vice versa. Thus in this case $\mathcal{C}$ is the set of pairs (c_1, c_2) with each c_i vanishing on $D^{(1)} \cap D^{(2)}$. If transversality does not hold then this may no longer be the case. For example, when $D^{(1)} = \{x_1 = 0\} \subset \mathbb{C}^2$ and $D^{(2)} = \{x_1 = x_2^2\} \subset \mathbb{C}^2$, $g^{(1)}$ is the square of the generator of $I_{D^{(1)} \cap D^{(2)}}$ in $\mathcal{O}_{D^{(2)}}$, and vice versa. ♢

To justify Step 5 in Algorithm 8.1, we need the following characterisation of the conductor ideal.

Proposition 8.4 *Suppose that $f : (\mathbb{C}^n_r, 0_r) \to (\mathbb{C}^{n+1}, 0)$ is $\mathcal{A}$-finite, with $n \geq 2$, and let $D^2_1(f)$ be the closure, in $\mathbb{C}^n_r$, of the set $\{x \in \mathbb{C}^n : |f^{-1}(f(x))| \geq 2\}$. Then*

(1) $D^2_1(f)$ is analytic, of pure dimension $n - 1$.
(2) As an ideal of $\mathcal{O}_{n,r}$, $\mathcal{C}(f)$ is the ideal $I(D^2_1(f))$ of all function germs vanishing on $D^2_1(f)$.
(3) As an ideal of $\mathcal{O}_{n+1}$, $\mathcal{C}$ is the ideal of function germs vanishing on D_{sing}, and thus equal to the radical of the Jacobian ideal J_g, where g is a reduced defining equation of D.

Proof By the Mather–Gaffney criterion for $\mathcal{A}$-finiteness, for $y \neq 0$, the multi-germ

$$f_y : (\mathbb{C}^n, f^{-1}(y)) \to (\mathbb{C}^{n+1}, y)$$

is stable if $y \neq 0$. In particular, if f takes the same value $y_0 \neq 0$ at k distinct immersive points, then by Theorem 3.3, the k distinct branches must meet in general position at y_0. Their k-fold intersection is analytic, with codimension (in $\mathbb{C}^{n+1}$) therefore equal to k. In particular, when $k = 2$ the set in question has pure dimension $n - 1$. As f is finite, its preimage also has pure dimension $n - 1$, and so also the closure of that preimage, which is evidently analytic.

By another consequence of the same criterion, Proposition A.1(2), the set Σ_f of points in the domain where f is not an immersion has codimension ≥ 2. It follows that there is a subset D' of the image D, of codimension ≥ 2 in D, such that at each point of $D \setminus D'$, the germ of D is either smooth, or is the transverse crossing of two immersed sheets. Once again, the complement S of $f^{-1}(D')$ has codimension ≥ 2 in $\mathbb{C}^n$. Now let $h \in \mathscr{O}_{n,r}$, and consider its restriction to S. By Example 8.6(3), a necessary and sufficient condition for $h|_S$ to be the composite with f of some analytic function p on $D \setminus D'$ is simply that for each pair of points x_1, x_2 in $D^2_1(f)$, $h(x_1) = h(x_2)$. Note that the function p on $D \setminus D'$ is uniquely determined by $h|_S$, so representatives of the germs at points of $D \setminus D'$ piece together to give a well-defined analytic function on $D \setminus D'$.

Suppose now that $h \in \mathscr{O}_n$ is a germ with the property that for each pair of points x_1, x_2, if $f(x_1) = f(x_2)$ then $h(x_1) = h(x_2)$. By the preceding paragraph, there is an analytic function r on $D \setminus D'$ such that $h|_S = r \circ f$. We claim that r extends to an analytic function on D. This follows by an extension theorem proved in Sect. D.4: as D is a hypersurface, and therefore Cohen–Macaulay, every function analytic outside a subset of codimension 2 extends to an analytic function on D.

Now let h be a member of the ideal described in (2). Then for every germ $h' \in \mathscr{O}_{n,r}$, hh' vanishes on $D^2_1(f)$. It follows that for every pair of distinct points (x_1, x_2) such that $f(x_1) = f(x_2)$, we have $hh'(x_1) = 0 = hh'(x_2)$. Therefore by the previous paragraph, $hh' \in \mathscr{O}_D$. So $h \in \mathscr{C}$. Conversely let $h \in \mathscr{C}$ and suppose $x_1 \neq x_2$ and $f(x_1) = f(x_2)$. Then $h(x_1) = h(x_2)$. Moreover there exists a germ h' such that $0 \neq h'(x_1) \neq h'(x_2) \neq 0$—for example, one of the coordinate functions, possibly plus a constant. In order that hh' should lie in $\mathscr{O}_D$, we must have $hh'(x_1) = hh'(x_2)$. The only possibility is that $h(x_1) = h(x_2) = 0$. This proves (2). For (3), we observe that D_{sing} is the closure of $f(D^2_1(f))$. □

Corollary 8.1 *Suppose that* $f : (\mathbb{C}^n_r, 0_r) \to (\mathbb{C}^{n+1}, 0)$ *is* $\mathscr{A}$*-finite, and let* g *be a reduced equation for the image. Then*

(1) The Jacobian ideal J_g *is contained in the conductor ideal* $\mathscr{C}$.
(2) The ring morphism $f^* : \mathscr{O}_{n+1} \to \mathscr{O}_n$ *induces a surjection* $(f^*)^{-1}\big(f^*(J_g)\mathscr{O}_n\big) \to f^*(J_g)\mathscr{O}_n$.
(3) The homomorphism (8.11) *induced by* f^* *is an isomorphism.*

Proof (1) is immediate from Proposition 8.4(3), and (2) follows, since by (1), $f^*(J_g)\mathscr{O}_n \subset \mathscr{C}$. The last statement holds because in addition, $f^* : \mathscr{O}_D \to \mathscr{O}_n$ is injective. □

The quotient λ that we introduced in the proof is of particular interest. In [Pie79], Ragni Piene shows that λ generates the conductor ideal $\mathscr{C}$ as an ideal in $\mathscr{O}_n$, for all germs $f : (\mathbb{C}^n, 0) \to (\mathbb{C}^{n+1}, 0)$ which are finite and generically 1-to-1. Proving this is beyond our reach here, but following Bruce and Marar in [BM96], we can easily prove it for $\mathscr{A}$-finite f.

Proposition 8.5 *For an* $\mathscr{A}$*-finite map-germ* $f : (\mathbb{C}^n_r, 0_r) \to (\mathbb{C}^{n+1}, 0)$*, with image defined by the reduced equation* g*,*

(1) the quotient λ *defined by*

$$\lambda(x) = (-1)^{i-1}\frac{\partial g}{\partial y_i}\Big/\frac{\partial(f_1,\ldots,\widehat{f_i},\ldots,f_{n+1})}{\partial(x_1,\ldots,x_n)}$$

is independent of i *and lies in* $\mathcal{O}_{n,0_r}$.
(2) λ *generates* $\mathcal{C}$ *over* $\mathcal{O}_{n,r}$.

Proof (1) was proved in the proof of Lemma 8.1. In that proof it was also shown that λ generates $\mathcal{C}$ in $\mathbb{C}^n \setminus \Sigma_f$. Let λ_1 be a generator of $\mathcal{C}$ in $\mathcal{O}_{n,r}$. Then it follows that the quotient λ_1/λ is regular outside Σ_f. Since Σ_f has codimension ≥ 2, λ_1/λ extends to an analytic function on all of $(\mathbb{C}^n_r, 0_r)$, by Hartogs' extension lemma, and thus λ generates $\mathcal{C}$ over $\mathcal{O}_{n,r}$.

In order to prove Theorem 8.2, we will need to know more about $\mathcal{C}$ as an ideal in $\mathcal{O}_D$. In Chap. 11 we develop a technique using Fitting ideals, which gives us a list of $\mathcal{O}_D$-generators for $\mathcal{C}$, and shows, in particular, that the preimage of $\mathcal{C}$ with respect to the quotient map $\mathcal{O}_{n+1} \to \mathcal{O}_D = \mathcal{O}_{n+1}/(g)\mathcal{O}_{n+1}$ is a determinantal ideal. We use this result later to prove one of the central results in the theory:

Theorem 8.5 ([dJvS91]) *Let* $f : (\mathbb{C}^n, S) \to (\mathbb{C}^{n+1}, 0)$ *be* $\mathcal{A}$*-finite, with* $n = 1$ *or* $n = 2$*, and let* g *be a reduced equation for its image. Then*

$$\operatorname{codim}_{\mathcal{A}_e}(f) \leq \mu_I(f)$$

with equality if f *is weighted homogeneous.*

We postpone the proof until Sect. 11.4, by which point we will have developed the algebra necessary to prove the necessary result on conservation of multiplicity. The first proof for $n = 2$ was found by de Jong and Pellikaan, and the first published proof is due to de Jong and van Straten [dJvS91]. This uses their theory of admissible deformations of pairs (space, subspace) and leads rather quickly to the result we want. However it demands a greater familiarity with the techniques of deformation theory than we want to assume or develop in this text.

Remark 8.1 The "Mond conjecture" A conjecture (or, more honestly, a question) of the first author asks whether if $f : (\mathbb{C}^n, S) \to (\mathbb{C}^{n+1}, 0)$ is a map-germ of finite codimension, and $(n, n+1)$ are in Mather's nice dimensions, then $\mu_I(f) \geq \mathcal{A}_e - \operatorname{codim}(f)$, with equality if f is weighted homogeneous. The answer Yes is supported by many examples, beyond the cases where it is proved: see, for instance, [Hou98, HK99, Hou98, CMWA02, Sha14, Sha17], but a general statement remains unproved. The supporting evidence is very strong: the paper of Sharland [Sha17] contains the germ

$$f(x,y,z) = (y^2+xz, x^5+yz+xy^2, x^6+y^3+z^2, x^7+x^4z+xz^2+y^2z) \tag{8.14}$$

It is weighted homogeneous of rank 0, and has $\mathscr{A}_e$-codimension $= \mu_I = 18,967$. Recent work of Fernández de Bobadilla, Nuño-Ballesteros and Peñafort-Sanchis in [FdBNnBPnS19] finds new means of calculating μ_I, and other modules whose flatness over the base of an unfolding, where it proved, would yield a proof of the conjecture.

Part of the difficulty in proving it lies in the fact that we do not have an effective method for computing image Milnor numbers. There is a conjectural formula (see Sect. 8.8 below) but without a proof that the relative module, $T^1_{\mathscr{A}_e}(F/d)$ is Cohen–Macaulay, it only gives an upper bound for μ_I. Other information on the image Milnor number comes from the image computing spectral sequence (see [GM93, Gor95, Hou99]), but so far this only yields an answer when f has corank 1 (see [GM93, Section 3]). In contrast, we do have a method for computing discriminant Milnor numbers, described in the next section. ◊

Exercises for Sect. 8.4

1

(i) Generalise the result of (2) in the Example 8.6 to the case of r germs of immersion $(\mathbb{C}^n, 0) \to (\mathbb{C}^{n+1}, 0)$ meeting at $0 \in \mathbb{C}^{n+1}$ with images defined by equations $g^{(1)}, \ldots, g^{(r)}$.

(ii) Show that in these circumstances, if the r germs of immersion meet in general position, then the conductor ideal is equal to the Jacobian ideal of a reduced defining equation of the image.

8.5 Finding $T^1_{\mathscr{A}_e} f$ in the Geometry of f: Sections of Stable Discriminants and Images

In this section we give an alternative description of $T^1_{\mathscr{A}_e} f$, due to Jim Damon in [Dam91], which is valid for all map-germs $(\mathbb{F}^n_r, S) \to (\mathbb{F}^p, 0)$, but which turns out to be particularly useful when $n \geq p$.

If $F : V \to W$ and $i : Y \to W$ are two maps, the *fibre product of V and Y over W*, denoted by $V \times_W Y$, is the space

$$V \times_W Y = \{(v, y) \in V \times Y : F(v) = i(y)\}.$$

A *fibre square* is the commutative diagram

$$\begin{array}{ccc} V & \xrightarrow{F} & W \\ \uparrow{\scriptstyle \pi_V} & & \uparrow{\scriptstyle i} \\ V \times_W Y & \xrightarrow[f:=\pi_Y]{} & Y \end{array} \tag{8.15}$$

which results, where π_Y and π_V are the restrictions to $V \times_W Y$ of the projections $V \times Y \to Y$ and $V \times Y \to V$. If V, W and Y are smooth spaces and $i \pitchfork F$, then $V \times_W Y$ is smooth also. We say that the map $\pi_Y : V \times_W Y \to Y$, which here we will denote by f, is the *pull-back of F by i*, or *transverse pull-back* in the case where $i \pitchfork F$, and write $f = i^*(F)$. The transversality of i to F guarantees that $V \times_W Y$ is smooth, but there is no canonical choice of coordinate system on $V \times_W Y$, so the map $i^*(F)$ is really defined only up to right equivalence.

By abuse of notation, we say that $f : (\mathbb{F}^n, S_0) \to (\mathbb{F}^p, 0)$ is obtained by transverse pull-back from $F : (\mathbb{F}^N, S_1) \to (\mathbb{F}^P, 0)$ if there is a diagram

$$\begin{array}{ccc}
(\mathbb{F}^N, S_1) & \xrightarrow{F} & (\mathbb{F}^P, 0) \\
\uparrow & & \uparrow{\scriptstyle i} \\
(\mathbb{F}^N \times_{\mathbb{F}^P} \mathbb{F}^p, S_1 \times \{0\}) & \longrightarrow & (\mathbb{F}^p, 0) \\
{\scriptstyle \simeq}\uparrow & \nearrow{\scriptstyle f} & \\
(\mathbb{F}^n, S_0) & &
\end{array} \tag{8.16}$$

in which $i \pitchfork F$.

In what follows, a *standard fibre square* is a fibre square of the form

$$\begin{array}{ccc}
(X \times U, (x_0, s_0)) & \xrightarrow{F} & (Y \times U, (y_0, s_0)) \\
{\scriptstyle j}\uparrow & & \uparrow{\scriptstyle i} \\
(X, x_0) & \xrightarrow{f} & (Y, y_0)
\end{array} \tag{8.17}$$

in which F is an unfolding of f, and i and j are given by $i(y) = (y, 0)$ and $j(x) = (x, 0)$. Every map-germ $f : (\mathbb{C}^n, S) \to (\mathbb{C}^p, 0)$ of finite singularity type can be obtained by transverse pull-back from a stable map-germ: simply construct a stable unfolding $F : (\mathbb{C}^n \times \mathbb{C}^d, 0) \to (\mathbb{C}^p \times \mathbb{C}^d, 0)$ and then recover f from F by means of a standard fibre square. This may not be the most economical construction, as the following example shows.

Example 8.7 In the situation of (8.15), take $V = \mathbb{F}^2$, $W = \mathbb{F}^3$, $Y = \mathbb{F}^3$, and let $F(v_1, v_2) = (v_1, v_2^2, v_1 v_2)$ and $i(y_1, y_2, y_3) = (p(y_1, y_2), y_2, y_3)$. Then $i \pitchfork F$ and

$$V \times_W Y = \{(v_1, v_2, y_1, y_2, y_3) : \ v_1 = p(y_1, y_2), \ v_2^2 = y_2, \ v_1 v_2 = y_3\}.$$

The three equations defining $V \times_W Y$ allow us to dispense with the coordinates v_1, y_2 and y_3, retaining y_1, v_2 as coordinates on $V \times_W Y$. With respect to these coordinates, the maps π_V and π_Y are then given by

$$\pi_Y(y_1, v_2) = (y_1, v_2^2, v_2 p(y_1, v_2^2))$$
$$\pi_V(y_1, v_2) = (p(y_1, v_2^2), v_2).$$

$\diamondsuit$

Thus every map germ of the form $f(x, y) = (x, y^2, yp(x, y^2))$ can be obtained by transverse pull-back from the map-germ F parameterising the Whitney umbrella. This has the following generalisation.

Proposition 8.6 *Suppose that* $f : (\mathbb{F}^n, S_0) \to (\mathbb{F}^p, 0)$ *and* $F : (\mathbb{F}^N, S_1) \to (\mathbb{F}^P, 0)$ *have isomorphic local algebras, that* $P - N = p - n$*, and that* F *is stable. Then* f *can be obtained by transverse pull-back from* F.

Proof See Exercise 4 below. □

Suppose that f is obtained from the stable map F by transverse pull-back by i. The main theorem of this section, Theorem 8.7, shows how to recover the module $T^1_{\mathscr{A}_e} f$ in terms of the interaction of F and i.

Before stating it, we need a definition.

Definition 8.4 If $D \subset W$ is an analytic subvariety of a smooth space W, $\mathrm{Der}(-\log D)$ is the sheaf of germs of vector fields on W tangent to D at its smooth points. It is in a natural way a sheaf of $\mathscr{O}_W$-modules, and, indeed, a subsheaf of θ_W.

It is easy to show that if D is the variety of zeros of an ideal I then

$$\mathrm{Der}(-\log D) = \{\chi \in \theta_W : \chi \cdot g \in I \text{ for all } g \in I\},$$

and in particular if D is a hypersurface with equation h then

$$\mathrm{Der}(-\log D) = \{\chi \in \theta_W : \chi \cdot h = \alpha h \text{ for some } \alpha \in \mathscr{O}_W\}.$$

For motivation, we mention a theorem of Ephraim. If D is any complex space and $x_0 \in D$, the *isosingular locus of* D *at* w, $\mathrm{Iso}(D)_{x_0}$, is the germ of the set of points

$$\{x \in (D, x_0) : \text{ the germs } (D, x_0) \text{ and } (D, x) \text{ are isomorphic}\}.$$

Theorem 8.6 (Ephraim [Eph78]) *Let* $D \subset W$. *Then* $Iso(D)_{x_0}$ *is a germ of smooth submanifold of* W *and* $T_{x_0} Iso(D)_{x_0} = \{\chi(x_0) : \chi \in \mathrm{Der}(-\log D)_{x_0}\}$. □

The inclusion of right-hand side in left in Theorem 8.6 is clear: if $\chi \in \mathrm{Der}(-\log D)_{x_0}$ then the integral flow of χ preserves D, and thus induces a family of isomorphisms of D. Evidently the integral curve of χ through w is contained in $\mathrm{Iso}(D)_{x_0}$, and so its tangent vector $\chi(x_0)$ is contained in $T_{x_0}\mathrm{Iso}(D)_{x_0}$.

The vector space $\{\chi(x_0) : \chi \in \text{Der}(-\log D)_{x_0}\}$ in Theorem 8.6 is known as the *logarithmic tangent space to D at w*; we denote it by $T_w^{\log} D$. If Y is a smooth space and $i : Y \to W$ a map, we say i is *logarithmically transverse to D at $y_0 \in Y$*, and write $i \pitchfork^{\log} D$, if

$$d_{y_0} i(T_{y_0} Y) + T_{i(y_0)}^{\log} D = T_{i(y_0)} W. \tag{8.18}$$

Each of the three vector spaces in (8.18) is the evaluation at y_0 of the stalk of a sheaf of $\mathcal{O}_Y$-modules: the three sheaves are, respectively, $ti(\theta_Y)$, $i^*(\text{Der}(-\log D))$ and $\theta(i)$.

Proposition 8.7 *The equality* (8.18) *holds if and only if*

$$\frac{\theta(i)}{ti(\theta_Y) + i^*(\text{Der}(-\log D))} = 0. \tag{8.19}$$

Proof This is just Nakayama's Lemma: we have

$$\frac{\theta(i)}{ti(\theta_Y) + i^*(\text{Der}(-log D)) + \mathfrak{m}_{Y,y_0}\theta(i)} \simeq \frac{T_{i(y_0)} W}{d_{y_0} i(T_{y_0} Y) + T_{i(y_0)}^{\log} D}.$$

□

The quotient module on the left-hand side of (8.19) thus measures the failure of logarithmic transversality of i to D. Its $\mathbb{C}$-vector-space dimension is finite if and only if i is logarithmically transverse to D outside y_0.

Let $F : (\mathbb{F}_r^N, S) \to (\mathbb{F}^P, 0)$ be a stable map-germ, and let D be its discriminant, if $N \geq P$, or image, if $N < P$.

Theorem 8.7 (J.N. Damon [Dam91]) *Suppose $f : (\mathbb{F}^n, S) \to (\mathbb{F}^p, 0)$ is obtained from the stable map-germ $F : X \to Z$ by transverse pull back by $i : Y \to Z$.*

(1) f is stable if and only if $i \pitchfork^{\log} D$.
(2) More generally,

$$T^1_{\mathscr{A}_e} f \simeq \frac{\theta(i)}{ti(\theta_Y) + i^*(\text{Der}(-\log D))}.$$

Before beginning the proof, note that the denominator $ti(\theta_Y) + i^*(\text{Der}(-\log D))$ is a sum of two terms, like the denominator in the original definition of $T^1_{\mathscr{A}_e} f$ as $\theta(f)/tf(\theta_n)+\omega f(\theta_p)$. However, there is an important difference: here, $\theta(i)$, $ti(\theta_p)$ and $i^*(\text{Der}(-\log D))$ are all finitely generated $\mathcal{O}_p$-modules, whereas in the original definition, only $\omega f(\theta_p)$ is guaranteed to be finitely generated over $\mathcal{O}_p$.

To prove the theorem, we need first to assemble some information about the discriminant.

8.5.1 Critical Space and Discriminant

Let $f : (\mathbb{C}^n, S) \to (\mathbb{C}^p, 0)$ be $\mathscr{A}$-finite, with $n \geq p$, and not a submersion. In Appendix D it is shown that the critical locus Σ_f is Cohen–Macaulay of pure dimension $p - 1$ (with the analytic structure given by the $p \times p$ minors of the Jacobian matrix J_f). Moreover, Σ_f is also reduced, provided that $p \geq 2$ (see Proposition D.4). When f is stable, then this is also true even for $p = 1$, since in that case Σ_f is smooth of dimension 0.

Proposition 8.8 *Let $F : (V, S) \to (W, w_0)$ be a stable map-germ and let D be its discriminant, when* $\dim V \geq \dim W$, *or image when* $\dim V < \dim W$. *Let $\chi \in \theta_{W,w_0}$. Then $\chi \in \mathrm{Der}(-\log D)_{w_0}$ if and only if it can be lifted to a vector field $\tilde{\chi}$ on (V, S)—that is, there exists $\tilde{\chi} \in \theta_{V,S}$ such that*

$$tF(\tilde{\chi}) = \omega F(\chi).$$

Proof We closely follow Looijenga's proof in [Loo84]. Suppose that $\chi \in \theta_W$ has lift $\tilde{\chi} \in \theta_V$. By integrating χ and $\tilde{\chi}$ we obtain flows Ψ_t, Φ_t on W and V, respectively, such that

$$F \circ \Phi_t = \Psi_t \circ F. \tag{8.20}$$

Suppose $y \in D$, and let $x \in \Sigma_F$ satisfy $y = F(x)$. For every t, (8.20) shows that the germs

$$F : (X, \Phi_t(x)) \to (Y, \Psi_t(y)) \quad \text{and} \quad F : (X, x) \to (Y, y)$$

are left-right equivalent. Since x is a critical point of F, so is $\Phi_t(x)$, and therefore $\Psi_t(y)$, which is equal to $F(\Phi_t(x))$, lies in D. That is, we have shown that the flows Φ_t and Ψ_t preserve Σ_F and D, respectively. It follows that the vector fields $\tilde{\chi}$ and χ are tangent to Σ_F and D respectively. In particular, $\chi \in \mathrm{Der}(-\log D)$.

Reciprocally, if $\chi \in \mathrm{Der}(-\log D)$ then we can certainly lift $\chi|_D$ to a vector field $\tilde{\chi}_0$ on Σ_F. For by Proposition D.4, Σ_F is the normalisation of D, and vector fields lift to the normalisation, by a theorem of Seidenberg (see Theorem D.11 in Appendix D). Suppose $\tilde{\chi}_0$ is the restriction to Σ_F of a vector field $\tilde{\chi}_1 \in \theta_X$. We have no guarantee that $\tilde{\chi}_1$ is a lift of χ—i.e. that $tF(\tilde{\chi}_1) = \omega F(\chi)$—only that this equality holds on Σ_F. But because J_F is radical, the fact that $\tilde{\chi}|_{\Sigma_F}$ is a lift of $\chi|_D$ means that $tF(\tilde{\chi}_1) - \omega F(\chi) \in J_F\theta(F)$. By Cramer's rule,

$$J_F\theta(F) \subset tF(\theta_V), \tag{8.21}$$

and thus there exists a vector field $\xi \in \theta_V$ such that

$$tF(\xi) = tF(\tilde{\chi}_1) - \omega F(\chi), \tag{8.22}$$

so that finally

$$tF(\tilde{\chi}_1 - \xi) = \omega F(\chi), \tag{8.23}$$

showing that χ is liftable. □

Remark 8.2 The "if" part of Proposition 8.8 holds without the stability assumption on the map-germ F. For the "only if" part, the stability of F is only necessary in order to ensure that the restriction $F : \Sigma_F \to D$ is the normalisation of D. But this is also true more generally when F is $\mathscr{A}$-finite and either $\dim V < \dim W$ or $\dim V \geq \dim W \geq 3$ (see Proposition D.4). ◊

Proof of Theorem 8.7 We show first that we can assume that f, i, j and F form a standard fibre square as in (8.17). To see this, we may suppose that i is an immersion, for if we replace i by the immersion $i_1(y) = (i(y), y) \in Z \times Y$, and F by $F_1 = F \times \mathrm{id}_Y : X \times Y \to Z \times Y$, then the discriminant of F_1 is equal to $D \times Y$, the pull-back map

$$i_1^*(F_1) : (X \times Y) \times_{Z \times Y} \to Y$$

is isomorphic (even the same as) the previous pull-back $i^*(F)$, and (by Exercise 8.5.10 below)

$$\frac{\theta(i_1)}{ti_1(\theta_Y) + i_1^*(\mathrm{Der}(-\log(D \times Y))} \simeq \frac{\theta(i)}{ti(\theta_Y) + i^*(\mathrm{Der}(-\log D)}.$$

With this assumption, by Theorem 1.2 we can choose coordinates $y_1, \ldots, y_p$ on Y and $y_1, \ldots, y_p, u_1, \ldots, u_d$ on $Z \times Y$ so that i becomes the standard immersion $i(y) = (y, 0)$. Of course this changes the map f, but the new $T^1_{\mathscr{A}_e} f$ is isomorphic to the old. As F is transverse to i, the map $(u_1 \circ F, \ldots, u_d \circ F)$ is a submersion, so its d component functions form part of a coordinate system on X. With respect to this new coordinate system, F is an unfolding of f.

Now that we are in the situation of the standard fibre square, we revert to the notation of (8.17) in which the parameter space is denoted by U. We denote by $\theta_{X \times U/U}$ the $\mathscr{O}_{X \times U}$-submodule of $\theta_{X \times U}$ consisting of vector fields on $X \times U$ with zero component in the U direction, and, similarly, by $\theta(F/U)$ the $\mathscr{O}_{X \times U}$-submodule of $\theta(F)$ consisting of vector fields along F with zero component in the U direction. We define $\theta_{Y \times U/U}$ and $\mathrm{Der}(-\log D)/U$ analogously.

Let $\pi : Y \times U \to U$ be projection. Consider the following diagram.

$$\begin{array}{ccccccccc}
 & & 0 & & 0 & & & & \\
 & & \uparrow & & \uparrow & & & & \\
0 & \longrightarrow & t\pi(\mathrm{Der}(-\log D)) & \longrightarrow & \theta(\pi) & \longrightarrow & 0 & & \\
 & & \uparrow{\scriptstyle t\pi} & & \uparrow{\scriptstyle t\pi} & & \uparrow & & \\
0 & \longrightarrow & \mathrm{Der}(-\log D) & \longrightarrow & \theta_{Y\times U} & \longrightarrow & \dfrac{\theta(F)}{tF(\theta_{X\times U})} & \longrightarrow & 0 \\
 & & \uparrow & & \uparrow & & \uparrow & & \\
0 & \longrightarrow & \mathrm{Der}(-\log D)/U & \longrightarrow & \theta_{Y\times U/U} & \longrightarrow & \dfrac{\theta(F/U)}{tF(\theta_{X\times U/U})} & \longrightarrow & 0 \\
 & & \uparrow & & \uparrow & & \uparrow & & \\
 & & 0 & & 0 & & 0 & &
\end{array} \tag{8.24}$$

Each column is exact. This is obvious for the first two columns; for the third, it is an easy calculation that the homomorphism

$$\frac{\theta(F/U)}{tF(\theta_{X\times U/U})} \to \frac{\theta(F)}{tF(\theta_{X\times U})},$$

induced by the inclusion $\theta(F/U) \hookrightarrow \theta(F)$, is an isomorphism.

Each row in the diagram is a complex, and thus we have a short exact sequence of complexes. Let us call the complexes, in ascending order, A, B and C, and give the columns from left to right indices 2, 1, 0. The resulting long exact sequence of homology contains the portion

$$\cdots \longrightarrow H_1(B_\bullet) \longrightarrow H_1(C_\bullet) \longrightarrow H_0(A_\bullet) \longrightarrow H_0(B_\bullet) \longrightarrow \cdots$$

However, $B_\bullet$ is exact, by Proposition 8.8, and thus $H_1(C_\bullet) \simeq H_0(A_\bullet)$. Evaluating these homology modules, we obtain

$$\frac{\theta(\pi)}{t\pi(\mathrm{Der}(-\log D))} \simeq \frac{\theta(F/U)}{tF(\theta_{X\times U/U}) + \omega F(\theta_{Y\times U/U})}. \tag{8.25}$$

Dividing each side by $\mathfrak{m}_U$ times itself gives

$$\frac{\theta(\pi)}{t\pi(\mathrm{Der}(-\log D)) + \mathfrak{m}_U\theta(\pi)} \simeq \frac{\theta(F/U)}{tF(\theta_{X\times U/U}) + \omega F(\theta_{Y\times U/U}) + \mathfrak{m}_U\theta(F/U)}; \tag{8.26}$$

the right-hand side in (8.26) is just $T^1_{\mathscr{A}_e} f$.

Now we show that the left-hand side in (8.26) is isomorphic to $\theta(i)/ti(\theta_Y) + i^*(\mathrm{Der}(-\log D))$. We have

$$\theta(\pi) = \sum_{j=1}^{d} \mathcal{O}_{Y\times U}\frac{\partial}{\partial u_j}$$

so

$$\frac{\theta(\pi)}{\mathfrak{m}_U\theta(\pi)} = \sum_{j=1}^{d} \mathcal{O}_Y\frac{\partial}{\partial u_i}.$$

Also

$$\theta(i) = \sum_{k=1}^{p} \mathcal{O}_Y\frac{\partial}{\partial y_k} \oplus \sum_{j=1}^{d} \mathcal{O}_Y\frac{\partial}{\partial u_j}$$

so

$$\frac{\theta(i)}{ti(\theta_Y)} = \sum_{j=1}^{d} \mathcal{O}_Y\frac{\partial}{\partial u_j}.$$

So $\theta(i)/ti(\theta_Y)$ can be identified with $\theta(\pi)/\mathfrak{m}_U\theta(\pi)$. Using this identification, we have

$$\frac{\theta(i)}{ti(\theta_Y) + i^*\,\mathrm{Der}(-\log D)} \simeq \frac{\theta(\pi)}{t\pi(\mathrm{Der}(-\log D)) + \mathfrak{m}_U\theta(\pi)} \simeq T^1_{\mathcal{A}_e} f.$$

□

The module in the denominator of (8.19) is in fact the (extended) tangent space to the orbit of i under a variant of contact equivalence introduced by Damon in [Dam87] and called $\mathcal{K}_D$*-equivalence*. We do not make use of $\mathcal{K}_D$-equivalence here, though it was the key to Damon's proof of Theorem 8.7 in [Dam91]. There, he showed that if i_t is a deformation of i then the family $i_t^*(F)$ is $\mathcal{A}$-trivial if and only if i_t is $\mathcal{K}_D$-trivial.

Definition 8.5 Let $f, g : (Y, y_0) \to (W, w_0)$ and let $(D, w_0) \subset (W, w_0)$. We say that f is $\mathcal{K}_D$-equivalent to g if there exists diffeomorphisms $\Phi : (Y \times W, (y_0, w_0)) \to (Y \times W, (y_0, w_0))$ and $\varphi : (Y, y_0) \to (Y, y_0)$ such that

1. Φ lifts φ, i.e. $\pi_Y \circ \Phi = \varphi \circ \pi_Y$;
2. $\Phi(Y \times D) = Y \times D$,
3. $\Phi(\mathrm{graph}(f)) = \mathrm{graph}(g)$.

The usual version of contact equivalence ($\mathcal{K}$-quivalence) is, in effect, $\mathcal{K}_D$-equivalence with $D = \{w_0\}$.

Definition 8.6 If D is a divisor (hypersurface) in W, we say D is a *free divisor* if $\mathrm{Der}(-\log D)$ is a locally free $\mathscr{O}_W$-module.

The definition is due to Kyoji Saito in [Sai80].

Proposition 8.9 ([Loo84, 6.13]) *If $F : (\mathbb{C}^n, S) \to (\mathbb{C}^p, y_0)$ $(n \geq p)$ is stable, then the discriminant D of F is a free divisor.*

Proof Let us write $(\mathbb{C}^n, S) =: X$, $(\mathbb{C}^p, y_0) =: Y$. The proof uses the short exact sequence

$$0 \longrightarrow \mathrm{Der}(-\log D) \longrightarrow \theta_Y \xrightarrow{\bar{\omega}F} \frac{\theta(F)}{tF(\theta_X} \longrightarrow 0 \tag{8.27}$$

which already appeared (in the special case that F is a parameterised unfolding) as the complex $B_\bullet$ of (8.24). We may assume that F is not a trivial unfolding of a lower dimensional germ; freeness of $\mathrm{Der}(-\log D)$ under this assumption implies freeness in general, since the discriminant of a trivial unfolding $F \times \mathrm{id}_U$ is equal to the product of U with the discriminant of F (see Exercise 8.5.1 below).

From the assumption, it follows that all of the members of $\mathrm{Der}(-\log D)$ vanish at y_0. For if $\chi \in \mathrm{Der}(-\log D)_{y_0}$ and $\chi(y_0) \neq 0$, then lifting χ to $\tilde{\chi}$ in θ_X (using Proposition 8.8), the Thom–Levine Lemma 2.6 implies that the integral flows of χ and $\tilde{\chi}$ give a 1-parameter trivialisation of F.

Let $\chi_1, \ldots, \chi_\ell$ be a minimal set of generators of $\mathrm{Der}(-\log D)$, with $\chi_i = \sum_j \chi_i^j \partial/\partial y_j$, and let χ be the $p \times \ell$ matrix of coefficients χ_i^j. Then

$$\mathscr{O}_Y^\ell \xrightarrow{\chi} \theta_Y \xrightarrow{\bar{\omega}F} \frac{\theta(F)}{tF(\theta_X)} \longrightarrow 0 \tag{8.28}$$

is exact. Since all of the entries in χ vanish at y_0, (8.28) is the right-hand end of a minimal free resolution of $\theta(F)/tF(\theta_Y)$. We claim that a minimal free resolution must have length 1; from this it follows that χ is injective, and thus an isomorphism onto its image, so that this image, $\mathrm{Der}(-\log D)$, is free. We prove that the length of a minimal free resolution of $\theta(F)/tF(\theta_n)$ is 1, in two steps as follows:

Step 1 $\theta(F)/tF(\theta_X)$ is Cohen–Macaulay of dimension $p-1$. This follows from a classical theorem of Buchsbaum and Rim [BR64, Corollary 2.7], which we quote as Theorem C.7 in Appendix C, via Theorem C.12. To apply it here, we need that the codimension of the support of $\theta(F)/tF(\theta_n)$ is the same as the codimension in the generic case, $p-n+1$. The support of $\theta(F)/tF(\theta_n)$ is just $V(J_F)$, whose codimension is shown to take the required value in the proof of Proposition D.4.

Step 2 Because $\theta(F)/tF(\mathrm{Der}_X)$ has depth $p-1$ over $\mathscr{O}_X$, and is finite over $\mathscr{O}_Y$, its $\mathscr{O}_Y$-depth is also $p-1$. Therefore by the Auslander–Buchsbaum Theorem C.4 its projective dimension is 1.

Exercises for Sect. 8.5

1 Show that if $D \subset W$ is a free divisor and U is smooth, then $D \times U \subset W \times U$ is also a free divisor.

2 In the situation of (8.15), show that if V, W and Y are smooth and $i \pitchfork F$, then $V \times_Y W$ is smooth of dimension $\dim V + \dim Y - \dim W$.

3 Show that if f is obtained by transverse pull-back from F, then

(i) the set of critical points of f is the preimage by π_X of the set of critical points of F;
(ii) the set of critical values of f is the preimage by i of the set of critical values of F;
(iii) the local algebras $Q(f)$ and $Q(F)$ are isomorphic.

4 Prove Proposition 8.6, by first obtaining f by transverse pull-back from a stable unfolding F_0, and then using the fact that because stable maps are classified by their local algebras (Theorem 7.1), F and F_0 are $\mathscr{A}$-equivalent, perhaps after taking a constant unfolding of one or other of them so as to equalise the dimension of their domains.

5 Let f be the germ of type H_2 given by $(x, y) \mapsto (x, y^3, xy + y^5)$ and let $F(a, b, c, y) = (a, b, c, y^3 + ay, by^2 + cy)$. Find $i : \mathbb{C}^3 \to \mathbb{C}^5$ such that $i^*(F) \simeq_{\mathscr{A}} f$.

6 Let $F(u, v, y) = (u, v, y^4 + uy^2 + vy)$. Find $i : (\mathbb{C}^2, 0) \to (\mathbb{C}^3, 0)$ such that $i^*(F) \simeq_{\mathscr{A}} f$ where $f(x, y) = (x, xy + y^4)$.

7 Let $f(x, y) = (x, y^3 + x^k y)$. Find a stable germ F and a germ i such that $f \simeq_{\mathscr{A}} i^*(F)$.

8 Find a stable bi-germ $F : (\mathbb{C}, \{0, 0'\}) \to (\mathbb{C}^2, 0)$ and $i : (\mathbb{C}^2, 0) \to (\mathbb{C}^2, 0)$ such that i^*F is $\mathscr{A}$-equivalent to

$$\begin{cases} s \mapsto (s, s^2) \\ t \mapsto (t, -t^2) \end{cases}$$

9 Show that, as stated in the proof of Theorem 8.7, the natural map

$$\frac{\theta(F/U)}{tF(\theta_{X \times U/U})} \to \frac{\theta(F)}{tF(\theta_{X \times U})}$$

is an isomorphism.

10 Suppose that $D = D_0 \times S \subset Z \times S$, let $i : Y \to Z \times S$ be any germ, and let $\pi : Z \times S \to Z$ be projection.

(i) Show that

$$\frac{\theta(i)}{ti(\theta_Y) + i^* \operatorname{Der}(-\log D)} \simeq \frac{\theta(\pi \circ i)}{t(\pi \circ i)(\theta_Y) + (\pi \circ i)^*(\operatorname{Der}(-\log D_0))}.$$

Hint: $\operatorname{Der}(-\log D)$ contains all the vector fields $\partial/\partial s_i$. You can choose the remaining generators for $\operatorname{Der}(-\log D)$ in $\theta_{Z\times U/U}$.

(ii) Let also $j_0 : Z \to Z \times S$ be the inclusion $z \mapsto (z, 0)$. Show that

$$\frac{\theta(i)}{ti(\theta_Y) + i^* \operatorname{Der}(-\log D_0)} \simeq \frac{\theta(j_0 \circ i)}{t(j_0 \circ i)(\theta_Y) + (j_0 \circ i)^*(\operatorname{Der}(-\log D))}.$$

(iii) Show that if $j : Z \to Z \times U$ is any inclusion of the form $z \mapsto (z, s(z))$ then

$$\frac{\theta(i)}{ti(\theta_Y) + i^* \operatorname{Der}(-\log D_0)} \simeq \frac{\theta(j \circ i)}{t(j \circ i)(\theta_Y) + (j \circ i)^*(\operatorname{Der}(-\log D))}.$$

11 Let $g : (\mathbb{F}^n, 0) \to (\mathbb{F}^p, 0)$ and $(D, 0) \subset (\mathbb{F}^p, 0)$.

(i) Show that

$$\left\{\frac{d\Phi_t \cdot f}{dt}|_{t=0} : \Phi_t \in \mathscr{K}_D \text{ is analytic }, \Phi_0 = \operatorname{id}\right\} = tf(\mathfrak{m}_n\theta_n) + f^* \operatorname{Der}(-\log D). \tag{8.29}$$

(ii) Show that if g is $\mathscr{K}_D$-finite then it is finitely determined for $\mathscr{K}_D$-equivalence. This can be proved by following the lines of the proof of finite determinacy for (ordinary) $\mathscr{K}$-equivalence given in Exercise 6.1.1.

12

(i) Suppose that the stable map $F : (\mathbb{C}^N, S) \to (\mathbb{C}^P, 0)$ is *minimal* in the sense that it is not $\mathscr{A}$-equivalent to a trivial unfolding of a germ $(\mathbb{F}^{N-1}, 0) \to (\mathbb{F}^{P-1}, 0)$. Show that all the vector fields in $\operatorname{Der}(-\log D(F))$ vanish at 0.

(ii) Suppose that $f : (\mathbb{F}^n, 0) \to (\mathbb{F}^p, 0)$ is induced from a mimimal stable germ $F : (\mathbb{F}^N, 0) \to (\mathbb{F}^P, 0)$ by transverse pull-back. Show that

$$\mathscr{A}_e - \text{codimension}(f) \geq P - p.$$

8.6 Bifurcation Sets

Let $f : (\mathbb{C}^n, S) \to (\mathbb{C}^p, 0)$ be an $\mathscr{A}$-finite map-germ, with $n \geq p$, and let $F : (\mathbb{C}^n \times \mathbb{C}^d, (S \times \{0\}) \to (\mathbb{C}^p \times \mathbb{C}^d, (0, 0))$ be an $\mathscr{A}_e$-versal unfolding. The isomorphism (8.25), together with the fact that the discriminant of a stable map is a free divisor, allows us to prove

Theorem 8.8 *Suppose that f does have a stable perturbation. Then the bifurcation set $\mathscr{B}(F)$ is a hypersurface in $\mathbb{C}^d$.*

Proof The locus of points $(y, u) \in \mathbb{C}^p \times \mathbb{C}^d$ such that the germ $f_u : (\mathbb{C}^p, f^{-1}(y) \cap \Sigma_{f_u}) \to (\mathbb{C}^p, y)$ is not stable is the support of the coherent sheaf

$$\mathscr{T}^1_{\mathscr{A}_e}(F/U) := \frac{F_*\left(\theta(F/U)/tF(\theta_{X\times U/U})\right)}{\omega F(\theta_{Y\times U/U})}$$

already studied in the proof of Lemma 5.3 and $\mathscr{B}(F)$ is the image of this set under the projection $\pi : \mathbb{C}^p \times \mathbb{C}^d \to \mathbb{C}^d$. This projection is finite on $\mathrm{supp}\mathscr{T}^1_{\mathscr{A}_e}(F/U)$, by the Mather–Gaffney criterion for $\mathscr{A}$-finiteness: the only point in this support lying over $0 \in \mathbb{C}^d$ is $0 \in \mathbb{C}^p$. Thus, the dimension of $\mathscr{B}(F)$ is equal to the dimension of $T^1_{\mathscr{A}_e}(F/U)$.

By (8.25), the sequence

$$\mathrm{Der}(-\log D) \xrightarrow{t\pi} \theta(\pi) \longrightarrow T^1_{\mathscr{A}_e}(F/U) \longrightarrow 0 \tag{8.30}$$

is a presentation. Since D is a free divisor, $\mathrm{Der}(-\log D)$ is a free $\mathscr{O}_{p+d}$-module of rank $p+d$, while $\theta(\pi)$ is free of rank d. Thus (8.30) can be written more abstractly as

$$\mathscr{O}^{p+d} \xrightarrow{K} \mathscr{O}^d \longrightarrow T^1_{\mathscr{A}_e}(F/U) \longrightarrow 0 \tag{8.31}$$

where K consists of the last d rows of the matrix of coefficients of a basis of $\mathrm{Der}(-\log D)$. By Theorem C.6, the codimension of $\mathrm{supp}T^1_{\mathscr{A}_e}(F/U)$ is at most $p+1$, and if it is $p+1$ then $T^1_{\mathscr{A}_e}(F/U)$ is a Cohen–Macaulay module over $\mathscr{O}_{p+d}$. In other words, $\dim \mathrm{supp}T^1_{\mathscr{A}_e}(F/U) \geq d-1$, and if it is $d-1$ then $T^1_{\mathscr{A}_e}(F/U)$ is Cohen–Macaulay. By Proposition 5.4, there are two possibilities as follows:

1. $\dim \mathrm{supp}T^1_{\mathscr{A}_e}(F/U) = d-1$; this holds if $\mathscr{B}(F)$ is a proper subset, and thus an analytic subvariety, of $\mathbb{C}^d$—in particular, if and only if f has a stable perturbation.
2. $\dim \mathrm{supp}T^1_{\mathscr{A}_e}(F/U) = d$, when f has no stable perturbation.

In the first case, $\mathscr{B}(F)$ is a hypersurface. □

In [Dam98], Jim Damon went on to give conditions under which $\mathscr{B}(F)$ is a free divisor. If $\mathscr{B}$ is a hypersurface then $T^1_{\mathscr{A}_e}(F/U)$ is Cohen–Macaulay of dimension $d-1$. Since π is finite on its support, $\mathscr{T}^1_{\mathscr{A}_e}(F/U)_{(0,0)}$ is Cohen–Macaulay over $\mathscr{O}_d$, with depth $d-1$ and therefore (by Theorem C.4) projective dimension 1. The $\mathscr{A}_e$-versality of F is equivalent to the surjectivity of the Kodaira–Spencer map

$$\theta_d \xrightarrow{\mathrm{KS}} T^1_{\mathscr{A}_e}(F/U)$$

given by $\mathrm{KS}(\vartheta) = \overline{\omega F}(\vartheta)$, where $\overline{\omega F} : \theta_d \to T^1_{\mathscr{A}_e}(F/U)$ is the composite of

$$\theta_d \hookrightarrow \theta_{p+d} \to \theta(F/U) \to T^1_{\mathscr{A}_e}(F/U) = \frac{\theta(F/U)}{tF(\theta_{X\times U/U}) + \omega F(\theta_{Y\times U/U})}$$

This follows from the versality criterion Theorem 5.1 by Nakayama's Lemma: since the module $\mathscr{T}^1_{\mathscr{A}_e}(F/U)_{(0,0)}$ is finite over $\mathscr{O}_{\mathbb{C}^d,0}$, by Nakayama's lemma surjectivity of KS is equivalent to surjectivity of

$$\theta_d/\mathfrak{m}_d\theta_d \xrightarrow{\mathrm{KS}_0} T^1_{\mathscr{A}_e}(F/U)/\mathfrak{m}_d T^1_{\mathscr{A}_e}(F/U) = T^1_{\mathscr{A}_e} f\,,$$

which is equivalent to versality by Theorem 5.1.

Because KS is surjective, we have a short exact sequence

$$0 \longrightarrow \ker\ \mathrm{KS} \hookrightarrow \theta_d \xrightarrow{\mathrm{KS}} T^1_{\mathscr{A}_e}(F/U) \longrightarrow 0\,.$$

From the depth lemma (Exercise C.4.5 Part (iii)), it follows that $\ker \mathrm{KS}$ is free. Now by (8.25), $\ker \mathrm{KS}$ is the $\mathscr{O}_d$ module of vector fields ϑ on $(\mathbb{C}^d, 0)$ which lift to a vector field $\eta \in \mathrm{Der}(-\log D(F))$. Every such ϑ is tangent to $\mathscr{B}(F)$, by an argument like in the first paragraph of the proof of Proposition 8.8. Hence

$$\ker \mathrm{KS} \subset \mathrm{Der}(-\log \mathscr{B}(F)).$$

If this inclusion is in fact an equality, then $\mathrm{Der}(-\log \mathscr{B}(F))$ is free and $\mathscr{B}(F)$ is a free divisor. To prove equality, it is enough to show that the set of points where they differ has codimension of at least 2. The argument to show this is a module-theoretic version of the proof of Proposition C.5. Let $M = \mathrm{Der}(-\log D)/\ker \mathrm{KS}$. Then

$$M \subset \theta_d/\ker \mathrm{KS} \simeq \pi_*(T^1_{\mathscr{A}_e}(F/U)),$$

and so $\mathrm{Ass}\ M \subset \mathrm{Ass}\ T^1_{\mathscr{A}_e}(F/U)$. Because $T^1_{\mathscr{A}_e}(F/U)$ is Cohen–Macaulay of dimension $d-1$, each of its associated primes has height 1, and so the same goes for the associated primes of M. But $\mathrm{Ass}\ M \subseteq \mathrm{supp}\, M$, and if $\ker \mathrm{KS}$ and $\mathrm{Der}(-\log D)$ coincide outside a codimension 2 set, then M is not supported at any prime ideal of height 1. This means that it has no associated primes at all—which means (see Sect. C.2) that $M = 0$.

In [Dam98], Damon gives conditions for $\mathrm{Der}(-\log D)$ and $\ker \mathrm{KS}$ to coincide outside a codimension 2 subset. The idea is that the codimension 1 components of $\mathrm{supp}(T^1_{\mathscr{A}_e}(F/U))$ correspond to codimension 1 singularities of germs of maps $f_u : (\mathbb{C}^n, S') \to (\mathbb{C}^p, y)$. So $\mathrm{Der}(-\log D) = \ker \mathrm{KS}$ globally if and only if the equality holds for each of the codimension 1 singularities appearing in the versal deformation of f.

One way to ensure this is to insist on quasihomogeneity. Let $g : (\mathbb{C}^n, S') \to (\mathbb{C}^p, 0)$ be an $\mathscr{A}_e$ codimension 1 singularity. If g is weighted homogeneous on weights $w_1, \ldots, w_n$ and degrees $d_1, \ldots, d_p$, and if it is possible to choose a generator h for $T^1_{\mathscr{A}_e} g$ of positive weight w_0, then g has a weighted homogeneous $\mathscr{A}_e$-versal unfolding $G(x, u) = (g(x) + uh(x), u)$, and now the Euler vector field $w_0\partial/\partial u$ on the base space $\mathbb{C}$ of G lifts to the Euler vector field $\sum_j d_j y_j \partial/\partial y_j + w_0 u\partial/\partial u$ in $\mathrm{Der}(-\log D(G))$, and thus lies in ker KS. Since $\mathscr{B}(G)$ is just the origin in the base space $\mathbb{C}$ of G, the Euler field generates $\mathrm{Der}(-\log \mathscr{B}(G))$, and thus $\ker \mathrm{KS} = \mathrm{Der}(-\log \mathscr{B}(G))$. This local calculation tells us what happens at a point $u \in \mathscr{B}(F)$ lying under a point $(u, y) \in \mathrm{supp} T^1_{\mathscr{A}_e}(F/U)$ where $f_u : (\mathbb{C}^p, \Sigma_{f_u} \cap f_u^{-1}(y)) \to (\mathbb{C}^p, y)$ is left-right equivalent to g. Thus, on this component of $\mathscr{B}(F)$, ker KS and $\mathrm{Der}(-\log \mathscr{B}(F))$ coincide generically.

If every codimension 1 singularity appearing in the versal deformation of f fulfils these conditions, it follows that $\ker \mathrm{KS} = \mathrm{Der}(-\log \mathscr{B}(F))$, and $\mathrm{Der}(-\log \mathscr{B}(F))$ is a free divisor. In [Dam98], Damon establishes the dimension range for which this is guaranteed.

Theorem 8.9 (J.N. Damon [Dam98, Theorem 3]) *If $f : (\mathbb{C}^n, S) \to (\mathbb{C}^p, 0)$ is $\mathscr{A}$-finite, with $n \geq p$, then provided (n, p) and f satisfy one of the numbered conditions below,* $\mathrm{Der}(-\log \mathscr{B}(F)) = \ker KS$, *and the bifurcation set in the base of an $\mathscr{A}_e$-versal unfolding F of f is a free divisor.*

1. $n > p + 1 \leq 5$
2. $n = p + 1 \leq 4$
3. *f has cokernel rank 1,* $n = p + 1 \leq 7$
4. *f has cokernel rank 1,* $n > p + 1 \leq 6$
5. *f has Boardman symbol* $\Sigma^{n-p+1,0}$ *or* $\Sigma^{2,1,0}$.

□

In Sect. 9.6 we study the bifurcation set in the base of a versal unfolding of an $\mathscr{A}$-finite germ $(\mathbb{C}^n, S) \to (\mathbb{C}^p, 0)$ with $n < p$. For $n = p - 1$ the conclusions are similar; when $n < p - 1$ they are very different. The methods also are quite different: when $n < p$ it is no longer true that the module of F-liftable vector fields is free, and the arguments we have used in this section do not apply in that case.

8.7 Calculating the Discriminant Milnor Number

Let h be an equation of the discriminant D of F, and define $\mathrm{Der}(-\log h)$ to be the $\mathscr{O}_Y$-module of germs of vector fields which annihilate h; that is, which are tangent not only to $D = h^{-1}(0)$, but to all level sets of h. Clearly $\mathrm{Der}(-\log h)$ is a submodule of $\mathrm{Der}(-\log D)$, but it depends on the choice of equation h, and is not determined by D alone. The following lemma gives a simple condition under which, if D is a free divisor, then not only $\mathrm{Der}(-\log D)$ but also $\mathrm{Der}(-\log h)$ is a locally free $\mathscr{O}_{\mathbb{C}^p}$-module.

Lemma 8.3 *If $h \in J_h$ (that is, if there exists a vector field χ such that $\chi \cdot h = h$) then*

$$\mathrm{Der}(-\log D) = \mathrm{Der}(-\log h) \oplus \mathscr{O}_Y \cdot \chi. \tag{8.32}$$

Proof Any vector field $\vartheta \in \mathrm{Der}(-\log D)$ can be written in the form

$$\vartheta = (\vartheta - (\vartheta \cdot h)\chi) + ((\vartheta \cdot h)\chi).$$

□

The lemma applies in case h is weighted homogeneous, taking as χ the Euler vector field. But even if no such vector field exists for h, we can find an equation for the discriminant of a constant unfolding $F_1 := F \times \mathrm{id}_{\mathbb{C}}$, for which the splitting (8.32) holds. Take t as the coordinate on the extra copy of $\mathbb{C}$; since $D(F_1) = D \times \mathbb{C}$, we can use $h_1(y, t) = e^t h(y)$ as equation for $D(F_1)$, and $\partial/\partial t$ as the vector field χ. We say that h is a *good defining equation* for D if (8.32) holds.

Lemma 8.4 *If $f : (\mathbb{C}^n, S) \to (\mathbb{C}^p, 0)$ is obtained from F by transverse pull-back, then it can also be obtained from $F \times id_{\mathbb{C}}$ by transverse pull back.* □

Lemma 8.5 *Suppose that the hypersurface germ $(D, 0) \subset (\mathbb{C}^q, 0)$ has a defining equation h which is weighted homogeneous with weight $\alpha \neq 0$ with respect to some set of weights $w_1, \ldots, w_q$ on $(\mathbb{C}^q, 0)$. If $i : (\mathbb{C}^p, 0) \to (\mathbb{C}^q, 0)$ is weighted homogeneous with degree $\beta \neq 0$ with respect to $w_1, \ldots, w_q$ and to weights $v_1 \ldots, v_p$ on $(\mathbb{C}^p, 0)$, then for any reduced defining equation h' of D,*

$$ti(\theta_p) + i^* \mathrm{Der}(-\log h') = ti(\theta_p) + i^* \mathrm{Der}(-\log D). \tag{8.33}$$

Proof Let E_q be the Euler vector field with respect to the weights on $\mathbb{C}^q$. Then $E_q \cdot h = \alpha h$. We have $h' = uh$, for some unit $u \in \mathscr{O}_q$. Then

$$E_q \cdot h' = (\alpha u + E_q \cdot u)h = (\alpha + u^{-1}(E_q \cdot u))h'.$$

Since E_q vanishes at 0, $\alpha + u^{-1}(E_q \cdot u)$ is a unit in $\mathscr{O}_q$. By (8.32), $\mathrm{Der}(-\log D) = \mathscr{O}_q \cdot E_q \oplus \mathrm{Der}(-\log h')$.

To prove (8.33), it is therefore enough to show that $E_q \circ i \in ti(\theta_p)$. Let E_p be the Euler vector field on $(\mathbb{C}^p, 0)$, with respect to $v_1, \ldots, v_p$. The fact that i is weighted homogeneous with degree β means that $ti(E_p) = \beta E_q \circ i$, so that if $\beta \neq 0$, $E_q \circ i \in ti(\theta_p)$ as required. □

Theorem 8.10 ([DM91]) *Suppose that $F : (\mathbb{C}^N, 0) \to (\mathbb{C}^P, 0)$ is stable, and that (8.32) holds. If $f : (\mathbb{C}^n, S) \to (\mathbb{C}^p, 0)$ has finite $\mathscr{A}_e$-codimension, with $n \geq p$ and (n, p) nice dimensions, and f is obtained from F by transverse pull-back by i, then*

$$\mu_\Delta(f) = dim_{\mathbb{C}} \frac{\theta(i)}{ti\big(\theta_{\mathbb{C}^p,0}\big) + i^*\big(\mathrm{Der}(-\log h)\big)}. \tag{8.34}$$

Proof The proof involves four steps

(1) Let h be a reduced defining equation of the discriminant D of F, and let

$$I : \big(\mathbb{C}^p \times \mathbb{C}, (0,0)\big) \to (\mathbb{C}^p, 0)$$

be a deformation of i, inducing a stable perturbation of f, that is, such that if $F(x,u) = (f_u(x), u)$ is the induced unfolding of f, then f_u is locally stable for $u \neq 0$. Such a deformation exists since the pair (n, p) is in the nice dimensions. Write $I(y,u) = i_u(y)$. Then for $u \neq 0$, i_u is logarithmically transverse to D. Also, for each $u \in (\mathbb{C}, 0)$, $h \circ i_u$ is a reduced defining equation for $D(f_u)$. We pick a representative for I whose image meets no critical point of h off $h^{-1}(0)$.

Lemma 8.6 *For each point y, the homomorphism* $i_u^*(th) : \theta(i_u)_y \to \theta(h \circ i_u) \simeq \mathscr{O}_{\mathbb{C}^p,y}$ *defined by*

$$i_u^*(th)\left(\sum_j \alpha_j \frac{\partial}{\partial z_j}\right) = \sum_j \alpha_j \left(\frac{\partial h}{\partial z_j} \circ i_u\right)$$

passes to the quotient to give rise to a homomorphism

$$\frac{\theta(i_u)_y}{ti_u(\theta_{\mathbb{C}^p,y}) + i_u^*(\mathrm{Der}(-\log h))_y} \to \frac{\mathscr{O}_{\mathbb{C}^p,y}}{J_{h\circ i_u}}. \tag{8.35}$$

If h is non-singular at $i_u(y)$*, then this is an isomorphism.* □

Proof of Lemma 8.6 The surjectivity of $i_u^*(th)$ follows from the fact that at least one partial derivative $\partial h/\partial Y_j$ is non-zero at $i_u(y)$.

That $i_u^*(th)$ passes to the quotient is also clear: if $\alpha \in i_u^*(\mathrm{Der}(-\log h))$, then evidently $i_u^*(th)(\alpha) = 0$. If $\alpha = ti_u(\beta)$, then $i_u^*(th)(\alpha) = t(h \circ i_u)(\beta) \in J_{h\circ i_u}$.

To see that the quotient morphism (8.35) is injective, suppose first that i_u is an immersion. If $i^*(th)(\alpha) \in J_{h\circ i_u}$, then there is a germ of vector field $\beta \in \theta_p$ such that $i^*(th)(\alpha) = t(h \circ i_u)(\beta)$. Both α and β may be extended to vector fields $\tilde{\alpha}$ and $\tilde{\beta}$ on $(\mathbb{C}^P, 0)$. The equality $i^*(th)(\alpha) = t(h \circ i_u)(\beta)$ implies that $(th(\tilde{\alpha} - \tilde{\beta})) \circ i_u = 0$. Because h is non-singular at $i_u(y)$, there is a germ of vector field $\chi \in \theta_{\mathbb{C}^P, i_u(y)}$ such that $dh(\chi) = 1$. Then the vector field

$$\eta := (\tilde{\alpha} - \tilde{\beta}) - (dh(\tilde{\alpha} - \tilde{\beta}))\chi$$

lies in $\mathrm{Der}(-\log h)$, and

$$\alpha - ti_u(\beta) = \eta \circ i_u \in i_u^*(\mathrm{Der}(-\log h)),$$

showing that (8.35) is injective and thus an isomorphism.

We complete the proof of the lemma by showing that any map i_u can be replaced by an immersion, without changing either the source or the target of (8.35). Very simply, we replace $\mathbb{C}^P$ by $\mathbb{C}^P \times \mathbb{C}^p$ and i_u by the immersion $\tilde{i}_u := (i_u, \mathrm{id}_{\mathbb{C}^p})$, and we extend h to the function $\tilde{h}$ on $\mathbb{C}^P \times \mathbb{C}^p$ defined by $\tilde{h}(z, y) = h(z)$. Then $\mathrm{Der}(-\log \tilde{h})$ is generated by the members of $\mathrm{Der}(-\log h)$ (regarded as vector fields on $\mathbb{C}^P \times \mathbb{C}^p$) together with $\partial/\partial y_1, \ldots, \partial/\partial y_p$. These latter cancel the extra generators of $\theta(\tilde{i}_u)$, and it follows that

$$\frac{\theta(\tilde{i}_u)}{t\tilde{i}_u(\theta_p) + i_u^*(\mathrm{Der}(-\log \tilde{h}))} \simeq \frac{\theta(i_u)}{ti_u(\theta_p) + i_u^*(\mathrm{Der}(-\log h))}.$$

On the other hand, $\tilde{h} \circ \tilde{i}_u = h \circ i_u$, so the right-hand side of (8.35) is also unchanged. □

By our choice of representative of I, from the lemma it follows that

$$\sum_{y \notin D(f_u)} \dim_{\mathbb{C}} \frac{\theta(i_u)_y}{ti_u(\theta_{\mathbb{C}^p,y}) + i_u^*(\mathrm{Der}(-\log h))_y} = \sum_{y \notin D(f_u)} \mu(h \circ i_u, y) \tag{8.36}$$

(2) The right-hand side in (8.36) is equal to $\mu_\Delta(f)$, by Theorem 8.4.
(3) Because f_u is stable, Theorem 8.7 tells us that at all points $y \in D(f_u)$,

$$\frac{\theta(i_u)_y}{ti_u(\theta_{\mathbb{C}^p,y}) + i_u^*(\mathrm{Der}(-\log D))_y} = 0.$$

We claim that in fact at all these points,

$$\frac{\theta(i_u)_y}{ti_u(\theta_{\mathbb{C}^p,y}) + i_u^*(\mathrm{Der}(-\log h))_y} = 0.$$

For this we use the hypothesis that (n, p) are nice dimensions. By Theorem 7.6, in the nice dimensions, all stable germs are quasihomogeneous in suitable coordinates. Thus, the germ of f_u at $f_u^{-1}(y) \cap C(f_u)$, being stable, is weighted homogeneous with respect to suitable coordinates around $f_u^{-1}(y) \cap C(f_u)$ in the source and around y in the target. Give or take smooth factors, the germ of F at $F^{-1}(i_u(y)) \cap C(F)$ is a trivial unfolding of the stable germ of f_u, and thus the good $\mathbb{C}^*$ action on $(\mathbb{C}^p, y)$ extends to a good $\mathbb{C}^*$ action on $D(F)$ at $i_u(y)$. The germ of i_u at y is now equivariant with respect to these actions, and thus Lemma 8.5 applies, proving the claim. Thus,

$$\sum_y \dim_{\mathbb{C}} \frac{\theta(i_u)_y}{ti_u(\theta_{\mathbb{C}^p,y}) + i_u^*(\mathrm{Der}(-\log h))_y} = \sum_{y \notin D(f_u)} + \sum_{y \in D(f_u)} = \sum_{y \notin D(f_u)} = \mu_\Delta(f).$$

(4) The final step is to show that

$$\sum_y \dim_{\mathbb{C}} \frac{\theta(i_u)_y}{ti_u(\theta_{\mathbb{C}^p,y}) + i_u^*(\mathrm{Der}(-\log h))_y} = \dim_{\mathbb{C}} \frac{\theta(i)}{ti\big(\theta_{\mathbb{C}^p,0}\big) + i^*\big(\mathrm{Der}(-\log h)\big)},$$

in other words, that we have conservation of multiplicity. To show this, consider the relative module,

$$T^1_{\mathscr{K}_{h,e}}(I/\mathbb{C}) := \frac{\theta(I)}{tI(\theta_{\mathbb{C}^p\times\mathbb{C}/\mathbb{C}}) + I^*(\mathrm{Der}(-\log h))}.$$

This has presentation

$$\theta_{\mathbb{C}^p\times\mathbb{C}/\mathbb{C}} \oplus I^*(\mathrm{Der}(-\log h)) \to \theta(I) \to T^1_{\mathscr{K}_{h,e}}(I/\mathbb{C}) \to 0. \tag{8.37}$$

Now $\theta_{\mathbb{C}^p\times\mathbb{C}/\mathbb{C}}$ is free of rank p, and because $\mathrm{Der}(-\log h)$ is a free $\mathscr{O}_{\mathbb{C}^P}$-module of rank $P-1$, $I^*(\mathrm{Der}(-\log h))$ is free of rank $P-1$ over $\mathscr{O}_{\mathbb{C}^p\times\mathbb{C}}$; thus (8.37) can be written in the form

$$\mathscr{O}^p \oplus \mathscr{O}^{P-1} \to \mathscr{O}^P \to T^1_{\mathscr{K}_{h,e}}(I/\mathbb{C}) \to 0, \tag{8.38}$$

where $\mathscr{O} = \mathscr{O}_{\mathbb{C}^p\times\mathbb{C},0}$. By the argument using the theorem of Buchsbaum–Rim [BR64, Corollary 2.7] used in the proof of Proposition 8.9,

$$\mathrm{codim}(\mathrm{supp}(T^1_{\mathscr{K}_h}(I/\mathbb{C})) \leq (p+P-1) - P + 1 = p,$$

and if equality holds then $T^1_{\mathscr{K}_h}(I/\mathbb{C})$ is a Cohen–Macaulay $\mathscr{O}$-module. Equality holds here because

$$\mathrm{supp}\left(T^1_{\mathscr{K}_{h,e}}(I/\mathbb{C})\right) \cap \left(\mathbb{C}^p \times \{0\}\right) = \mathrm{supp} T^1_{\mathscr{K}_{h,e}} i = \{0\}; \tag{8.39}$$

if the codimension of $\mathrm{supp}(T^1_{\mathscr{K}_{h,e}}(I/\mathbb{C}))$ was less than p, then its intersection with $\mathbb{C}^p \times \{0\}$ would have strictly positive dimension.

We conclude that $T^1_{\mathscr{K}_{h,e}}(I/\mathbb{C})$ is Cohen–Macaulay of dimension 1. Let $\pi : \mathbb{C}^p \times \mathbb{C} \to \mathbb{C}$ be projection; it is finite on $\mathrm{supp}\, T^1_{\mathscr{K}_{h,e}}(I/\mathbb{C})$ by (8.39), and it follows that the push-forward $\pi_*\left(T^1_{\mathscr{K}_{h,e}}(I/\mathbb{C})\right)$ to the base space $\mathbb{C}$ is a Cohen–Macaulay $\mathscr{O}_{\mathbb{C}}$-module of dimension 1, and therefore free. This is a standard argument: depth and dimension do not change under finite push-forward, so $\pi_*\left(T^1_{\mathscr{K}_{h,e}}(I/\mathbb{C})\right)$ is Cohen–Macaulay of dimension 1; since $\mathscr{O}_{\mathbb{C}}$ is regular, every finite module has a finite projective resolution; the Auslander–Buchsbaum formula (Theorem C.4 in Sect. C.4),

$$\mathrm{depth}_R M + \text{projective dimension}_R(M) = \mathrm{depth}_R R,$$

now tells us that the length of a minimal free resolution of $\pi_*\left(T^1_{\mathcal{K}_{h,e}}(I/\mathbb{C})\right)$ over $\mathcal{O}_{\mathbb{C}}$ is zero—i.e. that $\pi_*\left(T^1_{\mathcal{K}_{h,e}}(I/\mathbb{C})\right)$ is free. Finally

$$\sum_y \dim_{\mathbb{C}} \frac{\theta(i_u)_y}{ti_u(\theta_{\mathbb{C}^p,y}) + i_u^*(\mathrm{Der}(-\log h))_y} = \mathrm{rank}\,\pi_*\left(T^1_{\mathcal{K}_{h,e}}(I/\mathbb{C})\right)_u$$

$$= \mathrm{rank}\,\pi_*\left(T^1_{\mathcal{K}_{h,e}}(I/\mathbb{C})\right)_0 = \dim_{\mathbb{C}} \frac{\theta(i)}{ti\left(\theta_{\mathbb{C}^p,0}\right) + i^*\left(\mathrm{Der}(-\log h)\right)}.$$

Proof of Theorem 8.1 The inequality in Theorem 8.1 follows immediately from Theorems 8.10 and 8.7. Equality holds when f is weighted homogeneous, because then

$$\frac{\theta(i)}{ti(\theta_{\mathbb{C}^p,0}) + i^*(\mathrm{Der}(-\log h))} = \frac{\theta(i)}{ti(\theta_{\mathbb{C}^p,0}) + i^*(\mathrm{Der}(-\log D))}.$$

□

Exercises for Sect. 8.7

1 Let $f : (\mathbb{C}^2, 0) \to (\mathbb{C}^2, 0)$ be $\mathcal{A}$-finite, with critical set Σ_f and discriminant $D(f)$, and let c and d be the number of cusps and the number of nodes, respectively, on the discriminant of a stable perturbation of f. Show that

(i) $\mu_\Delta(f) = \mu(\Sigma_f) + d$, and
(ii) $\mu_\Delta(f) = \mu(D(f)) - 2c - d$.

2

(a) The proof of Lemma 8.6 is completely elementary, but is not necessarily what one would first think of when confronted with this kind of question. Another proof, possibly more obvious, though more sophisticated, would be to consider the short exact sequence

$$0 \longrightarrow \mathrm{Der}(-\log h)_z \hookrightarrow \theta_{\mathbb{C}^q,z} \xrightarrow{th} \mathcal{O}_{\mathbb{C}^q,z} \longrightarrow 0$$

where $z = i_u(y)$, and tensor over $\mathcal{O}_q$ with $\mathcal{O}_p$. Supply the details.

(b) When h is singular, the morphism (8.35) is neither injective nor surjective. Show that

1. it is not injective, and
2. the cokernel of (8.35) is $\mathscr{O}_p/i_u^*(J_h)$.

In fact (8.35) fits into a long exact sequence, constructed in [GM05], which in particular gives an alternative proof of Lemma 8.6.

8.8 Image Milnor Number and $\mathscr{A}_e$-Codimension

All except the last step in the proof just given work unchanged with the image Milnor number of a map-germ $(\mathbb{C}^n, S) \to (\mathbb{C}^{n+1}, 0)$ in place of the discriminant Milnor number of a map $(\mathbb{C}^n, S) \to (\mathbb{C}^p, 0)$ with $n \geq p$. However, without a proof of conservation of multiplicity, all that can be concluded is that

$$\mu_I(f) \leq \dim_{\mathbb{C}} \frac{\theta(i)}{ti(\theta_{\mathbb{C}^p,0}) + i^*(\mathrm{Der}(-\log h))}. \tag{8.40}$$

Attempts to prove the statement of Theorem 8.2 for $n > 2$ have generally focussed on proving this missing step. The crucial extra ingredient for discriminants of map-germs $(\mathbb{F}^n, 0) \to (\mathbb{F}^p, 0)$ with $n \geq p$, that the discriminant of a stable map-germ is a free divisor, is no longer true when $n < p$ and "discriminant" becomes "image".

In Sect. 11.4 we give *ad hoc* proofs that $\mu_I \geq \mathscr{A}_e$-codimension, with equality in the weighted homogeneous case, for map-germs $(\mathbb{C}^n, S) \to (\mathbb{C}^{n+1}, 0)$ for $n = 1$ and $n = 2$.

Remark 8.3 Despite the lack of a general theorem, the method used to calculate μ_Δ in Theorem 8.10 can be used successfully to calculate μ_I in examples. For it to yield the right answer, all that is necessary is that if the $\mathscr{A}$-finite germ $f : (\mathbb{C}^n, S) \to (\mathbb{C}^{n+1}, 0)$ is induced from a stable germ $F : (\mathbb{C}^N, S) \to (\mathbb{C}^{N+1}, 0)$ by transverse pull-back by $i : (\mathbb{C}^{n+1}, 0) \to (\mathbb{C}^{N+1}, 0)$, and if $I : (\mathbb{C}^{n+1} \times \mathbb{C}^d, (0, 0)) \to (\mathbb{C}^{N+1}, 0)$ is a deformation of i, then the relative T^1 for $\mathscr{K}_h$-equivalence, $T^1_{\mathscr{K}_h}(I/\mathbb{C}^d)$, should be Cohen–Macaulay of dimension d. This turns out to hold in all known examples, though we have no proof that it is always the case. Its truth in general is in fact equivalent to the conjecture of Remark 8.1. It is by this means that many examples supporting the general conjecture have been found, such as those in [Sha14, Sha17]. ◊

8.9 Further Developments

8.9.1 Almost Free Divisors

Theorem 8.1, showing the relation between $\mathscr{A}_e$-codimension and discriminant Milnor number, is part of a wider theory developed by Jim Damon in [Dam96, Dam98, Dam01] and later papers, which studies non-linear sections of free divisors. Recall that a map-germ $i : (V, p) \to (W, q)$ is *logarithmically transverse*, (or *algebraically transverse*) to the hypersurface D at p if

$$d_p i(T_p V) + T_q^{\log} D = T_q W$$

(see the discussion in Sect. 8.5). If $i : (V, p) \to (W, q)$ is logarithmically transverse to D at p, then it is easy to show that the germ of D at q is diffeomorphic to the product $(i^{-1}(D), p) \times (\mathbb{F}^c, 0)$ where $c = \dim W - \dim V$ (we leave the reader to interpret this in case $\dim V > \dim W$). Thus, if (D, q) is a free divisor, then so is $(i^{-1}(D), p)$. Suppose that $i : (V, 0) \to (D, 0)$. We say that $i^{-1}(D)$ is an *almost free divisor* if i is logarithmically transverse to D outside 0. This holds if and only if

$$\dim_{\mathbb{F}} \frac{\theta(i)}{ti(\theta_V) + i^* \operatorname{Der}(-\log D)} < \infty. \tag{8.41}$$

If it is possible to deform i in a 1-parameter family i_t so that for $t \neq 0$, i_t is logarithmically transverse to D, then $i_t^{-1}(D)$ is called the *singular Milnor fibre* of the almost free divisor $i^{-1}(0)$. Theorem 8.3 and the calculation of μ_Δ in Theorem 8.10 show that in this case the singular Milnor fibre has the homotopy type of a wedge of spheres of dimension $\dim_{\mathbb{C}} V - 1$, with the number of spheres equal to

$$\dim_{\mathbb{C}} \frac{\theta(i)}{ti(\theta_V) + i^* \operatorname{Der}(-\log h)},$$

where h is a good reduced defining equation for D. Damon, and later Damon and Pike, have used an ingenious technique based on this to gain information on the homology of some codimension 2 determinantal singularities, in [DP15, DP14]. The ease and success of these techniques have sparked a search for free divisors, which at one time were known only from discriminants of isolated complete intersection singularities. Now many other sources are known, including the discriminants in the base spaces of versal deformations of non-ICIS singularities [vS95, BEGvB09], equisingular deformations [Dam02], representations of Dynkin quivers [BM06], prehomogeneous vector spaces [GMNRS09] and explicit constructions of new free divisors from old [BC13].

8.9.2 Thom Polynomial Techniques

In [Ohm16], Toru Ohmoto uses Thom polynomial techniques to compute μ_I for quasihomogeneous map-germs $(\mathbb{C}^3, 0) \to (\mathbb{C}^4, 0)$ in terms of weights and degrees. His formula was used by A. Sharland to calculate the value of μ_I in her examples in [Sha17] (see Remarks 8.1 and 8.3). In [PTPS18], Guillermo Peñafort-Sanchis and Irma Pallarés-Torres have generalised this approach to give a formula for $\mu_I(f)$ as a rational function of weights and degrees, in the case of weighted homogeneous germs $f : (\mathbb{C}^5, 0) \to (\mathbb{C}^6, 0)$. They use a purely interpolative method to determine the coefficients of the rational function, together with theoretical considerations to bound the degrees of its numerator and denominator.

8.9.3 Does μ_Δ Constant Imply Topological Triviality?

The celebrated theorem of Lê and Ramanujan in [LêR76] states that a μ-constant family of isolated hypersurface singularities is topologically trivial, provided the ambient dimension is not 3. It is unknown whether this holds also in dimension 3. Do the image and discriminant Milnor numbers have an equally crucial role in determining the topology? For some results in this direction see [JPNnB09, FdBP08, NnBT12, NnBOOT18].

8.9.4 The Milnor–Tjurina Relation

The "Milnor–Tjurina relation" is a relation between topological complexity and deformation-theoretic codimension, which takes the form

$$\mu \geq \tau \quad \text{with equality in weighted homogeneous case.}$$

The classical example, from where the name originates, is for isolated hypersurface singularities, with the Milnor number (rank of the middle homology of the Milnor fibre) as the measure of topological complexity, and the Tjurina number (the $\mathscr{K}_e$-codimension of a defining equation) as deformation-theoretic codimension. As

$$\mu = \dim_{\mathbb{C}} \frac{\mathscr{O}_n}{J_f}, \qquad \tau = \dim_{\mathbb{C}} \frac{\mathscr{O}_n}{(f) + J_f},$$

the inequality is obvious, and the equality also, as soon as one observes that if f is weighted homogeneous then $f \in J_f$. The Milnor–Tjurina relation holds for isolated complete intersection singularities [Gre80, LS85]. Theorem 8.1 is also an example, now with μ_Δ and $\mathscr{A}_e$-codimension playing the rôle of μ and τ; so also is the conjectured relation between image Milnor number and $\mathscr{A}_e$-codimension of germs of maps $(\mathbb{C}^n, S) \to (\mathbb{C}^{n+1}, 0)$ discussed in Sect. 8.1 (and proved for $n = 1, 2$ in Chap. 10).

K. Saito showed in [Sai71] that for isolated hypersurface singularities, equality of μ and τ holds *only* if f is weighted homogeneous, and after much effort this was extended to the case of ICIS by Vosegaard in [Vos02]. Is it true, in the case of germs of mappings $(\mathbb{C}^n, S) \to (\mathbb{C}^p, 0)$ with $n \geq p$ and (n, p) nice dimensions, that the corresponding equality holds only for weighted homogeneous germs? If so, it would generalise the empirically observed fact that in the nice dimensions, all stable germs are weighted homogeneous in suitable coordinates, discussed in Sect. 7.4.

8.9.5 Augmentation and Concatenation: New Germs from Old

The Warwick PhD thesis of Tom Cooper [Coo94] showed how all $\mathscr{A}_e$-codimension 1 multi- and mono-germ singularities of maps $(\mathbb{F}^n, S) \to (\mathbb{F}^p, 0)$ with $n \geq p-1$ are generated from "primitive" $\mathscr{A}_e$-codimension 1 monogerms in lower dimensions by three operations: *augmentation, monic concatenation* and *binary concatenation.* Suppose that $f : (\mathbb{F}^n, S) \to (\mathbb{F}^p, 0)$ has $\mathscr{A}_e$-codimension 1 and that $F(x, u) = (f_u(x), u)$ is an $\mathscr{A}_e$-versal unfolding.

(i) The *augmentations* of f are the two map-germs $A_F^{\pm}(f) : (\mathbb{F}^{n+1}, S \times \{0\}) \to (\mathbb{F}^{p+1}, 0)$ defined by

$$A_F^{\pm}(f)(x, u) = (f_{\pm u^2}(x), u).$$

When $\mathbb{F} = \mathbb{C}$, the two are $\mathscr{A}$-equivalent. Cooper showed that the $\mathscr{A}_e$-codimension of each is 1, and that up to $\mathscr{A}$-equivalence, they depend only on f.

(ii) The monic concatenation of f is obtained by adding an extra branch to F: an immersion if $n = p - 1$ and a fold if $n \geq p$. The image or discriminant of the extra branch should be the space $\mathbb{C}^p \times \{0\}$. Once again, the new multi-germ is independent of the choice of versal unfolding F and has $\mathscr{A}_e$-codimension 1.

(iii) Binary concatenation requires a second codimension 1 germ $g : (\mathbb{C}^m, T) \to (\mathbb{C}^q, 0)$ with miniversal unfolding G, and forms a multi-germ with $|S| + |T|$ branches by putting together $\mathrm{id}_{\mathbb{C}^q} \times F$ and $G \times \mathrm{id}_{\mathbb{C}^p}$ in such a way that their analytic strata (whose dimensions sum to $p+q$) meet in $(\mathbb{C}^{p+q+1}, 0)$ in general position, in the sense that the intersection of their tangent spaces has dimension 0. Once again, the resulting germ has $\mathscr{A}_e$ codimension 1 and, over $\mathbb{C}$ at least, is independent of the choices of F and G. The bi-germ of Fig. 8.3 is an example of this construction.

Details can be found in [CMWA02].

In [Hou02], Kevin Houston generalised the notion of augmentation by replacing the function u^2 in the definition of augmentation by any analytic function $a : (\mathbb{C}^q, 0) \to (\mathbb{C}, 0)$ with isolated singularity, and gave formulae for the image/discriminant Milnor numbers of the germs so obtained.

Chapter 9
Multiple Points

The multiple-point spaces of a map-germ $(X, S) \to (Y, 0)$ with $\dim X \leq Y$ play an important rôle in the study of its geometry, as well as the topology of the image (or discriminant when $\dim X = \dim Y$) of a stable perturbation.

9.1 Introduction

We will study multiple-point spaces of maps $X \to Y$ in both the source and the target. Source k-tuple point spaces live in X, where they may be defined, roughly speaking, as the set of points x such that $|f^{-1}(f(x))| \geq k$, perhaps taking multiplicity into account, or in X^k, where they may be defined as the set of ordered k-tuples of points having the same image—though complications arise because of the presence of the diagonals (k-tuples of points of which some or even all are equal to one another), which are contained, willy-nilly, in the spaces defined in the most naive way. The advantage of locating the space in X^k is the presence of the symmetric group action, by permutation of the copies of X, which carries the gluing data. In Chap. 10 we will see how the homology of the image of a map can be recovered from information on the symmetric group actions on the homology of the various multiple point spaces in X^k.

Target multiple point spaces no longer carry gluing data but are integral to the understanding of the geometry of the image. They will be studied in Chap. 11.

D. Mond, J. J. Nuño-Ballesteros, *Singularities of Mappings*,
Grundlehren der mathematischen Wissenschaften 357,
https://doi.org/10.1007/978-3-030-34440-5_9

9.2 Choosing the Right Definition

Let $f : X \to Y$ be a map of complex spaces. For each natural number $k \geq 2$ the naive double point space is the fibre product $X \times_Y X$, and, analogously the naive k-point space is the k-fold fibre product of X over Y, denoted X^k/Y and defined by

$$X^k/Y = \{(x_1, \ldots, x_k) \in X^k : f(x_i) = f(x_j) \text{ for all } i, j\}. \tag{9.1}$$

One obvious property, which will be shared by all of the versions of multiple point spaces that we consider, is that the symmetric group S_k acts on X^k/Y by permuting the copies of X. We will see that the corresponding representation of the symmetric groups S_k on the homology of the multiple point spaces is crucial for relating this homology with the homology of the image of f.

The space X^k/Y is easy to define and to write down equations for, but suffers from many disadvantages. For every f and every k, it contains the small diagonal $\Delta_k = \{(x_1, \ldots, x_k) \in X^k : x_1 = \cdots = x_k\}$, which really does not deserve to be called a k-tuple point. Even when (x_1, x_2) is a genuine double point, with $x_1 \neq x_2$ and $f(x_1) = f(x_2)$, the bogus triple points (x_1, x_1, x_2) and (x_1, x_2, x_2) (and their permutations) lie in X^3/Y. So X^k/Y has many irreducible components of differing dimensions and differing degrees of naiveté, most of them there for reasons which are nothing to do with the behaviour of f itself. Clearly if we want to derive information about f from its multiple points we should find a way of removing, or disregarding, some of these components.

Let $X^{\langle k \rangle} := \{(x_1, \ldots, x_k) \in X^k : x_i \neq x_j \text{ if } i \neq j\}$. Then $(X^k/Y) \cap X^{\langle k \rangle}$ is the set of genuine multiple points. If X is a complex space and f is an analytic map, this is a constructible set—it is the intersection of the Zariski-closed set X^k/Y with the Zariski-open set $X^{\langle k \rangle}$.

Proposition 9.1 *The closure of $(X^k/Y) \cap X^{\langle k \rangle}$ with respect to the product Euclidean topology on X^k is also its Zariski closure, and is thus an analytic set.*

Proof X^k/Y is the union of finitely many analytically irreducible components $Z_i, i = 1, \ldots, m$. If Z is any irreducible analytic subset of a complex space X, and U is open in the Zariski topology on X (that is, $X \smallsetminus U$ is an analytic subset of X) then $Z \cap U$, if not empty, is dense in Z *with respect to the Euclidean topology* as well as with respect to the Zariski topology. For at each point, $Z \smallsetminus U$ is the zero set of some coherent analytic ideal, and if this ideal is equal to 0 on a Euclidean-open subset of Z then it is equal to 0 on all of Z, by the irreducibility of Z. Thus, the Euclidean topological closure of $Z \cap U$ is equal to its Zariski-topological closure, namely Z if $Z \cap U \neq \emptyset$ and $\emptyset$ otherwise. Now with respect to both topologies, the closure of $(X^k/Y) \cap X^{\langle k \rangle}$ is the union of the closures of $Z_i \cap X^{\langle k \rangle}$, and since these closures are the same in the two topologies, namely either Z_i or $\emptyset$, the Euclidean closure of $(X^k/Y) \cap X^{\langle k \rangle}$ is the union of the irreducible components Z_i for which $Z_i \cap X^{\langle k \rangle} \neq \emptyset$. □

Definition 9.1 Given an analytic map $f : X \to Y$ of complex spaces, we define $D^k_{\mathrm{cl}}(f)$, as a set, to be the closure of $(X^k/Y) \cap X^{\langle k \rangle}$.

Thus, $D^k_{\mathrm{cl}}(f)$ is the union of the irreducible components of X^k/Y which are not contained in any of the diagonals. To find its defining ideal, we have to remove some of the irreducible components X^k/Y, and thus remove primary components from its defining ideal. The standard way of doing this is to use *transporter ideals.*

Definition 9.2 Let I and J be ideals in the ring R.

(i) The transporter ideal (or colon ideal) $(I : J)$ is the annihilator in R of $(J+I)/I$. That is,

$$(I : J) = \{r \in R : I \supset rJ.\}$$

(ii) The saturation $(I : J^\infty)$ is defined by

$$(I : J^\infty) = \bigcup_{r \geq 1} (I : J^r).$$

Evidently $(I : J^r) \subset (I : J^{r+1})$, so we have an ascending chain of ideals. In a Noetherian ring R, it must terminate, so that for some finite r, $(I : J^\infty) = (I : J^r)$.

A clear treatment of transporter ideals can be found in the book [CLO15] of Cox, Little and O'Shea. They work in the context of polynomial rings, but the proofs work without change in $\mathcal{O}_n$.

Proposition 9.2 ([CLO15, Chapter 4 Theorem 10]) *Let I and J be ideals in $\mathcal{O}_n$. Then*

$$V(I : J^\infty) = \overline{V(I) \smallsetminus V(J)}.$$

□

That is, $V(I : J^\infty)$ is the union of the irreducible components of $V(J)$ which are not contained in I. In general it is not easy to determine *a priori* for which r the increasing sequence of ideals $(I : J^r)$ becomes stationary. However, it is easy to see that this is the first integer r for which $(I : J^r) = (I : J^{r+1})$. For then

$$(I : J^{r+2}) = \left((I : J^{r+1}) : J\right) = \left((I : J^r) : J\right) = \left(I : J^{r+1}\right) = \left(I : J^r\right),$$

etc., so that the sequence indeed stabilises at $(I : J^r)$.

This gives us a procedure for finding a defining ideal for $D^k_{\mathrm{cl}}(f)$. Suppose that $f : (\mathbb{C}^n, 0) \to (\mathbb{C}^p, 0)$ is an analytic map-germ. We continue to write X and Y for its domain and target. Let $x^{(i)}$ denote a point in the i'th copy of X, and $x^{(i)}_1, \ldots, x^{(i)}_n$

its coordinates. The union of all of the diagonals in X^k, the set $\{(x^{(1)}, \ldots, x^{(k)}) : x^{(i)} = x^{(j)} \text{ for some } i \neq j\}$ is defined by a product of ideals,

$$J_k := \prod_{i<j} \left(x_1^{(i)} - x_1^{(j)}, \ldots, x_n^{(i)} - x_n^{(j)} \right) \tag{9.2}$$

while the ideal defining X^k/Y is a sum of ideals,

$$I_k(f) := \sum_{j=2}^{k} \left(f_1(x^{(1)}) - f_1(x^{(j)}), \ldots, f_p(x^{(1)}) - f_p(x^{(j)}) \right). \tag{9.3}$$

The relevant transporter ideal is $(I_k(f) : J_k)$, with saturation $(I_k(f) : J_k^\infty)$. We denote this ideal by $\mathscr{I}_k^{\mathrm{s}}(f)$ (the upper index s to indicate "saturation").

Proposition 9.3 *If $f : (\mathbb{C}^n, S) \to (\mathbb{C}^p, 0)$ is finite, and*

$$\dim_{\mathbb{C}} \frac{\mathscr{O}_{\mathbb{C}^n, S}}{f^* \mathfrak{m}_{\mathbb{C}^p,0} \mathscr{O}_{\mathbb{C}^n, S}} = \ell,$$

then for $k > \ell$, $D_{cl}^k(f) = \emptyset$.

Proof For $k > \ell$, $(X^k/Y) \cap X^{\langle k \rangle} = \emptyset$; this follows from the upper semicontinuity of the number of generators of a coherent sheaf, applied to the sheaf $\tilde{f}_*(\mathscr{O}_U)$, where $\tilde{f} : U \to V$ is a good representative.[1] Thus, $V(I_k(f)) \subset V(J_k)$, and hence $V(I_k(f) : J_k^\infty) = \emptyset$. □

We will shortly consider two examples. Both are germs of rank > 0, which means that, written in linearly adapted coordinates, each can be regarded as an unfolding of a lower-dimensional germ. In preparation for the calculations, we note that if $X = \mathbb{C}^n \times \mathbb{C}^d$ and $Y = \mathbb{C}^p \times \mathbb{C}^d$, and $F : X \to Y$ is an unfolding on parameters $u_1, \ldots, u_d$, then any point $(x^{(1)}, u^{(1)}), \ldots, (x^{(k)}, u^{(k)})$ in $(X^k/Y) \cap X^{\langle k \rangle}$ must have $u^{(1)} = \cdots = u^{(k)}$. Thus, $D_{\mathrm{cl}}^k(F)$ ought to embed in a natural way in $(\mathbb{C}^n)^k \times \mathbb{C}^d$. Fortunately the procedure with transporter ideals easily takes account of this. Let $c : (\mathbb{C}^n)^k \times \mathbb{C}^d \to (\mathbb{C}^n \times \mathbb{C}^d)^k$ be the "diagonal embedding"

$$(x^{(1)}, \ldots, x^{(k)}, u) \mapsto \left((x^{(1)}, u), \ldots, (x^{(k)}, u) \right). \tag{9.4}$$

Lemma 9.1 *Let $F : (\mathbb{C}^n \times \mathbb{C}^d, (0,0)) \to (\mathbb{C}^p \times \mathbb{C}^d, (0,0))$ be an unfolding. Then for all k,*

$$\frac{\mathscr{O}_{(\mathbb{C}^n)^k \times \mathbb{C}^d, 0}}{(c^* I_k(F) : (c^* J_k)^\infty)} \simeq \frac{\mathscr{O}_{(\mathbb{C}^n \times \mathbb{C}^d)^k, 0}}{(I_k(F) : J_k^\infty)}. \tag{9.5}$$

[1] This argument is spelled out in more detail in the proof of Proposition 9.5 below.

Proof For brevity, write the ideals of differences in the form

$$\left(x^{(i)} - x^{(j)}\right), \left(u^{(i)} - u^{(j)}\right) \text{ and } \left(f\left(x^{(i)}, u^{(j)}\right) - f\left(x^{(j)}, u^{(j)}\right)\right),$$

collapsing the subindices.

It is obvious that for ideals I, J_α, P_α in any ring, and any natural number ℓ,

$$\left(I + \sum P_\alpha\right) : \left(\prod (J_\alpha + P_\alpha)\right)^\ell = \left(I + \sum P_\alpha\right) : \left(\prod J_\alpha\right).^\ell \tag{9.6}$$

We apply this equality with $I = \sum_{j=2}^{k} \left(f\left(x^{(1)}, u^{(1)}\right) - f\left(x^{(j)}, u^{(j)}\right)\right)$, $J_\alpha = (x^{(i)} - x^{(j)})$ and $P_\alpha = (u^{(i)} - u^{(j)})$. If

$$r\left(x^{(1)}, \ldots, x^{(k)}, u\right) \in \left(c^*(I) : c^*\left(\prod_\alpha J_\alpha\right)\right)$$

then $r\left(x^{(1)}, \ldots, x^{(k)}, u^{(1)}\right)$, regarded as a function of $x^{(1)}, \ldots, x^{(k)}, u^{(1)}, \ldots, u^{(k)}$, lies in $\left(I + \sum_\alpha J_\alpha : \prod_\alpha J_\alpha\right)$ since the difference between functions evaluated at $u^{(i)}$ and at $u^{(1)}$ lies in the ideal $\sum_\alpha J_\alpha$. This shows that

$$(c^* I_k(F) : c^* J_k^\infty) = c^*(I_k(F) : J_k^\infty). \tag{9.7}$$

The conclusion now follows from the exactness of the rows of the diagram

$$\begin{array}{ccccccccc}
0 & \longrightarrow & \sum_\alpha P_\alpha & \longrightarrow & \mathscr{O}_{(\mathbb{C}^n \times \mathbb{C}^d)^k} & \xrightarrow{c^*} & \mathscr{O}_{(\mathbb{C}^n)^k \times \mathbb{C}^d} & \longrightarrow & 0 \\
 & & \Vert & & \uparrow & & \uparrow & & \\
0 & \longrightarrow & \sum_\alpha P_\alpha & \longrightarrow & (I_k(F) : J_k^\infty) & \xrightarrow{c^*} & (c^* I_k(F) : c^* J_k^\infty) & \longrightarrow & 0
\end{array}$$

□

In the light of this lemma, for an unfolding $F : \mathbb{C}^n \times \mathbb{C}^d \to (\mathbb{C}^p \times \mathbb{C}^d)$, we will regard J_k, $I_k(F)$ and $\mathscr{I}_k^s(F)$ as ideals in $\mathscr{O}_{(\mathbb{C}^n)^k \times \mathbb{C}^d}$.

We will denote by i the inclusion $\mathbb{C}^n = \mathbb{C}^n \times \{0\} \hookrightarrow \mathbb{C}^n \times \mathbb{C}^d$, and by i_k the corresponding inclusion $(\mathbb{C}^n)^k = (\mathbb{C}^n \times 0)^k \hookrightarrow (\mathbb{C}^n \times \mathbb{C}^d)^k$.

Example 9.1

(1) Consider the stable germ $F : (\mathbb{C}^2, 0) \to (\mathbb{C}^3, 0)$ defined by

$$F(x, u) = (x^2, x^3 + ux, u).$$

This is a stable (indeed versal) unfolding of the cusp map $f(x) = (x^2, x^3)$. A *Macaulay 2* calculation shows

$$(I_2(F) : (J_2)^2) = (I_2(F) : J_2) = (x_1 + x_2, x_1 x_2 + u),$$

so this is the ideal $\mathscr{I}_2^{\mathrm{s}}(F)$; also $(I_3(F), J_3) = (1)$. Thus $D^2_{\mathrm{cl}}(F)$ is a curve and $D^3_{\mathrm{cl}}(F) = \emptyset$.

However, while $(I_2(f) : J_2)$ is equal to $i_2^*(I_2(F) : J_2) = (x_1 + x_2, x_1x_2)$, on the other hand $(I_2(f) : J_2^2) = (x_1, x_2)$ strictly includes $i_2^*(I_2(F) : J_2^2)$.

(2) Consider the stable germ $F\colon (\mathbb{C}^6, 0) \to (\mathbb{C}^7, 0)$ defined by

$$F(x_1, x_2, u_1, u_2, u_3, u_4) = (x_1^2+u_1x_2, x_1x_2+u_2x_1+u_3x_2, x_2^2+u_4x_1, u_1, u_2, u_3, u_4).$$

A *Macaulay 2* calculation shows that $(I_2 : J_2^2) = (I_2 : J_2)$, so that $\mathscr{I}_2(F) = (I_2 : J_2)$. It is easy to check that $(I_2 : J_2)$ is radical, and defines a Cohen–Macaulay variety of codimension 2 in $(\mathbb{C}^2)^2 \times \mathbb{C}^4$.

Likewise, $\mathscr{I}_3^{\mathrm{s}}(F)$ is defined by the transporter ideal $(I_3 : J_3)$, and is radical, and Cohen–Macaulay—even Gorenstein—of codimension 6 in $(\mathbb{C}^2)^3 \times \mathbb{C}^4$.

The germ F is a stable unfolding of the 2-to-1 germ $f\colon (\mathbb{C}^2, 0) \to (\mathbb{C}^3, 0)$ obtained by setting $u = 0$. Applying the above procedure to f, we obtain ideals $\mathscr{I}_2^{\mathrm{s}}(f)$ and $\mathscr{I}_3^{\mathrm{s}}(f)$ in $\mathscr{O}_{(\mathbb{C}^2)^2}$ and $\mathscr{O}_{(\mathbb{C}^2)^3}$. *Macaulay 2* calculations show that $\mathscr{I}_3^{\mathrm{s}}(f)$ is the maximal ideal in $\mathscr{O}_{(\mathbb{C}^2)^3,(0,0,0)}$, i.e. $D^3_{\mathrm{cl}}(f)$ is the reduced point $(0, 0, 0) \in (\mathbb{C}^2)^3$. On the other hand, $i_3^*(\mathscr{I}_3^{\mathrm{s}}(F))$ defines $(0, 0, 0)$ with non-reduced structure

$$\dim_{\mathbb{C}} \frac{\mathscr{O}_{(\mathbb{C}^2)^3,(0,0,0)}}{i_3^*(I_3(F))} = 6. \tag{9.8}$$

◇

Our examples show a general phenomenon: if $F\colon (X \times U, (x_0, 0)) \to (Y \times U, (y_0, 0))$ is an unfolding of $f\colon (X, x_0) \to (Y, y_0)$, and i_k is the inclusion $X^k \to (X \times \{0\})^k \subset (X \times U)^k$, then in general

$$i_k^{-1}(D^k_{\mathrm{cl}}(F)) \neq D^k_{\mathrm{cl}}(f).$$

This is an instance of a more general inequality: if $\varphi : R \to S$ is a ring homomorphism and I, J are ideals of R then in general $\varphi(I : J)$ is strictly contained in $(\varphi(I)S : \varphi(J)S)$.

The number 6 in (9.8) is significant. It is the number of elements in the symmetric group S_3, and, therefore, for any map g it is the number of points in $D^3_s(g)$ corresponding to a genuine triple point: if P, Q and R are pairwise distinct but share the same image under g, then

$$(P, Q, R), (P, R, Q), (Q, R, P), (R, Q, P), (R, P, Q) \text{ and } (Q, P, R)$$

all lie in $D^3_s(g)$. The map f in Example 9.1(1) has no genuine double point, though one emerges when we deform it (this is the first Reidemeister move of knot theory—see Example 3.4). Similarly, the map f in Example 9.1(2) has no genuine triple point, as is very easy to check. However it is possible to perturb it so that one, and only one genuine triple point emerges: for example, when $u \neq 0$

$$f_u(x_1, x_2) = (x_1^2 - ux_1, x_1x_2, x_2^2 - ux_2),$$

maps the three distinct points $(0,0)$, $(u,0)$, $(0,u)$ to $(0,0,0)$. Thus, the fact that here $i_3^*(\mathscr{I}_3^S(F))$ defines $0 \in (\mathbb{C}^2)^3$ as a fat point of length 6 suggests that in general, the multiple point space $i_k^{-1}(D_{\mathrm{cl}}^k(F))$, pulled back from a suitable unfolding, may behave better under deformation than the space $D_{\mathrm{cl}}^k(f)$ defined directly from f.

"Behave better under deformation" is an imprecise phrase which nevertheless should be sufficient to motivate the following definition, due to Terry Gaffney in [Gaf83]. The reader may recognise the notion of *flatness* lurking behind this discussion.

We work locally; that is, for each point $\mathbf{x} = (x_1, \ldots, x_k) \in N^k(f)$ we define the germ of $D^k(f)$ at $\mathbf{x}$ by means of an ideal in $\mathscr{O}_{X^k,\mathbf{x}}$.

Definition 9.3 Given a $\mathscr{K}$-finite germ $f: (X, S) \to (Y, 0)$ (with X and Y smooth), choose a germ of stable unfolding $F : (X \times P, S \times \{0\}) \to (Y \times P, (y_0, 0))$ and set

$$D^k(F) = D_{\mathrm{cl}}^k(F) \tag{9.9}$$

with structure defined by $\mathscr{I}_k^S(F)$. Then we define $D^k(f)$ as the fibre over $0 \in P$ of the projection $\pi_k : D^k(F) \to P$. That is, the defining ideal of $D^k(f)$ at S is the restriction to $X^k = (X \times \{0\})^k$ of the defining ideal $\mathscr{I}_k^S(F)$ of $D^k(F)$ at $S \times \{0\}$. We denote this defining ideal by $\mathscr{I}_k(f)$. Thus, if f is stable, then $\mathscr{I}_k(f) = \mathscr{I}_k^S(f)$.

Now the structure of $D^k(f)$ may not be reduced, of course. Here we include some basic properties of the multiple point spaces $D^k(f)$. Other properties can be found in [NnBPnS17].

Proposition 9.4

1. *The definition of $D^k(f)$ is independent of the choice of stable unfolding used in Definition 9.3.*
2. *If F is any unfolding of f on parameters $u_1, \ldots, u_a$, let $\pi_F : D^k(F) \to (\mathbb{F}^a, 0)$ be the projection. Then $D^k(f) = \pi_F^{-1}(0)$.*
3. *Suppose that X and Y are complex manifolds and $f : X \to Y$ is an analytic map. Then the sheaf of ideals on X^k by which we have defined $D^k(f)$ is coherent.*
4. *If $f: (X, 0) \to (Y, y_0)$ is $\mathscr{A}$-finite and $D_{cl}^k(f) \neq \emptyset$ then $D^k(f) = D_{cl}^k(f)$ away from $(0, \ldots, 0)$. The equality extends to all of $D^k(f)$ if, in addition, $D^k(f)$ is Cohen–Macaulay, or, more generally, unmixed.*
5. *$D^k(f) \setminus D_{cl}^k(f)$ consists of at most one point, and, if it is not empty, then $D_{cl}^k(f)$ is empty.*

Proof

1. Suppose that

$$F_1 \colon (\mathbb{F}^n \times \mathbb{F}^a, S \times \{0\}) \to (\mathbb{F}^p \times \mathbb{F}^a, (0,0)) \quad (x,u) \mapsto (f_1(x,u), u)$$

and

$$F_2 \colon (\mathbb{F}^n \times \mathbb{F}^b, S \times \{0\}) \to (\mathbb{F}^p \times \mathbb{F}^b, (0,0)) \quad (x,v) \mapsto (f_2(x,v), v)$$

are stable unfoldings of f. Let H be the direct sum unfolding of f:

$$H(x,u,v) = (f_1(x,u) + f_2(x,v) - f(x), u, v).$$

Thus $H(x,u,0) = (F_1(x,u), 0)$ and $H(x,0,v) = (f_2(x,v), 0, v)$. Then H is also a stable unfolding of F_1, on parameters v_j, $j = 1, \ldots, b$, and of F_2, on parameters u_i, $i = 1, \ldots, a$. Since F_1 and F_2 are stable, in each case H is a trivial unfolding. Considering first F_1, it follows that there is a diffeomorphism

$$\Phi_{k,1} : D^k_{\mathrm{cl}}(H) \to D^k_{\mathrm{cl}}(F_1) \times (\mathbb{F}^b, 0)$$

whose restriction to $D^k_{\mathrm{cl}}(H) \cap (\mathbb{F}^n \times \mathbb{F}^a \times 0)$ is the identity map, and such that the diagram

$$\begin{array}{ccc} D^k_{\mathrm{cl}}(H) & \xrightarrow{\Phi_{k,1}} & D^k_{\mathrm{cl}}(F_1) \times (\mathbb{F}^b, 0) \\ & {\scriptstyle \pi_H} \searrow \quad \swarrow {\scriptstyle \pi_{F_1} \times \mathrm{id}_{\mathbb{F}^b}} & \\ & (\mathbb{F}^a \times \mathbb{F}^b, (0,0)) & \end{array}$$

commutes. By commutativity, $\Phi_{k,1}$ restricts to a diffeomorphism $\pi_H^{-1}(0,0) \to (\pi_{F_1} \times \mathrm{id}_{\mathbb{F}^b})^{-1}(0,0)$. Evidently $(\pi_{F_1} \times \mathrm{id}_{\mathbb{F}^b})^{-1}(0,0) = \pi_{F_1}^{-1}(0)$. The domain of this diffeomorphism is contained in the region where $\Phi_{k,1}$ is the identity map. Hence the version of $D^k(f)$ we get using H is the same as the one we get using F_1. The same argument applies to F_2, showing that the definition is independent of the choice of stable unfolding.
2. Let G be a stable unfolding of F, on parameters v_j; G is also a stable unfolding of f, on parameters u_i, v_j. Then $D^k(F) = D^k(G) \cap \{v = 0\}$ and $D^k(f) = D^k(G) \cap \{u = v = 0\}$. Thus $D^k(f) = D^k(F) \cap \{u = 0\}$.
3. This follows from the fact that for a stable map-germ F, $D^k_{\mathrm{cl}}(F)$ is an analytic subspace, and the coherence of the ideal of functions vanishing on an analytic subspace.

4. Away from 0, f is stable, and so is its own stable unfolding. Thus, at all points of $D^k(f)$ outside $(0, \ldots, 0)$, $D^k(f) = D^k_{\mathrm{cl}}(f)$. In particular, $D^k(f)$ is generically reduced. If also $D^k(f)$ is Cohen–Macaulay, then by Proposition C.5 (a generically reduced Cohen–Macaulay space is reduced) it is reduced, and thus $D^k(f) = D^k_{\mathrm{cl}}(f)$ everywhere.
5. If $(x_1, \ldots, x_k) \in D^k(f) \smallsetminus \mathrm{closure}(D^k_S(f))$ then the multi-germ $f : (\mathbb{C}^n, \{x_1, \ldots, x_k\}) \to (\mathbb{C}^p, f(x_i))$ is not stable. If f is $\mathscr{A}$-finite then by the Mather–Gaffney criterion, $x_1 = \cdots = x_k = 0$, for the instability locus of f in $(\mathbb{C}^p, 0)$ is just 0, and $f^{-1}(0) = 0$. In this case the point $(0, \ldots, 0)$ is not in the closed set $D^k(f)$, and thus there is a neighbourhood of $(0, \ldots, 0)$ in $(\mathbb{C}^n)^k$ which does not meet $D^k(f)$.

□

9.2.1 Semi-Simplicial Spaces

The collection of spaces $D^k(f), k \geq 2$ has a rich structure. For each $k \geq 2$ and $1 \leq i \leq k$, the projection $X^k \to X^{k-1}$ which forgets the i'th copy of X restricts to a projection $\varepsilon^{i,k} : D^k(f) \to D^{k-1}(f)$. The $D^k(f)$ for different k and the maps $\varepsilon^{i,k}$ between them fit together to form a *semi-simplicial space* over Y—see, e.g. [Wei94]. In the next chapter we will use this structure, though we will not use the formalities of semi-simplicial objects.

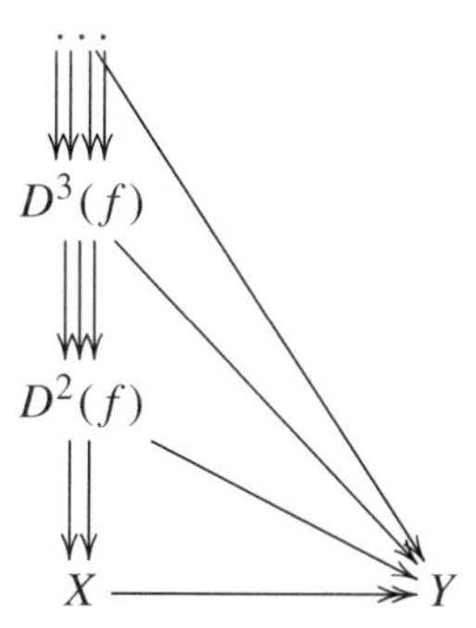

(9.10)

It will be useful to have a name for the image of $D^k(f)$ in $D^\ell(f)$ under the composite of the projections $D^k(f) \to D^{k-1}(f) \to \cdots \to D^\ell(f)$. We denote it by $D^k_\ell(f)$. Note that although there is a choice of projections, all have the same image.

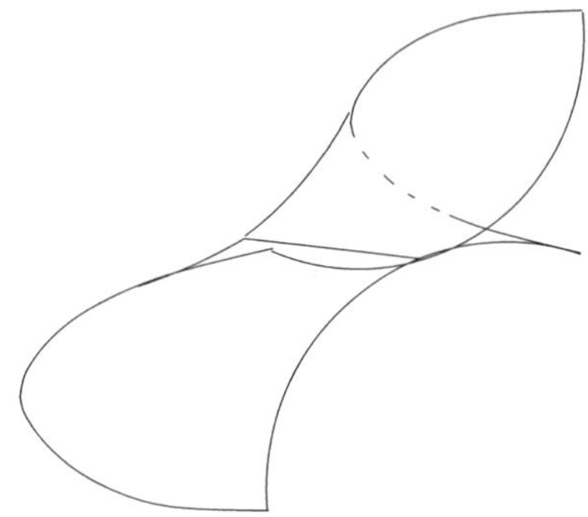

Fig. 9.1 Image of the germ $f(u, x) = (u, x^2, u^2x)$

9.2.2 *When Is $D^k_{cl}(f)$ Reduced?*

$\mathscr{I}^s_k(f)$ is not always a radical ideal. For $f(u, x) = (u, x^2, u^2x)$, $\mathscr{I}^s_2(f) = (x_1 + x_2, u^2)$. As Fig. 9.1 shows, the image has a line of points of non-transverse self-intersection. Each such point is unstable, so by the Mather–Gaffney criterion, f is not $\mathscr{A}$-finite.

Is it true that if f is $\mathscr{A}$-finite, then our definition gives $D^k(f)$ its reduced structure? Certainly not in case $D^k(f)$ is zero-dimensional—we have just seen an example in Example 9.1(1). But we have shown in Proposition 9.4 that positive dimensional multiple point spaces are generically reduced, and reduced if $D^k(f)$ is Cohen–Macaulay. To show that $D^k(f)$ is Cohen–Macaulay it will be enough to show that $D^k(F)$ is Cohen–Macaulay, for a stable unfolding F of f. For $\mathscr{A}$-finiteness tells us that the codimension of $D^k(f)$ in $(\mathbb{C}^n)^k$ is equal to the codimension of $D^k(F)$ in $(\mathbb{C}^n \times \mathbb{C}^d)^k$—see Sect. 9.3 below. From this it follows that the unfolding parameters form a regular sequence on $\mathscr{O}_{D^k(F)}$, and thus the Cohen–Macaulayness of$D^k(F)$ in turn implies that $D^k(f)$ is Cohen–Macaulay—see Exercise C.4.1.

9.2.3 *Irritating Notation, Occasionally Necessary*

To work with $D^k(f)$ in the case of a multi-germ, part of the difficulty we face is one of notation. To describe a multi-germ $f: (X, S) \to (Y, y_0)$, where $|S| = r$, we need labels for the r members of S and r systems of coordinates, one around each of the points of S. To talk about multiple points, we need multiple copies of each set of coordinates, which have to be distinguished from one another in some way. It is most natural to make each point of S the origin in the local coordinate system which describes its neighbourhood. We therefore adopt the following conventions:

1. Members of S are labelled $0^{(1)}, \ldots, 0^{(r)}$, and in this context S will often be denoted by $\mathbf{0}_r$

2. Points in the neighbourhood of $0^{(i)}$ are labelled $x^{(i)}$
3. The germ of f at $0^{(i)}$ is labelled $f^{(i)}$
4. Coordinates around $0^{(i)}$ are simply labelled $x_1, \ldots, x_n$, or as $x_1^{(i)}, \ldots, x_n^{(i)}$ when two sets of coordinates appear in the same expression;
5. Where we consider a point in $D^k(f)$ lying on one of the diagonals, points in the neighbourhood of the j'th copy of $0^{(i)}$ are labelled $x^{(ij)}$ and coordinates around the j'th copy of $0^{(i)}$ are labelled $x_1^{(ij)}, \ldots, x_n^{(ij)}$.

In what follows, we make use of this cumbersome notation to check that our definitions provide $D^k(f)$ with the right properties, and then try to simplify it wherever possible.

Consider a point $(x_1, \ldots, x_k) \in X^k$. After a permutation we may assume that

$$x_1 = \cdots = x_{k_1},\ x_{k_1+1} = \cdots = x_{k_1+k_2},\ \cdots, \cdots,\ x_{k_1+\cdots+k_{r-1}+1} = \cdots = x_k$$

and there are no further equalities among the x_j. Following our labelling convention, $(x_1, \ldots, x_k)$ becomes

$$(\underbrace{0^{(1)}, \ldots, 0^{(1)}}_{k_1}, \underbrace{0^{(2)}, \ldots, 0^{(2)}}_{k_2}, \ldots, \ldots, \underbrace{0^{(r)}, \ldots, 0^{(r)}}_{k_r}).$$

The following example tests our notation and probably the reader's patience.

Example 9.2 We consider the bi-germ $f\colon (\mathbb{F}^2, \mathbf{0}_2) \to (\mathbb{F}^3, 0)$ defined by

$$f : \begin{cases} f^{(1)}(x_1, x_2) = (x_1, x_1, x_2) \\ f^{(2)}(x_1, x_2) = (x_1, x_2^2, x_1x_2). \end{cases}$$

The two mono-germs making up this bi-germ are both stable. As $f^{(1)}$ is an immersion, its isosingular locus is its image, the plane $\{(X_1, X_2, X_3) : X_1 = X_2\}$; the isosingular locus of $f^{(2)}$ is just the origin $\{(0, 0, 0)\}$, and the instability of f is due to the failure of transversality of the two (see Theorem 3.3). In order to stabilise f we take the unfolding

$$F : \begin{cases} F^{(1)}(x_1, x_2, \lambda) = (x_1, x_1 + \lambda, x_2, \lambda) \\ F^{(2)}(x_1, x_2, \lambda) = (x_1, x_2^2, x_1x_2, \lambda) \end{cases} \tag{9.11}$$

of f. The isosingular loci of $F^{(1)}$ and $F^{(2)}$ are now $\{(X_1, X_2, X_3, \lambda) \in \mathbb{F}^4 : X_2 - X_1 - \lambda = 0\}$ and $\{(0, 0, 0)\} \times \mathbb{F}$. These meet transversely, so that F is indeed stable. Then $D^2(F)$ has three components:

$$\begin{aligned} D^{(2)(2)}(F) &:= D^2(F^{(2)}) = D^2(f^{(2)}) \times \mathbb{F} \\ D^{(1)(2)}(F) &:= \{\big((x^{(1)}, \lambda), (x^{(2)}, \lambda)\big) : F^{(1)}(x^{(1)}, \lambda) = F^{(2)}(x^{(2)}, \lambda)\} \\ D^{(2)(1)}(F) &:= \{\big((x^{(2)}, \lambda), (x^{(1)}, \lambda)\big) : F^{(1)}(x^{(1)}, \lambda) = F^{(2)}(x^{(2)}, \lambda)\} \\ &= \sigma\big(D^{(1)(2)}(F)\big), \end{aligned}$$

where σ is the involution on $D^2(F)$ reversing the order of the point pairs. We see directly from (5.1) that $D^{(1)(2)}(F)$ is defined by equations

$$\begin{cases} x_1^{(1)} = x_1^{(2)} \\ x_1^{(1)} + \lambda = (x_2^{(2)})^2 \\ x_2^{(1)} = x_1^{(2)} x_2^{(2)}. \end{cases}$$

Further calculation shows that $D_g^3(F)$ is a union of three disjoint smooth curves:

$$D^3(F) = D^{(1)(2)(2)}(F) \cup D^{(2)(1)(2)}(F) \cup D^{(2)(2)(1)}(F); \tag{9.12}$$

$D^{(1)(2)(2)}(F)$ consists of points $\big((x^{(1)}, \lambda), (x^{(2,1)}, \lambda), (x^{(2,2)}, \lambda)\big)$ such that

$$\begin{cases} \big((x^{(2,1)}, \lambda), (x^{(2,2)}, \lambda)\big) \in D^2(F^{(2)}) \\ F^{(1)}(x^{(1)}, \lambda) = F^{(2)}(x^{(2,1)}, \lambda) \end{cases}$$

and thus has equations

$$\begin{cases} x_1^{(2,1)} = x_1^{(2,2)} = 0 \qquad x_2^{(2,1)} + x_2^{(2,2)} = 0 \\ x_1^{(1)} = 0 \qquad x_1^{(1)} + \lambda = (x_2^{(2,1)})^2 \qquad x_2^{(1)} = 0. \end{cases}$$

The other two terms on the right-hand side of (9.12) are permutations of $D^{(1)(2)(2)}(F)$. It follows that the defining ideal of the germ of $D^3(f)$ at $\mathbf{0}_3$ is

$$\big(x_1^{(2,1)}, x_1^{(2,2)}, x_2^{(2,1)} + x_2^{(2,2)}, x_1^{(1)}, (x_2^{(2,1)})^2, x_2^{(1)}\big) \subset \mathcal{O}_{\left(\mathbb{F}^2\right)^3, \mathbf{0}_3};$$

this germ is thus a fat point of length 2. Altogether $D^3(f)$ consists of three fat points of length 2, forming a single orbit of S_3 (Fig. 9.2). ◊

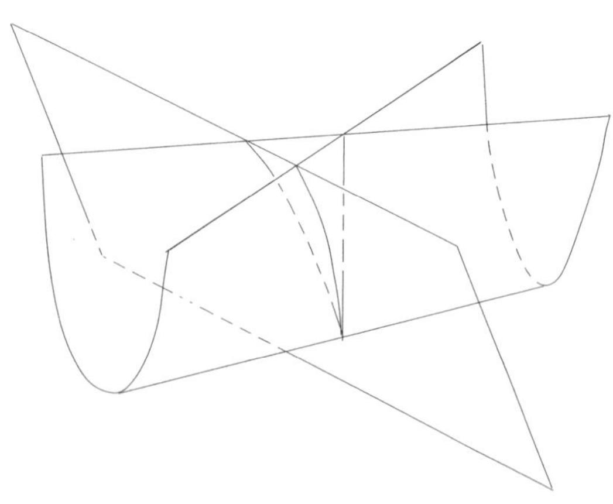

Fig. 9.2 Image of the bi-germ F

9.2.4 Equations or Procedures?

For $\mathscr{K}$-finite f, we have a procedure for finding $\mathscr{I}_k(f)$, but one which gives us little *a priori* information about $D^k(f)$. In a range of cases we are able to improve on this, in the sense that we can give explicit generators for $\mathscr{I}_k(f)$, at least for $\mathscr{A}$-finite f, directly from knowledge of f, without needing to first find a stable unfolding of f as Definition 9.3 prescribes. The cases are

1. For $k = 2$: we have a structure theorem for $D^2(f)$, for $\mathscr{A}$-finite f—see Theorem 9.9 below. It allows us to write down equations for $D^2(f)$ from the expression for f itself, and, in particular, shows that $D^2(f)$ is Cohen–Macaulay.
2. For germs of corank 1, we can do the same for $D^k(f)$ for all k. This allows us to characterise stability and $\mathscr{A}$-finiteness of a corank 1 germ f in terms of the geometry of the $D^k(f)$, and to describe the homology of $D^k(f_t)$ when f_t is a stable perturbation of f.
3. For a small range of stable germs F of corank ≥ 2, we can show by computer calculation that $D^k(F)$ is Cohen–Macaulay. This at least allows us to compute $D^k(f)$ as $D^k_{\mathrm{cl}}(f)$ directly from $\mathscr{I}^{\mathrm{s}}_k(f)$.

We describe these cases in detail after first deriving some estimates for the dimension of $D^k(f)$.

9.3 Expected Dimension

Let $M_k = \{y \in (\mathbb{C}^p, 0) : |f^{-1}(y)| \geq k\}$. This definition is meant precisely: $M_k(f)$ is the set of points with at least k distinct preimages. Here we are not counting the multiplicity of each preimage.

Lemma 9.2 *If $f: (\mathbb{C}^n, S) \to (\mathbb{C}^p, 0)$ is an $\mathscr{A}$-finite multi-germ then*

$$\dim M_k(f) = n - (k-1)(p-n) \tag{9.13}$$

provided $M_k(f)$ is not empty.

Proof Choose a good proper representative $\tilde{f} : U \to V$ for which every multi-germ $f_y: (U, f^{-1}(y)) \to (V, y)$ with $y \neq 0$ is stable (see Theorem 4.5) and let $U^* = U \smallsetminus S$. Consider the map $\tilde{f}^k: (U^*)^{\langle k\rangle} \to V^k$. Because each multi-germ is stable, this map is transverse to the small diagonal $\Delta_k := \{(y_1, \ldots, y_k) \in V^k : y_1 = \ldots = y_k\}$. The preimage $(\tilde{f}^k)^{-1}(\Delta_k)$ is therefore a manifold of dimension $kn - (k-1)p$, if it is not empty. We have $M_k(f) = f(\pi_i(\tilde{f}^k)^{-1}(\Delta_k))$, where $\pi_i: (U^*)^{\langle k\rangle} \to U$ is any one of the standard projections. Each projection $\pi_i|: (\tilde{f}^k)^{-1}(\Delta_k) \to U$ is finite, as is f, and so $f(\pi_i(\tilde{f}^k)^{-1}(\Delta_k))$ is a constructible set of dimension $kn - (k-1)p = n - (k-1)(p-n)$. □

Proposition 9.5 *Suppose X and Y are smooth, with* $\dim X \leq \dim Y$, *and* $f : X \to Y$ *is an analytic map. Let* $\Delta_k \subset X^k$ *denote the small diagonal, and, for* $x \in X$, *denote by* x^k *the point* $(x, \ldots, x) \in \Delta_k$.

1. Then

$$D^k(f) \cap \Delta_k \subseteq \{x^k : \dim_{\mathbb{F}} Q(f)_x \geq k\}. \tag{9.14}$$

2. If the local algebra $Q(f)_x$ *is of type* Σ^1 *or* Σ^2, *or of discrete algebra type, then equality holds in* (9.14). *In particular, equality in* (9.14) *holds for all stable germs in the nice dimensions.*

Proof

1. This is straightforward. Choose a proper representative $f : V \to W$ of the germ of f at x_0. Let $g_1, \ldots, g_\ell \in \mathcal{O}_{X,x_0}$ project to a basis of $Q(f)_x$. Then $\mathcal{O}_{X,x_0}$ is generated over $\mathcal{O}_{Y,y_0}$ by $g_1, \ldots, g_\ell$, and so there is a presentation

$$\mathcal{O}^m_{Y,y_0} \longrightarrow \mathcal{O}^\ell_{Y,y_0} \longrightarrow \mathcal{O}_{X,x_0} \longrightarrow 0\,.$$

By the finite coherence theorem, this extends to a presentation of $f_*\,\mathcal{O}_X$ on some neighbourhood U of y_0,

$$\mathcal{O}^m_U \longrightarrow \mathcal{O}^\ell_U \longrightarrow f_*\,\mathcal{O}_{V\cap f^{-1}(U)} \longrightarrow 0\,.$$

It follows that for each $y \in U$,

$$\sum_{x \in V} \dim_{\mathbb{C}} \frac{\mathcal{O}_{X,x}}{f^*\,\mathfrak{m}_{Y,y}\,\mathcal{O}_{X,x}} = \dim_{\mathbb{C}} \frac{f_*(\mathcal{O}_{V\cap f^{-1}(U)})_y}{\mathfrak{m}_{Y,y}\,f_*(\mathcal{O}_{V\cap f^{-1}(U)})_y} \leq \ell.$$

Thus, y can have no more than ℓ distinct preimages. Exactly the same argument applies to a stable unfolding F of f, showing that $(x,0)^k \notin D^k_{\mathrm{cl}}(F)$ if $k > \dim_{\mathbb{C}} Q(F)(x,0)$. Since $Q(F)_{(x,0)} \simeq Q(f)_x$, this shows that $x^k \notin D^k(f)$ if $k > \ell$.
2. The equality in (9.14) for the stated algebra types is the principal result of [DG76] and we do not give their argument here, though see Remark 9.2 for a discussion of the easiest case. Their statement is as follows: for a germ $f : (\mathbb{R}^n, 0) \to (\mathbb{R}^p, 0)$, choose a representative $\tilde{f} : U \to V$ and define its *topological multiplicity* $m(\tilde{f})$, as the greatest integer k such that in every neighbourhood V_1 of 0 in V and for every neighbourhood U_1 of 0 in U, there is a point $y \in V_1$ such that y has k distinct preimages in U_1. It is easy to see that $m(\tilde{f})$ is independent of the choice of representative, and thus that we have defined an invariant of the germ f itself, which we denote by $m(f)$. We will abbreviate this rather cumbersome definition of $m(f)$ by saying that it is the largest integer k such that there are points in $(\mathbb{F}^p, 0)$ with k distinct preimages in $(\mathbb{F}^n, 0)$. Let $\delta(f) = \dim_{\mathbb{R}} Q(f)$.

Theorem 9.1 ([DG76]) *Let* $f : (\mathbb{F}^n, 0) \to (\mathbb{F}^p, 0)$, *with* $n < p$, *be a stable map-germ of corank* ≤ 2 *or of discrete algebra type. Then* $m(f) = \delta(f)$. □

The proof in [DG76] is for $\mathbb{F} = \mathbb{R}$, but works without change over $\mathbb{C}$. In fact we can circumvent the need to check this for a great number of algebra types. A large number of local algebra types are represented over $\mathbb{Z}$—that is, by polynomial maps whose coefficients are integers. If $f\colon (\mathbb{C}^n, 0) \to (\mathbb{C}^p, 0$ is a stable complex germ whose local algebra is isomorphic to a local algebra represented over $\mathbb{R}$, then f is $\mathcal{A}$-equivalent to a polynomial germ $g\colon (\mathbb{C}^n, 0) \to (\mathbb{C}^p, 0)$ all of whose component functions are real polynomials, constructed by the standard procedure (*cf* Sect. 7.2) from a real representative of the algebra. The results of Damon and Galligo for g evidently apply also to f.

The statement about the nice dimensions follows simply by inspection of Mather's list in [Mat71] of stable germs in the nice dimensions: all have local algebras to which Damon and Galligo's theorem applies. □

Theorem 9.1 and its complex version easily extend to multi-germs. We first extend the definitions of $m(f)$ and $\delta(f)$ to multi-germs, in the obvious way: $\delta(f) = \sum_{0^{(i)}\in S} \delta(f^{(i)})$, and $m(f)$ is defined by replacing each occurrence of 0 (in $\mathbb{C}^n$), in the definition for mono-germs, by S.

Proposition 9.6 *Suppose that* $f\colon (\mathbb{C}^n, S) \to (\mathbb{C}^p, 0)$ *(*$n < p$*) is a stable multi-germ and that for each point* $0^{(i)} \in S$*, the mono-germ* $f^{(i)}\colon (\mathbb{C}^n, 0^{(i)}) \to (\mathbb{C}^n, 0)$ *satisfies* $m(f^{(i)}) = \delta(f^{(i)})$*. Then* $m(f) = \delta(f)$*.*

Proof Stability of f means that each mono-germ $f^{(i)}$ is stable and that the analytic strata $T^{(i)}, i = 1, \ldots, k$ of the $f^{(i)}$ meet in general position at $0 \in \mathbb{C}^p$. After a coordinate change in $(\mathbb{C}^p, 0)$ and in each of the $(\mathbb{C}^n, 0^{(i)})$ we may assume that each $Y^{(i)}$ is a linear subspace, and that $f^{(i)}$ is a product map of the form $\mathbb{C}^{n-d_i} \times T^{(i)} \xrightarrow{f_0^{(i)} \times \mathrm{id}_{T^{(i)}}} \mathbb{C}^{p-d_i} \times T^{(i)}$. By our hypothesis on the $f^{(i)}$, we can choose $y^{(i)} \in (\mathbb{C}^p, 0)$ with $\delta(f^{(i)})$ preimages under $f^{(i)}$ (in $(\mathbb{C}^n, 0^{(i)})$). Because $f^{(i)}$ is trivial in the $T^{(i)}$ direction, translation of $y(i)$ in a direction parallel to $T^{(i)}$ does not affect the number of its preimages. Thus for all points $y \in y^{(i)} + T^{(i)}$, we have $|\left(f^{(i)}\right)^{-1}(y)| = \delta(f^{(i)})$. The proposition now follows, because the general position of the $T^{(i)}$ implies that

$$\bigcap_{i=1}^{k} \left(y^{(i)} + T^{(i)}\right) \neq \emptyset;$$

any point in the intersection has $\delta(f^{(i)})$ preimages under $f^{(i)}$ and thus $\sum_i \delta(f^{(i)})$ preimages under f. To see that this intersection is non-empty, note that it is the preimage, under the epimorphism

$$\mathbb{C}^n \to \frac{\mathbb{C}^n}{T_1} \oplus \cdots \oplus \frac{\mathbb{C}^n}{T_k},$$

of the point $(y^{(1)} + T^{(1)}, \ldots, y^{(k)} + T^{(k)})$. □

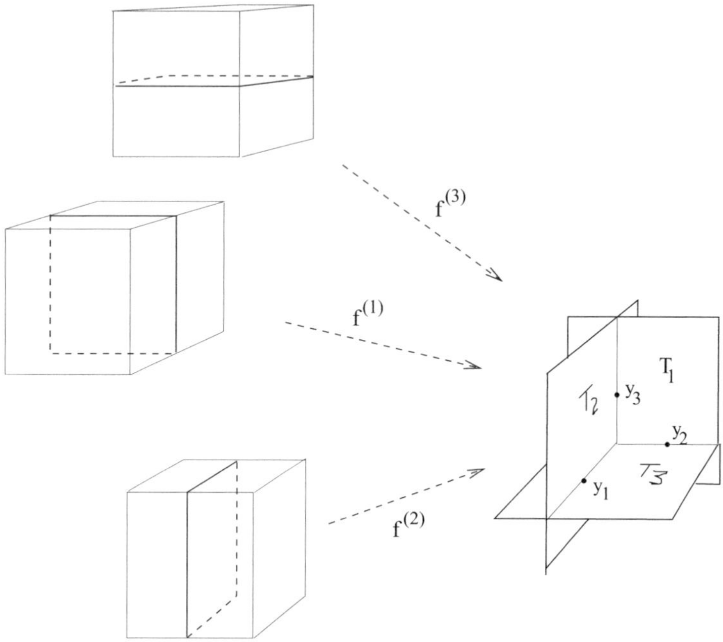

Fig. 9.3 Schematic diagram of multi-germ

Remark 9.1 If $f: (\mathbb{F}^n, S) \to (\mathbb{F}^p, 0)$ is a stable germ then all points y in $(\mathbb{F}^p, 0)$ satisfying $|f^{-1}(y)| = \delta(f)$ are normal crossing points. For by upper semicontinuity, $\sum_{x \in f^{-1}(y)} \delta(f)_x$ can be no greater than $\delta(f)$, and on the other hand $|f^{-1}(y)| = \delta(f)$, forcing $\delta(f)_x = 1$ for each $x \in f^{-1}(y)$. Thus $f: (\mathbb{F}^n, f^{-1}(y)) \to (\mathbb{F}^p, y)$ is a multi-germ of immersions. As it is also stable, it must be the germ of immersions with normal crossings (Fig. 9.3).

The hypothesis that the analytic strata of the $f^{(i)}$ be in general position is needed in Proposition 9.6. Example 9.2 shows an unstable bi-germ in which each of the component mono-germs is stable, and satisfies $m(f^{(i)}) = \delta(f^{(i)})$, but in which nevertheless $m(f) < \delta(f)$.

Remark 9.2 Theorem 9.1 is proved essentially by detailed calculation. One can get a tiny idea of the nature of the strategy from the following very easy case. Suppose $f: (\mathbb{F}^n, 0) \to (\mathbb{F}^p, 0)$ is of corank 1. Then it is a stable unfolding of $f_0: (\mathbb{F}, 0) \to (\mathbb{F}^{p-n+1}, 0)$ given by $f_0(x) = (x^k, 0, \ldots, 0)$. By the classification of stable germs by their local algebra, Theorem 7.1 above, f is right-left equivalent to a constant unfolding of the germ

$$(x, \mathbf{u}, \mathbf{v_1}, \ldots, \mathbf{v_{p-1}}) \mapsto (x^k + u_1 x^{k-2} + \cdots + u_{k-1}x, \sum_{i=1}^{k-1} v_{1i}x^{k-i}, \ldots, \sum_{i=1}^{k-1} v_{p-1,i}x^i). \tag{9.15}$$

We wish to show that in any neighbourhood of 0 in $\mathbb{R}^p$ there are points with k distinct preimages. But this is already true of the restriction of f to $\mathbb{R} \times \mathbb{R}^{k-1}$ (setting all the v_{ij} to 0). For the unfolding $G(x, u) = (\prod_{i=1}^{k}(x - ua_i), u)$, with the a_i pairwise distinct, has this property; since it is equivalent to an unfolding induced from F, F itself must have the same property. So there are points in $\mathbb{R}^p$ of the form $(0, \mathbf{u}, \mathbf{0}, \dots, \mathbf{0})$ with k distinct preimages.

The simplest type of multiple points is a normal crossing, where k immersed branches of $\mathbb{F}^n$ meet in general position. The multi-germ parameterising a normal crossing is stable—this is an easy consequence of Theorem 3.3; in fact

Proposition 9.7 *Germs of immersions with normal crossings are the only stable multi-germs of immersions.*

Proof We have to show

1. two multi-germs of immersions parameterising crossings, with one normal and the other not normal, are not right-left equivalent, and
2. if $f: (\mathbb{F}^n, S) \to (\mathbb{F}^p, 0)$ is a multi-germ of immersions parameterising a non-normal crossing, then an arbitrarily small perturbation of one or more of the $f^{(i)}$ places all crossings in general position.

The first is obvious; for the second, f can be unfolded by a multi-germ $F = (f_u, u)$ whose branches do meet in general position. The set of parameter values u for which the branches of f_u meet in general position has full measure, by the Basic Transversality Lemma 2.7. Thus, f can be perturbed by an arbitrarily small amount to a germ whose branches meet in general position. If f is stable, then its branches must already be in general position, by (i). □

Corollary 9.1 *Suppose that* $f: (\mathbb{C}^n, S) \to (\mathbb{C}^p, 0)$ *is* $\mathscr{A}$*-finite and has corank* ≤ 2*, or is of discrete algebra type. Suppose also that* $\dim D^k(f) > 0$*. Then*

1. *the set of points* $(x^{(1)}, \dots, x^{(k)}) \in D^k(f)$ *such that*
   (i) *the multi-germ of* f *at* $\{x^{(1)}, \dots, x^{(k)}\}$ *is an immersion with normal crossings, and*
   (ii) $f^{-1}(f(x^{(i)})) = \{x^{(1)}, \dots, x^{(k)}\}$

   is open and dense in $D^k(f)$.
2. $\dim D^k(f) = n - (k-1)(p-n)$.

Proof

1. Pick a good representative of f, $\tilde{f}: U \to V$ such that for every $y \in V \setminus \{0\}$, $\tilde{f}_y: (U, f^{-1}(y)) \to (V, y)$ is stable. Let $(x^{(1}, \dots, x^{(k)}) \in D^k(\tilde{f})$ and suppose $0 \neq y := \tilde{f}(x^{(i)})$. Note that the $x^{(i)}$ are not necessarily distinct from one another. The multi-germ $\tilde{f}_y: (\mathbb{C}^n, \tilde{f}^{-1}(y)) \to (\mathbb{C}^p, y)$ is stable, and all of its local algebras are of the type covered by Theorem 9.1. Evidently $\delta(\tilde{f}_y) \geq k$. By Proposition 9.6, for every collection of neighbourhoods $U^{(i)}$ of $x^{(i)}$ for $i = 1, \dots, k$, and every neighbourhood V_y of y there are points $y_1 \in V_y$ with

$\delta := \delta(\tilde{f}_y)$ distinct preimages $x_1^{(1)}, \ldots, x_1^{(\delta)}$ in $U_1 := \bigcup_i U^{(i)}$, and if y_1 is such a point then the germ $\tilde{f}_{y_1} : (U_1, f|_{U_1}^{-1}(y)) \to (V, y)$, is an immersion with normal crossings. Among these $\delta(\tilde{f}_y)$ preimages, choose k, say $x_1^{(i_1)}, \ldots, x_1^{(i_k)}$, with $x_1^{(i_j)} \in U^{(j)}$. The germ $\tilde{f} : (U, \{x_1^{(i_1)}, \ldots, x_1^{(i_k)}\}) \to (V, y_1)$ is an immersion with normal crossings, and $(x_1^{(i_1)}, \ldots, x_k^{(i_k)})$ lies in the neighbourhood $U^{(1)} \times \cdots \times U^{(k)} \cap D^k(\tilde{f})$ of $(x^{(1)}, \ldots, x^{(k)})$.

2. Provided $D^k(f)$ does not consist just of a single point, the dimension of the set of points with the two properties listed is $n - (k-1)(p-n)$.

□

Proposition 9.8 *If $n < p$ and $f : (\mathbb{C}^n, 0) \to (\mathbb{C}^p, 0)$ is stable and not an immersion at 0, then in every neighbourhood of 0 in $\mathbb{C}^n$ there are pairs of distinct points sharing the same image.*

Proof If f is not an immersion, then $\mathscr{I}_2(f) \neq (1)$, so by part 4 of Proposition 9.9 below, the codimension of $D^2(f)$ in $\mathbb{C}^n \times \mathbb{C}^n$ is no greater than p, so its dimension is at least $2n - p$. On the other hand, if f is stable and not an immersion, then the codimension of the non-immersive locus in $\mathbb{C}^n$ is $p - n + 1$, so that its dimension is $2n - p - 1$. Thus the complement of the non-immersive locus in $D^2(f)$ is non-empty. □

Let $d_k(n, p) = p - k(p-n)$. We will say that f is *dimensionally correct* if for each k for which $d_k(n, p) \geq 0$, the dimension of $D^k(f)$ is $d_k(n, p)$, and when $p - k(p-n) < 0$ then $D^k(f)$ is at most zero-dimensional.

9.4 Equations for $D^2(f)$

We find equations for $D^2(f)$ by finding a relatively Cohen–Macaulay ideal $\mathfrak{i}_2(f)$, with an explicit set of generators, which is contained in $(I_2(f) : J_2)$ and coincides with it generically. From this it follows by Proposition C.5 that the two ideals are equal.

Since $I_2(f) \subset J_2$, there exist α_{ij}, $1 \leq i \leq p$, $1 \leq j \leq n$, such that

$$f_i(x^{(1)}) - f_i(x^{(2)}) = \sum_{j=1}^{n} \alpha_{ij}(x^{(1)}, x^{(2)})(x_j^{(1)} - x_j^{(2)}) \tag{9.16}$$

for $i = 1, \ldots, p$. Let A denote the matrix $(\alpha_{ij})_{1 \leq i \leq p,\, 1 \leq j \leq n}$, and let $\min_n(A)$ denote the ideal of $\mathscr{O}_{\mathbb{C}^n \times \mathbb{C}^n, (0,0)}$ generated by its $n \times n$ minors. Equations (9.16) together give a matrix equation

$$f(x^{(1)}) - f(x^{(2)}) = A(x^{(1)}, x^{(2)})(x^{(1)} - x^{(2)}). \tag{9.17}$$

Define

$$\mathrm{i}^2(f) = \min_n(A) + I_2(f). \tag{9.18}$$

Proposition 9.9

1. *Let $i : \mathbb{C}^n \to \mathbb{C}^n \times \mathbb{C}^n$ be the diagonal embedding. Then $A \circ i$ is the Jacobian matrix of f, $[df]$, and $i^*(\mathrm{i}_2(f))$ is the ramification ideal $\mathcal{R}_f$, generated by the maximal minors of $[df]$.*
2. $V(\mathrm{i}_2(f)) = \big(D^2(f) \cap (\mathbb{C}^n)^{(2)}\big) \bigcup i\big($*non-immersive locus of* $f\big)$.
3. $$\mathrm{i}_2(f) = F_0\left(\frac{J_2}{I_2(f)}\right) + I_2(f)$$

   (where the first summand on the right means the zero'th Fitting ideal of $J_2/I_2(f)$ as an $\mathcal{O}_{\mathbb{C}^n\times\mathbb{C}^n}$-module[2]*) and is thus independent of the choice of the α_{ij} satisfying* (9.16).
4. *The codimension of $V(\mathrm{i}_2(f))$ is less than or equal to p; if it is equal to p, then $\mathcal{O}_{\mathbb{C}^n\times\mathbb{C}^n}/\mathrm{i}_2(f)$ is Cohen–Macaulay.*
5. *If $V(\mathcal{I}_2(f))$ has the expected dimension, $2n - p$, and the set of genuine double points (where $x^{(1)} \neq x^{(2)}$) is dense in $V(\mathcal{I}_2(f))$, then $\mathrm{i}_2(f) = (I_2(f) : J_2)$.*

Proof

1. We have

$$f_i(x^{(1)}) - f_i(x^{(2)}) = \sum_{j=1}^{n}(x_j^{(1)} - x_j^{(2)})\Big(\frac{\partial f_i}{\partial x_j}(x^{(1)}) + E_{ij}\Big),$$

where $E_{ij} \in I_{\Delta_n}$. Thus for each i the n-tuple

$$\left(\alpha_{i1} - \Big(\frac{\partial f_i}{\partial x_1} + E_{i1}\Big), \ldots, \alpha_{in} - \Big(\frac{\partial f_i}{\partial x_n} + E_{in}\Big)\right)$$

(with partials evaluated at $x^{(1)}$ and α_{ij} evaluated at $(x^{(1)}, x^{(2)})$) is a relation between the n linear forms $x_1^{(1)} - x_1^{(2)}, \ldots, x_n^{(1)} - x_n^{(2)}$. Since these form a regular sequence, all relations among them have coefficients in J_2. It follows that

$$\frac{\partial f_i}{\partial x_j}(x^{(1)}) = \alpha_{ij}(x^{(1)}, x^{(1)}) \mod J_2.$$

Thus, on the diagonal $V(J_2)$, α_{ij} coincides with $\partial f_i/\partial x_j$.

[2]Fitting ideals are introduced and described in more detail in Sect. 11.2.

2. If $x^{(1)} \neq x^{(2)}$ and $f(x^{(1)}) = f(x^{(2)})$, then the matrix A in Eq. (9.17) has rank less than n, so $(x^{(1)}, x^{(2)}) \in V(\mathrm{i}_2(f))$. This shows that $V(\mathrm{i}_2(f)) \smallsetminus V(J_2) = D^2(f) \smallsetminus V(J_2)$. The conclusion now follows from item 1.
3. Let M be the set of n-tuples $(\beta_1, \ldots, \beta_n)$ of germs in $\mathscr{O}_{X\times X,(0,0)}$ such that

$$\sum_i \beta_i \cdot (x_i^{(1)} - x_i^{(2)}) \in I_2(f),$$

let M_0 be the set of n-tuples such that

$$\sum_i \beta_i \cdot (x_i^{(1)} - x_i^{(2)}) = 0$$

and consider the ideal $\bigwedge^n M$ of determinants of n-tuples of elements of M. We claim that

$$\mathrm{i}^2(f) = \bigwedge^n M + I_2(f). \tag{9.19}$$

To see this, suppose that $\boldsymbol{\beta}_1, \ldots, \boldsymbol{\beta}_n$ is an n-tuple of elements of M, and let B be the $n \times n$ matrix wth rows $\boldsymbol{\beta}_1, \ldots, \boldsymbol{\beta}_n$. By definition of the matrix A, there exists a matrix C with entries in $\mathscr{O}_{X\times X,(0,0)}$ such that

$$B \cdot (x^{(1)} - x^{(2)}) = CA(x^{(1)} - x^{(2)})$$

so the rows of $B - CA$ are members of M_0. Thus

$$B - H = CA$$

where the rows of H lie in M_0. Now $x_1^{(1)} - x_1^{(2)}, \ldots, x_n^{(1)} - x_n^{(2)}$ is a regular sequence, so the members of M_0 are linear combinations of the trivial relations

$$(x_i^{(1)} - x_i^{(2)})\mathbf{e}_j - (x_j^{(1)} - x_j^{(2)})\mathbf{e}_i. \tag{9.20}$$

We wish to prove that $\det B \in \min_n A + I_2(f)$. Evidently $\det(B - H) \in \min_n A$. By multilinearity, $\det(B - H)$ is a sum of determinants of matrices B' consisting of k rows of B and $n - k$ rows of H. We claim that every such determinant with $k < n$ lies in $I_2(f)$.

To see this, observe that by rearranging rows we may assume the first k rows of B' come from B and the last $n-k$ come from H. Suppose first that the last row is the relation (9.20). Then

$$B'(x^{(1)}-x^{(2)}) = \begin{pmatrix} k \text{ rows of } B \\ \\ (n-k-1) \text{ rows of } H \\ \\ (x_i^{(1)}-x_i^{(2)})\mathbf{e}_j - (x_j^{(1)}-x_j^{(2)})\mathbf{e}_i \end{pmatrix} \begin{pmatrix} x_1^{(1)}-x_1^{(2)} \\ \vdots \\ \vdots \\ x_n^{(1)}-x_n^{(2)} \end{pmatrix} = \begin{pmatrix} c_1 \\ \vdots \\ c_k \\ 0 \\ \vdots \\ 0 \end{pmatrix} \tag{9.21}$$

where by definition of B, $c_1,\ldots,c_k \in I_2(f)$. By Cramer's rule, $(x_i^{(1)} - x_i^{(2)})\det B'$ is equal to the determinant of the matrix obtained from B' by replacing its i'th column by the right- hand side of (9.21). The only non-zero entry in the last row of this matrix is $x_i^{(1)} - x_i^{(2)}$ in the j'th place, and all the elements in its i'th column lie in $I_2(f)$; so its determinant lies in $(x_i^{(1)} - x_i^{(2)})I_2(f)$. We have shown that

$$(x_i^{(1)} - x_i^{(2)})\det B' \in (x_i^{(1)} - x_i^{(2)})I_2(f) \tag{9.22}$$

and it follows that $\det B' \in I_2(f)$. By multilinearity of the determinant, the same holds if the last row is any relation among the generators $x_i^{(1)} - x_i^{(2)}$ of J_2. This proves the inclusion of the right-hand side in the left-hand side in (9.19). The opposite inclusion is immediate from the definition of $\mathrm{i}_2(f)$.

Let $B_1,\ldots,B_r$ generate M and let B be the $r\times n$ matrix of their coefficients. Then B^t is the matrix of a presentation of $J_2/I_2(f)$ as $\mathcal{O}_{\mathbb{C}^n\times\mathbb{C}^n}$-module. The 0'th Fitting ideal of $J_2/I_2(f)$ as $\mathcal{O}_{\mathbb{C}^n\times\mathbb{C}^n}$-module is by definition the ideal generated by the $n\times n$ minors of B^t. By what we have just proved, $\min_n(B^t)\subseteq \mathrm{i}_2(f)$, so

$$\mathrm{i}_2(f)\subseteq F_0\big(J_2/I_2(f)\big)+I_2(f)\subseteq \mathrm{i}_2(f),$$

proving the required equality.

4. To prove that $\mathrm{i}_2(f)$ is relatively Cohen–Macaulay, we need to use an important theorem of G. Kempf. In [Kem76], Kempf proved a very general result concerning the union of the translates of a subspace $W\subset H$ of a k-vector space H, under the linear action of an algebraic group G on H. Denote this union by $G\cdot W$.

Theorem 9.2

1. *Suppose that there is a subgroup P of $Stab(W)\subset G$ such that G/P is complete. Then $G\cdot W$ is a closed subvariety of H.*
2. *Suppose in addition that $char(k)=0$, and the action of G on H, and of P on W, are both completely reducible. Then $G\cdot W$ is normal and Cohen–Macaulay.*

In [G.K75], he applied it to the Buchsbaum–Eisenbud *variety of complexes*, BE. For fixed integers $d_0, \ldots, d_n$, this is the variety of n-tuples of matrices

$$\left\{(A_1, \ldots, A_m) \in \prod_{i=1}^{n} \mathrm{Mat}_{d_{i-1}\times d_i} : A_{i+1}A_i = 0 \text{ for } i = 1, \ldots n\right\}. \tag{9.23}$$

In general BE is not Cohen–Macaulay (Exercise 9.4.5 below), but Kempf shows that if $(m_1, \ldots, m_n)$ are non-negative integers such that $m_i + m_{i+1} \leq d_i$ for $i = 1, \ldots, n$, and we define $\mathrm{BE}(m_1, \ldots, m_n)$ to be the set of $(A_1, \ldots, A_n)$ in BE such that $\mathrm{rank}(A_i) \leq m_i$ for $i = 1, \ldots, n$, then $\mathrm{BE}(m_1, \ldots, m_n)$ is Cohen–Macaulay.

To explain this consequence of Theorem 9.2, we introduce some notation. Let H be the product $\prod_{i=1}^{n} \mathrm{Mat}_{d_{i-1}\times d_i}$; we refer to an n-tuple of non-negative integers $(m_1, \ldots, m_n)$ as *admissible* if $m_i + m_{i+1} \leq d_i$ for $i = 1, \ldots, n$. One way of ensuring that $A_{i+1} \circ A_i = 0$, and that the rank of A_i is less than or equal to m_i, is to specify a linear subspace $L_i \subseteq k^{d_i}$ of dimension m_i such that $\mathrm{im}\, A_i \subseteq L_i \subseteq \ker A_{i+1}$. Write $m = (m_1, \ldots, m_n)$, let $\mathrm{Gr}(m_i, d_i)$ be the Grassmannian of m_i-planes in k^{d_i} and

$$\mathrm{Gr}(m) = \prod_{i=1}^{n} \mathrm{Gr}(m_i, d_i),$$

and given $L_i \in \mathrm{Gr}(m_i, d_i)$ for $i = 1, \ldots, n$ denote by $L \in \mathrm{Gr}(m_1, \ldots, m_n)$ the n-tuple $(L_1, \ldots, L_n)$. Given $L \in \mathrm{Gr}(m)$, let

$$W(L) = \{A \in H : \mathrm{im}\, A_i \subseteq L_i \subseteq \ker A_{i+1} \text{ for } i = 1, \ldots, n\}, \tag{9.24}$$

(where we take $A_{n+1} = 0$). Note that $W(L)$ is a linear subspace of H. Let $G = \mathrm{Gl}(k^{d_1}) \times \cdots \times \mathrm{Gl}(k^{d_n})$. Then G acts on H by

$$g \cdot (A_1, \ldots, A_n) = \left(g_1 \circ A_1, g_2 \circ A_2 \circ g_1^{-1}, \ldots, \; g_n \circ A_n \circ g_{n-1}^{-1}\right).$$

This action preserves $\mathrm{BE}(m)$, and moreover for any fixed choice of $L \in \mathrm{Gr}(m)$,

$$\mathrm{BE}(m) = G \cdot W(L).$$

Parts 1 and 2 of Theorem 9.2 apply, for we can take $P \subset G$ to be the product of the stabilisers of the L_i, for $i = 1, \ldots, n$. Then $G/P = \mathrm{Gr}(m_1, k^{d_1}) \times \cdots \times \mathrm{Gr}(m_n, k^{d_n})$

is complete, P stabilises $W(L)$, and both G and P are reductive groups, so that the actions of G on V and of P on W are both completely reducible. Thus, we have

Theorem 9.3 *Suppose that k has characteristic 0, and $(m_1, \ldots, m_n)$ is admissible. Then*

1. *$BE(m_1, \ldots, m_n)$ is normal and Cohen–Macaulay.*
2. *The dimension of $BE(m_1, \ldots, m_n)$ is equal to*

$$\sum_{i=1}^{n} \dim Gr(m_i, d_i) + \sum_{i=1}^{n} \dim \mathrm{Hom}\left(k^{d_{i-1}}/L_{i-1}, L_i\right), \tag{9.25}$$

(where we take $L_0 = 0$).

3. *The ideal of $BE(m)$ in H is generated by the entries in the composites $A_i A_{i-1}$ and by the minors of size $m_i + 1$ of the matrices A_i, for $i = 1, \ldots, n$.* □

The formula (9.25) does not appear in [G.K75], but is implicit in the proof. Exercise 9.4.4 leads to a proof of (9.25). The final statement of the theorem is only conjectured in [G.K75], but is proved in the rather technical paper [CS81], which in addition reproves Theorem 9.3 in arbitrary characteristic. Exercise 9.4.7 leads to a proof of part 3.

Now we return to the proof of part 4 of our Proposition 9.9. To derive the statement concerning the ideal $\mathrm{i}_2(f)$, we consider the case where the complex in question has just two maps:

$$\mathbb{C}^p \xrightarrow{A_1} \mathbb{C}^n \xrightarrow{A_2} \mathbb{C} . \tag{9.26}$$

By the formula (9.25), here BE$(n-1, 1)$ has dimension $n(p+1) - p$, and therefore codimension p in H. Given $f : (\mathbb{C}^n, 0) \to (\mathbb{C}^p, 0)$, pick a matrix $A = (\alpha_{ij})$ as defined in (9.16), and let $A_1 = A$ and $A_2 = (x_1^{(1)} - x_1^{(2)}, \ldots, x_n^{(1)} - x_n^{(2)})$. Define $\varphi : (\mathbb{C}^n \times \mathbb{C}^n, (0, 0)) \to H$ by $\varphi(x^{(1)}, x^{(2)}) = (A_1, A_2)$. Theorem 9.3 tells us that BE$(n - 1, 1)$ is Cohen–Macaulay of codimension p; thus by Proposition C.12 the codimension of $\varphi^{-1}(BE(n-1, 1))$ is no greater than p, and if it is equal to p, then φ^{-1}(BE$(n - 1, 1)$) is Cohen–Macaulay. Our $\mathrm{i}_2(f)$ is the pull-back of the ideal of $BE(n - 1, 1)$ by the map φ. Part 4 is proved.

5. Obviously $\mathrm{i}_2(f) \subset (I_2(f) : J_2)$. If the set $N_2(f) \cap X^{\langle 2\rangle}$ of genuine double points is dense $V(\mathrm{i}_2(f))$, then the two ideals coincide generically, since at genuine double points, both ideals are equal to $I_2(f)$. Now if $\mathrm{i}_2(f)$ is Cohen–Macaulay, then Proposition C.5 applies and $\mathrm{i}_2(f) = (I_2(f) : J_2)$.

□

Corollary 9.2 *If f is $\mathscr{A}$-finite, then $\mathscr{I}_2(f) = (I_2(f) : J_2) = \mathscr{I}_2^s(f)$ and $D^2(f)$ is Cohen–Macaulay.*

Proof All but the second equality follows immediately from the proposition. The second equality holds because for $\mathscr{A}$-finite f the genuine double points are transverse, so that $(I_2(f) : J_2)$ is reduced, and thus coincides with the potentially larger ideal $(I_2(f) : J_2^\infty)$. □

Exercises for Sect. 9.4

1 Show by an example that if we replace "stable" by "$\mathscr{A}$-finite" in Proposition 9.8, then the conclusion of the proposition does not hold. Where does the argument fail? Under which additional hypotheses does it hold?

2 If f has corank ≤ 2, then $I_2(f) \subset F_0(J_2/I_2(f))$ so that $\mathscr{I}_2(f) = F_0(J_2/I_2(f))$. In general if f has corank ≥ 3, then $I_2(f) \not\subseteq F_0(J_2/I_2(f))$.

3 If $f : (\mathbb{C}^n, 0) \to (\mathbb{C}^p, 0)$ has corank ≥ 2 (with $n < p$), then $D^2(f)$ is singular at $(0, 0) \in \mathbb{C}^n \times \mathbb{C}^n$.

4 In [G.K75], Kempf introduced the space

$$X := \{(L, A) \in \mathrm{Gr}(m) \times H : A \in W(L)\},$$

with maps $p : X \to \mathrm{Gr}(m)$ and $h : X \to H$ defined by $p(L, A) = L$, $h(L, A) = A$.

(a) Calculate the dimension of the fibres of $p : X \to \mathrm{Gr}(m)$, and so obtain a formula for the dimension of X.
(b) Show that the image of h is $\mathrm{BE}(m)$, and that $h : X \to \mathrm{BE}(m)$ is birational, by showing that at the open set of points (L, A) such that $\mathrm{rank}(A_i) = m_i$ (and therefore $\mathrm{im}(A_i) = L_i$) for $i = 1, \ldots, m$, h is an isomorphism.
(c) From (a) and (b) obtain the formula (9.25).

5

(a) Show that $p : X \to G/P$ is a vector bundle, and coincides with the homogeneous bundle over $X' \to G/P$ associated to the representation of P on $W(L)$, with total space X' equal to the quotient of $G \times W(L)$ by the equivalence relation $(g, A) \sim (gp^{-1}, p \cdot A)$ for $p \in P$, and projection $(g, A) \mapsto gP$.
(b) Show that $h : X \to \mathrm{BE}(m)$ is a resolution of singularities—i.e., X is smooth, h is proper, and h is an isomorphism over the smooth locus of $\mathrm{BE}(m)$.

6 Use *Macaulay 2* or *Singular* to check that when $n = 2$, $d_0 = d_2 = 3$ and $d_1 = 4$, then BE is not Cohen–Macaulay.

7 Use the Cohen–Macaulayness of $\mathrm{BE}(m)$ to prove part 3 of Theorem 9.3, by showing that there is a dense set of points in $\mathrm{BE}(m)$ where the ideal of germs of functions vanishing on $\mathrm{BE}(m)$ is equal to the ideal described in part 3.

9.5 Equations for $D^k(f)$ When f Is a Corank 1 Germ

If f has corank 1, we can choose linearly adapted coordinates $u, x := u_1, \ldots, u_{n-1}, x$ so that f takes the form

$$f(u, x) = (u, f_n(u, x), \ldots, f_p(u, x)). \tag{9.27}$$

That is, we write f explicitly as an unfolding of a map-germ in the single variable x. As $D^k(f)$ then embeds naturally in $\mathbb{C}^{n-1} \times \mathbb{C}^k$ (see Lemma 9.1), we take coordinates $u, x_1, \ldots, x_k$ on $\mathbb{C}^{n-1} \times \mathbb{C}^k$ and look for equations defining $D^k(f)$ in $\mathbb{C}^{n-1} \times \mathbb{C}^k$. The ideal J_k of (9.2) is now generated by the Vandermonde determinant vdM $:= \prod_{1 \le i < j \le k}(x_i - x_j)$ so $\mathscr{I}_k(f)$ consists of quotients of the form $g(u, x_1, \ldots, x_k)/\mathrm{vdM}$ where $g \in I_k(f) \cap J_k$. We will see that $\mathscr{I}_k(f)$ is generated by S_k-invariant quotients g/vdM, where g, like vdM, satisfies $\sigma^*(g) = \mathrm{sign}(\sigma) g$ for $\sigma \in S_k$.

9.5.1 *Generalities on Functions of One Variable*

The following analysis will be applied to each of the component functions f_j, $j = n \ldots, p$ of f. To spare notation for the moment, let h be any function of u, x. The map

$$(u, x_1, \ldots, x_k) \mapsto \big(h(u, x_1), \ldots, h(x, x_p)\big) \tag{9.28}$$

is equivariant with respect to the symmetric group actions on the source permuting the x_i and on the target permuting the $h(u, x_i)$. The set $\mathscr{E}^{S_k}$ of equivariant maps $\mathbb{C}^{n-1+k} \to \mathbb{C}^k$ is a module over the ring $\mathscr{O}^{S_k}$ of invariant functions on the source, and (although we will not *need* this fact) as such is generated by the gradient vectors of the generators of the ring $\mathscr{O}^{S_k}$ of invariant functions [Poé76]. The ring $\mathscr{O}^{S_k}$ is generated over $\mathscr{O}_{\mathbb{C}^{n-1},0}$ by the sums of powers $\rho_1 = x_1 + \cdots + x_k, \ldots, \rho_k = x_1^k + \cdots + x_k^k$, and so every equivariant mapping can be written as a linear combination, over $\mathscr{O}^{S_k}$, of the maps

$$\begin{aligned} m_1(x_1, \ldots, x_k) &= (1, \ldots, 1) \\ m_2(x_1, \ldots, x_k) &= (x_1, \ldots, x_k) \\ &\cdots\cdots\cdots \\ m_{k-1}(x_1, \ldots, x_k) &= (x_1^{k-1}, \ldots, x_k^{k-1}). \end{aligned} \tag{9.29}$$

Thus there exist invariant functions $\alpha_0^k, \alpha_1^k, \ldots, \alpha_{k-1}^k$ such that

$$\begin{pmatrix} h(u, x_1) \\ \vdots \\ h(u, x_k) \end{pmatrix} = \alpha_0^k \begin{pmatrix} 1 \\ \vdots \\ 1 \end{pmatrix} + \alpha_1^k \begin{pmatrix} x_1 \\ \vdots \\ x_k \end{pmatrix} + \cdots + \alpha_{k-1}^k \begin{pmatrix} x_1^{k-1} \\ \vdots \\ x_k^{k-1} \end{pmatrix}. \tag{9.30}$$

Solving for the α_i^k by Cramer's rule gives

$$\alpha_\ell^k(u, x_1, \ldots, x_k) = \frac{\begin{vmatrix} 1 & x_1 & \cdots & x_1^{\ell-1} & h(u,x_1) & x_1^{\ell+1} & \cdots & x_1^{k-1} \\ \vdots & \vdots & \vdots & \vdots & \vdots & \vdots & \vdots & \vdots \\ 1 & x_k & \cdots & x_k^{\ell-1} & h(u,x_k) & x_k^{\ell+1} & \cdots & x_k^{k-1} \end{vmatrix}}{\begin{vmatrix} 1 & x_1 & \cdots & x_1^{k-1} \\ \vdots & \vdots & \vdots & \vdots \\ 1 & x_k & \cdots & x_k^{k-1} \end{vmatrix}}. \tag{9.31}$$

In fact we do not need the statement of Poénaru referred to above, to see that the α_ℓ^k are regular (analytic): the numerator in (9.31) vanishes whenever $x_i = x_\ell$ for any $i \neq \ell$, and thus is divisible in $\mathcal{O}$ by the Vandermonde determinant, which is the denominator in (9.31). In other words the system of Eqs. (9.30) has analytic solutions. As can be seen from (9.31), they are S_k-invariant. They are also unique, since the Vandermonde determinant vanishes only along a hypersurface.

Recall that $I_k(h)$ is the ideal generated by the pairwise differences $h(u, x_i) - h(u, x_j)$. The determinant in the numerator of (9.31) lies in $I_k(h)$ (subtract $h(u, x_1)$ times column 1 from column $\ell + 1$). Hence

Lemma 9.3 $\alpha_\ell^k \in (I_k(h) : J_k)$ □

Let $\mathcal{A}_k(h)$ be the ideal generated by the α_ℓ^k for $\ell = 1, \ldots, k-1$.

Now in (9.30) subtract the first row from each of the others. Omitting the first row in the resulting equation gives

$$\begin{pmatrix} h(u,x_2) - h(u,x_1) \\ \vdots \\ h(u,x_k) - h(x,x_1) \end{pmatrix} = \begin{pmatrix} (x_2 - x_1) & \cdots & (x_2^{k-1} - x_1^{k-1}) \\ \vdots & \vdots & \vdots \\ (x_k - x_1) & \cdots & (x_k^{k-1} - x_1^{k-1}) \end{pmatrix} \begin{pmatrix} \alpha_1^k \\ \vdots \\ \alpha_k^{k-1}. \end{pmatrix} \tag{9.32}$$

It follows that

$$\mathcal{A}_k(h) \supseteq I_k(h). \tag{9.33}$$

Since the determinant of the $(k-1) \times (k-1)$ matrix on the right in (9.32) is still $\mathrm{vdM}(x_1, \ldots, x_k)$, we also have

$$x_1, \ldots, x_k \text{ are pairwise distinct} \implies \mathcal{A}_k(h) = I_k(h). \tag{9.34}$$

By contrast, when all the x_i are equal, we find that $\mathcal{A}_k(h)$ reduces to an ideal of partial derivatives.

Lemma 9.4 *$\alpha_0^k = h(u, x_i) \mod \mathcal{A}_k(h)$ and if $x_1 = \cdots = x_k =: x$, then for $1 \leq \ell \leq k-1$,*

$$\alpha_\ell^k + \binom{\ell+1}{\ell} x\alpha_{\ell+1}^k + \cdots + \binom{k-1}{\ell} x^{k-1-\ell}\alpha_{k-1}^k = \frac{1}{\ell!}\frac{\partial^\ell h}{\partial x^\ell} \tag{9.35}$$

and hence

$$(\alpha_\ell^k, \ldots, \alpha_{k-1}^k) + (x_1 - x_2, \ldots, x_1 - x_k) = \left(\frac{\partial^\ell h}{\partial x^\ell}, \ldots, \frac{\partial^{k-1} h}{\partial x^{k-1}}\right) + (x_1 - x_2, \ldots, x_1 - x_k). \tag{9.36}$$

Proof The first statement can be read off from the i'th row in (9.30). For the rest, write $x_i = x + \delta_i$ for $i = 1, \ldots, k$ and substitute this into (9.30). We get

$$\begin{pmatrix} h(u, x_1) \\ \vdots \\ h(u, x_k) \end{pmatrix} = (\alpha_0^k + x\alpha_1^k + \cdots + x^{k-1}\alpha_{k-1}^k) \begin{pmatrix} 1 \\ \vdots \\ 1 \end{pmatrix} + \cdots + \left(\alpha_\ell^k + \binom{\ell+1}{\ell} x\alpha_{\ell+1}^k + \cdots \right.$$
$$\left. + \binom{k-1}{\ell} x^{k-1-\ell}\alpha_{k-1}^k\right) \begin{pmatrix} \delta_1^\ell \\ \vdots \\ \delta_k^\ell \end{pmatrix} + \cdots + \alpha_{k-1}^k \begin{pmatrix} \delta_1^{k-1} \\ \vdots \\ \delta_k^{k-1} \end{pmatrix}. \tag{9.37}$$

In the left-hand side replace $h(u, x_i)$ by its Taylor expansion about (u, x). Writing $h^{(\ell)}$ for $\partial^\ell h/\partial x^\ell(u, x)$, and shifting powers of δ_i lower than k to the right-hand side, we obtain

$$\begin{pmatrix} \sum_{j \geq k} \left(\frac{h^{(j)}}{j!}\right) \delta_1^j \\ \vdots \\ \sum_{j \geq k} \left(\frac{h^{(j)}}{j!}\right) \delta_k^j \end{pmatrix} = (\alpha_0^k + x\alpha_1^k + \cdots + x^{k-1}\alpha_{k-1}^k - h) \begin{pmatrix} 1 \\ \vdots \\ 1 \end{pmatrix} + \cdots$$
$$+ \left(\alpha_\ell^k + \binom{\ell+1}{\ell} x\alpha_{\ell+1}^k + \cdots + \binom{k-1}{\ell} x^{k-1-\ell}\alpha_{k-1}^k - \frac{h^{(\ell)}}{\ell!}\right) \begin{pmatrix} \delta_1^\ell \\ \vdots \\ \delta_k^\ell \end{pmatrix} + \cdots$$
$$+ \left(\alpha_{k-1}^k - \frac{h^{(k-1)}}{(k-1)!}\right) \begin{pmatrix} \delta_1^{k-1} \\ \vdots \\ \delta_k^{k-1} \end{pmatrix}. \tag{9.38}$$

Divide both sides of the equation by the Vandermonde *matrix*

$$\begin{pmatrix} 1 & \delta_1 & \cdots & \delta_1^{k-1} \\ \vdots & \vdots & \vdots & \vdots \\ 1 & \delta_k & \cdots & \delta_k^{k-1} \end{pmatrix}; \tag{9.39}$$

it is obvious that we can do this on the right, while on the left it is possible by the statement of Poénaru already cited. All entries in the quotient on the left lie in the ideal $(\delta_1, \ldots, \delta_k)$, and so the same is true on the right. It follows that when the δ_i all vanish, the coefficients of the columns of the Vandermonde matrix that we see on the right in (9.37) must all vanish, and we have

$$\frac{h^{(\ell)}}{\ell!} = \alpha_\ell^k + \binom{\ell+1}{\ell} x\alpha_{\ell+1}^k + \cdots + \binom{k-1}{\ell} x^{k-1-\ell}\alpha_{k-1}^k. \tag{9.40}$$

This proves the second statement, and the third follows. □

What happens in situations intermediate between the situation of (9.34), where all the x_i are distinct from one another, and (9.35), where they are all equal? In other words, how does the ideal generated by the α_ℓ^k behave on the different diagonals (strata determined by S_k orbit-type) in $\mathbb{C}^{n-1} \times \mathbb{C}^k$? Each stratum in $\mathbb{C}^{n-1} \times \mathbb{C}^k$ corresponds to a partition of $\{1, \ldots, k\}$. By re-ordering the x_i, any partition can be brought to the form

$$\Big(\{1, \ldots, r_1\}, \{r_1+1, \ldots, r_1+r_2\}, \ldots, \{r_1+\cdots+r_{m-1}+1, \ldots, r_1+\cdots+r_m = k\}\Big).$$

The corresponding stratum $\Delta(\mathscr{P})$ consists of points $(u, x_1, \ldots, x_k) \in \mathbb{C}^{n-1} \times \mathbb{C}^k$ such that

$$x_j = x_{j+1} \text{ iff } r_1 + \ldots + r_{i-1} < j < r_1 + \cdots + r_i \text{ and } 1 \le i \le m, \tag{9.41}$$

whose closure has defining ideal

$$\mathrm{i}(\mathscr{P}) := \big(y_i - y_{i+1} : r_1 + \ldots + r_{i-1} < j < r_1 + \cdots + r_i \text{ for } 1 \le i \le m\}\big).$$

We call m the *length* of the partition. Observe that the dimension of $\Delta(\mathscr{P})$ is $n - 1 + m$.

We say the partition $\mathscr{P}'$ is *finer* than $\mathscr{P}$ if $\Delta(\mathscr{P}')$ is contained in the closure of $\Delta(\mathscr{P})$, and *strictly finer* if in addition they are not equal.

We need some notation. Define $s_i = r_1 + \cdots + r_i$ for $i = 1, \ldots, m$, and set $s_0 = 0$. Define $\pi_i^{\mathscr{P}} : \mathbb{C}^{n-1+k} \to \mathbb{C}^{n-1+r_i}$ and $\mathrm{pr}_i^{\mathscr{P}} : \mathbb{C}^{n-1+k} \to \mathbb{C}^n$ by

$$\pi_i^{\mathscr{P}}(u, x_1, \ldots, x_k) = (u, x_{s_{i-1}+1}, \ldots, x_{s_i}), \quad \mathrm{pr}_i^{\mathscr{P}}(, x_1, \ldots, x_k) = (u, x_{s_i}). \tag{9.42}$$

For the purpose for which we will use $\mathrm{pr}_i^{\mathscr{P}}$, we could just as well have defined it using x_j, for any j between $s_{i-1}+1$ and s_i, instead of x_{s_i}. The point is to select one of the x_j's in the i'th block of the partition. The purpose will become clear in the next proposition.

When $(u, x_1, \ldots, x_k) \in \Delta(\mathscr{P})$, it will sometimes it will be useful to denote the common value of the x_j for $r_1 + \cdots + r_{i-1} < j < r_1 + \cdots + r_i$ by $x^{(i)}$. With this notation, $(x_1, \ldots, x_k)$ becomes

$$\big(\underbrace{x^{(1)}, \ldots, x^{(1)}}_{r_1 \text{ times}}, \ldots, \underbrace{x^{(m)}, \ldots, x^{(m)}}_{r_m \text{ times}}\big).$$

Let $A_{\mathscr{P}}(h) = \mathcal{A}_k(h) + \mathrm{i}(\mathscr{P})$.

Proposition 9.10 *If (u, x) does not lie in $\Delta(\mathscr{P}')$ for any partition $\mathscr{P}'$ strictly finer than $\mathscr{P}$, then the stalks in $\mathscr{O}_{\mathbb{C}^{n-1+k},(u,\mathbf{x})}$ of $I_k(h)$ and $A_{\mathscr{P}}(h)$ are equal, respectively, to*

$$(\pi_1^{\mathscr{P}})^*(\mathrm{i}_{r_1}(h)) + \cdots + (\pi_m^{\mathscr{P}})^*(\mathrm{i}_{r_m}(h)) + \left(\left\{h \circ pr_i^{\mathscr{P}} - h \circ pr_1^{\mathscr{P}} : i = 2, \ldots, m\right\}\right) \tag{9.43}$$

and

$$\mathrm{i}(\mathscr{P}) + \left(\left\{\frac{\partial^\ell h}{\partial y^\ell} \circ pr_i^{\mathscr{P}} \,:\, \ell = 1, \ldots, r_i - 1, i = 1 \ldots, m\right\}\right)$$
$$+ \left(\left\{h \circ pr_1^{\mathscr{P}} - h \circ pr_i^{\mathscr{P}} : i = 1, \ldots, m\right\}\right) \tag{9.44}$$

Proof The subgroup $G = S_{r_1} \times \cdots \times S_{r_m}$ of S_k acts on $\mathbb{C}^{n-1} \times \mathbb{C}^k$ via the permutation representation of S_k. The ring of G-invariant functions $\mathscr{O}^G$ is generated by the power sums $g_{i,\ell} = (x_{r_1+\cdots+r_{i-1}+1})^\ell + \cdots + (x_{r_1+\cdots+r_i})^\ell$ for $0 \le j \le r_i$ and $i = 1, \ldots, m$. Let $\mathscr{E}^G$ be the $\mathscr{O}^G$-module of G-equivariant maps $\mathbb{C}^{n-1} \times \mathbb{C}^k \to \mathbb{C}^k$. By the theorem of Poénaru in *op cit*, $\mathscr{E}^G$ is generated over $\mathscr{O}^G$ by the gradients $e_{i,\ell}$ of the $g_{i,\ell+1}$; after dividing by constant coefficients, $e_{i,\ell}$ is the k-tuple with x_j^ℓ in the j'th place, for $r_1 + \cdots + r_{i-1} + 1 \le j \le r_1 + \cdots r_i$ and $0 \le \ell \le r_i - 1$, and with 0's elsewhere. Let $E(G)$ be the matrix with columns $e_{1,0}, \ldots, e_{1,r_1-1}, \ldots, e_{m,0}, \ldots, e_{m,r_m-1}$, and, as before, let $E(S_k)$ be the matrix whose columns are the generators of $\mathscr{E}(S_k)$ over $\mathscr{O}^{S_k}$. Because $\mathscr{E}(S_k) \subset \mathscr{E}(G)$, $E(S_k)$ is divisible by $E(G)$: there is a $k \times k$ matrix Q with entries in $\mathscr{O}^G$, such that $E(S_k) = E(G)Q$. The first column of Q has a 1 in the first, $r_1 + 1$'st, ..., $r_1 + \cdots + r_{m-1} + 1$'st places and 0's elsewhere.

Observe that

$$\det\ E(S_k) = \mathrm{Vdm}(x_1, \ldots, x_k)$$

$$\det\ E(G) = \mathrm{Vdm}(x_1, \ldots, x_{r_1}) \times \cdots \times \mathrm{Vdm}(x_{r_1+\cdots+r_{m-1}+1}, \ldots, x_k).$$

Since $(u, x) \notin \Delta(\mathscr{P}')$ for any $\mathscr{P}'$ strictly finer than $\mathscr{P}$,

$$\det Q = \det\ E(S_k)\big(\det\ E(G)\big)^{-1}$$

does not vanish at (u, x). We have $\big(h(u, x_1), \ldots, h(u, x_k)\big)^t \in \mathscr{E}(S_k) \subset \mathscr{E}(G)$ and so

$$\big(h(u, x_1), \ldots, h(u, x_k)\big)^t = E(G)\beta = E(S_k)\alpha = E(G)Q\alpha \tag{9.45}$$

for suitable column vectors

$$\beta = (\beta_{1,0}, \ldots, \beta_{1,r_1-1}, \beta_{2,0}, \ldots, \beta_{2,r_2-1}, \ldots, \beta_{m,0}, \ldots, \beta_{m,r_m-1})^t$$

and $\alpha = (\alpha_0, \ldots, \alpha_{k-1})^t$. Notice that

$$(\beta_{i,1}, \ldots, \beta_{i,r_i-1}) = \big(\pi_i^{\mathscr{P}}\big)^*\big(I_{r_i}(h)\big).$$

Cancelling $E(G)$ in (9.45) we obtain $\beta = Q\alpha$. In order to obtain an equation relating the $\beta_{i\ell}$ to the generators $\alpha_1^k, \ldots, \alpha_{k-1}^k$ of $\mathrm{i}_k(h)$, we reduce the first column of Q to a 1 in the first place followed by zeros elsewhere, by subtracting the first row of Q from the $r_1 + 1$'st, $r_1 + r_2 + 1$'st, ..., $r_1 + \cdots + r_{m-1} + 1$'st rows. We obtain

$$\begin{aligned}&\big(\beta_{1,0}, \ldots, \beta_{1,r_1-1}, \beta_{2,0} - \beta_{1,0}, \beta_{2,1}, \ldots, \beta_{2,r_2-1}, \ldots, \beta_{m,0} - \beta_{1,0}, \beta_{m,1} \ldots, \beta_{m,r_m-1}\big)^t \\ &\qquad = \left(\begin{array}{c|c} 1 & \cdots \\ \hline & \\ 0 & Q' \\ & \end{array}\right)\begin{pmatrix} \alpha_0^k \\ \alpha_1^k \\ \vdots \\ \alpha_{k-1}^k \end{pmatrix}.\end{aligned} \tag{9.46}$$

Notice that by Lemma 9.4, $\beta_{i,0} = h \circ \mathrm{pr}_i^{\mathscr{P}} \mod (\beta_{i,1}, \ldots, \beta_{i,r_i-1})$, so $\beta_{i,0} - \beta_{1,0} = h \circ \mathrm{pr}_i^{\mathscr{P}} - h \circ \mathrm{pr}_1^{\mathscr{P}} \mod (\beta_{i,1}, \ldots, \beta_{i,r_i-1}, \beta_{1,1}, \ldots, \beta_{1,r_1-1})$. Since $\det Q' = \det Q \neq 0$ at (u, x), the ideal $\mathscr{A}_k(h)_{(u,x)}$ generated by the germs at (u, x) of $\alpha_1^k, \ldots, \alpha_{k-1}^k$ is equal to the ideal generated by the entries in rows $2, \ldots, k$ of the left-hand side of (9.46); that is, to $\big(\{\beta_{i,\ell} : 1 \leq \ell \leq r_i - 1, 1 \leq i \leq m\}\big) + \big(h \circ \mathrm{pr}_2 - h \circ \mathrm{pr}_1, \ldots, h \circ \mathrm{pr}_m - h \circ \mathrm{pr}_1\big)$. Finally, by (9.35),

$$\big(\beta_{i,1}, \ldots, \beta_{i,r_i-1}\big) + \mathrm{i}(\mathscr{P}) = \left(\left\{\frac{\partial^\ell h}{\partial y^\ell} : 1 \leq \ell \leq r_i - 1\right\}\right),$$

and this completes the proof. □

Corollary 9.3 *Let* $(u, x_1, \ldots, x_k) \in \Delta(\mathscr{P})$ *with* $\mathscr{P}$ *as in* (9.41). *Then* $(u, x_1, \ldots, x_k) \in V(\mathcal{A}_k(h))$ *if and only if* $h(u, x_1) = \cdots = h(u, x_k)$ *and the ideal* $(u, h)\,\mathscr{O}_{\mathbb{C}^n,(u,x^{(i)})}$ *satisfies*

$$\dim_{\mathbb{C}} \frac{\mathscr{O}_{\mathbb{C}^n,(u,x^{(i)})}}{(u, h)} \geq r_i.$$

□

9.5.2 Application to Multiple Points

Now we use this construction to find defining equations for the multiple point spaces $D^k(f)$ for a corank 1 map-germ $f : (\mathbb{C}^n, S) \to (\mathbb{C}^p, 0)$ of the form (9.27). Apply (9.32) with f_j in place of h. Define

$$\mathcal{A}_k(f) = \mathcal{A}_k(f_n) + \cdots + \mathcal{A}_k(f_p) \tag{9.47}$$

and, for any partition $\mathscr{P}$ of $\{1, \ldots, k\}$,

$$\mathcal{A}_k(f, \mathscr{P}) = \mathcal{A}_k(f_n, \mathscr{P}) + \cdots + \mathcal{A}_k(f_p, \mathscr{P}) = A_k(f) + \mathrm{i}(\mathscr{P}). \tag{9.48}$$

We wish to compare $D^k(f)$ and $V(\mathcal{A}_k(f))$. By (9.34), in $\mathbb{C}^{n-1} \times \mathbb{C}^{\langle k \rangle}$ (i.e. where the x_j are pairwise different) $V(\mathcal{A}_k(f))$ is just the set of points where $f(u, x_1) = \cdots = f(u, x_k)$. It follows that

$$V(\mathcal{A}_k(f)) \supseteq D^k_{\mathrm{cl}}(f), \tag{9.49}$$

since $D^k_{\mathrm{cl}}(f)$ is the closure of points $(u, x_1 \ldots, x_k) \in \mathbb{C}^{n-1} \times \mathbb{C}^{\langle k \rangle}$ such that $f(u, x_i) = f(u, x_j)$ for all i, j (see Definition 9.1). Proposition 9.10 describes the intersection of $V(\mathcal{A}_k(f))$ with each stratum $\Delta(\mathscr{P})$ of the diagonal and enables us to turn this inclusion into an equality when f is stable. Suppose $(u, x) \in \Delta(\mathscr{P})$, with $\mathscr{P}$ as in (9.41); then by 9.10, $(u, x) \in D^k(f)$ if and only if $f(u, x_1) = \cdots = f(u, x_k)$ and the partials $\partial^\ell f_j / \partial x^\ell$ vanish at $(u, x^{(i)})$ for $j = n, \ldots, p$ and for $i = 1, \ldots, r_i - 1$. This latter condition is equivalent to

$$\dim_{\mathbb{C}} Q(f)_{(u, x^{(i)})} \geq r_i;$$

if f is stable then by the Damon–Galligo Theorem 9.1 (or, rather, its complex version), this implies that $m(f)_{(u, x^{(i)})} \geq r_i$ also; now by Proposition 9.6, it follows that (u, x) is in the closure of the set of k-tuples of pairwise distinct points sharing the same image. That is, $(u, x) \in D^k_{\mathrm{cl}}(f)$. Taking into account (9.49), we have therefore shown that for stable map-germs,

$$V(\mathcal{A}_k(f)) = D^k_{\mathrm{cl}}(f). \tag{9.50}$$

It remains to show that this is an equality of schemes—i.e. that $\mathcal{A}_k(f) = \mathscr{I}^{\mathrm{s}}_k(f)$.

By Lemma 9.3, $\mathcal{A}_k(f) \subset \mathscr{I}^{\mathrm{s}}_k(f)$. The ideal $\mathcal{A}_k(f)$ is generated by $(k-1)(p-n+1)$ elements. So its codimension in $\mathbb{C}^{n-1+k}$ is less than or equal to $(k-1)(p-n+1)$, and thus its dimension is at least $n - 1 + k - (k-1)(p-n+1)$, which simplifies to $p - k(p-n)$. By Corollary 9.1, this is the dimension of $D^k(f)$ for $\mathscr{A}$-finite f. Thus

Proposition 9.11 *If the $\mathscr{A}$-finite germ $f : (\mathbb{C}^n, 0) \to (\mathbb{C}^p, 0)$ $n \leq p$ has corank 1, and if $p - k(p-n) > 0$, then $\mathcal{A}_k(f)$ is a complete intersection ideal.* □

Corollary 9.4 *With the hypotheses of the proposition,* $\mathcal{A}_k(f) = \mathcal{I}_k^s(f)$.

Proof Complete intersections are Cohen–Macaulay. The two ideals have the same set of zeros. Since $\mathcal{A}_k(f) \subseteq \mathcal{I}_k^s(f)$, it is only necessary to show that $\mathcal{A}_k(f)$ is generically radical, for then C.5 shows that $\mathcal{A}_k(f)$ is radical, and hence contains $\mathcal{I}_k(f)$. In $D^k_{\mathrm{cl}}(f) \cap X^{\langle k \rangle}$, both ideals are indeed radical: both are equal to $I_k(f)$, and since this subset is dense in $D^k(f)$, the conclusion follows. □

In what follows it will be useful to denote by $\alpha^k_\ell(f_j)$ the germ defined by (9.31) with f_j in place of h.

Proposition 9.12 *Let* $f: (\mathbb{C}^n, 0) \to (\mathbb{C}^p, 0)$ *be a corank 1 map-germ of the form (9.27), and let* $\dim_{\mathbb{C}} Q(f) = k$. *Then* f *is stable if and only if* $\mathcal{O}_{\mathbb{C}^{n-1+k},0}/\mathcal{A}_k(f)$ *is regular of dimension* $d_k(n, p)$.

Proof Define $A_k(f): (\mathbb{C}^{n-1} \times \mathbb{C}^k, 0) \to (\mathbb{C}^{(k-1)(p-n)}, 0)$ by

$$\left(\alpha^k_1(f_n), \ldots, \alpha^k_{k-1}(f_n), \ldots, \alpha^k_1(f_p), \ldots, \alpha^k_{k-1}(f_p)\right). \tag{9.51}$$

That is, $A_k(f)$ is the map whose components are the generators of $\mathcal{A}_k(f)$; $\mathcal{O}_{\mathbb{C}^{n-1+k},0}$ is regular of dimension $d_k(n, p)$ if and only if $A_k(f)$ is a submersion. As $A_k(f)$ is invariant with respect to the S_k action on $\mathbb{C}^{n-1} \times \mathbb{C}^k$, it follows by Lemma 9.5 below, applied to the derivative $d_0 A$, that $A_k(f)$ is a submersion at 0 if and only if its restriction to the fixed point set of the group action is a submersion. The fixed point set is of course the small diagonal where all the x_i are equal. Here, the ideal $\mathcal{A}_k(f)$ is equal to the ideal $\left(\{\partial^\ell f_j/\partial y^\ell : n \leq j \leq p, 1 \leq \ell \leq k-1\}\right)$, by (9.35), so A_k is a submersion at $(0, 0) \in \mathbb{C}^{n-1} \times \mathbb{C}^k$ if and only if the map $B_k(f)$ with components

$$\left(\frac{\partial f_n}{\partial x}, \ldots, \frac{\partial^{k-1} f_n}{\partial x^{k-1}}, \ldots, \frac{\partial f_p}{\partial x}, \ldots, \frac{\partial^{k-1} f_p}{\partial x^{k-1}}\right) \tag{9.52}$$

is a submersion at $(0, 0)$. We can view $B_k(f)$ as the composite of the jet extension map, $j^{k-1}f : (\mathbb{C}^n, 0) \to J^{k-1}(n, p)$, with a projection $J^{k-1}(n, p) \to \mathbb{C}^{(k-1)(p-n)}$. In Theorem 4.6 we gave Mather's characterisation of the stability of a map-germ f by the transversality of its jet extension map $j^\ell f$, for sufficiently high ℓ, to its contact class $\mathcal{K} j^\ell f(0)$ in $J^\ell(n, p)$. Given that f has the form (9.27), its contact class is determined by the smallest ℓ such that $\partial^\ell f_j/\partial y^\ell(0) \neq 0$ for some j. Let $LJ^{k-1}(n, p)$ be the subspace of jets of maps in the form (9.27). Since $\dim_{\mathbb{C}} Q(f) > k-1$, $\mathcal{K} j^{k-1} f(0) \cap LJ^{k-1}(n, p)$ consists of $(k-1)$-jets $j^k g(0)$ such that $\partial^\ell g_j/\partial y^\ell(0) = 0$ for $j = n, \ldots, p$ and $\ell = 1, \ldots, k-1$. So $B_k(f)$ is the composite of $j^{k-1}f$ with the projection $\pi : LJ^{k-1}(n, p) \to \mathbb{C}^{(k-1)(p-n)}$ for which $\pi^{-1}(0) = \left(\mathcal{K} j^{k-1} f(0)\right) \cap LJ^{k-1}(n, p)$. It follows that

$$B_k(f) \text{ is a submersion if and only if } j^{k-1} f \pitchfork \mathcal{K} j^{k-1} f(0).$$

Now because $\dim_{\mathbb{C}} Q(f) = k$, it follows that for any $\ell \geq k-1$, $\mathscr{K} j^\ell f(0) \cap LJ^\ell(n,p)$ is the preimage, under the projection $\pi^{\ell,k-1} : LJ^\ell(n,p) \to LJ^{k-1}(n,p)$, of $\mathscr{K} j^{k-1} f(0) \cap LJ^{k-1}(n,p)$. Since $j^{k-1}f = \pi^{\ell,k-1} \circ j^\ell f$, we deduce that

$$j^{k-1} f \pitchfork \mathscr{K} j^{k-1} f(0) \text{ if and only if } j^\ell f \pitchfork \mathscr{K} j^\ell f(0) \text{ for every } \ell \geq k-1.$$

This proves the proposition. □

Lemma 9.5 *Let the finite group G act linearly on the vector space V and let $H : V \to W$ be a linear map. Then H is an epimorphism if and only if $H| : \operatorname{Fix} G \to W$ is an epimorphism.*

Proof Here Fix G is the fixed point set of G. Because H is G-invariant,

$$H(v) = H\Big(\frac{1}{|G|}\sum_{g\in G} gv\Big).$$

The argument of H on the right-hand side lies in Fix G. □

Because any regular ring is reduced, we deduce immediately

Corollary 9.5 *If f is stable then $\mathcal{A}_k(f)$ defines $D^k(f)$ with its reduced structure.*

Recall from Sect. 9.2 that when f is merely finite and not necessarily stable, we define $D^k(f)$ by taking a stable unfolding F of f on parameter space P; $D^k(f)$ is then the (scheme-theoretic) fibre of $D^k_{\mathrm{cl}}(F)$ over $0 \in P$. Now we can improve on Corollary 9.4.

Proposition 9.13 *If f is a finite corank 1 map-germ of the form* (9.27), *then $\mathcal{A}_k(f)$ is equal to $\mathscr{I}_k(f)$ for all k, and to $\mathscr{I}^s_k(f)$ for all k for which $d_k(n,p) > 0$.*

Proof If f has form (9.27), it is possible to choose a stable unfolding F of the same form. Then $\mathcal{A}_k(F)$ is generated by the $\alpha^k_\ell(F_j)$ for $j = n, \ldots, p$. It is clear that $\alpha^k_\ell(f_j) = \alpha^k_\ell(F_j)|_{\mathbb{C}^n \times \{0\}}$, so $\mathcal{A}_k(f)$ is the restriction to $\mathbb{C}^n \times \{0\}$ of $\mathcal{A}_k(F)$. The theorem now follows by Corollary 9.5. □

The proof of Proposition 9.12 reveals the slightly surprising fact that for a finite corank 1 map-germ, if $D^k(f)$ is smooth at 0, then so is $D^\ell(f)$ for all $\ell < k$. For after a change of coordinates we may assume that f is of the form (9.27), and then

$$\begin{array}{ccccc}
D^k(f) \text{ smooth at } 0 & \Longleftrightarrow & A_k(f) \text{ is a submersion} & \Longleftrightarrow & B_k(f) \text{ is a submersion} \\
 & & & & \Big\Downarrow \\
D^\ell(f) \text{ smooth at } 0 & \Longleftrightarrow & A_\ell(f) \text{ is a submersion} & \Longleftrightarrow & B_\ell(f) \text{ is a submersion}
\end{array} \tag{9.53}$$

The opposite implication is easily seen to be false.

Example 9.3 Following the procedure of Sect. 4.3, the stable unfolding F of $x \mapsto (x^{n+1}, 0)$ is given by

$$F(u_1, \ldots, u_{n-1}, v_1, \ldots, v_n, x) = \left(u, v, x^{n+1} + \sum_i u_i x^{n-i}, \sum_j v_j x^{n+1-j}\right).$$

Restricting to $\{v_n = 0\}$ in source and target, and adding x^{n+2} to the last component, we obtain a germ $f : (\mathbb{C}^{2n-1}, 0) \to (\mathbb{C}^{2n}, 0)$ for which

(a) $D^k(f)$ is smooth for $2 \le k \le n$, has an A_1 singularity for $k = n+1$, and is empty for $k > n+1$.
(b) f has $\mathscr{A}_e$-codimension 1.

◊

Example 9.4 Let $f(u, x) = (u, x^3, x^5 + ux)$. This is the simple singularity of type H_2 described in [Mon85]. Here $\mathcal{A}_2(f)$ is the ideal generated by

$$\frac{\begin{vmatrix} 1 & x_1^3 \\ 1 & x_2^3 \end{vmatrix}}{\text{vdM}}, \quad \frac{\begin{vmatrix} 1 & x_1^5 + ux_1 \\ 1 & x_2^5 + ux_2 \end{vmatrix}}{\text{vdM}} \tag{9.54}$$

$$= (x_1^2 + x_1x_2 + x_2^2, \quad u + x_1^4 + x_1^3x_2 + x_1^2x_2^2 + x_1x_2^3 + x_2^4)$$

so $D^2(f)$ is a plane curve singularity, of type A_1; $\mathcal{A}_3(f)$ is the ideal generated by

$$\frac{\begin{vmatrix} 1 & x_1^3 & x_1^2 \\ 1 & x_2^3 & x_2^2 \\ 1 & x_3^3 & x_3^2 \end{vmatrix}}{\text{vdM}}, \quad \frac{\begin{vmatrix} 1 & x_1 & x_1^3 \\ 1 & x_2 & x_2^3 \\ 1 & x_3 & x_3^3 \end{vmatrix}}{\text{vdM}}, \quad \frac{\begin{vmatrix} 1 & x_1^5 + ux_1 & x_1^2 \\ 1 & x_2^5 + ux_2 & x_2^2 \\ 1 & x_3^5 + ux_3 & x_3^2 \end{vmatrix}}{\text{vdM}}, \quad \frac{\begin{vmatrix} 1 & x_1 & x_1^5 + ux_1 \\ 1 & x_2 & x_2^5 + ux_2 \\ 1 & x_3 & x_3^5 + ux_3 \end{vmatrix}}{\text{vdM}},$$

$$= (P_2(x_1, x_2, x_3), \quad x_1 + x_2 + x_3, \quad x + P_4(x_1, x_2, x_3), \quad P_3(x_1, x_2, x_3),$$

where each of the P_i is a symmetric polynomial of the indicated degree. So $D^3(f)$ is a planar fat point singularity of length 6. ◊

The next proposition is close in appearance to (9.53), and is proved by a similar strategy. To put it into effect we need first to construct a different set of generators for $\mathcal{A}_k(f)$. Observe that for any function h,

$$\alpha_1^2(h) = \frac{\begin{vmatrix} 1 & h(u, x_1) \\ 1 & h(u, x_2) \end{vmatrix}}{\text{vdM}} = \frac{h(u, x_2) - h(u, x_1)}{x_2 - x_1} \tag{9.55}$$

and

$$\alpha_2^3(h) = \frac{\begin{vmatrix} 1 & x_1 & h(u,x_1) \\ 1 & x_2 & h(u,x_2) \\ 1 & x_3 & h(u,x_3) \end{vmatrix}}{\text{vdM}} = \frac{\begin{vmatrix} 1 & x_1 & h(u,x_1) \\ 0 & x_2 - x_1 & h(u,x_2) - h(u,x_1) \\ 0 & x_3 - x_1 & h(u,x_3) - h(u,x_1) \end{vmatrix}}{(x_3 - x_2)(x_3 - x_1)(x_2 - x_1)} = \frac{\begin{vmatrix} 1 & \frac{h(u,x_2)-h(u,x_1)}{x_2-x_1} \\ 1 & \frac{h(u,x_3)-h(u,x_1)}{x_3-x_1} \end{vmatrix}}{x_3 - x_2}$$

$$= \frac{(h(u,x_3) - h(u,x_1))/(x_3 - x_1) - (h(u,x_2) - h(u,x_1))/(x_2 - x_1)}{x_3 - x_2}. \tag{9.56}$$

The right-hand sides of (9.55) and (9.56) are the first and second *iterated divided differences* of h; we denote them by $d_{(1)}h$ and $d_{(2)}h$. Define the ℓ'th iterated divided difference of h inductively, by

$$d_{(\ell)}h(u, x_1, \ldots, x_{\ell+1}) = \frac{d_{(\ell-1)}h(u, x_1, \ldots, x_{\ell-1}, x_{\ell+1}) - d_{(\ell-1)}h(u, x_1, \ldots, x_{\ell-1}, x_\ell)}{x_{\ell+1} - x_\ell}. \tag{9.57}$$

In fact $d_{(\ell)}h(u, x_1, \ldots, x_{\ell+1})$ is symmetric in $x_1, \ldots, x_{\ell+1}$, and moreover the slightly unnatural selection of indices in the inductive definition (9.57), with a distinguished role being given to x_ℓ and $x_{\ell+1}$, can be replaced by any other. This will become clear in the following proof.

Lemma 9.6

$$\mathcal{A}_k(h) = \Big(d_{(1)}h(u, x_1, x_2),\ d_{(2)}h(u, x_1, x_2, x_3), \ldots, d_{(k-1)}h(u, x_1, \ldots, x_k)\Big)$$

Proof By subtracting row i_1 of (9.30) from row i_2 and dividing by $x_{i_1} - x_{i_2}$ we obtain

$$d_{(1)}h(u, x_{i_1}, x_{i_2}) = \alpha_1^k + \alpha_2^k(x_{i_1} + x_{i_2}) \cdots + \alpha_{k-1}^k \frac{x_{i_1}^{k-1} - x_{i_2}^{k-1}}{x_{i_1} - x_{i_2}}; \tag{9.58}$$

so the divided difference $d_{(1)}(h)(u, x_{i_1}, x_{i_2})$ lies in $\mathcal{A}_k(h)$. Now subtract (9.58), with i_3 in place of i_2, from (9.58); both sides can be divided by $x_{i_2} - x_{i_3}$, giving

$$d_{(2)}h(u, x_{i_1}, x_{i_2}, x_{i_3}) = \alpha_2^k + \alpha_3^k(x_{i_1} + x_{i_2} + x_{i_3}) + \cdots + \alpha_{k-1}^k P(x_{i_1}, x_{i_2}, x_{i_3}), \tag{9.59}$$

where P is a symmetric polynomial of degree $k - 3$. Continuing in this way, we see the ideal $\big(\{d_{(\ell)}h(u, x_{i_1}, \ldots, x_{i_{\ell+1}}), 1 \le \ell \le k-1\}\big)$ of iterated divided differences is contained in $\mathcal{A}_k(h)$. Moreover, after $k - 1$ subtractions and divisions, we are left with the equation

$$d_{(k-1)}h(u, x_{i_1}, \ldots, x_{i_k}) = \alpha_{k-1}^k$$

(and this proves the symmetry of the $d_{(\ell)}h$). So α^k_{k-1} is in the ideal of iterated divided differences. By downward induction on ℓ, we see that the ideals are in fact equal: for any ordering $i_1, \ldots, i_k$ of $1, \ldots, k$, we have

$$\big(\{d_{(\ell)}h(u, x_{i_1}, \ldots, x_{i_{\ell+1}}), 1 \le \ell \le k-1\}\big) = \mathcal{A}_k(h).$$

□

Proposition 9.14 *Suppose that $D^k(f)$ is smooth of dimension $d_k(n, p)$ at the point $(u, x_1, \ldots, x_k)$. Then for every $\ell < k$ and every projection $\varepsilon : D^k(f) \to D^\ell(f)$ (composite of the projections $\varepsilon^{i,j}$ described in Sect. 9.2.1), $D^\ell(f)$ is smooth of dimension $d_\ell(n, p)$ at $\varepsilon(u, x_1, \ldots, x_k)$.*

Proof This follows easily from the preceding lemma. After permuting coordinates (which does not affect smoothness or dimension), we may assume that $\varepsilon(u, x_1, \ldots, x_k) = (u, x_1, \ldots, x_\ell)$. Let $C_k(f)$ be the map $\mathbb{C}^{n-1+k} \to \mathbb{C}^{(k-1)(p-n+1)}$ with components

$$\big(d_{(1)} f_j(u, x_1, x_2),\ d_{(2)} f_j(u, x_1, x_2, x_3),\ \ldots,\ d_{(k-1)} f_j(u, x, \ldots, x_k)\big)_{j=n,\ldots,p}.$$

Then $D^k(f)$ is smooth of dimension $d_k(n, p)$ at $(u, x_1, \ldots, x_k)$ if and only if $C^k(f)$ is a submersion at $(u, x_1, \ldots, x_k)$. This implies that $C^\ell(f)$ is a submersion at $(u, x_1, \ldots, x_\ell)$, which is equivalent to that $D^\ell(f)$ is smooth of dimension $d_\ell(n, p)$ at $(u, x_1, \ldots, x_\ell)$. □

Suppose $f(u, x) = (u, f_n(u, x), \ldots, f_p(u, x))$ and $\mathscr{P}$ is a partition of $\{1, \ldots, k\}$ as in (9.41), of length m. We define $D^{\mathscr{P}}(f) = D^k(f) \cap \overline{\Delta(\mathscr{P})} = V\big(\mathrm{i}_k(f) + \mathrm{i}(\mathscr{P})\big)$. As $D^{\mathscr{P}}(f)$ is the zero locus, in $\overline{\Delta(\mathscr{P})}$, of an ideal generated by $(k-1)(n-p+1)$ elements, its dimension is at least

$$\dim \Delta(\mathscr{P}) - (k-1)(p-n+1) = n - 1 + m - (k-1)(p-n+1). \tag{9.60}$$

We denote this number by $d_{\mathscr{P}}(n, p)$.

Theorem 9.4 *Let $f : (\mathbb{C}^n, 0) \to (\mathbb{C}^p, 0)$ be a finite corank 1 map-germ and let $\ell = \dim_{\mathbb{C}} Q(f)$. Then the following are equivalent:*

1. *f is stable.*
2. *$D^k(f)$ is smooth of dimension $d_k(n, p)$, or empty, for all k such that $d_k(n, p) \ge 0$, and empty if $d_k(n, p) < 0$.*
3. *$D^\ell(f)$ is smooth of dimension $d_\ell(n, p)$.*
4. *$D^{\mathscr{P}}(f)$ is smooth of dimension $d_{\mathscr{P}}(n, p)$, or empty, for every k and $\mathscr{P}$ such that $d_{\mathscr{P}}(n, p) \ge 0$, and empty if $d_{\mathscr{P}}(n, p) < 0$.*

Proof 1⇔2 is Proposition 9.12. To see that 2⇔3, observe that by (9.53), statement 3 is simply a way of stating Proposition 9.12 without mentioning the value of $\dim_{\mathbb{C}} Q(f)$. For suppose that $\dim Q(f) = k$. We know by Proposition 9.12 that f

is stable if and only if $A_\ell(f)$ is a submersion, which holds, provided $d_k(n, p) \geq 0$, if and only if $D^\ell(f)$ is smooth of dimension $d_\ell(n, p)$. In this case, (9.53) shows that for all $k \leq \ell$, $D^k(f)$ is smooth of dimension $d_k(n, p)$. And for $k > \ell$, $D^k(f) = \emptyset$, since $A_k(f)(0) \neq 0$. If $d_\ell(n, p) < 0$, then $A_\ell(f)$ cannot be a submersion so f cannot be stable—unless $A_\ell(f)(0) \neq 0$, i.e. unless $D^\ell(f) = \emptyset$. But the fact that $\dim Q(f) = \ell$ implies, by the Damon–Galligo Theorem, that $D^\ell(f) \neq \emptyset$. So taken together, the facts that $d_\ell(n, p) < 0$ and that $\dim Q(f) = \ell$ imply that f is not stable.

(3⇔4): This is just the argument using Lemma 9.5 given in the proof of Proposition 9.10. Let $G = S_{r_1} \times \cdots \times S_{r_m}$. Then $V(\mathrm{i}(\mathscr{P}))$ is the fixed set of G. The map $A_k(f)$ is G-invariant, so $A_k(f)$ is a submersion if and only if $A_k(f)|_{V(\mathrm{d}(\mathscr{P}))}$ is a submersion. That is, $D^k(f)$ is smooth of dimension $d_k(n, p)$ if and only if $D^{\mathscr{P}}(f)$ is smooth of dimension $d_{\mathscr{P}}(n, p)$.

□

For a map-germ $f : (\mathbb{C}^n, x) \to (\mathbb{C}^p, y)$ let $\delta(f)_x = \dim_{\mathbb{F}} \mathscr{O}_{\mathbb{F}^n,x} / f^* \mathfrak{m}_{\mathbb{F}^p,y} \mathscr{O}_{\mathbb{F}^n,x}$.

Lemma 9.7 *Let $f(u, x) = \big(u, f_n(u, x), \ldots, f_p(u, x)\big)$, and suppose that*

$$f\left(u, x^{(1)}\right) = \cdots = f\left(u, x^{(m)}\right) =: z$$

with $x^{(i)} \neq x^{(j)}$ if $i \neq j$. Suppose that $\delta(f)_{(u,x^{(i)}))} = r_i$. Let $k = r_1 + \cdots + r_m$ and let $\mathscr{P} = (r_1, \ldots, r_m)$. Then if $\mathbf{x} = (x^{(1)}, \ldots, x^{(1)}, \ldots, x^{(m)}, \ldots, x^{(m)})$ (where $x^{(m)}$ is iterated r_i times) and $S = \{(u, x^{(1)}), \ldots, (u, x^{(m)})\}$, the following are equivalent:

1. *The multi-germ $f\colon (\mathbb{C}^n, S) \to (\mathbb{C}^p, z)$ is stable;*
2. *the map $A_k(f)$ defining $D^k(f)$ is a submersion at $(u, \mathbf{x})$;*
3. *the map $(A_k(f), I(\mathscr{P}))$ defining $D^{\mathscr{P}}(f)$ is a submersion at $(u, \mathbf{x})$.*

Proof (2⇔3) This is just an application of Lemma 9.5. If $G = S_{r_1} \times \cdots \times S_{r_m}$, then $A_k(f)$ is G invariant, so $A_k(f)$ is a submersion at $(u, \mathbf{x})$ if and only if its restriction to the fixed point set of G, $\Delta(\mathscr{P})$, is a submersion. This is the case if and only if the map $(A_k(f), I(\mathscr{P}))$ is a submersion.

(1⇔3) By Lemma 9.5, (3) is equivalent to the restriction to $\mathrm{d}(\mathscr{P})$ of $A_k(f)$ being a submersion. As in the proof of Proposition 9.12, we let $B_r(f)$ be the map with components

$$\frac{\partial^\ell f_j}{\partial x^\ell} \quad \text{for } n \leq p, \quad 1 \leq \ell \leq r - 1,$$

and we let $C^{\mathscr{P}}(f)$ be the map with components

$$f_j \circ \pi_i^{\mathscr{P}} - f_j \circ \pi_1^{\mathscr{P}} \quad \text{for } n \leq j \leq p \text{ and } 2 \leq i \leq m.$$

Then by Proposition 9.10, $A_k(f)|_{\mathrm{d}(\mathscr{P})}$ is a submersion if and only if the map $\mathrm{d}(\mathscr{P}) \to \mathbb{C}^{(l-1)(n-p+1)}$ defined by

$$(B_{r_1} \circ \pi_1^{\mathscr{P}}, \ldots, B_{r_m} \circ \pi_m^{\mathscr{P}}, C^{\mathscr{P}}) \tag{9.61}$$

is a submersion. As we saw in the proof of Proposition 9.12, submersiveness of $B_{r_i} \circ \pi_i^{\mathscr{P}}$ is equivalent to stability of the mono-germ $f: (\mathbb{C}^n, (u, x^{(i)}) \to (\mathbb{C}^p, z)$. The additional information contained in the submersiveness of (9.61) is that the images under f of the analytic strata of these germs meet in general position at z. Note that our embedding of $D^k(f)$ in $\mathbb{C}^{n-1} \times \mathbb{C}^k$ already takes into account the equality of the first $n-1$ components of these images. In any case, this general position, together with the stability of the mono-germs, is equivalent to the stability of the multi-germ $f: (\mathbb{C}^n, S) \to (\mathbb{C}^p, z)$. □

Theorem 9.5 *Let $f: (\mathbb{C}^n, 0) \to (\mathbb{C}^p, 0)$ be a finite corank 1 map-germ. Then the following are equivalent:*

1. *f is $\mathscr{A}$-finite.*
2. *$D^k(f)$ is an isolated complete intersection singularity (ICIS) of dimension $d_k(n, p)$, or empty, for all k such that $d_k(n, p) \geq 0$, and consists at most of $\{0\}$ if $d_k(n, p) < 0$.*
3. *$D^{\mathscr{P}}(f)$ is an ICIS of dimension $d_{\mathscr{P}}(n, p)$, or empty, for every ℓ and $\mathscr{P}$ such that $d_{\mathscr{P}}(n, p) \geq 0$, and consists at most of $\{0\}$ if $d_{\mathscr{P}}(n, p) < 0$.*

Proof (2⇔3) is just an application of Lemma 9.5.

(1⇔2). If f is $\mathscr{A}$-finite, there is a representative $\tilde{f}: U \to V$ with the property that for every $v \in V \setminus \{0\}$, the multi-germ $\tilde{f}: (U, \tilde{f}^{-1}(v)) \to (V, v)$, which we denote by $\tilde{f}_v$, is stable. After shrinking U and V if necessary, we may choose linearly adapted coordinates so $\tilde{f}$ takes the form (9.27) and each multiple point space $D^\ell(\tilde{f})$ embeds in $\mathbb{C}^{n-1} \times \mathbb{C}^\ell$. Suppose $(u, x_1, \ldots, x_\ell) \in D^\ell(\tilde{f}) \setminus \{0\}$, and let $v = f(u, x_j)$. Let $\tilde{f}^{-1}(v) = \{x^{(1)}, \ldots, x^{(m)}\}$, let $\delta(\tilde{f})_{(u, x^{(i)})} = r_i$, and let $k = \sum_i r_i$. Clearly $k \geq \ell$. We have

$$\big(u, \underbrace{x^{(1)}, \ldots, x^{(1)}}_{r_1 \text{ times}}, \ldots, \underbrace{x^{(m)}, \ldots, x^{(m)}}_{r_m \text{ times}}\big) \in D^k(\tilde{f}).$$

Denote this point by $(u, \mathbf{x})$. After permuting the x_j, $(u, x_1, \ldots, x_\ell) = \varepsilon(u, \mathbf{x})$ for some projection ε as in Proposition 9.14. So by Proposition 9.14 it is enough to show that $D^k(f)$ is smooth at $(u, \mathbf{x})$. But this is what we proved in Lemma 9.7.

The converse holds by the same chain of reasoning. If all the $D^k(f)$ have isolated singularities at 0 and are of the right dimension $d_k(n, p)$, then we can choose a representative of f whose multi-germs outside 0 are all stable. □

Exercises for Sect. 9.5

1 Prove the assertions of Example 9.3.

2 Find equations for $D^2(f)$ and $D^3(f)$ where $f(u, x) = (u, x^3, ux + x^{3k-1})$ (H_k in the classification of [Mon85]).

3 If $f(u,x) = (u, x^3, x^5 + ux)$, one calculates (e.g. by eliminating x_2 from the Eqs. (9.54) for $D^2(f)$) that the double point set in the source, $D_1^2(f)$, is the curve singularity of type A_7, with equation $u^2 + ux^4 + x^8 = 0$. This curve has two branches, $C_1 := \{(u,x) : x - \omega x^4 = 0\}$ and $C_2 = \{(u,x) : u - \omega^2 x^4 = 0\}$, where $\omega = \frac{-1+i\sqrt{3}}{2}$ is a cube root of unity. It is also easy to check that f has no genuine triple points: for each $(u,x) \in D_1^2(f) \smallsetminus \{(0,0)\}$ there exists a unique $(u',x') \neq (u,x)$ such that $f(u',x') = f(u,x)$. This determines an involution ι on $D_1^2(f) \smallsetminus \{(0,0)\}$, which evidently extends to $D_1^2(f)$ by setting $\iota(0,0) = (0,0)$.

To Do

(i) Find an expression for ι in coordinates u, x. Hint: Clearly $u' = u$; use the equation $x^2 + xx' + x'^2$ to find the relation between x' and x.
(ii) Show that ι is not holomorphic—i.e. is not the restriction to $D_1^2(f)$ of any holomorphic map-germ $(\mathbb{C}^2,0) \to (\mathbb{C}^2,0)$.

4 Calculate the number of transverse double points appearing in the image of a stable perturbation of the germ

$$f(x_1,x_2) = (x_1, x_1x_2 + x_2^3, x_1x_2^2 + x_2^5, x_2^4).$$

Recall that we calculated the $\mathscr{A}_e$-codimension of this germ in Example 4.2. In [KPR07] it is shown that there exists a real stable perturbation in which all of the double points are real.

9.6 Bifurcation Sets for Germs of Corank 1

In Sect. 8.6 we showed that the bifurcation set in the base of a versal unfolding of a germ $(\mathbb{C}^n, S) \to (\mathbb{C}^p, 0)$ with $n \geq p$ is a hypersurface. Crucial for the analysis leading to this result is the fact that the set of liftable vector fields on the target is a free module. This is no longer true when $n < p$, and the method of proof used when $n \geq p$ yields nothing when $n < p$. Here we use a quite different approach, based on the characterisation of stability in terms of non-singularity of the multiple point spaces. We show that when $p = n + 1$ and f has corank 1, $\mathscr{B}(F)$ is a hypersurface, though not in general a free divisor, and when $p > n + 1$ then $\mathscr{B}(F)$ may have components of codimension 2 or more, as well as hypersurface components. In all of this section we assume $n < p$, unless explicitly stated otherwise.

First, bifurcation sets need not be hypersurfaces.

Example 9.5 Let $f\colon (\mathbb{C}^2, S) \to (\mathbb{C}^4, 0)$ be a tri-germ of immersions, with each pair of branches meeting transversely. This has $\mathscr{A}_e$ codimension 2 (Exercise 3.4.9). Let $F : (\mathbb{C}^2 \times \mathbb{C}^2, S \times \{0\}) \to (\mathbb{C}^4 \times \mathbb{C}^2, (0,0))$ be an $\mathscr{A}_e$-miniversal unfolding. By the transversality assumption, $D^2(f)$ is smooth, and it follows that $D^2(F)$ is smooth also, and moreover that the projection $D^2(F) \to \mathbb{C}^2$ is a submersion. The bifurcation set consists only of the point 0—it is the only value of the parameters for which f_u has a triple point. This singularity can be understood as a higher-dimensional version of Reidemeister III: here two immersed planes P_1 and P_2 meet transversely; a third plane, P_{u_1,u_2}, transverse to each of them, moves in a 2-parameter family and for a single value of the parameter passes through their point of intersection. ◊

Suppose that $f\colon(\mathbb{C}^n,S)\to(\mathbb{C}^p,0)$ is a multi-germ of corank ≤ 1, i.e. each of its branches is either an immersion or a germ of corank 1. An $\mathscr{A}_e$- versal unfolding $F:\big(\mathbb{C}^n\times\mathbb{C}^d, S\times\{0\}\big)\to\big(\mathbb{C}^p\times\mathbb{C}^d,(0,0)\big)$ of f is in particular a stable germ, and also of corank ≤ 1. It follows by Theorem 9.4 that $D^k(F)$ is either smooth of dimension $d_k(n+d,p+d)$ or empty. For each k, let $\pi_k : D^k(F)\to\mathbb{C}^d$ be the projection. For each k-tuple $(x^{(1)},u^{(1)}),\ldots,(x^{(k)},u^{(k)})\in D^k(F)$, we have $u^{(1)}=\cdots=u^{(k)}$ so π_k is well defined. Let $\mathscr{C}_k$ and $\mathscr{B}_k$ be the set of critical points of π_k and the set of critical values, respectively. Then by Theorem 9.4 again, f_u is stable if and only if $u\notin\mathscr{B}_k$ for any k. Thus

$$\mathscr{B}(F)=\bigcup_{k\geq 2}\mathscr{B}_k.$$

Each projection $\pi_k| : \mathscr{C}_k \to \mathbb{C}^d$ is finite, for the original germ f that we are unfolding is $\mathscr{A}$-finite, and so by the Mather–Gaffney criterion, Theorem 4.5, $\pi_k^{-1}(0)\cap\mathscr{C}_k$ consists of at most a single point. Thus dim $\mathscr{B}_k=$ dim $\mathscr{C}_k$. There are two cases to consider: $d_k(n,p)\geq 0$ and $d_k(n,p)<0$.

1. We have $d_k(n+d,p+d)=d+d_k(n,p)$. If $d_k(n,p)\geq 0$ then dim $D^k(F)\geq d$. By Sard's theorem, $\mathscr{B}_k$ is nowhere dense in $\mathbb{C}^d$, and if we choose a proper representative then it follows that $\mathscr{B}_k$ is a proper analytic subset of $\mathbb{C}^d$, and thus of dimension $\leq d-1$. On the other hand, dim $\mathscr{B}_k\geq d-1$, by the standard argument: using local coordinate on $D^k(F)$, $\mathscr{C}_k$ is the zero set of the maximal minors of the $d\times(d+d_k(n,p))$ Jacobian matrix of $d\pi_k$; if this space is not empty, it has codimension no greater than $d_k(n,p)-1$ —see Theorem C.6. So its dimension must be at least $d-1$. Combined with the previous opposite inequality, this shows that dim $\mathscr{B}_k=d-1$.
2. If $d_k(n,p)<0$ then $\mathscr{C}_k=D^k(F)$ and $\mathscr{B}_k=\pi_k(D^k(F))$. Let $k_0=k_0(n,p)$ be the least value of k for which $d_k(n,p)<0$. For $k>k_0$, $\mathscr{B}_k=\pi_k(D^k(F))\subset\pi_{k_0}(D^{k_0}(F))=\mathscr{B}_{k_0}$.

It follows that

$$\mathscr{B}(F)=\mathscr{B}_2\cup\cdots\cup\mathscr{B}_{k_0-1}\cup\pi_{k_0}(D^{k_0}(F)). \tag{9.62}$$

Proposition 9.15 *Let $f\colon(\mathbb{C}^n,0)\to(\mathbb{C}^{n+r},0)$ be an $\mathscr{A}$-finite germ of corank 1, and let $F\colon(\mathbb{C}^n\times\mathbb{C}^d,(0,0)\to(\mathbb{C}^{n+r}\times\mathbb{C}^d,(0,0))$ be an $\mathscr{A}_e$-versal unfolding.*

(i) If $r = 1$, *then* $\pi_{k_0}(D^{k_0}(F))$ *is itself a hypersurface, and hence* $\mathscr{B}(F)$ *is a hypersurface.*

(ii) Suppose that $r > 1$ *and* $n \equiv s \mod r$ *with* $0 \leq s \leq r-1$. *If* $s = r-1$, *then* $\mathscr{B}(F)$ *is a hypersurface. If* $s < r-1$ *and* $D^{k_0}(F) \neq \emptyset$, *then* $\pi_{k_0}(D^{k_0}(F))$ *has codimension* > 1 *in* $\mathbb{C}^d$ *and so* $\mathscr{B}(F)$ *has component(s) of codimension greater than 1.*

Proof All that remains to do is some arithmetic. As $d_k(n, n+r) = n - (k-1)r$, k_0 is the least integer greater than $(n/r) + 1$.

If $r = 1$, then $k_0 = n + 2$, and dim $D^{k_0}(F) = d - 1$, so $\pi_{k_0}(D^{k_0}(F)$ is a hypersurface.

Suppose $r > 1$. Let $n = qr+s$, with $q, s \in \mathbb{N}$ and $0 \leq s < r$. Then $n-(k-1)r = (q-k+1)r + s$. This is less than 0 if $k > q+1$. So the least integer k for which dim $D^k(F) < d$ is $k_0 := q+2$. Then $n - (k_0 - 1)r = s - r$. If $s - r = -1$, then $\pi_{k_0}(D^{k_0}(F)$, if not empty, is a hypersurface, and hence so is $\mathscr{B}(F)$. If $s - r < -1$, then $\pi_{k_0}(D^{k_0}(F))$, if not empty, has codimension greater than 1, and $\mathscr{B}(F)$ is not a hypersurface. □

The description (9.62) of $\mathscr{B}(F)$ does not give us complete information about the irreducible components of $\mathscr{B}(F)$. In general $\mathscr{B}_k$ has more than one component. Each corresponds to a distinct $\mathscr{A}_e$- codimension 1 singularity, and these are indexed by the partition $\mathscr{P}$ of k determined by the multiplicities of the distinct branches, which sum to k, by the Damon–Galligo Theorem 9.1.

Example 9.6 There are five codimension 1 singularities of maps from 2-space to 3-space. Figures 5.5 and 8.2 in Sect. 8.1 show the images of their stable perturbations. They are:

Name		Which $\mathscr{B}_k$?	$\mathscr{P}$
S_1	$(u, x) \mapsto (u, x^2, x^3 + u^2 x)$	$\mathscr{B}_2$	(2)
Immersions separated by A_1	$\begin{cases} (x_1, x_2) \mapsto (x_1, x_2, 0) \\ (x_1, x_2) \mapsto (x_1, x_2, x_1^2 \pm x_2^2) \end{cases}$	$\mathscr{B}_2$	(1, 1)
Birth of two triple points	$\begin{cases} (x_1, x_2) \mapsto (x_1, x_2, 0) \\ (x_1, x_2) \mapsto (0, x_1, x_2) \\ (x_1, x_2) \mapsto (x_1 + x_2^2, x_2, x_1 + x_2^2) \end{cases}$	$\mathscr{B}_3$	(1, 1, 1)
Umbrella + plane	$\begin{cases} (x_1, x_2) \mapsto (x_1, x_2^2, x_1 x_2) \\ (x_1, x_2) \mapsto (x_1, x_2, x_2) \end{cases}$	$\mathscr{B}_3$	(2, 1)
Quadruple point	$\begin{cases} (x_1, x_2) \mapsto (0, x_1, x_2) \\ (x_1, x_2) \mapsto (x_1, 0, x_2) \\ (x_1, x_2) \mapsto (x_1, x_2, 0) \\ (x_1, x_2) \mapsto (x_1, x_2 - x_1, -x_2) \end{cases}$	$\mathscr{B}_4$	(1, 1, 1, 1)

(9.63)

◊

Remark 9.3 We do not know whether the bifurcation set is a hypersurface when $f\colon (\mathbb{C}^n, S) \to (\mathbb{C}^p, 0)$ has corank greater than 1, even in the case $p = n + 1$. It does not seem possible to deduce this from Proposition 9.15, even though "corank 1 singularities are dense".

Example 9.7 The lowest dimension stable corank 2 germ is the germ F of Example 9.1,

$$F(x_1, x_2, u_1, u_2, u_3, u_4) = (x_1^2+u_1x_2, x_1x_2+u_2x_1+u_3x_2, x_2^2+u_4x_1, u_1, u_2, u_{,}u_4).$$

If $i\colon (\mathbb{C}^6, 0) \to (\mathbb{C}^7, 0)$ is a generic linear injection, then $i^*(F)$, the map obtained from F by transverse pull-back using i, as described in Sect. 8.5, has $\mathscr{A}_e$-codimension 1. For example, we restrict to the subspace of the source and target where $u_1 = u_2$, obtaining the map

$$g(x_1, x_2, v_1, v_2, v_3) = (x_1^2 + v_1x_2, x_1x_2 + v_2(x_1 + x_2), x_2^3 + v_3x_1, v_1, v_2, v_3).$$

This has versal unfolding

$$G(x, v, u) = (x_1^2+v_1x_2, x_1x_2+v_2(x_1+x_2)+u(x_1-x_2), x_2^3+v_3x_1, v_1, v_2, v_3, u).$$

A calculation shows that $D^2(G)$ and $D^3(G)$ have isolated singularity at 0, and their projections to the parameter space $\mathbb{C}$ are non-singular outside 0. ♢

In this example, the bifurcation set is just the single point $0 \in \mathbb{C}$, and thus does indeed have codimension 1. Since there are no corank 1 bifurcations in the vicinity, it seems that our understanding of the corank 1 case has no bearing on what happens here. ♢

Exercises for Sect. 9.6

1 Show that there is no $\mathscr{A}_e$-codimension 1 germ $(\mathbb{F}^3, 0) \to \mathbb{F}^4, 0)$ or $(\mathbb{F}^4, 0) \to (\mathbb{F}^5, 0)$ of corank > 1 (see Exercise 8.5.11).

2 By considering the distinct partitions of k for $k \leq k_0$ (where k_0 is defined just before Proposition 9.15), give geometrical descriptions of the $\mathscr{A}_e$-codimension 1 singularities of maps from 3-space to 4-space.

9.7 Disentangling a Singularity: The Geometry of a Stable Perturbation

The term "disentanglement", for the image of a stable perturbation f_t of an $\mathscr{A}$-finite map-germ f, was introduced by de Jong and van Straten in [dJvS91]. Our description of the multiple point spaces $D^k(f)$ and $D^k(f_t)$ contributes to understanding the disentanglement.

As a more or less immediate corollary of Theorems 9.4 and 9.5, we have

Corollary 9.6 *Suppose f_{t_0} is a stable perturbation of an $\mathscr{A}$-finite germ $f\colon (\mathbb{C}^n, 0) \to (\mathbb{C}^p, 0)$ of corank 1, with $n \le p$. Then for each k for which the expected dimension $d_k(n, p)$ of $D^k(f)$ is non-negative, $D^k(f_{t_0})$ is a Milnor fibre of the ICIS $D^k(f)$, and if $d_k(n, p)$ is negative, then $D^k(f_{t_0}) = \emptyset$.*

Proof Since f is $\mathscr{A}$-finite, its instability is concentrated at 0 (see Theorem 4.5). Thus, near the boundary of the domain U of a suitable representative, f is stable, and all the strata of the stratification by stable type (the "strata of f") are transverse to the boundary ∂U. The multiple point space $D^k(f)$ is a complete intersection in U^k, with isolated singularity at $(0, \ldots, 0)$, by Theorem 9.5, and it follows from the transversality of the strata of f to ∂U that $D^k(f)$ is transverse to ∂U^k. If $f_{t_0} : U \to V$ is a stable perturbation of f, then as t varies from 0 to t_0, transversality of $D^k(f_t)$ to ∂U^k is not lost. Since $D^k(f_{t_0})$ is smooth, again by Theorem 9.5, it is a Milnor fibre of $D^k(f)$. □

Remark 9.4 Let F be an unfolding of the $\mathscr{A}$-finite germ $f\colon (\mathbb{C}^n, 0) \to (\mathbb{C}^p, 0)$, on (smooth) parameter space S, and suppose that $D^k(f) \neq \emptyset$. Then by definition of multiple point space (*cf* Sect. 9.2) there is a fibre square

$$\begin{array}{ccc} (D^k(f), 0) & \longrightarrow & (D^k(F), (0, 0)) \\ \big\downarrow & & \big\downarrow \pi \\ \{0\} & \longrightarrow & (S, 0) \end{array} \tag{9.64}$$

where π is the coordinate projection. If f has corank 1 then so also does F. Now suppose that $d_k(n, p) \ge 0$. Then $D^k(f)$ is a complete intersection; it follows that the same is true for $D^k(F)$. Moreover, the fibre of π over 0, $D^k(f)$, has codimension in $D^k(F)$ equal to the dimension of S. As $D^k(F)$ is Cohen–Macaulay, we deduce that the map π is flat. Here we are using the following theorem: if X is Cohen–Macaulay and S is smooth then $\pi\colon (X, x_0) \to (S, s_0)$ is flat if and only if $\dim \pi^{-1}(0) + \dim S = \dim X$. See, e.g. [Mat89, Theorem 23.1]. Thus the diagram (9.64) is a bona fide "deformation" of the isolated singularity $D^k(f)$—flatness is part of the standard definition of deformation (see, e.g. [GLS07]). We have recovered the very well known fact that flatness is automatic for deformations of complete intersection singularities. For singularities which are not complete intersections, *flatness is no longer automatic*, and hence if f has corank greater than 1, it is not automatic that

(9.64) is a deformation. In fact if $D^2(f)$ has the expected dimension, then, with the structure described in Sect. 9.4, it is Cohen–Macaulay (see Theorem 9.9), and the same therefore goes for $D^2(F)$. It follows, again from [Mat89, 23.1], that π in (9.64) is flat, and once again we do have a deformation. For $k > 2$ this is not clear, since we do not yet have a structure theory for the higher multiple point spaces. $\Diamond$

Suppose that $f: (\mathbb{C}^n, S) \to (\mathbb{C}^p, 0)$ is a finitely determined germ, and let $f_t : U \to V$ be a stable perturbation of f. Suppose that f_t has singularities of corank at most 1. Since the codimension of $\overline{\Sigma^2}$ in $J^1(n, p)$ is $2(p-n+2)$, and $j^1 f_t$ is transverse to every $\mathscr{A}$-invariant stratum, this holds if $n < 2(p-n+2)$ and in particular, for maps from n-space to $n+1$-space if $n < 6$, and for maps from n-space to $n+2$-space if $n < 8$. By Theorem 9.4, the multiple point spaces $D^k(f_t)$ are all smooth. It turns out that the projections $\varepsilon^{k,i} : D^k(f_t) \to D^{k-1}(f_t)$ are also stable maps. Different projections $\varepsilon^{k,i}$ and $\varepsilon^{k,j}$ are left-right equivalent via the symmetric group actions, so we consider only $\varepsilon^{k,k}$, which we denote by ε^k. To see that ε^k is stable, we invoke the *principle of iteration*—the observation, due to Salomonsen, that

$$D^\ell(\varepsilon^k) \simeq D^{k+\ell-1}(f_t). \tag{9.65}$$

We will prove this in a moment, but let us first apply it to prove that ε^k is stable. By Theorem 9.4, $D^k(f_t)$ is smooth of dimension $d_k(n, p)$, unless it is empty. Again by Theorem 9.4, it is enough to prove that $D^\ell(\varepsilon^k)$ is smooth and of dimension $d_\ell(d_k(n, p), d_{k-1}(n, p))$, or empty. The smoothness, or emptiness, follow immediately from the principle of iteration, and as for the dimension, we have

$$\begin{aligned} d_\ell\big(d_k(n, p), d_{k-1}(n, p)\big) &= d_\ell\big(n - (k-1)(p-n), n - (k-2)(p-n)\big) \\ &= n - (k-1)(p-n) - (\ell-1)((p-n) \\ &= n - (k+\ell-2)(p-n) \\ &= d_{k+\ell-1}(n, p). \end{aligned} \tag{9.66}$$

Thus ε^k is stable.

The collection of spaces and maps $(D^\bullet(f_t), \varepsilon^\bullet)$ already described in Sect. 9.2.1 is a “semi-simplicial stable map”, although since the composite of two stable maps is not in general stable, stable maps do not form a category, and the term “semi-simplicial stable map” , which ought to mean “semi-simplicial object in the category of smooth spaces and stable maps”, is not quite legitimate. We could recover legitimacy at the price of clarity by considering the category *generated by* smooth spaces and stable maps—that is, which allows *composites* of stable maps.

Whatever we call it, the spaces $D^\bullet(f_t)$ and maps $\varepsilon^\bullet$, together with the symmetric group actions, together make up an intricate structure. Although our explicit description of multiple point spaces is only available when the maps in question have corank 1 singularities, this assumption need not be made for the map-germ f

whose stable perturbation we are considering. All that is needed is to be sure that its stable perturbation f_t has only corank 1 singularities. And this can be guaranteed by assumptions on the dimension of domain and target of f.

To prove (9.65), let us first define obvious morphisms in each direction. A point of $D^\ell(\varepsilon^k)$ is an ℓ-tuple of points in $D^k(f)$ sharing the same image under ε^k. Since ε^k simply forgets the last point in each k-tuple, a point in $D^\ell(\varepsilon^k)$ must take the form

$$\Big((x^{(1)},\dots,x^{(k-1)},x^{(k,1)}),\ (x^{(1)},\dots,x^{(k-1)},x^{(k,2)}),\dots,(x^{(1)},\dots,x^{(k-1)},x^{(k,\ell)})\Big) \tag{9.67}$$

– ℓ k-tuples of points in X differing only in the k'th spot. Define $\varphi : D^\ell(\varepsilon^k) \to D^{k+\ell-1}(f)$ by sending the point (9.67) to the $k+\ell-1$-tuple

$$\Big(x^{(1)},\dots,x^{(k-1)},x^{(k,1)},\dots,x^{(k,\ell)}\Big) \tag{9.68}$$

which certainly lies in $D^{k+\ell-1}(f)$; equally simply, we define $\varphi^{-1} : D^{k+\ell-1}(f) \to D^\ell(\varepsilon)$ by $\varphi^{-1}\Big(x^{(1)},\dots,x^{(k+\ell-1)}\Big) =$

$$\Big((x^{(1)},\dots,x^{(k-1)},x^{(k)}),((x^{(1)},\dots,x^{(k-1)},x^{(k+1)}),\dots,(x^{(1)},\dots,x^{(k-1)},x^{(k+\ell-1)})\Big).$$

If we knew that both $D^\ell(\varepsilon)$ and $D^{k+\ell-1}(f)$ were reduced, this would be enough to prove that φ is an isomorphism. However, although stability of f means that $D^{k+\ell-1}(f)$ is reduced, we do not know yet that $D^\ell(\varepsilon)$ is reduced, since our only evidence that ε is stable relies on the map (9.65) which we are trying to prove an isomorphism. So we have to supplement this construction with a more painstaking analysis. The argument comes from [Gor95], where it is a little more succinct than here. To begin, note that in case $\ell = 2$, the morphism φ of (9.68) fits into a commutative diagram

$$\begin{array}{ccc} D^2(\varepsilon^k) & \xrightarrow{\ \varphi\ } & D^{k+1}(f) \\ & {\scriptstyle\varepsilon^2}\searrow \quad \swarrow{\scriptstyle\varepsilon^{k+1}} & \\ & D^k(f) & \end{array} \tag{9.69}$$

in which the map ε^2 is the projection associated to the double point space of the map $\varepsilon^k : D^k(f) \to D^{k-1}(f)$. We use this observation as the basis of an argument by induction on k. The argument makes use of one *ad hoc* calculation, for which we require an explicit formula for a stable map-germ of corank 1 and multiplicity $n+1$ from $(\mathbb{C}^N, 0)$ to $(\mathbb{C}^{N+r}, 0)$. The smallest such N is equal to rn, and with respect to coordinates $x, u_1, \ldots, u_{n-1}$ and $v_{i,1}, \ldots, v_{i,n}$ for $i = 1, \ldots, r-1$ the map-germ we want takes the form

$$(u, v, x) \mapsto \Big(u, v, x^{n+1} + \sum_{i=1}^{n-1} u_i x^{n-i}, \sum_{j=1}^{n} v_{1,j} x^{n-j+1}, \ldots, \sum_{j=1}^{n} v_{r-1,j} x^{n-j+1}\Big). \tag{9.70}$$

Denote this map by $F_{n,r}$. To simplify notation, denote the source and target of $F_{n,r}$ as $D^1(F_{n,r})$ and $D^0(F_{n,r})$, respectively.

Lemma 9.8 *The projection $\varepsilon^2 : D^2(F_{n,r}) \to D^1(F_{n,r})$ is left-right-equivalent to $F_{n-1,r} \times id_{\mathbb{C}}$.*

Proof The defining equations for $D^2(F_{n,r})$,

$$\frac{x_1^{n+1} - x_2^{n+1} + \sum_{j=1}^{n-1} u_j (x_1^{n-j} - x_2^{n-j})}{x_1 - x_2} = 0$$

$$\frac{\sum_{i=1}^{n} v_{i,j} (x_1^{n-j+1} - x_2^{n-j+1})}{x_1 - x_2} = 0 \text{ for } i = 1, \ldots, r-1$$

allow us to express u_{n-1} and the $v_{i,n}$, for $i = 1, \ldots, r-1$, in terms of the other coordinates. Thus we can take $x_1, x_2, u' := u_1 \ldots, u_{n-2}$ and $v_i' := v_{i,1} \ldots v_{i,n-1}$, for $i = 1, \ldots, r-1$, as coordinates on $D^2(F_{n,r})$. Consider the projection

$$D^2(F_{n,r}) \cap \{x_1 = 0\} \to D^1(F_{n,r}) \cap \{x_1 = 0\}.$$

It takes the form

$$(x_2, u', v') \mapsto -\left(-u', -v', x_2^n + \sum_{i=1}^{n-2} u_i x_2^{n-i-1}, \sum_{i=1}^{n-1} v_{1,i} x_2^{n-i}, \ldots, \sum_{i=1}^{n-1} v_{r-1,i} x_2^{n-i}\right)$$

with respect to coordinates $(u', v', u_{n-1}, v_{1,n}, \ldots, v_{r-1,n})$ on $\{x_1 = 0\} \subset D^1(F_{n,r})$. This is precisely (9.70) with $n-1$ in place of n. So $\varepsilon^2|_{\{x_1=0\}}$ is stable. It follows that ε^2, as an unfolding of $\varepsilon^2|_{\{x_1=0\}}$ on parameter x_1, is a trivial unfolding, equivalent to $F_{n-1,r} \times \mathrm{id}_{\mathbb{C}}$. □

Proposition 9.16 *The projection $\varepsilon^k : (D^k(F_{n,r}), 0) \to (D^{k-1}(F_{n,r}), 0)$ is left-right equivalent to the germ at 0 of the map $F_{n-k+1,r} \times id_{\mathbb{C}^{k-1}}$.*

Proof We have just shown that this is true for $k = 2$. Suppose by induction that it is true for k: that there are isomorphisms making the diagram

$$\begin{array}{ccc} D^k(F_{n,r}) & \xrightarrow{\simeq} & D^1(F_{n-k+1,r} \times \mathrm{id}_{\mathbb{C}^{k-1}}) \\ {\scriptstyle \varepsilon^k}\downarrow & & \downarrow{\scriptstyle F_{n-k+1,r} \times \mathrm{id}_{\mathbb{C}^{k-1}}} \\ D^{k-1}(F_{n,r}) & \xrightarrow[\simeq]{} & D^0(F_{n-r+k,r} \times \mathrm{id}_{\mathbb{C}^{k-1}}) \end{array} \tag{9.71}$$

commute. Then by Lemma 9.8 and the induction hypothesis, there is a commutative diagram

$$\begin{array}{ccccc} D^2(\varepsilon^k) & \xrightarrow{\simeq} & D^2(F_{n-k+1,r} \times \mathrm{id}_{\mathbb{C}^{k-1}}) & \xrightarrow{\simeq} & D^1(F_{n-k,r} \times \mathrm{id}_{\mathbb{C}^k}) \\ \downarrow{\scriptstyle \varepsilon^2(\varepsilon^k)} & & \downarrow{\scriptstyle \varepsilon^2} & & \downarrow{\scriptstyle F_{n-k,r}\times \mathrm{id}_{\mathbb{C}^k}} \\ D^1(\varepsilon^k) & \xrightarrow[\simeq]{} & D^1(F_{n-k+1,r} \times \mathrm{id}_{\mathbb{C}^{k-1}}) & \xrightarrow[\simeq]{} & D^0(F_{n-k+1,r}) \end{array} \tag{9.72}$$

where the spaces in the top left and bottom left are the double-point space and domain of the map $\varepsilon^k : D^k(F_{n,r}) \to D^{k-1}(F_{n,r})$ on the left of (9.71). It follows that the left-hand vertical map is stable. This was all that was missing from our earlier argument that the map φ of (9.68) and (9.69) is an isomorphism: so we conclude that $\varepsilon^{k+1} : D^{k+1}(F_{n,r}) \to D^k(F_{n,r})$ is stable. Now juxtaposing (9.69) and (9.72) gives a commutative diagram

$$\begin{array}{ccc} D^{k+1}(F_{n,r})) & \xrightarrow{\simeq} & D^2(F_{n-k+1,r} \times \mathrm{id}_{\mathbb{C}^{k-1}}) . \\ \downarrow{\scriptstyle \varepsilon^{k+1}} & & \downarrow{\scriptstyle \varepsilon^2} \\ D^k(F_{n,r}) & \xrightarrow[\simeq]{} & D^1(F_{n-k+1,r} \times \mathrm{id}_{\mathbb{C}^{k-1}}) \end{array} \tag{9.73}$$

The induction step is now proved by juxtaposing *this* diagram with the diagram

$$\begin{array}{ccc} D^2(F_{n-k+1} \times \mathrm{id}_{\mathbb{C}^{k-1}}) & \xrightarrow{\simeq} & D^1(F_{n-k,r} \times \mathrm{id}_{\mathbb{C}^k}) \\ \downarrow{\scriptstyle \varepsilon^2} & & \downarrow{\scriptstyle F_{n-k,r}\times \mathrm{id}_{\mathbb{C}^k}} \\ D^1(F_{n-k+1,r} \times \mathrm{id}_{\mathbb{C}^{k-1}}) & \xrightarrow[\simeq]{} & D^0(F_{n-k,r} \times \mathrm{id}_{\mathbb{C}^k}) \end{array} \tag{9.74}$$

provided by Lemma 9.8. This completes the induction. □

Corollary 9.7 (The Principle of Iteration for Stable Maps of Corank 1) *If $f : U \to V$ is a stable map with only corank 1 singularities then all germs of projection $\varepsilon^k : D^k(f) \to D^{k-1}(f)$ are stable, and moreover the map $\varphi : D^\ell\big(\varepsilon^k : D^k(f) \to D^{k-1}(f)\big) \to D^{k+\ell-1}(f)$ of (9.68) is an isomorphism.*

Proof This is now immediate from Proposition 9.16 and Lemma 9.9. □

A stable singularity of corank 1 map $(\mathbb{C}^n, S) \to (\mathbb{C}^p, y)$, $n \leq p$ with $|S| = k$ is characterised, up to isomorphism, by the k-tuple of its multiplicities at the k distinct points of S. This is simply because each of the mono-germs $f^{(i)}: (\mathbb{C}^n, x^{(i)}) \to (\mathbb{C}^p, y)$, for $x^{(i)} \in S$, is determined, up to isomorphism, by its local algebra, and thus by its mutiplicity, and because stability of f means that the analytic strata of the distinct mono-germs making up f are in general position in $\mathbb{C}^p$.

Lemma 9.9 *Up to isomorphism and trivial unfolding, all stable multi-germs of corank 1 maps* $(M^m, S) \to (N^{m+r}, y)$ *appear in the germs* $F_{n,r}$, *for n sufficiently large.*

Proof Suppose that $(M^m, S) \to (N^p, y)$, $n \leq p$ is stable, $|S| = k < \infty$, and the germ of f at $x^{(i)} \in S$ is of corank ≤ 1, with multiplicity n_i, for $i = 1, \ldots, k$. Let $n = \sum_i n_i$. Then a (necessarily stable) multi-germ $(\mathbb{C}^n, S') \to (\mathbb{C}^p, y)$ with multiplicities $n_1, \ldots, n_k$ appears in every representative of $F_{n,r}$. To see this, identify the $n-1$-dimensional space of the parameters u_i with the space P_{n+1} of monic polynomials of degree $n+1$ with root sum and product both equal to 0. In P_n there is a stratum consisting of polynomials with k distinct roots with multiplicities $n_1, \ldots, n_k$. Let $u(t) = (u_1(t), \ldots, u_{n-1}(t))$, with $u(0) = 0$, parameterise an analytic curve which lies in in this stratum for $t \neq 0$. Let $x^{(1)}(t), \ldots, x^{(k)}(t)$ be the corresponding roots. They depend analytically on t, for $t \neq 0$. Then the multi-germ of $F_{n+1,r}$ at $S'_t = \{(u(t), 0, x^{(1)}t)), \ldots, (u(t), 0, x^{(k)}(t))\}$ has the required multiplicities. It must be stable, as the germ at 0 of $F_{n,r}$ is stable, stability is an open condition, and the weighted homogeneity of $F_{n+1,r}$ means that its stability in some neighbourhood of 0 is propagated to all of its domain. □

Question Consider the full disentanglement $D^\bullet(f_t) := (D^\bullet(f_t), \varepsilon^{\bullet,\bullet})$ of a map-germ $f: (\mathbb{C}^n, S) \to (\mathbb{C}^p, 0)$, and suppose that f_t has only corank 1 singularities. To what extent do invariants of $D^\bullet(f_t)$ detect invariants of f? In particular, can the rank of $d_0 f$ be recovered from $D^\bullet(f_t)$?

Exercises for Sect. 9.7

1 Find equations for $D^2(f)$ and $D^3(f)$ when f is the map-germ given by

(i) $f(u_1, u_2, u_3, x) = (u_1, u_2, u_3, x^3 + u_1 x, u_2 x^2 + u_3 x)$ (stable map-germ of type $\sum^{1,1,0}$).
(ii) $f(u, x) = (u, x^2, x^3 + u^{k+1} x)$ (type S_k in [Mon85]—here $D^3(f) = \emptyset$).
(iii) $f(u, x) = (u, x^3, ux + x^5)$ (type H_2 in [Mon85]).
(iv) $f(u, x) = (u, x^3, ux + x^{3k-1})$ (type H_k in [Mon85]).

and use them to find their Milnor numbers.

2 In 1(i), check that $D^k(f)$ is smooth whenever non-empty.

3 For 1(ii),1(iii) and 1(iv), check that $D^2(f)$ has isolated singularity.

4 For 1(iii) and 1(iv), check that $D^3(f)$ is zero-dimensional. What is the complex vector space dimension of $\mathscr{O}_{\mathbb{C}\times\mathbb{C}^3,0}/\mathcal{I}_3(f)$ in these two cases? Your answer should be divisible by 6.

5 Suppose that $f\colon (\mathbb{C}^n, 0) \to (\mathbb{C}^p, 0)$, with $n < p$, has corank 1 and is $\mathscr{A}$-finite, and suppose that $d_k(n, p) = 0$. Use conservation of mutiplicity to show that

(i) $\dim_{\mathbb{C}} \mathscr{O}_{D^k(f),0}$ is divisible by $(n+1)!$, and
(ii) if f_t is a stable perturbation of f, then the image of f_t contains $\frac{1}{k!}\dim_{\mathbb{C}} \mathscr{O}_{D^k(f),0}$ k-tuple points, at each of which it is locally isomorphic to the union of the k coordinate hyperplanes in $(\mathbb{C}^{n+1}, 0)$.

6 Let $f\colon (\mathbb{F}, 0) \to (\mathbb{F}^2, 0)$ be 1-to-1. Show that

$$\delta(f) = \frac{1}{2}\dim_{\mathbb{F}} \frac{(x - x')\,\mathscr{O}_{\mathbb{C}^2,0}}{(f_1(x) - f_1(x'), f_2(x) - f_2(x'))}.$$

7 Deduce the principle of iteration for corank 1 $\mathscr{A}$-finite multi-germs $(\mathbb{C}^n, S) \to (\mathbb{C}^p, 0)$ from the principle of iteration for stable germs.

8 Does the principle of iteration hold for germs of corank > 1? For germs which are not $\mathscr{A}$-finite?

9.8 Blowing-Up Multiple Points

We have seen in Sects. 9.4 and 9.5 that we have explicit equations for $D^2(f)$ and also for higher order multiple points $D^k(f)$ when f has singularities of corank one. That is, we can compute these multiple spaces directly from f, without having to take a stable unfolding F of f and use the procedure of Definition 9.3. In the presence of singularities of corank ≥ 2, we cannot yet ensure in general that $D^k_{\mathrm{cl}}(F)$ is Cohen–Macaulay, so we cannot expect to find explicit equations for $D^k(f)$ in the above sense.

Another unpleasant phenomenon at the points of corank ≥ 2 is that $D^2(f)$ is not smooth, even when f is stable (see Exercise 3). As a consequence, the important results of Theorems 9.4 and 9.5, which characterise the stability and the finite determinacy of map-germs in terms of the multiple point spaces $D^k(f)$, do not hold if we allow singularities of corank ≥ 2.

In order to overcome these two problems we introduce another class of multiple point spaces, K_k in place of D^k, due originally to Kleiman [Kle81] (though the notation K_k is ours). Kleiman's original construction is more general than we need: in fact it works for general separated maps between not necessarily smooth schemes. We follow a different approach based on the results of [NBPS18], which works for mappings $f : X \to Y$ between smooth manifolds and which generalises the notion of divided differences used to define $D^k(f)$ in the corank one case. In fact, this alternative construction coincides with that of Ronga [Ron73] in the case of double points. Another classical reference for double points is that of Fulton and Laksov [FL77].

Our construction of the K_k takes place in two stages. We first construct a smooth ambient space $B_k(X)$ in which we will embed $K_k(f)$.

9.8.1 Construction of an Ambient Space for K_k

Let X be a manifold and $Y \subset X$ a closed submanifold. The *blow-up* of X along Y is a manifold $\mathrm{Bl}_Y X$ together with a surjective proper smooth map $b : \mathrm{Bl}_Y X \to X$ with the property that b is a diffeomorphism on $X \setminus Y$ and $E := b^{-1}(Y)$ is a hypersurface in $\mathrm{Bl}_Y X$, called the *exceptional divisor*.

In a neighbourhood of a point $x_0 \in X \setminus Y$, $\mathrm{Bl}_Y X$ is equal to X and b is just the identity. When $x_0 \in Y$, by taking local coordinates, we can assume that $X \subset \mathbb{C}^n$ is an open subset and that Y is given by equations $g_1(x) = \cdots = g_m(x) = 0$, where $g = (g_1, \ldots, g_m) : X \to \mathbb{C}^m$ is a submersion. Then $\mathrm{Bl}_Y X$ is the submanifold of $X \times \mathbb{P}^{m-1}$ of pairs $(x, [v])$ such that $g(x) \wedge v = 0$. The map b is then the restriction of the projection onto the first factor, and the exceptional divisor is $E = Y \times \mathbb{P}^{m-1}$ which has codimension one in $\mathrm{Bl}_Y X$. It is not difficult to see that this construction can be extended globally to give a well-defined structure of manifold on $\mathrm{Bl}_Y X$ and a surjective proper map $b : \mathrm{Bl}_Y X \to X$ (see, for instance, [Dem12] for more details).

Definition 9.4 Given a manifold X, we recursively define manifolds $B_k := B_k(X)$ and submersions $p : B_k \to B_{k-1}$ as follows: $B_0 = *$ (a one point manifold), $B_1 = X$ and $p : B_1 \to B_0$ is the constant map; for $k \geq 2$,

$$B_k = \mathrm{Bl}_{\Delta B_{k-1}}(B_{k-1} \times_{B_{k-2}} B_{k-1})$$

and $p : B_k \to B_{k-1}$ is the composition $p_1 \circ b$, where $b : B_k \to B_{k-1} \times_{B_{k-2}} B_{k-1}$ is the blow-up map and $p_1 : B_{k-1} \times_{B_{k-2}} B_{k-1} \to B_{k-1}$ is the projection onto the first component.

We recall that $B_{k-1} \times_{B_{k-2}} B_{k-1}$ is the fibre product, given by the following fibre square:

$$\begin{array}{ccc} B_{k-1} & \xrightarrow{\;p\;} & B_{k-2} \\ {\scriptstyle p_1}\uparrow & & \uparrow{\scriptstyle p} \\ B_{k-1} \times_{B_{k-2}} B_{k-1} & \xrightarrow{\;p_2\;} & B_{k-1} \end{array}$$

The fact that $p : B_{k-1} \to B_{k-2}$ is a submersion implies that $B_{k-1} \times_{B_{k-2}} B_{k-1}$ is a submanifold of $B_{k-1} \times B_{k-1}$ which contains the diagonal ΔB_{k-1}. Thus, it makes sense to define B_k as the blow-up of this manifold along ΔB_{k-1}. As the new map $p : B_k \to B_{k-1}$ turns out to be again a submersion, we can proceed to the next step.

By construction, $B_k(X)$ is a manifold of dimension nk, where $n = \dim X$. We now define an adapted atlas for B_k which will be necessary to define the generalised divided differences. Let X be any manifold and fix an atlas $\mathcal{A}$ for X. We recursively construct atlases $\mathcal{A}_k$ for the manifolds $B_k(X)$ with the following property:

(*) For each chart $\varphi\colon V \to A$ in $\mathcal{A}_k$, there is a unique chart $\varphi'\colon V' \to A'$ in $\mathcal{A}_{k-1}$ such that $p(V) = V'$ and the diagram

$$\begin{array}{ccc} V & \xrightarrow{p} & V' \\ \downarrow{\scriptstyle\varphi} & & \downarrow{\scriptstyle\varphi'} \\ A & \xrightarrow{\pi_1} & A' \end{array}$$

commutes, where $p\colon B_k(X) \to B_{k-1}(X)$ and $\pi_1 : \mathbb{C}^{kn} \to \mathbb{C}^{(k-1)n}$ is the projection onto the first coordinates.

The construction is obvious for $k = 0, 1$, so we set $k \geq 2$ and assume that we have already constructed atlases $\mathcal{A}_0, \dots, \mathcal{A}_{k-1}$ with property (*). We define the atlas $\mathcal{A}_k$ by constructing two types of charts:

(A) *Diagonal charts.* Let $\varphi'\colon V' \to A'$ be any chart in $\mathcal{A}_{k-1}$. By induction hypothesis, there exists a unique chart $\varphi''\colon V'' \to A''$ in $\mathcal{A}_{k-2}$ such that $p(V') = V''$ and we have a commutative diagram

$$\begin{array}{ccc} V' & \xrightarrow{p} & V'' \\ \downarrow{\scriptstyle\varphi'} & & \downarrow{\scriptstyle\varphi''} \\ A' & \xrightarrow{\pi_1} & A'' \end{array}$$

Then $W = V' \times_{V''} V'$ is a coordinate open subset of $B_{k-1} \times_{B_{k-2}} B_{k-1}$. For all $(x^{(1)}, x^{(2)}) \in W$, the first $n(k-2)$ coordinates of $\varphi'(x^{(1)})$ and $\varphi'(x^{(2)})$ are equal. By construction,

$$b^{-1}(W) = \left\{(x^{(1)}, x^{(2)}, [v]) \in W \times \mathbb{P}^{n-1} : \left(\pi_2(\varphi'(x^{(2)})) - \pi_2(\varphi'(x^{(1)}))\right) \wedge v = 0\right\},$$

where $\pi_2\colon \mathbb{C}^{n(k-1)} \to \mathbb{C}^n$ is now the projection onto the last coordinates. For each $i = 1, \dots, n$, let $\omega_i : \mathbb{C}^{n-1} \to \mathbb{C}^n$ be the embedding $a \mapsto (a_1, \dots, a_{i-1}, 1, a_i, \dots, a_n)$. We define:

- V as the open subset of $b^{-1}(W)$ of triples $(x^{(1)}, x^{(2)}, [v])$ such that $v_i \neq 0$.
- A as the open subset of $(\mathbb{C}^n)^k$ of k-tuples $(x, \gamma^1, \dots, \gamma^{k-1})$ such that

$$(x, \gamma^1, \dots, \gamma^{k-2}) \in A'$$

and if $\gamma^{k-1} = (\lambda, a)$ with $\lambda \in \mathbb{C}$ and $a \in \mathbb{C}^{n-1}$, then

$$(x, \gamma^1, \ldots, \gamma^{k-2} + \lambda\omega_i(a)) \in A'.$$

- $\varphi\colon V \to A$ as the map sending $(x^{(1)}, x^{(2)}, [v]) \mapsto (\varphi'(x^{(1)}), \gamma^{k-1})$ such that $\gamma^{k-1} = (\lambda, a)$, with $[v] = [\omega_i(a)]$ and

$$\pi_2(\varphi'(x^{(2)})) - \pi_2(\varphi'(x^{(1)})) = \lambda\omega_i(a).$$

(B) *Non-diagonal charts.* Let $\varphi_1'\colon V_1' \to A_1'$ and $\varphi_2'\colon V_2' \to A_2'$ be charts in $\mathcal{A}_{k-1}$ such that $V_1' \cap V_2' = \emptyset$. Again, there exist unique charts $\varphi_1''\colon V_1'' \to A_1''$ and $\varphi_2''\colon V_2'' \to A_2''$ in $\mathcal{A}_{k-2}$ such that for $i = 1, 2$, $p(V_i') = V_i''$ and we have a commutative diagram

$$\begin{array}{ccc} V_i' & \xrightarrow{p} & V_i'' \\ \downarrow{\scriptstyle \varphi_i'} & & \downarrow{\scriptstyle \varphi_i''} \\ A_i' & \xrightarrow{\pi_1} & A_i'' \end{array}$$

In this case, we define:

- $V = (B_{k-1} \times_{B_{k-2}} B_{k-1}) \cap (V_1' \times V_2')$. Since $V \cap \Delta B_{k-1} = \emptyset$, we have $b^{-1}(V) \equiv V$. The elements of V are pairs $(x^{(1)}, x^{(2)})$ such that $b(x^{(1)}) = b(x^{(2)})$ and thus,

$$\varphi_2'' \circ (\varphi_1'')^{-1}\left(\pi_1(\varphi_1'(x^{(1)}))\right) = \pi_1(\varphi_2'(x^{(2)})).$$

- A as the open subset of $(\mathbb{C}^n)^k$ of k-tuples $(x, \gamma^1, \ldots, \gamma^{k-1})$ such that

$$(x, \gamma^1, \ldots, \gamma^{k-2}) \in A_1'$$

and

$$\left(\varphi_2'' \circ (\varphi_1'')^{-1}(x, \gamma^1, \ldots, \gamma^{k-3}), \gamma^{k-1}\right) \in A_2'.$$

(We follow the convention that this $(k-1)$-tuple is just γ^{k-1} when $k = 2$).
- $\varphi\colon V \to A$ as the map given by $(x^{(1)}, x^{(2)}) \mapsto (\varphi_1'(x^{(1)}), \pi_2(\varphi_2'(x^{(2)})))$.

Proposition 9.17 *With the above notation, we have:*

1. *$\varphi\colon V \to A$ is a biholomorphism, so we take it as a chart in $\mathcal{A}_k$.*
2. *Property (*) holds.*
3. *Let E be the exceptional divisor of $B_k(X)$. If $\varphi\colon V \to A$ is a diagonal chart and $\gamma^{k-1} = (\lambda, a)$, then $\varphi(E \cap V) = \{\lambda = 0\} \cap A$.*
4. *The family of all diagonal and non-diagonal charts defines an atlas for $B_k(X)$.*

Proof To see (1) we use induction on k. Suppose first that $\varphi\colon V \to A$ is a diagonal chart, then φ is well defined and holomorphic. In fact, $\varphi(x^{(1)}, x^{(2)}, [v]) = (\varphi'(x^{(1)}), \lambda, a)$, where

$$a = \frac{1}{v_i}(v_1, \ldots, v_{i-1}, v_{i+1}, \ldots, v_n)$$

and λ is the i-th coordinate of $\pi_2(\varphi'(x^{(2)})) - \pi_2(\varphi'(x^{(1)}))$. The inverse of φ is also well defined and holomorphic: it is given by $(x, \gamma^1, \ldots, \gamma^{k-1}) \mapsto (x^{(1)}, x^{(2)}, [v])$, where $\gamma^{k-1} = (\lambda, a)$ and

$$x^{(1)} = (\varphi')^{-1}(x, \gamma^1, \ldots, \gamma^{k-2}),$$
$$x^{(2)} = (\varphi')^{-1}(x, \gamma^1, \ldots, \gamma^{k-2} + \lambda\omega_i(a)), \qquad v = \omega_i(a).$$

If φ is a non-diagonal chart, it is clear that it is well defined and holomorphic. Its inverse is $(x, \gamma^1, \ldots, \gamma^{k-1}) \mapsto (x^{(1)}, x^{(2)})$, where now

$$x^{(1)} = (\varphi_1')^{-1}(x, \gamma^1, \ldots, \gamma^{k-2}),$$
$$x^{(2)} = (\varphi_2')^{-1}\left(\varphi_2'' \circ (\varphi_1'')^{-1}(x, \gamma^1, \ldots, \gamma^{k-3}), \gamma^{k-1}\right),$$

which is also well defined and holomorphic. Statements (2), (3) and (4) are obvious from the construction. □

The notation $(x, \gamma^1, \ldots, \gamma^{k-1})$ for the local coordinates of $B_k(X)$ has been chosen in such a way that $(x, \gamma^1, \ldots, \gamma^{r-1})$ are local coordinates of $B_r(X)$, for each $r \leq k$. In particular x are the local coordinates of X. At each step r, the last coordinates γ^r may represent either a pair (λ, a) of projective coordinates or the last n coordinates of another point in the corresponding $B_{r-1}(X)$, depending on whether or not the chart is diagonal.

Example 9.8 Assume for simplicity that $X \subset \mathbb{C}^n$ is an open subset, with atlas $\mathcal{A}_1$ consisting just of the identity map, and consider the case $k = 2$. The atlas $\mathcal{A}_2$ of $B_2(X)$ has only diagonal charts $\varphi_i : V_i \to A_i$, for $i = 1, \ldots, n$, where

- V_i is the open subset of $B_2(X)$ of triples $(x^{(1)}, x^{(2)}, [v])$ such that $v_i \neq 0$;
- A_i is the open subset of $X \times \mathbb{C} \times \mathbb{C}^{n-1}$ of triples (x, λ, a) such that $x + \lambda\omega_i(a) \in X$;
- $\varphi_i : V_i \to A_i$ is given by $(x^{(1)}, x^{(2)}, [v]) \mapsto (x, \lambda, a)$ such that $x^{(1)} = x$, $x^{(2)} = x + \lambda\omega_i(a)$ and $[v] = [\omega_i(a)]$.

◊

9.8.2 Construction of $K_k(f)$ as Subspace of $B_k(X)$

Let X and Y be closed complex subspaces of a complex space Z and assume that $X = V(\mathcal{I})$ and $Y = V(\mathcal{J})$, for some coherent ideal sheaves $\mathcal{I}$ and $\mathcal{J}$ on Z. We set $X : Y = V(\mathcal{I} : \mathcal{J})$, which is also a closed complex subspace of Z. Notice that here we do not assume that X, Y or $X : Y$ are reduced.

Definition 9.5 Given a smooth map $f : X \to Y$, we recursively define complex spaces $K_k := K_k(f)$ and maps $K_k \to K_{k-1}$ as follows: $K_0 = Y$, $K_1 = X$ and $K_1 \to K_0$ is equal to f; for $k \geq 2$, $K_k \to K_{k-1}$ is the restriction of $p : B_k \to B_{k-1}$, and

$$K_k = b^{-1}(K_{k-1} \times_{K_{k-2}} K_{k-1}) : E,$$

where E is the exceptional divisor of the blow-up $b : B_k \to B_{k-1} \times_{B_{k-2}} B_{k-1}$.

As we pointed out at the beginning of this section, this definition of the multiple point spaces $K_k(f)$ is not the original definition of Kleiman in [Kle81], which is more generally applicable. In [Kle81], the double point space $K_2(f)$ is defined as the residual scheme of $X \times_Y X$ along the diagonal ΔX. When ΔX is regularly embedded in $X \times_Y X$, the residual scheme is equal to the blow-up, but in general, these two spaces may be different. Higher order multiple point spaces are now defined by the iteration principle: the triple point space $K_3(f)$ is the double point space of the projection $K_2(f) \to K_1(f) = X$ and so on. The proof that the two definitions of $K_k(f)$ coincide when X and Y are smooth is not easy and we will not include it. Details can be found in the paper [NBPS18].

Definition 9.6 Let $h : X \to \mathbb{C}$ be any smooth function on a manifold X. Consider in $B_k(X)$ the atlas $\mathcal{A}_k$ of Proposition 9.17. We define the *generalised divided differences* recursively as follows: for $k = 1$ and for each $\varphi : V \to A$ a chart of $B_1 = X$, we set $h_\varphi[x] = h(\varphi^{-1}(x))$, $x \in A$. Let $k \geq 2$ and let $\varphi : V \to A$ be a chart of B_k. If φ is a diagonal chart, we set

$$h_\varphi[x, \gamma^1, \dots, \gamma^{k-1}] = \frac{h_{\varphi'}[x, \gamma^1, \dots, \gamma^{k-2} + \lambda\omega_i(a)] - h_{\varphi'}[x, \gamma^1, \dots, \gamma^{k-2}]}{\lambda},$$

where $\gamma^{k-1} = (\lambda, a)$. Otherwise, if φ is non-diagonal, we set

$$h_\varphi[x, \gamma^1, \dots, \gamma^{k-1}] = h_{\varphi_2'}[\varphi_2'' \circ (\varphi_1'')^{-1}(x, \gamma^1, \dots, \gamma^{k-3}), \gamma^{k-1}] - h_{\varphi_1'}[x, \gamma^1, \dots, \gamma^{k-2}].$$

For simplicity, we omit the subindex φ in h_φ, whenever there is no risk of confusion. It is clear that the generalised divided differences $h[x, \gamma^1, \dots, \gamma^{k-1}]$ are smooth functions on each open set A for a chart $\varphi : V \to A$. Due to the choice of coordinates, the functions $h[x, \gamma^1, \dots, \gamma^r]$ are also well defined and smooth on A, for all $r = 1, \dots, k-1$.

Example 9.9 As in Example 9.8, we assume that $X \subset \mathbb{C}^n$ is open and consider $k = 2$. Then the generalised divided differences of $h : X \to \mathbb{C}$ in $B_2(X)$ are

$$h[x, \lambda, a] = \frac{h(x + \lambda\omega_i(a)) - h(x)}{\lambda},$$

for $(x, \lambda, a) \in A_i$ and $i = 1, \dots, n$. ◊

For simplicity, the following theorem is stated only for open subsets $X \subset \mathbb{C}^n$ and $Y \subset \mathbb{C}^p$, where we begin with the atlas $\mathcal{A}_1$ on X consisting just of the identity map. The general case where X and Y are manifolds can be adapted easily, by taking convenient atlases on X and Y. Details are left for the reader.

Theorem 9.6 ([NBPS18]) *Let $f: X \to Y$ be a smooth map between open subsets $X \subset \mathbb{C}^n$ and $Y \subset \mathbb{C}^p$. Let $k \geq 2$ and take a chart $\varphi: V \to A$ of the atlas $\mathcal{A}_k$ of $B_k(X)$. Then $\varphi(K_k(f) \cap V)$ is the complex subspace of A given by the vanishing of the generalised divided differences*

$$f_j[x, \gamma^1], \dots, f_j[x, \gamma^1, \dots, \gamma^{k-1}],$$

of the components $f_j : X \to \mathbb{C}$, $j = 1, \dots, p$ with respect to φ.

Proof We proceed by induction on k. The result is clear for $k = 1$ and we assume it is true for $k - 1$. By construction, $K_k = b^{-1}(K_{k-1} \times_{K_{k-2}} K_{k-1}) : E$. If φ is a diagonal chart, then $\varphi'(K_{k-1} \cap V')$ is given in A' by the vanishing of

$$f_j[x, \gamma^1], \dots, f_j[x, \gamma^1, \dots, \gamma^{k-2}], \; j = 1, \dots, p.$$

We denote by I the ideal in $\mathcal{O}_A$ generated by these functions. The defining ideal of $\varphi(b^{-1}(K_{k-1} \times_{K_{k-2}} K_{k-1}) \cap V)$ is now $I + J$, where J is generated by

$$f_j[x, \gamma^1, \dots, \gamma^{k-2} + \lambda\omega_i(a)], \; j = 1, \dots, p,$$

where $\gamma^{k-1} = (\lambda, a)$. But obviously $I + J = I + J'$, where J' is generated by

$$f_j[x, \gamma^1, \dots, \gamma^{k-2} + \lambda\omega_i(a)] - f_j[x, \gamma^1, \dots, \gamma^{k-2}], \; j = 1, \dots, p.$$

Since $\varphi(E \cap V)$ is the hypersurface $\lambda = 0$ in A, it follows that $\varphi(K_k \cap V)$ is defined in A by the ideal $(I + J') : \lambda$. But the generators of I do not depend on λ and the generators of J' are all divisible by λ, so $(I + J') : \lambda$ is generated by

$$f_j[x, \gamma^1], \dots, f_j[x, \gamma^1, \dots, \gamma^{k-1}], \; j = 1, \dots, p.$$

The case where φ is non-diagonal is easier, since $E \cap V = \emptyset$ and hence $\varphi(K_k \cap V)$ is equal to $\varphi(b^{-1}(K_{k-1} \times_{K_{k-2}} K_{k-1}) \cap V)$. As above, this space is defined by $I + J$, where I is generated in $\mathcal{O}_A$ by

$$f_j[x, \gamma^1], \dots, f_j[x, \gamma^1, \dots, \gamma^{k-2}], \; j = 1, \dots, p,$$

and J by

$$f_j[\varphi_2'' \circ (\varphi_1'')^{-1}(x, \gamma^1, \dots, \gamma^{k-3}), \gamma^{k-1}], \; j = 1, \dots, p.$$

But $I + J = I + J'$, where J' is generated by

$$f_j[x, \gamma^1, \ldots, \gamma^{k-1}],\ j = 1, \ldots, p.$$

□

Corollary 9.8 *Let $f: X \to Y$ be a smooth map between open subsets $X \subset \mathbb{C}^n$ and $Y \subset \mathbb{C}^p$ and let $k \geq 2$. Then* $\dim K_k(f) \geq d_k(n, p)$ *and if* $\dim K_k(f) = d_k(n, p)$ *then it is a local complete intersection in $B_k(X)$.*

Example 9.10 We explore the link between $K_2(f)$ and $D^2(f)$. Let $f : X \to Y$ be as in Theorem 9.6 and let $k = 2$. We follow the notation of Examples 9.8 and 9.9. For each $i = 1, \ldots, n$, $K_2(f)$ is given in the chart φ_i by

$$f_j[x, \lambda, a] = 0,\ j = 1, \ldots, p.$$

Let $A = (\alpha_{ij})$ be the matrix used in Sect. 9.4 to define $D^2(f)$ (see displayed equation (9.16)). We have

$$\begin{aligned} f_j[x, \lambda, a] &= \frac{f_j(x + \lambda\omega_i(a)) - f_j(x)}{\lambda} \\ &= \frac{A_j(x, x + \lambda\omega_i(a))(x + \lambda\omega_i(a) - x)}{\lambda} \\ &= A_j(x, x + \lambda\omega_i(a))\omega_i(a) = A_j(x^{(1)}, x^{(2)})v, \end{aligned}$$

where $\varphi_i(x^{(1)}, x^{(2)}, [v]) = (x, \lambda, a)$ and A_j is the j-th column of A. We deduce that $K_2(f)$ is the complex subspace of $B_2(X)$ given by the vanishing of $A_j(x^{(1)}, x^{(2)})v$, $j = 1, \ldots, p$.

Set theoretically, $K_2(f)$ is the subset of $X \times X \times \mathbb{P}^{n-1}$ of points $(x^{(1)}, x^{(2)}, [v])$ such that either $x^{(1)} \neq x^{(2)}$, $f(x^{(1)}) = f(x^{(2)})$ and $[v] = [x^{(2)} - x^{(1)}]$ or $x^{(1)} = x^{(2)}$ and $d_{x^{(1)}} f(v) = 0$. Consequently, $K_2(f)$ coincides with Ronga's definition of double point space (denoted by $\widetilde{M}_2(f)$ in [Ron73]). Moreover, if $\dim D^2(f) = 2n - p$ and f has corank $\leq 2n - p + 1$, then $\dim K_2(f) = 2n - p$ and hence is a local complete intersection in $B_2(X)$. Observe that the above inequality on the corank is always satisfied when $p \leq n + 1$. ◊

From the above description of $K_2(f)$ it follows that the blow-up map $b : B_2(x) \to X \times X$ restricts to a proper surjective map $K_2(f) \to D^2(f)$. What is not so obvious is that this map is, in fact, a morphism of complex spaces.

Proposition 9.18 *The blow-up map $b : B_2(X) \to X \times X$ restricts to a proper surjective morphism of complex spaces $K_2(f) \to D^2(f)$.*

Proof We must show that $b^*(\mathcal{I}^2(f)) \subset J$, where $b^*: \mathcal{O}_{X \times X} \to \mathcal{O}_{B_2(X)}$ is the pull-back of b and J is generated by $A_j v$, with $j = 1, \ldots, p$. We recall that $\mathcal{I}^2(f)$

is generated by the functions $f_j(x^{(2)}) - f_j(x^{(1)})$ and by the $n \times n$-minors of A. On the one hand,

$$f_j(x^{(2)}) - f_j(x^{(1)}) = \lambda A_j v,$$

and thus $b^*(f_j(x^{(2)}) - f_j(x^{(1)})) \in J$. On the other hand, let A' be an $n \times n$ submatrix obtained by choosing some rows $j_1, \dots, j_n$ of A. By Cramer's rule, for all $s = 1, \dots, n$,

$$|A'|\lambda v_s = |A'_s|,$$

where A'_s is the matrix obtained by substitution of the s-th column of A' by the elements $f_{j_\ell}(x^{(2)}) - f_{j_\ell}(x^{(1)})$, $\ell = 1, \dots, n$. In particular, for $s = i$, $v_i = 1$, so $\lambda|A'| = |A'_i|$. Expanding the determinant $|A'_i|$ by the i-th column, we obtain

$$\lambda|A'| = \sum_{\ell=1}^{n} (f_{j_\ell}(x^{(2)}) - f_{j_\ell}(x^{(1)}))|A''_\ell| = \lambda \sum_{\ell=1}^{n} (A_{j_\ell} v)|A''_\ell|,$$

for the corresponding submatrices A''_ℓ of A'_i. This implies that

$$|A'| = \sum_{\ell=1}^{n} (A_{j_\ell} v)|A''_\ell|,$$

and hence $b^*\left(|A'|\right) \in J$. □

An important consequence of Theorem 9.6 is that $K_k(f)$ is well behaved under unfoldings, just as is $D^k(f)$. Let $X \subset \mathbb{C}^n$ and $U \subset \mathbb{C}^r$ be open subsets. Then $B_1(X \times U) = X \times U = B_1(X) \times U$. For $k = 2$, $B_2(X \times U)$ is the submanifold of $X \times U \times X \times U \times \mathbb{P}^{n+r-1}$ of tuples $(x^{(1)}, u^{(1)}, x^{(2)}, u^{(2)}, [w])$ such that

$$((x^{(2)}, u^{(2)}) - (x^{(1)}, u^{(1)})) \wedge w = 0.$$

Then there is an embedding $i : B_2(X) \times U \hookrightarrow B_2(X \times U)$ given by

$$(x^{(1)}, x^{(2)}, [v]) \mapsto (x^{(1)}, u, x^{(2)}, u, [v, 0]).$$

In local coordinates, the embedding can be read as $(x, \gamma, u) \mapsto (x, u, \gamma, 0)$. Let $F : X \times U \to Y \times U$ be a smooth map given by $F(x, u) = (f_u(x), u)$, where $Y \subset \mathbb{C}^p$ is also open. We denote the coordinates in $B_2(X \times U)$ by (x, u, γ, b), where $\gamma = (\lambda, a)$, with $\lambda \in \mathbb{C}$, $a \in \mathbb{C}^{n-1}$ and $b \in \mathbb{C}^r$. For $j = 1, \dots, r$, the divided difference is

$$\begin{aligned} F_{p+j}[x, u, \gamma, b] &= \frac{F_{p+j}((x, u) + \lambda\omega_i(a, b)) - F_{p+j}(x, u)}{\lambda} \\ &= \frac{u_j + \lambda(\omega_i(a, b))_{p+j}}{\lambda} = (\omega_i(a, b))_{p+j} = b_j, \end{aligned}$$

when $i < p + j$ (otherwise we have 1 or b_{j+1}). We deduce that $b = 0$ and hence $K_2(F) \subset i(B_2(X) \times U)$. Moreover, for $j = 1, \ldots, p$,

$$F_j[x, u, \gamma, 0] = \frac{F_j((x, u) + \lambda\omega_i(a, 0)) - F_j(x, u)}{\lambda} = (f_u)_j[x, \gamma].$$

Thus, $(i_u)^{-1}(K_2(F)) = K_2(f_u)$, for all $u \in U$, where $i_u : B_2(X) \hookrightarrow B_2(X \times U)$ is the induced embedding $i_u(z) = i(z, u)$.

In general, there is an embedding $i : B_k(X) \times U \hookrightarrow B_k(X \times U)$ given in local coordinates by

$$(x, \gamma^1, \ldots, \gamma^{(k-1)}, u) \mapsto (x, u, \gamma^1, 0, \ldots, \gamma^{(k-1)}, 0).$$

If we denote by $(x, u, \gamma^1, b^1, \ldots, \gamma^{k-1}, b^{k-1})$ the coordinates of $B_k(X \times U)$, then $i(B_k(X) \times U)$ corresponds to $b^1 = \cdots = b^{k-1} = 0$.

Proposition 9.19 *With the above notation we have:*

1. $K_k(F) \subset i(B_k(X) \times U)$ *and* $(i_u)^{-1}(K_k(F)) = K_k(f_u)$, *for all* $u \in U$.
2. *For all* $j = 1, \ldots, r$ *and for all charts which cover* $i(B_k(X) \times U)$,
$$F_{p+j}[x, u, \gamma^1, 0, \ldots, \gamma^{k-2}, 0, \gamma^{k-1}, b^{k-1}] = b_j^{k-1}.$$
3. *For all* $j = 1, \ldots, p$ *and for all charts which cover* $i(B_k(X) \times U)$,
$$F_j[x, u, \gamma^1, 0, \ldots, \gamma^{k-1}, 0] = (f_u)_j[x, \gamma^1, \ldots, \gamma^{k-1}].$$

Proof Assume the three properties are true for $k - 1$. If we use a diagonal chart φ in $B_k(X \times U)$, then for all $j = 1, \ldots, r$ we get

$$F_{p+j}[x, u, \gamma^1, 0, \ldots, \gamma^{k-2}, 0, \gamma^{k-1}, b^{k-1}] = (\omega_i(a, b^{k-1}))_{p+j} = b_j^{k-1},$$

if $i < p + j$ (or 1 or b_{j+1}^{k-1} if $i \geq p + j$). Moreover, for all $j = 1, \ldots, p$,

$$F_j[x, u, \gamma^1, 0, \ldots, \gamma^{k-2}, 0, \gamma^{k-1}, 0] = (f_u)_j[x, \gamma^1, \ldots, \gamma^{k-1}].$$

Otherwise, if φ is non-diagonal, we have that

$$\varphi_2'' \circ (\varphi_1'')^{-1}(x, u, \gamma^1, 0, \ldots, \gamma^{k-3}, 0) = (x, u, \gamma_1^1, 0, \ldots, \gamma_1^{k-3}, 0),$$

for some $\gamma_1^1, \ldots, \gamma_1^{k-3}$. For all $j = 1, \ldots, r$,

$$\begin{aligned} F_{p+j}[x, u, \gamma^1, 0, \ldots, \gamma^{k-2}, 0, \gamma^{k-1}, b^{k-1}] =& F_{p+j}[x, u, \gamma_1^1, 0, \ldots, \gamma_1^{k-3}, 0\gamma^{k-1}, b^{k-1}] \\ &- F_{p+j}[x, u, \gamma_1^1, 0, \ldots, \gamma_1^{k-2}, 0] \\ =& b_j^{k-1}, \end{aligned}$$

and for all $j = 1, \ldots, p$,

$$F_j[x, u, \gamma^1, 0, \ldots, \gamma^{k-2}, 0, \gamma^{k-1}, 0] = (f_u)_j[x, \gamma^1, \ldots, \gamma^{k-1}].$$

In both cases we deduce the three properties for k. □

The multiple point spaces $K_k(f)$ are also invariant under $\mathscr{A}$-equivalence. The proof is not difficult and is left as an exercise (see Exercises 9.8.1 and 9.8.2). Another consequence of Theorem 9.6, together with the fact that $K_k(f)$ are well behaved under unfoldings and $\mathscr{A}$-equivalence, is that in the corank one case $K_k(f)$ coincides with $D^k(f)$, up to isomorphism of complex spaces.

From now on, we assume that dim $X = n$ and dim $Y = p$, with $n \leq p$.

Corollary 9.9 *Let $f : X \to Y$ be a smooth mapping with only corank one singularities. Then $K_k(f)$ is isomorphic, as a complex space, to $D^k(f)$, for all $k \geq 2$.*

Proof In a neighbourhood of each point, we can choose coordinates in the source and target such that f can be seen as an unfolding of a 1-variable mapping. That is, we can assume that $X = X' \times U$ and $Y = Y' \times U$, where $X' \subset \mathbb{C}$, $U \subset \mathbb{C}^{n-1}$ and $Y' \subset \mathbb{C}^{p-n+1}$ and $f(x, u) = (g_u(x), u)$. By Proposition 9.19, $K_k(f) \subset i(B_k(X') \times U)$ and $(i_u)^{-1}(K_k(f)) = K_k(g_u)$, for all $u \in U$. Since $X' \subset \mathbb{C}$, we have a canonical isomorphism $B_k(X') \equiv (X')^k$ and the generalised divided differences of g_u correspond exactly to the equations of $D^k(g_u)$ in $(X')^k$. □

The following theorem was proved by Ronga in [Ron73] under some genericity assumptions. Here we give a proof in the particular case that the mapping is stable.

Theorem 9.7 *If $f : X \to Y$ is locally stable, then $K_2(f)$ is smooth of dimension $2n - p$ and transverse to the exceptional divisor of the blow-up $b: B_2(X) \to X \times X$.*

Proof We assume $X \subset \mathbb{C}^n$ and $Y \subset \mathbb{C}^p$ are open subsets. We claim that it is enough to prove that there exists an unfolding F such that $K_2(F)$ is smooth of codimension p and transverse to the exceptional divisor. In fact, since f is stable, after shrinking the open sets if necessary, F is trivial, which implies that $K_2(F)$ is isomorphic to $K_2(f) \times U$ (where U is the parameter space), by a diffeomorphism which preserves the parameters and commutes with the blow-up maps. Hence, $K_2(f)$ is also smooth of codimension p and transverse to the exceptional divisor.

Denote by H the space of linear maps from $\mathbb{C}^n$ to $\mathbb{C}^p$. We consider H as the parameter space and define the unfolding $F(x, L) = (f_L(x), L)$ given by

$$f_L = f + L, \ L \in H.$$

By Proposition 9.19, we can embed $K_2(F)$ in $B_2(X) \times H$, with local equations given by the divided differences of f_L. That is, after taking a chart $\varphi_i : V_i \to A_i$ in $B_2(X)$, the equations of $K_2(F)$ are the components of the divided differences $f_L[x, \gamma]$. For the linear map L, the divided difference is

$$L[x, \gamma] = \frac{L(x + \lambda\omega_i(a)) - L(x)}{\lambda} = L(\omega_i(a)),$$

where $\gamma = (\lambda, a)$, $\lambda \in \mathbb{C}$ and $a \in \mathbb{C}^{n-1}$. By the linearity of the divided differences, we have

$$f_L[x, \gamma] = f[x, \gamma] + L(\omega_i(a)).$$

Since $\omega_i(a) \neq 0$, $f_L[x, \gamma]$ has rank p when we consider the coefficients of L as variables. Hence, $K_2(F)$ is smooth of codimension p. Moreover, since $L(\omega_i(a))$ is independent of λ, $K_2(F)$ is transverse to the exceptional divisor given by $\lambda = 0$. □

It follows from Theorem 9.7 that $K_2(f)$ is a desingularisation of the double point space $D^2(f)$ when f is stable. If Z is any analytic space, a *resolution* of Z is a proper surjective map $\pi : Z' \to Z$ such that Z' is smooth, $\dim(\pi^{-1}(\text{Sing } Z)) < \dim Z$ and the restriction $\pi : Z' \setminus \pi^{-1}(\text{Sing } Z) \to Z \setminus \text{Sing } Z$ is a diffeomorphism.

Corollary 9.10 *If $f : X \to Y$ is locally stable, then $b : K_2(f) \to D^2(f)$ is a resolution of $D^2(f)$ and $K_2(f)$ is the closure of the inverse image of the strict double points.*

Proof We already know that $b : K_2(f) \to D^2(f)$ is a proper surjective mapping. If f is stable, $K_2(f)$ is smooth of dimension $2n - p$ by Theorem 9.7. Also by Theorem 9.4, $\text{Sing}(D^2(f)) = \Delta\Sigma$, where Σ is the set of points $x \in X$ such that f has corank ≥ 2 at x and the restriction of b to $D^2(f) \setminus \Delta\Sigma$ is a diffeomorphism by Corollary 9.9. To prove that it is a resolution it only remains to see that $\dim(b^{-1}(\Delta\Sigma)) < 2n - p$.

Write $\Sigma = \Sigma^2 \cup \cdots \cup \Sigma^n$, where Σ^r is the set of points $x \in X$ such that f has corank r at x. We have

$$b^{-1}(\Delta\Sigma) = b^{-1}(\Delta\Sigma^2) \cup \cdots \cup b^{-1}(\Delta\Sigma^n),$$

with $\dim b^{-1}(\Delta\Sigma^r) = \dim \Sigma^r + r - 1$, for all $r = 2, \ldots, n$ such that $\Sigma^r \neq \emptyset$. If $\Sigma^r \neq \emptyset$, then necessarily $r \geq n - p$ and thus,

$$\dim b^{-1}(\Delta\Sigma^r) = n - r(p - n + r) + r - 1 < 2n - p,$$

since

$$2n - p - (n - r(p - n + r) + r - 1) = (p - n + r)(r - 1) + 1 > 0.$$

Finally, $K_2(f)$ is transverse to the exceptional divisor E of the blow-up $b : B_2(X) \to X \times X$. Hence, $E \cap K_2(f)$ is a smooth hypersurface of $K_2(f)$ and $K_2(f)$ is equal to the closure of the complement of $E \cap K_2(f)$. □

The analogues of Theorem 9.7 and Corollary 9.10 for triple points are also true. The proofs use the same techniques as Theorem 9.7 and Corollary 9.10 and can be found in [NBPS19].

Theorem 9.8 ([NBPS19]) *If $f : X \to Y$ is locally stable, then $K_3(f)$ is smooth of dimension $3n - 2p$ and transverse to the exceptional divisor of the blow-up, the projection $K_3(f) \to D^3(f)$ is a resolution of $D^3(f)$ and $K_3(f)$ is the closure of the inverse image of the strict triple points.*

However, for $k \geq 4$, it is not true that $K_k(f)$ is smooth for every stable map $f : X \to Y$. In fact, we will see that if $n < p$ and f has a singularity $x_0 \in X$ of corank ≥ 2, then for all $k \geq 2$ $K_k(f)$ is non-empty. Worse, its dimension increases with k. Since $d_k(n, p)$ is a decreasing function of k, $\dim K_k(f) > d_k(n, p)$ for sufficiently high k. In particular, if f is stable, there are corank one singularities near x_0, over which $D^k(f)$ has dimension $d_k(n, p)$, and it follows that $K_k(f)$ has components of distinct dimensions on each neighbourhood of x_0.

Proposition 9.20 *Let $f : X \to Y$ be a smooth mapping with singularities of corank $r \geq 2$. Then $K_k(f) \neq \emptyset$, for all $k \geq 2$ and there exists k_0 such that for all $k \geq k_0$, $\dim K_k(f) > d_k(n, p)$.*

Proof Let $x_0 \in X$ be a singularity of corank $r \geq 2$ of f. We will show first that the fibre of $(x_0, \dots, x_0) \in X^k$ by the map $K_k \to X^k$ has dimension $\geq (r - 1)(k - 1)$. In fact, in a neighbourhood of x_0, we can assume that f is an unfolding $f : X' \times U \to Y' \times U$, given by $f(x', u) = (g_u(x'), u)$, for some open subsets $X' \subset \mathbb{C}^r$, $U \subset \mathbb{C}^{n-r}$ and $Y' \subset \mathbb{C}^{p-n+r}$. Moreover, if $x_0 = (x'_0, u_0)$, then $d_{x'_0} g_{u_0} = 0$. By Proposition 9.19, we can consider $K_k(f)$ embedded in $B_k(X') \times U$, where it is given by the divided differences of g_u. When $k = 2$, we have $g_{u_0}[x'_0, 0, a] = 0$, for all $a \in \mathbb{C}^{r-1}$. Since $r \geq 2$, we can choose a pair $a, a' \in \mathbb{C}^{r-1}$ such that $a \neq a'$. We have $g_{u_0}[x'_0, 0, a] = 0 = g[x'_0, 0, a']$ and taking limits when $a' \to a$ we deduce that

$$g_u[x'_0, 0, a^1, 0, a^2] = 0,$$

for all $a^1, a^2 \in \mathbb{C}^{r-1}$. We can proceed analogously by induction and prove that

$$g_{u_0}[x'_0, 0, a^1, \dots, 0, a^{k-1}] = 0,$$

for all $a^1, \dots, a^{k-1} \in \mathbb{C}^{r-1}$. All the points in $K_k(f)$ with coordinates

$$(x'_0, 0, a^1, \dots, 0, a^{k-1}, u_0)$$

are contained in the fibre of $(x_0, \dots, x_0)$, which implies that the fibre has dimension $\geq (r - 1)(k - 1)$.

Since the set $\Sigma^r(f)$ of points of corank r has dimension $\geq n - (p - n + r)r$ we have

$$\dim K_k(f) \geq (n - (p - n + r)r) + (r - 1)(k - 1).$$

Thus, $K_k(f) \neq \emptyset$, for all $k \geq 2$ and there exists k_0 such that for all $k \geq k_0$, $\dim K_k(f) > d_k(n, p)$. □

The fact that the dimension of $\dim K_k(f) > d_k(n, p)$ for all $k \geq k_0$ when f has corank 2 singularities is a pathology of the multiple point spaces $K_k(f)$. However, we can use this property to give a geometric characterisation of $\mathscr{A}$-finiteness of map-germs $f : (\mathbb{C}^n, S) \to (\mathbb{C}^p, 0)$ if (n, p) are dimensions such that the corank 2 singularities of stable maps are at most isolated points. If f is stable, then $\Sigma^2(f)$, if not empty, has dimension $n - 2(p - n + 2)$ so we assume that $n - 2(p - n + 2) \leq 0$. We cannot hope that $K_k(f)$ should have isolated singularity. For unlike in the case of $D^k(f)$, the preimage of S in $K_k(f)$ has positive dimension, due to the blowing up involved. The analogue of the condition of isolated singularity here is that $K_k(f)$ should be non-singular away from the preimage of S.

Theorem 9.9 ([NBPS19]) *Let $f : (\mathbb{C}^n, S) \to (\mathbb{C}^p, 0)$ be a finite map-germ, with $n < p$ and $n - 2(p - n + 2) \leq 0$. Then f is $\mathscr{A}$-finite if and only if all the spaces $K_k(f)$ are smooth of dimension $d_k(n, p)$ away from the preimage of S.*

Proof Let f be $\mathscr{A}$-finite. By the Mather–Gaffney criterion Theorem 4.5 there exists a representative $f : X \to Y$ such that $f^{-1}(0) = S$ and the restriction $f : X \setminus S \to Y \setminus \{0\}$ is stable. The condition on dimensions means that f has at most isolated singularities of corank 2 on $X \setminus S$. After shrinking X and Y if necessary, we can assume that f has only corank 1 singularities on $X \setminus S$. Thus, $K_k(f)$ is isomorphic to $D^k(f)$, which is smooth of dimension $d_k(n, p)$ on $X \setminus S$, by Theorem 9.4 and Corollary 9.9.

Conversely, assume there is a representative $f : X \to Y$ such that $f^{-1}(0) = S$ and $K_k(f_{X \setminus S})$ are smooth of dimension $d_k(n, p)$. By Proposition 9.20, $f|_{X \setminus S}$ has only corank one singularities. It follows that $K_k(f|_{X \setminus S}) \equiv D^k(f|_{X \setminus S})$ are smooth of dimension $d_k(n, p)$ by Corollary 9.9. Thus, $f|_{X \setminus S}$ is stable by Theorem 9.4 and f is $\mathscr{A}$-finite by the Mather–Gaffney criterion Theorem 4.5. □

Example 9.11 ([NBPS18]) Let $f : (\mathbb{C}^3, 0) \to (\mathbb{C}^4, 0)$ be the corank 2 germ

$$f(x, y, z) = (x^2 + zy, y^2 - zx, x^3 + y^3 + xy, z).$$

We can see this as an unfolding $f(x, y, z) = (g_z(x, y), z)$. Figure 9.4 shows real pictures of the mappings g_0 and g_z, with $z \neq 0$. We compute $K_k(f)$ as the subspace of $B_2(\mathbb{C}^2) \times \mathbb{C}$ given by the divided differences of g_z. In the chart φ_1 with coordinates (x, y, λ, a, z), the first divided differences of g_z are

$$g_z[x, y, \lambda, a] = \frac{g_z(x + \lambda, y + \lambda a) - g_z(x, y)}{\lambda},$$

which give three equations $az + \lambda + 2x$, $a^2\lambda + 2ay - z$ and

$$a^3\lambda^2 + 3a^2\lambda y + a\lambda + ax + 3ay^2 + \lambda^2 + 3\lambda x + 3x^2 + y.$$

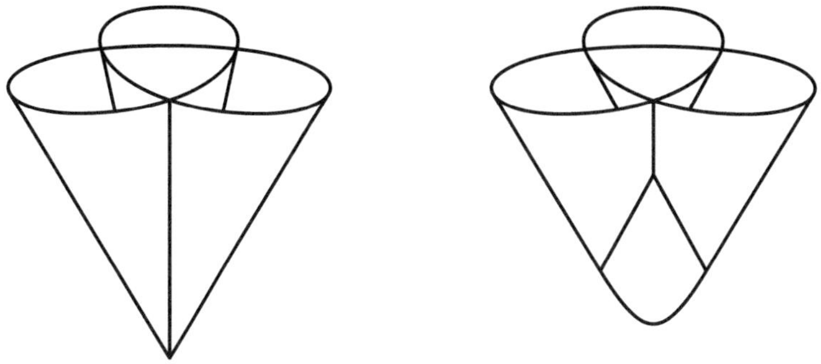

Fig. 9.4 The mappings g_0 (left-hand side) and g_z, $z \neq 0$ (right-hand side)

With the aid of a computer it is easy to see that $K_2(f)$ is a surface in $B_2(\mathbb{C}^2) \times \mathbb{C}$ with only three singular points corresponding to $x = y = z = \lambda = 0$ and $a \in \mathbb{C}^3$ such that $a^3 = 1$. The computations with the chart φ_2 are analogous and give the same results.

We now compute $K_3(f)$. Assume we have used the chart φ_1 in the first divided differences and take also φ_1 for the second ones. Thus,

$$g_z[x, y, \lambda, a, \mu, b] = \frac{g_z[x, y, \lambda + \mu, a + \mu b] - g_z[x, y, \lambda, a]}{\mu},$$

which gives three new equations $bz + 1$, $a^2 + 2ab\mu + 2ab\lambda + b^2\mu^2 + b^2\mu\lambda + 2by$ and

$$\begin{aligned} &a^3\mu + 2a^3\lambda + 3a^2b\mu^2 + 6a^2b\mu\lambda + 3a^2b\lambda^2 + 3a^2y + 3ab^2\mu^3 \\ &+ 6ab^2\mu^2\lambda + 3ab^2\mu\lambda^2 + 6ab\mu y + 6ab\lambda y + a + b^3\mu^4 + 2b^3\mu^3\lambda \\ &+ b^3\mu^2\lambda^2 + 3b^2\mu^2 y + 3b^2\mu\lambda y + b\mu + b\lambda + bx + 3by^2 + \mu + 2\lambda + 3x. \end{aligned}$$

Again it is not difficult to see (with the aid of the computer) that these three equations together with the three equations above define a curve with a finite number of singularities. Observe that since f has corank 2 at the origin, $K_3(f)$ must have a 2-dimensional component which does not appear in this chart.

The result is different if we take the chart φ_2 instead of φ_1 for the second divided differences. In this case, the three new equations are $b + z$, $a^2b + 2ab\mu + 2a\lambda + b\mu^2 + \mu\lambda + 2y$ and

$$\begin{aligned} &a^3b^2\mu + 2a^3b\lambda + 3a^2b^2\mu^2 + 6a^2b\mu\lambda + 3a^2by + 3a^2\lambda^2 + 3ab^2\mu^3 \\ &+ 6ab\mu^2\lambda + 6ab\mu y + ab + 3a\mu\lambda^2 + 6a\lambda y + b^2\mu^4 + b^2\mu + 2b\mu^3\lambda \\ &+ 3b\mu^2 y + b\mu + 2b\lambda + 3bx + \mu^2\lambda^2 + 3\mu\lambda y + \lambda + x + 3y^2. \end{aligned}$$

The 2-dimensional component appears now with equations $x = y = z = \lambda = b = 0$, but when $(x, y, z) \neq 0$, $K_3(f)$ has dimension 1 with a finite number of singularities. The results are analogous if we start with φ_2 in the first step. In all cases, we see that $K_3(f)$ is smooth of dimension 1 on $X \setminus \{0\}$ for some open neighbourhood X of 0 in $\mathbb{C}^3$.

We know that f has only corank 1 singularities when $(x, y, z) \neq 0$. Since f has multiplicity 3, we have $D^k(f) = \emptyset$ for all $k \geq 4$ and thus $K_k(f)$ is contained in the fibre of the origin, for all $k \geq 4$. It follows that f is $\mathscr{A}$-finite by Theorem 9.9. $\diamondsuit$

Exercises for Sect. 9.8

1 Let $f : X \to Y$ be a smooth mapping and $\varphi : X \to X'$ be a diffeomorphism. Show that for each k, φ induces a diffeomorphism $\tilde{\varphi} : B_k(X) \to B_k(X')$ which commutes with the projection maps such that $\tilde{\varphi}(K_k(f)) = K_k(f \circ \varphi^{-1})$, as complex spaces.

2 Let $f : X \to Y$ be a smooth mapping and $\psi : Y \to Y'$ an embedding. Show that $K_k(f) = K_k(\psi \circ f)$, for all k.

3 Let $f\colon (\mathbb{C}^n, S) \to (\mathbb{C}^p, 0)$ be a smooth germ such that $K_2(f)$ is smooth of dimension $2n - p$ and with no triple points. Show that f is stable.

4 Let $f : \mathbb{C}^4 \to \mathbb{C}^4$ be the mapping $f(x, y, u, v) = (x^2 + uy, y^2 + vx, u, v)$. Find the equations of $K_2(f)$ and $K_3(f)$ and show that they are smooth of dimension 4.

9.9 What Remains To Be Done

1. In Chap. 10, we apply the results of this chapter to calculate the homology of the image of a stable perturbation of an $\mathscr{A}$-finite germ. Crucial to this, when the germ f has corank 1, is the fact that $D^k(f)$ is an isolated complete intersection singularity (ICIS). For this means that when f_t is a stable perturbation of f, then $D^k(f_t)$, as a Milnor fibre of an ICIS, has reduced homology only in middle dimension—see, e.g. [Loo84, Chapter 5]. This homological information on $D^k(f_t)$ is turned into homological information on the image of f_t by the *image computing spectral sequence*, which we introduce in the next chapter.

   A lot remains to be done here. For germs f of corank > 1, we do not have explicit equations for $D^k(f)$ when $k > 2$, and even the equations for $D^2(f)$ that we do have do not give us any insight into the homology of $D^2(f_t)$, since we do not have a description of a Milnor fibre for singularities of the type to which $D^2(f)$ belongs—of the variety of complexes. Indeed, in high dimensions a stable perturbation of a germ of corank greater than 1 may not even result in a smoothing of its multiple point spaces, as noted in Sect. 9.7. So from the point of view of the light they might shed on the geometry and topology of a stable perturbation, our explicit equations are not, as yet, of very much help.

We have defined $D^k(f)$, for stable f, using transporter ideals. If L and J are ideals and $L = (I : J)$ for some ideal I, then L is said to be a *residual intersection* of J. It is an *s-residual intersection* if I is generated by s elements, where $s \geq \text{height}(J)$, and $\text{height}(L) \geq s$. If $f : (\mathbb{C}^n, 0) \to (\mathbb{C}^p, 0)$ with $n \leq p$ is $\mathscr{A}$-finite, then the codimension of the set $D^k(f)$ in $(\mathbb{C}^n)^k$ is $p(k-1)$, equal to the number of generators of $I_k(f)$, and greater than the height, n, of J_k, so the ideal $(I_k(f) : J_k)$ is a $p(k-1)$-residual intersection of J_k.

So it is desirable to understand the properties of the residual intersections involved. Many authors, beginning with Artin and Nagata in [AN72], have given conditions which guarantee that residual intersections are Cohen–Macaulay. Unfortunately it seems that none of them seem to cover the cases that interest us here. The most widely applicable result, due to Hassanzadeh in [Has12], uses the notion, due to Herzog, Vasconcelos and Villareal in [HVV85], of *ideals of sliding depth*. An ideal J generated by r elements $g_1, \ldots, g_r$ in a local ring R is said to have sliding depth if for each i,

$$\text{depth}\, H_i\big(K_\bullet(g_1, \ldots, g_r)\big) \geq \dim\ R - r + i,$$

where $K_\bullet(g_1, \ldots, g_r)$ is the Koszul complex on the generators of the ideal, and H_i is its i'th homology. By convention, $\text{depth}(0) = \infty$. Hassanzadeh shows that if R is a Cohen–Macaulay ring and J is an ideal with sliding depth, then every residual intersection of J is Cohen–Macaulay. Unfortunately, as one can check with a *Macaulay 2* calculation, our ideal J_k does not have sliding depth if $k > 2$. The ideal J_2 is generated by a regular sequence, so does indeed have sliding depth—more precisely, all of its Koszul homology modules vanish. Thus from Hassanzadeh's result we get only another proof that $(I_2(f) : J_2)$ defines a Cohen–Macaulay scheme.

Despite the lack of a general theorem, at least for those stable germs of corank 2 for which computer calculations terminate, $D^k(f)$ is Cohen–Macaulay, and the top multiple point space $D^m(f)$ is Gorenstein. It would be very interesting to discover how widely this holds.

2. A second task is to describe the homology of $D^k(f_t)$ when f_t is a stable perturbation of a germ of corank > 1. In contrast to the corank 1 case, $D^k(f_t)$ may have reduced homology outside middle dimension—calculations with an $\mathscr{A}_e$-codimension 1 germ in [Mon18] show this. A remarkable theorem of Kevin Houston in [Hou97] states that despite this, the so-called *alternating homology* of $D^k(f_t)$, which we study in Chap. 10, is confined to middle dimension. The alternating homology is, roughly speaking, the subspace of the homology of $D^k(f_t)$ on which the symmetric group S_k acts by its sign representation. It is the part of the homology that contributes to the spectral sequence described in the next chapter, and it is thanks to the fact that the alternating homology is confined to middle dimension that the spectral sequence collapses at E^1, giving a simple

way to read off the homology of the image of f_t from the alternating homology of the $D^k(f_t)$.

3. The recent paper [NP15] of Némethi and Pinter identifies the Whitney umbrella number W of a map-germ $f : (\mathbb{C}^2, 0) \to (\mathbb{C}^3, 0)$ (i.e. the number of Whitney umbrellas on the image of a stable perturbation—see Exercise E.4.2) with the Smale invariant of the immersion $S^3 \to S^5$ obtained by restricting f to the boundary of a small ball around 0. This shows in particular that W is a topological invariant of f. Is the triple point number of f also a topological invariant? Which other analytic invariants are in fact topological invariants?

Chapter 10
Calculating the Homology of the Image

Let $f : X \to Y$ be a finite map. For each $k \geq 2$ and $1 \leq i \leq k$ there is a projection $\varepsilon^{k,i} : D^k(f) \to D^{k-1}(f)$ defined by forgetting the i'th copy of X. These projections give rise to maps on the homology of the spaces $D^k(f_t)$ when f_t is a stable perturbation of f; there is thus a rich structure of homology groups and homomorphisms associated to a stable perturbation. It turns out that from this, one can obtain information about the homology of the image of the stable perturbation. The action of the symmetric group S_k on $D^k(f)$ determines the glueing which takes place when the domain of f_t is mapped to the image, and it is therefore no surprise that in the computation of the homology of the image, this action should play a rôle. In fact it is the *alternating* part of the homology which enters into the calculation of $H_*(\text{image}(f_t))$. This was first observed in [GM93] at the level of rational homology. For any map $f : X \to Y$, we define

$$H_q^{\text{Alt}}(D^k(f); \mathbb{Q}) = \{[c] \in H_q(D^k(f); \mathbb{Q}) : \sigma_*([c]) = \text{sign}(\sigma)[c] \text{ for all } \sigma \in S_k\},$$

and refer to it as the *alternating part* of $H_q(D^k(f); \mathbb{Q})$. Later the construction was greatly clarified by Goryunov in [Gor95], by the introduction of the *alternating chain complex*, which we now describe. The description here differs from Goryunov's only in that it uses singular homology, and later simplicial homology, in place of cellular homology.

Before we do this, we simplify our notation slightly. The projections $\varepsilon^{k,i} : D^k(f) \to D^{k-1}(f)$ for different i are all left-right equivalent to one another thanks to the symmetric group actions on $D^k(f)$ and $D^{k-1}(f)$. We will use only $\varepsilon^{k,k}$, which we refer to from now on as ε^k.

Remark 10.1 In all of what follows, we use the purely topological version of the multiple point space, namely $D^k_{\text{cl}}(f)$, the closure in X^k of the set of strict k-tuple points. When X is simply a topological space, with no analytic structure assumed, no other definition is available to us. However, it is worth noting that when X is a

D. Mond, J. J. Nuño-Ballesteros, *Singularities of Mappings*,
Grundlehren der mathematischen Wissenschaften 357,
https://doi.org/10.1007/978-3-030-34440-5_10

complex manifold, then if we assume that all the germs of f are $\mathscr{A}$-finite, and use Definition 9.3 to obtain a possibly larger space, the substitution of $D^k(f)$ by $D^k_{\mathrm{cl}}(f)$ will make no difference to the alternating homology groups. For by Proposition 9.4, any point of $D^k(f) \smallsetminus D^k_{\mathrm{cl}}$ is isolated in $D^k(f)$. Moreover, it involves fewer than k distinct points of X, and so does not contribute to the alternating homology which we will now define.

However, *because the subscript* "cl" *interferes with other subscripts, we will drop it, and refer only to* $D^k(f)$. ◊

10.1 The Alternating Chain Complex

Let D be any space on which the symmetric group S_k acts, and let $C_\ell(D)$ be the usual free abelian group of singular ℓ-chains in D. A chain $c \in C_\ell(D)$ is *alternating* if for each $\sigma \in S_k$, $\sigma_\#(c) = \operatorname{sign}(\sigma)c$. We denote the set of all alternating ℓ-chains with integer coefficients on D by $C_\ell^{\mathrm{Alt}}(D)$. It is, evidently, a subgroup of $C_\ell(D)$, and therefore free abelian. The $C_\ell^{\mathrm{Alt}}(D)$ form a complex under the usual boundary map; we call its homology the *alternating homology of* D, and denote it by $AH_*(D)$. The next proposition shows that if one replaces coefficients in $\mathbb{Z}$ with coefficients in $\mathbb{Q}$, then AH_* coincides with H_*^{Alt}; the next example shows that over $\mathbb{Z}$ the two need not coincide.

Proposition 10.1 *For any space D on which S_k acts,*

$$H_q^{Alt}(D; \mathbb{Q}) \simeq AH_q(D; \mathbb{Q}).$$

Proof The operator $\mathrm{Alt}_{\mathbb{Q}} : C_\ell(D; \mathbb{Q}) \to C_\ell(D; \mathbb{Q})$ defined by

$$\mathrm{Alt}_{\mathbb{Q}}(c) = \frac{1}{k!} \sum_{\sigma \in S_k} \operatorname{sign}(\sigma)\sigma_\#(c)$$

is idempotent ($\mathrm{Alt}_{\mathbb{Q}} \circ \mathrm{Alt}_{\mathbb{Q}} = \mathrm{Alt}_{\mathbb{Q}}$) and has image $C_\ell^{\mathrm{Alt}}(D; \mathbb{Q})$. It gives a left inverse to the inclusion of chain complexes $i : C_\bullet^{\mathrm{Alt}}(D; \mathbb{Q}) \hookrightarrow C_\bullet(D; \mathbb{Q})$. Suppose $[c_\ell] \in AH_\ell(D; \mathbb{Q})$ and $i_*([c_\ell]) = 0$ in $H_\ell(D; \mathbb{Q})$. Then $i_\#(c_\ell) = \partial c_{\ell+1}$ in $C_\bullet(D; \mathbb{Q})$; thus $c_\ell = \mathrm{Alt}_{\mathbb{Q}}(c_\ell) = \mathrm{Alt}_{\mathbb{Q}}\partial c_{\ell+1} = \partial \mathrm{Alt}_{\mathbb{Q}} c_{\ell+1}$, so that $[c_\ell] = 0$ in $AH_\ell(D; \mathbb{Q})$, and i_* is injective.

Clearly $i_* AH_\ell(D; \mathbb{Q}) \subset H_\ell^{\mathrm{Alt}}(D; \mathbb{Q})$. If $[c_\ell] \in H_\ell^{\mathrm{Alt}}(D; \mathbb{Q})$ then for each $\sigma \in S_k$, $\sigma_*[c_\ell] = \operatorname{sign}(\sigma)[c_\ell]$, and so $c_\ell - \operatorname{sign}(\sigma)\sigma_\#(c_\ell) = \partial \bar{c}_\sigma$ for some $\bar{c}_\sigma \in C_{\ell+1}(D; \mathbb{Q})$. Then

$$c_\ell - \mathrm{Alt}_{\mathbb{Q}}\, c_\ell = \frac{1}{k!} \sum_{\sigma \in S_k} (c_\ell - \operatorname{sign}(\sigma)\sigma_\#(c_\ell)) = \partial\Big(\frac{1}{k!} \sum_{\sigma \in S_k} \bar{c}_\sigma\Big)$$

and thus $[c_\ell] = i_*[\mathrm{Alt}_{\mathbb{Q}}\, c_\ell] \in i_* AH_\ell(D; \mathbb{Q})$. □

Example 10.1 With integer coefficients, $AH_\ell^{\text{Alt}}(D^k(f)) \to H_\ell^{\text{Alt}}(D^k(f))$ is not always an isomorphism. Let B^2 be the closed unit ball in $\mathbb{R}^2$ and consider the quotient surjection $q : B^2 \to \mathbb{RP}^2$ under which opposite points on the boundary of B^2 are identified. Then

$$D^2(q) = \{(x_1, x_2) \in \partial B^2 \times \partial B^2 : x_2 = -x_1\}$$

and the projection ε^2 identifies $D^2(q)$ with $S^1 = \partial B^2$, with the involution $(1, 2)$ on $D^2(q)$ descending to the antipodal map a on S^1. Thus we can identify $C_j^{\text{Alt}}(D^2(q))$ with $\{c \in C_j(S^1) : a_\#(c) = -c\}$. For any $x \in S^1$, $x - a_\#(x)$ is an alternating 0-chain, but is not an alternating boundary (see Exercise 10.1.1 below), although it is evidently a boundary in $C_\bullet(S^1)$. Thus $AH_0(D^2(q)) \neq 0 = H_0^{\text{Alt}}(D^2(q))$, and $AH_0(D^2(q)) \to H_0^{\text{Alt}}(D^2(q))$ is not injective. On the other hand, if c parameterises the anticlockwise arc running from $a(x)$ to x, then $c - a_\# c$ is an alternating 1-chain whose boundary is $2([x] - [a(x)])$, so the class of $[x] - a_\#[x]$ in $AH_0(D^2(q))$ is 2-torsion. It turns out that in fact this class generates $AH_0(D^2(q))$ and $AH_0(D^2(q)) \simeq \mathbb{Z}_2$.

As we will see, it is no coincidence that $AH_0(D^2_{\text{cl}\,1}(q)) \simeq H_1(\mathbb{RP}^2)$.

Lemma 10.1 $\varepsilon_\#^k(C_n^{Alt}(D^k(f)) \subset C_n^{Alt}(D^{k-1}(f))$.

Proof Consider the standard inclusion $\{1, \ldots, k-1\} \hookrightarrow \{1, \ldots, k\}$. Any $\sigma \in S_{k-1}$ extends to $\bar{\sigma} \in S_k$ by setting $\bar{\sigma}(k) = k$; then $\text{sign}(\bar{\sigma}) = \text{sign}(\sigma)$. For any chain c,

$$\sigma_\#(\varepsilon_\#^k(c)) = \varepsilon_\#^k(\bar{\sigma}_\#(c)).$$

If c is alternating then the right-hand side is $\varepsilon_\#^k(\text{sign}(\bar{\sigma})c)$ and thus equal to $\text{sign}(\sigma)\varepsilon_\#^k(c)$. □

It is convenient to include X and Y in the sequence of multiple points spaces, as $D^1(f)$ and $D^0(f)$, and in this context to regard $f : X \to Y$ as $\varepsilon^1 : D^1(f) \to D^0(f)$. We obtain the sequence

$$\cdots \longrightarrow D^{k+1}(f) \xrightarrow{\varepsilon^{k+1}} D^k(f) \xrightarrow{\varepsilon^k} D^{k-1}(f) \xrightarrow{\varepsilon^{k-1}}$$

$$\cdots \longrightarrow D^2(f) \xrightarrow{\varepsilon^2} X \xrightarrow{f} Y \tag{10.1}$$

In view of the lemma, for each fixed $n \in \mathbb{N}$ this gives rise to a sequence of chain maps

$$\cdots \longrightarrow C_n^{\mathrm{Alt}}(D^{k+1}(f)) \xrightarrow{\varepsilon_\#^{k+1}} C_n^{\mathrm{Alt}}(D^k(f)) \xrightarrow{\varepsilon_\#^{k}}$$

$$\cdots \xrightarrow{\varepsilon_\#^{3}} C_n^{\mathrm{Alt}}(D^2(f)) \xrightarrow{\varepsilon_\#^{2}} C_n(X) \xrightarrow{f_\#} C_n(Y) \longrightarrow 0, \tag{10.2}$$

Lemma 10.2 *For each $n \geq 0$, (10.2) is a chain complex, i.e. for $k \geq 2$, $\varepsilon_\#^{k-1} \circ \varepsilon_\#^k = 0$ on $C_\bullet^{Alt}(D^k(f))$.*

Proof Consider the action of the transposition $(k-1, k)$. We have

$$\varepsilon^{k-1} \circ \varepsilon^k \circ (k-1, k) = \varepsilon^{k-1} \circ \varepsilon^k,$$

since $\varepsilon^{k-1} \circ \varepsilon^k$ forgets both of the last two components in X^k. So

$$\varepsilon_\#^{k-1} \circ \varepsilon_\#^k \circ (k-1, k)_\# = \varepsilon_\#^{k-1} \circ \varepsilon_\#^k.$$

But if c is an alternating chain,

$$\varepsilon_\#^{k-1} \circ \varepsilon_\#^k \circ (k-1, k)_\#(c) = \varepsilon_\#^{k-1} \circ \varepsilon_\#^k(-c) = -\varepsilon_\#^{k-1} \varepsilon_\#^k(c).$$

Thus $2\varepsilon_\#^{k-1} \circ \varepsilon_\#^k(c) = 0$, and since we are working with integer coefficients, $\varepsilon_\#^{k-1} \circ \varepsilon_\#^k(c) = 0$. □

There is more: since $\varepsilon_\#$ is a chain morphism, the diagram of free abelian groups and homomorphisms given by the two-index array $(C_n^{\mathrm{Alt}}(D^k(f)), \partial, \varepsilon_{\#,n}^k)_{n,k}$ commutes, and each row and column is a chain complex.

The key step in our construction of the image computing spectral sequence is to show that if f is a simplicial map, and we replace singular chains by simplicial chains, with respect to triangulations of the $D^k(f)$ which we will construct, then the complex (10.2) is exact.

A doubly indexed array $C_{j,k}$ with differentials $d_h : C_{j,k} \to C_{j-1,k}$ and $d_v : C_{j,k} \to C_{j,k-1}$ is a *double complex* if each row $C_{\bullet,k}$ and each column $C_{j,\bullet}$ is a complex, and if each square

$$\begin{array}{ccc} C_{j,k} & \xrightarrow{d_h} & C_{j-1,k} \\ {\scriptstyle d_v}\downarrow & & \downarrow{\scriptstyle d_v} \\ C_{j,k-1} & \xrightarrow{d_h} & C_{j-1,k-1} \end{array}$$

anticommutes: $d_h \circ d_v = -d_v \circ d_h$. Anti-commutation is required so that the direct sums $T_n := \bigoplus_{j+k} = n$ form a complex, with differential $T_n \to T_{n-1}$ defined by applying $d_v + d_h$ to each summand in T_n. The complex $(T_\bullet, d_h + d_v)$ is known as the *total complex* associated to the double complex. In our case, with $C_{j,k} = C_j^{\mathrm{Alt}}(D^k(f))$, $d_h = \partial$ (the usual chain boundary operator) and $d_v = \varepsilon_{\#}^k$, we have commutation rather than anti-commutation; to remedy this, we use a standard trick [Rot09, Example 10.4] and define

$$\epsilon : C_j^{\mathrm{Alt}}(D^k(f)) \to C_j^{\mathrm{Alt}}(D^{k-1}(f)) = (-1)^j \varepsilon_{\#}^k,$$

to get a (first quadrant) double complex $(C_j^{\mathrm{Alt}}(D^k(f)), \partial, \epsilon)$ with $j, k \geq 0$.

This double complex, and its total complex, will play a central role in the rest of this chapter. For the sake of later developments we draw a diagram.

$$\begin{array}{ccccccccccc}
 & & \downarrow & & \downarrow & & & & \downarrow & & \\
0 & \longleftarrow & C_0^{\mathrm{Alt}}(D^k) & \longleftarrow & C_1^{\mathrm{Alt}}(D^k) & \longleftarrow & \cdots & \longleftarrow & C_r^{\mathrm{Alt}}(D^k(f)) & \longleftarrow & \cdots \\
 & & \downarrow & & \downarrow & & & & \downarrow & & \\
0 & \longleftarrow & C_0^{\mathrm{Alt}}(D^{k-1}) & \longleftarrow & C_1^{\mathrm{Alt}}(D^{k-1}) & \longleftarrow & \cdots & \longleftarrow & C_r^{\mathrm{Alt}}(D^{k-1}) & \longleftarrow & \cdots \\
 & & \downarrow & & \downarrow & & & & \downarrow & & \\
 & & \vdots & & \vdots & & \cdots & & \vdots & & \cdots \\
 & & \downarrow & & \downarrow & & & & \downarrow & & \\
 & \longleftarrow & C_0^{\mathrm{Alt}}(D^2(f) & \overset{\partial}{\longleftarrow} & C_1^{\mathrm{Alt}}(D^2)f) & \overset{\partial}{\longleftarrow} & \cdots & \overset{\partial}{\longleftarrow} & C_r^{\mathrm{Alt}}(D^2(f)) & & \cdots \\
 & & \downarrow & & \downarrow & & & & \downarrow & & \\
0 & \longleftarrow & C_0(X) & \overset{\partial}{\longleftarrow} & C_1(X) & \overset{\partial}{\longleftarrow} & \cdots & \overset{\partial}{\longleftarrow} & C_r(X) & \longleftarrow & \cdots \\
 & & \downarrow & & \downarrow & & & & \downarrow & & \\
 & & 0 & & 0 & & \cdots & & 0 & & \cdots
\end{array} \qquad (10.3)$$

10.1.1 Motivation

We deviate briefly from this account in order to motivate the definition of the alternating chain complex. Let $f : X \to Y$ be a finite map. The significance of the alternating homology groups $AH_\ell(D^k(f))$ for the homology of the image of f is suggested by the following two sample calculations.

Example 10.2

(i) Suppose that $z_\ell^2 \in C_\ell^{\mathrm{Alt}}(D^2(f))$ represents a homology class in $H_\ell^{\mathrm{Alt}}(D^2(f))$, and that $\varepsilon_\#^2 z_\ell^2$ is a boundary in X. This is certainly the case if, for example, X is the (contractible) domain of a stable perturbation of a mono-germ. Then there exists $c_{\ell+1}^1 \in C_{\ell+1}(X)$ such that $\partial c_{\ell+1}^1 = \varepsilon_\#^2(z_\ell^2)$. Moreover $f_\#(c_{\ell+1}^1)$ is a cycle in the image of f, for $\partial f_\#(c_{\ell+1}^1) = f_\#(\partial c_{\ell+1}^1) = f_\# \varepsilon_\#^2(z_\ell^2)$, and this is equal to 0 by Lemma 10.2. The argument is summarised by the following diagram, which should be read from top to bottom.

$$\begin{array}{ccccccc}
D^2(f) & \qquad & 0 & \xleftarrow{\ \partial\ } & z_\ell^2 & & \\
\downarrow{\scriptstyle \varepsilon^2} & & & & \downarrow & & \\
X & & 0 & \longleftarrow & \varepsilon_\#^2(z_\ell^2) & \xleftarrow{\ \partial\ } & \exists\, c_{\ell+1}^1 \\
\downarrow{\scriptstyle f} & & & & \downarrow{\scriptstyle f_\#} & & \downarrow \\
Y & & & & f_\#\varepsilon_\#^2(z_\ell^2) = 0 & \xleftarrow{\ \partial\ } & f_\#(c_{\ell+1}^1)
\end{array}$$

Conclusion From an alternating ℓ-cycle z_ℓ^2 in $D^2(f)$, we get an $\ell + 1$ cycle on Y—provided $\varepsilon_\#^2(z_\ell^2)$ is a boundary in X, i.e. provided $\varepsilon_*^2[z_\ell^2] = 0$ in $H_\ell(X)$. Note that

- If $z_\ell^2 = \varepsilon_\#^3(c_\ell^3)$ for some $c_\ell^3 \in C_\ell^{\mathrm{Alt}}(D^3(f))$ then $\varepsilon_\#^2(z_\ell^2) = 0$, and as $c_{\ell+1}^1$ we can simply pick 0.
- If $z_\ell^2 = \partial c_{\ell+1}^2$ for some $c_{\ell+1}^2 \in C_{\ell+1}^{\mathrm{Alt}}(D^2(f))$, then can take $c_{\ell+1}^1 = \varepsilon_\#^2(c_{\ell+1}^2)$ so the homology class we get in $H_{\ell+1}(Y)$ is zero.

So we are really interested in

$$\frac{\ker \varepsilon_*^2 : H_\ell^{\mathrm{Alt}}(D^2(f)) \to H_\ell(X)}{\mathrm{im}\ \varepsilon_*^3 : H_\ell^{\mathrm{Alt}}(D^3(f)) \to H_\ell^{\mathrm{Alt}}(D^2(f))},$$

i.e. in *the homology of the vertical complex* $(H_\ell^{\mathrm{Alt}}(D^\bullet(f_t)), \varepsilon_*^\bullet)$.

(ii) let $z^3_\ell \in Z_\ell(D^3(f))$.

$$
\begin{array}{ccccccccc}
D^3(f) & & 0 & \xleftarrow{\partial} & c^3_\ell & & & & \\
\downarrow{\scriptstyle \varepsilon^3} & & & & \downarrow & & & & \\
D^2(f) & & 0 & \xleftarrow{\partial} & \varepsilon^3_\#(z^3_\ell) & \xleftarrow{\partial} & \exists\, c^2_{\ell+1} & & \\
\downarrow{\scriptstyle \varepsilon^2} & & & & \downarrow & & \downarrow & & \\
X & & & & 0 & \xleftarrow{\partial} & \varepsilon^2_\#(c^2_{\ell+1}) & \xleftarrow{\partial} & \exists\, c^1_{\ell+2} \\
\downarrow{\scriptstyle f} & & & & & & \downarrow & & \downarrow \\
Y & & & & & & 0 & \xleftarrow{\partial} & f_\#(c^1_{\ell+2})
\end{array}
$$

where the element $c^2_{\ell+1}$ in the second row exists if $\varepsilon^3_*[z^3_\ell] = 0 \in H^{\mathrm{Alt}}_\ell(D^2(f))$ and the element $c^1_{\ell+2}$ in the third row exists if $\varepsilon^2_*[c^2_{\ell+1}] = 0 \in H_{\ell+1}(X)$.

Here, an ℓ-dimensional alternating homology class on $D^3(f)$ leads to a $\ell+2$-dimensional class in Y, once again provided certain homology classes vanish.

Remark 10.2 In Example (i), we did not quite construct a map from the second homology of the complex $(H^{\mathrm{Alt}}_\ell(D^\bullet(f)), \varepsilon^\bullet)$ to $H_{\ell+1}(Y)$; there was an element of arbitrariness in the choice of $c^1_{\ell+1}$. In fact if $c'^1_{\ell+1}$ is any other choice of $\ell+1$-chain on X such that $\partial c'^1_{\ell+1} = \varepsilon^2_\#(c^2_\ell)$ then $c^1_{\ell+1} - c'^1_{\ell+1}$ represents a homology class in $H_{\ell+1}(X)$, and thus the homology classes of $f_\#(c^1_{\ell+1})$ and $f_\#(c'^1_{\ell+1})$ in $H_{\ell+1}(Y)$ differ by an element of $f_* H_{\ell+1}(X)$. Thus our construction yields a map from the second homology of the complex $(H^{\mathrm{Alt}}_\ell(D^\bullet(f)), \varepsilon^\bullet)$ to $H_{\ell+1}(Y)/f_* H_{\ell+1}(X)$.

The next two examples follow the schema of the calculations in Example 10.2 in concrete cases.

Example 10.3 A 2-cycle in the image from an alternating 1-cycle in D^2: Consider the stable perturbation $f_t : \mathbb{R}^2 \to \mathbb{R}^3$, defined by $f_t(x, y) = (x, y^2, y^3 + x^2 y + ty)$, of the singularity $f = f_0$ of type S_1. By the calculations described in Sect. 9.5, we have

$$D^2(f_t) = \{(x, y_1, y_2) : y_1 + y_2 = 0 = x^2 + y_1^2 + y_1 y_2 + y_2^2 + t\}.$$

The projection $\varepsilon^2(x, y_1, y_2) = (x, y_1)$ (with inverse $(x, y) \mapsto (x, y, -y)$) maps this isomorphically to the conic

$$D^2_1(f_t) := \{(x, y) \in \mathbb{C}^2 : x^2 + y^2 + t = 0\},$$

with the involution $\sigma(x, y_1, y_2) = (x, y_2, y_1)$ now induced by $(x, y) \mapsto (x, -y)$.

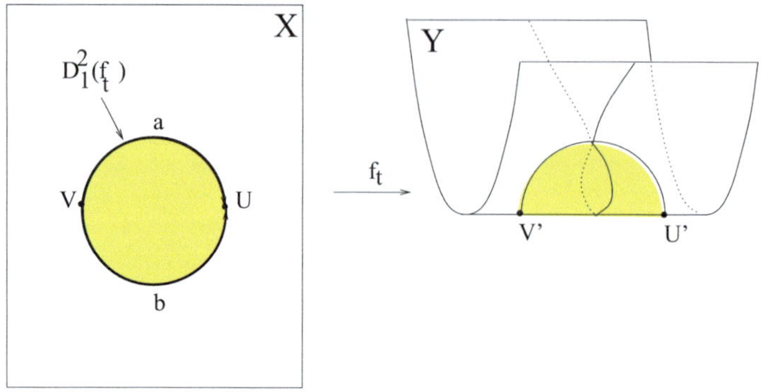

Fig. 10.1 The creation of a 2-cycle in the target from an alternating 1-cycle in the source

Figure 10.1 shows 1-simplices a and b running from V to U, a on the upper arc and b on the lower arc of $D^2_1(f_t)$, such that $\sigma \circ a = b$. Then $AH_1(D^2(f_t)$ is generated by $a - b$. Taking as c_2 a suitable triangulation of the interior of the shaded disc, we have $\partial c_2 = a - b$. As can be seen in the picture, $f_{\#}(c_2)$ forms a bubble, whose homology class generates $H_2(Y)$.

In fact this picture shows all the important features of the complexified map $\mathbb{C}^2 \to \mathbb{C}^3$. Here $D^2(f_t) \simeq D^2_1(f_t)$ is the complex Milnor fibre of an A_1 singularity, and is diffeomorphic to a cylinder. However, its alternating homology is generated by the cycle shown in the real picture, and from there on the construction is the same.

Example 10.4 Homology of the image of a stable perturbation of the germ $f(x, y) = (x, y^3, xy + y^5)$ *of type* H_2. Here we obtain a 2-cycle in the image from an alternating 0-cycle in D^3, and a second 2-cycle in the image from an alternating 1-cycle in D^2. The germ f has $\mathscr{A}_e$-codimension and image Milnor number 2. As shown in Example 9.4, $D^2(f)$ is defined in $\mathbb{C} \times \mathbb{C}^2$ with coordinates x, y_1, y_2 by the vanishing of

$$y_1^2 + y_1 y_2 + y_2^2, \quad x + y_1^4 + y_1^3 y_2 + y_1^2 y_2^2 + y_1 y_2^3 + y_2^4$$

and $D^3(f)$ is defined in $\mathbb{C} \times \mathbb{C}^3$ with coordinates x, y_1, y_2, y_3 by the vanishing of

$$P_2(y_1, y_2, y_3), \quad y_1 + y_2 + y_3, \quad x + P_4(y_1, y_2, y_3), \quad P_3(y_1, y_2, y_3),$$

where each P_j is a symmetric polynomial of degree j. Thus $D^2(f)$ is an A_1 curve singularity and $D^3(f)$ is a non-reduced point of multiplicity 6. By Corollary 9.6, if f_t is a stable perturbation then $D^2(f_t)$ is a Milnor fibre of the A_1 singularity, homotopy-equivalent to a circle, and $D^3(f_t)$ consists of 6 points forming a single S_3-orbit. By eliminating y_2 from the equations for $D^2(f)$ we find that $D^2_{\mathrm{cl}\,1}(f_0)$ has equation $x^2 + xy^4 + y^8$, and is thus a curve singularity of type A_7. By judicious choice of parameter values u and v in the miniversal deformation $f_{u,v}(x, y) =$

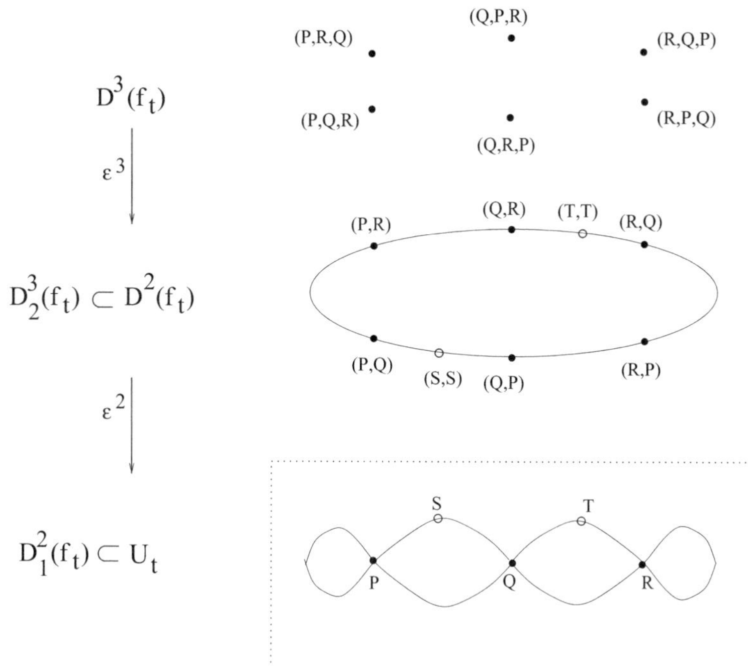

Fig. 10.2 Projections of $D^3(f_t)$ and $D^2(f_t)$

$(x, y^3 + uy, xy + y^5 + vy^2)$, one can arrange that the triple point and the two Whitney umbrellas are real. Let us denote the chosen perturbation by f_t. It follows that $D^2_1(f_t)$ has three nodes and that the involution on $D^2(f_t)$ has two fixed points. Note that the real part of $D^2_1(f_0)$ is just the single point $(0, 0)$. Thus $D^2_1(f_t)$ is also a compact curve, and in fact must be a real morsification of A_7. A real picture of $D^2(f_t)$ and $D^3(f_t)$, and their projections $D^3_1(f_t)$, $D^2_1(f_t)$, is shown in Fig. 10.2.

Here S and T in U_t are the non-immersive points of f_t. At each, the germ of f_t is equivalent to the parameterisation of the Whitney umbrella, $(x, y) \mapsto (x, y^2, xy)$, since this is the only stable non-immersive germ in this dimension range. The non-strict double points (S, S) and (T, T) are the fixed points of the involution $(1, 2)$ on $D^2(f_t)$, which, in our picture, is induced by a reflection in the straight line joining them. Thus $D^2_1(f_t)$ is a real morsification of $D^2_1(f_0)$.

As a single faithful S_3-orbit, $D^3(f_t)$ carries an alternating 0-cycle,

$$c_0 = (P, Q, R) - (P, R, Q) + (R, P, Q) - (R, Q, P) + (Q, R, P) - (Q, P, R).$$

The projection of this cycle to $D^2(f_t)$, $\varepsilon^3_\#(c_0)$ is an alternating boundary:

$$\varepsilon^3_\#(c_0) = (P, Q) - (P, R) + (R, P) - (R, Q) + (Q, R) - (Q, P) = \partial(c_1)$$

where c_1 is the alternating 1-chain

$$[(T,T)(Q,R)] - [(T,T)(R,Q)] + [(Q,P)(R,P)] - [(P,Q)(P,R)]$$

(here, for any two non-antipodal points $A, B \in D^2(f_t)$, $[A, B]$ denotes the singular 1-simplex parameterising the shorter arc from A to B). The projection of c_1 to U_t is a 1-cycle in U_t, and is the boundary of the 2-chain c_2 consisting of a suitable triangulation of the first and third (counting from the left) of the four regions enclosed by the curve $D_1^2(f_t)$ in U_t. And by the argument above, $f_{t\#}(c_2)$ is a cycle in the image X_t, indeed one of the two generators of $H_2(X_t)$. The other generator comes from the alternating 1-cycle c_1' on $D^2(f_t)$ consisting of the anticlockwise arc $[(S,S)(T,T)]$ minus the clockwise arc $[(S,S)(T,T)]$. We encourage the reader to find a 2-chain c_2' on U_t such that $\partial c_2' = \varepsilon_\#^2(c_1')$.

Exercises for Sect. 10.1

1 In the context of Example 10.1,

(i) show that for any $x \in S^1$, $x - a_\# x$ is not a boundary in the alternating chain complex. Hint: counting (the absolute value of) multiplicities, the number of points in the support of the boundary of any alternating 1-chain is divisible by 4;

(ii) show for any two points $x, x' \in S^1$, the alternating 0-chain $(x - a_\# x) - (x' - a_\# x')$ *is* a boundary in the alternating chain complex.

2 The map-germ of type H_3, $f_0(x, y) = (x, y^3, xy + y^8)$, has two triple points and two Whitney umbrellas in its disentanglement.

(i) Find equations for $D^2(f_0)$ and use them to find an equation for $D_1^2(f_0)$.

(ii) Find equations for $D^3(f_0)$ and use them to check that f_t does indeed have two triple points.

(iii) Assuming that it is possible to find a stable perturbation of f_t whose real part has two triple points and two Whitney umbrellas, make a drawing like the one above for H_2. How are the two S_3-orbits in $D_2^3(f_t)$ distributed on $D^2(f_t)$?

(iv) Show that the stable perturbation

$$f_t(x, y) = \left(x, y^3 - uy, xy + y^2 \prod_{i=1}^{2} (y^3 - uy - c_i)\right)$$

with $u > 0$, $|c_i| > \frac{2}{3} u^{\frac{3}{2}}$ and $c_1 \neq c_2$, does indeed have two real triple points and two real Whitney umbrellas.

(v) Follow the procedure used for H_2 to obtain a 2-cycle in the real image of f_t for each S_3-orbit in $D^3(f_t)$.

3 *Three classical examples:*

(i) Consider the well-known immersion of the Klein bottle, $f : K \to \mathbb{R}^3$, whose image is shown in Fig. 10.3. Find the non-empty multiple point spaces of f, and calculate the non-trivial alternating homology groups $AH_\ell(D^k(f))$.

(ii) Figure 10.4 shows two Whitney umbrellas, with instructions for joining them. Joining the domains of their parameterisations gives a Möbius strip. Gluing a disc D to the conjoined umbrellas along the boundary (shown in bold), we get a surface Y in $\mathbb{R}^3$ known as the *cross cap*. The same gluing, back in their conjoined domains, gives a real projective plane. The parameterisations of the

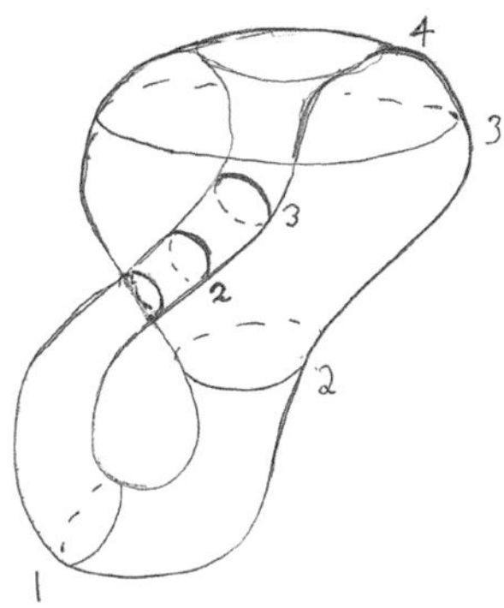

Fig. 10.3 Immersion of the Klein bottle in $\mathbb{R}^3$

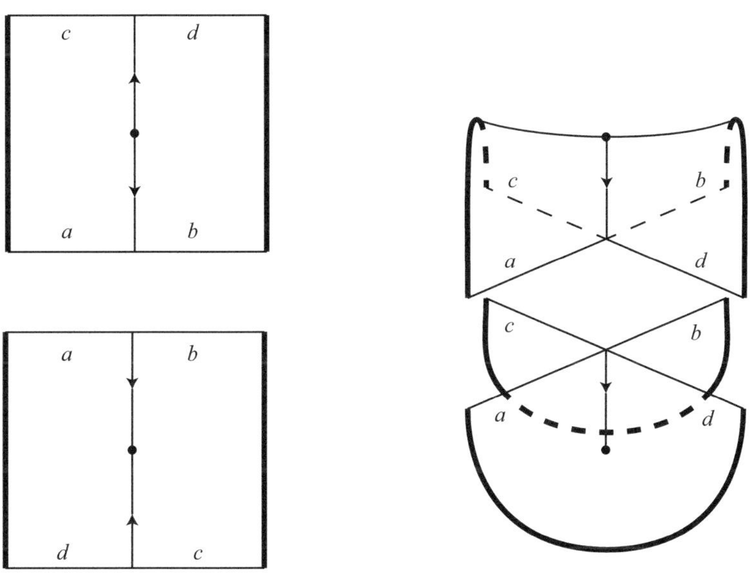

Fig. 10.4 Image of the Möbius strip in $\mathbb{R}^3$ with two Whitney umbrellas

two umbrellas, and the identification of the two discs, piece together to give a map $f : \mathbb{RP}^2 \to Y$. **To do** Describe the multiple point spaces $D^k(f)$, and compute their alternating homology groups.

(iii) Do the same for the immersion $b : \mathbb{RP}^2 \to \mathbb{R}^3$ parameterising *Boy's surface*. Information about Boy's surface can be found in its Wikipedia entry.

4 Let $f : X \to Y$ be a continuous map. Show that the inclusion $D^k_{\mathrm{cl}}(f) \hookrightarrow (X^k/Y)$ induces an isomorphism of chain complexes

$$C^{\mathrm{Alt}}_{\bullet}(D^k_{\mathrm{cl}}(f)) \to C^{\mathrm{Alt}}_{\bullet}(X^k/Y)$$

and therefore also an isomorphism $AH_*(D^k_{\mathrm{cl}}(f)) \simeq AH_*(X^k/Y)$. Thus, alternating homology can also be seen as a way of removing the unwanted diagonals from X^k/Y at the level of homology.

5 Is it true that $AH_*(D^k_{\mathrm{cl}}(f) \to AH_*(D^k(f))$ is always an isomorphism, whether or not the singularities of f are $\mathscr{A}$-finite?

10.2 The Image Computing Spectral Sequence

The combinatorics of the previous constructions are all bundled up together in a spectral sequence, the *image computing spectral sequence*, or ICSS, which was first described in [GM93] and later developed and extended in [Gor95, Hou97] and [Hou99]. The main theorem of [Gor95] on this topic is the following.

Theorem 10.1 *Let $f : X \to Y$ be a finite surjective algebraic map of compact semialgebraic spaces. Then*

1. *There is a spectral sequence with $E^1_{pq} = AH_{q+1}(D^p(f))$, converging to $H_{p+q}(Y)$.*
2. *The differential $d^1 : E^1_{p,q} = AH_q(D^{p+1}(f)) \to AH_q(D^p(f)) = E^1_{p,q-1}$ is the morphism ε^{p+1}_* induced by the projection $\varepsilon^{p+1} : D^{p+1}(f) \to D^p(f)$.*

This means, in particular, that the homology of the image that does not come from the homology of X comes from the alternating homology of the multiple point spaces.

For those not familiar with them (and even for those who are) spectral sequences can seem rather formidable objects. Certainly the detailed definition can be more than one can take in without motivating examples, as the existence of numerous expository papers demonstrates. We believe that the examples provided on the previous pages do provide good motivation for the image computing spectral sequence, and indeed help to make this a reasonably comprehensible first encounter with spectral sequences in general.

Let us now integrate the experience of the examples discussed above into the rather technical structure of a spectral sequence.

First, a spectral sequence consists of a sequence (with index r) of two-index collections of modules (with indices p, q), $E^1_{p,q}, E^2_{p,q}, \ldots, E^r_{p,q}, \ldots$. Each can be displayed as a two dimensional array, and one often speaks of the E^1 *page*, the E^2 *page* etc.. Each page $E^r_{p,q}$ is equipped with arrows, whose slope and length depends on r. The composite of any two composable arrows is required to be 0, and for each r, $E^{r+1}_{p,q}$ is the homology of the resulting complex at $E^r_{p,q}$. The E^1 page in the ICSS is

$$\begin{array}{ccccc}
\downarrow & \downarrow & & \downarrow & \\
H_0^{\text{Alt}}(D^p) & H_1^{\text{Alt}}(D^p) & \cdots & H_r^{\text{Alt}}(D^p(f)) & \cdots \\
\downarrow & \downarrow & & \downarrow & \\
H_0^{\text{Alt}}(D^{p-1}) & H_1^{\text{Alt}}(D^{p-1}) & \cdots & H_r^{\text{Alt}}(D^{p-1}) & \cdots \\
\downarrow & \downarrow & & \downarrow & \\
\vdots & \vdots & \cdots & \vdots & \cdots \\
 & & \cdots & & \cdots \\
\downarrow & \downarrow & & \downarrow & \\
H_0(X) & H_1(X) & \cdots & H_r(X) & \cdots \\
\downarrow & \downarrow & & \downarrow & \\
0 & 0 & \cdots & 0 & \cdots
\end{array} \tag{10.4}$$

with the arrows given by the ε^p_*. We have already seen, in Lemma 10.2, that $\varepsilon^{p-1}_* \circ \varepsilon^p_* = 0$.

The statement "the spectral sequence converges to $H_{p+q}(Y)$" means the following:

1. If we calculate the E^2 page, the E^3 page, etc., then for each pair (p, q), the result eventually stabilises: for some r (which may depend on p and q), $E^r_{p,q} = E^{r+1}_{p,q} = \cdots$. This stable value of $E^r_{p,q}$ is denoted by $E^\infty_{p,q}$.
2. For each n, $H_n(Y)$ is, roughly speaking, the direct sum of all the $E^\infty_{p,q}$ with $p + q = n$. More precisely, there is an increasing filtration $F_\bullet(H_n(Y))$ on $H_n(Y)$ whose associated graded module is isomorphic to this direct sum:

$$F_q(H_n(Y))/F_{q-1}(H_n(Y)) \simeq E^\infty_{n-q,q}. \tag{10.5}$$

It seems remarkable that such a baroque structure should have been codified at all. More remarkable still is that the definition perfectly fits the structure we observe.

We will explore the filtration in the case of the ICSS later. For now, note that Remark 10.2 gives a glimpse of a filtration on $H_{\ell+1}(Y)$, whose first term is $f_*(H_{\ell+1}(X))$, for which (10.5) holds.

Theorem 10.1 above appears to leave out crucial information. It tells us what are the arrows at E^1, but leaves us in the dark regarding the definition of the arrows at later pages. This is, in fact, typical for theorems of this type—see, e.g., [Loo84, Section 8.C]. Part of the reason for the incompleteness is that in very many cases, the spectral sequence arises from a filtered complex, and all the differentials arise from the differential of the complex in a standard way. This is the case with the ICSS. The other reason, which also applies to the ICSS, is that spectral sequences are most useful, and most likely to be the subject of theorems, when they converge quickly. If $E^\infty_{p,q} = E^r_{p,q}$ for all p and q—i.e. if all differentials at E^r, and at all later pages, are equal to 0—then the sequence is said to *collapse* at E^r. As we will see, when f_t is a stable perturbation of an $\mathscr{A}$-finite map-germ $(\mathbb{C}^n, 0) \to (\mathbb{C}^p, 0)$ with $n < p$, the ICSS for f_t collapses at E^1 for a very simple reason—either the source or the target of every differential is zero—and thus we do not need to worry about differentials at all. When f_t is a stable perturbation of an $\mathscr{A}$-finite multi-germ, the ICSS collapses at E^2. Thus for multi-germs it is important to understand the differential at E^2. We will leave this until later.

Remark 10.3 In the ICSS, $E^2_{p,q}$ is the homology of the vertical complex

$$\left(AH_q(D^\bullet(f)), \varepsilon^\bullet_*\right)$$

at $AH^{\mathrm{Alt}}_q(D^{p+1}(f))$. The alternating homology class $[c^{p+1}_q]$ only contributes to $H_{p+q}(Y)$ if it survives to E^2. After supposing that $\varepsilon^{p+1}_\#(c^{p+1}_q) = \partial c^p_{q+1}$ for some alternating p-chain on $D^p(f)$, we then asked that $\varepsilon^p_\#(c^p_{q+1})$ should be the boundary of an alternating $q+2$ chain on $D^{p-1}(f)$. This second check essentially determines the definition of the differential d^2 at E^2: the class of c^{p+1}_q in $E^2_{p,q}$ is now sent to the class of $\varepsilon^p_\#(c^p_{q+1})$ in $E^2_{p-2,q+1}$. Thus, c^{p+1}_q will only contribute to the homology of Y if this too is 0.

Theorem 10.1 holds also for homology with coefficients in $\mathbb{Q}$. As we saw in Proposition 10.1, $AH^{\mathrm{Alt}}_q(D^p(f); \mathbb{Q})$ is isomorphic to the subspace $H^{\mathrm{Alt}}_q(D^p(f); \mathbb{Q})$ of $H_q(D^p(f); \mathbb{Q})$. If $f\colon (\mathbb{C}^n, 0) \to (\mathbb{C}^{n+r}, 0)$ is $\mathscr{A}$-finite and has corank 1, and f_t is a stable perturbation, then each non-empty $D^p(f_t)$ is a Milnor fibre of the corresponding ICIS $\Big((D^p(f), (0, \ldots, 0)\Big)$, by Corollary 9.6, and therefore has reduced homology only in middle dimension, $n - (r-1)(p-r)$. This means that on the E^1 page of the spectral sequence, the possibly non-zero modules are arrayed along a line of negative slope of absolute value ≤ 1. The arrows on the E^1 page are all vertical, and therefore each one either begins or ends at 0. Therefore for all p, q, $E^2_{p,q} = E^1_{p,q}$. The differential at E^2 runs two down and one to the right. Therefore an arrow from a non-zero module ends at 0. So $E^3_{p,q} = E^2_{p,q}$. In fact all arrows on later pages have negative slope of absolute value greater than 1; it follows that at every page, every arrow either begins or ends at 0. So the spectral sequence collapses at E^1, and (modulo replacing each $H_*(Y)$ by the associated graded module coming from the filtration) we can read off the homology of Y from the E^1 page. In fact if $\ell > 1$ then the filtration is trivial, as we will see.

Corollary 10.1 *Suppose that* $f_t : X_t \twoheadrightarrow Y_t$ *is a stable perturbation of a corank 1 map-germ* $f : (\mathbb{C}^n, 0) \to (\mathbb{C}^{n+\ell}, 0)$.

1. *If* $\ell \geq 2$, *then*

$$H_q(Y_t; \mathbb{Q}) = \begin{cases} H^{Alt}_{n-(k-1)\ell}(D^k(f_t); \mathbb{Q}) & \text{if } q = n - (k-1)(\ell-1) \text{ for some } k \\ \mathbb{Q} & \text{if } q = 0 \\ 0 & \text{otherwise} \end{cases}$$

2. *If* $\ell = 1$, *then* $H_q(Y) = 0$ *if* $q \neq 0, n$, *and*

$$H_n(Y; \mathbb{Q}) \simeq \bigoplus_{k=2}^{n+1} H^{Alt}_{n-k+1}(D^k(f_t); \mathbb{Q}).$$

If D^k is an S_k-invariant ICIS of dimension r with S_k-invariant Milnor fibre D^k_t, let us refer to the rank of $H^{\text{Alt}}_r(D^k_t)$ as the *alternating Milnor number* of D^k. Then we have

Corollary 10.2 *In the situation of Corollary 10.1(2), the image Milnor number of* f *is the sum of the alternating Milnor numbers of the ICISs* $D^k(f)$ *for* $k = 2, \ldots, n+1$.

Warning! Corollaries 10.1 and 10.2 hold only for stable perturbations of mono-germs. For multi-germs the situation is more complicated, and the spectral sequence may not collapse at E^1. Kevin Houston shows in [Hou97] that if it does not collapse at E^1 then it collapses at E^2, and gives general formulae for the homology of the image in this case.

Remark 10.4 When one works with another coefficient ring R—e.g. $R = \mathbb{Z}$—then the second part of Corollary 10.1 would be replaced by

2'. There is a filtration on $H_n(Y_t; R)$ such that the associated graded module is isomorphic to the direct sum

$$\bigoplus_{k=2}^{n+1} AH_{n-k+1}(D^k(f_t); R).$$

In the situation of Corollary 10.1, with coefficients in a field, the associated graded object is isomorphic to the original vector space.

See also Sect. 10.7.8 for a stronger version of Corollary 10.1 in which $\mathbb{Q}$ is replaced by $\mathbb{Z}$.

If $f: (\mathbb{C}^n, S) \to (\mathbb{C}^{n+r}, 0)$ is no longer assumed to have corank 1, then we know very little about its multiple point spaces $D^k(f)$ and those of a stable perturbation f_t. In particular, in general $D^k(f)$ is not an ICIS, and its homology is not concentrated in middle dimension. Moreover, if the dimensions $(n, n+\ell)$ are such that there exist stable singularities of maps $\mathbb{C}^n \to \mathbb{C}^{n+\ell}$ with corank 2, then $D^k(f_t)$ is not in general a smoothing of $D^k(f)$. Nevertheless, in [Hou97], Kevin Houston proved the remarkable theorem that the conclusion of Corollary 10.1 still holds. The main step in the proof is the following.

Theorem 10.2 *Let f_t be a stable perturbation of an $\mathscr{A}$—finite map-germ $(\mathbb{C}^n, 0) \to (\mathbb{C}^{n+\ell}, 0)$. Then $AH_q(D^k(f_t)) = 0$ if $q \neq \dim D^k(f_t)$.* □

Thus,

Corollary 10.3 *The formulae of Corollary 10.1 hold without the hypothesis that f have corank 1.* □

An example in [Mon18] shows that $D^k(f_t)$ may have vanishing homology outside middle dimension. It is only the alternating homology that is confined to middle dimension.

Unfortunately the proof of Houston's theorem is beyond the scope of this book.

10.2.1 Towards the ICSS

We return to the double complex $\left(C_n^{\mathrm{Alt}}(D^k(f)), (-1)^{k-1}\partial, \varepsilon_{\#}^k\right)_{k\geq 1, n\geq 0}$. Given any double complex, $(C_{p,q}, d_v, d_h)$ with $d_h : C_{p,q} \to C_{p-1,q}$ and $d_v : C_{p,q} \to C_{p,q-1}$, and $d_v \circ d_h + d_h \circ d_v = 0$, the *total complex* is the complex of direct sums $T_n = \bigoplus_{p+q=n} C_{p,q}$ with differential $d : T_n \to T_{n-1}$ given by $d = d_v + d_h$. It is a complex because of the anti-commuting of d_v and d_h. A double complex with $C_{p,q} = 0$ for p and q less than 0 is a *first quadrant double complex*, and in this case two standard spectral sequences ${}^{I}E$ and ${}^{II}E$ converge to the homology of the total complex. These are explained below. In our case, one of these two is the ICSS. Our proof that the ICSS converges to the homology of the image uses the standard strategy of showing that the same is true of the *other* spectral sequence arising from the double complex.

Now we briefly explain the two spectral sequences ${}^{I}E$ and ${}^{II}E$ for the general case. More complete explanations can be found e.g. in [Rot09, Chapter 10] or in [McC01, Chapter 3], so we restrict ourselves to the following partial explanation:

1. We point out the two natural filtrations on the total complex of a double complex.
2. We state without proof the main theorem on the existence of a spectral sequence for a filtered complex.

3. We identify the modules ${}^{I}E^{1}_{p,q}$ and ${}^{II}E^{1}_{pq}$ in the spectral sequences arising from the two filtrations, and the differentials in the E^1 pages of these two spectral sequences.

We leave the reader to learn elsewhere about the spectral sequence of a filtered complex; especially recommended is [Rot09, Chapter 10], which moreover gives a very clear explanation of the spectral sequences arising from the two filtrations on the total complex of a double complex.

10.2.2 The Filtrations

Given a first quadrant double complex $(C_{p,q}, d_h, d_v)$, with $d_h : C_{p,q} \to C_{p-1,q}$ and $d_v : C_{p,q} \to C_{p,q-1}$, the spectral sequences ${}^{I}E$ and ${}^{II}E$ arise from two natural increasing filtrations on the total complex $T_\bullet$. The first, $\{{}^{I}F_p\}_{p\in\mathbb{N}}$, is defined by

$$ {}^{I}F_pT_n = \bigoplus_{i+j=n,\ i\leq p} C_{i,j}; \tag{10.6} $$

the second is defined by

$$ {}^{II}F_pT_n = \bigoplus_{i+j=n,\ j\leq p} C_{i,j}. \tag{10.7} $$

10.2.3 The Spectral Sequence of a Filtered Complex

Suppose that $(C_\bullet, d)$ is a complex with an increasing filtration F_p compatible with the differential d—i.e. so that $d(F_pC_n) \subset F_p(C_{n-1})$, and thus $(F_pC_\bullet, d)$ is a subcomplex for each p. The spectral sequence of the filtered complex combines the following ingredients:

1. The inclusion of complexes $F_pC_\bullet \hookrightarrow C_\bullet$ induces a morphism of homology $H_n(F_pC_\bullet) \to H_n(C_\bullet)$.
2. The quotients $F_pC_n/F_{p-1}C_n$ also form a complex, since the original differential d passes to the quotient to give a differential

$$ d_p : \left(F_pC_n/F_{p-1}C_n\right) \to \left(F_pC_{n-1}/F_{p-1}C_{n-1}\right). $$

The information we will use on the spectral sequence is summarised in the following theorem, which combines Theorem 10.11 and Exercise 10.4 of [Rot09].

Theorem 10.3 *In these circumstances, there is a spectral sequence in which*

- $E^1_{p,q}$ *is equal to the* $(p+q)'th$ *homology of the quotient complex* $(F_pC_\bullet/F_{p-1}C_\bullet)$, *and*
- $d^1 : E^1_{p,q} \to E^1_{p-1,q}$ *is the connecting homomorphism in the long exact sequence of homology arising from the short exact sequence of complexes*

$$0 \to \frac{F_{p-1}C_\bullet}{F_{p-2}C_\bullet} \to \frac{F_pC_\bullet}{F_{p-2}C_\bullet} \to \frac{F_pC_\bullet}{F_{p-1}C_\bullet} \to 0. \tag{10.8}$$

This spectral sequence converges to the homology of $C_\bullet$, *in the sense that* $E^\infty_{p,q}$ *is isomorphic to*

$$\frac{\textit{image of } H_{p+q}(F_pC_\bullet) \to H_{p+q}(C_\bullet)}{\textit{image of } H_{p+q}(F_{p-1}C_\bullet) \to H_{p+q}(C_\bullet)}. \tag{10.9}$$

□

Observe that the filtration on the homology $H_{p+q}(C_\bullet)$ is implicitly defined by the last line of the theorem: the p'th term in the filtration is the image of $H_{p+q}(F_pC_\bullet)$ in $H_{p+q}(C_\bullet)$.

10.2.4 The Spectral Sequences Arising from the Two Filtrations on the Total Complex of the Double Complex

Now let us consider the filtration ${}^IF$. We have

$$\frac{{}^IF_pT_n}{{}^IF_{p-1}T_n} = \frac{\bigoplus_{i+j=n,\ i\le p} C_{i,j}}{\bigoplus_{i+j=n,\ i\le p-1} C_{i,j}} \simeq C_{p,n-p}$$

$$\frac{{}^IF_pT_{n-1}}{{}^IF_{p-1}T_{n-1}} = \frac{\bigoplus_{i+j=n-1,\ i\le p} C_{i,j}}{\bigoplus_{i+j=n-1,\ i\le p-1} C_{i,j}} \simeq C_{p,n-p-1}$$

and $d = d_h + d_v : F_pT_n \to F_pT_{n-1}$ passes to the quotient as $d_v : C_{p,n-p} \to C_{p,n-p-1}$. Thus, in the first spectral sequence,

$${}^IE^1_{p,n-p} = \frac{\ker d_v : C_{p,n-p} \to C_{p,n-p-1}}{\operatorname{im} d_v : C_{p,n-p+1} \to C_{p,n-p}}. \tag{10.10}$$

Analogously, in the second spectral sequence,

$${}^{II}E^1_{n-p,p} = \frac{\ker d_h : C_{n-p,p} \to C_{n-p-1,p}}{\operatorname{im} d_h : C_{n-p,p+1} \to C_{n-p,p}}. \tag{10.11}$$

Both filtrations give rise to spectral sequences which converge to the homology of the total complex. A number of important theorems follow from this simple fact. For example, this is the standard proof of the isomorphism between de Rham cohomology and Cech cohomology for a smooth manifold (see, e.g., [BT82]).

The ICSS is the spectral sequence resulting from the second filtration of the total complex of the double complex $\left(C_n^{\mathrm{Alt}}(D^k(f)), (-1)^{k-1}\partial, \varepsilon_\#^k\right)_{k\geq 1, n\geq 0}$. That is, we first apply the differential $d_h = \partial$ to get $E^1_{p,q} = AH_q(D^{p+1}(f))$, equipped with differential $d^1_{p,q} : E^1_{p,q} \to E_{p,q-1}$ equal to ε_*^p.

The spectral sequence arising from the first filtration, in which one first applies the differential $\varepsilon_\#^q$, is hard to make sense of with alternating singular homology. In particular, in general if f is not injective then it is not clear what is the image of $f_\# : C_n(X) \to C_n(Y)$; it is not surjective for $n > 0$ (see Exercise 10.2.8 below). On the other hand, if the map f is triangulated—i.e. f is a simplicial map with respect to some triangulations of X and Y— and one replaces singular with simplicial homology, then $f_\# : C_n(X) \to C_n(Y)$ is surjective for all ℓ, and, as we shall see, there are natural triangulations of $D^k(f)$ for all k, making the complex

$$\cdots \longrightarrow C_\ell^{\mathrm{Alt}}(D^3(f)) \xrightarrow{\varepsilon_\#^3} C_\ell^{\mathrm{Alt}}(D^2(f)) \xrightarrow{\varepsilon_\#^2} C_\ell(X) \xrightarrow{f_\#} C_\ell(Y) \longrightarrow 0 \tag{10.12}$$

exact for all ℓ. This means that each of the columns in the double complex (see (10.4)) has homology only in degree 0, and the sequence collapses onto the p-axis:

$${}^{I}E^1_{p,q} = \begin{cases} C_p(Y) & \text{if } q = 0 \\ 0 & \text{if } q > 0 \end{cases}$$

and $d^1 : E^1_{p,0} \to E^1_{p-1,0}$ is the usual boundary operator. From this it follows that that

$$H_n(T_\bullet) = H_n(Y) \text{ for all } n. \tag{10.13}$$

Since both spectral sequences converge to the same homology, that of $T_\bullet$, the main theorem on the ICSS, Theorem 10.1, follows immediately from (10.13) and the fact that ${}^{II}E^1_{p,q} = AH_q(D^{p+1}(f))$, together with the fact that the different versions of alternating homology (singular and simplicial) coincide.

We record the exactness of (10.12) as a theorem:

Theorem 10.4 *Let $f : X \to Y$ be a finite surjective simplicial map. Then there is a triangulation of each space $D^k(f)$ such that each projection $\varepsilon^k : D^k(f) \to D^{k-1}(f)$ is a simplicial map, and for each $n \in \mathbb{N}$, the resulting complex $C_n^{Alt}(D^{\bullet+1}(f))$ is an acyclic resolution of $C_n(Y)$.*

Here, and in what follows, X and Y are geometric simplicial complexes, and C_n and C_n^{Alt} denote groups of simplicial chains, and alternating simplicial chains.

Corollary 10.4 *The second spectral sequence* ${}^{II}E^1_{p.q}$ *of the double complex*

$$(C_n^{Alt}(D^{k+1}(f)), \partial_n^k, \epsilon_n^k)$$

converges to $H_\bullet(Y)$*, i. e.,*

$${}^{II}E^1_{p.q} \cong AH_q D^{p+1}(f)) \Longrightarrow H_{p+q}(Y), \tag{10.14}$$

with differentials at E^1 *given by* $(\epsilon_q^{p+1})_*$

Proof By (10.11), the first page of the second spectral sequence is given by taking the homology with respect to the horizontal differentials ∂ in diagram (10.3), so we get the alternating homology of the multiple point spaces. It converges to the homology of the total complex, which by Proposition 10.4 is isomorphic to the homology $H_*(Y)$. □

The second spectral sequence of the double complex $(C_n^{\text{Alt}}(D^{k+1}(f)), \partial_n^k, \epsilon_n^k)$ given in (10.14) is *the Image Computing Spectral Sequence, ICSS.*

Exercises for Sect. 10.2

1 Check that $d^2 : E^2_{p,q} \to E^2_{p-2,q+1}$, as defined in Remark 10.3, is well defined, and that $d^2 \circ d^2 = 0$.

2 Example 10.1 computes the alternating homology of $D^2(q)$, where q is the quotient map from the unit disc in the real plane to $\mathbb{RP}^2$. In view of Theorem 10.1, what does this say about the homology of $\mathbb{RP}^2$?

3 Use Theorem 10.1 to calculate the homology of $\mathbb{RP}^3$ as image of a suitable finite quotient map.

4 Suppose that $X \to S^1$ is a locally trivial fibre bundle with fibre X_1 and monodromy homeomorphism $h : X_1 \to X_1$. Calculate the homology of X, by considering it as the image of $[0, 1] \times X_1$ under the map identifying all points $(0, x)$ and $(1, h(x))$.

5 Let X be the disjoint union of 3 real lines (see Fig. 10.5) and $f : X \to \mathbb{R}^2$ be the map

$$\begin{cases} u \mapsto (u, 0) \\ v \mapsto (0, v) \\ w \mapsto (w, 1 - w) \end{cases}$$

(i) Where in the ICSS does the 1-cycle in the image of f come from?
(ii) Does complexifying f into a map from the disjoint union of three complex lines into $\mathbb{C}^2$ make any difference to the spectral sequence?

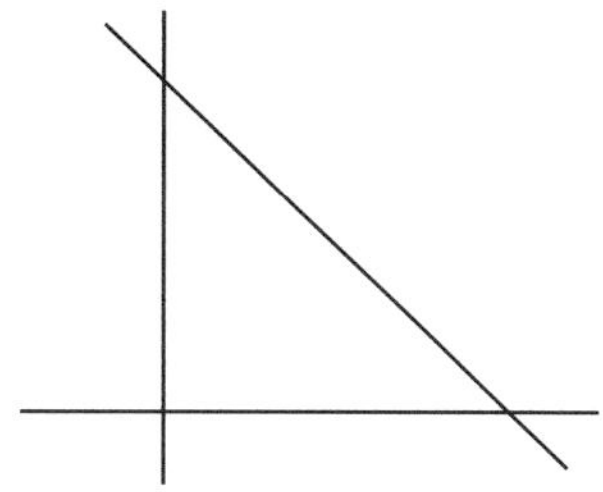

Fig. 10.5 The union of three real lines

6 Generalising the previous exercise, consider the map from the disjoint union of $n+2$ copies of $\mathbb{R}^n$ to $\mathbb{R}^{n+1}$, mapping the j'th copy of $\mathbb{R}^n$ to the coordinate plane $\{x_j = 0\}$ for $j = 1, \ldots, n+1$ and mapping the last copy of $\mathbb{R}^n$ by

$$(x_1, \ldots, x_n) \mapsto (x_1, \ldots, x_n, (1 - \sum_i x_i)).$$

The image, Y, contracts to the boundary of an $n+1$ simplex, which is topologically a sphere. Where in the ICSS does the n- cycle generating $H_n(Y)$ come from?

7 *More on the three classical examples*: Following on from Exercise 10.1.3: find $\varepsilon_*^k : AH_j(D^k(f)) \to AH_j(D^{k-1}(f))$ for each of the non-trivial alternating homology groups you found, and calculate the modules in the E^2 page in the image computing spectral sequence. Go on to use the ICSS to calculate the homology of the immersed Klein bottle, the cross cap, and Boy's surface.

8 Let Y be the figure 8, and let $f : S^1 \to Y$ be the usual parameterisation, with one double point. Show that the homomorphism of singular chain groups, $f_\# : C_1(S^1) \to C_1(Y)$, is not surjective.

9 Deduce the statement in Corollary 10.4 about $d^1 : E^1_{p,q} \to E^1_{p-1,q}$ from the definition of the second filtration.

10.3 Finite Simplicial Maps

Here we prove the main result on the ICSS, Theorem 10.4. The reader who is concerned more with *using* the ICSS than with its construction can skip this section and go on to Sects. 10.4 and 10.7.

We suppose that X and Y are geometric simplicial complexes embedded in $\mathbb{R}^N, \mathbb{R}^P$, respectively, and that $f : X \to Y$ is simplicial, surjective, and finite-to-one. "Simplicial" means that on each simplex $[v_0, \ldots, v_n]$ of X, f is given by

$$f(\sum_i t_i v_i) = \sum_i t_i f(v_i). \tag{10.15}$$

We begin by establishing some notation and proving some elementary lemmas. An *ordered simplex* in X or Y is an affine homeomorphism

$$(t_0, \ldots, t_n) \mapsto \sum_i t_i v_i$$

from the standard n-simplex into X or Y which forms part of its simplicial structure; the image of an ordered simplex is a *geometric simplex*. Thus, an ordered simplex of X is determined by a geometric k-simplex together with an ordering of its vertices. We will use the symbol Δ to denote an ordered simplex, and $|\Delta|$ to denote its image. When we need to speak of a geometric simplex without specifying an ordering of its vertices, we will use the symbol Γ. We write $C_k(X)$ for the free abelian group generated by the ordered k-simplices of X. If f is a simplicial map and Δ an ordered simplex of X, then $f(|\Delta|)$ is a geometric simplex of Y and $f_{\#}(\Delta)$ an ordered simplex. Thus $f(|\Delta|) = |f_{\#}(\Delta)|$. If $\Delta = [v_0, \ldots, v_n]$ then $|\mathring{\Delta}|$ denotes its *interior*:

$$|\mathring{\Delta}| = \{\sum_i t_i v_i : \sum_i t_i = 1, t_i > 0, i = 0, \ldots, n\}.$$

The following properties of any simplicial complex K are well known:

1. Each point $x \in K$ lies in the interior of a unique geometric simplex of K, which we denote by Γ_x.
2. We write $|\Delta| \leq |\Delta'|$ when $|\Delta|$ is a face of $|\Delta'|$ (or equal to Δ'), and note that if $|\Delta'| \cap |\mathring{\Delta}| \neq 0$ then $|\Delta| \leq |\Delta'|$. For two simplices meet only along a common face of each, so $\Delta = \overline{\mathring{\Delta}}$ is a face of Δ'.
3. For each point x,

$$\mathrm{St}(x) := \bigcup_{\Gamma_x \leq |\Delta|} |\mathring{\Delta}|$$

is an open neighbourhood of x. Suppose that $(x_n) \to x$. By local finiteness of K we can suppose that all x_n lie in the interior of one simplex Δ. Then $x \in \Delta \cap \mathring{\Gamma}_x$, so by the preceding statement, $\Gamma_x \leq \Delta$, and thus all the x_n lie in $\mathrm{St}(x)$.

Recall that we are assuming that f is finite-to-one.

Lemma 10.3 *f maps each geometric n-simplex of X isomorphically to a geometric n-simplex of Y.*

Proof By definition of simplicial map, on the simplex $\Delta := [v_0, \ldots, v_n]$, we have

$$f(\sum_i t_i v_i) = \sum_i t_i f(v_i).$$

If the vertices $f(v_0), \ldots, f(v_n)$ were affinely dependent, then this map would not be finite to one. Thus $\Delta' := [f(v_0), \ldots, f(v_n)]$ is an n-simplex of Y and $f : |\Delta| \to |\Delta'|$ is an isomorphism. □

Lemma 10.4 *If $f(x_1) = f(x_2)$ then the barycentric coordinates of x_1 in Γ_{x_1}, with respect to a suitable ordering of the vertices, are the same as those of x_2 in Γ_{x_2}.*

Proof Write $y := f(x_1) = f(x_2)$. If $\Gamma_y = |[w_1, \ldots, w_n]|$ then for $i = 1, 2$ order the vertices $v_j^{(i)}$ of Γ_{x_i} so that $f(v_j^{(i)}) = w_j$. Since f is linear on Γ_{x_i}, the barycentric coordinates of x_i in Γ_{x_i} are the same as those of y in Γ_y. □

Recall that we are assuming that $f : X \to Y$ is surjective.

Lemma 10.5 *$f_\# : C_n(X) \to C_n(Y)$ is surjective.*

Proof Let Γ' be a k-simplex in Y, and $y \in \mathring{\Gamma}'$. Pick $x \in X$ such that $f(x) = y$. By Lemma 10.3, $f : \Gamma_x \to \Gamma'$ is an isomorphism. It follows that for each Δ' such that $\Gamma' = |\Delta'|$, there is an ordered k-simplex Δ in X such that $f_\#(\Delta) = \Delta'$, so that $f_\#$ is surjective. □

The statement of Lemma 10.5 is false in singular homology when $n > 0$: for example, the figure 8 is the image of the circle S^1 under a generically 1-to-1 immersion with one double point. A 1-simplex in the figure 8 which turns a corner at the vertex of the 8 is not the image of any 1-simplex in X.

10.3.1 Triangulating $D^k(f)$

Recall first that since $X \subset \mathbb{R}^N$, we have $D^k(f) \subset (\mathbb{R}^N)^k$. Recall also that $D^k(f)$ is the closure of the set $D_S^k(f)$ of strict k-tuple points: k-tuples $(x_1, \ldots, x_k) \in X^k$ sharing the same image, but pairwise distinct.

Let $y \in Y$, suppose that Γ_y is an n-simplex, and suppose $\{x_1, \ldots, x_r\} \subset f^{-1}(y)$ with $x_i \neq x_{i'}$ if $i \neq i'$. By Lemma 10.3, the preimage of Γ_y contains the geometric n-simplices Γ_{x_i}, each of which is mapped isomorphically to Γ_y. Suppose that $\Gamma_y = |[w_0, \ldots, w_n]|$, and that $y = \sum_j t_j w_j$. Let $v_j^{(i)}$ be the unique vertex of Γ_{x_i} such that $f(v_j^{(i)}) = w_j$, for $i = 1, \ldots, r$ and $j = 0, \ldots, n$. Note that it may happen that $v_j^{(i)} = v_j^{(i')}$ for some $i \neq i'$. By Lemma 10.4, the barycentric coordinates of each point x_i in Γ_{x_i}, with the vertices ordered as described, are the same as those of y in Γ_y. Thus, the geometric n-simplex

$$\left|\left[\left(v_0^{(1)}, \ldots, v_0^{(r)}\right), \ldots, \left(v_n^{(1)}, \ldots, v_n^{(r)}\right)\right]\right| \tag{10.16}$$

is contained in $D^r(f)$, and

$$(x_1, \ldots, x_r) = \sum_j t_j \left(v_j^{(1)}, \ldots, v_j^{(r)}\right)$$

lies in its interior: all of the t_{j_i} are strictly positive since $y \in \mathring{\Gamma}_y$.

We denote the geometric simplex (10.16) by

$$\left(\Gamma_{x_1} \times \cdots \times \Gamma_{x_r}\right)/Y.$$

Its interior is contained in $D^r_S(f)$, but some of its faces may not be. When $r = 2$, it is more commonly written as $\Gamma_{x_1} \times_Y \Gamma_{x_2}$.

Let us consider now more general points $\mathbf{x} \in D^k(f)$, where there may be repetitions in the k components. Such points appear in the closure of $D^k_S(f)$. The components of $\mathbf{x}$ may be denoted by $\pi_\ell(\mathbf{x})$, $\ell = 1, \ldots, k$, where $\pi_\ell : D^k(f) \to X$ is the restriction of the ℓ'th standard projection $X^k \to X$. Denote the common image $f(\pi_\ell(\mathbf{x}))$ by y. Picking an order for the vertices of Γ_y, say $\Gamma_y = ||[w_0, \ldots, w_n]||$, determines an order for the vertices of each simplex $\Gamma_{\pi_\ell(\mathbf{x})}$,

$$\Gamma_{\pi_\ell(\mathbf{x})} = \left|\left[v_0^{(\ell)}, \ldots, v_n^{(\ell)}\right]\right|,$$

where $f(v_j^{(\ell)}) = w_j$ for $j = 0, \ldots, n$. We claim that the simplex

$$\Gamma_{\mathbf{x}}^{(k)} := \left|\left[\left(v_0^{(1)}, \ldots, v_0^{(k)}\right), \ldots, \left(v_n^{(1)}, \ldots, v_n^{(k)}\right)\right]\right| \tag{10.17}$$

is contained in $D^k(f)$, and contains $\mathbf{x}$ in its interior.

To see this, observe that by definition of $D^k(f)$ as the closure of $D^k_S(f)$, there is a sequence $\left(x_n^{(1)}, \ldots, x_n^{(k)}\right)_{n\in\mathbb{N}}$ in $D^k_S(f)$ converging to $\mathbf{x}$. By local finiteness of the triangulation of X, after passing to a subsequence we may suppose each individual sequence $(x_n^{(\ell)})_{n\in\mathbb{N}}$ is contained in the interior of a single simplex, which we denote by $\Gamma^{(\ell)}$. Note that $\Gamma^{(\ell)} \neq \Gamma^{(\ell')}$ if $\ell \neq \ell'$ since f is one-to-one on each simplex. Then by the argument of the first part of the construction,

$$\left(\Gamma^{(1)} \times \cdots \times \Gamma^{(k)}\right)/Y$$

is contained in $D^k(f)$. Since $\Gamma_{\pi_\ell(\mathbf{x})} \leq \Gamma^{(\ell)}$ for $\ell = 1, \ldots, k$, it follows that the simplex (10.17) is contained in $D^k(f)$. Again, $\mathbf{x}$ is contained in its interior, since the barycentric coordinates of each point $\pi_\ell(\mathbf{x})$ in $||[v_0^{(\ell)}, \ldots, v_n^{(\ell)}]||$ are equal to those of y in $[w_0, \ldots, w_n]$, which are all strictly positive.

We denote the simplex of (10.17) by $\Gamma_{\mathbf{x}}^{(k)}$. The "$k$" here refers to the fact that we are speaking of a simplex in $D^k(f)$; it is not the dimension of the simplex. Consistently with our previous notation, this simplex can also be referred to as

$$\left(\Gamma_{\pi_1(\mathbf{x})} \times \cdots \times \Gamma_{\pi_k(\mathbf{x})}\right)/Y. \tag{10.18}$$

We will show shortly that $\Gamma_{\mathbf{x}}^{(k)}$ is the unique geometric simplex of $D^k(f)$ containing $\mathbf{x}$ in its interior, so that the notation $\Gamma_{\mathbf{x}}^{(k)}$ is consistent with the notation Γ_x for the unique geometric simplex of X containing x in its interior.

Notation As described in the construction, each of the simplices $\Delta_\ell := [v_0^{(\ell)}, \ldots, v_n^{(\ell)}]$ appearing in (10.17) satisfies $f_\#(\Delta_\ell) = [w_0, \ldots, w_n]$. It will also be useful to refer to the ordered n simplex of $D^k(f)$ appearing in (10.17) as $(\Delta_1 \times \cdots \times \Delta_k)/Y$.

Proposition 10.2

1. *For each k, this construction gives $D^k(f)$ the structure of a simplicial complex. In particular, $\Delta_{\mathbf{x}}^{(k)}$ is the unique simplex containing* **x** *in its interior.*
2. *For each k, the map $\varepsilon^k : D^k(f) \to D^{k-1}(f)$ is a simplicial map, with respect to these structures.*

Proof

1. It is necessary to show
   (a) that if we omit any vertex of a simplex of $D^k(f)$ then the resulting simplex is still a simplex of $D^k(f)$;
   (b) that each point of $D^k(f)$ lies in the interior of exactly one of the geometric simplices we have constructed.

   The first statement is obvious, since by omitting a vertex of a simplex $\Gamma^{(k)}$ we obtain a geometric simplex in the topological closure of $\Gamma^{(k)}$. For the second, suppose $\mathbf{x}'$ lies in the interior of the n-simplex $\Gamma_{\mathbf{x}}^{(k)}$. Then, with the notation of the construction,

$$\mathbf{x}' = \sum_j t'_j \left(v_j^{(1)}, \ldots, v_j^{(k)}\right),$$

   for some t'_j, with $t'_j > 0$ for $j = 0, \ldots, n$ and $\sum_j t'_j = 1$. Then $\pi_\ell(\mathbf{x}') = \sum_j t'_j v_j^{(\ell)}$, and $\pi_\ell(\mathbf{x}') \in \mathring{\Gamma}_{\pi_\ell(\mathbf{x})}$, so that $\Gamma_{\pi_\ell(\mathbf{x}')} = \Gamma_{\pi_\ell(\mathbf{x})}$. It follows that $\Gamma_{\mathbf{x}'}^{(k)} = \Gamma_{\mathbf{x}}^{(k)}$. Thus each point of $D^k(f)$ lies in the interior of a unique geometric simplex, proving (b).
2. It is clear from (10.17) that $\varepsilon^k(\Gamma_{\mathbf{x}}^{(k)}) = \Gamma_{\varepsilon^k(\mathbf{x})}^{(k-1)}$. The two geometric simplices have the same dimension, and with respect to the barycentric coordinates defined above, ε^k is the identity map.

□

Remark 10.5 Consider the map $f^{(k)} : D^k(f) \to Y$ defined by $f^{(k)}(\mathbf{x}) = f\left(\pi_\ell(\mathbf{x})\right)$ for some ℓ with $1 \le \ell \le k$. Evidently this is independent of the choice of ℓ, so for convenience we take $\ell = 1$. We have

$$\begin{aligned} f^{(k)}\left(\sum_j t_j \left(v_j^{(1)}, \ldots, v_j^{(1)}, v_j^{(2)}, \ldots, v_j^{(2)}, \ldots, \ldots, v_j^{(r)}, \ldots, v_j^{(r)}\right)\right) &= f\Big(\sum_j t_j v_j^{(1)}\Big) \\ &= \sum_j t_j f(v_j^{(1)}), \end{aligned}$$

the last equality because f is linear on each simplex of X. Thus $f^{(k)}$ defines an isomorphism $\Gamma_{\mathbf{x}}^{(k)} \to \Gamma_{f(\pi_\ell(\mathbf{x}))}$. Conversely, for any n-simplex Γ in Y, each component of any point $\mathbf{x}$ in $D^k(f)$ such that $f^{(k)}(\Gamma_{\mathbf{x}}^{(k)}) = \Gamma_y$, lies in the interior of a simplex which is mapped isomorphically to Γ_y.

Let $\Delta = [w_0, \ldots, w_n]$ be an ordered n-simplex of Y. Denote by $C_n(D^k(f))|_\Delta$ and $C_n^{\mathrm{Alt}}(D^k(f))|_\Delta$ the collection of n-chains, and alternating n-chains, respectively, consisting of linear combinations of ordered n-simplices $\Delta^{(k)}$ such that $f_\#^{(k)}(\Delta^{(k)}) = \Delta$. Every such simplex is of the form $(\Delta_1 \times \cdots \times \Delta_k)/Y$, where each Δ_ℓ is a simplex of X such that $f_\#(\Delta_\ell) = \Delta$. Recall that by $D^1(f)$ we mean simply X, so that both $C_n(D^1(f))|_\Delta$ and $C_n^{\mathrm{Alt}}(D^1(f))|_\Delta$ consist of the free abelian group generated by the n-simplices Δ' of X such that $f_\#(\Delta') = \Delta$. We have

$$\varepsilon_\#^k\big((\Delta_1 \times \cdots \times \Delta_k)/Y\big) = (\Delta_1 \times \cdots \times \Delta_{k-1})/Y,$$

which implies the following simple but important lemma.

Lemma 10.6 *$C_n(D^\bullet(f))|_\Delta$ and $C_n^{Alt}(D^\bullet(f))|_\Delta$ are subcomplexes of the complexes $C_n(D^\bullet(f))$ and $C_n^{Alt}(D^\bullet(f))$.* □

Recall that $\epsilon^k : C_j^{\mathrm{Alt}}(D^k(f)) \to C_j^{\mathrm{Alt}}(D^{k-1}(f)) = (-1)^j \varepsilon_\#^k$.

10.3.2 $(C_n^{Alt}(D^\bullet(f)), \epsilon_\#^\bullet)$ Is a Resolution of $C_n(Y)$

Lemma 10.7 $\varepsilon_\#^2\left(C_n^{Alt}(D^2(f))\right) = \ker\left(f_\# : C_n(X) \to C_n(Y)\right)$.

Proof We have already seen in Lemma 10.2 that

$$\varepsilon_\#^2\left(C_n^{\mathrm{Alt}}(D^2(f))\right) \subseteq \ker\left(f_\# : C_n(X) \to C_n(Y)\right).$$

Now we prove the opposite inclusion.

Suppose that $c := \sum_i n_i \Delta_i$ is a chain in $\ker f_\# : C_k(X) \to C_k(Y)$:

$$\sum_i n_i f_\#(\Delta_i) = 0.$$

Write $\|c\| := \sum_i |n_i|$. We can assume that for each $i \neq j$ with $n_i \neq 0 \neq n_j$, we have $\Delta_i \neq \Delta_j$. For each simplex Δ_{i_1} with $n_{i_1} \neq 0$, let $\Delta_{i_2}, \ldots, \Delta_{i_r}$ be the simplices with non-zero coefficients in c such that $f_\#(\Delta_{i_j}) = f_\#(\Delta_{i_1})$. We can assume that $n_{i_1} > 0$. Thus $\sum_{j=1}^r n_{i_j} = 0$. Hence, perhaps after a reordering, we can assume that $n_{i_2} < 0$. Let $\Delta_{i_1} = [v_0, \ldots, v_n]$ and $\Delta_{i_2} = [v_0', \ldots, v_n']$; our assumptions means that $f(v_j) = f(v_j')$ for $j = 0, \ldots, n$. Thus, as explained in the construction of the triangulation of the $D^k(f)$ preceding Proposition 10.2, $\Delta_{i_1} \times_Y \Delta_{i_2} = [(v_0, v_0'), \ldots, (v_n, v_n')]$ is an n-simplex in the triangulation of $D^2(f)$. Then

$$c' := \Delta_{i_1} \times_Y \Delta_{i_2} - \Delta_{i_2} \times_Y \Delta_{i_1} = [(v_0, v_0'), \ldots, (v_k, v_k')] - [(v_0', v_0), \ldots, (v_k', v_k)]$$

is an alternating k-chain on $D^2(f)$ and

$$\varepsilon^2_\#(c') = \Delta_{i_1} - \Delta_{i_2}.$$

The chain $c - \varepsilon^2_\#(c')$ lies in $\ker f_\#$, and moreover $\|c - \varepsilon^2_\#(c')\| < \|c\|$. By induction on $\|c\|$, we conclude that

$$\ker f_\# \subset \varepsilon^2_\# \left(C_k^{\mathrm{Alt}}(D^2(f)) \right).$$

Since the opposite inclusion always holds, this is an equality. □

For any simplex $\Delta^{(k)}$ in the triangulation of $D^k(f)$, let

$$\mathrm{Alt}_{\mathbb{Z}}(\Delta^{(k)}) = \sum_{\sigma \in S_k} \mathrm{sign}\,(\sigma)\sigma_\#(\Delta^{(k)}). \tag{10.19}$$

Clearly $\mathrm{Alt}_{\mathbb{Z}}(\Delta^{(k)})$ is an alternating chain. We use the subindex $\mathbb{Z}$ to distinguish the operator $\mathrm{Alt}_{\mathbb{Z}}$ from the more familiar projection operator $\mathrm{Alt}_{\mathbb{Q}} = \frac{1}{k!}\sum_{\sigma \in S_k} \mathrm{sign}\,(\sigma)\sigma_\#$.

Lemma 10.8

1. *If the simplex $\Delta^{(k)}$ in $D^k(f)$ is invariant under some non-trivial permutation $\sigma \in S_k$, then $\Delta^{(k)}$ does not appear in any irredundant expression of any alternating chain c, and moreover $Alt_{\mathbb{Z}}(\Delta^{(k)}) = 0$.*
2. *Let Δ_y be an n-simplex in Y, and let $\Delta_1, \ldots, \Delta_N$ be the simplices Δ in X for which $f_\#(\Delta) = \Delta_y$, with $\Delta_i \neq \Delta_j$ for $i \neq j$. For $1 \leq i_1 < \cdots < i_k \leq N$, denote the sequence $i_1, \ldots, i_k$ by I, and let $\Delta_I^{(k)}$ be the n-simplex $(\Delta_{i_1} \times \cdots \times \Delta_{i_k})/Y$ in $D^k(f)$. Then $C_n^{Alt}(D^k(f))$ is freely generated over $\mathbb{Z}$ by the alternating chains $Alt_{\mathbb{Z}}(\Delta_I^{(k)})$ for such I.*

Proof

1. If $\Delta^{(k)}$ is invariant under some non-trivial permutation $\sigma \in S_k$, pick $i \neq j$ such that $\sigma(i) = j$. The fact that $\Delta^{(k)}$ is invariant under σ means that $|\Delta^{(k)}|$ is contained in the set $\{x_i = x_j\} \subset X^k$, and hence $\Delta^{(k)}$ is invariant under (i, j). Suppose that the coefficient of $\Delta^{(k)}$ in some alternating chain c on $D^k(f)$ is m. Because (i, j) leaves $\Delta^{(k)}$ fixed, its coefficient in $(i, j)_\#(c)$ is also m. But because c is alternating and sign $(i, j) = -1$, we must have $m = -m$. Thus $m = 0$.

   If $\Delta^{(k)}$ is invariant under some non-trivial permutation, so is $\sigma_\#(\Delta^{(k)})$ for each $\sigma \in S_k$. It follows that $\mathrm{Alt}_{\mathbb{Z}}(\Delta_i^{(k)}) = 0$.

2. By part 1 of the lemma, every simplex $\Delta^{(k)}$ appearing non-trivially in an alternating chain is in the S_k orbit of some simplex $\Delta_I^{(k)}$ with I as described. The argument just given shows that for any such simplex $\Delta_I^{(k)}$, the simplices $\sigma_\#(\Delta_I^{(k)})$, $\sigma \in S_k$, are pairwise distinct, and thus that the coefficient of $\Delta_I^{(k)}$ in $\mathrm{Alt}_{\mathbb{Z}}(\Delta_I^{(k)})$ is equal to 1. Now let c be an alternating simplicial n-chain on $D^k(f)$. We claim that c is a linear combination of chains $\mathrm{Alt}_{\mathbb{Z}}(\Delta_I^{(k)})$. To see this, suppose that $m \neq 0$ is the coefficient of $\Delta_I^{(k)}$ in an irredundant expression of c as linear combination of ordered simplices. Let $\|c\|$ be the sum of the absolute values of the coefficients of the simplices in such an expression. Because c is alternating, for each $\sigma \in S_k$, the coefficient of $\sigma_\#(\Delta_I^{(k)})$ is sign $(\sigma)m$. It follows that

$$\|c - m\mathrm{Alt}_{\mathbb{Z}}(\Delta_{\mathbf{x}}^{(k)})\| < \|c\|.$$

Hence, by induction on $\|c\|$, c is a linear combination of alternating chains $\mathrm{Alt}_{\mathbb{Z}}(\Delta_I^{(k)})$.

That there are no relations among distinct chains $\mathrm{Alt}_{\mathbb{Z}}\Delta(k)_I$ is obvious: if $I = \{i_1, \ldots, i_k\}$ and $I' \neq I$ then the simplex $(\Delta_{i_1} \times \cdots \times \Delta_{i_k})/Y$ does not appear in $\mathrm{Alt}_{\mathbb{Z}}\Delta_{I'}^{(k)}$. □

Remark 10.6 The lemma shows that $C_n^{\mathrm{Alt}}(D^k(f))$ is equal to the image of the "Alternation Operator" $\mathrm{Alt}_{\mathbb{Z}} : C_n(D^k(f)) \to C_n^{\mathrm{Alt}}(D^k(f))$. We learned this from Kevin Houston's paper [Hou97]. This is a special feature of this S_k-action. In contrast, consider the action of S_3 on $\mathbb{R} \times \mathbb{R}^3$ defined by

$$\sigma(t, x_1, x_2, x_3) = ((\mathrm{sign}\, \sigma)t, x_{\sigma(1)}, x_{\sigma(2)}, x_{\sigma(3)}).$$

The point $(1, x, x, x)$ is then invariant only under even permutations. If we apply $\mathrm{Alt}_{\mathbb{Z}}$ to the 0-simplex $[(1, x, x, x)]$, we get $3([(1, x, x, x)] - [(-1, x, x, x)])$. The alternating 0-chain $[(1, x, x, x)] - [(-1, x, x, x)]$ is not in the image of $\mathrm{Alt}_{\mathbb{Z}}$.

Lemma 10.9 *For each n, the complex $(C_n^{Alt}(D^\bullet(f)), \epsilon_\#^\bullet)$ splits as a direct sum of complexes:*

$$C_n^{Alt}(D^\bullet(f)) = \bigoplus C_n^{Alt}(D^\bullet(f))|_\Delta. \tag{10.20}$$

Here the direct sum is over ordered n-simplices Δ in Y.

Proof We have already seen, in Lemma 10.6, that each of the $(C_n^{\mathrm{Alt}}(D^\bullet(f))|_{\Delta_y}, \varepsilon^\bullet)$ is a subcomplex. If $\Delta \neq \Delta'$, then

$$C_n^{\mathrm{Alt}}(D^k(f))|_\Delta \cap C_n^{\mathrm{Alt}}(D^k(f))|_{\Delta'} = 0.$$

The splitting (10.20) now follows from Lemma 10.8. □

Now we prove the main theorem.

Proposition 10.3 *For each $n \geq 0$, the complex of alternating simplicial chains $(C_n^{\mathrm{Alt}}(D^\bullet(f)), \epsilon^\bullet)$ is an acyclic resolution of $C_n(Y)$.*

Proof We have $C_n(Y) = \bigoplus \mathbb{Z} \cdot \Delta$, where the direct sum is over n-simplices of Y. We claim that for each n-simplex Δ of Y, the complex

$$0 \longleftarrow \mathbb{Z} \cdot \Delta \xleftarrow{f_\#} C_n(X)|_\Delta \xleftarrow{\epsilon_\#^2} C_n^{\mathrm{Alt}}(D^2(f))|_\Delta \longleftarrow$$

$$\xleftarrow{\epsilon_\#^3} \cdots \xleftarrow{\epsilon_\#^k} C_n^{\mathrm{Alt}}(D^k(f))|_\Delta \longleftarrow \tag{10.21}$$

is exact. Because of the splitting of $(C_n^{\mathrm{Alt}}(D^\bullet(f)), \epsilon^\bullet)$ as a direct sum of subcomplexes $(C_n^{\mathrm{Alt}}(D^\bullet(f))|_\Delta, \epsilon_n^\bullet)$, this will prove the proposition. Let $\Delta_1, \ldots, \Delta_N$ be the distinct simplices Δ' of X such that $f_\#(\Delta') = \Delta$. To prove exactness of (10.21), we show it is, up to some signs, chain-isomorphic to a well-known exact complex, namely the augmented oriented simplicial chain complex of the standard $N-1$-simplex $\mathbf{\Delta^{N-1}} := [e_1, \ldots, e_N]$, with degree shifted by 1, $C_\bullet^{\mathrm{or}}(\mathbf{\Delta^{N-1}})[1]$,

$$0 \longleftarrow \mathbb{Z} \xleftarrow{\varepsilon} C_0^{\mathrm{or}}(\mathbf{\Delta^{N-1}}) \longleftarrow C_1^{\mathrm{or}}(\mathbf{\Delta^{N-1}}) \longleftarrow$$

$$\longleftarrow \cdots \longleftarrow C_{k-1}^{\mathrm{or}}(\mathbf{\Delta^{N-1}}) \longleftarrow \cdots \tag{10.22}$$

where $\varepsilon(\sum_i m_i e_i) = \sum_i m_i$. To obtain a *bona fide* chain isomorphism, we would have to modify the sign of the boundary operators in (10.22). Since our aim is simply to prove exactness of (10.21), it is enough instead to construct an isomorphism $\varphi_k : C_n^{\mathrm{Alt}}(D^k(f))|_\Delta \to C_{k-1}^{\mathrm{or}}(\mathbf{\Delta^{N-1}})$ for each k such that

$$\varphi_{k-1} \circ \epsilon_n^k = (-1)^{k+n-1} \partial \circ \varphi_k. \tag{10.23}$$

Recall from Lemma 10.8 that $C_n^{\mathrm{Alt}}(D^k(f))|_\Delta$ is freely generated by the n-chains $\mathrm{Alt}_{\mathbb{Z}}(\Delta_{i_1} \times \cdots \times \Delta_{i_k})/Y$, where $1 \leq i_1 < \cdots < i_k \leq N$. When $k = 1$, this of course just reduces to the single simplex Δ_{i_1}. We map each such generator to the oriented $k-1$ face $[e_{i_1}, \ldots, e_{i_k}]$ of $\mathbf{\Delta^{N-1}}$, obtaining in this way an isomorphism

$$\varphi_k : C_n^{\mathrm{Alt}}(D^k(f))|_\Delta \to C_{k-1}^{\mathrm{or}}(\mathbf{\Delta^{N-1}}).$$

And we define $\varphi_0(m\Delta) = m$. To prove (10.23), recall first that $\epsilon_n^k = (-1)^n \varepsilon_\#^k$. Since $\varepsilon_\#^k$ maps alternating chains to alternating chains, so does ϵ^k; moreover since

$$\varepsilon_\#^k((\Delta_{i_1} \times \cdots \times \Delta_{i_k})/Y) = (\Delta_{i_1} \times \cdots \times \Delta_{i_{k-1}})/Y,$$

it follows that $\epsilon^k(\mathrm{Alt}_{\mathbb{Z}}(\Delta_{i_1} \times \cdots \times \Delta_{i_k})/Y)$ must be a linear combination of chains

$$\mathrm{Alt}_{\mathbb{Z}}((\Delta_{i_1} \times \cdots \times \widehat{\Delta_{i_\ell}} \times \cdots \times \Delta_{i_k})/Y)$$

with $1 \leq \ell \leq k$, and it is necessary only to determine the coefficient of each. The simplex $(\Delta_{i_1} \times \cdots \times \widehat{\Delta_{i_\ell}} \times \cdots \times \Delta_{i_k})/Y$ itself appears only once in the expression of $\varepsilon^k_{\#}(\mathrm{Alt}_{\mathbb{Z}}((\Delta_{i_1} \times \cdots \times \Delta_{i_k})/Y)$ as a linear combination of simplices, namely as

$$\varepsilon^k_{\#}\mathrm{sign}\,(\sigma_{k,\ell})(\sigma_{k,\ell})_{\#}\left((\Delta_{i_1} \times \cdots \times \Delta_{i_k})/Y\right),$$

where $\sigma_{k,\ell}$ is the permutation taking the ordered set $(0, \ldots, k)$ to $(0, \ldots, \hat{\ell}, \ldots, k, \ell)$. This permutation has sign $(-1)^{k-\ell}$. It follows that this, multiplied by $(-1)^n$, is the coefficient of $\mathrm{Alt}_{\mathbb{Z}}((\Delta_{i_1} \times \cdots \times \widehat{\Delta_{i_\ell}} \times \cdots \times \Delta_{i_k})/Y)$ in $\epsilon^k(\mathrm{Alt}_{\mathbb{Z}}(\Delta_{i_1} \times \cdots \times \Delta_{i_k})/Y)$. The coefficient of $[e_{i_1}, \ldots, \widehat{e_{i_\ell}}, \ldots, e_{i_k}]$ in $\partial\varphi_k(\mathrm{Alt}_{\mathbb{Z}}((\Delta_{i_1} \times \cdots \times \Delta_{i_k})/Y)$ is $(-1)^{\ell-1}$, thus showing that (10.23) holds, and completing the proof. □

Remark 10.7 Simplicial vs Singular Our theorem relates the alternating simplicial homology of the $D^k(f)$ to the simplicial homology of the image, Y. It is well known that singular and simplicial homology coincide. In fact a variant of the standard proof shows that the same goes for alternating simplicial homology and alternating singular homology.

10.4 Finite Complex Maps Are Triangulable

Applicability of our version of the ICSS relies on the triangulability of the map $f : X \to Y$. Fortunately, all of the maps we deal with have this property:

Theorem 10.5 ([Har77, Theorem 3])

1. *Let $P \subset \mathbb{R}^n$ be a closed finite-dimensional subanalytic set and $N \subset \mathbb{R}^m$ a real analytic space, and let $f : P \to N$ be a proper light subanalytic map. Then there exist simplicial complexes $\mathscr{G}$ and $\mathscr{H}$ and homeomorphisms $g : \mathscr{G} \to P$ and $h : \mathscr{H} \to f(P)$, and a simplicial map $p : \mathscr{G} \to \mathscr{H}$ such that $f = h \circ p \circ g^{-1}$.*
2. *Suppose, in the same situation, that $\mathscr{Q}$ and $\mathscr{R}$ are locally finite families of closed subanalytic sets in P and $f(P)$, respectively. Then the simplicial complexes $\mathscr{G}$ and $\mathscr{H}$, and the homeomorphisms g and h, can all be chosen so that $g^{-1}(Q)$ is a subcomplex of $\mathscr{G}$ and $h^{-1}(R)$ is a subcomplex of $\mathscr{H}$ for all $Q \in \mathscr{Q}$ and $R \in \mathscr{R}$.* □

A map is *light* if the preimage of a discrete set is discrete. A subset of a Euclidean space is *subanalytic* if locally it is the image of a semianalytic set under a linear projection, and *semianalytic* if it can be defined by the vanishing and/or positivity of real analytic functions—see, e.g., [BM88]. Evidently every real analytic space (and hence every complex analytic space) is subanalytic. A *map* is subanalytic if its graph is a subanalytic set. Thus, complex and real analytic maps are subanalytic.

Our version of the ICSS applies, for example, to a stable perturbation f_t of an $\mathscr{A}$-finite map-germs $(\mathbb{F}^n, S) \to (\mathbb{F}^p, 0)$ when $n \leq p$, and, when $n \geq p$, to the restriction to the critical space of a stable perturbation of an $\mathscr{A}$-finite germ. The ICSS calculates the homology of the image in the former case, and of the discriminant (set of critical values) in the latter.

10.5 Other Proofs

The ICSS was introduced in [GM93] in a version for rational cohomology. It was based on the exactness of the sequence of sheaves associated with a surjection $f : X \to Y$,

$$0 \longrightarrow \mathbb{Q}_Y \longrightarrow f_*\mathbb{Q}_X \longrightarrow \pi^2_*\mathbb{Q}^{\mathrm{Alt}}_{D^2(f)} \longrightarrow \pi^3_*\mathbb{Q}^{\mathrm{Alt}}_{D^3(f)} \longrightarrow \cdots \tag{10.24}$$

where $\pi^k_*\mathbb{Q}^{\mathrm{Alt}}_{D^k(f)}$ is the S_k—alternating part of the push-forward of the constant sheaf $\mathbb{Q}_{D^k(f)}$ by $\pi^k : D^k(f) \to Y$. Exactness is proved by exactly the same argument as used here in the proof of Proposition 10.3.

In [Gor95], Victor Goryunov introduced the alternating chain complex. He used a geometric realisation Y_G of the semisimple object $D^\bullet(f)$ due to Vassiliev and described in [Vas01]: it is formed by beginning with X embedded in some high-dimensional Euclidean space, and then adding, for each pair of points x_1, x_2 with $f(x_1) = f(x_2)$, the 1-simplex $[x_1, x_2]$, for each triple x_1, x_2, x_3 with $f(x_1) = f(x_2) = f(x_3)$ the 2-simplex $[x_1, x_2, x_3]$, etc. The embedding of X in $\mathbb{R}^N$ is chosen so general that none of these added simplices meet one another except by the standard face inclusions. There is then a natural surjection $Y_G \to Y$, and for each $y \in Y$ the preimage in Y_G is a simplex. From this it follows that $Y_G \to Y$ is a homotopy equivalence, so that the homology of Y may be computed as the homology of Y_G. Goryunov showed in [Gor95] that the alternating homology $AH_q^(D^k(f))$ is isomorphic to the relative homology $H_{q+k-1}(Y_G^k, Y_G^{k-1})$, where Y_G^k is X together with the added simplices of dimension $\leq k$, and the ICSS is thus reduced to the spectral sequence for the homology of the filtered space Y_G.

Our argument here is clearly closely related to Goryunov's proof. It seems likely that the total complex of the double complex $(C_\bullet^{\mathrm{Alt}}(D^\bullet(f)), \partial, \varepsilon_\#^\bullet)$ is isomorphic to the simplicial chain complex of some triangulation of Y_G. It would be interesting to prove this.

10.6 Cohomology

Dualising all of the modules in the homological double complex (10.3) with respect to $\mathbb{Z}$, we obtain a second double complex with all arrows reversed, which we will refer to as the *cohomological double complex*. The columns are still exact except in degree 0, for $\left(C_n^{\mathrm{Alt}}(D^{\bullet+1}(f)), \varepsilon_{\#}^{\bullet}\right)$ is a resolution of the free $\mathbb{Z}$-module, $C_n(Y)$ (see Exercise 10.6.1 below) and thus

$$H^q\Big(\mathrm{Hom}_{\mathbb{Z}}\left(C_n^{\mathrm{Alt}}(D^{\bullet+1}(f))\right), \mathbb{Z}\Big) = \begin{cases} \ker(\varepsilon^2)^{\#} & q = 0 \\ 0 & q > 0 \end{cases},$$

where $(\varepsilon^2)^{\#} : C^n(X) \to \mathrm{Hom}_{\mathbb{Z}}(C_n^{\mathrm{Alt}}(D^2(f), \mathbb{Z})$. By exactness of (10.2), the kernel of $(\varepsilon^2)^{\#}$ is just $\mathrm{Hom}_{\mathbb{Z}}(C_n(Y), \mathbb{Z})$, i.e. $C^n(Y)$.

Just as for the homological double complex, the cohomology of the total complex of the cohomological double complex can be calculated by the spectral sequences arising from two standard filtrations—see, e.g., [McC01, Theorem 2.15]. The exactness of the columns means that, as in the homological case, the first of these spectral sequences collapses onto the p-axis, and the E_1 differential $C^n(Y) \to C^{n+1}(Y)$ is the usual coboundary operator. The cohomology of the total complex is therefore that of Y. The second spectral sequence has

$$E_1^{p,q} = H^q\left(Hom_{\mathbb{Z}}(C_\bullet^{\mathrm{Alt}}(D^{q+1}(f)), \mathbb{Z})\right),$$

with the differential d_1^{pq} inherited from the vertical differential $(\varepsilon^{q+1})^{\#}$. Thus we have

Proposition 10.4 *Let $f : X \to Y$ be a finite surjective triangulable map of triangulable topological spaces.*

1. *There is a spectral sequence with $E_1^{pq} = H^q\left(Hom_{\mathbb{Z}}(C_\bullet^{Alt}(D^{q+1}(f)), \mathbb{Z})\right)$, converging to $H^{p+q}(Y)$.*
2. *The differential $d^1 : E_1^{p,q} \to E_1^{p+1,q}$ is the dual of the morphism $\varepsilon_{\#}^{p+2}$ induced by the projection $\varepsilon^{p+2} : D^{p+2}(f) \to D^{p+1}(f)$.*

□

Now, by analogy with the definition of C_n^{Alt}, we define alternating cochains:

$$C^n_{\mathrm{Alt}}(D^k(f)) := \{\psi \in C^n(D^k(f)) : \sigma^{\#}(\psi) = \mathrm{sign}\,(\sigma)\psi \ \text{ for all } \sigma \in S_k\}.$$

Like $C_\bullet^{\mathrm{Alt}}(D^k(f))$, $C^\bullet_{\mathrm{Alt}}(D^k(f))$, with the usual coboundary operator is a complex, and gives rise to *alternating cohomology*, $AH^*(D^k(f))$.

In fact, thanks to Lemma 10.8, $C^\bullet_{\mathrm{Alt}}(D^k(f))$ can be identified with the complex $\mathrm{Hom}_{\mathbb{Z}}(C_\bullet^{\mathrm{Alt}}(D^{q+1}(f)), \mathbb{Z})$, and $AH^q(D^k(f))$ with $H^q\left(\mathrm{Hom}_{\mathbb{Z}}(C_\bullet^{\mathrm{Alt}}(D^{q+1}(f)), \mathbb{Z})\right)$. Recall the operator $\mathrm{Alt}_{\mathbb{Z}} : C_n(D^k(f)) \to C_n^{\mathrm{Alt}}(D^k(f))$, defined by $\mathrm{Alt}_{\mathbb{Z}}(c) = \sum_{\sigma \in S_k} \mathrm{sign}(\sigma)\sigma_{\#}(c)$. Given $\psi \in \mathrm{Hom}(C_n^{\mathrm{Alt}}(D^k(f)), \mathbb{Z})$, we define a cochain $\mathrm{Alt}_{\mathbb{Z}}^{*}\psi \in C^n(D^k(f))$ by composing with $\mathrm{Alt}_{\mathbb{Z}} : C_n(D^k(f)) \to C_n^{\mathrm{Alt}}(D^k(f))$.

Proposition 10.5 *$Alt_{\mathbb{Z}}^*$ defines an isomorphism of complexes*

$$\mathrm{Hom}_{\mathbb{Z}}(C_n^{Alt}(D^\bullet(f)), \mathbb{Z}) \to C^n_{Alt}(D^\bullet(f)).$$

Proof We have to check first that for $\psi \in \mathrm{Hom}(C_n^{\mathrm{Alt}}(D^k(f)), \mathbb{Z})$, $\mathrm{Alt}_{\mathbb{Z}}^*\psi \in C^n_{\mathrm{Alt}}(D^k(f))$. Suppose $\sigma \in S_k$ and c is any chain in $C_n(D^k(f))$. Then

$$\begin{aligned}\left(\sigma^\#(\mathrm{Alt}_{\mathbb{Z}}^*\psi)\right)(c) = (\mathrm{Alt}_{\mathbb{Z}}^*\psi)(\sigma_\# c) = \psi(\mathrm{Alt}_{\mathbb{Z}}(\sigma_\#(c)) &= \psi\left(\sum_{\rho\in S_k} \mathrm{sign}\,(\rho)\rho_\#\sigma_\#(c)\right)\\ &= \mathrm{sign}\,(\sigma)\psi\left(\sum_{\rho\in S_k}\mathrm{sign}\,(\rho)\mathrm{sign}\,(\sigma)(\rho\circ\sigma)_\#(c)\right)\\ &= \mathrm{sign}\,(\sigma)(\psi)(\mathrm{Alt}_{\mathbb{Z}}(c)) = \mathrm{sign}\,(\sigma)\mathrm{Alt}_{\mathbb{Z}}^*\psi(c).\end{aligned}$$

Thus $\mathrm{Alt}_{\mathbb{Z}}^*\psi \in C^n_{\mathrm{Alt}}(D^k(f))$. Because for any $\sigma \in S_k$, $\varepsilon_\#^{p+1}\circ\sigma_\# = \sigma_\#\circ\varepsilon_\#^{p+1}$, $\mathrm{Alt}_{\mathbb{Z}}^*$ gives a homomorphism of cochain complexes

$$\mathrm{Hom}_{\mathbb{Z}}(C_\bullet^{\mathrm{Alt}}(D^k(f)), \mathbb{Z}) \to C^\bullet_{\mathrm{Alt}}(D^k(f)).$$

It has an inverse $\theta : C^\bullet_{\mathrm{Alt}}(D^k(f)) \to \mathrm{Hom}_{\mathbb{Z}}(C_\bullet^{\mathrm{Alt}}(D^k(f)), \mathbb{Z})$, defined as follows. Suppose that $\varphi \in C^n_{\mathrm{Alt}}(D^k(f))$ and $c \in C_n^{\mathrm{Alt}}(D^k(f))$. By Lemma 10.8, we can write $c = \sum_i m_i\mathrm{Alt}_{\mathbb{Z}}(c_i)$, where each c_i is a simplex. This representation is not unique, because for any $\sigma \in S_k$, the alternating chain $\mathrm{Alt}_{\mathbb{Z}}(c_i)$ can also be written as $\mathrm{sign}\,(\sigma)\mathrm{Alt}_{\mathbb{Z}}(\sigma_\#(c_i))$. However, because φ is alternating, we have

$$\varphi\Big(\mathrm{sign}\,(\sigma)\sigma_\#(c_i)\Big) = \sigma^\#(\varphi)\big(\mathrm{sign}\,(\sigma)c_i\big) = \varphi(c_i),$$

so that $\theta(\varphi)$ is well-defined by the formula

$$\theta(\varphi)\Big(\sum_i m_i\mathrm{Alt}_{\mathbb{Z}}(c_i)\Big) = \sum_i m_i\varphi(c_i).$$

Once again, θ is a homomorphism of cochain complexes.

Since, for any simplex c,

$$\mathrm{Alt}_{\mathbb{Z}}^*(\theta(\varphi))(c) = \theta(\varphi)(\mathrm{Alt}_{\mathbb{Z}}(c)) = \varphi(c),$$

and for any alternating chain $c = \sum_i m_i\mathrm{Alt}_{\mathbb{Z}}(c_i)$,

$$\theta(\mathrm{Alt}_{\mathbb{Z}}^*(\psi))(c) = \sum_i m_i\mathrm{Alt}_{\mathbb{Z}}^*(\psi)(c_i) = \sum_i m_i\psi(\mathrm{Alt}_{\mathbb{Z}}(c_i)) = \psi(c),$$

θ and $\mathrm{Alt}_{\mathbb{Z}}^*$ are mutually inverse. □

Now Proposition 10.4 becomes

Theorem 10.6 *Let $f : X \to Y$ be a finite surjective triangulable map of triangulable topological spaces. Then*

1. *there is a spectral sequence with $E_1^{pq} = AH^q(D^{p+1}(f))$, converging to $H^{p+q}(Y)$.*
2. *The differential $d^1 : E_1^{p,q} \to E_1^{p+1,q}$ is the morphism $(\varepsilon^{p+2})^*$ induced by the projection $\varepsilon^{p+2} : D^{p+2}(f) \to D^{p+1}(f)$.*

□

Exercises for Sect. 10.6

1 This exercise is concerned with the assertion in the first line of this section, that the dual of a free resolution of a free abelian group is exact.

Prove the following sequence of statements.

(i) If F is a free abelian group then any exact sequence

$$0 \longrightarrow A \xrightarrow{i} B \xrightarrow{j} F \to 0$$

splits (i.e. there is an isomorphism $B \xrightarrow{(\alpha, j)} A \oplus F$ such that $i \circ \alpha = 1_A$).

(ii) Any subgroup of a free abelian group is also free.

(iii) If $F_1 \longrightarrow F_0 \xrightarrow{\alpha} F \longrightarrow 0$ is exact, with all three groups free abelian, then $F_1 \simeq \ker\alpha \oplus F_1'$ for some subgroup F_1'.

(iv) The dual of any exact sequence $\cdots \to F_n \to \cdots \to F_0 \to F \to 0$ of free abelian groups is exact.

(v) For comparison, the dual of the sequence

$$0 \longrightarrow \mathbb{Z} \xrightarrow{n} \mathbb{Z} \longrightarrow (\mathbb{Z}/n\mathbb{Z}) \longrightarrow 0$$

is not exact.

10.7 Examples and Applications of the ICSS

We have already described what are perhaps the most striking applications of the ICSS in Corollaries 10.1 and 10.3. Here we discuss some further applications. We begin with some simple calculations which provide useful illustrations.

10.7.1 The Reidemeister Moves

The Reidemeister moves of knot theory are versal deformations of the three $\mathcal{A}_e$-codimension 1 singularities of mappings from the line to the plane. The ICSSs resulting from their disentanglements are good examples through which to familiarise oneself with the spectral sequence. The codimension 1 germs are shown in the middle column of the table below, and the right-hand column shows a 1-parameter versal deformation, which, fixing $t \neq 0$, gives a stable perturbation.

I	$f_0 : x \mapsto (x^2, x^3)$	$f_t : x \mapsto (x^2, x^3 - tx)$
II	$f_0 : \begin{cases} x \mapsto (x,\ x^2) \\ y \mapsto (y, -y^2) \end{cases}$	$f_t : \begin{cases} x \mapsto (x, x^2 - t) \\ y \mapsto (y, -y^2) \end{cases}$
III	$f_0 : \begin{cases} x \mapsto (x,\ x) \\ y \mapsto (y,\ 0) \\ z \mapsto (z, -z) \end{cases}$	$f_t : \begin{cases} x \mapsto (x,\ x) \\ y \mapsto (y,\ t) \\ z \mapsto (z, -z) \end{cases}$

(10.25)

For all three cases, the non-trivial modules in the E^1 page of the ICSS for f_t are contained in the single column

$$\begin{array}{c} 0 \\ \downarrow \\ H_0^{\mathrm{Alt}}(D^3(f_t)) \\ \downarrow \varepsilon_*^3 \\ H_0^{\mathrm{Alt}}(D^2(f_t)) \\ \downarrow \varepsilon_*^2 \\ H_0(U_t) \\ \downarrow \\ 0 \end{array} \tag{10.26}$$

10.7.2 Reidemeister I

Take $F\colon (x, t) \mapsto (t, f_t(x))$ as stable unfolding. As usual, we embed $D^2(F)$ in $\mathbb{C}^3$ with coordinates t, x_1, x_2. There, following the recipe in Sect. 9.5, we find that

$D^2(F)$ is defined by the equations

$$\frac{x_1^2 - x_2^2}{x_1 - x_2} = \frac{x_1^3 - tx_1 - (x_2^3 - tx_2)}{x_1 - x_2} = 0. \tag{10.27}$$

Simplifying, this gives

$$x_1 + x_2 = 0 \quad x_1^2 + x_1x_2 + x_2^2 = t. \tag{10.28}$$

Thus $D^2(f_0)$ is a 0-dimensional A_1 singularity. For $t \neq 0$,denote the two complex square roots of t by P and Q. Then $D^2(f_t)$ is the point pair $\{(P, Q), (Q, P)\}$, and $H_0^{\mathrm{Alt}}(D^2(f_t)) \simeq \mathbb{Z}$, generated by the class of $[(P, Q)] - [(Q, P)]$. Note that $H_0^{\mathrm{Alt}}(D^2(f_0)) = 0$, since when $t = 0$, $P = Q$. For both f_0 and f_t, $D^3 = \emptyset$.

In the E^1 page (10.26), we have $\varepsilon_*^2 = 0$, for $\varepsilon_*^2\big([(P, Q)]-[(Q, P)]\big) = [P]-[Q]$, and U_t is connected, so that $[P] = [Q]$. Hence the spectral sequence collapses at E^1, and for $t \neq 0$

$$H_0(X_t) = H_0(U_t) = \mathbb{Z}, \quad H_1(X_t) = H_0^{\mathrm{Alt}}(D^2(f_t)) = \mathbb{Z}.$$

The same groups and homomorphisms arise in the real case when we take $t > 0$.

10.7.3 Reidemeister II

Here both branches of the bi-germ f_0 are immersions, so all multiple points are strict. Denote by 0_x and 0_y the origins of the coordinate systems with coordinates x and y respectively. The domain of the stable perturbation f_t is a disjoint union $U_t = U_{x,t} \cup U_{y,t}$, where $U_{x,t}$ and $U_{y,t}$ are contractible neighbourhoods of 0_x and 0_y. Thus $H_0(U_t) \simeq \mathbb{Z}^2$. There are no triple points, and $D^2(f_t)$ consists of

$$\{(x, y) \in (\mathbb{C}, 0_x) \times (\mathbb{C}, 0_y) : x = y, x^2 - t = -y^2\} \tag{10.29}$$

together with its image under the involution $(1, 2)$ sending (x, y) to (y, x). When $t = 0$ this is a pair of 0-dimensional A_1 singularities, interchanged by $(1, 2)$. To describe $D^2(f_t)$ for $t \neq 0$, denote the points in $(\mathbb{C}, 0_x)$ with x coordinates $\sqrt{t/2}$ and $-\sqrt{t/2}$ by P_x and Q_x respectively, and the points in $(\mathbb{C}, 0_y)$ with y coordinates $\sqrt{t/2}$ and $-\sqrt{t/2}$ by P_y and Q_y. Then for $t \neq 0$,

$$D^2(f_t) = \{(P_x, P_y), (P_y, P_x), (Q_x, Q_y), (Q_y, Q_x)\}, \tag{10.30}$$

with the involution $(1, 2)$ interchanging the first and second points, and the third and fourth. For $t = 0$, this collapses just to

$$D^2(f_0) = \{(0_x, 0_y), (0_y, 0_x)\}.$$

Thus for $t \neq 0$, $H_0^{\mathrm{Alt}}(D^2(f_t))$ has basis $[(P_x, P_y)] - [(P_y, P_x)]$, $[(Q_x, Q_y)] - [(Q_y, Q_x)]$, and for $t = 0$, $H_0^{\mathrm{Alt}}(D^2(f_0))$ has basis $[(0_x, 0_y)] - [(0_y, 0_x)]$. With respect to the basis of $H_0^{\mathrm{Alt}}(D^2(f_t))$ described above, and the basis $[P_x], [P_y]$ for $H_0(U_t)$, ε_*^2 has matrix $\begin{pmatrix} 1 & 1 \\ -1 & -1 \end{pmatrix}$ when $t \neq 0$, and thus has 1-dimensional kernel and cokernel. The spectral sequence collapses at E^2, and

$$H_1(X_t) = E_{1,0}^2 = \ker \varepsilon_*^2 \simeq \mathbb{Z}, \quad H_0(X_t) = E_{0,0}^2 = \operatorname{coker} \varepsilon_*^2 \simeq \mathbb{Z}.$$

10.7.4 Reidemeister III

We use the same conventions as for Reidemeister II. Let P_x and P_y denote the points in $U_{x,t}$ and $U_{y,t}$ with x and y coordinate t, and let Q_y and Q_z denote the points in $U_{y,t}$ and $U_{z,t}$ with y and z coordinate $-t$. Note that when $t = 0$, then $P_x = 0_x$, etc. Then

$$\begin{aligned} D^2(f_t) \cap (U_{x,t} \times U_{y,t}) &= \{(P_x, P_y)\} \\ D^2(f_t) \cap (U_{x,t} \times U_{z,t}) &= \{(0_x, 0_y)\} \\ D^2(f_t) \cap (U_{y,t} \times U_{z,t}) &= \{(Q_x, Q_z)\} \end{aligned} \tag{10.31}$$

and

$$D^3(f_0) \bigcap U_{x,t} \times U_{y,t} \times U_{z,t} = \{(0_x, 0_y, 0_z)\}.$$

Thus

$$\begin{aligned} &H_0^{\mathrm{Alt}}(D^3(f_0)) \simeq \mathbb{Z} \quad H_0^{\mathrm{Alt}}(D^3(f_t)) = 0 \\ &H_0^{\mathrm{Alt}}(D^2(f_0)) \simeq \mathbb{Z}^3 \quad H_0^{\mathrm{Alt}}(D^2(f_t)) \simeq \mathbb{Z}^3 \\ &H_0(U_0) \simeq \mathbb{Z}^3 \qquad\quad H_0(U_t) \simeq \mathbb{Z}^3 \end{aligned} \tag{10.32}$$

with bases shown in the following table.

Module	Basis
$H_0^{\mathrm{Alt}}(D^3(f_0))$	$[(0_x, 0_y, 0_z)] - [(0_x, 0_z, 0_y)] + [(0_z, 0_x, 0_y)]$ $-[(0_z, 0_y, 0_x)] + [(0_y, 0_z, 0_x)] - [(0_y, 0_x, 0_z)]$
$H_0^{\mathrm{Alt}}(D^2(f_t))$	$[(P_x, P_y)] - [(P_y, P_x)]$, $[(0_z, 0_x)] - [(0_x, 0_z)]$, $[(Q_y, Q_z)] - [(Q_z, Q_y)]$
$H_0(U_t)$	$[0_x] = [P_x]$, $[P_y] = [Q_y]$, $[Q_z] = [0_z]$

With respect to these bases, the differentials ε_*^k have the following matrices (with the first only for $t = 0$):

$\varepsilon_*^3 = \begin{pmatrix} 1 \\ 1 \\ 1 \end{pmatrix}$	$\varepsilon_*^2 = \begin{pmatrix} 1 & -1 & 0 \\ -1 & 0 & 1 \\ 0 & 1 & -1 \end{pmatrix}$

In the spectral sequence for f_0, the image of ε_*^3 kills the kernel of ε_*^2. When $t \neq 0$, D^3 vanishes, along with its homology, while $H_0^{\mathrm{Alt}}(D^2(f_t))$ remains unchanged. The spectral sequence collapses at E^2, and

$$H_1(X_t) = E_{1,0}^2 = \ker \varepsilon_*^2 \simeq \mathbb{Z}, \quad H_0(X_t) = E_{0,0}^2 = \operatorname{coker} \varepsilon_*^2 \simeq \mathbb{Z}.$$

10.7.5 Map-Germs of Multiplicity 2

Recall that the *multiplicity* of a finite map-germ $f\colon (\mathbb{F}^n, 0) \to (\mathbb{F}^p, 0)$ is by definition $\dim_{\mathbb{F}} \mathscr{O}_n / f^* \mathfrak{m}_p \mathscr{O}_n$. A germ of multiplicity 1 is an immersion, and thus germs of multiplicity 2 are by this measure the simplest non-immersive germs.

1. Two Examples Consider the map-germ $f(x, y) = (x, y^2, y^3 + x^{k+1} y)$. Because f has corank 1 and is written in linearly adapted form, $D^2(f)$ embeds naturally in $\mathbb{F}^3$, where, with respect to coordinates x, y_1, y_2, it is defined by the equations

$$\begin{aligned} y_1 + y_2 &= 0 \\ y_1^2 + y_1 y_2 + y_2^2 + x^{k+1} &= 0. \end{aligned} \tag{10.33}$$

Since f has multiplicity 2, $D^3(f) = \emptyset$ and $\varepsilon^2 : D^2(f) \to D_1^2(f)$ is an isomorphism. An easy calculation shows that

$$D_1^2(f) = \{(x, y) : y^2 + x^{k+1} = 0\},$$

a curve singularity of type A_k. By the calculation of Exercise 3.3.1, f has $\mathscr{A}_e$-codimension k. If f_u is a stable perturbation of f, then $D^2(f_u)$ is a Milnor fibre of a curve singularity of type A_k. This has first homology of rank k, and either one or two boundary components, depending on whether k is even or odd. Thus, when $\mathbb{F} = \mathbb{C}$, the genus g of $D^2(f_u)$ satisfies

$$k = \begin{cases} 2g & k \text{ even} \\ 2g - 1 & k \text{ odd} \end{cases}$$

2. A General Result We follow Kevin Houston's argument in [Hou98], to prove that the relation

$$\mathscr{A}_e - \text{codimension} \leq \mu_I$$

with equality in case f is quasihomogeneous, holds for *all* germs $(\mathbb{C}^n, 0) \to (\mathbb{C}^{n+1}, 0)$ of multiplicity 2. In coordinates $(x, x_n) = (x_1, \ldots, x_{n-1}, x_n)$ on $(\mathbb{C}^n, 0)$ and $(y_1, \ldots, y_{n+1})$ on $(\mathbb{C}^{n+1}, 0)$, such a germ takes the form

$$f(x, x_n) = (x, x_n^2, x_n p(x, x_n^2)).$$

Write $h(x, x_n) = (x, x_n^2)$, and, as in Example 3.5,write $\mathscr{O}^e$ for the subring of germs in $\mathscr{O}_{\mathbb{C}^n,0}$ invariant under reflection $(x, x_n) \mapsto (x, -x_n)$, and $\mathscr{O}^o$ for the anti-invariants. Note that $\mathscr{O}^o = x_n\, \mathscr{O}^e$. By the calculations of Exercises 3.3.2, we have

$$T\mathscr{A}_e f = \Big(\bigoplus_{i=1}^{n-1} \mathscr{O}_n \frac{\partial}{\partial y_i}\Big) \oplus \mathscr{O}_n \frac{\partial}{\partial y_n} \oplus \Big(\mathscr{O}^e + x_n\, \mathscr{O}^e \left(x_n^2 \frac{\partial p}{\partial x_n} \circ h,\ \frac{\partial p}{\partial x} \circ h,\ p \circ h\right)\Big) \frac{\partial}{\partial y_{n+1}},$$

where for brevity we write $\frac{\partial p}{\partial x}$ in place of $\frac{\partial p}{\partial x_1}, \ldots, \frac{\partial p}{\partial x_{n-1}}$. Thus

$$\begin{aligned} \operatorname{codim}_{\mathscr{A}_e}(f) &= \dim_{\mathbb{C}} \frac{\mathscr{O}^e}{\left(x_n^2 \frac{\partial p}{\partial x_n} \circ h,\ \frac{\partial p}{\partial x} \circ h \circ h,\ p \circ h\right)} \\ &= \dim_{\mathbb{C}} \frac{\mathscr{O}_{\mathbb{C}^n,0}}{\left(x_n \frac{\partial p}{\partial x_n},\ \frac{\partial p}{\partial x}\right)}, \end{aligned} \tag{10.34}$$

where the last equality is because h^* defines an isomorphism $\mathscr{O} \to \mathscr{O}^{\mathrm{e}}$. Now we use the ICSS to calculate $\mu_I(f)$. In this case the calculation is straightforward since there is only one non-empty multiple point space. Because f has multiplicity 2, $D^3(f) = \emptyset$, and the projection $\varepsilon^2\colon (D^2(f), (0, 0)) \to (\mathbb{C}^n, 0)$ is an embedding. The equations defining $D^2(f)$ in $\mathbb{C}^{n-1} \times \mathbb{C}^2$, with coordinates x, x_n, x_n', are

$$x_n + x_n' = 0, \qquad \frac{x_n p(x, x_n^2) - x_n' p(x, x_n^2)}{x_n - x_n'} = 0.$$

In the presence of the first equation, the second is equivalent to $p(x, x_n^2) = 0$, and this is the equation of $D^2(f)$ in $\mathbb{C}^n$. The involution is now induced by reflection in the hyperplane $\{x_n = 0\}$. If f has finite codimension then $D^2(f)$ has isolated singularity, by Theorem 9.5. Let f_t be a stable perturbation of f and let Y_t be the image of f_t. We will calculate the rank of $H_n(Y_t)$ and compare it to the $\mathscr{A}_e$-codimension of f. By Corollary 10.1,

$$H_n(Y_t) \simeq AH_{n-1}(D^2(f_t)).$$

We may assume that f_t is of the form

$$(x, y) \mapsto (x, x_n^2, x_n p_t(x, x_n^2)).$$

As $D^2(f_t)$ is a Milnor fibre of $D^2(f)$, the vanishing homology of $D^2(f_t)$ has rank equal to the Milnor number of $D^2(f)$. The involution $(x, x_n) \mapsto (x, -x_n)$ acts on $D^2(f_t)$ and induces an eigenspace decomposition

$$H_{n-1}(D^2(f_t); \mathbb{Q}) = H_{n-1}^{\mathrm{Alt}}(D^2(f_t); \mathbb{Q}) \oplus H_{n-1}(D^2(f_t); \mathbb{Q})^{\mathbb{Z}/2\mathbb{Z}}.$$

The invariant part of the homology is the homology of the quotient of $D^2(f_t)$ by the involution. This quotient has equation $p_t(x, x_n) = 0$, and so the rank of $H_{n-1}(D^2(f_t); \mathbb{Q})^{\mathbb{Z}/2\mathbb{Z}}$ is equal to the Milnor number of the singularity of p at (0, 0). Thus

$$\mu_I(f) = \mu(p \circ h) - \mu(p).$$

We have

$$\begin{aligned} \mu(p \circ h) &= \dim_{\mathbb{C}} \frac{\mathcal{O}_{\mathbb{C}^n,0}}{\left(x_n \frac{\partial p}{\partial x_n} \circ h, \frac{\partial p}{\partial x} \circ h\right)} \\ &= \dim_{\mathbb{C}} \frac{\mathcal{O}^e}{\mathcal{O}^e \bigcap \left(x_n \frac{\partial p}{\partial x_n} \circ h, \frac{\partial p}{\partial x} \circ h\right)} + \dim_{\mathbb{C}} \frac{\mathcal{O}^o}{\mathcal{O}^o \bigcap \left(x_n \frac{\partial p}{\partial x_n} \circ h, \frac{\partial p}{\partial x} \circ h\right)}. \end{aligned} \tag{10.35}$$

We consider the two summands in (10.35) separately. For the first we have

$$\dim_{\mathbb{C}} \frac{\mathcal{O}^e}{\mathcal{O}^e \bigcap \left(x_n \frac{\partial p}{\partial x_n} \circ h, \frac{\partial p}{\partial x} \circ h\right)} = \dim_{\mathbb{C}} \frac{\mathcal{O}^e}{\left(x_n^2 \frac{\partial p}{\partial x_n} \circ h, \frac{\partial p}{\partial x} \circ h\right) \mathcal{O}^e}.$$

For the second,

$$\begin{aligned} \dim_{\mathbb{C}} \frac{\mathcal{O}^o}{\mathcal{O}^o \bigcap \left(x_n \frac{\partial p}{\partial x_n} \circ h, \frac{\partial p}{\partial x} \circ h\right)} &= \dim_{\mathbb{C}} \frac{x_n \mathcal{O}^e}{\left(x_n \frac{\partial p}{\partial x_n} \circ h, x_n \frac{\partial p}{\partial x} \circ h\right) \mathcal{O}^e} \\ &= \dim_{\mathbb{C}} \frac{\mathcal{O}^e}{\left(\frac{\partial p}{\partial x_n} \circ h, \frac{\partial p}{\partial x} \circ h\right) \mathcal{O}^e}. \end{aligned}$$

The last quotient is the isomorphic image under pull-back by h of the Jacobian algebra $\mathcal{O}_{\mathbb{C}^n,0} / J_p$, and has length $\mu(p)$. So

$$\mu_I(f) = \mu(p \circ h) - \mu(p) = \dim_{\mathbb{C}} \frac{\mathcal{O}^e}{\left(x_n^2 \frac{\partial p}{\partial x_n} \circ h, \frac{\partial p}{\partial x} \circ h\right)} = \dim_{\mathbb{C}} \frac{\mathcal{O}_{\mathbb{C}^n,0}}{\left(x_n \frac{\partial p}{\partial x_n}, \frac{\partial p}{\partial x}\right)}.$$

By comparing this with (10.34) we see that $\mathscr{A}_e -$ codimension $(f) \leq \mu_I(f)$ with equality if and only if $p \in J_p$. By a well-known theorem of K.Saito, $p \in J_p$ if and only if p is quasihomogeneous in suitable coordinates. Quasihomogeneity of p is equivalent to quasihomogeneity of f, so we conclude that the $\mathscr{A}_e$-codimension and the image Milnor number coincide precisely when f is quasihomogeneous.

10.7.6 Codimension 1 Corank 1 Germs

For each $n \geq 1$, the germ $f^n\colon (\mathbb{C}^{2n-1}, 0) \to (\mathbb{C}^{2n}, 0)$ given by

$$(u, v, x) \mapsto (u, v, x^{n+1} + u_1 x^{n-1} + \cdots + u_{n-1}x, x^{n+2} + v_1 x^{n-1} + \cdots + v_{n-1}x),$$

where $u = (u_1, \ldots, u_{n-1})$ and $v = (v_1, \ldots, v_{n-1})$, has $\mathscr{A}_e$-codimension 1. This can be checked quite easily: denote the coordinates in the target by u, v, Y_1, Y_2; a straightforward calculation using Theorem 4.2 shows that

$$T\mathscr{A}_e f^n = \theta(f^n) \smallsetminus \left\{x^n \frac{\partial}{\partial Y_2}, x^{\ell-1}\frac{\partial}{\partial v_1}, \ldots, x\frac{\partial}{\partial v_{\ell-1}}\right\}$$
$$+ \mathrm{Sp}_{\mathbb{F}}\left\{x^{\ell-1}\frac{\partial}{\partial v_1} + x^{\ell}\frac{\partial}{\partial Y_2}, \ldots, x\frac{\partial}{\partial v_{\ell-1}} + x^{\ell}\frac{\partial}{\partial Y_2}\right\},$$

where we use the notation explained in Sect. 3.3.

Every $\mathscr{A}_e$-codimension 1 corank 1 germ $(\mathbb{C}^n, 0) \to (\mathbb{C}^{n+1}, 0)$ is $\mathscr{A}$-equivalent to one of these germs, or to an iterated augmentation of one of them.

Let us denote the last two component functions of f^n by f_{2n-1} and f_{2n}. It is easy to check that $D^k(f^n)$ is smooth for $2 \leq k \leq n$. Recall that $D^k(f^n)$ is defined in $\mathbb{C}^{2n-2+k} = u, v, x_1, \ldots, x_k$ – space by the Vandermonde quotients $\alpha^k_\ell(f_{2n-1}), \alpha^k_\ell(f_{2n})$ for $1 \leq \ell \leq k-1$ of Proposition 9.12; an easy calculation shows that

$$\alpha^k_\ell(f_{2n-1})(u, v, x_1, \ldots, x_k) = u_\ell + \text{ function of } x,$$
$$\alpha^k_\ell(f_{2n})(u, v, x_1, \ldots, x_k) = v_\ell + \text{ function of } x \tag{10.36}$$

for $1 \leq \ell \leq k-1$ provided $k \leq n$.

Another calculation shows that $D^{n+1}(f^n)$ has a non-degenerate hypersurface singularity at 0. For with $k = n+1$, (10.36) still holds for $1 \leq \ell \leq n-1$, and

$$\alpha^{n+1}_n(f_{2n-1})(u, v, x) = x_1 + \cdots + x_{n+1}, \quad \alpha^{n+1}_n(f_{2n})(u, v, x) = s_2(x),$$

where $s_2(x)$ is the quadratic function of $x_1, \ldots, x_{n+1}$ in which every quadratic monomial appears, each with coefficient 1. Note that

$$s_2(x_1, \ldots, x_{n+1}) = \frac{1}{2}(x_1 + \cdots + x_{n+1})^2 + \frac{1}{2}(x_1^2 + \cdots + x_{n+1}^2).$$

Now consider the stable perturbation f_t^n of f^n given by

$$f_t^n(u, v, x) = (u, v, f_{2n-1}, f_{2n} + tx^n).$$

Here the only change to any of the multiple point spaces is that $D^{n+1}(f^n)$ is smoothed: on the hyperplane $\{\alpha_n^{n+1}(f_{2n-1}) = 0\}$,

$$\alpha_n^{n+1}(f_{2n} + tx^n) = \frac{1}{2}(x_1^2 + \cdots + x_{n+1}^2 + t).$$

Now it is easy to see that for real $t < 0$, the real part of this space is an $n-1$-sphere $S^{n-1}_{\mathbb{R},t}$, and that on $H_{n-1}(S^{n-1}_{\mathbb{R},t}) \simeq \mathbb{Z}$, the symmetric group S_{n+1} acts by its sign representation. That is,

$$\mathrm{Alt}H_{n-1}(D^{n+1} f^n_{\mathbb{R},t}) = H_{n-1}(D^{n+1} f^n_{\mathbb{R},t}) \simeq \mathbb{Z}.$$

Since the remaining multiple point spaces are contractible or empty, we conclude from the ICSS that the image of $f^n_{\mathbb{R},t}$ has $2n-1$'st Betti number equal to 1, and the remaining Betti numbers equal to 0.

What about the image of the *complex* map f_t^n? For real $t < 0$, the inclusion $D^n(f^n_{\mathbb{R},t}) \hookrightarrow D^n(f_t^n)$ is a homotopy equivalence—the same homotopy equivalence familiar from elementary Morse theory, where the complex level set $\sum_j z_j^2 = t$ is diffeomorphic to the tangent bundle of the real level set. Thus at least for real $t < 0$, the homology of $D^{n+1}(f_t^n)$ is alternating, and the image of f_t^n again has $2n-1$'st Betti number 1. However, the image of f_t^n for (suitably small) complex value $t \neq 0$ is independent of t, up to homeomorphism, so our calculation of the homology of the image is independent of the assumption that t is real and negative.

10.7.7 Generalised Mayer–Vietoris

Suppose that $\{U_\alpha\}$ is an open cover of a topological space X with the property that each U_α, and each non-empty intersection of U_α's, is contractible. Consider the map from the disjoint union $\coprod_\alpha U_\alpha$ to X given simply by the inclusion of each U_α in X. Then the E_1 page of the ICSS reduces to the Čech complex of sections of the constant sheaf on the acyclic cover $\{U_\alpha\}$, and the convergence of the ICSS to the cohomology of X simply becomes the statement that the cohomology of the Čech complex of the constant sheaf, with respect to an acyclic cover, is isomorphic to the cohomology of X. This is referred to as the the generalised Mayer–Vietoris principle in [BT82].

10.7.8 Relation Between AH_* and H_*

As we have seen in Example 10.1, for a space X on which S_k acts, the morphism $AH_*(X) \to H_*^{\mathrm{Alt}}(X)$ induced by the inclusion of chain complexes $C_\bullet^{\mathrm{Alt}}(X) \to C_\bullet(X)$ may not be injective. In principle it may not even be surjective—there may exist an alternating homology class which is not represented by an alternating cycle. On the other hand, since the inclusion of chain complexes $C_\bullet^{\mathrm{Alt}}(X) \otimes_{\mathbb{Z}} \mathbb{Q} \hookrightarrow C_\bullet(X) \otimes_{\mathbb{Z}} \mathbb{Q}$ induces an isomorphism of $AH_*(X;\mathbb{Q}) \to H_*^{\mathrm{Alt}}(X;\mathbb{Q})$ (Proposition 10.1), we know that both the kernel K_ℓ and cokernel L_ℓ of the morphism $AH_\ell(X) \to H_\ell^{\mathrm{Alt}}(X)$ are made up of torsion classes. In fact, using the operator Alt, we can slightly improve on this statement.

Proposition 10.6 $k!K_\ell = 0 = k!L_\ell$.

Proof See Exercise 10.7.3 □

Much more significantly, Goryunov showed in [Gor95, Theorem 2.1.2] that if X is an S_k-invariant Milnor fibre of an isolated complete intersection singularity defined in some ambient $\mathbb{C}^N$ by S_k-invariant analytic functions, and if the representation of the symmetric group on $\mathbb{C}^N$ has the property that for all $\sigma \in S_k$, $\det \rho(\sigma) = \mathrm{sign}(\sigma)$, then for all ℓ, $K_\ell = 0 = L_\ell$. This applies to the spaces $D^k(f_t)$ where f_t is a stable perturbation of a finitely determined corank 1 germ $f_0\colon (\mathbb{C}^n, 0) \to (\mathbb{C}^p, 0)$ with $n \le p$—see Sect. 9.7 of Chap. 9. Thus, a stronger version of Corollary 10.1 holds, in which rational coefficients are replaced by integer coefficients.

10.7.9 Exercises for Sect. 10.7

1 For the map-germ $f(x, y) = (x, y^2, y^3 + x^k y)$ and its stable perturbation f_u,

a) Draw topologically accurate pictures of $D^2(f_u)$ for the cases $k = 3$ and $k = 4$.
b) How many Whitney umbrellas appear on the image of f_u when u is stable? Hint: let $\mathscr{R}_f$ and $\mathscr{R}_{f_u}$ denote the ramification ideals of f and f_u, generated by the 2×2 minors of the matrix of the derivative. By conservation of multiplicity,

$$\sum_x \dim_{\mathbb{C}} \mathscr{O}_{\mathbb{C}^2,x}/\mathscr{R}_{f_u} = \dim_{\mathbb{C}} \mathscr{O}_2/\mathscr{R}_f.$$

If f_u is stable then the only non-immersive points it has are Whitney umbrella points, each of which contributes 1 to the sum on the left here.
c) Each Whitney umbrella point is a fixed point of the involution on $D^2(f_u)$. Use this, the calculation of b) and the Lefschetz fixed point formula, to show that $AH_1(D^2(f_u);\mathbb{Q}) = H_1(D^2(f_u);\mathbb{Q})$.
d) Combine the results of (c), and of Exercise 3.3.2, to show that for these germs, $\mu_I(f) = \mathscr{A}_e$-codimension(f).
e) In each case find an involution i on the curve $D_1^2(f)$ such that $i_* = -1$ on $H_1(D^2(f_u)$.

2 how that the conclusion of 1d) holds also for the germs $f(x, y) = (x, y^2, x^2y + y^{2k+1})$ for $k \geq 2$—even though here it is not true that $AH_1(D^2(f_u); \mathbb{Q}) = H_1(D^2(f_u); \mathbb{Q})$.

3 Prove Proposition 10.6—use the fact that if z is an alternating chain then $\text{Alt}z = k!z$, and if $[z]$ is an alternating homology class then $\text{Alt}[z] = k![z]$.

10.8 Open Questions

(i) Theorem 10.2 is proved by a rather complicated argument using equivariant stratified Morse theory. This remarkable theorem has not received the attention it deserves, in part because the published version is hard to read and suffers from some unfortunate typography. It would be a worthwhile project to write a clearer account. Houston's philosophical motivation for the theorem is worth describing because it is simple and illuminating. The difficulty in describing $D^k(f)$ is entirely due to the need to remove the diagonals, by which $D^k(f)$ differs from the simple minded scheme

$$\begin{aligned}(X/Y)^k &:= X \times_Y X \times_Y \cdots \times_Y X \\ &:= \{(x_1, \ldots, x_k) \in X^k : f(x_i) = f(x_j) \text{ for all } i, j\}.\end{aligned}$$

Away from these diagonals, $(X/Y)^k$ is a complete intersection, defined in X^k by the $(k-1)p$ equations $f_k(x_1) = f_k(x_i)$ for $1 \leq k \leq p$ and $2 \leq i \leq k$. Indeed, if f is $\mathscr{A}$-finite, then $(X/Y)^k$ is non-singular away from the diagonals, since at all genuine k-tuple points, which by the conic structure theorem occur away from 0, the corresponding multi-germ of f is stable. Now in the alternating chain complex $C_\bullet^{\text{Alt}}(D^k(f_t))$, the support of no chain can contain any simplex c lying entirely in any diagonal $\{x_i = x_j\}$, since, evidently, the transposition (i, j) leaves c fixed. It follows that for the alternating homology, $D^k(f)$ ought to behave like a complete intersection with isolated singularity, and new cycles should appear only in middle dimension. The extent to which this argument can be turned into a proof is not clear!

(ii) How can one compute the "alternating Milnor number" of $D^k(f)$ when f has corank > 1?

(iii) How can one compute the image Milnor number of a map-germ $(\mathbb{C}^n, S) \to (\mathbb{C}^{n+1}, 0)$? An answer to (i), together with Corollary 10.2, would provide a method; beyond this, there is only a conjectural method which is part of the "Mond conjecture," that

$$\mu_I(f) = \dim_{\mathbb{C}} \frac{\theta(i)}{ti(\theta_p) + i^*(\text{Der}(-\log h))}$$

—see Remark 8.3.

Chapter 11
Multiple Points in the Target: The Case of Parameterised Hypersurfaces

By the preparation theorem, if $f\colon (\mathbb{C}^n, S) \to (\mathbb{C}^p, 0)$ is a finite map-germ, then $\mathscr{O}_{\mathbb{C}^n,S}$ is a finite (i.e. finitely generated) module over $\mathscr{O}_{\mathbb{C}^{n+1},0}$. For any ring R, a *presentation* of a finite R module M is an exact sequence

$$R^p \xrightarrow{\lambda} R^q \xrightarrow{\alpha} M \longrightarrow 0\,. \tag{11.1}$$

Thus, if $m_i = \alpha(e_i)$, where e_i is the i'th standard generator of R^q, then $m_1, \ldots, m_q$ generate M over R, the columns of λ are relations among the m_i, and, because λ maps onto the kernel of α, these relations generate the R-module of all such relations. The presentation (11.1) expresses M as the 'abstract' R-module coker λ: the morphism sending e_i to m_i for $i = 1, \ldots, q$ passes to the quotient to give an isomorphism of R-modules

$$R^q/\text{image}(\lambda) \simeq M. \tag{11.2}$$

Although α maps R^q to the 'concrete' incarnation of this R-module, namely M, (11.2) shows that the crucial information about the structure of M as R-module is contained in the matrix λ.

In this chapter we study the case where $p = n + 1$. For $\mathscr{A}$-finite non-immersive germs $(\mathbb{C}^n, 0) \to (\mathbb{C}^p, 0)$ with $p > n + 1$, the image is never a Cohen–Macaulay space: if it were, then so would be the image of every nearby perturbation, but such a germ can be perturbed to a bi-germ consisting of two immersions meeting with intersection of dimension less than $n - 1$, whose image is not Cohen–Macaulay (Exercise C.3.3 below). In this case, finding a presentation and deriving information from it is harder than the case we consider here. In contrast, when $p = n + 1$, from the matrix λ of a presentation of $\mathscr{O}_n$ as $\mathscr{O}_{n+1}$-module, one can easily learn a great deal about the geometry of f. In [Tei76], Bernard Teissier showed that in this case the matrix λ of a minimal presentation is square, and its determinant

D. Mond, J. J. Nuño-Ballesteros, *Singularities of Mappings*,
Grundlehren der mathematischen Wissenschaften 357,
https://doi.org/10.1007/978-3-030-34440-5_11

is an equation for the image. From this equation one can, in principle, determine the germ f itself, up to isomorphism: it is the normalisation of its image, and the normalisation is unique up to isomorphism (see Definition D.5 and the text following it). Other information, in the form of the Fitting ideals, can be derived more immediately. We return to this after first developing an algorithm for finding a presentation. This algorithm has been implemented recently in *Singular* [GP08] by Hernandes, Miranda and Peñafort [HMPnS18].

The theory we develop here for map-germs $(\mathbb{C}^n, S) \to (\mathbb{C}^{n+1}, 0)$ is valid with a Cohen–Macaulay complex space-germ (X, S) in place of $(\mathbb{C}^n, 0)$, and we will develop it in this generality. Note that if $(X, S) = \coprod_{i=1}^r (X_i, x_i)$, then $\mathscr{O}_{X,S} = \oplus_{i=1}^r \mathscr{O}_{X_i,x_i}$. So if λ_i is the matrix of a presentation of $\mathscr{O}_{X_i,x_i}$ over $\mathscr{O}_{\mathbb{C}^{n+1},0}$, then the block diagonal matrix $\oplus_{i=1}^r \lambda_i$ presents $\mathscr{O}_{X,S}$ over $\mathscr{O}_{\mathbb{C}^{n+1},0}$. Hence it is enough to develop a procedure to find each local presentation λ_i. In what follows we consider a finite map-germ $f\colon (X, x_0) \to (\mathbb{C}^{n+1}, 0)$, where (X, x_0) is an n-dimensional Cohen–Macaulay complex space-germ.

11.1 Finding a Presentation

Nakayama's lemma tells us that if $g_1, \ldots, g_m \in \mathscr{O}_{X,x_0}$ project to a $\mathbb{C}$-basis for $\mathscr{O}_{X,x_0}/f^*\mathfrak{m}_{n+1}\mathscr{O}_{X,x_0}$, then $g_1, \ldots, g_m$ form a minimal set of generators for $\mathscr{O}_{X,x_0}$ over $\mathscr{O}_{n+1}$ (see Corollary C.2). The structure of $\mathscr{O}_{X,x_0}$ as $\mathscr{O}_{n+1}$-module is determined by the relations between these generators. The fact that the g_i generate $\mathscr{O}_{X,x_0}$ over $\mathscr{O}_{n+1}$ is equivalent to the surjectivity of

$$\mathscr{O}_{n+1}^m \xrightarrow{\mathbf{g}} \mathscr{O}_{X,x_0},$$

where $\mathbf{g}$ sends the i-th basis vector e_i to g_i. The module of relations between the g_i is the kernel of $\mathbf{g}$, and because $\mathscr{O}_{n+1}$ is Noetherian, it is finitely generated. Thus there is an $m \times r$ matrix λ over $\mathscr{O}_{n+1}$ such that

$$\mathscr{O}_{n+1}^r \xrightarrow{\lambda} \mathscr{O}_{n+1}^m \xrightarrow{\mathbf{g}} \mathscr{O}_{X,x_0} \longrightarrow 0 \tag{11.3}$$

is exact. Because the g_i form a minimal generating set for $\mathscr{O}_{X,x_0}$, all entries in λ lie in the maximal ideal of $\mathscr{O}_{n+1}$. Thus (11.3) is the beginning of a minimal free resolution of $\mathscr{O}_{X,x_0}$ over $\mathscr{O}_{n+1}$. The Auslander–Buchsbaum formula (see Theorem C.4) tells us that if p is the length of such a free resolution (the projective dimension of $\mathscr{O}_{X,x_0}$ as $\mathscr{O}_{n+1}$-module), then

$$p + \operatorname{depth}_{\mathscr{O}_{n+1}} \mathscr{O}_{X,x_0} = \operatorname{depth}_{\mathscr{O}_{n+1}} \mathscr{O}_{n+1} = n + 1.$$

On the other hand, by Lemma C.4, $\mathscr{O}_{X,x_0}$ has depth n not only as a ring, but also as $\mathscr{O}_{n+1}$-module. It follows that $p = 1$. In other words, λ may be chosen injective.

This forces r to be equal to m; for tensoring the exact sequence

$$0 \longrightarrow \mathcal{O}^r_{n+1} \xrightarrow{\lambda} \mathcal{O}^m_{n+1} \xrightarrow{g} \mathcal{O}_{X,x_0} \longrightarrow 0$$

with the field of fractions of $\mathcal{O}_{n+1}$ (the field $\mathcal{M}_{n+1} = \mathcal{M}_{\mathbb{C}^{n+1},0}$ of meromorphic functions), we retain exactness and thus get an exact sequence

$$0 \longrightarrow \mathcal{M}^r_{n+1} \longrightarrow \mathcal{M}^m_{n+1} \longrightarrow \mathcal{M}_{n+1} \otimes \mathcal{O}_{X,x_0} \longrightarrow 0\,.$$

But $\mathcal{O}_{X,x_0}$ is a torsion $\mathcal{O}_{n+1}$-module: if $h \in \mathcal{O}_{n+1}$ defines the image of f, then $ha = 0$, for all $a \in \mathcal{O}_{X,x_0}$. This implies that $\mathcal{M}_{n+1} \otimes \mathcal{O}_{X,x_0} = 0$. It follows that $r = m$.

To find a matrix λ, one can use the following procedure:

1. Choose a projection $\pi : \mathbb{C}^{n+1} \to \mathbb{C}^n$ such that $\pi \circ f$ is finite. A suitable projection always exists: by Theorem D.6 the image $(f(X), 0)$ is analytic of dimension n and as π we can take any generic projection to an n-plane (see Corollary D.4). In practice this usually means selecting n of the $n+1$ component functions of f, though in principle it may be that none of these coordinate projections is finite. *In what follows we will assume that coordinates are chosen so that* $\pi(y_1, \ldots, y_{n+1}) = (y_1, \ldots, y_n)$.
2. Then $\mathcal{O}_{X,x_0}$ is free over $\mathcal{O}_n$. This follows again from the Auslander–Buchsbaum formula since now its projective dimension is zero. Let $g_0, \ldots, g_m$ be a basis. Once again, by Nakayama's lemma it is sufficient that the g_i form a $\mathbb{C}$-vector-space basis for $\mathcal{O}_{X,x_0}/(\pi \circ f)^* \mathfrak{m}_n$, which is finite-dimensional by finiteness of $\pi \circ f$. *One of the* g_i *at least must be a unit in* $\mathcal{O}_{X,x_0}$; *we take* $g_0 = 1$.
3. Find the unique $\lambda^i_j \in \mathcal{O}_n$ such that

$$\begin{aligned}
f_{n+1} &= \lambda^0_0 g_0 + \cdots + \lambda^m_0 g_m \\
g_1 f_{n+1} &= \lambda^0_1 g_0 + \cdots + \lambda^m_1 g_m \\
\cdots &= \cdots \;\cdots\cdots\cdots \;\cdots \\
g_m f_{n+1} &= \lambda^0_m g_0 + \cdots + \lambda^m_m g_m .
\end{aligned} \tag{11.4}$$

Since $f_{n+1} = y_{n+1} \circ f$, (11.4) can be rewritten as

$$
\begin{array}{lcccccc}
0 &=& (\lambda_0^0 - y_{n+1})g_0 & + & \cdots & + \cdots + & \lambda_0^m g_m \\
0 &=& \lambda_1^0 g_0 & + & (\lambda_1^1 - y_{n+1})g_1 & + \cdots + & \lambda_1^m g_m \\
\cdots &=& \cdots & \cdots & \cdots & \cdots\cdots\cdots & \cdots \\
0 &=& \lambda_m^0 g_0 & + & \cdots & + \cdots + & (\lambda_m^m - y_{n+1})g_m .
\end{array}
\tag{11.5}
$$

Thus the columns of the matrix

$$
\begin{pmatrix}
\lambda_0^0 - y_{n+1} & \lambda_1^0 & \cdots & \lambda_m^0 \\
\lambda_0^1 & \lambda_1^1 - y_{n+1} & \cdots & \lambda_m^1 \\
\cdots & \cdots & \cdots & \cdots \\
\lambda_0^m & \lambda_1^m & \cdots & \lambda_m^m - y_{n+1}
\end{pmatrix}
\tag{11.6}
$$

are relations between the g_i.

Proposition 11.1 (11.6) *is the matrix of a presentation of* $\mathcal{O}_{X,x_0}$ *over* $\mathcal{O}_{n+1}$. *In other words, the columns of* (11.6) *generate all the relations among the* g_i *over* $\mathcal{O}_{n+1}$.

Proof A useful trick is described in [MP89, 2.2]: embed X as the hyperplane $\{t = 0\}$ in $X \times \mathbb{C}$, and define $F\colon (X \times \mathbb{C}, (x_0, 0)) \to (\mathbb{C}^{n+1}, 0)$ by

$$F(x,t) = (f_1(x), \ldots, f_n(x), f_{n+1}(x) - t).$$

Write S for $(X \times \mathbb{C}, (x_0, 0))$ and T for $(\mathbb{C}^{n+1}, 0)$. Then

$$\mathcal{O}_S / F^* \mathfrak{m}_T = \frac{\mathcal{O}_S}{(f_1, \ldots, f_n, f_{n+1} - t)} \simeq \frac{\mathcal{O}_{X,x_0}}{(f_1, \ldots, f_n)}$$

so that by the preparation theorem, $g_0, \ldots, g_m$ form a minimal generating set for $\mathcal{O}_T$ over $\mathcal{O}_S$. Since, by Auslander–Buchsbaum again, $\mathcal{O}_S$ is free over $\mathcal{O}_T$, it follows that $g_0, \ldots, g_m$ are a basis for $\mathcal{O}_S$ over $\mathcal{O}_T$, and thus determine an $\mathcal{O}_T$-isomorphism $\mathcal{O}_T^{m+1} \xrightarrow{\varphi} \mathcal{O}_S$. In the diagram

$$
\begin{array}{ccccccccc}
0 & \longrightarrow & \mathcal{O}_S & \xrightarrow{t} & \mathcal{O}_S & \longrightarrow & \mathcal{O}_n & \longrightarrow & 0 \\
 & & \uparrow{\scriptstyle\varphi} & & \uparrow{\scriptstyle\varphi} & & & & \\
 & & \mathcal{O}_T^{m+1} & \xrightarrow{[t]_G^G} & \mathcal{O}_T^{m+1} & & & &
\end{array}
\tag{11.7}
$$

$\left[t\right]_G^G$ denotes the matrix of the $\mathcal{O}_T$-linear map $\mathcal{O}_S \xrightarrow{t} \mathcal{O}_S$ (multiplication by t), with respect to the basis $g_0, \ldots, g_m$ of $\mathcal{O}_S$. We have

$$tg_i = (f_{n+1} - y_{n+1})g_i = \lambda_i^0 g_0 + \cdots + (\lambda_i^i - y_{n+1})g_i + \cdots + \lambda_i^m g_m$$

and thus $\left[t\right]_G^G$ is equal to the matrix (11.6). From the commutativity of (11.7) it follows that the cokernel of (11.6) is indeed isomorphic to $\mathcal{O}_{X,x_0}$ as claimed. □

The presentation obtained above is not necessarily minimal, since in general

$$\dim_{\mathbb{C}} \frac{\mathcal{O}_{X,x_0}}{f^* \mathfrak{m}_{n+1}} < \dim_{\mathbb{C}} \frac{\mathcal{O}_{X,x_0}}{(\pi \circ f)^* \mathfrak{m}_{\mathbb{C}^n,0}}.$$

Nevertheless it is always injective, since the determinant of (11.6) is not zero—as can easily be seen, it is a monic polynomial of degree $m+1$ in $\mathbb{C}\{y_1, \ldots, y_n\}[y_{n+1}]$.

Example 11.1

1. Let $f\colon (\mathbb{C}^2, 0) \to (\mathbb{C}^3, 0)$ be defined by

$$f(x, y) = (x, y^3, xy + y^5)$$

(this is the simple singularity of type H_2 in the classification [Mon85]). Take $\pi(Y_1, Y_2, Y_3) = (Y_1, Y_2)$; then $\mathcal{O}_2$ (source) is generated over $\mathcal{O}_2$ (target) by the classes of $1, y, y^2$. We have

$$\begin{aligned} f_3 &= xy + y^5 &&= 01 + Y_1 y + Y_2 y^2 \\ g_1 f_3 &= xy^2 + y^6 &&= Y_2^2 1 + 0y + Y_1 y^2 \\ g_2 f_3 &= xy^3 + y^7 &&= Y_1 Y_2 1 + Y_2^2 y + 0y^2 \end{aligned}$$

so as matrix of the presentation we obtain

$$\begin{pmatrix} -Y_3 & Y_2^2 & Y_1 Y_2 \\ Y_1 & -Y_3 & Y_2^2 \\ Y_2 & Y_1 & -Y_3 \end{pmatrix}.$$

2. Let $f\colon (\mathbb{C}^2, 0) \to (\mathbb{C}^3, 0)$ be defined by $f(x_1, x_2) = (x_1^2, x_2^2, x_1 x_2)$, and as before take $\pi(Y_1, Y_2, Y_3) = (Y_1, Y_2)$. Then $\mathcal{O}_2$ (source) is generated over $\mathcal{O}_2$ (target) by $1, x_1, x_2, x_1 x_2$. We have

$$\begin{aligned} f_3 &= x_1 x_2 &&= 01 + 0x_1 + 0x_2 + 1x_1 x_2 \\ g_1 f_3 &= x_1^2 x_2 &&= 01 + 0x_1 + Y_1 x_2 + 0x_1 x_2 \\ g_2 f_3 &= x_1 x_2^2 &&= 01 + Y_2 x_1 + 0x_2 + 0x_1 x_2 \\ g_3 f_3 &= x_1^2 x_2^2 &&= Y_1 Y_2 1 + 0x_1 + 0x_2 + 0x_1 x_2 \end{aligned}$$

giving presentation matrix

$$\begin{pmatrix} -Y_3 & 0 & 0 & Y_1Y_2 \\ 0 & -Y_3 & Y_2 & 0 \\ 0 & Y_1 & -Y_3 & 0 \\ 1 & 0 & 0 & -Y_3 \end{pmatrix}.$$

Row and column operations transform this to

$$\begin{pmatrix} 0 & 0 & 0 & Y_3^2 - Y_1Y_2 \\ 0 & -Y_3 & Y_2 & 0 \\ 0 & Y_1 & -Y_3 & 0 \\ 1 & 0 & 0 & 0 \end{pmatrix}.$$

This is now the matrix of a presentation with respect to different set of generators (**Exercise:** which?), of which, according to the first column, the last is superfluous. Deleting it gives the minimal presentation

$$\begin{pmatrix} 0 & 0 & Y_3^2 - Y_1Y_2 \\ -Y_3 & Y_2 & 0 \\ Y_1 & -Y_3 & 0 \end{pmatrix}.$$

The determinant here is a square: as we will see, this corresponds to the fact that f is a double covering of its image.

Every presentation can be reduced to a minimal presentation by row and column operations followed by the deletion of superfluous rows and columns, as in this example.

11.1.1 Using Macaulay2 to Find a Presentation

We continue with Example 4.3. To find a presentation using *Macaulay2* we have to define weights for the variables—in common with other computer algebra systems, *Macaulay2* can only compute a presentation of the push-forward in the weighted homogeneous case, to which, fortunately, this example belongs. The stable germ

$$F(x, y, a, b, c, d, e) = (x^2 + ay + by^2, y^3 + cx + dy, xy + ex + fy, a, b, c, d, e, f) \tag{11.8}$$

is weighted homogeneous with respect to the following weights:

x	y	a	b	c	d	e	f	X	Y	Z	A	B	C	D	E	F
2	1	3	2	1	2	1	2	4	3	3	3	2	1	2	1	2

The following sequence returns the matrix of a presentation:

```
S=QQ[x,y,a,b,c,d,e,g,Degrees=>{{2},{1},{3},{2},{1},{2},{1},{2}}]
T=QQ[X,Y,Z,A,B,C,D,E,F,Degrees=>{{4}},{3},{3},{3},{2},{1}{2}{1}
{2}}]
F=map(S,T,matrix{{x^2+a*y+b*y^2, y^3+c*x+d*y, x*y+e*x+f*y,a,b,c,
d,e,f}})
M=module S
P=mingens image presentation pushForward(G,M).
```

We will return to this example in Sect. 11.7 below.

Algorithm 11.1 To find a presentation of $f_* \mathscr{O}_n$ over $\mathscr{O}_{n+1}$ one can also use the following procedure to partially automate the algorithm described in Proposition 11.1. It yields a result in some cases where the calculation is too big for the *Macaulay2* procedure just described.

Step 1. Pick a projection $p : \mathbb{C}^{n+1} \to \mathbb{C}^n$ so that $p \circ f$ is finite. As in Proposition 11.1, we assume coordinates $y_1, \ldots, y_{n+1}$ on $\mathbb{C}^{n+1}$ are chosen so that $p(y_1, \ldots, y_{n+1}) = (y_1, \ldots, y_n)$. Denote $\mathscr{O}_n$ (source) by S, and $\mathscr{O}_n$ (target of $p \circ f$) by T. Note that this is not quite the same notation as in Proposition 11.1.

Step 2. Find a *monomial* basis G for S over T, say $G = 1, g_1, \ldots, g_m$. One always exists, of course.

Step 3. For $i = 1, \ldots, n$, find the matrix M_i with respect to G of the T-linear map $S \to S$ given by multiplication by the i'th coordinate function x_i. If the basis is chosen intelligently, this can be relatively straightforward, and M_i will have many zeros.

Step 4. Find the coefficients $\alpha_0^j \in T$ in the expression

$$f_{n+1} = \sum_{j=0}^{m} \alpha_0^j g_j.$$

Let P_0 be the column vector formed by these coefficients. Once again, a wise choice of basis will ensure that the coefficients in P_0 are simple, and in particular that all coefficients which are units in T are in fact equal to 1.

Step 5. Suppose that $g_i = x_1^{r_1} \cdots x_n^{r_n}$. Then

$$P_i := M_1^{r_1} \cdots M_n^{r_n} P_0$$

is the column vector of coefficients α_i^j in the expression

$$f_{n+1} g_i = \sum_{j=0}^{m} \alpha_i^j g_j.$$

In practice, g_i will usually be obtained from some g_ℓ with $\ell < i$ by multiplying by some x_j, so $P_i = M_j P_\ell$, and beginning with P_0 one can find the P_i cumulatively.

Step 6. Let N be the $(m+1) \times (m+1)$ matrix with y_{n+1} along the anti-diagonal and zeros elsewhere. Then the $(m+1) \times (m+1)$ matrix

$$P := \begin{pmatrix} P_m & P_{m-1} & \cdots & P_1 & P_0 \end{pmatrix} - N$$

is the presentation matrix (11.6) described in Sect. 11.1.

Step 7. Unless $f_{n+1} \in (f_1, \ldots, f_n)$, the presentation so obtained will not be minimal. To minimise it, one has to perform row and column operations. These can all be done easily in *Macaulay2*, and can be partially automated as follows. Suppose that the matrix P obtained by this procedure has a 1 in position (s, t). To perform column operations it is necessary first to create a "mutable matrix".

```
PP=mutable matrix P
```

Then the commands

```
for i from 0 to t-1 do columnAdd(PP, i, -PP_(s,i),t)
for i from t+1 to m do columnAdd(PP, i, -PP_(s,i),t)
```

clear all the non-zero entries in row s, except for the 1 in position (s, t). Once this has been done for each row containing a 1, say rows $s_1, \ldots, s_a$, do

```
P=matrix PP
```

and then delete from P the columns containing the 1's, and the resulting rows which have been reduced to zero. This is best done using the `submatrix'` command,

```
P=submatrix'(P,{s1,...,sa},{t1,...,ta})
```

where t_i is the column with a 1 in row s_i. This yields a minimal presentation.

Exercises for Sect. 11.1

1 Given a commutative diagram of R-linear maps,

$$\begin{array}{ccc} R^p & \xrightarrow{\lambda_1} & R^q \\ \downarrow{\scriptstyle P} & & \downarrow{\scriptstyle Q} \\ R^p & \xrightarrow{\lambda_2} & R^q \end{array}$$

with P and Q isomorphisms, show that coker λ_1 and coker λ_2 are isomophic as R-modules.

2 Find a presentation for $\mathcal{O}_n$ over $\mathcal{O}_{n+1}$ when

(i) $f(x) = (x^2, x^5)$;
(ii) $f(x = (x^2, x^{2k+1})$;
(iii) $f(u, x) = (u, x^2, xp(u, x^2))$;
(iv) $f(u, x) = (u, x^3, ux + x^{3k-2})$;
(v) $f(u_1, u_2, u_3, u_4, x_1, x_2) = (u_1, u_2, u_3, u_4, x_1^2+u_1x_2, x_2^2+u_2x_1, x_1x_2+u_3x_1+u_4x_2)$.

3 Find a minimal presentation for a stable unfolding of each of the rank 0 map-germs

(i) $f(x_1, x_2) = (x_1^2, x_2^5, x_1x_2)$,
(ii) $f(x_1, x_2) = (x_1^3, x_1x_2, x_2^3)$,
(iii) $f(x_1, x_2, x_3) = (x_1^2, x_2^2, x_3^2, x_2x_3 + x_1x_3 + x_1x_2)$.

4 Use *Macaulay* or *Singular* to find presentations for $\mathcal{O}_1$ over $\mathcal{O}_3$ when

(i) $f(x) = (x^3, x^4, x^5)$.
(ii) $f(u, x) = (u, x^2, x^3 + ux, ux)$.

5 Find a stable unfolding F of the germ $x \mapsto (x^3, x^4, x^5)$, and then a presentation of $F_*(\mathcal{O}_{d+1})$ over $\mathcal{O}_{d+3}$ where d is the number of parameters.

11.2 Fitting Ideals and Multiple Points in the Target

From the square matrix λ one can extract a great deal of information about the geometry of f. In general, given any $q \times p$ matrix λ with entries in a ring R, we adopt the following conventions:

$$\min_\ell(\lambda) = \begin{cases} R, & \text{if } \ell \leq 0; \\ \text{ideal in } R \text{ generated by the } \ell \times \ell \text{ minors of } \lambda, & \text{if } 1 \leq \ell \leq \min\{p, q\}; \\ 0, & \text{if } \ell > \min\{p, q\}. \end{cases}$$

Definition 11.1 Let $R^p \xrightarrow{\lambda} R^q \xrightarrow{g} M \longrightarrow 0$ be a presentation of the R-module M. The k'th *Fitting ideal of M as R-module*, $\mathrm{Fitt}_k^R(M)$, or simply $\mathrm{Fitt}_k(M)$ if it is clear which ring we are talking about, is the ideal $\min_{q-k}(\lambda)$.

The above definition makes sense only for modules of finite presentation, but this is always the case when R is Noetherian and M is finitely generated over R. By construction, we have the following chain of Fitting ideals:

$$0 = \mathrm{Fitt}_{-1}(M) \subseteq \mathrm{Fitt}_0(M) \subseteq \cdots \subseteq \mathrm{Fitt}_{q-1}(M) \subseteq \mathrm{Fitt}_q(M) = R.$$

The next proposition says that Fitting ideals are preserved under change base. More precisely:

Proposition 11.2 *Let M be an R-module of finite presentation and let S be an R-algebra. Then:*

$$Fitt_k^S(M \otimes_R S) = Fitt_k^R(M) \cdot S.$$

Proof Given a presentation of M over R

$$R^p \xrightarrow{\lambda} R^q \longrightarrow M \longrightarrow 0,$$

by tensoring with S we get a presentation of $M \otimes_R S$ over S

$$S^p \xrightarrow{\lambda \otimes \mathrm{id}} S^q \longrightarrow M \otimes_R S \longrightarrow 0\ .$$

This is due to the fact that the product tensor is right exact. But $\lambda \otimes \mathrm{id}$ has the same matrix as λ, which implies that $\mathrm{Fitt}_k^S(M \otimes_R S) = \mathrm{Fitt}_k^R(M) \cdot S$. □

This property has several important consequences. For instance, we will see later in Proposition 11.6 that it can be used to compute the Fitting ideals of a transverse fibre product. Another consequence is that Fitting ideals commute with localisation.

Corollary 11.1 *Let M be an R-module of finite presentation. Then,*

$$Fitt_k^{R_{\mathfrak{p}}}(M_{\mathfrak{p}}) = Fitt_k^R(M) \cdot R_{\mathfrak{p}},$$

for any prime ideal $\mathfrak{p} \subset R$.

The 0th-Fitting ideal is related to the annihilator of a module, as we can see in the next proposition.

Proposition 11.3 *Let M be an R-module generated by q elements, then*

$$\mathrm{Ann}(M)^q \subset Fitt_0(M) \subset \mathrm{Ann}(M).$$

In particular, $\sqrt{Fitt_0(M)} = \sqrt{\mathrm{Ann}(M)}$.

Proof See, e.g., [Eis95, Proposition 20.7]. □

Recall that when R is a local ring and M is a finitely generated R-module, then the dimension of $M/\mathfrak{m}M$ as $R/\mathfrak{m}$-vector space is equal to the minimal number of generators of M (see Corollary C.2). As a consequence, if $r = \dim_{R/\mathfrak{m}} M/\mathfrak{m}M$, given a presentation

$$R^p \xrightarrow{\lambda} R^q \longrightarrow M \longrightarrow 0,$$

then $q \geq r$ and we say that the presentation is *minimal* if $q = r$. A minimal presentation always exists if R is Noetherian. Moreover, if we tensor with $k = R/\mathfrak{m}$, we get an exact sequence

$$k^p \xrightarrow{\overline{\lambda}} k^q \longrightarrow M/\mathfrak{m}M \longrightarrow 0,$$

where $\overline{\lambda} = (\overline{\lambda}_{ij})$ is the matrix of residue classes of the λ_{ij}. It follows that the presentation is minimal if and only if $\lambda_{ij} \in \mathfrak{m}$, for all i, j. In particular, we have

$$0 \subseteq F_0(M) \subseteq \cdots \subseteq F_{r-1}(M) \subseteq \mathfrak{m}, \quad F_r(M) = R. \tag{11.9}$$

The Fitting ideals tell us a great deal about the geometry of f. We give two versions of this, first, one from algebraic geometry:

Proposition 11.4 *Let M be an R-module of finite presentation. Then,*

$$V(Fitt_k(M)) = \{\mathfrak{p} \in \operatorname{Spec} R : M_{\mathfrak{p}} \textit{ needs more than } k \textit{ generators over } R_{\mathfrak{p}}\}.$$

Proof By Corollary 11.1, it is enough to prove that if R is a local ring, then M can be generated by k elements over R if and only if $\mathrm{Fitt}_k(M) = R$. But this follows directly from (11.9). □

In analytic geometry there are always two ways of looking at the same object. Let $\mathscr{S}$ be a coherent sheaf on the complex space X. Define the ideal sheaf $\mathscr{F}_k(\mathscr{S})$ as the sheaf associated to the presheaf

$$U \mapsto \mathrm{Fitt}_k^{\mathscr{O}_X(U)}(\mathscr{S}(U)).$$

Proposition 11.5 *Let $\mathscr{S}$ be a coherent sheaf on the complex space X. Then:*

1. *$Fitt_k^{\mathscr{O}_{X,x}}(\mathscr{S}_x) = \big(\mathscr{F}_k(\mathscr{S})\big)_x$, for all $x \in X$.*
2. *Set theoretically, $V(\mathscr{F}_k(\mathscr{S}))$ is the set of points $x \in X$ such that $\mathscr{S}_x$ needs more than k generators over $\mathscr{O}_{X,x}$. In particular, $V(\mathscr{F}_0(\mathscr{S})) = supp(\mathscr{S})$.*

Proof By coherence, for each $x \in X$, there exists an open neighbourhood U of x in X and an exact sequence of sheaves:

$$\mathscr{O}^p_{X|U} \xrightarrow{\lambda} \mathscr{O}^q_{X|U} \longrightarrow \mathscr{S}_{|U} \longrightarrow 0.$$

Taking stalks at x, we get an exact sequence of $\mathcal{O}_{X,x}$-modules

$$\mathcal{O}^p_{X,x} \xrightarrow{\lambda_x} \mathcal{O}^q_{X,x} \longrightarrow \mathcal{S}_x \longrightarrow 0\,,$$

where λ_x is the induced morphism by λ. But this implies that $\left(\mathcal{F}_k(\mathcal{S})\right)_x = \left(\mathcal{F}_k(\mathcal{S}_{|U})\right)_x = \mathrm{Fitt}_k^{\mathcal{O}_{X,x}}(\mathcal{S}_x)$ and we have item 1. Item 2 follows directly from item 1 and (11.9), since $\mathcal{O}_{X,x}$ is a local ring. □

Definition 11.2 Let $f : X \to Y$ be a finite morphism of complex spaces. By the Finite Mapping Theorem E.2, $f_* \mathcal{O}_X$ is a coherent sheaf of $\mathcal{O}_Y$-modules. We define the *k'th Fitting ideal sheaf of f* as $\mathcal{F}_k(f) = \mathcal{F}_k(f_* \mathcal{O}_X)$. We also define the *k'th target multiple point space of f* as the closed complex subspace of Y given by $M_k(f) = V(\mathcal{F}_{k-1}(f))$ with structure sheaf $\mathcal{O}_{M_k(f)} = \mathcal{O}_Y / \mathcal{F}_{k-1}(f)$.

There is a local version of the above definitions:

Definition 11.3 Let $f : (X, S) \to (Y, y)$ be a finite morphism of complex space-germs. We define $\mathcal{F}_k(f) = \mathrm{Fitt}_k^{\mathcal{O}_{Y,y}}(\mathcal{O}_{X,S})$ and $M_k(f) = V(\mathcal{F}_{k-1}(f))$ (so it is a complex subspace germ of (Y, y) with $\mathcal{O}_{M_k(f),y} = \mathcal{O}_{Y,y} / \mathcal{F}_{k-1}(f))$.

The relationship between the global and local definitions as well as the set theoretical description of $M_k(f)$ is given in the next corollary:

Corollary 11.2 *Let $f : X \to Y$ be a finite morphism of complex spaces.*

1. *For all $y \in Y$, $(M_k(f), y) = M_k(f_S)$, where $f_S : (X, S) \to (Y, y)$ is the germ of f at $S = f^{-1}(y)$.*
2. *As a set, $M_k(f)$ is the set of points $y \in Y$ with at least k preimages, counting multiplicity. More precisely:*

$$M_k(f) = \left\{ y \in Y : \sum_{f(x)=y} \dim_{\mathbb{C}} \frac{\mathcal{O}_{X,x}}{f^*\mathfrak{m}_{Y,y}} \geq k \right\}.$$

In particular, $M_1(f) = f(X)$.

Proof

1. By item 1 of Proposition 11.5,

$$\left(\mathcal{F}_k(f)\right)_y = \mathrm{Fitt}_k^{\mathcal{O}_{Y,y}}\left((f_* \mathcal{O}_X)_y\right) = \mathrm{Fitt}_k^{\mathcal{O}_{Y,y}}\left(\bigoplus_{f(x)=y} \mathcal{O}_{X,x}\right)$$

$$= \mathrm{Fitt}_k^{\mathcal{O}_{Y,y}}(\mathcal{O}_{X,S}) = \mathcal{F}_k(f_S).$$

2. This is a consequence of item 2 of Proposition 11.5, since for each $y \in Y$, the minimal number of generators of $(f_* \mathcal{O}_X)_y = \bigoplus_{f(x)=y} \mathcal{O}_{X,x}$ over $\mathcal{O}_{Y,y}$ is equal to $\sum_{f(x)=y} \dim_{\mathbb{C}} \mathcal{O}_{X,x} / f^* \mathfrak{m}_{Y,y}$.

□

An important property of the multiple point spaces is that they behave well under deformations. This means that if $f\colon (\mathbb{C}^n, S) \to (\mathbb{C}^p, 0)$ is finite and $F(x, u) = (f_u(x), u)$ is an unfolding of f, then $M_k(f)$ is obtained from $M_k(F)$ just by taking the intersection (as complex space-germs) with the subspace $u = 0$. We will prove this property in a more general setting, namely for any fibre product of complex space-germs.

Proposition 11.6 *Suppose we have a fibre-product diagram*

$$\begin{array}{ccc} (X, S) & \xrightarrow{\;F\;} & (Y, y_0) \\ {\scriptstyle j}\uparrow & & \uparrow{\scriptstyle i} \\ (X \times_Y Z, S_0) & \xrightarrow{\;f\;} & (Z, z_0) \end{array} \tag{11.10}$$

where F is finite. Then:

1. *f is also finite.*
2. *For each matrix presentation of $\mathcal{O}_{X,S}$ over $\mathcal{O}_{Y,y_0}$*

$$\mathcal{O}^r_{Y,y_0} \xrightarrow{\;\lambda\;} \mathcal{O}^s_{Y,y_0} \xrightarrow{\;\mu\;} \mathcal{O}_{X,S} \longrightarrow 0\,,$$

there is an induced matrix presentation of $\mathcal{O}_{X\times_Y Z, S_0}$ over $\mathcal{O}_{Z,z_0}$

$$\mathcal{O}^r_{Z,z_0} \xrightarrow{\;i^*\lambda\;} \mathcal{O}^s_{Z,z_0} \xrightarrow{\;i^*\mu\;} \mathcal{O}_{X\times_Y Z, S_0} \longrightarrow 0\,.$$

3. *$M_k(f) = i^{-1}(M_k(F))$ as complex space-germs.*

Proof The fact that f is also finite is clear from the definition of fibre product (see Exercise 11.2.6 below); item 3 is a direct consequence of item 2. To prove item 2 we proceed in three steps:

Step 1. Assume that $(Z, z_0) = (Y \times \mathbb{C}^q, (y_0, 0))$ and that i is the projection $i(y, t) = t$. Then we have $X \times_Y Z = X \times \mathbb{C}^q$ and $f = F \times \mathrm{id}$, that is, our diagram looks as follows:

$$\begin{array}{ccc} (X, S) & \xrightarrow{\;F\;} & (Y, y_0) \\ \uparrow & & \uparrow{\scriptstyle i} \\ (X \times \mathbb{C}^q, S_0) & \xrightarrow{\;F\times \mathrm{id}\;} & (Y \times \mathbb{C}^q, z_0) \end{array}$$

We obtain a presentation of $\mathcal{O}_{S\times\mathbb{C}^q} = \mathcal{O}_S \hat{\otimes}_{\mathbb{C}} \mathcal{O}_{\mathbb{C}^q}$ over $\mathcal{O}_{Y\times\mathbb{C}^q} = \mathcal{O}_Y \hat{\otimes}_{\mathbb{C}} \mathcal{O}_{\mathbb{C}^q}$ by taking the analytic tensor product with $\mathcal{O}_{\mathbb{C}^q,0}$ of a presentation

$$\mathcal{O}_Y^r \xrightarrow{\lambda} \mathcal{O}_Y^s \xrightarrow{\mu} \mathcal{O}_X \longrightarrow 0\,, \tag{11.11}$$

to get

$$\mathcal{O}_y^r \hat{\otimes}_{\mathbb{C}} \mathcal{O}_{\mathbb{C}^q} \xrightarrow{\lambda\hat{\otimes}_{\mathbb{C}}1} \mathcal{O}_Y^s \hat{\otimes}_{\mathbb{C}} \mathcal{O}_{\mathbb{C}^q} \xrightarrow{\mu\hat{\otimes}_{\mathbb{C}}1} \mathcal{O}_X \hat{\otimes}_{\mathbb{C}} \mathcal{O}_{\mathbb{C}^q} \longrightarrow 0\,, \tag{11.12}$$

which is still exact by the right exactness of analytic tensor product (though see Exercise 11.2.7 below).

Step 2. Suppose that $i\colon (Z, z_0) \to (Y, y_0)$ is an embedding. Without loss of generality, we can assume that (Z, z_0) is a closed subspace germ of (Y, z_0) and that i is the inclusion. Then we can take $X = F^{-1}(Z)$ and f as the restriction of F. That is, we now have a diagram:

$$\begin{array}{ccc} (X, S) & \xrightarrow{F} & (Y, z_0) \\ \uparrow & & \uparrow{\scriptstyle i} \\ (F^{-1}(Z), S) & \xrightarrow{f} & (Z, z_0) \end{array}$$

As $\mathcal{O}_{Z,z_0} = \mathcal{O}_{Y,z_0}/I$, for some ideal $I \subset \mathcal{O}_{Y,z_0}$ hence, $\mathcal{O}_{F^{-1}(Z),S} = \mathcal{O}_{X,S}/F^*(I)\,\mathcal{O}_{X,S}$. There is isomorphism of $\mathcal{O}_{Y,y_0}$-modules:

$$\mathcal{O}_{X,S} \otimes_{\mathcal{O}_{Y,y_0}} \mathcal{O}_{Z,z_0} = \mathcal{O}_{X,S} \otimes_{\mathcal{O}_{Y,z_0}} \frac{\mathcal{O}_{Y,z_0}}{I} \cong \frac{\mathcal{O}_{X,S}}{I\cdot\mathcal{O}_{X,S}} = \mathcal{O}_{F^{-1}(Z),S}\,.$$

We obtain a presentation of $\mathcal{O}_{F^{-1}(Z),S}$ over $\mathcal{O}_{Z,z_0}$ by applying the functor $\otimes_{\mathcal{O}_{Y,z_0}} \mathcal{O}_{Z,z_0}$ to a presentation

$$\mathcal{O}_{Y,z_0}^r \xrightarrow{\lambda} \mathcal{O}_{Y,z_0}^s \xrightarrow{\mu} \mathcal{O}_{X,S} \longrightarrow 0$$

of $\mathcal{O}_{X,S}$. Since the product tensor is right exact, the resulting sequence

$$\mathcal{O}_{Z,z_0}^r \xrightarrow{i^*\lambda} \mathcal{O}_{Z,z_0}^s \xrightarrow{i^*\mu} \mathcal{O}_{F^{-1}(Z),S} \longrightarrow 0\,,$$

is exact.

Step 3. Finally, we consider the general case. Suppose that $Z \subset \mathbb{C}^q$ for some q. We split our fibre product as a composition of three fibre products as follows:

$$\begin{array}{ccc}
(X, S) & \xrightarrow{\quad F \quad} & (Y, y_0) \\
\uparrow & & \uparrow \alpha_3 \\
(X \times \mathbb{C}^q, S_0') & \xrightarrow{F \times \mathrm{id}_{\mathbb{C}^q}} & (Y \times \mathbb{C}^q, (y_0, z_0)) \\
\uparrow & & \uparrow \alpha_2 \\
(X \times Z, S_0') & \xrightarrow{F \times \mathrm{id}_Z} & (Y \times Z, (y_0, z_0)) \\
\uparrow & & \uparrow \alpha_1 \\
(X \times_Y Z, S_0) & \xrightarrow{\quad f \quad} & (Z, z_0)
\end{array}$$

where α_1 is the immersion $\alpha_1(z) = (i(z), z)$, α_2 is the inclusion $\alpha_2(y, z) = (y, z)$ and α_3 is the projection $\alpha_3(y, z) = y$. The result now follows from steps 1 and 2. □

We know from Corollary 11.2 that set-theoretically the space $M_1(f)$ is equal to the image of f. But we would like to have an explicit description of its analytic structure. In fact, $M_1(f)$ may be non-reduced in general, as we will see later. We will restrict ourselves to the case of a finite germ $f\colon (X, S) \to (\mathbb{C}^{n+1}, 0)$, where (X, S) is a Cohen–Macaulay analytic space of dimension n. We saw in Sect. 11.1 that in this situation, $\mathcal{O}_{X,S}$ admits a presentation over $\mathcal{O}_{n+1}$ by a square matrix λ and hence, $\mathcal{F}_0(f)$ is generated by the determinant of λ. Thus, $M_1(f)$ is always a hypersurface of $(\mathbb{C}^{n+1}, 0)$. In particular, $M_1(f)$ is also Cohen–Macaulay.

We first consider the mono-germ case $S = \{x_0\}$ and also assume that (X, x_0) is irreducible. This implies that $f\colon (X, x_0) \to (\mathbb{C}^{n+1}, 0)$ has a well-defined degree onto its image. Roughly speaking, the degree is the number of preimages of a generic value in the image of f. The precise definition is given in Definition D.2.

Proposition 11.7 *Let (X, x_0) be an n-dimensional irreducible Cohen–Macaulay analytic germ and let $f\colon (X, x_0) \to (\mathbb{C}^{n+1}, 0)$ be finite. Then $\mathcal{F}_0(f)$ is generated by h^k, where $h \in \mathcal{O}_{n+1}$ is a reduced equation for the image of f and k is the degree of f onto its image.*

Proof Since $M_1(f)$ is equal to the image of f as a set-germ, $\mathcal{F}_0(f)$ is generated by h^r, for some $r \geq 1$. The image is irreducible, so by Lemma D.3 there exists $(D, 0)$ a hypersurface in the image of f such that f is k-to-1 in the complement to $(D, 0)$. Let (Σ, x_0) and $(\Sigma', 0)$ denote the singular loci of (X, x_0) and $(f(X), 0)$, respectively. Both (Σ, x_0) and $(\Sigma', 0)$ are analytic germs of dimension $\leq n - 1$. Moreover, since f is finite, $(f(\Sigma), 0)$ is also analytic of dimension $\leq n - 1$.

We choose small enough representatives $f : X \to V$, $h : V \to \mathbb{C}$, D, S and Σ of all the germs, such that $f(X) = h^{-1}(0)$, $\mathcal{F}_0(f)$ is locally generated by h^r and D, $f(\Sigma)$ and Σ' are closed analytic subsets of V. Let $y \in$

$f(X) \setminus (D \cup f(\Sigma) \cup \Sigma')$ and assume $S = f^{-1}(y) = \{x_1, \dots, x_k\}$. For each $i = 1, \dots, k$, the mono-germ $\tilde{f}_{x_i} : (X, x_i) \to (f(X), y)$ is a finite germ between smooth germs of the same dimension. By Corollary D.5, each $\tilde{f}_{x_i}$ is surjective and hence each point in a neighbourhood of y has at least one preimage. But the total number of such preimages is equal to k, so each $\tilde{f}_{x_i}$ must be one-to-one. By the inverse function theorem, $\tilde{f}_{x_i}$ is a biholomorphism. In other words, the germ $f_{x_i} : (X, x_i) \to (\mathbb{C}^{n+1}, y)$ is an embedding. It follows that $\mathcal{F}_0(f_{x_i})$ is generated by h (see Exercises 3(iii) and 3(iv) of Sect. 11.2 and Corollary 11.2),

$$\langle h^r \rangle = \left(\mathcal{F}_0(f)\right)_y = \mathcal{F}_0(f_S) = \mathcal{F}_0(f_{x_1}) \dots \mathcal{F}_0(f_{x_k}) = \langle h^k \rangle,$$

hence, $k = r$. □

Corollary 11.3 *Let* $f : (X, S) \to (\mathbb{C}^{n+1}, 0)$ *be finite, where* (X, S) *is an* n*-dimensional Cohen–Macaulay analytic germ. Then* $M_1(f)$ *is reduced if and only if* f *is generically one-to-one.*

We have seen that $M_1(f)$ has a nice analytic structure when we consider a finite germ $f : (X, S) \to (\mathbb{C}^{n+1}, 0)$, where (X, S) is Cohen–Macaulay of dimension n. In fact, $M_1(f)$ is always a hypersurface in $(\mathbb{C}^{n+1}, 0)$ and hence, is Cohen–Macaulay. The situation is much worse if we look at finite germs $f : (X, S) \to (\mathbb{C}^p, 0)$, with $p > n + 1$, as the following example shows.

Example 11.2 Let $f : (\mathbb{C}, 0) \to (\mathbb{C}^3, 0)$ be given by $f(x) = (x^3, x^4, x^5)$. It is easy to see by means of *Macaulay* or *Singular* that $\mathcal{F}_0(f)$ has a primary component of dimension 0, and so $M_1(f)$ is neither reduced nor Cohen–Macaulay.

11.2.1 Are the Fitting Ideal Spaces $M_k(f)$ Cohen–Macaulay?

In the remainder of this chapter, we consider whether Fitting ideals are relatively Cohen–Macaulay (i.e. give a Cohen–Macaulay structure to the multiple point spaces $M_k(f)$). Proposition 11.6 shows that they behave well under deformation. If they are relatively Cohen–Macaulay, then by conservation of multiplicity, they give information on the number of ordinary double points (nodes) which emerge when we deform a singular parameterisation $f : (\mathbb{C}, 0) \to (\mathbb{C}^2, 0)$ of a plane curve, or the number of triple points which emerge when we deform an unstable map-germ $g : (\mathbb{C}^2, 0) \to (\mathbb{C}^3, 0)$:

$$\#\,\text{nodes} = \dim_{\mathbb{C}} \frac{\mathcal{O}_2}{\mathcal{F}_1(f)}$$

$$\#\text{triple points} = \dim_{\mathbb{C}} \frac{\mathcal{O}_3}{\mathcal{F}_2(g)}.$$

Both of these statements hold: they will follow from Proposition 11.6 once we show that given unfoldings F of f and G of g, the corresponding multiple point spaces $M_2(F)$ and $M_3(G)$ are Cohen–Macaulay. This we will do in the next two sections. However, beyond M_3, the general proof of Cohen–Macaulayness runs out. We do not know whether in general it is true that for map-germs $h\colon (\mathbb{C}^n, 0) \to (\mathbb{C}^{n+1}, 0)$ with $n > 3$, the number of $n+1$-tuple points appearing in a generic deformation is equal to $\dim_{\mathbb{C}} \mathscr{O}_{n+1}/\mathcal{F}_n(h)$, because we do not know in full generality whether the multiple point spaces $M_k(F)$ for $k > 3$ are Cohen–Macaulay. We *do* know this for certain classes of germs (for example, germs of corank 1), but the general question remains open. The direct approach, exemplified below by Corollaries 11.5 and 11.6, valid for all $\mathscr{A}$-finite germs, seems not to have progressed since [MP89] and [KU97], even though computer algebra software like *Macaulay 2* and *Singular* allows us to check a wide range of examples. On the other hand, there is a piecemeal approach which can quickly yield complete information about germs of low multiplicity. It relies on the following proposition.

Proposition 11.8 *Let $f\colon (\mathbb{C}^n, 0) \to (\mathbb{C}^{n+1}, 0)$ be $\mathscr{K}$-finite, and let $F\colon (\mathbb{C}^N, 0) \to (\mathbb{C}^{N+1}, 0)$ be a stable germ whose local algebra is isomorphic to that of f. If*

(1) $M_k(F)$ is Cohen–Macaulay, and
(2) codim $M_k(f) = k$
then $M_k(f)$ is Cohen–Macaulay.

Proof Let $F_1\colon (\mathbb{C}^n \times \mathbb{C}^d, (0,0)) \to (\mathbb{C}^{n+1} \times \mathbb{C}^d, (0,0))$ be a stable unfolding of f; F_1 and F are stable and have isomorphic local algebras, so one is $\mathscr{A}$-equivalent to a trivial unfolding of the other, by Theorem 7.1. It follows from (1) that $M_k(F_1)$ is Cohen–Macaulay. By Proposition 11.6, $M_k(f) = i^{-1}(M_k(F_1))$, where $i\colon \mathbb{C}^{n+1} \to \mathbb{C}^{n+1} \times \mathbb{C}^d$ is given by $i(y) = (y, 0)$. Most k-tuple points of F are the intersection of k immersed sheets, which, by stability, are in general position. Thus the codimension of $M_k(F)$ is k, so Cohen–Macaulayness of $M_k(f)$ now follows from (2) by Proposition C.12. □

Proposition 11.9 *For all stable germs $f\colon (\mathbb{C}^n, 0) \to (\mathbb{C}^{n+1}, 0)$ of multiplicity ≤ 5 and for each k with $2 \leq k \leq 5$, $M_k(f)$ is Cohen–Macaulay.*

Here to avoid listing cases, we allow the empty set to be Cohen–Macaulay.

Proof There are only finitely many $\mathscr{K}$-equivalence classes of rank 0 germs $(\mathbb{C}^n, 0) \to (\mathbb{C}^{n+1}, 0)$ of multiplicity ≤ 5. For each one, a stable map-germ F with isomorphic local algebra can be constructed following the recipe of Sect. 4.1, and the Fitting ideals $\mathcal{F}_k(F)$ computed from a presentation. Each is relatively Cohen–Macaulay. For $k = 1$ and 2 this holds by general theorems (Corollaries 11.5 and 11.6) which we prove below. For $k = 3$ and 4, it is possible to use, e.g. *Macaulay2* to check for Cohen–Macaulayness. □

In fact the minimal stable maps of multiplicity 5 have domains of rather high dimension, and the *pushForward* command in *Macaulay2* may not return an answer on a laptop. The procedure described in Algorithm 11.1 can be used instead. However, deriving conclusions from the presentation is itself computationally intensive, and for this reason we have not checked systematically beyond multiplicity 5.

From the last two propositions we deduce

Corollary 11.4 *For all $\mathscr{A}$-finite germs $f: (\mathbb{C}^n, 0) \to (\mathbb{C}^{n+1}, 0)$ of multiplicity $\leq$ 5, all of the multiple point spaces $M_k(f)$ are Cohen–Macaulay.*

Proof $\mathscr{A}$-finite germs are stable outside the base-point. Most k-tuple points are the intersection of k immersed sheets, which, by stability, are in general position. Thus, the codimension of $M_k(f)$ is k, as required by Proposition 11.9. □

It is not difficult to extend the range of local algebra types to which the arguments here apply.

Exercises for Sect. 11.2

1 The Fitting ideals are independent of the choice of presentation of M. Prove this by showing

(i) If $R^a \xrightarrow{\alpha} R^q \xrightarrow{g} M \longrightarrow 0$ and $R^b \xrightarrow{\beta} R^q \xrightarrow{g} M \longrightarrow 0$ are presentations of the same module with respect to the same set of generators, then

$$\min_{q-k}(\alpha) = \min_{q-k}(\beta).$$

Suggestion: lift the identity map $M \to M$ to a chain morphism from one presentation to the other (see Exercise 3).

(ii) If $R^s \xrightarrow{\mu} R^t \xrightarrow{h} M \longrightarrow 0$ is another presentation of the same module M, then $g + h : R^{q+t} \to M$ is surjective. For each basis vector e_i in R^t there exists $c_i \in R^q$ such that $g(c_i) = h(e_i)$, and thus $(c_i, -e_i) \in \ker(g + h)$. Show that the kernel of $g + h$ is generated by such pairs $(c_i, -e_i)$ together with pairs $(c, 0)$ with $c \in \ker g$, so that there is a presentation of the form

$$R^{p+t} \xrightarrow{\nu} R^{q+t} \xrightarrow{g+h} M \longrightarrow 0 \tag{11.13}$$

with

$$\nu = \begin{pmatrix} \lambda & -c \\ 0 & I_t \end{pmatrix}.$$

Clearly

$$\min_{q+t-k}(\nu) = \min_{q-k}(\lambda).$$

By symmetry, the kernel of $g+h$ is also generated by pairs $(0, d)$ with $d \in \ker h$ and pairs (e_j, d_j) where e_j is the j'th basis vector of R^p and $g(e_j) = -h(d_j)$. By 1, the ideals of $(q-k+t)$-minors are the same.

2 Let M and M' be R-modules of finite presentation. Show that

$$\mathrm{Fitt}_k(M \oplus M') = \sum_{r+s=k} \mathrm{Fitt}_r(M)\mathrm{Fitt}_s(M').$$

3 Let $f\colon (\mathbb{C}^n, S) \to (\mathbb{C}^p, 0)$ be finite with $S = \{x_1, \dots, x_r\}$. Show that

(i) If $\phi\colon (\mathbb{C}^n, S) \to (\mathbb{C}^n, S)$ is a diffeomorphism, then $\mathcal{F}_k(f \circ \phi) = \mathcal{F}_k(f)$.
(ii) If $\psi\colon (\mathbb{C}^p, 0) \to (\mathbb{C}^p, 0)$ is a diffeomorphism, then $\psi^*\mathcal{F}_k(f) = \mathcal{F}_k(\psi \circ f)$.
(iii) $\mathcal{F}_0(f) = \mathcal{F}_0(f_1)\dots\mathcal{F}_0(f_r)$, where f_i is the mono-germ of f at each x_i.
(iv) Let $n < p$. Then $M_1(f)$ is smooth if and only if f is an embedding (i.e. an isomorphism onto its image).

4 Let $f\colon (X, S) \to (\mathbb{C}^{n+1}, 0)$ be finite, where (X, S) is an n-dimensional Cohen–Macaulay analytic germ. Assume that the image of f has r irreducible components $(Y_1, 0), \dots, (Y_r, 0)$ and that f has degree k_i onto each $(Y_i, 0)$. Show that $\mathcal{F}_0(f)$ is generated by

$$h = h_1^{k_1} \dots h_r^{k_r},$$

where $h_i \in \mathcal{O}_{n+1}$ is a reduced equation for each $(Y_i, 0)$, $i = 1, \dots, r$.

5

(i) Check the assertions of Example 11.2.
(ii) Show that in contrast, if F is a stable unfolding of $f(x) = (x^3, x^4, x^5)$ then $M_1(F)$ is reduced, though not Cohen–Macaulay.

6 In the fibre diagram (11.10), show that the local algebras of f and F are isomorphic. Suggestion: since $\mathcal{O}_{X,S} = \bigoplus_{x_i \in S} \mathcal{O}_{X,x}$, this reduces immediately to the case where $|S| = 1$.

7 Prove directly from first principles that the sequence (11.12) is exact.

11.3 Double Points in the Target

In this section we focus on local properties of the space $M_2(f)$ of double points in the target for mappings from an n-dimensional Cohen–Macaulay analytic space X to $\mathbb{C}^{n+1}$. For simplicity, we restrict ourselves to the case of mono-germs

$f\colon (X, x_0) \to (\mathbb{C}^{n+1}, 0)$, although most of the results can be adapted with a little effort to the general case of multi-germs. From now on in this section, and unless otherwise specified, we will assume that (X, x_0) is an n-dimensional Cohen–Macaulay analytic germ.

The main result will be that $M_2(f)$ is Cohen–Macaulay, in fact, a determinantal variety, when f is finite and generically 1–1. This is shown by proving first that $\mathcal{F}_1(f)$ is generated by the maximal minors of the matrix of a presentation *with its first row, corresponding to the generator 1, deleted*, even when f is many-to-one. When f is generically 1-to-1, then the codimension of $M_2(f)$ is 2, which is equal to the codimension of the variety of maximal minors of a generic $k \times (k+1)$ matrix, and it follows that $M_2(f)$ is a determinantal variety.

The following result follows directly from Exercise 3(iv) of Sect. 11.2.

Proposition 11.10 *Suppose that X is smooth and let $f\colon (X, x_0) \to (\mathbb{C}^{n+1}, 0)$ be finite and generically 1–1. Then $M_2(f)$ coincides, as a set, with the singular locus of $M_1(f)$.* □

If f is generically k-to-1, with $k \geq 2$, then $M_2(f) = M_1(f)$, as set-germs. Otherwise, if f is generically 1-to-1 and $h \in \mathcal{O}_{n+1}$ is a reduced equation of the image, then $M_2(f) = V(J_h + \langle h\rangle)$, as set germs. These two spaces may be different as complex space-germs.

Example 11.3 Consider the cusp $x \mapsto (x^2, x^3)$. Then $h = X^3 - Y^2$ and $J_h + \langle h\rangle = \langle X^2, Y\rangle$, but $\mathcal{F}_1(f) = \langle X, Y\rangle$.

The next result strengthens theorems of [Cat84] and [MP89] and is the key to understanding the determinantal nature of $M_2(f)$.

Proposition 11.11 *Let $f\colon (X, x_0) \to (\mathbb{C}^{n+1}, 0)$ be finite and let λ be the $(m+1) \times (m+1)$ matrix of a presentation of $\mathcal{O}_{X,x_0}$ over $\mathcal{O}_{n+1}$, with respect to generators $g_0 = 1, g_1, \ldots, g_m$. Then the ideal $\mathcal{F}_1(f)$ is generated by the $m \times m$ minors of the matrix λ' obtained from λ by deleting its first row.*

Proof We first show that the property is true for the particular matrix presentation λ obtained in Proposition 11.1. We continue with the notation of the proof of Proposition 11.1: $\mathcal{O}_S$ is free over $\mathcal{O}_T$ with basis $G = \{g_0, \ldots, g_m\}$ and $\lambda = [t]_G^G$. Let $e_0, \ldots, e_m$ denote the standard basis of $\mathcal{O}_T^{m+1}$. Recall that for each $a \in \mathcal{O}_S$, $[a]_G^G = [a]$ is the matrix of $\mathcal{O}_S \xrightarrow{a} \mathcal{O}_S$ (multiplication by a), with respect to G. If $\varphi : \mathcal{O}_T^{m+1} \to \mathcal{O}_S$ is the $\mathcal{O}_T$-isomorphism sending e_i to g_i for $i = 0, \ldots, m$, then for $x \in \mathcal{O}_T^{m+1}$,

$$a\varphi(x) = \varphi([a]x). \tag{11.14}$$

In particular, $a = \varphi([a]e_0)$.

Instead of $\mathcal{O}_S$ and $\mathcal{O}_T$, we work now over the respective rings of fractions K_S and K_T. Let $\overline{\varphi} : K_T^{m+1} \to K_S$ be the K_T-morphism extending φ. This is obviously a monomorphism, since if there were a non-trivial relation among the g_i over K_T, after multiplication by a common denominator, we would obtain a non-trivial relation over $\mathcal{O}_T$. In fact $\overline{\varphi}$ is also surjective, and therefore an isomorphism. For let $a/b \in K_S$, with a, b in $\mathcal{O}_S$ and $b \neq 0$. Note that (11.14) holds also over K_T, that is,

$$b\overline{\varphi}(x) = \overline{\varphi}([b]x), \quad \text{for all } x \in K_T^{m+1}.$$

Since $b \neq 0$ and K_T is a field, $[b]$ has an inverse $[b]^{-1}$ over K_T and thus,

$$b\overline{\varphi}([b]^{-1}[a]e_0) = \overline{\varphi}([b][b]^{-1}[a]e_0) = \overline{\varphi}([a]e_0) = a.$$

This implies $\overline{\varphi}([b]^{-1}[a]e_0) = a/b$, proving surjectivity. It follows that K_S also has K_T–basis G. As before, for $a \in K_S$ we denote by $[a]$ the matrix of multiplication by a on K_S with respect to basis G. Note that if $a \in K_T$ then $[a]$ is the scalar matrix with diagonal entries equal to a.

Now consider again the presentation matrix $\lambda = [t]$. By construction, $\mathcal{F}_1(f)$ is generated by the submaximal minors of λ, which are the entries of the adjugate matrix λ^{adj}, and the maximal minors of λ' are the entries of the first column of λ^{adj}. Now

$$\lambda^{\text{adj}} = [ht^{-1}],$$

where $h = \det \lambda$. Thus for $j = 0, \ldots, m$, the j'th column of λ^{adj} is equal to $[ht^{-1}]e_j$, so

$$[ht^{-1}]e_j = [ht^{-1}][g_j]e_0 = [ht^{-1}g_j]e_0 = [g_j][ht^{-1}]e_0.$$

But the entries of $[g_j]$ are in $\mathcal{O}_T$ (since $g_j \in \mathcal{O}_S$) and thus each entry of $[ht^{-1}]e_j$ is a linear combination of the entries of $[ht^{-1}]e_0$ over $\mathcal{O}_T$.

To finish the proof, it only remains to show that if the property is true for a particular matrix presentation λ, then it is also true for every matrix presentation. Let D be the image of f. We can consider its local ring $\mathcal{O}_D$ as a subring of $\mathcal{O}_{X,x_0}$. It is the image of the ring homomorphism $f^* : \mathcal{O}_{n+1} \to \mathcal{O}_{X,x_0}$, and so as an $\mathcal{O}_{n+1}$-submodule, it is generated by $g_0 = 1$. As a consequence, from the presentation of $\mathcal{O}_{X,x_0}$, we get a presentation of the quotient $\mathcal{O}_{X,x_0} / \mathcal{O}_D$ by deleting the first row of λ:

$$\mathcal{O}_{n+1}^{m+1} \xrightarrow{\lambda'} \mathcal{O}_{n+1}^{m} \longrightarrow \frac{\mathcal{O}_{X,x_0}}{\mathcal{O}_D} \longrightarrow 0\,.$$

The ideal generated by the maximal minors of λ' is therefore $\mathrm{Fitt}_0^{\mathcal{O}_{n+1}}(\mathcal{O}_{X,x_0} / \mathcal{O}_D)$. We know that both this ideal and $\mathcal{F}_1(f)$ are independent of the choice of presentations. This proves our claim. □

Corollary 11.5 *For any finite and generically one-to-one germ $f\colon (X, x_0) \to (\mathbb{C}^{n+1}, 0)$, $M_2(f)$ is a determinantal variety of dimension $n-1$ and hence, Cohen–Macaulay.*

Proof By Proposition 11.11, $F_1(f)$ is defined by the maximal minors of an $m \times (m+1)$ matrix, hence $M_2(f)$ has dimension $\geq n-1$. But f is generically one-to-one, which implies that $M_2(f)$ has dimension $\leq n-1$. Thus, $M_2(f)$ has dimension $n-1$ and is a determinantal variety (see Theorem C.6). □

Corollary 11.6 *Let $f\colon (\mathbb{C}^n, 0) \to (\mathbb{C}^{n+1}, 0)$ be $\mathscr{A}$-finite, with $n \geq 2$. Then $M_2(f)$ is Cohen–Macaulay and reduced of dimension $n-1$.*

Proof Since f is $\mathscr{A}$-finite, it is finite and stable in a punctured neighbourhood of the origin in $\mathbb{C}^{n+1}$ by the Mather–Gaffney Criterion Theorem 4.5. In particular, f is generically one-to-one and hence, $M_2(f)$ is Cohen–Macaulay of dimension $n-1$. Moreover, for $n \geq 2$, the multi-germ of f at a generic point y of $M_2(f)$ is a transverse double point. It follows that $M_2(f)$ is smooth at y (see Example 8.6). Thus, $M_2(f)$ is Cohen–Macaulay and generically smooth, and hence is reduced, by Proposition D.2. □

As a byproduct of the proof of Proposition 11.11, we get the following lemma, which gives the relationship between $\mathcal{F}_1(f)$ and the conductor ideal. Recall from Subsection 8.4.1 that the conductor ideal of a finite germ $f\colon (X, x_0) \to (\mathbb{C}^{n+1}, 0)$ is $\mathscr{C} := \{h \in \mathcal{O}_{X,x_0} : h\,\mathcal{O}_{X,x_0} \subset \mathcal{O}_D\}$, where D is the image of f (see Definition 8.3 in the case $X = \mathbb{C}^n$). Regarding $\mathcal{O}_D$ as a subring of $\mathcal{O}_{X,x_0}$, $\mathscr{C}$ is evidently contained in $\mathcal{O}_D$, where it is also an ideal.

Lemma 11.1 *If $f\colon (X, x_0) \to (\mathbb{C}^{n+1}, 0)$ is finite and generically 1-to-1, then*

1. $(f^*)^{-1}(\mathscr{C}) = \mathcal{F}_1(f)$.
2. $\mathscr{C} = \mathcal{F}_1(f)\,\mathcal{O}_D$.

Proof The two statements are equivalent to one another; the first views $\mathscr{C}$ as an ideal of $\mathcal{O}_{X,x_0}$, and the second as an ideal of $\mathcal{O}_D$.

We saw in the proof of Proposition 11.11 that $\mathcal{O}_{X,x_0} / \mathcal{O}_D$ has a presentation

$$\mathcal{O}_{n+1}^{m+1} \xrightarrow{\lambda'} \mathcal{O}_{n+1}^{m} \longrightarrow \frac{\mathcal{O}_{X,x_0}}{\mathcal{O}_D} \longrightarrow 0$$

and that $\mathcal{F}_1(f) = \mathrm{Fitt}_0^{\mathcal{O}_{n+1}}(\mathcal{O}_{X,x_0} / \mathcal{O}_D)$. In addition, here f is generically 1-to-1, hence $\mathcal{O}_{n+1} / \mathcal{F}_1(f)$ and $\mathcal{O}_{X,x_0} / O_D$ both have dimension $n-1$. This implies that $\mathcal{O}_{X,x_0} / \mathcal{O}_D$ is a Buchsbaum–Rim module (or determinantal module, in Damon's terminology). This means a module which admits a matrix presentation of size $r \times s$, with $r \leq s$ and has codimension $s - r + 1$. But such modules M have the property that $\mathrm{Fitt}_0(M) = \mathrm{Ann}(M)$ (see [BE77b]), so $\mathcal{F}_1(f) = \mathrm{Ann}_{\mathcal{O}_{n+1}}(\mathcal{O}_{X,x_0} / \mathcal{O}_D) = (f^*)^{-1}(\mathcal{C})$, by definition of $\mathcal{C}$. □

As a consequence of Lemma 11.1, $f^*\mathcal{F}_1(f) = \mathcal{C}$. We have already seen, in Proposition 8.5 that provided X is smooth and f is generically 1-to-1, $\mathcal{C}$ is a principal ideal, generated by $\big((\partial h/\partial Y_i) \circ f\big)/\Delta_i$, where h is an equation for the image and

$$\Delta_i := (-1)^{i+1}\frac{\partial(f_1, \ldots, \widehat{f_i}, \ldots, f_{n+1})}{\partial(x_1, \ldots, x_n)}$$

is the minor determinant of the matrix of the derivative df obtained by omitting row i.

Exercises for Sect. 11.3

1 We continue with the notation of the proof of Proposition 11.11. Let m^i_j be the $m \times m$ minor determinant of λ obtained by omitting row i and column j. Writing out columns $0, \ldots, \hat{j}, \ldots, m$ of λ as relations between $g_0, \ldots, g_m$, subtract $\lambda^i_k\, g_i$ from both sides of the k'th relation, to obtain a set of m equations with an $m \times m$ matrix of coefficients.

(i) Use Cramer's rule to show that for all i, j, k,

$$m^i_j g_k = \pm m^k_j g_i \tag{11.15}$$

and in particular

$$m^i_j = \pm m^0_j g_i. \tag{11.16}$$

(ii) Because $g_i g_j$ lies in $\mathcal{O}_{X,x_0}$ and $\mathcal{O}_{X,x_0}$ is generated over $\mathcal{O}_{n+1}$ by the g_k, there exist $\Gamma^k_{ij} \in \mathcal{O}_{n+1}$ such that $g_i g_j = \sum_k \Gamma^k_{ij} g_k$, with $\Gamma^k_{ij} \in \mathcal{O}_{n+1}$. Use (i) to show that

$$m^i_j g_k = \sum_\ell \pm\Gamma^\ell_{ik} m^\ell_j.$$

From this it follows that $\mathcal{F}_1(f)\,\mathcal{O}_D$ is contained in the conductor ideal $\mathcal{C}$.

2

(i) Find equations for the double point locus, C, of the image of the map-germ f of type H_2, given by $f(x, y) = (x, y^3, xy + y^5)$.
(ii) Show that C is the image of the map $t \mapsto (t^4, t^3, t^5)$.

(iii) Check that $f^*(\mathcal{F}_1(f))$ really is a principal ideal in $\mathcal{O}_2$.
(iv) Find the preimage in $\mathbb{C}^2$ of C, and show that it has a singularity of type A_6 at 0.
(v) Show that the set of real points on this curve is just 0.
(vi) Can you reconcile the conclusions of (ii) and (v)?

3 Find a generator for the conductor when f is the stable map-germ

$$(x, u_1, u_2, u_3) \mapsto (x^3 + u_1 x, u_2 x^2 + u_3 x, u_1, u_2, u_3)$$

(see Exercise 4.3.2). Show that $D_1^2(f)$ is isomorphic to the product of a Whitney umbrella and a line. What is the explanation for this? What is the generalisation of this fact, for stable map-germs of corank 1 from $\mathbb{F}^n$ to $\mathbb{F}^{n+1}$?

4 The δ-invariant of a curve singularity (X, x_0) is defined as follows. There exists a parameterisation $\alpha\colon (\mathbb{C}, S) \to (X, x_0)$; this follows from the existence of a normalisation (see Sect. D.4) and the fact that any normal 1-dimensional space is smooth. Since α is surjective, we can consider $\mathcal{O}_{X,x_0}$ as a subring of $\mathcal{O}_{\mathbb{C},S}$. Then

$$\delta(X, x_0) = \dim_{\mathbb{C}} \frac{\mathcal{O}_{\mathbb{C},S}}{\mathcal{O}_{X,x_0}}.$$

To do: show that if (X, x_0) is a Cohen–Macaulay curve germ and $f\colon (X, x_0) \to (\mathbb{C}^2, 0)$ is 1-to-1, and has image Y, then

$$\dim_{\mathbb{C}} \mathcal{O}_{\mathbb{C}^2,0}/\mathcal{F}_1(f) = \delta(Y, 0) - \delta(X, x_0).$$

Suggestion: by composing f with α we get a parametrisation of $(Y, 0)$. In the proof of Proposition 11.11 we have seen that $\mathcal{O}_{X,x_0}/\mathcal{O}_{Y,0}$ is presented by the matrix of a presentation of $\mathcal{O}_{X,0}$ over $\mathcal{O}_{\mathbb{C}^2,0}$ with its first row deleted. Now use Theorem C.8.

11.4 $\mathcal{A}_e$-Codimension and Image Milnor Number of Map-Germs$(\mathbb{C}^n, S) \to (\mathbb{C}^{n+1}, 0)$

In Chap. 8 we discussed the relation between these two invariants of $\mathcal{A}$-finite germs, but postponed the proof of the theorem which relates them for $n = 1$ and $n = 2$. Having proved that the double point space $M_2(f)$ is a determinantal variety, we are now able to give a proof. First let us re-state the result.

Theorem 11.1 ([dJvS91]) *Let $f\colon (\mathbb{C}^n, S) \to (\mathbb{C}^{n+1}, 0)$ be $\mathscr{A}$-finite, with $n = 1$ or $n = 2$, and let g be a reduced equation for its image. Then*

$$\operatorname{codim}_{\mathscr{A}_e}(f) \leq \mu_I(f)$$

with equality if f is weighted homogeneous.

Proof Recall that in Lemma 8.1 we showed that there is an epimorphism

$$\frac{\theta(f)}{T\mathscr{A}_e f} \to \frac{J_g\, \mathscr{O}_{\mathbb{C}^n,S}}{J_g\, \mathscr{O}_D}$$

which is injective when $n = 2$, and whose kernel in case $n = 1$ is described in Lemma 8.2. We use the natural short exact sequence

$$0 \to \frac{J_g\, \mathscr{O}_{\mathbb{C}^n,S}}{J_g\, \mathscr{O}_D} \to \frac{\mathscr{C}}{J_g\, \mathscr{O}_D} \to \frac{\mathscr{C}}{J_g\, \mathscr{O}_{\mathbb{C}^n,S}} \to 0 \tag{11.17}$$

to relate the $\mathscr{A}_e$-codimension of f to the number of critical points of the defining function g of the image which move off the image as f is deformed. In both cases, we will use conservation of multiplicity. To do this we have to first replace $\mathscr{C}/J_g\, \mathscr{O}_D$ by a quotient of ideals in $\mathscr{O}_{n+1}$. Recall from Lemma 11.1 that the preimage of $\mathscr{C}$ under the passage to the quotient $\mathscr{O}_{n+1} \to \mathscr{O}_D$ is the first Fitting ideal of $\mathscr{O}_{\mathbb{C}^n,S}$ as $\mathscr{O}_{n+1}$-module, $\mathcal{F}_1(f)$. Moreover, by Proposition 11.11, this ideal is determinantal: it is generated by the maximal minors of the matrix λ' obtained by deleting the first row of the matrix of a presentation of $\mathscr{O}_{\mathbb{C}^n,S}$ as $\mathscr{O}_{n+1}$-module. By the Hilbert–Burch Theorem C.5, this means that it has a free resolution

$$0 \longrightarrow \mathscr{O}_{n+1}^m \xrightarrow{(\lambda')^t} \mathscr{O}_{n+1}^{m+1} \xrightarrow{\Delta} \mathcal{F}_1(f) \longrightarrow 0 \tag{11.18}$$

Lifting the inclusion $J_g \to \mathcal{F}_1(f)$ to a morphism $\alpha : \theta_{n+1} \to \mathscr{O}_{n+1}^{m+1}$ and combining this with the presentation (11.18), we obtain a diagram

$$\begin{array}{ccccccc}
 & & \theta_{n+1} = \mathscr{O}_{n+1}^{n+1} & \longrightarrow & J_g & \longrightarrow & 0 \\
 & & \downarrow{\scriptstyle \alpha} & & \downarrow & & \\
\mathscr{O}_{n+1}^m & \xrightarrow{\beta} & \mathscr{O}_{n+1}^{m+1} & \longrightarrow & \mathcal{F}_1(f) & \longrightarrow & 0
\end{array} \tag{11.19}$$

(where $\beta = (\lambda')^t$) from which it follows by a diagram chase that $\mathcal{F}_1(f)/J_g$ has a presentation

$$\mathscr{O}_{n+1}^{n+1} \oplus \mathscr{O}_{n+1}^m \xrightarrow{\alpha+\beta} \mathscr{O}_{n+1}^{n+1} \longrightarrow \mathcal{F}_1(f)/J_g \longrightarrow 0 \tag{11.20}$$

Now in the two cases $n = 1$ and $n = 2$, the quotient $\mathcal{F}_1(f)/J_g$ is supported only at $0 \in \mathbb{C}^{n+1}$; for since the image of f is non-singular outside 0 (in case $n = 1$), and has normal crossings outside 0 (in case $n = 2$), J_g is radical outside 0. Since $J_g \subset \mathcal{F}_1(f)$, by Corollary 8.1, the two ideals coincide outside 0.

Now suppose that $F\colon (\mathbb{C}^n \times \mathbb{C}^d, S \times \{0\}) \to (\mathbb{C}^{n+1} \times \mathbb{C}^d, (0,0))$ is a versal unfolding of f, and let G denote a reduced equation of the image of F which specialises to g when the parameter $u \in \mathbb{C}^d$ is equal to 0. We form the relative version of the quotient $\mathcal{F}_1(f)/J_g$, namely $\mathcal{F}_1(F)/J_G^{\text{rel}}$, where the upper index "rel" means that we take the ideal generated by the partials $\partial G/\partial y_1, \ldots, \partial G/\partial y_{n+1}$, omitting the partials with respect to the unfolding parameters. By the argument above, this has a presentation

$$\mathcal{O}^{n+1}_{n+d+1} \oplus \mathcal{O}^{m}_{n+d+1} \xrightarrow{A+B} \mathcal{O}^{m+1}_{n+d+1} \longrightarrow \mathcal{F}_1(F)/J_G^{\text{rel}} \longrightarrow 0 \tag{11.21}$$

The intersection of the support of $\mathcal{F}_1(F)/J_G^{\text{rel}}$ with $\mathbb{C}^{n+1} \times \{0\} \subset \mathbb{C}^{n+1} \times \mathbb{C}^d$ is just the support of $\mathcal{F}(f)/J_g$. Since this has dimension 0, the dimension of the support of $\mathcal{F}_1(F)/J_G^{\text{rel}}$ can be no greater than d. By Theorem C.7, it can be no less than d, and it follows, also by Theorem C.7, that $\mathcal{F}_1(F)/J_G^{\text{rel}}$ is Cohen–Macaulay, evidently of dimension d.

Let π denote the projection $\mathbb{C}^{n+1} \times \mathbb{C}^d \to \mathbb{C}^d$. Because $\operatorname{supp}(\mathcal{F}_1(f)/J_g)$ is just $\{0\}$, π is finite on $\operatorname{supp}\mathcal{F}_1(F)/J_G^{\text{rel}}$. It follows from the finite coherence theorem that $\pi_*(\mathcal{F}_1(F)/J_G^{\text{rel}})$ is coherent. By Proposition C.10, its stalk at $0 \in \mathbb{C}^d$, equal to the stalk of $\mathcal{F}_1(F)/J_G^{\text{rel}}$ at $(0,0) \in \mathbb{C}^{n+1} \times \mathbb{C}^d$, is free, and thus $\pi_*(\mathcal{F}_1(F)/J^{\text{rel}})$ is a locally free sheaf. Its rank is $\dim_{\mathbb{C}}(\mathcal{F}_1(f)/J_g)$. To show this, one can compute

$$\dim_{\mathbb{C}} \frac{\mathcal{F}_1(F)/J_G^{\text{rel}}}{\mathfrak{m}_d\, \mathcal{F}_1(F)/J_G^{\text{rel}}}$$

directly, but this raises problems of the intersection of ideals; it is simpler to use the presentation (11.21), which gives

$$\begin{aligned}
\frac{\mathcal{F}_1(F)/J_G^{\text{rel}}}{\mathfrak{m}_d\, \mathcal{F}_1(F)/J_G^{\text{rel}}} &\cong \frac{\mathcal{O}^{m+1}_{n+d+1}}{A(\mathcal{O}^{m}_{n+d+1}) + B(\mathcal{O}^{m}_{n+d+1}) + \mathfrak{m}_d\, \mathcal{O}^{m+1}_{n+d+1}} \\
&\cong \frac{\mathcal{O}^{m+1}_{n+1}}{\alpha(\mathcal{O}^{n+1}_{n+1}) + \beta(\mathcal{O}^{m}_{n+1})} \\
&\cong \mathcal{F}(f)/J_g.
\end{aligned}$$

Now let u be a parameter value for which $f_u : U \to V$ is a stable perturbation of f, and let $g_u = G(\_, u)$ be the equation of its image obtained by restricting G. By the freeness of $\pi_*(\mathcal{F}_1(F)/J_G^{\mathrm{rel}})$ and the above calculation using presentations,

$$\sum \dim_{\mathbb{C}}(\mathcal{F}_1(f_u)/J_{g_u})_y = \dim_{\mathbb{C}} \mathcal{O}_{n+1}/J_g, \tag{11.22}$$

where the sum is over points y in the support of the quotient. As f_u is stable, we can easily describe all of these points.

Case 1 $n = 1$ The image of f_u is an immersed curve with normal crossings. At normal crossing points, $J_{g_u} = \mathcal{F}_1(f_u)$ (Exercise 8.6). Thus the only points of the support of $\mathcal{F}_1(f_u)/J_{g_u}$ lie off the image. At such a point y, $(\mathcal{F}_1(f_u)/J_{g_u})_y = \mathcal{O}_{\mathbb{C}^2,y}/J_{g_u}$. Hence, by (11.22),

$$\sum_y \dim_{\mathbb{C}}(\mathcal{O}_{\mathbb{C}^2,y}/J_{g_u}) = \dim_{\mathbb{C}} \mathcal{F}_1(f)/J_g.$$

In Theorem 8.4 we showed that the sum on the left is equal to the image Milnor number $\mu_I(f)$.

Our proof for $n = 1$ is nearly complete. Suppose first that $g \in J_g$, so that $\mathscr{C}/J_g\mathcal{O}_D = \mathcal{F}_1(f)/J_g$. From the calculation we have just completed, and the exact sequence (11.17), we have

$$\mu_I(f) = \dim_{\mathbb{C}}(\mathcal{F}_1(f)/J_g) = \dim_{\mathbb{C}} \frac{J_g\,\mathcal{O}_{\mathbb{C},S}}{J_g\,\mathcal{O}_D} + \dim_{\mathbb{C}} \frac{\mathscr{C}}{J_g\,\mathcal{O}_{\mathbb{C},S}}. \tag{11.23}$$

Now recall from Lemmas 8.1 and 8.2 that

$$\operatorname{codim}_{\mathscr{A}_e}(f) = \dim_{\mathbb{C}} \frac{J_g\,\mathcal{O}_{\mathbb{C},S}}{J_g\,\mathcal{O}_D} + \sum_{x\in S} \dim_{\mathbb{C}} \frac{\mathcal{O}_{\mathbb{C},x}}{(f_1', f_2')}. \tag{11.24}$$

Two anomalies stand in the way of the desired equality of μ_I and $\mathscr{A}_e$-codimension—the second summands on the right-hand side of (11.23) and (11.24). Remarkably, they are equal, as we show in the following lemma, which holds for all $n \geq 1$:

Lemma 11.2 *Let c be an $\mathcal{O}_{\mathbb{C}^n,S}$-generator of the conductor ideal $\mathscr{C}$, let $\mathscr{R}_f$ be the ramification ideal, generated by the $n \times n$ minors of the matrix of df, and let D be the image of f, with defining equation g. Then multiplication by c gives an $\mathcal{O}_{\mathbb{C}^m,S}$-linear isomorphism*

$$\mathcal{O}_{\mathbb{C}^n,S}/\mathscr{R}_f \to \mathscr{C}/J_g\,\mathcal{O}_D.$$

Proof This is an immediate consequence of the fact, recorded in Proposition 8.5, that for each maximal minor $m_j := \partial(f_1, \ldots, \widehat{f_j} \ldots, f_{n+1})/\partial(x_1, \ldots, x_n)$ of the matrix of df, the quotient $(\partial g/\partial y_j)/m_j$ is independent of j and is an $\mathscr{O}_{\mathbb{C}^n,S}$ generator of the conductor. □

The lemma, together with (11.23) and (11.24), shows that when $g \in J_g$, $\mu_I(f) = \text{codim}_{\mathscr{A}_e} f$. If $g \notin Jg$, then this becomes $\mu_I(f) > \text{codim}_{\mathscr{A}_e} f$. Finally, it was shown by K.Saito in [Sai71] that for a function g with isolated singularity, $g \in J_g$ if and only if g is weighted homogeneous with respect to some coordinate system. This completes the proof of Theorem 11.1 for $n = 1$.

Case 2 $n = 2$. Now the support of $\mathcal{F}_1 f_u)/J_{g_u}$ consists of both points on the image of f_u and points off the image. The image of f_u has points of four types: regular (smooth) points, transverse double points, triple points and Whitney umbrella points. At points of the first three types, $J_{g_u} = \mathcal{F}_1(f)$ (see Exercise 11.4.5). At a Whitney umbrella point, let us take local coordinates x_1, x_2 in the source and y_1, y_2, y_3 in the target, with respect to which the germ of f takes the form

$$(x_1, x_2) \to (x_1, x_2^2, x_1x_2).$$

One calculates that $\mathcal{F}_1(f) = (y_1, y_3)$, and taking $g = y_1^2y_2 - y_3^2$ we have $J_g = (y_1y_2, y_1^2, y_3)$, so that the quotient $\mathcal{F}_1(f)/J_g$ has $\mathbb{C}$-basis y_1. Thus, the contribution of points on the image of f_u to the sum on the left-hand side of (11.22) is simply the number of Whitney umbrella points on the image of f_u. By a standard argument using conservation of multiplicity, this number is equal to $\dim_{\mathbb{C}} \mathscr{O}_{\mathbb{C}^2,S}/\mathscr{R}_f$.

As in the previous case, at a point y off the image of f_u, $\mathcal{F}_1(f_u)/J_{g_u} = \mathscr{O}_{\mathbb{C}^3,y}/J_{g_u}$, and thus the contribution of points off the image of f_u to the sum on the left-hand side of (11.22) is equal to the image Milnor number $\mu_I(f)$.

Again, let us suppose first that $g \in J_g$. Then $\mathscr{C}/J_g\mathscr{O}_D = \mathcal{F}_1(f)/J_g$, and from the exactness of (11.17) and the isomorphism $\theta(f)/T\mathscr{A}_e f \cong J_g\mathscr{O}_{\mathbb{C}^2,S}/J_g\mathscr{O}_D$ we obtain

$$\text{codim}_{\mathscr{A}_e} f = \dim_{\mathbb{C}} \mathcal{F}_1(f)/J_g - \dim_{\mathbb{C}} \mathscr{C}/J_g\mathscr{O}_D\,.$$

Conservation of multiplicity (11.22), and our calculations of the last paragraphs, then give

$$\text{codim}_{\mathscr{A}_e} f = \#\text{Whitney umbrellas} + \mu_I(f) - \dim_{\mathbb{C}} \mathscr{C}/J_g\mathscr{O}_D\,.$$

But #Whitney umbrellas $= \dim_{\mathbb{C}} \mathscr{O}_{\mathbb{C}^2,S}/\mathscr{R}_f$, and by Lemma 11.2,

$$\dim_{\mathbb{C}} \mathscr{O}_{\mathbb{C}^2,S}/\mathscr{R}_f = \dim_{\mathbb{C}} \mathscr{C}/J_g\mathscr{O}_D,$$

and we conclude that

$$\text{codim}_{\mathscr{A}_e} f = \mu_I(f).$$

Finally, if $g \notin J_g$ then $\dim_{\mathbb{C}} \mathscr{C}/J_g\, \mathscr{O}_D < \dim_{\mathbb{C}} \mathcal{F}_1(f)/J_g$, and it follows that

$$\operatorname{codim}_{\mathscr{A}_e} f < \mu_I(f).$$

Theorem 11.1 was first proved, for $n = 2$, by T. de Jong and R. Pellikaan (unpublished) and then by de Jong and van Straten in [dJvS91]. The proof given here is from [Mon91]; the key idea, Lemma 8.1, was suggested by de Jong and van Straten.

Example 11.4 (Looking at Bent Wires V) If $f: (\mathbb{C}, S) \to (\mathbb{C}^2, 0)$ is real (defined by power-series with real coefficients) then it has a real stable perturbation f_u for which all of the nodes are real, i.e. are crossing points of two real branches (see [GZ74]) or [A'C75]). Now suppose that C is a space curve, and consider the family of radial projections of C from points of $\mathbb{R}^3 \setminus C$ to a plane. Because the curve can be perturbed to have all its nodes real, it follows that if X is an $\mathscr{A}$-equivalence class of singularity which is versally unfolded in the family of projections, then each time X appears in the family, by an arbitrarily small shift of the viewpoint (point of projection) we can split the singularity of projected curve into $\delta(X)$ real nodes. If X is versally unfolded in the family of projections, then in particular the codimension of View(X) in $\mathbb{R}^3$ is equal to $\mathscr{A}_e$– codim(X). For a generic curve, all of the singularities of projection are weighted homogeneous in suitable coordinates [Dav83]. It follows by Theorem 11.1 that *for each $\mathscr{A}$-equivalence class X which is versally unfolded in the family of projections,*

$$\begin{aligned}\operatorname{codim} \operatorname{View}(X) = {} & \text{max. \# nodes appearing after small shift of the viewpoint} \\ & - \text{\# branches} + 1.\end{aligned}$$

As explained in Example 5.2, it can be shown that for a generic space curve, the hypothesis of versatility holds for all of the simple singularities of projection. The only one for which it does not hold is the quadruple point.

Exercises for Sect. 11.4

Suppose that $f_u: (U, S) \to V$ is a stable perturbation of an $\mathscr{A}$-finite germ $f: (\mathbb{C}, S) \to (\mathbb{C}^2, 0)$.

1 Show that $\mu_I(f) = \#$ nodes in image of $f_u - |S| + 1$.

2 Use the result of Exercise 11.3.4 and conservation of multiplicity to show that

$$\delta(f) = \#\text{ nodes in image of } f_u.$$

3 The proof of Theorem 11.1 shows that $\mu_I(f) = \dim_{\mathbb{C}} \mathcal{F}_1(f)/J_g$. Use this and the results of earlier exercises to prove Milnor's formula

$$\mu + 2\delta - r + 1$$

for a plane curve germ $\{g = 0\}$, where r is the number of branches.

4 Show that if $f\colon (\mathbb{C}, 0) \to (\mathbb{C}^2, 0)$ is $\mathscr{A}$-finite then $\mu_I(f) = \delta(f)$.

5 Use suitable local coordinate to check the assertion, used in the proof of Theorem 11.1 for $n = 2$, that at transverse points and triple points of f_u, $J_{g_u} = \mathcal{F}_1(f_u)$.

11.5 The Rank Condition

We have seen in Proposition 11.11 that if λ is the presentation matrix of a finite map-germ $f\colon (X, x_0) \to (\mathbb{C}^{n+1}, 0)$ with respect to generators $g_0 = 1, g_1, \ldots, g_m$, then it has the property that $\min_m(\lambda) = \min_m(\lambda')$, where λ' is the matrix obtained from λ by deleting its first row. This property is called the rank condition (RC) in [Cat84] and [MP89], and the ring condition (RC) in [dJvS90]. In this section, we will prove a converse: if λ is a matrix over $\mathcal{O}_{n+1}$ satisfying RC and with $\det \lambda$ not identically zero (briefly "$\det \lambda \neq 0$"), then it is the matrix presentation of some finite germ $f\colon (X, S) \to (\mathbb{C}^{n+1}, 0)$, where (X, S) is a multi-germ of Cohen–Macaulay complex space of dimension n.

Definition 11.4 Let λ be an $(m + 1) \times (m + 1)$-matrix with entries in $\mathcal{O}_{n+1}$ and λ' the matrix obtained from λ by deleting its first row. If $\min_m(\lambda) = \min_m(\lambda')$, then we say that λ satisfies the *rank condition* (RC).

As in Sect. 11.3, we index the rows and columns of the matrix λ by $\{0, \ldots, m\}$ and write m^i_j for the $m \times m$-minor of λ, obtained by omitting row i and column j. We also denote by $c^i_j = (-1)^{i+j} m^i_j$ the corresponding cofactor, and by Δ the $1 \times (m+1)$ matrix $(c^0_0, \ldots, c^0_m)$. Note that $\Delta(\lambda^0_0, \ldots, \lambda^0_m)^t = \det \lambda$, and $\Delta(\lambda^i_0, \ldots, \lambda^i_m)^t = 0$ for $i = 1, \ldots, m$; one can see the latter equality by viewing this product as the cofactor expansion, by the 0'th row, of the determinant of the matrix with a repeated row obtained from λ by replacing its zero'th row by its i'th row.

Lemma 11.3 *Suppose λ is an $(m + 1) \times (m + 1)$-matrix with entries in $\mathcal{O}_{n+1}$ satisfying* RC. *Let $I = \min_m(\lambda)$ and suppose $\mathcal{O}_{n+1}/I$ has codimension 2. Then I admits a presentation as $\mathcal{O}_{n+1}$-module:*

$$\mathcal{O}^m_{n+1} \xrightarrow{(\lambda')^T} \mathcal{O}^{m+1}_{n+1} \xrightarrow{\Delta} I \longrightarrow 0$$

where $\Delta = (c^0_0, \ldots, c^0_m)$.

Proof The ideal I is generated by the $m \times m$-minors of the $m \times (m+1)$ matrix λ', and $V(I)$ has codimension 2. The sequence

$$0 \longrightarrow \mathscr{O}^m_{n+1} \xrightarrow{(\lambda')^T} \mathscr{O}^{m+1}_{n+1} \xrightarrow{\Delta} \mathscr{O}_{n+1} \longrightarrow \frac{\mathscr{O}_{n+1}}{I} \longrightarrow 0$$

is a complex, by what we have just observed. The Hilbert–Burch Theorem C.5 says that in fact it is exact (though see [Eis95, §20.3] for a more generally applicable argument). Thus

$$\mathscr{O}^m_{n+1} \xrightarrow{(\lambda')^T} \mathscr{O}^{m+1}_{n+1} \xrightarrow{\Delta} I \longrightarrow 0$$

is also exact. □

With the notation of Lemma 11.3, let $h := \det(\lambda) \neq 0$ and assume $h \neq 0$. Write D for the hypersurface in $(\mathbb{C}^{n+1}, 0)$ defined by h.

Definition 11.5 In anticipation of its future rôle, let us call the ideal $\mathscr{C} = I\,\mathscr{O}_D$ in $\mathscr{O}_D$ the *conductor ideal* of λ.

Given any element $a \in \mathscr{O}_{n+1}$, we write $\overline{a}$ for its residue class in $\mathscr{O}_D$. Analogously, if $x = (x_{ij})$ is any matrix with entries in $\mathscr{O}_{n+1}$, we denote by $\overline{x} = (\overline{x_{ij}})$ the corresponding matrix over $\mathscr{O}_D$.

Lemma 11.4 *With the hypothesis of Lemma 11.3, there is an exact sequence*

$$0 \longrightarrow \mathrm{Hom}_{\mathscr{O}_D}(\mathscr{C}, \mathscr{O}_D) \longrightarrow \mathscr{O}^{m+1}_D \xrightarrow{\overline{\lambda}} \mathscr{O}^{m+1}_D$$

Proof We first prove that the conductor ideal $\mathscr{C}$ admits a presentation as $\mathscr{O}_D$-module:

$$\mathscr{O}^{m+1}_D \xrightarrow{\overline{\lambda}^T} \mathscr{O}^{m+1}_D \xrightarrow{\overline{\Delta}} \mathscr{C} \longrightarrow 0\,, \tag{11.25}$$

That $\bar{\Delta}\bar{\lambda}^T = 0$ follows from the remarks just before Lemma 11.3. Conversely, suppose $\overline{a} \in \mathscr{O}^{m+1}_D$ and $\overline{\Delta a} = 0$. Then for some $\alpha \in \mathscr{O}_{n+1}$,

$$\Delta a = \alpha h = \alpha \sum_{i=0}^{m} \lambda^0_i c^0_i = \Delta(\alpha\lambda^0)^T.$$

Therefore, $\Delta(a - (\alpha\lambda^0)^T) = 0$ and by Lemma 11.3,

$$a - \alpha(\lambda^0)^T = (\lambda')^T b,$$

for some $b \in \mathscr{O}^m_{n+1}$. Thus, $\overline{a} \in \mathrm{im}\,((\overline{\lambda^0})^T, (\overline{\lambda'})^T) = \mathrm{im}\,\overline{\lambda}^T$.

Now we know that (11.25) is exact, we can apply the right-exact functor $\mathrm{Hom}_{\mathscr{O}_D}(-, \mathscr{O}_D)$ and identify $\mathrm{Hom}_{\mathscr{O}_D}(\mathscr{O}_D^{m+1}, \mathscr{O}_D)$ with $\mathscr{O}_D^{m+1}$ to get the exact sequence:

$$0 \longrightarrow \mathrm{Hom}_{\mathscr{O}_D}(\mathscr{C}, \mathscr{O}_D) \longrightarrow \mathscr{O}_D^{m+1} \xrightarrow{\overline{\lambda}} \mathscr{O}_D^{m+1}$$

□

Lemma 11.5 *With the hypothesis of Lemma 11.3, the periodic sequence*

$$\cdots \longrightarrow \mathscr{O}_D^{m+1} \xrightarrow{\overline{\lambda}} \mathscr{O}_D^{m+1} \xrightarrow{\overline{\lambda}^{adj}} \mathscr{O}_D^{m+1} \xrightarrow{\overline{\lambda}} \mathscr{O}_D^{m+1} \xrightarrow{\overline{\lambda}^{adj}} \cdots$$

is exact.

Proof It is complex, because $\lambda \cdot \lambda^{\mathrm{adj}} = \lambda^{\mathrm{adj}} \cdot \lambda = hI_{m+1}$ and hence $\overline{\lambda} \cdot \overline{\lambda}^{\mathrm{adj}} = \overline{\lambda}^{\mathrm{adj}} \cdot \overline{\lambda} = 0$. It is acyclic, because if $\overline{\lambda}(\overline{a}) = 0$, then

$$\lambda(a) = hb = \lambda \cdot \lambda^{\mathrm{adj}}(b),$$

for some $b \in \mathscr{O}_{n+1}^{m+1}$ and so $\lambda(a - \lambda^{\mathrm{adj}}(b)) = 0$. Since $h = \det \lambda \neq 0$, λ is injective, and hence $a = \lambda^{\mathrm{adj}}(b)$. That is, $\ker \overline{\lambda} = \mathrm{im}\, \overline{\lambda}^{\mathrm{adj}}$. A similar argument shows that $\ker \overline{\lambda}^{\mathrm{adj}} = \mathrm{im}\, \overline{\lambda}$. □

The following theorem is the promised converse to Proposition 11.11: if the matrix λ has non-zero determinant and satisfies condition RC, then it presents a ring. The proof here (from [MP89, 3.14]) is due to Theo de Jong and Duco van Straten.

Theorem 11.2 *Suppose λ is an $(m+1) \times (m+1)$-matrix with entries in $\mathscr{O}_{n+1}$ satisfying* RC, *and such that* $\det \lambda$ *is not identically* 0. *Let $I = \min_m(\lambda)$, $h = \det(\lambda)$ and $D := \{h = 0\}$. Suppose that $V(I)$ has codimension 2 in $(\mathbb{C}^{n+1}, 0)$. If $R =$* $\mathrm{coker}(\lambda)$, *then:*

1. *$R \cong \mathrm{Hom}_{\mathscr{O}_{X,0}}(\mathscr{C}, \mathscr{C})$, where $\mathscr{C} := I\,\mathscr{O}_D$ is the conductor ideal,*
2. *R is a finite direct sum of analytic $\mathbb{C}$-algebras,*
3. *R is generated by $g_0, \ldots, g_m$ over $\mathscr{O}_{n+1}$, where g_i is multiplication by $\overline{c}_i^0/\overline{c}_0^0$ on $\mathscr{C}$.*

Recall that an *analytic algebra* is a local ring of the form $\mathscr{O}_k / I$ for some k and some ideal I [dJP00, Definition 3.2.8].

Proof Let $\mathbf{1}_{m+1}$ denote the identity matrix of size $m+1$. The identity $h\mathbf{1}_{m+1} = \lambda \cdot \lambda^{\mathrm{adj}}$ implies that $h\,\mathscr{O}_{n+1}^{m+1} \subset \mathrm{im}\, \lambda$, so

$$\mathrm{coker}\, \overline{\lambda} = \frac{\mathscr{O}_D^{m+1}}{\mathrm{im}\, \overline{\lambda}} = \frac{\mathscr{O}_{n+1}^{m+1} / h\, \mathscr{O}_{n+1}^{m+1}}{\mathrm{im}\, \lambda / h\, \mathscr{O}_{n+1}^{m+1}} \cong \frac{\mathscr{O}_{n+1}^{m+1}}{\mathrm{im}\, \lambda} = \mathrm{coker}\, \lambda = R.$$

By Lemmas 11.4 and 11.5,

$$\operatorname{coker}\overline{\lambda} = \frac{\mathscr{O}_D^{m+1}}{\operatorname{im}\overline{\lambda}} = \frac{\mathscr{O}_D^{m+1}}{\ker\overline{\lambda}^{\mathrm{adj}}} \cong \operatorname{im}\overline{\lambda}^{\mathrm{adj}} = \ker\overline{\lambda} \cong \operatorname{Hom}_{\mathscr{O}_D}(\mathscr{C}, \mathscr{O}_D).$$

Hence, $R \cong \operatorname{Hom}_{\mathscr{O}_D}(\mathscr{C}, \mathscr{O}_D)$. The module R is generated by the classes of the standard basis elements $e_0, \dots, e_m$ of $\mathscr{O}_D^{m+1}$ and we can find the generators of the module $\operatorname{Hom}_{\mathscr{O}_D}(\mathscr{C}, \mathscr{O}_D)$ by following these generators in the above isomorphisms. We see that $\operatorname{im}\overline{\lambda}^{\mathrm{adj}}$ is generated by $\overline{\lambda}^{\mathrm{adj}}(e_0), \dots, \overline{\lambda}^{\mathrm{adj}}(e_m)$, where

$$\overline{\lambda}^{\mathrm{adj}}(e_i) = \sum_{j=0}^{m} \overline{c}^i_j e_j.$$

Now the exact sequence in Lemma 11.4 says that the image of $\overline{\lambda}^{\mathrm{adj}}(e_i)$ in $\operatorname{Hom}_{\mathscr{O}_D}(\mathscr{C}, \mathscr{O}_D)$ is the homomorphism $\varphi_i : \mathscr{C} \to \mathscr{O}_D$ such that

$$\varphi_i(\overline{c}^0_j) = \overline{c}^i_j.$$

In particular, $\varphi_i(\mathscr{C}) \subset \mathscr{C}$, for all i, and thus $\operatorname{Hom}_{\mathscr{O}_D}(\mathscr{C}, \mathscr{O}_D) = \operatorname{Hom}_{\mathscr{O}_D}(\mathscr{C}, \mathscr{C})$. This shows item 1.

Now we show item 2. Obviously, $\operatorname{Hom}_{\mathscr{O}_D}(\mathscr{C}, \mathscr{C})$ is an $\mathscr{O}_{n+1}$-algebra with unit under composition. What is not so obvious is that it is commutative. To see that it is, suppose $\varphi \in \operatorname{Hom}_{\mathscr{O}_D}(\mathscr{C}, \mathscr{C})$ and $a, b \in \mathscr{C}$; we have

$$\varphi(ab) = a\varphi(b) = b\varphi(a).$$

If a is not a zero divisor in $\mathscr{O}_D$, then in the total ring of fractions $Q(\mathscr{O}_D)$,

$$\varphi(b) = \frac{\varphi(a)}{a} b.$$

This means that φ is given by multiplication by an element $\varphi(a)/a \in Q(\mathscr{O}_D)$ and that this element $\varphi(a)/a$ is independent of the choice of a. Thus, we have a ring embedding $\operatorname{Hom}_{\mathscr{O}_D}(\mathscr{C}, \mathscr{C}) \hookrightarrow Q(\mathscr{O}_D)$ which maps φ to $\varphi(a)/a$. This implies that $\operatorname{Hom}_{\mathscr{O}_D}(\mathscr{C}, \mathscr{C})$ is commutative and hence, a finite $\mathscr{O}_{n+1}$-algebra. By [dJP00, Corollary 3.3.26],

$$R \cong \bigoplus_{i=1}^{k} R_i,$$

where the R_i are analytic $\mathbb{C}$-algebras.

Finally, item 3 follows from items 1 and 2. In fact, by item 1, $\mathrm{Hom}_{\mathcal{O}_D}(\mathcal{C}, \mathcal{C})$ is generated by $g_0, \ldots, g_m$ such that $g_i(\overline{c}^0_j) = \overline{c}^i_j$. By item 2, g_i is given by multiplication by $\overline{c}^i_j / \overline{c}^0_j$, for all j. In particular, it is given by multiplication by $\overline{c}^0_i / \overline{c}^0_0$. □

Corollary 11.7 *With the hypothesis of Theorem 11.2, there exists a multi-germ of Cohen–Macaulay complex space (X, S) of dimension n and a finite map $f: (X, S) \to (\mathbb{C}^{n+1}, 0)$ such that λ is the matrix of a presentation of $\mathcal{O}_{X,S}$ over $\mathcal{O}_{n+1}$.*

Proof By Theorem 11.2, $R \cong \bigoplus_{i=1}^k R_i$, where R_i are analytic $\mathbb{C}$-algebras. For each i, $R_i = \mathcal{O}_{n_i}/I_i$, for some ideal $I_i \subset \mathcal{O}_{n_i}$. Take N big enough and embeddings $(\mathbb{C}^{n_i}, 0) \to (\mathbb{C}^N, x_i)$, with $x_i \neq x_j$ if $i \neq j$. So, we can assume that each I_i is in fact an ideal of $\mathcal{O}_{\mathbb{C}^N, x_i}$ and that $R_i = \mathcal{O}_{\mathbb{C}^N, x_i}/I_i$. Then, $R \cong \mathcal{O}_{\mathbb{C}^N, S}/I$, where $I = \bigoplus_{i=1}^k I_i$ and $S = \{x_1, \ldots, x_k\}$. By construction, $(X, S) = V(I)$ is a multi-germ of complex space such that $\mathcal{O}_{X,S} \cong R$. Furthermore, the morphism of analytic $\mathbb{C}$-algebras $\mathcal{O}_{n+1} \to R \cong \mathcal{O}_{X,S}$ also induces a finite morphism of complex space-germs $f: (X, S) \to (\mathbb{C}^{n+1}, 0)$ with the property that λ is a matrix presentation of $\mathcal{O}_{X,S}$ over $\mathcal{O}_{n+1}$. Since $h = \det(\lambda) \neq 0$ and h defines the image of f, (X, S) has dimension n. Finally, observe that we have an exact sequence

$$0 \longrightarrow \mathcal{O}_{n+1}^{m+1} \xrightarrow{\lambda} \mathcal{O}_{n+1}^{m+1} \longrightarrow \mathcal{O}_{X,S} \longrightarrow 0$$

which implies that $\mathrm{pd}(\mathcal{O}_{X,S}) \leq 1$. By the Auslander–Buchsbaum Formula Theorem C.4, $\mathrm{depth}(\mathcal{O}_{X,S}) \geq n$ and since $\dim \mathcal{O}_{X,S} = n$, $\mathcal{O}_{X,S}$ is Cohen–Macaulay. □

Remark 11.1 In [dJvS90] de Jong and van Straten give a cleaner treatment of the rank condition by reformulating it as a condition on fractional ideals M in a Gorenstein ring R.[1] Suppose that R is a Gorenstein ring and $Q(R)$ is its total ring of fractions. A *fractional ideal* is a finitely generated R-module M contained in $Q(R)$ which contains a non-zero divisor. For each fractional ideal M, let $M^\vee$ denote its dual $\mathrm{Hom}_R(M, R)$. If M is a fractional ideal containing R, then the inclusion $R \hookrightarrow M$ leads to a monomorphism $\mathrm{Hom}_R(M, R) \hookrightarrow \mathrm{Hom}_R(R, R) = R$, so that $\mathrm{Hom}_R(M, R)$ can be viewed as an ideal of R. In fact

Proposition 11.12 ([dJvS90, Proposition 1.7]) *Let R be a Gorenstein ring. The duality functor $M \mapsto M^\vee$ on the category of R modules has the following properties:*

1. *It converts factional ideals to fractional ideals.*
2. *It converts maximal Cohen–Macaulay modules (MCMs) to MCMs.*
3. *It is an inclusion-reversing involution on the category of fractional MCMs.* □

[1] We define Gorenstein rings in the preamble to Theorem 11.8 in Sect. 11.7 below. It is enough, to motivate the argument here, to think of R as the ring of functions on a hypersurface in $\mathbb{C}^{n+1}$: all hypersurface rings are Gorenstein.

A Cohen–Macaulay module on a ring R is *maximal* if $\dim M = \dim R$.

Proposition 11.13 ([dJvS90, Proposition 1.8]) *Let $\tilde{R}$ be a fractional ideal containing R and let $C \subset R$ be its dual. The following statements are equivalent.*

1. *$\tilde{R}$ is a subring of $Q(R)$.*
2. *(RC) The natural inclusion* $\mathrm{Hom}_R(C, C) \hookrightarrow \mathrm{Hom}_R(C, R)$ *is an equality.* □

Proof Suppose that RC holds. By involutivity of the functor $M \mapsto M^\vee$, $\tilde{R} = \mathrm{Hom}_R(C, R)$. By RC, $\mathrm{Hom}_R(C, R) = \mathrm{Hom}_R(C, C)$. The latter has a ring structure given by composition. The converse follows from the "change of rings isomorphism"

$$\mathrm{Hom}_{\tilde{R}}(M, \mathrm{Hom}_R(\tilde{R}, N)) \cong \mathrm{Hom}_R(M, N),$$

where M is a finitely generated $\tilde{R}$-module, and N is an R-module. For the ideal C is easily seen to be an $\tilde{R}$-module, and thus taking $M = C$ and $N = R$ we get

$$\mathrm{Hom}_{\tilde{R}}(C, C) = \mathrm{Hom}_R(C, R);$$

thus $\mathrm{Hom}_R(C, R) = \mathrm{Hom}_{\tilde{R}}(C, C) \subseteq \mathrm{Hom}_R(C, C) \subseteq \mathrm{Hom}_R(C, R)$. Evidently these containments must be equalities, and RC holds. □

The second statement in Proposition 11.13 is what de Jong and van Straten call the Ring Condition, RC. It has already made an appearance in the proof of Theorem 11.2.

11.6 Corank 1 Mappings: Cyclic Extensions

Let $(D, 0)$ be a hypersurface in $\mathbb{C}^{n+1}$ and $f : (X, S) \to (D, 0)$ a finite analytic map-germ, with S a finite set. We say that the resulting algebra morphism $\mathcal{O}_{D,0} \to \mathcal{O}_{X,S}$ is a *cyclic extension* if there is an element $g \in \mathcal{O}_{X,S}$ such that the powers $1 = g^0, g, g^2, \ldots, g^m$ generate $\mathcal{O}_{X,S}$ over $\mathcal{O}_D$. In this case we call the ordered $m + 1$-tuple $(1, g, \ldots, g^m)$ a *cyclic generating set.*

Example 11.5

(i) The definition is motivated by the case of finite corank 1 maps $(\mathbb{C}^n, 0) \to (\mathbb{C}^{n+1}, 0)$; in suitable ("linearly adapted") coordinates, such maps take the form

$$(x, y) \mapsto (x, f_n(x, y), f_{n+1}(x, y)),$$

where $x = (x_1, \ldots, x_{n-1})$, $f_n, f_{n+1} \in y\,\mathfrak{m}_n$, and $f_n(0, y) = y^{m+1}$ for some $m \in \mathbb{N}$. In this case,

$$\mathcal{O}_n / f^* \mathfrak{m}_{n+1} \mathcal{O}_n = \mathrm{Sp}\{1, y, \ldots, y^m\},$$

so that the extension $\mathcal{O}_D \to \mathcal{O}_n$ is cyclic, with $g = y$.

(ii) If $(D, 0)$ is the union of m hypersurfaces $(D_i, 0), 0 = 1, \ldots, m$ in $(\mathbb{C}^{n+1}, 0)$ and X is their disjoint union, then the morphism $\mathcal{O}_{D,0} \to \mathcal{O}_{X,\mathbf{0}_{m+1}}$ (where $\mathbf{0}_{m+1}$ is the set of base-points in the disjoint union of the hypersurfaces) is a cyclic extension. For since each ring $\mathcal{O}_{D_i,0}$ is a quotient of $\mathcal{O}_{D,0}$, it is generated over $\mathcal{O}_{D,0}$ by its multiplicative unit. Then the powers of the element $g = (1, 2, \ldots, m+1) \in \mathcal{O}_{X,\mathbf{0}_m}$ generate $\mathcal{O}_{X,\mathbf{0}_{m+1}}$ over $\mathcal{O}_{D,0}$, for the van der Monde matrix

$$\begin{pmatrix} 1 & 2 & \cdots & m+1 \\ 1^2 & 2^2 & \cdots & (m+1)^2 \\ \vdots & \vdots & \cdots & \vdots \\ 1^m & 2^m & \cdots & (m+1)^m \end{pmatrix}$$

is invertible.

The following proposition is due to Gruson and Peskine, [GP82, Prop. 2.6].

Proposition 11.14 *Suppose that $f\colon (X, S) \to (\mathbb{C}^{n+1}, 0)$ is a finite map-germ with image D, and that the resulting extension $\mathcal{O}_{D,0} \to \mathcal{O}_{X,S}$ is cyclic. Let λ be the $(m+1) \times (m+1)$ matrix of a presentation of $\mathcal{O}_{X,S}$ over $\mathcal{O}_{D,0}$ with respect to a cyclic generating set $(1, g, \ldots, g^m)$. Let $\lambda^{\widehat{[0,k-1]}}$ be the matrix λ with rows $0, \ldots, k-1$ deleted, and let $\mathcal{F}'_k$ be the ideal of maximal minors of $\lambda^{\widehat{[0,k-1]}}$. Then:*

(i) For $k = 0, \ldots, n-1$, $V(\mathcal{F}_k(f)) = V(\mathcal{F}'_k)$.
(ii) Suppose $\operatorname{codim} M_{k+1}(f) = k+1$, *and* $\operatorname{codim} M_{k+2}(f) > k+1$, *then*

$$\mathcal{F}_k(f) = \mathcal{F}'_k. \tag{11.26}$$

Observe that Corollary 11.5 says that $\mathcal{F}_1(f) = \mathcal{F}'_1(f)$, without any hypothesis on the codimension.

Proof We may choose a representative of f, say $f : U \to V$, such that $f_* \mathcal{O}_U$ has presentation

$$\mathcal{O}_V^{m+1} \xrightarrow{\lambda} \mathcal{O}_V^{m+1} \longrightarrow f_* \mathcal{O}_U \longrightarrow 0\,.$$

Each column $(\lambda_j^0, \dots, \lambda_j^m)^t$ of λ is a relation among the g^i:

$$\lambda_j^0 \cdot g^0 + \cdots + \lambda_j^m \cdot g^m = 0.$$

Now $V(\mathscr{F}_k(f))$ is the set of points $z \in V$ such that $f_*(\mathscr{O}_U)_z$ requires at least $k+1$ generators over $\mathscr{O}_{V,z}$. Suppose that $k < m$. We claim that if $z \notin V(\mathscr{F}_k(f))$, then $f_*(\mathscr{O}_U)_z$ is generated by $g^0, \dots, g^{k-1}$.

To see this, note first that since no more than k generators are needed, the $\mathbb{C}$-dimension of the quotient

$$(f_* \mathscr{O}_U)_z / \mathfrak{m}_{V,z}(f_* \mathscr{O}_U)_z$$

is no greater than k, and thus there are non-trivial $\mathbb{C}$-linear relations between the classes of $g^0, \dots, g^m$ in the quotient. Let s be the lowest power such that there is a relation

$$g^s = \lambda^0 g^0 + \cdots + \lambda^{s-1} g^{s-1} \mod \mathfrak{m}_{V,z}(f_* \mathscr{O}_U)_z.$$

By repeatedly multiplying this relation by g, we see that all g^t, for $t \geq s$, can be expressed as $\mathbb{C}$-linear combinations of $g^0, \dots, g^{s-1}$ modulo $\mathfrak{m}_{V,z}(f_* \mathscr{O}_U)_z$. Hence

$$f_*(\mathscr{O}_U)_z = \sum_{i=0}^{m} \mathscr{O}_V \cdot g^i = \sum_{i=0}^{s-1} \mathscr{O}_V \cdot g^i + \mathfrak{m}_{V,z}\, f_*(\mathscr{O}_U)_z$$

from which it follows by Nakayama's lemma that $g^0, \dots, g^{s-1}$ generate $f_*(\mathscr{O}_U)_z$ over $\mathscr{O}_{V,z}$. By our assumption on s, the classes of $g^0, \dots, g^{s-1}$ in $(f_* \mathscr{O}_U)_z / \mathfrak{m}_{V,z}(f_* \mathscr{O}_U)_z$ are linearly independent. Hence $s \leq k$.

It follows that

$$V(\mathscr{F}_k(f)) = \operatorname{supp}\left\{ f_*(\mathscr{O}_U) / \left(\sum_{i=0}^{s-1} \mathscr{O}_V \cdot g^i \right) \right\}.$$

The right-hand side here is also equal to $V(\mathcal{F}'_k)$; for $\lambda^{\widehat{[0,k-1]}}$ is a presentation of $f_*(\mathscr{O}_U) / \left(\sum_{i=0}^{s-1} \mathscr{O}_V \cdot g^i \right)$, and thus

$$\mathcal{F}'_k = \operatorname{Fitt}_0^{\mathscr{O}_V} \left\{ f_*(\mathscr{O}_U) / \left(\sum_{i=0}^{s-1} \mathscr{O}_V \cdot g^i \right) \right\}.$$

The 0'th Fitting ideal always defines the support of a sheaf. This proves (i).
(ii) Now suppose that the codimension of $M_{k+1}(f)$ is $k+1$. Then $\mathcal{F}'_k$ is determinantal, and hence unmixed. To prove that it coincides with $\mathcal{F}_k$, by Proposition C.5 it is

therefore enough to prove that they coincide generically. Here we make use of the hypothesis that codim $M_{k+2}(f) \geq k+2$. This implies that no irreducible component of $M_{k+1}(f)$ is contained in $M_{k+2}(f)$. At points $z \in M_{k+1}(f) \smallsetminus M_{k+2}(f)$, $f_*(\mathcal{O}_U)_z$ is generated by $g^0, \ldots, g^k$, by the argument of the first part of the proof, so for each $\ell \geq k+1$, there is a relation

$$g^{\ell} = \sum_{i=0}^{k} \beta_{\ell}^{i} g^{i}, \quad \beta_{\ell}^{i} \in \mathcal{O}_{V,z} \text{ for } i = 0, \ldots, k. \tag{11.27}$$

It follows that by column operations we can bring the germ at z of the matrix λ to the form

$$\left(\begin{array}{c|c} A & B \\ \hline 0 & C \end{array}\right)$$

with C invertible and of size $m-k$. The ideal $\mathcal{F}_k$ is then generated at z by the entries of the matrix A, whereas the ideal $\mathcal{F}_k'$ is generated by the entries in the last row of A.

We can also characterise $\mathcal{F}_k'$ as the ideal generated by the coefficients α_k in all relations

$$\alpha_0 + \alpha_1 g + \cdots + \alpha_k g^k = 0 \tag{11.28}$$

in $f_*(\mathcal{O}_U)_z$, for all such relations are linear combinations of the columns of A. We complete the proof by showing that *all* of the coefficients α_i in such a relation in fact lie in $\mathcal{F}_k'$. We show this inductively. Suppose it is true that for some j with $1 \leq j < k+1$, we have $\alpha_{j+1}, \ldots, \alpha_k \in \mathcal{F}_k'$. Multiplying the relation (11.28) by g^{k-j} we get

$$0 = \alpha_0 g^{k-j} + \cdots + \alpha_j g^k + \alpha_{j+1} g^{k+1} + \cdots + \alpha_k g^{2k-j}.$$

By using the relations (11.27) we can bring this to the form

$$0 = \gamma_0 + \gamma_1 g + \cdots + (\alpha_j + \beta_{k+1}^k \alpha_{j+1} + \cdots + \beta_{2k-j}^k \alpha_k) g^k$$

for some coefficients $\gamma_0, \ldots, \gamma_{k-1} \in \mathcal{O}_{V,z}$. It follows that $\alpha_j + \beta_1^k \alpha_{j+1} + \cdots + \beta_{k-j}^k \alpha_k \in \mathcal{F}_k'$, and thus, by induction hypothesis, that $\alpha_j \in \mathcal{F}_k'$.

This shows that $\mathcal{F}_k = \mathcal{F}_k'$ at z, and thus that $\mathcal{F}_k$ and $\mathcal{F}_k'$ are generically equal. Because $\mathcal{F}_k'$ is unmixed, this completes the proof of (ii). □

Note Proposition 11.14 was first proved by Gruson and Peskine in [GP82, Prop. 2.6] by a different argument. The proof we have given here is a simplification on the proof in [MP89].

11.7 Duality and Symmetric Presentations

We review in this section some results of Scheja and Storch in [SS75] about duality in relative complete intersections, and then go on to generalise to the case of Gorenstein spaces. Scheja and Storch consider at the same time the global case (of polynomial rings) and the local case (of analytic algebras). For simplicity, we will restrict ourselves here to the local case only.

Let $\phi : A \to B$ be a morphism of analytic $\mathbb{C}$-algebras A and B, such that B is A-finite and non-degenerate (that is, $\dim B = \dim A + \dim B/\mathfrak{m}_A B$). Phrased in geometric rather than algebraic terms, this says that if $A = \mathcal{O}_X$ and $B = \mathcal{O}_Y$, then the fibre of the morphism $Y \to X$ inducing ϕ has dimension equal to $\dim Y - \dim X$. Choose generators $g_0, g_1, \ldots, g_m$ of B as A-module such that $g_0 = 1$ and $g_i \in \mathfrak{m}_B$ for $i > 0$. Then there is an induced epimorphism of A-algebras $\rho : A\{z\} \to B$, where

$$A\{z\} = A\{z_1, \ldots, z_m\} := A\widehat{\otimes}_{\mathbb{C}}\mathbb{C}\{z_1, \ldots, z_m\}$$

and $\rho(z_i) = g_i$, for $i = 1, \ldots, m$.

Since $A\{z\}$ has dimension $\dim A + m$, at least m elements are needed to generate the kernel of ρ as an ideal in $A\{z\}$ (see the definition of *slicing dimension* and Theorem C.2 in Appendix C).

Definition 11.6 In these circumstances, we say that $\rho : A\{z\} \to B$ is a representation of B as a *complete intersection over A* if the kernel of ρ is generated by m elements.

If A and B are analytic algebras, and the algebra homomorphism $\phi : A \to B$ makes B into a finite A-algebra, then [SS75, §2] give the following simple criterion for ρ to be a representation of B as complete intersection over A.

Theorem 11.3 *Let $\phi : A \to B$ be a morphism of analytic $\mathbb{C}$-algebras A and B, such that B is A-finite. Then B a complete intersection over A if the following two conditions hold:*

1. *B is A-free,*
2. *$B/\mathfrak{m}_A B$ is a 0-dimensional complete intersection in the usual sense.* □

Example 11.6 Let $F\colon (X, x_0) \to (\mathbb{C}^n, 0)$ be a finite germ, where (X, x_0) is an n-dimensional complete intersection (in the usual sense). Then $\mathcal{O}_{X,x_0}$ is a complete intersection over $\mathcal{O}_n$ with respect to the induced morphism $F^* : \mathcal{O}_n \to \mathcal{O}_{X,x_0}$. In fact, $\mathcal{O}_n$ and $\mathcal{O}_{X,x_0}$ are both Cohen–Macaulay and by the Auslander-Buchsbaum Formula Theorem C.4,

$$\mathrm{pd}_{\mathcal{O}_n}(\mathcal{O}_{X,x_0}) = \mathrm{depth}(\mathcal{O}_n) - \mathrm{depth}_{\mathcal{O}_n}(\mathcal{O}_{X,x_0}) = n - n = 0,$$

so $\mathcal{O}_{X,x_0}$ is $\mathcal{O}_n$-free. Moreover, $\mathcal{O}_{X,x_0}/\mathfrak{m}_n\,\mathcal{O}_{X,x_0} = \mathcal{O}_{X,x_0}/(F_1, \ldots, F_n)$, which is a complete intersection in the usual sense, by the finiteness of F.

The main result of [SS75] is the following duality theorem for relative complete intersections.

Theorem 11.4 ([SS75]) *If B is a complete intersection over A, then there is an isomorphism of B-modules* $\Theta : \mathrm{Hom}_A(B, A) \to B$.

The B-module structure on $\mathrm{Hom}_A(B, A)$ is given by $(b_1 \cdot \psi)(b_2) = \psi(b_1 b_2)$.

The isomorphism Θ described in [SS75] is not canonical; as we will see below, its construction depends on the choice of the A-basis $g_0, \ldots, g_m$ of B and the generators $t_1, \ldots, t_m$ of $\ker \rho$. However, since every B-linear automorphism of B is simply multiplication by a unit, any two isomorphisms $\Theta : \mathrm{Hom}_A(B, A) \to B$ are similarly related, and we will speak of *the* isomorphism Θ.

The isomorphism Θ induces a perfect pairing[2] $B \times B \to A$, given by

$$\langle b|c\rangle := (\Theta^{-1}(b))(c) = (b\eta)(c) = \eta(bc), \qquad \text{for } b, c \in B.$$

Besides A-bilinearity, it has the following two properties:

1. $\langle b|c\rangle = \langle c|b\rangle$, for all $b, c \in B$ (symmetry),
2. $\langle ab|c\rangle = \langle b|ac\rangle$, for all $a, b, c \in B$ (compatibility with multiplication).

Given the basis $g_0, \ldots, g_m$ of B over A, the symmetric $(m+1) \times (m+1)$ matrix $(\langle g_i|g_j\rangle)$ is invertible. It is the transition matrix between the basis $g_0, \ldots, g_m$ and the dual basis $\check{g}_0, \ldots, \check{g}_m$ with respect to $\langle\cdot|\cdot\rangle$. This basis is uniquely determined by the property that $\langle \check{g}_i|g_j\rangle = \delta_{ij}$, so that

$$g_j = \sum_{i=0}^{m} \langle g_i|g_j\rangle \check{g}_i.$$

Scheja and Storch describe the following construction for the isomorphism Θ, or equivalently, for an A-linear homomorphism $\eta : B \to A$ generating $\mathrm{Hom}_A(B, A)$ over B, using the *trace form* of B over A, $\mathrm{Tr} \in \mathrm{Hom}_A(B, A)$, defined by

$$\mathrm{Tr}(b) = \text{trace of the } A\text{-endomorphism of } B \text{ given by multiplication by } b.$$

If η is to be a generator of $\mathrm{Hom}_A(B, A)$, Tr must be a multiple of η; in fact the coefficient is precisely the Jacobian determinant of the generators $t_1, \ldots, t_m$ of $\ker(\rho)$ with respect to the variables $z_1, \ldots, z_m$.

$$J = \rho\left(\frac{\partial(t_1, \ldots, t_m)}{\partial(z_1, \ldots, z_m)}\right).$$

[2] That is, an A-bilinear form $\langle\cdot|\cdot\rangle : B \times B \to A$ with the property that the map $B \to \mathrm{Hom}_A(B, A)$ given by $b \mapsto \langle b|\cdot\rangle$ is an isomorphism of A-modules.

To state this correctly, we have to work with K and L, the total rings of fractions of A and B, respectively. Let $\overline{\mathrm{Tr}} \in \mathrm{Hom}_K(L, K)$ be the corresponding trace form of L over K. Then

$$\eta(b) := \overline{\mathrm{Tr}}(b/J)$$

defines an A-linear homomorphism $B \to K$, which we denote by Tr/J.

Theorem 11.5 ([SS75, 4.2]) *With the above notation, $Tr/J \in \mathrm{Hom}_A(B, A)$, and generates $\mathrm{Hom}_A(B, A)$ as B-module.*

So to find a generator η, one computes the trace Tr and then obtains η by dividing by J. In particular, we always have

$$\eta(J) = \overline{\mathrm{Tr}}(J/J) = \overline{\mathrm{Tr}}(1) = m + 1$$

and thus

Corollary 11.8 *Any $\eta \in \mathrm{Hom}_A(B, A)$ such that $\eta(J)$ is a unit in A, is a generator of $\mathrm{Hom}_A(B, A)$ as B-module, and induces a perfect A-bilinear pairing $\langle \_|\_ \rangle : B \times B \to A$.*

Example 11.7 Let $f \in \mathscr{O}_n$ have isolated singularity at 0, so $\mathscr{O}_n / J_f$ is a 0-dimensional complete intersection ring. Let H be the Hessian determinant $\det(\partial^2 f/\partial x_i \partial x_j)_{1\le i,j\le n}$. Then any linear form $\ell : \mathscr{O}_n / J_f \to \mathbb{C}$ such that $\ell(H) \neq 0$ induces a perfect $\mathbb{C}$-valued pairing on $\mathscr{O}_n / J_f$, $\langle a|b\rangle = \ell(ab)$. ◊

Example 11.8 Let $F\colon (\mathbb{C}^2, 0) \to (\mathbb{C}^2, 0)$ be the finite germ $F(x, y) = (x, xy + y^3)$ (the Whitney cusp map). Let $B = \mathbb{C}\{x, y\}$ and $A = \mathbb{C}\{X, Y\}$ be the local rings of the domain and target of F. We take $1, y, y^2$ as basis for B as A-module. With respect to this basis, multiplication by 1, y and y^2 have matrices

$$\begin{pmatrix} 1 & 0 & 0 \\ 0 & 1 & 0 \\ 0 & 0 & 1 \end{pmatrix}, \quad \begin{pmatrix} 0 & 0 & Y \\ 1 & 0 & -X \\ 0 & 1 & 0 \end{pmatrix} \quad \begin{pmatrix} 0 & Y & 0 \\ 0 & -X & Y \\ 1 & 0 & -X \end{pmatrix}$$

respectively, so $\mathrm{Tr}(1) = 3$, $\mathrm{Tr}(y) = 0$, $\mathrm{Tr}(y^2) = -2X$. Multiplication by J has matrix

$$[J] := \begin{pmatrix} X & 3Y & 0 \\ 0 & -2X & 3Y \\ 3 & 0 & -2X \end{pmatrix}$$

so the matrix of multiplication by J^{-1} is the inverse of this,

$$\frac{1}{4X^3+27Y^2}\begin{pmatrix}4X^2 & 6XY & 9Y^2\\ 9Y & -2X^2 & -3XY\\ 6X & 9Y & -2X^2\end{pmatrix}. \tag{11.29}$$

Thus

$$\begin{aligned}
\eta(1) &= \mathrm{Tr}[J^{-1}][1] = 0\\
\eta(y) &= \mathrm{Tr}[J^{-1}][y] = \mathrm{Tr}\left(\frac{1}{4X^3+27Y^2}\begin{pmatrix}4X^2 & 6XY & 9Y^2\\ 9Y & -2X^2 & -3XY\\ 6X & 9Y & -2X^2\end{pmatrix}\begin{pmatrix}0 & 0 & Y\\ 1 & 0 & -X\\ 0 & 1 & 0\end{pmatrix}\right)\\
&= 0\\
\eta(y^2) &= \mathrm{Tr}[J^{-1}][y^2] = \mathrm{Tr}\left(\frac{1}{4X^3+27Y^2}\begin{pmatrix}4X^2 & 6XY & 9Y^2\\ 9Y & -2X^2 & -3XY\\ 6X & 9Y & -2X^2\end{pmatrix}\begin{pmatrix}0 & Y & 0\\ 0 & -X & Y\\ 1 & 0 & -X\end{pmatrix}\right)\\
&= \frac{1}{4X^3+27Y^2}(4X^3+27Y^2) = 1.
\end{aligned}$$

If $a, b \in \mathbb{C}\{x, y\}$, then $\langle a|b\rangle = \eta(ab)$ is thus the coefficient of y^2 in the expression of ab with respect to the basis $1, y, y^2$. The matrix of the pairing $\langle\cdot|\cdot\rangle$ with respect to this basis is therefore

$$\begin{pmatrix}0 & 0 & 1\\ 0 & 1 & 0\\ 1 & 0 & -X\end{pmatrix}$$

and the basis dual to $\{1, y, y^2\}$ is $\{y^2 + x, y, 1\}$.

The duality theorem 11.4 allows us to adapt the argument in the proof of Proposition 11.1 to prove the following result (see [Cat84, MP89]):

Theorem 11.6 *Let* $f\colon (X, x_0) \to (\mathbb{C}^{n+1}, 0)$ *be finite, where* (X, x_0) *is an* n*-dimensional complete intersection. Then the matrix* λ *of a presentation of* $\mathscr{O}_{X,x_0}$ *over* $\mathscr{O}_{n+1}$ *can be chosen symmetric.*

Proof We replace the diagram (11.7) by a second diagram in which the two isomorphisms of $\mathscr{O}_S = \mathscr{O}_{X\times\mathbb{C},(x_0,0)}$ with $\mathscr{O}_T^{m+1} = \mathscr{O}_{n+1}^{m+1}$ are no longer assumed to be the same. By Example 11.6, $\mathscr{O}_S$ is a complete intersection over $\mathscr{O}_T$ with respect to the induced morphism $F^*: \mathscr{O}_T \to \mathscr{O}_S$. By Theorem 11.4, there exists a perfect symmetric $\mathscr{O}_T$-bilinear pairing $\langle\cdot|\cdot\rangle : \mathscr{O}_S \times \mathscr{O}_S \to \mathscr{O}_T$.

For each basis $G := \{g_0, \dots, g_m\}$ for $\mathcal{O}_S$ as $\mathcal{O}_T$-module there is a dual basis $\check{G} := \{\check{g}_0, \dots, \check{g}_m\}$ with the property that $\langle \check{g}_i | g_j \rangle = \delta_{ij}$. Let $\lambda := [t]_G^{\check{G}}$ denote the matrix of multiplication by t with respect to $\check{G}$ in the source and G in the target. By the argument of the proof of Proposition 11.1, λ is the matrix of a presentation of $\mathcal{O}_{X,x_0}$ over $\mathcal{O}_{n+1}$. We claim that it is symmetric. In fact, if $\lambda = (\lambda^i_j)$, then by construction

$$t\check{g}_j = \sum_{j=0}^{m} \lambda^i_j g_i,$$

which implies

$$\langle t\check{g}_j | \check{g}_k \rangle = \sum_{j=0}^{m} \lambda^i_j \langle g_i | \check{g}_k \rangle = \lambda^k_j.$$

Since $\langle t\check{g}_j | \check{g}_k \rangle = \langle t\check{g}_k | \check{g}_j \rangle$, it follows that $\lambda^k_j = \lambda^j_k$ for all $j, k = 0, \dots, m$. □

The symmetric presentation matrix we have just obtained is minimal only when $f_{n+1} \subset (f_1, \dots, f_n)\,\mathcal{O}_n$. We can reduce it to a minimal symmetric presentation by means of symmetric row and column operations, as follows. Each column operation is effected by post-multiplying λ by the "editing matrix" E obtained from the identity matrix by performing this same operation. In order to preserve the symmetry, for each column operation, we carry out the dual row operation, pre-multiplying by E^t. The same goes for row operations, *mutatis mutandis*. Suitably chosen symmetric operations bring λ to the form

$$\begin{pmatrix} I_\ell & 0 \\ 0 & \tilde{\lambda} \end{pmatrix}$$

with $\tilde{\lambda}$ symmetric and with all entries in $\mathfrak{m}_{n+1}$. Then $\tilde{\lambda}$ is a symmetric minimal presentation matrix.

11.7.1 Gorenstein Rings and Symmetric Presentations

The conclusion of Theorem 11.6 holds in greater generality than just stated; in fact there is a symmetric presentation if (and only if) $\mathcal{O}_{X,x_0}$ is *Gorenstein*. We give only a minimal discussion of this property here. There are several equivalent ways of defining Gorenstein rings. For our purposes, it is most convenient to use the characterisation as those Cohen–Macaulay local rings R for which R itself is a *canonical module*. The notion of canonical module is easiest to define (though this is only to scratch the surface of the topic) in the 0-dimensional case. Suppose that R is a local ring of dimension 0, containing a field k such that the composite $k \hookrightarrow R \to R/\mathfrak{m}$ is an isomorphism. Consider the dual $D(R) := \mathrm{Hom}_k(R, k)$. This is an R-module with the action of $r \in R$ on a homomorphism $\varphi \in D(R)$ defined by $(r \cdot \varphi)(r_1) = \varphi(rr_1)$. We declare $D(R)$, with this R-module structure, to be a canonical module for R. In [Eis95], Eisenbud defines a canonical module for a Cohen–Macaulay ring R of dimension $n > 0$ by induction on n: M is a canonical module for R if M/rM is a canonical module for R/rR, where $r \in \mathfrak{m}$ is a regular element. In [Eis95, Corollary 21.14] he shows that any two canonical modules for the same ring are isomorphic, so we may speak of "the canonical module" of the ring R. It is usually denoted ω_R. It follows from the definition that an n-dimensional Cohen–Macaulay local ring R is Gorenstein if and only if $R/(r_1, \ldots, r_n)R$ is a zero-dimensional Gorenstein ring, where $r_1, \ldots, r_n$ is a maximal regular sequence in $\mathfrak{m}$. Evidently, a field k is a zero-dimensional Gorenstein ring (with maximal ideal (0)), and from this it follows (using Eisenbud's definition) that a regular local ring is Gorenstein (Exercise11.7.3 below).

Recall that for R-modules M and N, $\mathrm{Ext}^k_R(M, N)$ is the k'th homology of a free resolution $F_\bullet$ of M, dualised with respect to N:

$$\mathrm{Ext}^k(M, N) = H^k(\mathrm{Hom}_R(F_\bullet, N)).$$

Here it is conventional to use the upper index k because by dualising we reverse the arrows. The key to the proof of Theorem 11.6, when $\mathcal{O}_{X,x_0}$ is Gorenstein, is the following theorem (see, e.g., [Eis95, Theorem 21.15]).

Theorem 11.7 *Let A be a local Cohen–Macaulay ring with canonical module ω_A. If B is a Cohen–Macaulay local A-algebra that is finitely generated as A-module, with* $\dim A - \dim B = d$, *then* $\mathrm{Ext}^d(B, \omega_A)$ *is a canonical B-module.* □

Suppose that $S = (Z, z_0)$ is Gorenstein (i.e. $\mathcal{O}_{Z,z_0}$ is a Gorenstein ring) of dimension $n+1$ and that $f : S \to T = (\mathbb{C}^{n+1}, 0)$ is a finite map-germ. We know that $\mathcal{O}_{n+1}$ is Gorenstein. Now we apply Theorem 11.7. Since here $d = 0$, we consider $\mathrm{Ext}^0_T(\mathcal{O}_S, \omega_T)$, i.e. $\mathrm{Hom}_T(\mathcal{O}_S, \mathcal{O}_T)$. We get

$$\mathcal{O}_S \simeq \omega_S \cong \mathrm{Hom}_T(\mathcal{O}_S, \omega_T) \cong \mathrm{Hom}_T(\mathcal{O}_S, \mathcal{O}_{n+1}).$$

The composite of these isomorphisms is exactly what is needed to construct the perfect pairing

$$\mathcal{O}_S \times \mathcal{O}_S \to \mathcal{O}_T$$

which gives rise to the symmetry of the presentation, in the argument of Theorem 11.6. As in the proof of Theorem 11.6, given Gorenstein (X, x_0) we take $S = (X \times \mathbb{C}, (x_0, 0))$, which is also Gorenstein. Thus, we have

Theorem 11.8 *Suppose that (X, x_0) is a germ of n-dimensional Cohen–Macaulay space, and $f: (X, x_0) \to (\mathbb{C}^{n+1}, 0)$ is a finite map. If $\mathcal{O}_{X,x_0}$ is a Gorenstein ring, then $\mathcal{O}_{X,x_0}$ has a symmetric presentation matrix over $\mathcal{O}_{n+1}$.* □

For an illuminating discussion of Gorenstein rings, see [Eis95]. Even the zero-dimensional case is surprisingly rich.

As we already mentioned in Sect. 11.3, if $f : (\mathbb{C}^n, 0) \to (\mathbb{C}^{n+1}, 0)$ is finite and generically 1–1, then $f^*\mathcal{F}_1(f) = \mathcal{C}$, the conductor ideal, and we know from Proposition 8.5 that this ideal is principal if f is $\mathcal{A}$-finite. Now we can obtain this fact as a corollary of the symmetry of λ, without the $\mathcal{A}$-finiteness assumption.

Corollary 11.9 *Let $f: (X, x_0) \to (\mathbb{C}^{n+1}, 0)$ be finite and generically 1–1, where (X, x_0) is an n-dimensional Gorenstein space-germ. Then $f^*\mathcal{F}_1(f)$ is a principal ideal in $\mathcal{O}_{X,x_0}$.*

Proof Choose a symmetric presentation λ, with respect to generators $g_0 = 1, \ldots, g_m$. Then in the language of the proof of Proposition 11.11, $\mathcal{F}_1(f)$ is generated by $(m_0^0, \ldots, m_m^0)$, and so $f^*\mathcal{F}_1(f)$ is generated by $f^*(m_0^0), \ldots, f^*(m_m^0)$. Since $m_j^i = m_j^0 g_i$ (Exercises 11.3.1) it follows, by the symmetry of λ, that $f^*\mathcal{F}_1(f)$ is generated by $f^*(m_0^0)$. □

Example 11.9

1. Let $f: (\mathbb{C}^2, 0) \to (\mathbb{C}^3, 0)$ be defined by

$$f(x, y) = (x, xy + y^3, xy^2 + cy^4).$$

(this is a member of the unimodal family P_3 of [Mon85]). Take $\pi(Y_1, Y_2, Y_3) = (Y_1, Y_2)$; then $\mathcal{O}_2$ (source) is generated over $\mathcal{O}_2$ (target) by the classes of $1, y, y^2$. We have

$$\begin{aligned} f_3 &= xy^2 + cy^4 = 01 + (1-c)Y_1Y_2y + cY_2^2y^2 \\ yf_3 &= xy^3 + cy^5 = cY_21 + (c-1)Y_1^2y + (1-2c)Y_1Y_2y^2 \\ y^2f_3 &= xy^4 + cy^6 = (1-c)Y_11 + cY_2y + (c-1)Y_1^2y^2 \end{aligned}$$

so as matrix of the presentation we obtain

$$\begin{pmatrix} -Y_3 & (1-c)Y_1Y_2 & cY_2^2 \\ cY_2 & (c-1)Y_1^2 - Y_3 & (1-2c)Y_1Y_2 \\ (1-c)Y_1 & cY_2 & (c-1)Y_1^2 - Y_3 \end{pmatrix}.$$

By adding $Y_1\times$(column 1) to (column 3) and then interchanging columns 1 and 3, we obtain the symmetric matrix λ

$$\begin{pmatrix} cY_2^2 - Y_1Y_3 & (1-c)Y_1Y_2 & -Y_3 \\ (1-c)Y_1Y_2 & (c-1)Y_1^2 - Y_3 & cY_2 \\ -Y_3 & cY_2 & (1-c)Y_1 \end{pmatrix}.$$

2. In Example 11.8 we gave an example of a weighted homogeneous stable map-germ. The following is a symmetric presentation:

$$\begin{pmatrix} a_{11} & a_{12} & a_{13} & a_{14} \\ a_{12} & a_{22} & a_{23} & a_{24} \\ a_{13} & a_{23} & a_{33} & a_{34} \\ a_{14} & a_{24} & a_{34} & a_{44} \end{pmatrix}, \tag{11.30}$$

where

$$\begin{aligned}
a_{11} &= -XY - ZAC - Z^2E + YAE - XDE + XCF, \\
a_{12} &= XC + ZD + ZE^2 - YF, \\
a_{13} &= Z^2 + YA - ZBC + YBE + ZEF, \\
a_{14} &= YB - XE - ZF, \\
a_{22} &= -Y - DE - E^3 + CF, \\
a_{23} &= -AC - ZE - E^2F, \\
a_{24} &= Z - BC + EF, \\
a_{33} &= YB - AD - XE - BDE - 2ZF + BCF - EF^2, \\
a_{34} &= -X - BD - AE + F^2, \\
a_{44} &= A + BE.
\end{aligned}$$

11.7.2 Geometrical Interpretation of the Trace Homomorphism

Let $F\colon (S,0) \to (T,0)$ be finite with $\mathcal{O}_{S,0}$ free over $\mathcal{O}_{T,0}$, and let $F: S \to T$ be a proper representative. Let $g_0, \ldots, g_m$ minimally generate $F_*(\mathcal{O}_S)$ over $\mathcal{O}_T$. For $b \in \mathcal{O}_S$, let $[b]_G^G$ denote the matrix of the $\mathcal{O}_T$-linear endomorphism of $\mathcal{O}_S$ given by multiplication by b.

Lemma 11.6 *Let $b \in \mathcal{O}_{S,0}$. Suppose that $y_0 \in T$ is a regular value of F and let $x_0, \ldots, x_m$ be its preimages in S. Then $b(x_0), \ldots, b(x_m)$ are the eigenvalues of the matrix $[b]_G^G(y_0)$ (i.e. $[b]_G^G$ evaluated at y_0).*

Proof By the coherence of $F_*\mathcal{O}_S$, after possibly shrinking S and T, the $\mathcal{O}_{T,0}$-isomorphism

$$\mathcal{O}_{T,0}^{m+1} \xrightarrow{G} \mathcal{O}_{S,0}$$

defined by the basis $g_0 \dots, g_m$ extends to an isomorphism

$$\mathscr{O}_T^{m+1} \xrightarrow{G} F_* \mathscr{O}_S \,. \tag{11.31}$$

By Exercise E.1.4(ii), $(F_* \mathscr{O}_S)_{y_0} = \bigoplus_{i=0}^m \mathscr{O}_{S,x_i}$. For $i = 0, \dots, m$, let $q_i \in (F_* \mathscr{O}_S)_{y_0}$ be the multi-germ of function whose value in a small neighbourhood of each point x_j is δ_{ij}. Then there exist $a_{0\,i}, \dots, a_{m\,i} \in \mathscr{O}_{T,y_0}$ such that $q_i = \sum_j (a_{j\,i} \circ F) g_j$. We claim that the vector $(a_{0\,i}(y_0), \dots, a_{m\,i}(y_0))^t$ is an eigenvector of $[b]_G^G(y_0)$ with eigenvalue $b(x_i)$. Since y_0 is a regular value of F, the germ $F\colon (S, x_i) \to (T, y_0)$ is an isomorphism, and therefore there exists $c_i \in \mathscr{O}_{T,y_0}$ such that $b = F^* c_i$ in $\mathscr{O}_{S,x_i}$. Since $q_i \equiv 0$ in $\mathscr{O}_{S,x_j}$ for $j \neq i$, we also have $bq_i = (F^* c) q_i$ in $\mathscr{O}_{S,x_j}$ for $j = 0, \dots, m$. It follows that

$$bq_i = [g_0, \dots, g_m][b]_G^G \begin{pmatrix} a_0 \\ \vdots \\ a_m \end{pmatrix} = [g_0, \dots, g_m] \begin{pmatrix} c_i a_{0\,i} \\ \vdots \\ c_i a_{m,i} \end{pmatrix}$$

in $(F_* \mathscr{O}_S)_{y_0}$. Because there are no relations among the g_i over $\mathscr{O}_T$, we must have

$$[b]_G^G \begin{pmatrix} a_{0\,i} \\ \vdots \\ a_{m\,i} \end{pmatrix} = c_i \begin{pmatrix} a_{0\,i} \\ \vdots \\ a_{m\,i} \end{pmatrix}$$

and thus $(a_{0\,i}(y_0), \dots, a_{m\,i}(y_0))^t$ is an eigenvector of $[b]_G^G(y_0)$ with eigenvalue $b(x_i)$, as claimed. □

We have already defined the trace homomorphism $\mathrm{Tr} : \mathscr{O}_S \to \mathscr{O}_T$; the *norm* is a multiplicative homomorphism $n : \mathscr{O}_S \to \mathscr{O}_T$ defined by $n(b) = \det[b]_G^G$. According to Exercise 2(ii), $n(b)$ defines the image under F of $\{b = 0\} \subset S$.

Corollary 11.10 *Let $F\colon (\mathbb{C}^n, 0) \to (\mathbb{C}^n, 0)$ be finite, and let $F : S \to T$ be a proper representative. Let $g_0, \dots, g_m$ be a basis for $\mathscr{O}_{S,0}$ over $\mathscr{O}_{T,0}$ and, for $b \in \mathscr{O}_{S,0}$, let $[b]_G^G$ be the matrix of multiplication by b with respect to the basis $g_0, \dots, g_m$. Then*

(i) $\mathrm{Tr}\,[b]_G^G(y) = \sum_{x \in F^{-1}(y)} \mathrm{mult}_x(F) b(x)$;
(ii) $n(b)(y) = \prod_{x \in F^{-1}(y)} (b(x))^{\mathrm{mult}_x(F)}$.

Proof The trace of a square matrix is the sum of its eigenvalues, and the determinant is their product. Thus, by the lemma, the equalities hold on the set of regular values of F. To conclude that they hold everywhere it is necessary only to show that the functions on the right-hand side are continuous at critical values of F. This holds by continuity of b, conservation of multiplicity, and properness of F: suppose x_0 is a critical value of F with $F(x_0) = y_0$, and let $\varepsilon > 0$. There exists a neighbourhood

$B_\rho(y_0)$ and a neighbourhood $B_\delta(x_0)$ such that for $y \in B_\rho(y_0)$,

$$\sum_{x \in B_\delta(x_0) \cap F^{-1}(y)} \text{mult}_x(F) = \text{mult}_{x_0}(F).$$

By reducing ρ we can reduce δ until for $x \in B_\delta(x_0)$, $|b(x) - b(x_0)| < \varepsilon / \text{mult}_{x_0}(F)$. Then for $y \in B_\rho(y_0)$,

$$\begin{aligned} \left| \text{mult}_{x_0}(F) b(x_0) - \sum_x \text{mult}_x(F) b(x) \right| &\leq \left| \text{mult}_{x_0}(F) - \sum_x \text{mult}_x(F) \right| |b(x_0)| \\ &\quad + \sum_x \text{mult}_x(F) |b(x) - b(x_0)| \\ &< \varepsilon \end{aligned}$$

where the sums are over $x \in B_\delta(x_0) \cap F^{-1}(y)$. Conservation of multiplicity means that the first sum in the middle term of this chain of inequalities is equal to 0. Because f is proper, all preimages of points y near y_0 are near to preimages of y_0. We can argue similarly in the neighbourhood of the other preimages of y_0, replacing ε in the above argument by $\varepsilon / |f^{-1}(y_0)|$. This proves continuity of the right-hand side in the statement of the corollary concerning the trace. The argument for the norm is similar but the estimates are a little more complicated. □

Exercises for Sect. 11.7

1 Show that a 0-dimensional complete intersection ring $R := \mathscr{O}_n(f_1, \ldots, f_n)$ is Gorenstein by using Corollary 11.8 to give an R-linear isomorphism $R \to \text{Hom}_{\mathbb{C}}(R, \mathbb{C})$.

2 Let $F\colon (\mathbb{C}^n, 0) \to (\mathbb{C}^n, 0)$ be finite and consider $B = \mathbb{C}\{x\}$ as a complete intersection over $A = \mathbb{C}\{y\}$.

(i) Show that the Jacobian J of Theorem 11.5 coincides with the Jacobian determinant of F as a map, up to multiplication by a unit.

(ii) Let $b \in B$. Show that the determinant of the A-linear map $B \xrightarrow{b} B$ is an equation for the image under F of the set $\{b = 0\}$. Suggestion: see displayed equation (11.7) above. Explain why this accounts for the appearance of the familiar equation $4X^3 + 27Y^2$ for the discriminant in the denominator in (11.29).

3 Show that a regular local ring is Gorenstein by using the fact that in a regular local ring, the maximal ideal is generated by a regular sequence.

4 Suppose that (X, x_0) is a germ of complete intersection of dimension n defined in $\mathbb{C}^{n+k}$ by equations $h_1, \ldots, h_k$. Use Theorem 11.7 and the free resolution given by the Koszul complex (constructed in Example C.11) to show that a complete intersection is Gorenstein. See also Exercise 5 for an alternative proof.

5 Suppose that (X, x_0) is a germ of n-dimensional Cohen–Macaulay space, and $f\colon (X, x_0) \to (\mathbb{C}^{n+1}, 0)$ is a finite map. Show that if $\mathcal{O}_{X,x_0}$ has a symmetric presentation matrix over $\mathcal{O}_{n+1}$ then $\mathcal{O}_{X,x_0}$ is a Gorenstein ring. Thus, given a finite map $(X, x_0) \to (\mathbb{C}^{n+1}, 0)$, the existence of a symmetric presentation is equivalent to (X, x_0) being Gorenstein.

6 Suppose that A is a quotient of dimension n of a local Gorenstein ring R of dimension $n + k$, and has minimal free resolution

$$0 \longrightarrow R^p \xrightarrow{d_k} R^q \longrightarrow \cdots \longrightarrow R \longrightarrow A \longrightarrow 0$$

of length k. Show that if $p = 1$ then A is itself Gorenstein. Hint: $\mathrm{Ext}^k(A, R)$ is an A-module. It follows that the ideal generated by the entries of d_k contains the ideal defining A. Dualise to prove the opposite inclusion (and see [Mat89, Theorem 17.1] for the exactness of the dualised resolution).

7 Use Exercise 6 to show that a complete intersection is Gorenstein.

8 Let $S := \mathrm{Sym}_n$ be the space of symmetric $n \times n$ matrices, and let I_{n-2} be the ideal generated by the $(n-2) \times (n-2)$ minors of the generic symmetric matrix. Use computer calculations to show, using Exercise 5, that $\mathcal{O}_{S,0}/I_{n-2}$ is a Gorenstein ring for $n = 3, 4$ and 5. Which of these rings is a complete intersection ring?

9 Let T be the space of 5×5 skew-symmetric matrices, and let P_4 be the ideal generated by the 4×4 Pfaffians of the generic matrix in T. Use a computer calculation to show that $\mathcal{O}_{T,0}/P_4$ is a Gorenstein ring (see [BE77a] and [KL80]).

10 Prove the version of the first statement of Corollary C.12 obtained by replacing "Cohen–Macaulay" by "Gorenstein".

11 Show that the symmetric matrix λ obtained in Example 11.9 corresponds to the presentation taken with dual bases $\{1, y, y^2\}$ and $\{y^2 - x, y, 1\}$ (see Example 11.8).

12 Symmetrise the presentations you found in Exercise 11.1 by means of column operations.

13 Suppose that $f\colon (\mathbb{C}^n, 0) \to (\mathbb{C}, 0)$ has isolated singularity. Use Theorem 11.5 to show that if $\Phi : \mathscr{O}_n / J_f \to \mathbb{C}$ is $\mathbb{C}$-linear, then the pairing

$$\frac{\mathscr{O}_n}{J_f} \times \frac{\mathscr{O}_n}{J_f} \to \mathbb{C},\ \langle p, q\rangle = \Phi(pq)$$

is non-degenerate if and only if $\Phi(H) \neq 0$, where H is the class in $\mathscr{O}_n / J_f$ of the Hessian determinant of f. Evidently, it follows that $H \notin J_f$. Now suppose that f is weighted homogeneous with respect to strictly positive weights $w_1, \ldots, w_n$.

(i) Suppose that $E = \{H, g_2, \ldots, g_\mu\}$ is a $\mathbb{C}$-basis for $\mathscr{O}_n / J_f$, with all of the g_i weighted homogeneous. Show that for every non-zero $g \in \mathscr{O}_n / J_f$, there exists $g' \in \mathscr{O}_n / J_f$ such that the coefficient of H in the expression of gg' with respect to the basis E is non-zero.
(ii) Show that for all $g \in \mathfrak{m}_n$, $gH \in J_f$.
(iii) Show that the weighted degree of H is greater than the weight of any other homogeneous member of $\mathscr{O}_n / J_f$.
(iv) Deduce that a stable unfolding of a germ of function $f\colon (\mathbb{C}^3, 0) \to (\mathbb{C}, 0)$ for which $J_f \subset \mathfrak{m}_3^2$ cannot be quasihomogeneous.

In fact (ii) holds even without the hypothesis of weighted homogeneity. The Hessian determinant generates the "socle"—the annihilator of the maximal ideal—of the Milnor algebra $\mathscr{O}_n / J_f$.

11.8 Triple Points in the Target

We have already seen in the previous sections that $M_1(f)$ and $M_2(f)$ are Cohen–Macaulay spaces of codimension 1 and 2, respectively, provided that our map-germ $f\colon (X, x_0) \to (\mathbb{C}^{n+1}, 0)$ is finite and generically one-to-one. We will see in this section that the same is true for triple points, provided that (X, x_0) is Gorenstein, and $M_3(f)$ has the right codimension, namely codimension 3. In fact, we will prove that $M_3(f)$ is again a determinantal variety, not in the usual sense, but when we consider the space of square symmetric matrices. This will be possible because we have seen in Sect. 11.7 that one can choose the matrix presentation λ to be symmetric.

We start with a technical lemma, known as the Laplace identity. Suppose λ is an $(m + 1) \times (m + 1)$ matrix with entries in a commutative ring R. Set $h := \det \lambda$, denote by m^i_j the minor of λ obtained by deleting row i and column j, and denote by $m^{i,k}_{j,\ell}$ that obtained by deleting rows i and k and columns j and ℓ, where $m^{j,\ell}_{i,k} = 0$ if $i = k$ or $j = \ell$ and where the determinant of the empty matrix is 1.

Lemma 11.7 (Laplace Identity) *For $0 \leq i \leq k \leq m$ and $0 \leq j \leq \ell \leq m$,*

$$m^{i,k}_{j,\ell} h = m^i_j m^k_\ell - m^i_\ell m^k_j.$$

Proof Without loss of generality, we may assume that $0 = i < k$ and $0 = j < \ell$. Denote by v_i the column vectors of λ and by e_i the column vectors of the standard basis of R^{m+1}, with $i = 0, \ldots, m$. Then,

$$h = \det(v_0, \ldots, v_m),$$
$$m_{0,\ell}^{0,k} = (-1)^{k+\ell+1} \det(e_k, v_1, \ldots, v_{\ell-1}, e_1, v_{\ell+1}, \ldots, v_m),$$

and for all $s, t = 1, \ldots, n$,

$$m_t^s = (-1)^{s+t} \det(v_0, \ldots, v_{t-1}, e_s, v_{t+1}, \ldots, v_m).$$

Our assertion now translates into

$$\begin{aligned}
&\det(v_0, \ldots, v_m) \det(e_k, v_1, \ldots, v_{\ell-1}, e_1, v_{\ell+1}, \ldots, v_m)\\
&= \det(e_1, v_1, \ldots, v_m) \det(v_0, \ldots, v_{\ell-1}, e_k, v_{\ell+1}, \ldots, v_m)\\
&+ \det(e_k, v_1, \ldots, v_m) \det(v_0, \ldots, v_{\ell-1}, e_1, v_{\ell+1}, \ldots, v_m).
\end{aligned}$$

But this is just a particular case of the following determinantal identity:

$$\begin{aligned}
&\det(x_1, \ldots, x_n) \det(y_1, \ldots, y_n)\\
&= \sum_{s=1}^{n} \det(y_s, x_2, \ldots, x_n) \det(y_1, \ldots, y_{s-1}, x_1, y_{s+1}, \ldots, y_n),
\end{aligned}$$

for any column vectors $x_1, \ldots, x_n, y_1, \ldots, y_n$. This identity is just a particular case of [Ful97, Lemma 2, §8]. We give here a proof adapted to our case. To prove this identity it suffices to prove that the difference of the two sides is an alternating function of the $n + 1$ vectors $y_1, \ldots, y_n, x_1$, since any such function must vanish (as $\bigwedge^{n+1} R^n = 0$). For this it suffices to show that both sides are equal when two successive vectors y_t, y_{t+1} are equal and when $y_n = x_1$. When $y_t = y_{t+1}$, the left-hand side is zero and the right-hand side gives

$$\begin{aligned}
&\det(y_t, x_2, \ldots, x_n)\{\det(y_1, \ldots, y_{t-1}, x_1, y_t, y_{t+2}, \ldots, y_n)\\
&\quad + \det(y_1, \ldots, y_t, x_1, y_{t+2}, \ldots, y_n)\} = 0.
\end{aligned}$$

When $y_n = x_1$, both sides give

$$\det(y_n, x_2, \ldots, x_n) \det(y_1, \ldots, y_n).$$

□

In a similar vein to Proposition 11.11, the next theorem shows that if λ is a symmetric presentation matrix, then $\mathcal{F}_2(f)$ can be computed from the matrix obtained from λ by deleting its first row and column. This theorem appears in [MP89, Theorem 4.1], although the proof here follows the arguments of [KU97, Proposition 3.6].

Theorem 11.9 *Let $f\colon (X, x_0) \to (\mathbb{C}^{n+1}, 0)$ be finite and generically 1-to-1, where (X, x_0) is an n-dimensional Gorenstein germ. Let λ be a symmetric presentation of $\mathcal{O}_{X,x_0}$ over $\mathcal{O}_{n+1}$, with respect to generators $g_0 = 1, g_1, \dots, g_m$. Then $\mathcal{F}_2(f)$ is generated by the $(m-1)\times(m-1)$ minors of the matrix obtained from λ by deleting its first row and column.*

Proof Let λ' be the matrix obtained from λ by deleting the first row and let λ'' be the matrix obtained from λ' by deleting the first column. For simplicity, we put $F_i := \mathcal{F}_i(f)$, for all i, and $F_2'' := \min_{m-1}(\lambda'')$, so we need to show that $F_2 = F_2''$.

By Proposition 11.11, we have $F_1 = \min_m(\lambda')$. Since λ' is a matrix of size $m \times (m+1)$ and $\operatorname{codim} \mathcal{O}_{n+1}/F_1 = 2$, by the Hilbert–Burch Theorem C.5, F_1 admits the following presentation as $\mathcal{O}_{n+1}$-module:

$$\mathcal{O}_{n+1}^m \xrightarrow{(\lambda')^T} \mathcal{O}_{n+1}^{m+1} \xrightarrow{\Delta} F_1 \longrightarrow 0\,,$$

where $\Delta = (c_0^0, \dots, c_m^0)$. This implies $\operatorname{Fitt}_2(F_1) = \min_{m-1}(\lambda')$ and hence

$$F_2'' \subset \operatorname{Fitt}_2(F_1) \subset F_2.$$

To see the other inclusion, we will show that

$$F_2 \subset F_1^2 : (h) \subset F_2''.$$

It follows directly from Laplace identity (Lemma 11.7) that

$$(h)F_2 \subset F_1^2. \tag{11.32}$$

Furthermore, for $i = \ell = 0$, the Laplace identity gives

$$m_{j,0}^{0,k} h = m_0^j m_0^k - m_0^0 m_j^k,$$

(here we use the symmetry of λ). Hence,

$$F_1^2 \subset hF_2'' + m_0^0 F_1. \tag{11.33}$$

Putting together (11.32) and (11.33) we obtain

$$F_2 \subset F_1^2 : h \subset (hF_2'' + (m_0^0)) : h. \tag{11.34}$$

On the other hand, (11.33) also implies that

$$F_1^2 \subset (h, m_0^0) \subset F_1.$$

In particular, F_1 and (h, m_0^0) have the same radical and thus, $\mathcal{O}_{n+1}/(h, m_0^0)$ has codimension 2. This implies that h, m_0^0 form a regular sequence in $\mathcal{O}_{n+1}$.

By definition, for all $b \in (hF_2'' + (m_0^0) : h)$, we have $bh = ch + dm_0^0$, for some $c \in F_2''$ and $d \in \mathcal{O}_{n+1}$. This gives $(b - c)h = dm_0^0$ and since h, m_0^0 form a regular sequence, we must have $d = \beta h$, for some $\beta \in O_{n+1}$. So, $bh = ch + \beta hm_0^0$ and hence, $b = c + \beta m_0^0 \in F_2''$. By (11.34),

$$F_2 \subset F_1^2 : h \subset (hF_2'' + (m_0^0)) : h \subset F_2''.$$

□

By a theorem of Kutz [Kut74] (but see also [Joz78], and see also Theorem C.12 in the Appendix), the variety of zeros of the ideal of submaximal minors of a symmetric $m \times m$ matrix can have codimension no greater than 3, and if the codimension is 3 then the variety in question is Cohen–Macaulay. Thus

Corollary 11.11 *In the circumstances of Theorem 11.9,* $\operatorname{codim} M_3(f) \leq 3$. *Moreover, if* $\operatorname{codim} M_3(f) = 3$*, then* $M_3(f)$ *is Cohen–Macaulay.*

Corollary 11.12 *Let* $f : (\mathbb{C}^2, 0) \to (\mathbb{C}^3, 0)$ *be finite and generically* 1*-to-*1*. If* $M_3(f) = \{0\}$*, then the number of triple points in the image of a stable perturbation of* f *is equal to* $\dim_{\mathbb{C}} \mathcal{O}_{\mathbb{C}^3,0}/\mathcal{F}_2(f)$.

Proof Let $F : (\mathbb{C}^2 \times \mathbb{C}, 0) \to (\mathbb{C}^3 \times \mathbb{C}, 0)$ be a stabilisation of f, with $F(x, s) = (f_s(x), s)$. On the one hand, we know that $M_3(F)$ has codimension ≤ 3. On the other hand, $M_3(F) \cap \{s = 0\} = M_3(f) \times \{0\}$ by Proposition 11.6, so

$$4 = \operatorname{codim}(M_3(f) \times \{0\}) \leq \operatorname{codim}(M_3(F)) + 1$$

and hence $\operatorname{codim} M_3(F) = 3$. Thus, $M_3(F)$ is Cohen–Macaulay by Corollary 11.11. We consider the projection $\pi : M_3(F) \to (\mathbb{C}, 0)$ given by $\pi(y, s) = s$, which is finite. By conservation of multiplicity (see Corollary E.6), we have

$$\dim_{\mathbb{C}} \frac{\mathcal{O}_{\mathbb{C}^3,0}}{\mathcal{F}_2(f)} = \sum_{y \in B_\epsilon(0)} \dim_{\mathbb{C}} \frac{\mathcal{O}_{\mathbb{C}^3,y}}{\mathcal{F}_2(f_s)_y}, \tag{11.35}$$

for all $s \in B_\eta(0)$. For $s \neq 0$, f_s is stable and the only singularities which can appear are simple points, Whitney umbrellas, transverse double points or transverse triple points. All of them, except the transverse triple points, have multiplicity ≤ 2, which implies that $\mathcal{F}_2(f_s)_y = \mathcal{O}_{\mathbb{C}^3,y}$. Otherwise, if $y \in B_\epsilon(0)$ is a transverse triple point of f_s, after $\mathcal{A}$-equivalence, we can assume that $y = 0$ and that $f_s : (\mathbb{C}_3^2, 0) \to (\mathbb{C}^3, 0)$

is given by

$$f_s(x_1, y_1) = (x_1, y_1, 0), \quad f_s(x_2, y_2) = (x_2, 0, y_2), \quad f_s(x_3, y_3) = (0, x_3, y_3).$$

A matrix presentation of this multi-germ is

$$\begin{pmatrix} X & 0 & 0 \\ 0 & Y & 0 \\ 0 & 0 & Z \end{pmatrix}.$$

Hence, $\dim_{\mathbb{C}} \mathcal{O}_{\mathbb{C}^3,y}/\mathcal{F}_2(f_s)_y = 1$ and we deduce that the right-hand side of (11.35) is equal to the number of triple points of f_s in $B_\epsilon(0)$, for all $s \in B_\eta(0) \setminus \{0\}$. □

We end this section with another result of [KU97], which proves a weaker version of Theorem 11.9 for higher order multiple points.

Lemma 11.8 *Let R be a local ring, B a finite R-algebra. Assume that, with respect to generators $g_0 = 1, g_1, \ldots, g_m$, the R-module B can be presented by a symmetric matrix λ with regular determinant. Set $F_i = \mathrm{Fitt}_i^R(B)$. Let λ'' be the matrix obtained from λ by deleting its first row and column and set $F_i'' = \min_{m+1-i}(\lambda'')$. Then, for $i \geq 2$,*

$$\sqrt{F_i} = \sqrt{F_i''}.$$

Proof Evidently $\sqrt{F_i''} \subset \sqrt{F_i}$. To see the opposite inclusion, let $\mathfrak{p}$ be a prime ideal of R containing F_i''; we need to show that $\mathfrak{p}$ contains F_i. Localising at $\mathfrak{p}$, we may assume that $\mathfrak{p}$ is the maximal ideal $\mathfrak{m}$ of R. If any entries of λ'' lie outside of $\mathfrak{m}$, then we can add multiples of rows and columns $1, \ldots, m$ to the other rows and columns, until λ takes the form

$$\lambda = \begin{pmatrix} \lambda_0^0 & 0 & * \\ 0 & I_\ell & 0 \\ * & 0 & \tilde{\lambda}'' \end{pmatrix},$$

where $\tilde{\lambda}''$ has all its entries in $\mathfrak{m}$. Removing from λ rows and columns $1, \ldots, \ell$ we get a new presentation matrix $\tilde{\lambda}$ of size $m' + 1$ where $m' = m - \ell$, with the property that $\tilde{\lambda}''$ has all its entries in $\mathfrak{m}$. The ideals F_i and F_i'' are now the ideals of minors of size $m' + 1 - i$ of the new matrices $\tilde{\lambda}$ and $\tilde{\lambda}''$.

Since $F_i'' \neq R$, $m' + 1 - i \geq 1$. Hence $m' + 1 \geq i + 1 \geq 3$. Therefore every monomial in the expansion of $\det \tilde{\lambda}$ has an entry in $\tilde{\lambda}''$ as a factor, and thus $F_0 \subset \mathfrak{m}$, or equivalently, $B \neq 0$ (recall that we have localised at p, so this is a non-trivial deduction).

Now we show that the entries $\tilde{\lambda}_j^0$ and $\tilde{\lambda}_0^j$ lie in $\mathfrak{m}$. Suppose that $\tilde{\lambda}_j^0 \notin \mathfrak{m}$ for some j with $1 \leq j \leq m$. Then $1 \in \mathfrak{m}B$ because the columns of $\tilde{\lambda}$ yield relations between $1, g_1, \ldots, g_{m'}$ and because $\tilde{\lambda}''$ has all its entries in $\mathfrak{m}$. Hence $B = \mathfrak{m}B$, so $B = 0$ by Nakayama's Lemma C.1. Therefore, $\tilde{\lambda}_j^0 \in \mathfrak{m}$, for $1 \leq j \leq m$. Since $\tilde{\lambda}$ is symmetric, we also have $\tilde{\lambda}_0^j \in \mathfrak{m}$, for $1 \leq j \leq m$. Finally, repeating the argument now with the 0'th column of $\tilde{\lambda}$, we see that $\tilde{\lambda}_0^0 \in \mathfrak{m}$ as well. Thus, $\tilde{\lambda}_i^j \in \mathfrak{m}$, for all i, j and $F_i \subset \mathfrak{m}$, since $m + 1 - i \geq 1$. □

Kleiman and Ulrich point out that this proof works also in the case that λ is skew-symmetric.

Corollary 11.13 *Let $f\colon (X, x_0) \to (\mathbb{C}^{n+1}, 0)$ be finite and generically 1–1, where (X, x_0) is an n-dimensional Gorenstein germ. Let λ be a symmetric presentation of $\mathcal{O}_{X,x_0}$ over $\mathcal{O}_{n+1}$, with respect to generators $g_0 = 1, g_1, \ldots, g_m$. Then, set theoretically,*

$$M_i(f) = V(\min_{m+1-i}(\lambda'')),$$

for $i \geq 2$, where λ'' is the matrix obtained from λ by deleting its first row and column.

It is interesting to note that although the statement is false for $i = 1$, nevertheless there is a related result, valid also in that case:

Proposition 11.15 *In the circumstances of Lemma 11.8, $\sqrt{F_1} = \sqrt{F_1'' + F_0}$.*

Proof Clearly $F_1 \supset F_1'' + F_0$, so we need only show $\sqrt{F_1} \subset \sqrt{F_1'' + F_0}$. The only point where the hypothesis $i \geq 2$ is used in the proof of the corresponding inclusion in Lemma 11.8 is to conclude that $F_0 \subset \mathfrak{m}$. But now if $F_1'' + F_0 \subset \mathfrak{m}$, this is automatic. □

The resulting set-theoretic equality is already well known to us, since, as we have observed in Corollary 11.9, the symmetry of λ means that $F_1 B = F_1'' B$. In this case the inclusion $F_1 \supset F_1'' + F_0$ is always strict. However, computer calculation for stable germs, and an argument like in Corollary 11.4, show

Proposition 11.16 *For all $\mathcal{A}$-finite germs $(\mathbb{C}^n, 0) \to (\mathbb{C}^{n+1}, 0)$ of multiplicity ≤ 6, the ideals F_i'' and F_i are equal for $i \geq 2$.* □

Conjecture 11.1 If $f\colon (\mathbb{C}^n, 0) \to (\mathbb{C}^{n+1}, 0)$ is an $\mathcal{A}$-finite germ, then $F_i = F_i''$ for $i \geq 2$.

Remark 11.2 It can be shown that the matrix λ'' is the matrix of a presentation of the module $\mathrm{Ext}^1_{\mathcal{O}_D}(\mathcal{O}_X, \mathcal{O}_X)$, under the assumption that X is Gorenstein and $X \to D$ is the normalisation of D.

Exercises for Sect. 11.8

1 Compute the number of triple points that appear in the image of a stable perturbation of the germ H_2 of Exercises 11.3 number 2, and, more generally, in the image of a stable perturbation of the germ H_k given by $(x, y) \mapsto (x, y^3, xy + y^{3k-1})$ for $k \geq 2$.

2 Consider the germ P_3 of Example 11.9. For which values of $c \in \mathbb{C}$ is it true that $M_3(f)$ has $\mathscr{A}_e$-codimension 3? Find the number of triple points in the image of a stable perturbation of f for these values of c.

Aguas de março—Tom Jobim

Appendix A
Jet Spaces and Jet Bundles

We summarise information on jet spaces and jet bundles, in particular the Thom Transversality Theorem, and state a number of its consequences.

We denote by $J^k(n, p)$ the space of p-tuples of polynomials of degree $\leq k$ in n variables with no constant term. A map-germ : $(\mathbb{F}^n, 0) \to (\mathbb{F}^p, 0)$ determines a germ of map $j^k f\colon (\mathbb{F}^n, 0) \to J^k(n, p)$, the k-jet extension of f, defined by

$$j^k f(x) = \text{degree k Taylor polynomial of } f \text{ at } x\text{, without its constant term.}$$

The Taylor polynomial of f is determined by partial derivatives of order $\leq k$ of the component functions of f at x, so the k-jet can be thought of as simply recording these partial derivatives. There is also a *jet bundle* $J^k(X, Y)$ over any pair of manifolds X and Y, whose fibre over $(x_0, y_0) \in X \times Y$, which we denote by $J^k(X, Y)_{(x,y)}$, is the set of k-jets of germs of maps $(X, x_0) \to (Y, y_0)$. Two such map-germs determine the same k-jet at x if they have the same partials of order $\leq k$ at x, with respect to some, and therefore to any, local coordinate systems on X and Y. So in coordinate free terms, a k-jet is an equivalence class of map-germs $(X, x) \to (Y, y)$.

Although $J^k(n, p)$ is a vector space, the fibre $J^k(X, Y)_{(x_0,y_0)}$ is not a vector space in any natural way; for the identifications between the two spaces depend on a choice of coordinate system, and when we change coordinates the higher derivatives of f change in a non-linear way. Thus, there is no natural way of providing $J^k(X, Y)_{(x_0,y_0)}$ with the operations of a vector space, and $J^k(X, Y)$ is not a vector bundle over $X \times Y$.

Nevertheless, $J^k(X, Y)$ is a locally trivial fibre bundle over $X \times Y$.

D. Mond, J. J. Nuño-Ballesteros, *Singularities of Mappings*,
Grundlehren der mathematischen Wissenschaften 357,
https://doi.org/10.1007/978-3-030-34440-5

Its importance for us is because of its role as a kind of Platonic Third Realm[1] which houses ideal versions of all of the singularities which appear in mappings. We will spend the rest of this section justifying this metaphysical remark.

Consider first the 1-jet-bundle $J^1(X, Y)$. By the choice of local coordinates on $U_X \subset X$ and $U_Y \subset Y$ we can identify $\pi^{-1}(U_X \times U_Y)$ with a product $V_X \times V_Y \times J^1(n, p)$ where $V_X \subset \mathbb{F}^n$, $V_Y \subset \mathbb{F}^p$ are open sets. The information contained in the 1-jet $j^1 f(x)$ is just the values of the first order partials of f, so we can think of $j^1 f$ as the map

$$x \mapsto (x, f(x), [d_x f]) \in \mathbb{F}^n \times \mathbb{F}^p \times \mathrm{Mat}_{p\times n}(\mathbb{F})$$

where $[d_x f]$ is the $n \times p$ jacobian matrix of f at x. Let us suppose, to fix ideas, that $n \leq p$, and define $\Sigma^k(n, p)$ (or Σ^k when the dimensions are clear from the context) to be the set of $p \times n$ matrices over $\mathbb{F}$, of kernel rank k.

Let $f : X \to Y$ be a mapping, and denote now by $\Sigma^k(f)$ the set of points in X where $d_x f$ has kernel rank k. Then $\Sigma^k(f) = (j^1 f)^{-1}(\Sigma^k)$. Note, incidentally, that if we change coordinates on X then of course $j^1 f$ also changes, but $(j^1 f)^{-1}(\Sigma^k)$ is, evidently, unchanged. This is because Σ^k has the important property that it is preserved by the action of coordinate changes on X (and on Y).

Observation: suppose $x_0 \in \Sigma^k(f)$ and $j^1 f \pitchfork \Sigma^k$ at x_0. Then

- $\Sigma^k(f)$ is a smooth submanifold of X of codimension $k(p - n + k)$.
- Slightly less obvious: for $\ell < k$, $j^1 f \pitchfork \Sigma^\ell$ also, on some neighbourhood of x_0.
- Indeed, writing $m_0 := j^1 f(x_0)$, there is a local diffeomorphism of germs of filtered spaces

$$(\mathrm{Mat}_{p\times n}, m_0) \supset (\overline{\Sigma^1}, m_0) \supset \cdots \supset (\overline{\Sigma^{k-1}}, m_0) \supset (\Sigma^k, m_0)$$

  and

$$\Big((X, x_0) \supset (\overline{\Sigma^1(f)}, x_0) \supset \cdots \supset (\overline{\Sigma^{k-1}(f)}, x_0) \supset (\Sigma^k(f), x_0)\Big) \times \text{ smooth factor}$$

The second statement is a consequence of the fact that the corresponding stratification

$$\mathrm{Mat}_{p\times n}(\mathbb{F}) \supset (\Sigma^1 \smallsetminus \overline{\Sigma^2}) \supset \cdots \supset (\Sigma^\ell \smallsetminus \overline{\Sigma^{\ell+1}}) \cdots$$

is Whitney regular. We do not dwell on this now. The aim is simply to make clear that the transversality of $j^1 f$ to certain submanifolds of the jet bundle $J^k(X, Y)$ gives us a lot of information about submanifolds (subsets) of X determined by the geometry of f. The submanifolds of the jet bundle that we are interested in are those which are preserved by the action of the group of diffeomorphisms of X and

[1] See, e.g., https://plato.stanford.edu/entries/platonism-mathematics/.

Y—the so-called left-right invariant subsets of $J^k(X, Y)$. The hypothesis on the transversality of $j^1 f$ to Σ^k that we invoked in our observation is motivated by the following statement.

Proposition A.1 *Let $W \subset J^k(X, Y)$ be a left-right invariant submanifold. Then*

1. *If $f : X \to Y$ is a stable*[2] *map, then $j^k f \pitchfork W$.*
2. *If $f : (X, x_0) \to (Y, y_0)$ is an $\mathscr{A}$-finite germ (a map-germ of finite $\mathscr{A}_e$-codimension), then $j^k f \pitchfork W$ on $X \smallsetminus \{x_0\}$.*
3. *If $F : (X \times U, (x_0, 0)) \to (Y \times U, (y_0, 0))$ is an $\mathscr{A}_e$-versal unfolding of f then $j^k F/U \pitchfork W$.*

Here $j^k F/U : X \times U \to J^k(X, Y)$ is the map $(x, u) \mapsto j^k f_u(x)$.

Proof Suppose f is stable.

Step 1 Suppose that $j^k f(x_0) \in W$. There exists a germ of unfolding $F : (X \times U, (x_0, 0)) \to (Y \times U, (f(x_0), 0))$ of f such that the relative jet extension map $j^k F/U : X \times U \to J^k(X, Y)$ is transverse to W at $(x_0, 0)$. This can be arranged by choosing coordinates on X and Y around x_0 and y_0, and then taking as parameter space $U = J^k(n, p)$, and regarding its members as polynomial maps, which can be added to f. The resulting family is defined by $F(x, u) = f(x) + u(x)$, and $j^k F/U|_{\{x_0\} \times U} \to J^k(X, Y)_{(x_0, y_0)}$ is the identity map. It is thus transverse to W.

Step 2 f is stable, so F is a trivial unfolding. Thus, there exist germs of diffeomorphisms Φ of $(X \times U, (x_0, 0))$ with $\Phi(x, u) = (\varphi_u(x), u)$ and Ψ of $(Y \times U, (y_0, 0))$ with $\Psi(y, u) = \psi_u(y), u)$ such that $\Psi \circ (f \times \mathrm{id}_U) \circ \Phi = F$. As $j^k F/U \pitchfork W$, we have $j^k(\Psi \circ F \circ \Phi)/U \pitchfork W$. As W is left-right invariant, it follows that $j^k f \pitchfork W$ (Exercise).

The second statement follows by the Mather–Gaffney criterion for finite $\mathscr{A}$-codimension, Theorem 4.3. Since f is stable outside x_0, $j^k f$ is transverse to W outside x_0.

The third statement is a consequence of "openness of versality", Theorem 5.6. It is proved in Chap. 5 as Proposition 5.3. □

Using an auxiliary map such as $j^k f$ to pull back a universal object from jet space can give useful information. Provided the codimension of the pulled back object is the same as the codimension of the universal object, much of the associated algebraic structure pulls back also. We will see this in Sect. C.5.

A second important application of jet-space is through the Thom transversality theorem, which concerns the behaviour of smooth maps between smooth manifolds. A *residual* subset of a topological space is the intersection of a countable number of dense open sets, and a property is *generic* if it is held by all members of a residual subset. If M and N are smooth manifolds, the *Whitney C^k Topology* on the space

[2] The notions of stability and $\mathscr{A}_e$-codimension are discussed in Chap. 3.2. See also Theorem 4.3 for the geometrical importance of finite codimension—essentially it means "isolated instability".

$C^\infty(M, N)$ of smooth maps from M to N has as base the collection of subsets modelled on open sets $U \subset J^k(M, N)$:

$$C_U = \{f \in C^\infty(M, N) : j^k f(M) \subset U\}$$

and the Whitney C^∞ topology allows such sets for all values of k. We will always consider $C^\infty(M, N)$ with this topology. It is a Baire space—residual sets are dense. A property of mappings $M \to N$ is said to be *generic* if it is held by the members of a residual subset of $C^\infty(M, N)$. We discuss the Whitney C^∞ topology in more detail in Sect. 5.2.

Theorem A.1 (Thom Transversality Theorem) *Let M and N be C^∞ manifolds, let $W \subset J^k(M, N)$ be a smooth submanifold, and let $T(W)$ be the set of smooth maps $f : M \to N$ such that $j^k f \pitchfork W$. Then*

1. *$T(W)$ is residual in $C^\infty(M, N)$.*
2. *If W is closed in $J^k(M, N)$, then $T(W)$ is open in $C^\infty(M, N)$.*

Note that if codim $W >$ dim M, the only way that $j^k f : M \to J^k(M, N)$ can be transverse to W is if $(j^k f)^{-1}(W) = \emptyset$. This is often the way that one proves that sets of mappings with certain properties are residual.

An immersion is an *embedding* if it is a diffeomorphism onto its image.

It is just a short step to prove Whitney's "easy" embedding theorem from Theorem A.1:

Theorem A.2 *Let M be an n-dimensional smooth manifold. If $p \geq 2n + 1$ then the set of embeddings $M \to \mathbb{R}^p$ is residual in $C^\infty(M, \mathbb{R}^p)$.*

If the domain M is compact, one has only to prove that immersions are residual, and that injective maps are residual. Properness (that the preimage of every compact set is compact) is a global property with some subtlety, and we will not discuss it except to say that it is automatic if the domain is compact. Injectivity, on the other hand, is a property of jets, and can be arranged, if the dimensions are right, by requiring transversality to a suitable submanifold of the *multi-jet space* ${}_rJ^k(M, N)$, which is defined as follows: there is a natural map $p : J^k(M, N) \to M$ giving the source of each jet; ${}_rJ^k(M, N)$ is the preimage in $\left(J^k(M, N)\right)^r$ of the set

$$M^{(r)} = \{(x_1, \ldots, x_r) \in M^r : x_i \neq x_j \text{ if } i \neq j\},$$

under the r-fold Cartesian product map $p^r : \left(J^k(M, N)\right)^r \to M^r$. Each map $f : M \to N$ gives rise to a natural map ${}_rj^k f : M^{(r)} \to_r J^k(M, N)$. The following is Mather's multi-jet transversality theorem.

Theorem A.3 *Let M and N be C^∞ manifolds, and let $W \subset {}_rJ^k(M, N)$ be a smooth submanifold. Then the set of smooth maps $f : M \to N$ such that ${}_rj^k f \pitchfork W$ is residual in $C^\infty(M, N)$.*

Exercises for Appendix A

1 Show that if $p \geq n$ then the set $\Sigma^k(n, p)$ of matrices with kernel rank k is a submanifold of $\mathrm{Mat}_{p\times n}(\mathbb{F})$ of codimension $k(p-n+k)$. The formula for the codimension can be recalled as follows: a $p \times n$ matrix of the form

$$\begin{pmatrix} I_{n-k} & B \\ 0 & D \end{pmatrix}$$

has kernel rank k if and only if $D = 0$. The same is true if we have an invertible $(n-k) \times (n-k)$ matrix A in place of I_{n-k}. A more general matrix

$$\begin{pmatrix} A & B \\ C & D \end{pmatrix}$$

in which A is of size $(n-k) \times (n-k)$ and invertible can be brought to this form by left-multiplying by

$$\begin{pmatrix} I_{n-k} & 0 \\ -CA^{-1} & I_{p-n+k} \end{pmatrix}$$

The matrix is in Σ^k if all entries in the transformed D are equal to zero. This gives $(p-n+k)k$ equations.

2 If A is a residual subset of a Baire space B, can $B \smallsetminus A$ contain a residual subset of B?

3 The "elementary transversality theorem" says that if W is a smooth submanifold of N then the set $\{f \in C^\infty(M, N) : f \pitchfork W\}$ is residual. Show how to deduce this from the Thom transversality Theorem A.1.

4 Show that an immersion which is a homeomorphism onto its image is a diffeomorphism.

5 Show that if M is compact then an injective immersion is an embedding, and give an example of an injective immersion of $\mathbb{R}$ in $\mathbb{R}^2$ which is not an embedding.

6 Prove Whitney's easy embedding Theorem A.2 for compact manifolds M. The theorem does not *require* the hypothesis of compactness, but the proof in the general case has an added complication we prefer to avoid.

7 Whitney's hard embedding theorem says that an n-dimensional smooth manifold M can be embedded in $\mathbb{R}^{2n}$. Show that

(i) the set of immersions is residual in $C^\infty(M, \mathbb{R}^{2n})$, but
(ii) the set of embeddings is *not* residual in $C^\infty(M, \mathbb{R}^{2n})$. Suggestion: begin with $M = S^1$.

Proof of the hard embedding theorem requires clever arguments to eliminate singularities.

8

(i) Let $W = \{(x, 0, 0) \in \mathbb{R}^3 : -1 < x < 1\}$. Show that the set $\{f \in C^\infty(\mathbb{R}, \mathbb{R}^3) : f \pitchfork W\}$ is *not* open. Hint: consider $f(t) = (-1, t, 0)$.
(ii) Let $W = \{(x, 0) \in \mathbb{R}^2 : -1 < x < 1\}$. Show that the set $\{f \in C^\infty(S^1, \mathbb{R}^3) : f \pitchfork W\}$ is not open.

9

(i) Show that if $n < 6$, the set of mappings $M^n \to N^{n+1}$ for which all singularities have corank 1 (i.e. the kernel of the derivative has dimension 1) is residual. Is it open?
(ii) What is the smallest value of n for which a stable map $M^n \to N^{n+1}$ can have a corank 2 singularity? A corank 3 singularity?

10 A critical point x_0 of a smooth function $f : M^m \to \mathbb{R}$ is *non-degenerate* if the Hessian matrix $\det([\frac{\partial^2 f}{\partial x_i x_j}(x_0)]_{1 \le i,j \le m}])$ (with respect to some, and hence any, set of local coordinate) is invertible. A function $M \to \mathbb{R}$ is a *Morse function* if all of its critical points are non-degenerate and no two critical points share the same critical value. Show that for any smooth manifold M, Morse functions form a residual set in $C^\infty(M, \mathbb{R})$.

11 A fixed point x_0 of a smooth map $f : M \to M$ is *non-degenerate* if $d_{x_0} f$ does not have 1 as an eigenvalue. Show that this condition can be expressed in terms of the transversality of some jet extension map to a suitable submanifold of jet space, and deduce that the set of maps $f : M \to M$ with only non-degenerate fixed points is residual in $C^\infty(M, M)$.

12 René Thom showed that stable mappings are not always dense in $C^\infty(N, P)$ by means of an example of a singularity $(\mathbb{F}^{n^2}, 0) \to (\mathbb{F}^{n^2}, 0)$ with no stable perturbation. This exercise reconstructs the example, though it does not complete the proof that it has no stable perturbation. For this, see Sect. 5.2.

(i) Let M denote the $m \times m$ matrix $\begin{pmatrix} I_{m-r} & 0 \\ 0 & 0 \end{pmatrix}$. It lies in the subspace $\Sigma^r(m, m)$ of $\mathrm{Mat}_{m\times m}(\mathbb{F})$ consisting of matrices of kernel rank r. Let e_{ij} denote the $r \times r$ elementary matrix with a 1 in row i and column j and zeroes elsewhere, and let E_{ij} denote the $m \times m$ matrix $\begin{pmatrix} 0 & 0 \\ 0 & e_{ij} \end{pmatrix}$. Show that $\mathrm{Sp}_{\mathbb{F}}\{E_{ij} : 1 \le i, j \le r\}$ is a minimal transversal to $\Sigma^r(m, m)$ at M.
(ii) Let $m = 4$ and $r = 2$, and define $f : (\mathbb{F}^4, 0) \to (\mathbb{C}^4, 0)$ by

$$f(x_1, x_2, x_3, x_4) = (x_1, x_2, x_3^2 + x_1 x_4, x_4^2 + x_2 x_3).$$

Show that $j^1 f\colon (\mathbb{F}^4, 0) \to J^1(4, 4)$ meets $\Sigma^2(4, 4)$ transversely (we identify $J^1(4, 4)$ with $\mathrm{Mat}_{4\times 4}(\mathbb{F})$).

(iii) Let $m = n^2$ and $r = n$. Construct a polynomial germ $f\colon (\mathbb{F}^{n^2}, 0) \to (\mathbb{F}^{n^2}, 0)$ which meets $\Sigma^n(n^2, n^2)$ transversely.

Further Reading Chapter II of the textbook [GG73] of Guillemin and Golubitsky gives a thorough and illuminating treatment of jet bundles and the Thom transversality theorem.

Appendix B
Stratifications

We summarise the theory of stratifications, concluding with Thom's two Isotopy Lemmas, which are necessary to prove the fibration theorems for mappings. We also present some basic information on semi-algebraic sets and mappings.

The first Isotopy Lemma is a generalisation of the well-known fibration theorem of Ehresmann:

Theorem B.1 *If $f : N \to P$ is a proper submersion of smooth manifolds, then it is a locally trivial fibre bundle. In particular, when P is connected, the diffeomorphism type of the fibre $N_y := \phi^{-1}(y)$ is independent of $y \in P$.* □

Here we consider instead of a smooth manifold N, a more general class of sets, namely Whitney stratified sets. The main reference for this section is the book [GWdPL76].

B.1 Stratification of Sets

Definition B.1 Let Z be any subset of a smooth manifold M. A *stratification* of Z is a locally finite partition $\mathcal{A}$ of Z into submanifolds of M, called *strata*. The pair $(Z, \mathcal{A})$ is called a *stratified set*.

The local finiteness condition means that any point in Z has a neighbourhood in M which meets only finitely many strata. When the stratification $\mathcal{A}$ is understood, we write just Z for the stratified set instead of $(Z, \mathcal{A})$.

Given integers $1 \leq p \leq m$, we denote by $\mathrm{Gr}(m, p)$ the Grassmannian manifold of p-dimensional subspaces of $\mathbb{R}^m$.

Definition B.2 Let A^p, B^q be submanifolds of a smooth manifold M^m and let $y \in B$.

D. Mond, J. J. Nuño-Ballesteros, *Singularities of Mappings*,
Grundlehren der mathematischen Wissenschaften 357,
https://doi.org/10.1007/978-3-030-34440-5

1. Assume first that $M \subseteq \mathbb{R}^m$ is an open set. We say that A is *Whitney regular* (or *Whitney B-regular*) over B at y if for any pair of sequences $\{x_i\} \subset A$ and $\{y_i\} \subset B$, with $x_i \neq y_i$, both converging to y, and such that the sequence of tangent spaces $\{T_{x_i}A\}$ converges to a subspace T in $\mathrm{Gr}(m, p)$ and the sequence of secant lines $\{\overline{x_i y_i}\}$ converges to a line L in $\mathrm{Gr}(m, 1)$, then $L \subseteq T$. There is also a weaker condition, *Whitney A-regularity*: if x_n is a sequence in A converging to $b \in B$ and the sequence of tangent planes $T_{x_n}A$ converges, in the Grassmannian $\mathrm{Gr}(\dim A, m)$ to some subspace T, then $T \subset T_b B$. It is well known that Whitney B implies Whitney A.
2. Now we consider the general case where M is any smooth manifold. We say that A is *Whitney regular* over B at y if for some (and hence any) coordinate system (φ, U) of M at y, then $\varphi(U \cap A)$ is Whitney regular over $\varphi(U \cap B)$.
3. We say that A is *Whitney regular* over B when it is so at every point in B.

Definition B.3 A *Whitney stratification* of $Z \subseteq M$ is a stratification $\mathcal{A}$ such that for any pair of strata $A, B \in \mathcal{A}$, A is Whitney regular over B. The pair $(Z, \mathcal{A})$ is called a *Whitney* stratified set.

Example B.1 Assume that Z is any semialgebraic subset of $\mathbb{R}^m$ of dimension d. The subset $A_0 \subseteq Z$ of regular points of dimension d is a semialgebraic d-dimensional submanifold of $\mathbb{R}^m$ and its complement $Z' = Z \setminus A_0$ is again semialgebraic of dimension $< d$. By iterating this process, we obtain a finite stratification $\mathcal{A} = \{A_0, \ldots, A_\ell\}$ of Z. But in general, this stratification is not Whitney regular.

For instance, let $Z \subseteq \mathbb{R}^3$ be the Whitney umbrella, given by the equation $z^2 = x^2 y$. Then Z' is the y-axis which is smooth. Thus, the stratification $\mathcal{A}$ has only two strata $A = Z \setminus Z'$ and $B = Z'$. But A is not Whitney regular over B at the origin. We leave this as an exercise, and urge the reader to draw a picture of Z before starting it. ◊

If M is a complex manifold and $A, B \subseteq M$ are complex submanifolds, then we can define the complex version of Whitney regularity in a similar way to that of Definition B.2. The only precaution is that we have to consider the complex secant line $\overline{x_i y_i}$ instead of the real one.

Definition B.4 Let $(Z, \mathcal{A})$ be a stratified subset of a smooth manifold M. Given a smooth mapping $f : N \to M$, we say that f is *transverse to* Z (or *transverse to* $\mathcal{A}$) if it is transverse to all the strata of $\mathcal{A}$. When f is transverse to Z, then $f^{-1}\mathcal{A} = \{f^{-1}(A) : A \in \mathcal{A}\}$ is obviously a stratification of $f^{-1}(Z)$ in N.

It is common to include in the definition of Whitney stratified set the so-called *frontier condition*: let A, B be strata with $A \cap \overline{B} \neq \emptyset$; then $A \subseteq \overline{B}$ (in other words, the frontier of a stratum is a union of strata). Following [GWdPL76], we have not included this condition in our definition of Whitney stratified set, since it is not necessary in order to prove the isotopy lemmas. Nevertheless, it can be shown that if Z is any Whitney stratified set, then the stratification given by the connected components of all the strata of Z satisfies the frontier condition (see [GWdPL76, II.5.7]).

Definition B.5 A *stratified homeomorphism* between two stratified sets Z and W is a homeomorphism $h : Z \to W$ that takes strata into strata and is a diffeomorphism on each stratum.

The requirement of smoothness in the definition of stratified homeomorphism is surprising but well established.

Definition B.6 Let $(Z, \mathcal{A})$ be a stratified subset of a smooth manifold M and let $\phi : M \to P$ be a smooth map such that for all $A \in \mathcal{A}$, $\phi|_A$ is a submersion. Then the restriction $\phi : Z \to P$ is called a *submersion*.

For each $y \in P$, the *fibre* of the submersion is the stratified set $(Z_y, \mathcal{A}_y)$, where $Z_y := Z \cap \phi^{-1}(y)$ and $\mathcal{A}_y := \{A \cap \phi^{-1}(y) : A \in \mathcal{A}\}$.

We say that Z is *locally C^0-trivial* over P if for any $y \in P$ there exist an open neighbourhood $V \subseteq P$ of y and a commutative diagram

$$\begin{array}{ccc} Z \cap \phi^{-1}(V) & \xrightarrow{\;h\;} & Z_y \times V \\ \big\downarrow \pi & \swarrow \pi_2 & \\ V & & \end{array}$$

where h is a stratified homeomorphism and π_2 is the projection onto the second factor.

In other words, $\phi : Z \to P$ is a locally trivial fibre bundle with fibre Z_y.

If Z is locally C^0-trivial over P and P is connected, then the topological type of the fibre Z_y is independent of $y \in P$. If, in addition, P is contractible, then it is well known that any fibre bundle is trivial (see, e.g., [Hus94]). That is, there exists a global homeomorphism $h : Z \to Z_y \times P$ such that $\pi_2 \circ h = \phi$.

Now we can state precisely Thom's first isotopy lemma, which can be seen as a stratified version of Ehresmann's fibration theorem (the proof can be found in [GWdPL76, Theorem II.5.2]).

Theorem B.2 (Thom's First Isotopy Lemma) *Let $(Z, \mathcal{A})$ be a locally closed Whitney stratified subset of a smooth manifold M and let $\phi : Z \to P$ be a submersion such that for all $A \in \mathcal{A}$, $\phi|_{\overline{A} \cap Z}$ is proper. Then, $(Z, \mathcal{A})$ is locally C^0-trivial over P.*

Example B.2 We consider again the Whitney umbrella $Z \subset \mathbb{R}^3$ with the stratification $\{A, B\}$, where B is the z-axis and $A = Z \setminus B$ (see Example B.1). Let $\phi : Z \to \mathbb{R}$ be the projection onto the y-axis. It is obvious that the restrictions $\phi|_A$ and $\phi|_B$ are submersions and also that $\phi|_{\overline{A} \cap Z}$ and $\phi|_{\overline{B} \cap Z}$ are proper. However, Z is not locally C^0-trivial over the y-axis. In fact, Z_y is a pair of lines when $y > 0$, a line when $y = 0$ and a point for $y < 0$. ◊

Besides the *existence* of Whitney stratifications, it is also desirable to have some notion of uniqueness. If Z is a Whitney stratified subset of a manifold M, we can obtain a new Whitney stratification by adding unnecessary lower-dimensional strata in the interiors of existing strata, or by declaring different connected components of the same stratum to be distinct strata. To avoid such complications, consider the *associated filtration by dimension* of Z, denoted by $\{Z_i\}$, where Z_i the union of all strata of dimension $\leq i$.

Definition B.7 We say that a Whitney stratification of Z is *canonical* if for each i, Z_i is the largest subset of Z such that $Z_i \smallsetminus Z_{i-1}$ is a submanifold of M and $Z_j \smallsetminus Z_{j-1}$ is Whitney regular over $Z_i \smallsetminus Z_{i-1}$ for all $j > i$.

Clearly, any canonical Whitney stratification of Z is minimal if we consider the following partial order $\leq$ in the set of all Whitney stratifications of Z: given two stratifications $\mathcal{A}$ and $\mathcal{A}'$ with associated filtrations $\{Z_i\}$ and $\{Z_i'\}$, respectively, we say that $\mathcal{A} < \mathcal{A}'$ when there exists an integer i such that $Z_i \subset Z_i'$ but $Z_j = Z_j'$, for all $j > i$.

The existence of a canonical Whitney stratification plays an important role in Mather's work on topological stability (see his Harvard preprint *Notes on Topological Stability*, and also [GWdPL76] and [dPW95]).

Exercises for Sect. B.1

1 Suppose that N, P and U are smooth manifolds, with N compact, that $F : N \times U \to P$ is smooth, and that Z is a closed submanifold of P. Define $f_u : N \to P$ by $f_u(x) = F(x, u)$. Use Ehresmann's fibration theorem to show that if $f_u \pitchfork Z$ for all $u \in U$, then the projection $f^{-1}(Z) \to U$ is a locally trivial fibre bundle.

2 Let $Z \subseteq \mathbb{R}^3$ be the Whitney umbrella $\{z^2 = x^2 y\}$ and take B as the y-axis and $A = Z \setminus B$ (see Example B.1). Show that A is not Whitney regular over B at the origin.

3 Show that $\mathcal{A} = \{A, B \setminus \{0\}, \{0\}\}$ is a Whitney stratification of the Whitney umbrella.

4 Show that the definition of Whitney regularity at a point in a smooth manifold is independent of the chosen coordinate system.

5 Let A and B be submanifolds of a smooth manifold M. Let $x \in A \cap \overline{(B \smallsetminus A)}$ and assume B is Whitney regular over A at x. Show that $\dim A < \dim B$.

6 Let A and B be complex submanifolds of a complex manifold M. Show that A is Whitney regular over B at y in the complex sense if and only if it is so in the real sense.

7 Show that the product $Z_1 \times Z_2$ of two Whitney stratifed sets Z_1 and Z_2 is also a Whitney stratified set, with the natural induced stratification.

8 Let Z be a Whitney stratified set and let $U \subset Z$ be an open subset. Show that U is also a Whitney stratified set with the restricted stratification.

9 Show that if Z is a Whitney stratified subset of M and $f : N \to M$ is transverse to Z, then $f^{-1}(Z)$ is also a Whitney stratified subset of N.

B.2 Stratification of Mappings

Thom's second isotopy lemma is about stratifications of mappings. We include the basic definitions in order to understand its statement.

Definition B.8 Let $f : M \to N$ be a smooth mapping and let $Z \subseteq M$ and $W \subseteq N$ such that $f(Z) \subseteq W$. A *stratification* of $f : Z \to W$ is a pair $(\mathcal{A}, \mathcal{B})$ with $\mathcal{A}$ and $\mathcal{B}$ Whitney stratifications of Z and W, respectively, which satisfy the following two conditions:

1. f maps strata into strata,
2. for any $A \in \mathcal{A}$ and $B \in \mathcal{B}$ such that $f(A) \subseteq B$, the restriction $f : A \to B$ is a submersion.

The triple $(f, \mathcal{A}, \mathcal{B})$ is called a *stratified mapping*.

As in the case of stratified sets, we just write f for a stratified mapping instead of $(f, \mathcal{A}, \mathcal{B})$ when the stratifications $\mathcal{A}$ and $\mathcal{B}$ are understood.

Definition B.9 Let $f : M \to N$ be a smooth mapping, let $A, B \subseteq M$ be submanifolds such that the restrictions $f|_A$ and $f|_B$ have constant rank and let $y \in B$.

1. We say that A is *Thom regular* over B at $y \in B$ relative to f if for any sequence $\{x_i\} \subseteq A$ converging to y such that the sequence of subspaces $\{\ker d_{x_i}(f|_A)\}$ converges to a subspace $T \subseteq T_y M$ in the appropriate Grassmannian bundle, then $\ker d_y(f|_B) \subseteq T$.
2. We say that A is *Thom regular* over B relative to f if it is so at every point of B.

Definition B.10 A *Thom stratification* of a mapping $f : Z \to W$ is a stratification $(\mathcal{A}, \mathcal{B})$ such that for any pair of strata $A, B \in \mathcal{A}$, A is Thom regular over B relative to f. The triple $(f, \mathcal{A}, \mathcal{B})$ is called a *Thom* stratified mapping.

Definition B.11 Let $(f, \mathcal{A}, \mathcal{B})$ be a stratified mapping between subsets of smooth manifolds $Z \subseteq M$ and $W \subseteq N$ and let $\phi : W \to P$ be a submersion. For each $y \in P$, the *fibre* is the stratified map $(f_y, \mathcal{A}_y, \mathcal{B}_y)$ constructed as follows:

1. $Z_y = (\phi \circ f)^{-1}(y) \cap Z$, $W_y = \phi^{-1}(y) \cap W$ and $f_y : Z_y \to W_y$ is the restriction of f;
2. $\mathcal{A}_y := \{A \cap (\phi \circ f)^{-1}(y) : A \in \mathcal{A}\}$ and $\mathcal{B}_y := \{B \cap \phi^{-1}(y) : B \in \mathcal{B}\}$.

We say that f is *locally C^0-trivial* over P if for any $y \in P$ there exist an open neighbourhood $V \subseteq P$ of y and a commutative diagram

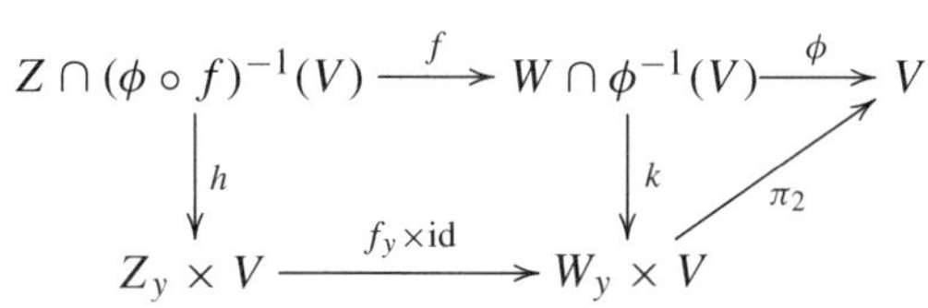

where h and k are stratified homeomorphisms and π_2 is the projection onto the second factor.

As in the case of stratified sets, if f is locally C^0-trivial over P and P is connected, then the topological type of the fibre f_y (up to homeomorphisms in the source and the target) is independent of $y \in P$. We can now state Thom's second isotopy lemma which is about topological triviality of Thom stratified mappings (see the proof in [GWdPL76, Theorem II.5.8]).

Theorem B.3 (Thom's Second Isotopy Lemma) *Let $(f, \mathcal{A}, \mathcal{B})$ be a Thom stratified mapping between locally closed subsets of smooth manifolds $Z \subseteq M$ and $W \subseteq N$ and let $\phi : W \to P$ be a submersion. Suppose that for all $A \in \mathcal{A}$ and $B \in \mathcal{B}$, the restrictions $f|_{\overline{A}\cap Z}$ and $\phi|_{\overline{B}\cap W}$ are proper. Then, $(f, \mathcal{A}, \mathcal{B})$ is locally C^0-trivial over P.*

Finally, we introduce the concept of partial stratification of a smooth mapping $f : M \to N$ between smooth manifolds. We see that this is a powerful tool in order to obtain a Thom stratification.

Definition B.12 Let $f : M \to N$ be a smooth mapping with critical set $C = C(f)$ and whose discriminant $\Delta = f(C)$ is closed in N. A *partial stratification* of f is a Whitney stratification $\mathcal{B}$ of Δ which satisfies the following three conditions for any pair of strata $A, B \in \mathcal{B}$ (possibly equal):

1. $f^{-1}(A) \cap C$ is a submanifold of M and the restriction $f : f^{-1}(A) \cap C \to A$ is a local diffeomorphism;
2. $f^{-1}(B) \cap C$ is Whitney regular over $f^{-1}(A) \cap C$;
3. $f^{-1}(B) \setminus C$ is Whitney regular over $f^{-1}(A) \cap C$.

The discriminant Δ is closed in N when the restriction to the critical set $f : C \to N$ is a closed mapping. But this is a generic condition which is satisfied, for instance, when $f : M \to N$ is locally stable. The next property shows how to produce a Thom stratification of f from a partial stratification (see [GWdPL76, I.3.1] for a proof).

Proposition B.1 *Let $f : M \to N$ be a smooth mapping whose discriminant Δ is closed. If $\mathcal{B}$ is a partial stratification of f, then $(\mathcal{A}, \mathcal{B}')$ is a Thom stratification of f, where $\mathcal{B}' = \mathcal{B} \cup \{N \setminus \Delta\}$ and the strata of $\mathcal{A}$ are either $M \setminus f^{-1}(\Delta)$, $f^{-1}(A) \cap C$ or $f^{-1}(A) \setminus C$, with $A \in \mathcal{B}$.*

Exercises for Sect. B.2

1 Let $Z \subseteq \mathbb{R}^3$ be the Whitney umbrella $\{z^2 = x^2 y\}$, let B be the y-axis and let $A = Z \setminus B$ (see Example B.1). Consider the projection to the y-axis, $f : Z \to \mathbb{F}$, $f(x, y, z) = y$. Is there a stratification of the y-axis making f into a stratified submersion? Is there a stratification making f into a Thom stratified mapping?

B.3 Semialgebraic Sets

Here we study the useful class of *semialgebraic sets*. These sets appear many times in the context of singularities of mappings. A key property is that they always admit a Whiney stratification. We present in this section the main definitions and properties. At the end of the section we will mention other larger classes of sets which also have more or less the same properties (real and complex analytic, semianalytic or subanalytic sets).

Definition B.13 The class of *semialgebraic sets* in $\mathbb{R}^n$ is the smallest Boolean algebra of subsets of $\mathbb{R}^n$ which contain all subsets of the form

$$\{x \in \mathbb{R}^n : p(x) > 0\},$$

for some polynomial function $p: \mathbb{R}^n \to \mathbb{R}$. Given $S \subset \mathbb{R}^n$, a mapping $f: S \to \mathbb{R}^m$ is called *semialgebraic* if its graph is semialgebraic in $\mathbb{R}^n \times \mathbb{R}^m$.

By construction, the semialgebraic sets are closed under finite unions, finite intersections, complements and products. It follows that $S \subset \mathbb{R}^n$ is semialgebraic if and only if it is defined by a finite number of equalities and inequalities given by polynomial functions. In particular, any algebraic set (defined by a finite number of polynomial equations) is semialgebraic. The notion of semialgebraic mapping is a little bit less intuitive (for instance, a semialgebraic mapping is not necessarily continuous). It is easy to prove that the inverse image of a semialgebraic set by a semialgebraic mapping is semialgebraic. However, the first non-obvious result, known as the Tarski–Seidenberg theorem, is that the image of a semialgebraic mapping is also semialgebraic.

Theorem B.4 ([Ło65]) *The image of any semialgebraic mapping $f: S \to \mathbb{R}^m$ is semialgebraic in $\mathbb{R}^m$.*

The proof of the Tarski–Seidenberg theorem is strongly based on the following result, which generalises a theorem of Whitney about finiteness of algebraic sets.

Proposition B.2 ([Ło65]) *A semialgebraic set has a finite number of connected components, which are also semialgebraic.*

Another important result is about the existence of regular points in semialgebraic sets.

Definition B.14 Given a subset $S \subset \mathbb{R}^n$, a point $x \in S$ is called *regular* (of dimension d) if there exists an open neighbourhood V of x in $\mathbb{R}^n$ such that $V \cap S$ is a submanifold of $\mathbb{R}^n$ (of dimension d).

Theorem B.5 ([Ło65]) *The set of regular points of a semialgebraic set S is also semialgebraic and dense in S.*

The *dimension* of a semialgebraic set S is defined as the maximal dimension of S at a regular point. This always makes sense because of the above theorem. The set of points of S which are not regular of maximal dimension is denoted by ΣS. It follows that ΣS is semialgebraic of dimension $< \dim S$. Another deeper result of the same nature is that if S and T are semialgebraic disjoint submanifolds, then the set of points $x \in S$ where T fails to be Whitney regular over S at x is also semialgebraic of dimension $< \dim S$. Based on this fact, it is possible to prove the following (see [GWdPL76, I.2.7] for details):

Theorem B.6 *Any semialgebraic set admits a canonical Whitney stratification having finitely many semialgebraic strata.*

The following easy application can be found in Milnor's book [Mil68] for algebraic sets:

Corollary B.1 *Let S be semialgebraic and let M be the subset of regular points of S. Any polynomial function g on M can have at most a finite number of critical values.*

Proof The set Σ of critical points of g on M is semialgebraic and hence admits a stratification with finitely many semialgebraic strata $\{A_1, \dots, A_r\}$. Given $x \in A_i$, x is a critical point of $g|_M$ and hence, it is also a critical point of $g|_{A_i}$. Since all points of A_i are critical, g must be constant on each connected component of A_i, so the image $g(A_i)$ is a finite set. Therefore, $g(\Sigma) = g(A_1) \cup \cdots \cup g(A_r)$ is also finite. □

Another important application of the existence of a Whitney stratification is that the semialgebraic sets are "locally conical". This is particularly important in the definition of the vanishing homology, which we referred to briefly in the Introduction.

If X is any topological space, the *cone on* X, which we denote by $C(X)$, is obtained by forming the Cartesian product $X \times [0, 1]$ and then identifying all of the points of $X \times \{1\}$ with one another. One writes $C(X) = (X \times [0, 1])/(X \times \{1\})$, where the notation B/A, for A a subset of B, means the quotient of B by the equivalence relation which identifies all the points of A to one another. The cone on the empty set is, by definition, a single point. If X is embedded in some $\mathbb{R}^n$, then the cone $C(X)$ can be described more concretely as follows: if v is an (arbitrary) point in $\mathbb{R}^n \times \{1\}$, then $C(X)$ is homeomorphic to the union of all of the line segments in $\mathbb{R}^n \times [0, 1]$ joining v to a point $(x, 0)$, for $x \in X$ (see Fig. B.1).

Because cones are contractible, their homology is equal to that of a point.

For $x_0 \in \mathbb{R}^n$, let $S_\varepsilon(x_0)$ be the sphere of radius ε centred at x_0, and let $B_\varepsilon(x_0)$ be the open ball of (Euclidean) radius ε centred at x_0.

Theorem B.7 (Local Conical Structure) *Let $X \subset \mathbb{R}^n$ be semialgebraic and let $x_0 \in X$. There exists $\varepsilon > 0$ such that $X_{\leq\varepsilon} := X \cap \overline{B}_\varepsilon(x_0)$ is homeomorphic to the cone on its boundary $X_\varepsilon := X \cap S_\varepsilon(x_0)$.*

Proof Let $\mathcal{A} = \{A_1, \dots, A_r\}$ be a Whitney stratification of X with semialgebraic strata. Let $\rho : \mathbb{R}^n \to \mathbb{R}$ be the distance-squared function $\rho(x) = \|x - x_0\|^2$, which is polynomial and proper. By Corollary B.1, each $g|_{A_i}$ has most a finite number of

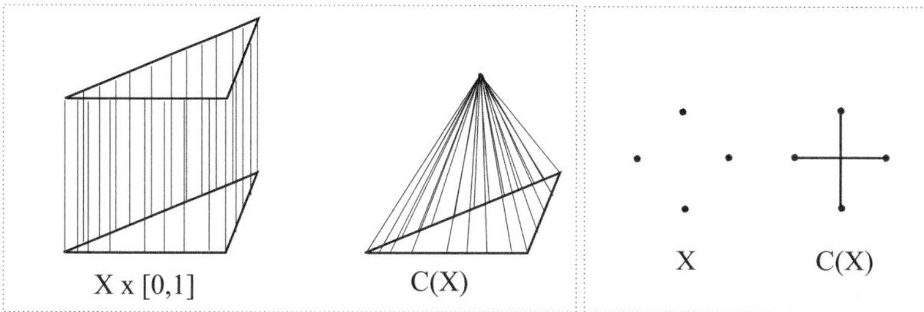

Fig. B.1 Some examples of cones

critical values. Hence, there exists $\delta > 0$ such that for each i, $g|_{A_i}$ has no critical values on the interval $(0, \delta^2)$.

Consider $Z := X \cap B_\delta(x_0) \setminus \{x_0\} = X \cap \rho^{-1}(0, \delta^2)$ with the Whitney stratification $\mathcal{A}|_Z$. Since X is semialgebraic, it is locally closed. After shrinking δ if necessary, we can assume that Z is closed in $B_\delta(x_0) \setminus \{x_0\}$. Hence, the restriction of $\rho\colon Z \to (0, \delta^2)$ is a proper submersion.

By Theorem B.2, Z is locally C^0-trivial over $(0, \delta^2)$. As $(0, \delta^2)$ is contractible, the fibre X_ε is independent of ε, for all $0 < \varepsilon < \delta$ and there exists a global homeomorphism $h\colon Z \to X_\varepsilon \times (0, \delta^2)$ such that $\pi_2 \circ h = \rho$. The restriction of h gives a homeomorphism between $X_{\leq\varepsilon} \setminus \{x_0\}$ and $X_\varepsilon \times (0, \varepsilon^2]$, which extends (automatically) to a homeomorphism between $X_{\leq\varepsilon}$ and the cone on X_ε. □

A proof without using the first isotopy lemma can be found in [Mil68] in the particular case that X is algebraic and x_0 is an isolated singularity. Milnor's proof is based on the construction of a "radial" tangent vector field v on X, pointing in towards x_0, on a neighbourhood of x_0 in X, and adjust the length of the vectors so that for each point $x \in X_\varepsilon$, the trajectory $\varphi_t(x)$ starting at x arrives at x_0 at time $t = 1$. The theorem does not hold in general for the zero loci of C^∞ functions. For example, it is possible to construct a C^∞ function f on $\mathbb{R}^n$ vanishing exactly on the points of a sequence converging to a point x_0. The zero locus of f in any ball $B_\varepsilon(x_0)$ is then not homeomorphic to a cone on its boundary.

Finally, yet another important property of the semialgebraic sets, which is very useful in singularity theory. It is called the curve selection lemma and was proved for semialgebraic sets by Milnor in [Mil68]. Roughly speaking, it says that any sequence in S converging to a point $x_0 \in \overline{S}$ can be approximated by an analytic curve.

Lemma B.1 *Let $S \subset \mathbb{R}^n$ be semialgebraic and $x_0 \in \overline{S}$ (the topological closure of S). Then there exists a real analytic curve $\gamma : [0, \epsilon) \to \mathbb{R}^n$ such that $\gamma(0) = x$ and $\gamma(t) \in S$, for all $t > 0$.*

The class of semialgebraic subsets can be extended conveniently to include sets defined by analytic equalities or inequalities. Given $x_0 \in \mathbb{R}^n$, we denote by $\mathscr{S}_{x_0}$

the smallest Boolean algebra of germs of sets (S, x_0) in $(\mathbb{R}^n, x_0)$ which contains all germs of sets of the form

$$\{x \in U : f(x) > 0\},$$

for some U open neighbourhood of x_0 and $f: U \to \mathbb{R}$ analytic. A subset $S \subset \mathbb{R}^n$ is called *semianalytic* if for each $x_0 \in \mathbb{R}^n$, $(S, x_0) \in \mathscr{S}_{x_0}$ (here we do not assume that $x_0 \in S$). This class includes not only the semialgebraic sets, but also the analytic and the complex analytic ones (when $n = 2m$ and we identify $\mathbb{R}^{2m} \equiv \mathbb{C}^m$). However, the class of semianalytic subsets is not good enough, for the image of a semianalytic set by a semianalytic mapping is not in general semianalytic. Thus, it is better to increase this class in order to include also projections of semianalytic sets. Subsets obtained as projections of semianalytic sets are called *subanalytic*. All definitions and properties of this section can be extended conveniently for subanalytic sets, including the existence of a Whitney stratification by subanalytic strata (not necessarily finite), the local conical structure theorem and the curve selection lemma. We refer to the Łojasiewicz notes [Ło65] and the paper by E. Bierstone and P. Milman [BM88] for details.

Exercises for Sect. B.3

1 Show that in order to prove the Tarski–Seidenberg Theorem B.4, it is enough to prove that for any $S \subset \mathbb{R}^n \times \mathbb{R}^m$, the projection $\pi_1(S)$ is semialgebraic in $\mathbb{R}^n$.

2 Show that the topological closure $\overline{S}$ of any semialgebraic set S is also semialgebraic.

3 Divide up the objects pictured below into subsets which are homeomorphic to cones on their boundary (Fig. B.2).

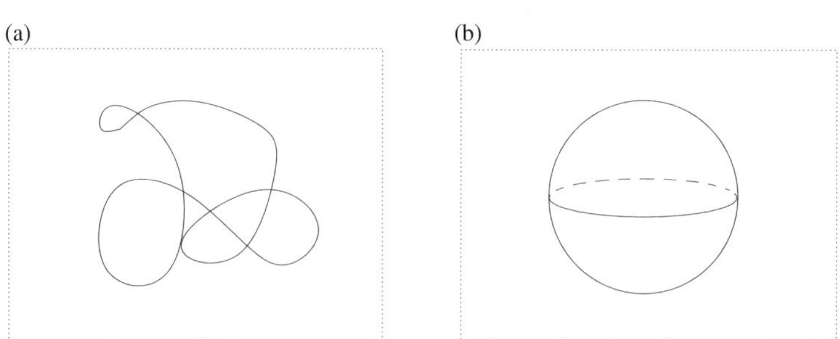

Fig. B.2 (**a**) Planar projection of a knot. (**b**) Sphere

4 Let X be a (real or complex) analytic set and let $x_0 \in X$ be an isolated singular point. Assume there exists an analytic function $\rho : X \to \mathbb{R}_{\geq 0}$ such that $\rho^{-1}(0) = \{x_0\}$ and ρ has no critical point in $X_{\leq \varepsilon} \setminus \{x_0\}$, for some $\varepsilon > 0$. Use the gradient vector of ρ to construct a tangent vector field v on X, pointing in towards x_0, on a neighbourhood of x_0 in X, and such that for each point $x \in X_\varepsilon$, the trajectory $\varphi_t(x)$ starting at x arrives at x_0 at time $t = 1$.

5 Let X and x_0 be as in Exercise 4. Use the curve selection lemma to show that if $\rho : X \to \mathbb{R}_{\geq 0}$ is a real analytic function such that $\rho^{-1}(0) = \{x_0\}$, then there exists $\varepsilon > 0$ such that ρ has no critical point in $X_{\leq \varepsilon} \setminus \{x_0\}$. In particular, one can use the Euclidean distance-squared function $\rho_E(x) := \|x - x_0\|^2$.

6 Let X and x_0 be as in Exercise 4 and let $\rho_E : X \to \mathbb{R}_{\geq 0}$ be Euclidean distance-squared function. Show that ρ has no critical point in $X_{\leq \varepsilon} \setminus \{x_0\}$ if and only if for all ε' with $0 < \varepsilon' \leq \varepsilon$, $X \pitchfork S_{\varepsilon'}(x_0)$.

Appendix C
Background in Commutative Algebra

We summarise the commutative algebra leading up to the theory of Cohen Macaulay modules and the phenomenon of conservation of multiplicity.

Like algebraic geometry, singularity theory applies commutative algebra to understand the structure of the ring of functions on a space and the modules over it. Below, we give an outline of some of the algebra we use in this book, assuming a basic background in the theory of rings and ideals, and sketching some of the material that we believe illuminates the arguments. The aim is to orient the reader rather than fully inform them. The book *Local Analytic Geometry* of de Jong and Pfister [dJP00]) develops this material in a geometrical context, and we frequently refer to it, and to *Commutative Rings* of Matsumura [Mat89], and *Commutative Algebra* of Eisenbud [Eis95]. We have one central objective: to introduce Cohen–Macaulay modules and review some of their properties. Cohen–Macaulayness is key to "conservation of multiplicity"—for example, the fact that when an isolated critical point x_0 of a function f with

$$\dim_{\mathbb{C}} \mathscr{O}_n / \left(\frac{\partial f}{\partial x_1}, \dots, \frac{\partial f}{\partial x_n}\right) = \mu < \infty$$

splits under perturbation into non-degenerate critical points, there are exactly μ of them.

C.1 Spaces and Functions on Spaces

Most of singularity theory is concerned with local pictures—germs and multi-germs of spaces and maps. Unlike closed complex varieties, these can all be described by embeddings in (real or complex) Euclidean spaces, and as maps between embedded subspaces, so this is where we locate them. In this book, unless otherwise stated,

D. Mond, J. J. Nuño-Ballesteros, *Singularities of Mappings*,
Grundlehren der mathematischen Wissenschaften 357,
https://doi.org/10.1007/978-3-030-34440-5

all analytic spaces are embedded in some $\mathbb{C}^n$. The bridge between geometry and algebra is the ring of functions on a space. Algebraic geometry has the coordinate ring of an affine variety, and its localisations. Analytic geometry, which we are more interested in, *begins* with local rings, since complex analytic functions defined in a neighbourhood of a point do not in general extend to any given open neighbourhood of the point. The ring of analytic germs on $\mathbb{C}^n$ at $x \in \mathbb{C}^n$, $\mathscr{O}_{\mathbb{C}^n,x}$, is not the localisation of any naturally defined global ring.

This ring, with its abundant supply of prime ideals, forms the basis for local analytic geometry. It is worth noting that this approach to geometry completely fails in the C^∞ case. The ring of smooth germs $C^\infty(\mathbb{R}^n)_{x_0}$ is not Noetherian and is not an integral domain, and moreover the ideal I_X of germs vanishing on the set-germ $X \subset (\mathbb{R}^n, x_0)$ is only prime when X consists of the single point x_0, so that the only prime ideal in $C^\infty(\mathbb{R}^n)_{x_0}$ is the maximal ideal. All of this follows from the existence of partitions of unity in the C^∞ category.

A closed analytic subspace X of an open subset U of $\mathbb{C}^n$ is a subset defined by the zeros of analytic functions $g_1, \ldots, g_k$ on U. At each point x_0, the set-germ (X, x_0) is called *analytic* and it has a ring of germs of analytic functions,

$$\mathscr{O}_{X,x_0} = \mathscr{O}_{\mathbb{C}^n,x_0}/(g_1, \ldots, g_k).$$

The pair $((X, x_0), \mathscr{O}_{X,x_0})$ is called a *complex space germ*. When there is no risk of confusion, it is usual to denote a complex space germ simply by (X, x_0). Given a complex space $(X, \mathscr{O}_X)$, the rings $\mathscr{O}_{X,x}$ are tied together in a *sheaf* of rings, $\mathscr{O}_X$—see Appendix E below. Here we discuss only the rings $\mathscr{O}_{X,x_0}$ rather than the sheaf.

We do not assume that the ideal $(g_1, \ldots, g_k)$ of $\mathscr{O}_{X,x_0}$ is radical, so the same point set can have many different analytic structures. Its *reduced structure* is the one given by the (unique) radical ideal defining it.

The ring $\mathscr{O}_{X,x_0}$ is a *local ring*—it has a unique maximal ideal,

$$\mathfrak{m}_{X,x_0} := \{f \in \mathscr{O}_{X,x_0} : f(x_0) = 0\}.$$

When $X = \mathbb{C}^n$ and $x_0 = 0$, it is common to abbreviate $\mathscr{O}_{\mathbb{C}^n,0}$ to $\mathscr{O}_n$ and $\mathfrak{m}_{\mathbb{C}^n,0}$ to $\mathfrak{m}_n$ or $\mathfrak{m}$. The functions in $\mathscr{O}_{X,x_0} \setminus \mathfrak{m}_{X,x_0}$, are *units*—they have a multiplicative inverse in $\mathscr{O}_{X,x_0}$.

A *semi-local ring* is a ring with a finite number of maximal ideals—for example, the ring $\mathscr{O}_{\mathbb{C}^n,S} := \bigoplus_{x\in S} \mathscr{O}_{\mathbb{C}^n,x}$, where S is a finite point set, or, less trivially, the ring $\mathbb{C}[x_1, \ldots, x_n]/I$ where I is an ideal with a finite number of zeros.

If $x_1, \ldots x_n$ are coordinates on $\mathbb{C}^n$ around x_0, and $x_0 = (p_1, \ldots, p_n)$ in these coordinates, then every germ $f \in \mathscr{O}_{X,x_0}$ can be written as a convergent power series in $x_1 - p_1, \ldots, x_n - p_n$. It follows that

$$\mathfrak{m}_{X,x_0} = (x_1 - p_1, \ldots, x_n - p_n). \tag{C.1}$$

Proposition C.1

1. *$\mathcal{O}_{X,x_0}$ is Noetherian: every strictly ascending chain of ideals is finite.*
2. *When $X = \mathbb{C}^n$, then for $k = 1, 2, \ldots,$*

$$\mathfrak{m}^k_{\mathbb{C}^n,x_0} = \{f \in \mathcal{O}_{\mathbb{C}^n,x_0} : f(x_0) = \partial^\alpha f/\partial x_\alpha(x_0) = 0 \text{ for } |\alpha| \le k-1\}, \quad \text{(C.2)}$$

where α is a multi-index $\alpha = (\alpha_1, \ldots, \alpha_n)$ and $|\alpha| = \alpha_1 + \cdots + \alpha_n$.

Besides the local ring of an analytic space germ, we will be interested in modules over this ring.

In the C^∞ category, (C.1) and (C.2) also hold. However they are no longer completely obvious; (C.1) is known as *Hadamard's lemma*; it may be proved as follows.

Assume for simplicity that $x_0 = 0$, so we have to show that $\mathfrak{m}_n$ is generated by the coordinate functions $x_1, \ldots, x_n$. If $f \in \mathfrak{m}_n$, then

$$\begin{aligned} f(x) = f(x) - f(0) &= \int_0^1 \frac{d}{dt}\{f(tx)\}dt = \int_0^1 \sum_{i=1}^n \frac{\partial f}{\partial x_i}(tx)x_i dt \\ &= \sum_{i=1}^n x_i \int_0^1 \frac{\partial f}{\partial x_i}(tx)dt = \sum_{i=1}^n x_i g_i(x), \end{aligned}$$

where $g_i(x) = \int_0^1 \frac{\partial f}{\partial x_i}(tx)dt$.

Note, however, that the ring of C^∞ germs is not Noetherian: the ideal $\mathfrak{m}_n^\infty = \cap_{n=1}^\infty \mathfrak{m}_n^k$ is not empty—for example, when $n = 1$ it contains the germ

$$f(x) = \begin{cases} e^{-1/x^2} & x \neq 0 \\ 0 & x = 0 \end{cases}$$

—and not finitely generated, for one can show that f/x is also in $\mathfrak{m}_1^\infty$ (see Exercise 1).

We will frequently use the following statement, known as *Nakayama's lemma*, though see [Mat89, Page 8] for Nakayama's disclaimer as well for a proof.

Lemma C.1 *Let M be a finitely generated module over a local ring R with maximal ideal $\mathfrak{m}$. If $\mathfrak{m} M = M$, then $M = 0$.* □

Corollary C.1 *Let M and N be submodules of an R-module P, with M finitely generated, and suppose that*

$$M \subset N + \mathfrak{m} M. \quad \text{(C.3)}$$

Then $M \subset N$.

Proof of Corollary Let $M_0 = (M+N)/N$. The hypothesis $M \subset N + \mathfrak{m} M$ implies that $M_0 = \mathfrak{m} M_0$. It follows by the Lemma that $M_0 = 0$, so that $M \subset N$. □

Nakayama's lemma holds for finitely generated modules over a semi-local ring R, if one replaces $\mathfrak{m}$ by the *Jacobson radical* of R, the intersection of all of its maximal ideals.

Corollary C.2 *Let M be a finite R-module over a local ring R and let $k = R/\mathfrak{m}$. Then $M/\mathfrak{m}M$ is a finite-dimensional k-vector space and its dimension is equal to the minimal number of generators of M over R.*

Proof Let r be the minimal number of generators of M over R and take $m_1, \dots, m_r$ which generate M over R. Their classes $\overline{m_1}, \dots, \overline{m_r}$ obviously generate $M/\mathfrak{m}M$ over k. This shows that $M/\mathfrak{m}M$ is a finite-dimensional k-vector space and that $\dim_k M/\mathfrak{m}M \leq r$. Conversely, if $\overline{v_1}, \dots, \overline{v_s}$ is any basis of $M/\mathfrak{m}M$ over k, then

$$M = \sum_{i=1}^{s} Rv_i + \mathfrak{m}M.$$

By Corollary C.1, $M \subset \sum_{i=1}^{s} Rv_i$, so $v_1, \dots, v_s$ generate M over R and hence, $s \geq r$. □

Besides the local ring $\mathcal{O}_{X,x_0}$, we will be interested in modules over $\mathcal{O}_{X,x_0}$. Every submodule of a finitely generated module over a Noetherian ring is itself finitely generated, and in particular this applies to the ideals of $\mathcal{O}_{X,x_0}$.

Definition C.1 If M is an R module and S is a multiplicatively closed subset of R containing 1, then the *localisation* of M at S, M_S, is the set

$$M_S = \{\frac{m}{s} : m \in M, s \in S\}$$

subject to the relation $m_1/s_1 = m_2/s_2$ if there exists $t \in S$ such that $t(s_2m_1 - s_1m_2) = 0$. It is an R module, and, indeed, an R_S-module. The localisation of M *at a prime ideal* p is the module M_S where $S = R \setminus p$. It is denoted by M_p. The passage from R to R_p simply makes every element not in p invertible.

The *support* of the R-module M, supp$M$, is the set of prime ideals p of R such that the localisation M_p is distinct from 0.

The set of all prime ideals in R is called Spec R. The following property is frequently useful:

Proposition C.2 *If*

$$0 \to M'' \to M \to M' \to 0$$

is an exact sequence of R-modules and $p \in \operatorname{Spec} R$*, then*

$$0 \to M''_p \to M_p \to M'_p \to 0$$

is also exact. □

Example C.1 $R = \mathscr{O}_n$, $M = \mathscr{O}_n/I$, where I is an ideal of $\mathscr{O}_n$. Then $(\mathscr{O}/I)_p = 0$ if and only if for all $r + I \in \mathscr{O}/I$, there exists $t \notin p$ such that $tr \in I$. This holds if and only if $I \not\subset p$. So

$$\operatorname{supp}(\mathscr{O}/I) = \{p \in \operatorname{Spec} \mathscr{O} : I \subset p\}.$$

◊

The set on the right here is better known as $V(I)$. Similarly, for $a \in R$, $V(a) = \{p \in \operatorname{Spec} R : a \in p\}$.

Exercises for Sect. C.1

1 Use Lemma C.1 to show that in the ring $R = \mathscr{O}_{\mathbb{R}^n,0}$ of C^∞ function germs, the ideal $\mathfrak{m}_n^\infty = \bigcap_{n=1}^\infty \mathfrak{m}_n^k$ is not finitely generated.

2 Show that for any ring R and R-module M,

(i) $\operatorname{supp} M = V(\operatorname{Ann}_R M)$.
(ii) The prime ideals of R_p are of the form qR_p where q is a prime ideal contained in p [Mat89, Theorem 4.1]. The prime ideals of R/p are of the form q/p, where q is a prime ideal containing p.
(iii) For $a \in R$, $\operatorname{supp} M/aM = \operatorname{supp} M \cap V(a)$. Hint: use Proposition C.2.

C.2 Associated Primes

Let M be an R-module. An *associated prime* of M is a prime ideal p such that $p = \operatorname{Ann}_R(m)$ for some $m \in M$. The set of all associated primes of M is denoted $\operatorname{Ass}_R M$, or $\operatorname{Ass} M$ when the ring R is clear from the context. If $p = \operatorname{Ann}_R(m)$, then $r \mapsto rm$ defines an injection $R/p \hookrightarrow M$, so $p \in \operatorname{Ass}(M)$ if M contains a submodule isomorphic to R/p.

Every ideal can be written as an intersection of primary ideals—proper ideals q with the property that if $fg \in q$ and $f \notin q$ then $g^n \in q$ for some n. This is the familiar *primary decomposition* of I. The radical of a primary ideal is prime, and the associated primes of R/I are the radicals of the primary ideals q_i appearing in a primary decomposition $I = \bigcap_i q_i$. Thus, the minimal associated primes define the irreducible components of $V(I)$. Non-minimal associated primes are described as *embedded.*

Example C.2

(i) In $\mathscr{O}_2$ we have

$$(x^2, xy) = (x) \cap (x, y).$$

Both ideals on the right are prime. The first is a minimal associated prime, and the second is an embedded associated prime.

(ii) In $\mathscr{O}_3$, let $I = (xy, yz, zx)$ and $J = (xyz, xy + yz + zx)$. Then

$$\begin{aligned} I =& (y, z) \cap (x, z) \cap (x, y) \\ J =& (y + z, z^2) \cap (x + z, x^2) \cap (x + y, y^2) \\ \operatorname{Ass} \mathscr{O}_3 / I = \operatorname{Ass} \mathscr{O}_3 / J =& \{(y, z), (x, z), (x, y)\} \end{aligned}$$

The $\mathfrak{m}$-primary ideal $\mathfrak{m}^2$ is in $V(I)$ and $V(J)$, but $\mathfrak{m}$ is not an associated prime of either quotient ring. It is an associated prime of $\mathscr{O}_3 / L$, where $L = (x^2yz, xy^2z, xyz^2, xy + yz + zx)$. And

$$V(L) = V(I) = V(J) = \{(y, z), (x, z), (x, y), (x, y, z)\}.$$

◊

Proposition C.3 *Let R be Noetherian and let M be an R-module. For every m in M there is a $p \in \operatorname{Ass} M$ such that $\operatorname{Ann}_R(m) \subset p$. Thus, the union of the associated primes is equal to the set of all zero-divisors on M.*

Proof Every maximal element in the set $\{\operatorname{Ann}_R(m) : m \in M\}$, ordered by inclusion, is prime. □

Note that in particular, $\operatorname{Ass} M \neq \emptyset$.

An element $r \in R$ is *regular* on M if r is not a zero-divisor on M and $aM \neq M$.

Exercises for Sect. C.2

1 For each of the ideals in Example C.2, and for each of the associated primes p of their quotient rings, find a monomorphism from $\mathscr{O}_3 / p$ to the quotient ring.

2 Let M be an R-module. Then

(i) Every associated prime of M contains $\operatorname{Ann} M$.
(ii) $\sqrt{\operatorname{Ann} M} = \bigcap_{p \in \operatorname{Ass} M} p$.

(iii) Ass $M \subset \text{supp}M$.
(iv) Harder: the minimal members of supp$M$ are in Ass M (see [Mat89, Theorem 6.5]).

3 Let M and N be R-modules. Show that

$$\text{supp}(M \otimes_R N) \subseteq \text{supp}(M) \cap \text{supp}(N).$$

The opposite inclusion does not always hold. For example, take $R = \mathbb{Z}$, $M = \mathbb{Q}$ and $N = \mathbb{Z}/2\mathbb{Z}$.

C.3 Dimension, Depth and Cohen–Macaulay Modules

There are many ways of characterising the dimension of Noetherian rings and modules. Here we describe four, which coincide, and complement one another, in the cases we are interested in.

C.3.1 Krull Dimension

The *height* of a prime ideal p of R is the maximum of the length, n, of a chain of strict inclusions

$$p_0 \subset p_1 \subset \cdots \subset p_n = p$$

where all of the p_i are prime ideals of R. The *Krull dimension* of a ring R is the maximum of the heights of prime ideals in R. Usually this is simply called the dimension of R.

Example C.3 The dimension of $\mathcal{O}_n$ is n. It is easy to check that this is so in the case $n = 1$: the only prime ideals are 0 and the maximal ideal (x). For $n > 1$, each of the ideals in the chain

$$0 \subset (x_1) \subset (x_1, x_2) \subset \cdots \subset (x_1, \ldots, x_n)$$

is prime, and there can be no prime strictly between any two of them, since in each case the successive quotients are isomorphic to $\mathcal{O}_1$. ◊

In a *catenary ring*, such as $\mathcal{O}_n$ and all of its quotients, the lengths of all maximal chains are the same.

The height of a not necessarily prime ideal I is defined to be the minimum of the heights of the prime ideals containing I. Geometrically, this says that the dimension of a reducible space or germ of space is the maximum of the dimensions of its irreducible components.

Definition C.2 The (Krull) dimension of a module, M, $\dim M$, is the Krull dimension of the ring $R/\operatorname{Ann}_R(M)$.

C.3.2 *Slicing Dimension*[1]

The *length* of an R- module M is the supremum of the lengths of chains of strict inclusions of R-modules

$$M_n \subset M_{n-1} \cdots \subset M_0 = M.$$

For example, the length of $\mathscr{O}_n$ is infinite, witness the chain

$$\cdots \subset \mathfrak{m}^n \subset \mathfrak{m}^{n-1} \subset \cdots \subset \mathfrak{m} \subset \mathscr{O}_n$$

whereas the length of the quotient ring $R := \mathscr{O}_n/(\mathfrak{m}^k)$ is finite: it is equal to the number of distinct monomials of degree $< k$ in n variables, since by adding one monomial at a time, one can form an increasing chain, of this length, of ideals, with each successive quotient having vector space dimension over $\mathbb{F}$ equal to 1. In fact,

Lemma C.2 *For any local ring R containing a field k such that the composite $k \to R \to R/\mathfrak{m}$ is an isomorphism, the length of an R-module M is equal to its dimension as a k-vector space.* □

The relation to dimension begins with the following proposition. The *radical* of an ideal I, usually denoted $\sqrt{I}$, is the set

$$\sqrt{I} = \{r \in R : r^k \in I \text{ for some } k > 0\}.$$

Proposition C.4 *Let R be a ring.*

1. *Let I be an ideal of R. Then*

$$\sqrt{I} = \bigcap_{p \in V(I)} p. \tag{C.4}$$

2. *Suppose that R is a local ring whose maximal ideal $\mathfrak{m}$ is finitely generated, and let $K = R/\mathfrak{m}$. Let M be a finitely generated R-module. Then*

$$\dim_K M < \infty \quad \Leftrightarrow \quad suppM = \{\mathfrak{m}\}.$$

[1]This is called *Chevalley dimension* in [dJP00]

Proof

1. That the left-hand side is contained in the right-hand side follows easily from the definitions. The opposite inclusion is less trivial; it involves showing that if $x \notin \sqrt{I}$ then a maximal element in the set of ideals containing I and not meeting $\{x, x^2, \ldots\}$ is prime. See [Mat89, p. 3].
2. Suppose that $\dim_K M < \infty$. Then the descending chain of inclusions

$$M \supset \mathfrak{m} M \supset \mathfrak{m}^2 M \supset \cdots$$

must be eventually constant. If $\mathfrak{m}^{k+1} M = \mathfrak{m}^k M$, then by Nakayama's lemma, $\mathfrak{m}^k M = 0$—for the hypothesis implies that $\mathfrak{m}^k M$ is finitely generated. As remarked above, the prime ideal p belongs to supp$M$ if and only if $p \supset \text{Ann}\, M$. The only prime ideal containing $\mathfrak{m}^k$ is $\mathfrak{m}$ itself.
Conversely, suppose that supp$M = \{\mathfrak{m}\}$. Then by 1.,

$$\text{radical}(\text{Ann}\, M) = \mathfrak{m}$$

and thus for each generator x_i of $\mathfrak{m}$, some finite power $x_i^{k_i}$ belongs to Ann M. Since $\mathfrak{m}$ is finitely generated, it follows that $\mathfrak{m}^k \subset \text{Ann}\, M$ for some finite k. Suppose that M is generated by n elements. Then there is an epimorphism

$$R^{\times n} \to M.$$

Since $\mathfrak{m}^k$ annihilates M, this epimorphism passes to the quotient to give an epimorphism

$$(R/\mathfrak{m}^k)^{\times n} \to M.$$

The left-hand term here is finite-dimensional over K (see [Mat89, page 12]). □

Definition C.3 Let R be a local ring and M an R-module. The *slicing dimension* $\delta(M)$ is defined to be the smallest integer n such that there exist $r_1, \ldots, r_n \in \mathfrak{m}$ such that

$$\text{length}(M/(r_1, \ldots, r_n)M) < \infty.$$

Each r_i slices the support of M, and $\delta(M)$ is the least number of slices that can reduce supp$M$ to a single point.

Example C.4 $\delta(\mathcal{O}_n) \leq n$, obviously, for $\dim_{\mathbb{C}} (\mathcal{O}_n/(x_1, \ldots, x_n)) < \infty$. That equality holds is less obvious; it follows from Theorem C.1 below. ◊

C.3.3 Hilbert–Samuel Dimension

There are $\binom{n+k-1}{k}$ monomials of degree k in n variables. Taking $\mathfrak{m} = \mathfrak{m}_{\mathbb{C}^n,0}$, this is therefore the $\mathbb{C}$-vector-space-dimension of $\mathfrak{m}^k / \mathfrak{m}^{k+1}$. It is a polynomial in k of degree n.

Definition C.4

1. Let M be a module over the local ring R . The *Hilbert–Samuel function* of M is the function $\chi_M : \mathbb{N} \to \mathbb{N}$ defined by

$$\chi_M(t) = \text{length}(\mathfrak{m}^t\, M/\, \mathfrak{m}^{t+1}\, M). \tag{C.5}$$

   For t large enough, it is a polynomial in t [Mat89, p. 97].
2. The *Hilbert–Samuel dimension* of M, $d(M)$, is the degree of this polynomial.

Remark C.1 For any ideal I such that $\mathfrak{m}^k \subset I$ for some finite k, one can similarly define

$$\chi_M^I(t) = \text{length}(I^t M/I^{t+1} M).$$

This too is a polynomial in the integer t for t large enough, and though its values may depend strongly on the choice of I, its degree does not [Mat89, pp 97–98]. ◊

C.3.4 Weierstrass Dimension

If $(X, x_0) \subset (\mathbb{C}^n, x_0)$ is a germ of analytic space and $\pi : X \to L^d$ is the linear projection to a d-dimensional coordinate plane L^d, then for a suitable choice of L, $\mathcal{O}_{X,x_0}$ is a finite module over $\mathcal{O}_{L^d,\pi(x_0)}$. The *Weierstrass dimension* of $d_W(X, x_0)$ is the smallest d for which this holds. Note that $\mathcal{O}_{X,x_0}$ is obviously finite over $\mathcal{O}_{\mathbb{C}^n,x_0}$, so the Weierstrass dimension of (X, x_0) is well defined, and $\leq n$. In fact the formal definition [dJP00, Definition 4.1.3] allows any finite map $X \to \mathbb{C}^d$ rather than just a coordinate projection, but this does not alter the value of d_W.

C.3.5 The Hauptidealsatz

The central result in the dimension theory of Noetherian local rings is the following.

Theorem C.1

1. Let R be a local Noetherian ring and M be a finitely generated R-module. Then

$$\dim M = \delta(M) = d(M).$$

Proof See [Mat89, Theorem 13.4]. □

For analytic spaces we also have

Theorem C.2 *If* (X, x_0) *is a germ of complex analytic space, then*

$$\dim \mathcal{O}_{X,x_0} = \delta(\mathcal{O}_{X,x_0}) = d(\mathcal{O}_{X,x_0}) = d_W(\mathcal{O}_{X,x_0}).$$

Proof See, e.g., [dJP00, Theorem 4.1.9]. □

From Theorem C.1 the celebrated *Hauptidealsatz* (principle theorem of ideal theory) of Wolfgang Krull follows:

Theorem C.3 *Let* R *be a Noetherian ring and* I *a proper ideal generated by* s *elements* $a_1, \ldots, a_s$ *. Then*

$$\text{height}(I) \leq s.$$

Proof Let p be a minimal overprime of I. In the local ring R_p, $\sqrt{IR_p} = pR_p$, for pR_p is the only prime ideal containing IR_p. Hence IR_p contains some power of pR_p and therefore $\text{length}(R_p/IR_p) < \infty$; that is, by slicing by the s elements $a_1, \ldots, a_s$ we reduce R_p to an R_p-module of finite length. Thus,

$$\text{height}\, I \leq \text{height}\, p = \dim R_p = \delta(R_p) \leq s.$$

□

Lemma C.3 ("Active Lemma", [dJP00, Corollary 4.1.10]) *Let* R *be a Noetherian local ring. If* $a \in R$ *is not in any of the minimal primes of* R*, then* $\dim R/(a) = \dim R - 1$.

Proof Any chain of prime ideals

$$\bar{p}_1 \subsetneq \bar{p}_2 \subsetneq \cdots \subsetneq \bar{p}_n$$

in $R/(a)$ corresponds to a chain $(a) \subset p_1 \subsetneq p_2 \subsetneq \cdots \subsetneq p_n$ in R. Because a is not in any minimal prime of R, p_1 is not minimal, so there is a longer chain in R. Hence $\dim R/(a) \leq \dim R - 1$. By Theorem C.3, the height of (a) is no greater than 1, so $\dim R/(a) \geq \dim R - 1$. □

An ideal is said to be *generically radical* if the ideal it generates in R_p is radical, for each of the minimal members p of $V(I)$.

Example C.5 The ideal $(xz, yz, z^2) = (x, y, z)^2 \cap (z) \subset \mathcal{O}_3$ is generically radical but not radical. ◊

Proposition C.5 *Let* $I \subset R$ *be an ideal. If* I *is unmixed, or more generally if* R/I *has no embedded associated primes, then* I *is radical if and only if it is generically radical.*

Proof I has an irredundant primary decomposition $I = q_1 \cap \cdots \cap q_n$ where each q_i is primary; the ideals $p_i := \sqrt{q_i}$ are the associated primes of R/I. The fact that I is generically radical means that $q_i = p_i$ for each minimal p_i. If there are no embedded associated primes, then all of the p_i are minimal, so I is an intersection of prime ideals, and thus is radical. □

Proposition C.6

(i) If (X, x_0) and (Y, x_0) are irreducible analytic subspaces of $(\mathbb{C}^n, x_0)$, then $\operatorname{codim} X \cap Y \leq \operatorname{codim} X + \operatorname{codim} Y$.

(ii) If $\varphi\colon (\mathbb{C}^n, x_0) \to (\mathbb{C}^p, y_0)$ is a germ of analytic map and $(Z, y_0) \subset (\mathbb{C}^p, y_0)$ is irreducible, then

$$\operatorname{codim}(\varphi^{-1}(Z), \mathbb{C}^n) \leq \operatorname{codim}(Z, \mathbb{C}^p).$$

Proof

(i) Let Δ be the diagonal in $\mathbb{C}^n \times \mathbb{C}^n$. Then $X \cap Y \simeq (X \times Y) \cap \Delta$. The codimension of $X \times Y$ in $\mathbb{C}^n \times \mathbb{C}^n$ is $\operatorname{codim} X + \operatorname{codim} Y$. The ideal of Δ is defined by n functions, and so $X \cap Y \simeq (X \times Y) \cap \Delta$ is obtained from $X \times Y$ by slicing by these n functions. Hence the dimension of $X \cap Y$ is no less than $\dim X + \dim Y - n$, by Theorem C.2. From this the formula for the codimension follows.

(ii) The graph embedding $\mathbb{C}^n \to \mathbb{C}^n \times \mathbb{C}^p$ sending x to $(x, \varphi(x))$ determines an isomorphism $\mathbb{C}^n \to \operatorname{graph}(\varphi)$ sending $\varphi^{-1}(Z)$ to $\operatorname{graph}(\varphi) \cap (\mathbb{C}^n \times Z)$. Hence the statement follows from (i).

□

Remark C.2

(i) The generalisation of Proposition C.6 to subspaces of a possibly singular variety W is false. For example, the two 2-planes $V(x_1, x_2)$ and $V(x_3, x_4)$ each have codimension 1 in $W := V(x_1x_3 - x_2x_4)$, but their intersection is just the single point $(0, 0, 0, 0)$, which has codimension 3 in W.

(ii) The difference between analytic and C^∞ geometry is exemplified by the fact that the intersection of smooth submanifolds of $\mathbb{R}^n$ of codimensions a and b may have codimension greater than $a + b$. ◊

C.3.6 Depth and Cohen–Macaulay Modules

The *depth* of an R-module is a more subtle measure than any of the notions of dimension we have considered. It is bounded above by dimension (Proposition C.7 below) but may diverge from it significantly.

Definition C.5 Let R be a Noetherian ring and M be an R-module.

1. A sequence $r_1, \ldots, r_s$ of elements of R is an *M-sequence*, or a *regular sequence for M*, if for each $i = 1, \ldots, s$, r_i is regular on $M/(r_1, \ldots, r_{i-1})M$, and $M/(r_1, \ldots, r_n)M \neq 0$.
2. If M is finitely generated and I is an ideal of R, all maximal M-sequences in I have the same length [Mat89, Theorem 16.7], which is called the *I-depth* of M. If R is a local ring with maximal ideal $\mathfrak{m}$, the $\mathfrak{m}$-depth of M is usually referred to simply as the *depth* of M. The I-depth of R itself is called simply depth I.

We remark that depth I is not the same as the depth of I as R-module: for example, consider $R = \mathcal{O}_n$: the depth of the maximal ideal $\mathfrak{m}$ as $\mathcal{O}_n$-module is 1, while the $\mathfrak{m}$-depth of the ring $\mathcal{O}_n$ (the depth of $\mathcal{O}_n$ as $\mathcal{O}_n$-module) is n (see Exercise 2 below). Nevertheless it generally held that this should not cause confusion, since "the meaning will be clear from the context". Some authors (e.g., Matsumura - see, e.g., [Mat89, page 132], but not Eisenbud—see [Eis95, page 425]) speak of "grade I" instead of "depth I" to avoid the problem. The risk of confusion only arises when I is an ideal of R; if M is an R-module but not an ideal, then there is only one meaning to the term "depth(M)".

Proposition C.7 *Let M be a finitely generated module over a Noetherian local ring R. Then*

$$\textit{depth}\ M \leq \dim M.$$

Proof If $a_1, \ldots, a_n$ is an M-sequence, then for each i we have a short exact sequence

$$0 \longrightarrow \frac{M}{(a_1,\ldots,a_{i-1})M} \xrightarrow{a_i} \frac{M}{(a_1,\ldots,a_{i-1})M} \longrightarrow \frac{M}{(a_1,\ldots,a_i)M} \longrightarrow 0\,.$$

Hence, by repeated applications of Exercise C.3.1 below, $\dim M/(a_1, \ldots, a_n)M = \dim M - n$. So $\dim M \geq n$. □

Now we can give one of the key definitions.

Definition C.6 A finitely generated module M over a Noetherian local ring R is *Cohen–Macaulay* if depth $M = \dim M$. In particular, R is called a *Cohen–Macaulay ring* if it is a Cohen–Macaulay R-module. A germ $(X, x_0) \subset (\mathbb{C}^n, x_0)$ is *Cohen–Macaulay* if its local ring O_{X,x_0} is Cohen–Macaulay.

Example C.6 $\mathcal{O}_n$ is Cohen–Macaulay (Exercise C.3.2 below). ◊

Proposition C.8 *If M is Cohen–Macaulay of dimension n, then* $\dim R/p = n$ *for all $p \in$* Ass M. *In particular, any Cohen–Macaulay module is unmixed.*

Proof See, e.g., [Mat89, Theorem 17.3]. □

Proposition C.9 *Let R be a Cohen–Macaulay local ring. For any ideal $I \subset R$,*

$$\operatorname{depth}(I, R) = \operatorname{height} I$$

and

$$\operatorname{height} I + \dim \frac{R}{I} = \dim R.$$

Proof See, e.g., [Mat89, Theorem 17.4]. □

Exercises for Sect. C.3

1 Let R be a Noetherian ring. Show

(i) If $a \in R$ is a regular element, then $\dim R/(a) = \dim R - 1$.
(ii) If M is a finitely generated R-module, and I is an ideal of R, then

$$\sqrt{\operatorname{Ann}(M/IM))} = \sqrt{\operatorname{Ann}(M) + I}.$$

Hint: suppose $m_1, \ldots, m_n$ generate M. If $r \in \operatorname{Ann}(M/IM)$, then for each i, $rm_i = \sum_j \alpha_{ij} m_j$ for some $\alpha_{ij} \in I$. Use Cramer's rule to show that $r^n \in \operatorname{Ann} M + I$.
(iii) If M is a finitely generated R-module and a is not in any of the minimal overprimes of $\operatorname{Ann} M$, then $\dim M/aM = \dim M - 1$.
(iv) If a is regular on M, then $\dim M/aM = \dim M - 1$.
(v) The converse to the Active Lemma is false. It is possible to reduce the dimension of a space by slicing by an element which is regular on a high-dimensional component but a zero divisor on a lower-dimensional component. Check this with the example of $\mathcal{O}_3 /(x, y) \cap (z)$.
(vi) However, there is a partial converse to the Active Lemma. An ideal $I \subset R$ is said to be *unmixed* if all of the associated primes of R/I have the same height. Suppose that I is unmixed and $\dim R/((a) + I) < \dim R/I$. Then a is regular on R/I. Hint: use Exercise 2.
(vii) A module M is called *unmixed* if $\operatorname{Ann} M$ is unmixed. Show that if M is unmixed and $\dim M/aM < \dim M$, then a is regular on M.

2 Show

(i) The R-depth of any R-module of finite length is 0.
(ii) The R-depth of any domain R is at least 1.
(iii) The sequence $x_1, \ldots, x_n$ is $\mathcal{O}_n$-regular; the depth of $\mathcal{O}_n$ as a module over itself is n.

(iv) The depth of the ideal $\mathfrak{m}_n \subset \mathcal{O}_n$, considered as an $\mathcal{O}_n$-module, is equal to 1. That is, whatever element $x \in \mathfrak{m}_n$ is chosen, every element of $\mathfrak{m}_n$ is a zero-divisor on $\mathfrak{m}/x\,\mathfrak{m}$.
(v) The depth of the $\mathcal{O}_4$-module $\mathcal{O}_4/(x_1x_3, x_1x_4, x_2x_3, x_2x_4)$ (the local ring of the union of two 2-planes meeting transversely) is 1. Its dimension is 2.

3 Show that the following spaces are *not* Cohen–Macaulay:

(i) The space consisting of the union of two 2-planes in 4-space, meeting transversely.
(ii) The space consisting of the union of two 2-planes in n-space, with $n > 4$, meeting in a single point. (Use (i)).
(iii) The space consisting of two three-planes meeting in general position in $\mathbb{C}^5$. Hint: slice this configuration by a hyperplane to get (i).
(iv) The space consisting of two k-planes in $\mathbb{C}^n$ with non-empty intersection of dimension less than $k-1$.

4 Show that the space consisting of two k-planes in n-space, with intersection of dimension $k-1$, is Cohen–Macaulay.

C.4 Free Resolutions

Let R be a local ring and M an R-module generated by $m_1, \ldots, m_n$. By mapping the i'th basis vector of $F_0 := R^{\times n}$ to m_i we obtain an epimorphism $\alpha_0 : F_0 \to M$. If in addition R is Noetherian, then $\ker \alpha_0$ is finitely generated and by the same procedure we obtain an epimorphism $F_1 \to \ker \alpha_0$, for some free module F_1. Continuing in this way we get an exact sequence

$$\cdots \longrightarrow F_k \xrightarrow{\alpha_k} F_{k-1} \xrightarrow{\alpha_{k-1}} \cdots \longrightarrow F_1 \xrightarrow{\alpha_1} F_0 \xrightarrow{\alpha_0} M \longrightarrow 0 \tag{C.6}$$

This terminates at some finite stage k if the kernel of α_k is free; if this does not happen, then it continues infinitely, in fact becoming periodic after a finite number of steps. In either case, the complete sequence is a *free resolution* of M. A free resolution is *minimal* if the image of each free module F_i is contained in $\mathfrak{m}\,F_{i-1}$. Any two minimal free resolutions have the same length; if this is finite, it is the shortest possible. If R is a regular local ring of dimension n, then every finitely generated R-module has a free resolution of length $\leq n$ [dJP00, Theorem 6.5.16]. If an R-module M has a finite length free resolution, the minimal length is called the *projective dimension* of M, and denoted by $\operatorname{pd}_R(M)$. If there is no finite-length free resolution, then $\operatorname{pd}_R(M) = \infty$.

The following theorem plays an important role in conservation of multiplicity.

Theorem C.4 (Auslander–Buchsbaum Formula) *Suppose that M is a finitely generated module over a Noetherian local ring R. If $\mathrm{pd}_R(M) < \infty$, then*

$$\mathrm{pd}_R(M) + \mathrm{depth}_R(M) = \mathrm{depth}_R(R).$$

Proof See, e.g., [dJP00, Theorem 6.5.20]. □

C.4.1 Cohen–Macaulay Modules and Freeness

The Auslander–Buchsbaum formula is useful for proving that a module is free, the key step in proving conservation of multiplicity. Typically, the situation is the following: M is a Cohen–Macaulay module over $\mathcal{O}_{X,x_0}$, $f\colon (X, x_0) \to (\mathbb{C}^d, y_0)$ is an analytic map, and $f|\colon (\mathrm{supp} M, x_0) \to (\mathbb{C}^d, y_0)$ is a finite map. This often occurs when $\mathbb{C}^d$ is the base of some deformation of a space or a singular mapping. Regarding M as an $\mathcal{O}_{\mathbb{C}^d,y_0}$-module via f, M is still Cohen–Macaulay, thanks to the following lemma, and if $\dim M = d$ then it will follow by the Auslander–Buchsbaum formula that M is free over $\mathcal{O}_{\mathbb{C}^d,y_0}$.

Lemma C.4 *Let $\varphi : R \to S$ be a homomorphism of Noetherian local rings, and M a finite S-module which is also finite when considered as an R-module via φ. Then*

$$\mathrm{depth}_R M = \mathrm{depth}_S M \quad \textit{and} \quad \dim_R M = \dim_S M.$$

Proof Neither depth nor dimension changes if we replace R and S by $R/\mathrm{Ann}_R M$ and $S/\mathrm{Ann}_S M$. Moreover φ passes to the quotient to define an injective homomorphism $R/\mathrm{Ann}_R M \to S/\mathrm{Ann}_S M$. So we may replace φ by this homomorphism, regard R as a subring of S, and suppose that $\mathrm{Ann}_S M = 0$. For each $s \in S$, consider the R-endomorphism φ_s of M given by multiplication by s. As M is finite over R, this satisfies a non-zero monic polynomial over R, by the Cayley Hamilton theorem: there exist $r_1, \ldots, r_n \in R$ such that

$$\varphi_s^n + r_1\varphi_s^{n-1} + \cdots + r_n = 0, \quad \text{i.e. for all } m \in M \quad (s^n + r_1 s^{n-1} + \cdots + r_n)m = 0.$$

As $\mathrm{Ann}_S M = 0$, this means that $s^n + r_1 s^{n-1} + \cdots + r_n = 0$. So S is an integral extension of R and therefore a finite extension (see, e.g., [Mat89, Theorem 9.1]). It follows that $\mathfrak{m}_S^k \subseteq \mathfrak{m}_R S$ for some k.

To show that $\mathrm{depth}_R M = \mathrm{depth}_S M$, let $r_1, \ldots, r_n \in \mathfrak{m}_R$ be a maximal M-sequence. Then the depth of $M/(r_1, \ldots, r_n)M$ is zero, and so $\mathfrak{m}_R$ annihilates some non-zero element of $M/(r_1, \ldots, r_n)M$. But then so must $\mathfrak{m}_S^k$, and thus $\mathrm{depth}_S(M/(r_1, \ldots, r_n)M) = 0$ and $r_1, \ldots, r_n$ is maximal as an M-sequence in S.

Because S is finite over R, $\dim R = \dim S$ (this is clear from the Weierstrass definition of dimension). Thus,

$$\dim_R M = \dim R/\operatorname{Ann}_R M = \dim S/\operatorname{Ann}_S M = \dim_S M.$$

□

Proposition C.10 *Let M be a Cohen–Macaulay $\mathcal{O}_{X,x_0}$-module of dimension d, and let $f\colon (X, x_0) \to (\mathbb{C}^d, y_0)$ be an analytic map such that M is a finite $\mathcal{O}_{\mathbb{C}^d,y_0}$-module via f^*. Then M is free over $\mathcal{O}_{\mathbb{C}^d,y_0}$.*

Proof Because M is Cohen–Macaulay over $S = \mathcal{O}_{X,x_0}$, it follows from the lemma that it is Cohen–Macaulay over $R = \mathcal{O}_{\mathbb{C}^d,y_0}$, of depth and dimension d. As R is a regular local ring, $\operatorname{pd}_R M < \infty$. Since $\operatorname{depth}_R M = d = \operatorname{depth} R$, by the Auslander–Buchsbaum formula we have $\operatorname{pd}_R M = 0$ —i.e. M is free over R. □

The Cohen–Macaulay property is essential here: if $d := \operatorname{depth} M < \dim M$ then $\operatorname{supp} M$ cannot be finite over $\mathcal{O}_{\mathbb{C}^d,0}$ (here we use the Weierstrass dimension of $\operatorname{supp} M$). So is the regularity of $\mathcal{O}_{\mathbb{C}^d,y_0}$: for example, there are modules of dimension d over the local ring of a singular hypersurface in $\mathbb{C}^{d+1}$, which are Cohen–Macaulay but not free—see, e.g., Lemma 11.5.

C.4.2 Examples of Cohen–Macaulay Spaces

Example C.7 A reduced curve-germ $(X, x_0) \subset (\mathbb{C}^n, x_0)$ is Cohen–Macaulay. Let R be a reduced 1-dimensional Noetherian local ring. Then the intersection of its prime ideals (equal to the set of nilpotents) is equal to 0. We may replace "all primes" by "all minimal primes". The maximal ideal $\mathfrak{m}$ is not minimal since R is 1-dimensional. By prime avoidance, there exists an $r \in \mathfrak{m}$ which does not lie in any minimal prime. This element is regular. ◊

Definition C.7 A germ $(X, x_0) \subset (\mathbb{C}^n, x_0)$ is a *complete intersection* if its radical ideal $I_X \subset \mathcal{O}_{\mathbb{C}^n,x_0}$ is generated by a regular sequence. Equivalently, I_X is generated by k elements and $\operatorname{codim} X = k$. (X, x_0) is a *set-theoretic complete intersection* if some possibly non-radical defining ideal is generated by $\operatorname{codim} X$ elements.

Example C.8 The union of the three coordinate axes in $\mathbb{C}^3$ has radical ideal generated by xy, yz, zx. Thus, it is not a complete intersection. Nevertheless, $V(xyz, xy+yz+zx) = V(xy, yz, zx)$, so it is a set-theoretic complete intersection. ◊

Exercise 1 proves the following result.

Proposition C.11 *Complete intersections are Cohen–Macaulay.* □

When R is a Cohen–Macaulay local ring and $I \subset R$ is an ideal, then R/I is Cohen–Macaulay of codimension 1 if and only if there exists a non-zero divisor $a \in R$ such that $I = (a)$ and in that case, R/I has a free resolution of the form

$$0 \longrightarrow R \xrightarrow{\ a\ } R \longrightarrow R/I \longrightarrow 0$$

This is an easy consequence of the Auslander–Buchsbaum formula.

The *Hilbert–Burch theorem* classifies all the ideals $I \subset R$ such that R/I is Cohen–Macaulay of codimension 2.

Theorem C.5 (Hilbert–Burch) *Let R be a Noetherian local ring.*

1. *Let $I \subset R$ be an ideal such that R/I has a free resolution of length 2*

$$0 \longrightarrow R^m \xrightarrow{\ \lambda\ } R^n \xrightarrow{\ \mu\ } R \longrightarrow R/I \longrightarrow 0$$

   Then, $n = m + 1$ and there exists a non-zero divisor $a \in R$ such that $I = a\,\mathrm{min}_m(\lambda)$ and for each $i = 1, \ldots, m+1$, $\mu_i = (-1)^i a M_i$, where M_i is the minor of λ obtained by omitting the ith row. Moreover, $\mathrm{depth}(\mathrm{min}_m(\lambda)) = 2$.
2. *Conversely, let λ be an $(m+1) \times m$ matrix with entries in R such that $grade(\mathrm{min}_m(\lambda)) \geq 2$, let $a \in R$ be a non-zero divisor and let $I = a\,\mathrm{min}_m(\lambda)$. Then R/I admits a free resolution of length 2 as in item 1, in which $n = m + 1$ and $\mu_i = (-1)^i a M_i$, for all $i = 1, \ldots, m+1$.*

Proof See [Eis95, Theorem 20.15]. □

The Cohen–Macaulay codimension 2 rings R/I are a particular case of the so-called determinantal rings. The definition is based on the following result.

Theorem C.6 *Let R be a Cohen–Macaulay local ring and let λ be an $m \times n$ matrix with entries in R. Let $I = \mathrm{min}_t(\lambda)$, with $1 \leq t \leq \min\{m, n\}$. Then:*

1. $\mathrm{codim}(R/I) \leq (m-t+1)(n-t+1)$, *and*
2. *if* $\mathrm{codim}(R/I) = (m-t+1)(n-t+1)$, *then R/I is Cohen–Macaulay.*

This is a classical theorem which was the result of work by many authors. The first complete statement appeared in [HE71]. The rings R/I obtained as in Theorem C.6 and such that $\mathrm{codim}(R/I) = (m-t+1)(n-t+1)$ are called *determinantal rings*. Thus, the theorem says that determinantal rings are Cohen–Macaulay. When $n = m + 1$ and $t = m$ we are exactly in the situation of the Hilbert–Burch theorem. Although this is not the original proof of the theorem, it is in fact a special case of the theorems of Kempf and de Concini–Strickland on the "variety of complexes"—see Theorems 9.2 and 9.3 in Sect. 9.4.

The analogue of determinantal rings for modules is due to Buchsbaum and Rim:

Theorem C.7 *Let φ be an $m \times n$ matrix of elements of a Cohen–Macaulay local ring R, with $n \geq m$, and let M be the cokernel of the map $R^n \to R^m$. Then* depth $\min_m(\varphi) \leq n - m + 1$, *and if it takes this value then* M *is a Cohen–Macaulay* R*-module of dimension* $\dim R - (n - m + 1)$.

This is due originally to Buchsbaum and Rim [BR64]. The statement here is a special case of [Eis95, Theorem A2.13]. The following statement can also be found in [BR64], though with difficulty—it's on page 224—and seems to have escaped the notice of some later writers (see [BV86, page 145 and page 157]).

Theorem C.8 *If, in the situation of Theorem C.7,* $\dim R = n - m + 1$, *and* $\dim M$ *takes its minimum possible value,* 0, *then*

$$\text{length } M = \text{length } R/\min_m(\varphi).$$

Exercises for Sect. C.4

1 Let M be a finitely generated module over the Noetherian local ring R, and suppose that $a_1, \ldots, a_n$ is a regular sequence for M. Show that $M/(a_1, \ldots, a_n)M$ is Cohen–Macaulay if and only if M is Cohen–Macaulay. Hint: A regular sequence of length n reduces dimension by n, by Exercise C.3.1. Every maximal M sequence has the same length, namely depth M.

2 A germ of an analytic space, (X, x_0), is Cohen–Macaulay if and only if $(X \times \mathbb{C}, (x_0, 0))$ is Cohen–Macaulay.

3

(i) Suppose that M and N are finitely generated modules over the Noetherian local ring R, and that

$$\cdots \longrightarrow F_n \longrightarrow F_{n-1} \longrightarrow \cdots \longrightarrow F_0 \longrightarrow M \longrightarrow 0 \qquad \text{(C.7)}$$

and

$$\cdots \longrightarrow G_n \longrightarrow G_{n-1} \longrightarrow \cdots \longrightarrow G_0 \longrightarrow N \longrightarrow 0 \qquad \text{(C.8)}$$

are free resolutions. Show that an R-linear map $\varphi : M \to N$ lifts to a morphism of complexes from (C.7) to (C.8).

(ii) Use (i) to show that any two minimal resolutions of the same module have the same length, so that $\text{pd}_R M$ is well defined.

(iii) Show that any two lifts of a morphism $M \to N$, as in (i), are chain homotopy equivalent.

4 The functor Ext can be defined as follows: if M and P are R-modules, then $\text{Ext}^k(M, P)$ is the k'th homology of the complex $\text{Hom}(F_\bullet, P)$, where $F_\bullet$ is the complex (C.7) with M replaced by 0. Use the previous exercise to show that for each k,

(i) $\mathrm{Ext}^k(M, P)$ is annihilated by $\mathrm{Ann}_R\, M$ and $\mathrm{Ann}_R\, P$.
(ii) $\mathrm{Ext}^k(\_, P)$ is a contravariant functor from R-modules to R-modules.

5

(i) Show that if M is a finite module of finite projective dimension over a local ring R then $\mathrm{pd}_R\, M = \max\{k : \mathrm{Ext}^k(M, R) \neq 0\}$.
(ii) Use the long exact sequence of the contravariant functor $\mathrm{Ext}(\_, R)$ arising from a short exact sequence, to prove that if

$$0 \to M'' \to M \to M' \to 0 \tag{C.9}$$

is a short exact sequence of R-modules of finite projective dimension, then

$$\mathrm{pd}\ M \leq \max\{\mathrm{pd}\ M'', \mathrm{pd}\ M'\}$$

and if the inequality is strict then

$$\mathrm{pd}\ M'' = \mathrm{pd}\ M' - 1.$$

(iii) Use the Auslander–Buchsbaum formula to deduce the *depth lemma*: given a short exact sequence (C.9), then $\mathrm{depth}\, M \geq \min\{\mathrm{depth}\, M'', \mathrm{depth}\, M'\}$, and if the inequality is strict then $\mathrm{depth}\, M'' = \mathrm{depth}\, M' + 1$. The statement and proof of the depth lemma without the hypothesis of finite projective dimension can be found in [dJP00, Lemma 6.5.18].

C.5 Pulling Back Algebraic Structures

In singularity theory, the objects we study are often the pull-backs of ideal objects located in jet spaces and bundles; for example, the set of critical points of a mapping $f : \mathbb{C}^n \to \mathbb{C}^p$, $n \geq p$ is the pull back of the set of matrices $\overline{\Sigma^{p-1}} \subset J^1(\mathbb{C}^n, \mathbb{C}^p)$, via the 1-jet extension map $j^1 f : \mathbb{C}^n \to J^1(\mathbb{C}^n, \mathbb{C}^p)$. Let $Z \subset J^k(\mathbb{C}^n, \mathbb{C}^p)$. Information about Z, together with information about the way that $j^k f$ meets Z, can give us information about $(j^k f)^{-1}(Z)$. An obvious example is transversality: if Z is smooth of codimension d in $J^k(\mathbb{C}^n, \mathbb{C}^p)$ and $j^k f$ is transverse to Z, then $(j^k f)^{-1}(Z)$, if not empty, is smooth of codimension d in X.

By Proposition C.6, we know that

$$\mathrm{codim}(j^k f^{-1}(Z), \mathbb{C}^n) \leq \mathrm{codim}(Z, J^k(\mathbb{C}^n, \mathbb{C}^p)).$$

For an $\mathscr{A}$-finite germ f and a geometrically meaningful subspace Z, this is an equality:

Lemma C.5 *Let $f\colon (\mathbb{C}^n, 0) \to (\mathbb{C}^p, 0)$ be an $\mathscr{A}$-finite germ, and $Z \subset J^k(\mathbb{C}^n, \mathbb{C}^p)$ a right-left invariant subspace of codimension $\leq n$. If $j^k f(0) \in Z$, then*

$$\operatorname{codim}(j^k f^{-1}(Z), \mathbb{C}^n) = \operatorname{codim}(Z, J^k(\mathbb{C}^n, \mathbb{C}^p)).$$

Proof Z has a right-left-invariant stratification $Z = \bigcup_i Z_i$. By Proposition A.1(2), for each i, $j^k f \pitchfork Z_i$ outside 0, and it follows that $\operatorname{codim}((j^k f)^{-1}(Z) \setminus \{0\}, \mathbb{C}^n) = \operatorname{codim}(Z, J^k(\mathbb{C}^n, \mathbb{C}^p))$. The addition of 0 does not alter the codimension if it is less than n; if it is n, then the equality holds already. □

This is important because of the following result.

Proposition C.12 *Suppose that $\varphi\colon (\mathbb{C}^n, x_0) \to (\mathbb{C}^N, z_0)$ is analytic, and that $Z \subset (\mathbb{C}^N, z_0)$ is Cohen–Macaulay. If* $\operatorname{codim}(\varphi^{-1}(Z), \mathbb{C}^n) = \operatorname{codim}(Z, \mathbb{C}^N)$*, then*

1. *$\varphi^{-1}(Z)$ is Cohen–Macaulay, and*
2. *The pull-back to $(\mathbb{C}^n, x_0)$ of a minimal free resolution of $\mathscr{O}_{Z,z_0}$ over $\mathscr{O}_{\mathbb{C}^N,z_0}$ is a minimal free resolution of $\mathscr{O}_{\varphi^{-1}(Z),x_0}$ over $\mathscr{O}_{\mathbb{C}^n,x_0}$.*

See Exercise C.5.2 below for an important application of this result. Proposition C.12 is a special case of a slightly more general result, which extends a result (Proposition 11.6) that we use in Chap. 11. In order to state it, we introduce the notion of fibre product. The set-theoretical version of it already was used in Chap. 8, but here we give a more general definition for complex space germs.

Let $f\colon (X, S) \to (Y, y_0)$ and $g\colon (Z, S') \to (Y, y_0)$ be two analytic maps between complex space germs. Assume that their local rings are $\mathscr{O}_{X,S} = \mathscr{O}_n/I$, $\mathscr{O}_{Z,S'} = \mathscr{O}_m/J$ and $\mathscr{O}_{Y,y_0} = \mathscr{O}_p/L$, for some ideals $I \subset \mathscr{O}_n$, $J \subset \mathscr{O}_m$ and $L \subset \mathscr{O}_p$. The *fibre product over Y* is defined as the complex space germ $(X \times_Y Z, S \times_Y S')$ whose local ring is $\mathscr{O}_{X\times_Y Z, S\times_Y S'} = \mathscr{O}_{m+n}/K$, where $K \subset \mathscr{O}_{m+n}$ is the ideal generated by the functions

$$f_i(x) - g_i(z),\ i = 1, \ldots, p.$$

As a set-germ, $(X \times_Y Z, S \times_Y S')$ is given by the pairs $(x, z) \in X \times Z$ such that $f(x) = g(z)$. It is easy to see that the restriction of the projections onto the factors are in fact analytic maps of complex space germs, so we have a commutative diagram

$$\begin{array}{ccc} (X, S) & \xrightarrow{\ f\ } & (Y, y_0) \\ \Big\uparrow{\scriptstyle \pi_1} & & \Big\uparrow{\scriptstyle g} \\ (X \times_Y Z, S \times_Y S') & \xrightarrow{\ \pi_2\ } & (Z, S') \end{array}$$

When (Y, y_0) is a single point, f and g are constant maps and in that case the fibre product is called the *Cartesian product* of two complex space germs $(X \times Z, S \times S')$.

In algebraic geometry, the structural ring of the Cartesian product of two schemes is given by the tensor product of their corresponding rings. But in analytic geometry, the usual tensor product has to be substituted by the analytic tensor product. Let R and S be two analytic $\mathbb{C}$-algebras given by $R = \mathcal{O}_n/I$ and $S = \mathcal{O}_m/J$ for some ideals $I \subset \mathcal{O}_n$, $J \subset \mathcal{O}_m$. The *analytic tensor product* is defined as

$$R \widehat{\otimes} S = \frac{\mathcal{O}_{m+n}}{I\mathcal{O}_{m+n} + J\mathcal{O}_{m+n}}.$$

It is obvious from the definitions that if (X, x_0) and (Z, z_0) are complex space germs, then

$$\mathcal{O}_{X \times Z,(x_0,z_0)} = \mathcal{O}_{X,x_0} \widehat{\otimes} \mathcal{O}_{Z,z_0}.$$

Proposition C.13 *Suppose given a fibre product of complex space germs as follows:*

$$\begin{array}{ccc} (X, S) & \xrightarrow{F} & (Y, y_0) \\ \uparrow & & \uparrow i \\ (X \times_Y Z, S_0) & \xrightarrow{f} & (Z, z_0) \end{array} \tag{C.10}$$

where F is finite. Suppose also that (X, S) is Cohen–Macaulay, (Y, y_0) and (Z, z_0) are smooth and that $\operatorname{codim}(F(X), Y) = \operatorname{codim}(f(X \times_Y Z), Z)$. *Then:*

1. *$(X \times_Y Z, S_0)$ is also Cohen–Macaulay, and*
2. *The pull-back to (Z, z_0) of a minimal resolution of $\mathcal{O}_{X,S}$ over $\mathcal{O}_{Y,y_0}$ is a minimal free resolution of $\mathcal{O}_{X \times_Y Z, S_0}$ over $\mathcal{O}_{Z,z_0}$.*

We need the following lemma, which can be proved by an elementary diagram chase:

Lemma C.6 *If*

$$\cdots \longrightarrow M_{n+1} \xrightarrow{\alpha_{n+1}} M_n \xrightarrow{\alpha_n} M_{n-1} \xrightarrow{\alpha_{n-1}} \cdots \tag{C.11}$$

is an exact sequence of R-modules and $x \in R$ is regular on each M_i, then

$$\cdots \longrightarrow M_{n+1}/xM_{n+1} \xrightarrow{\bar{\alpha}_{n+1}} M_n/xM_n \xrightarrow{\bar{\alpha}_n} M_{n-1}/xM_{n-1} \xrightarrow{\bar{\alpha}_{n-1}} \cdots \tag{C.12}$$

is exact. □

Proof (Proof of Proposition C.13) The proof follows the same strategy as in the proof of Proposition 11.6. We first observe that item 1 follows from item 2, by the Auslander–Buchsbaum formula Theorem C.4. To prove item 2 we proceed in three steps:

Step 1 Let i be a submersion. By taking coordinates in Y and Z we may assume that $(Y, y_0) = (\mathbb{C}^n, 0)$, $(Z, z_0) = (\mathbb{C}^n \times \mathbb{C}^q, 0)$ and $i(y,t) = y$. But this implies that $(X \times_Y Z, S_0) = (X \times \mathbb{C}^q, S_0)$ and that $f = F \times \mathrm{id}_{\mathbb{C}^q}$. By induction on q, it suffices to prove item 2 in the case $q = 1$.

Let

$$0 \longrightarrow \mathcal{O}_n^{d_k} \xrightarrow{\alpha_k} \mathcal{O}_n^{d_{k-1}} \xrightarrow{\alpha_{k-1}} \cdots \longrightarrow \mathcal{O}_n^{d_0} \xrightarrow{\alpha_0} \mathcal{O}_{X,S} \longrightarrow 0$$

be a minimal free resolution of $\mathcal{O}_{X,S}$. Then

$$0 \longrightarrow \mathcal{O}_{n+1}^{d_k} \xrightarrow{i^*\alpha_k} \mathcal{O}_{n+1}^{d_{k-1}} \xrightarrow{i^*\alpha_{k-1}} \cdots \longrightarrow \mathcal{O}_{n+1}^{d_0} \xrightarrow{i^*\alpha_0} \mathcal{O}_{X\times\mathbb{C},S_0} \longrightarrow 0$$

is obviously a complex. For simplicity, we put $\alpha_\ell = 0$, when $\ell > k$ or $\ell < 0$. For each ℓ, let $a \in \ker(i^*\alpha_\ell)$, with $a \in \mathcal{O}_{n+1}^{d_\ell}$. Write $a = b + tc$, where $b \in \mathcal{O}_n^{d_\ell}$ and $c \in \mathcal{O}_{n+1}^{d_\ell}$. Taking $t = 0$, we get $b \in \ker(\alpha_\ell) = \mathrm{im}\,(\alpha_{\ell+1}) \subset \mathrm{im}\,(i^*\alpha_{\ell+1})$ and $c \in \ker(i^*\alpha_\ell)$. This shows that

$$\ker(i^*\alpha_\ell) \subset \mathrm{im}\,(i^*\alpha_{\ell+1}) + \mathfrak{m}_{n+1}\ker(i^*\alpha_\ell),$$

and hence $\ker(i^*\alpha_\ell) \subset \mathrm{im}\,(i^*\alpha_{\ell+1})$, by Nakayama's lemma C.1. Thus, the induced complex is also exact. Moreover, it is minimal again by the Auslander–Buchsbaum formula.

Step 2 Let i be an embedding. In this case we choose the coordinates in such a way that $(Y, y_0) = (\mathbb{C}^n \times \mathbb{C}^q, 0)$, $(Z, z_0) = (\mathbb{C}^n, 0)$ and $i(z) = (z, 0)$. In this case, $(X \times_Y Z, S)$ is the complex subspace of (X, S) given by the vanishing of $F_{n+1}, \ldots, F_{n+q}$ and f is the restriction of F. As in Step 1, we can proceed by induction on q, so it is enough to prove it for $q = 1$.

Since (X, S) is Cohen–Macaulay and $(X \cap F_{n+1}^{-1}(0), S)$ has dimension equal to $\dim(X, S) - 1$, it follows that $y_{n+1} \in \mathcal{O}_{n+1}$ is regular on $\mathcal{O}_{X,S}$. Moreover, for each free module $\mathcal{O}_{n+1}^{d_\ell}$ in the minimal resolution, y_{n+1} is also regular on $\mathcal{O}_{n+1}^{d_\ell}$. Hence, the result is a consequence of Lemma C.6.

Step 3 In the general case, just write the map i as a composition $i = i_2 \circ i_1$, where i_1 is the embedding $(Z, z_0) \to (Z \times Y, (z_0, y_0))$ given by $z \mapsto (z, i(z))$ and i_2 is the projection $(Z \times Y, (z_0, y_0)) \to (Y, y_0)$ onto the second factor. Now, apply the steps 1 and 2.

□

Example C.9 Let $f\colon (\mathbb{C}^n, 0) \to (\mathbb{C}^p, 0)$ be an analytic germ, with $n < p$, and denote by Σ_f the non-immersive locus of f. Then

$$\operatorname{codim}(\Sigma_f) \le p - n + 1,$$

and in case of equality, Σ_f is Cohen–Macaulay. In fact, Σ_f is defined by the maximal ($= n \times n$) minors of the Jacobian matrix

$$\left(\frac{\partial f_i}{\partial x_j}\ 1 \le i \le p,\ 1 \le j \le n\right)$$

of f. So the claim is just an application of Theorem C.6. ◊

Example C.10 The conclusion of Theorem C.12(ii) may be false if Z is not Cohen–Macaulay, even when the codimension of $f^{-1}(Z)$ in $\mathbb{C}^n$ is equal to that of Z in $\mathbb{C}^N$. Here is an example. Let $n = 2$ and $N = 4$, let $Z \subset \mathbb{C}^4$ be the union of the 2-planes $\{w = x = 0\}$ and $\{y = z = 0\}$, and let $\varphi : \mathbb{C}^2 \to \mathbb{C}^4$ be defined by $\varphi(u, v) = (u, v, u, v)$. Then $\varphi^{-1}(Z) = \{(0, 0)\}$, and so the codimension condition holds. It is well known that W is not Cohen–Macaulay: $\mathcal{O}_{Z,0}$ has dimension 2 and depth 1. Thus, its minimal free resolution over $\mathcal{O}_Y$ is of length 3. The ideal of Z is minimally generated by wy, wz, xy, xz; when these four generators are composed with φ, the second and third generators coincide. This relation is not in the image of the first differential of the pulled back resolution, and thus the pull-back is no longer exact. It is worth noting that because it is zero-dimensional, $f^{-1}(Z)$ is automatically Cohen–Macaulay. ◊

Example C.11 It can be proved that the following is a minimal free resolution of the quotient $\mathbb{C} = \mathcal{O}_k / \mathfrak{m}_k$ as $\mathcal{O}_k$-algebra (we write $\mathcal{O}$ in place of $\mathcal{O}_k$):

$$0 \longrightarrow \mathcal{O} \xrightarrow{d_k} \bigwedge^{k-1} \mathcal{O}^k \xrightarrow{d_{k-1}} \bigwedge^{k-2} \mathcal{O}^k \xrightarrow{d_{k-2}}$$

$$\cdots \longrightarrow \bigwedge^2 \mathcal{O}^k \xrightarrow{d_2} \mathcal{O}^k \xrightarrow{d_1} \mathcal{O} \longrightarrow \mathbb{C} \longrightarrow 0 \qquad \text{(C.13)}$$

where $d_\ell : \bigwedge^\ell \mathcal{O}^k \to \bigwedge^{\ell-1} \mathcal{O}^k$ is given by

$$e_{i_1} \wedge \cdots \wedge e_{i_\ell} \mapsto \sum_j (-1)^{j-1} x_j e_{i_1} \wedge \cdots \wedge \widehat{e_{i_j}} \wedge \cdots \wedge e_{i_\ell}$$

and in particular d_1 sends e_i to x_i. By Proposition C.12, it follows from this that $f_1, \ldots, f_k \in \mathcal{O}_n$ form a regular sequence (and thus define a complete intersection) if and only if the *Koszul complex*, namely the complex obtained from (C.13) by replacing $\mathcal{O}_k$ by $\mathcal{O}_n$, and x_i by f_i for $i = 1, \ldots, k$, is exact. ◊

Exercises for Sect. C.5

1 State and prove a version of Proposition C.12 for modules in place of rings.

2 Use Proposition C.12 to deduce Theorem C.6 from the special case of "generic determinantal varieties", where the local ring R of the theorem is replaced by its residue field $R/\mathfrak{m}$.

C.6 Samuel Multiplicity

Let R be a Noetherian local ring of dimension d. An ideal $\mathfrak{q} \subset R$ is $\mathfrak{m}$-*primary* if $\mathfrak{m}^k \subset \mathfrak{q} \subset \mathfrak{m}$ for some $k \geq 1$. This is equivalent to having $\sqrt{\mathfrak{q}} = \mathfrak{m}$ and to having $\text{length}(R/\mathfrak{q}) < \infty$, by Proposition C.4. If, in addition, $\mathfrak{q}$ is generated by d elements, then $\mathfrak{q}$ is called a *parameter ideal*.

Definition C.8 The *Samuel function* of an R-finite module M with respect to an $\mathfrak{m}$-primary ideal $\mathfrak{q}$ is defined as

$$\chi_M^{\mathfrak{q}}(t) = \text{length}\left(\frac{M}{\mathfrak{q}^{t+1}M}\right).$$

For large enough values of t, this function is in fact a polynomial in $\mathbb{Q}[t]$ of degree equal to $\dim M \leq d$ (see [Mat89, pp. 97–98]). Moreover, $\chi_M^{\mathfrak{q}}(t)$ can only take integer values, so it is easy to see by induction on d (using the fact that $\chi(t+1)-\chi(t)$ has the same property) that

$$\chi_M^{\mathfrak{q}}(t) = \frac{e}{d!}t^d + \text{terms of lower order}, \tag{C.14}$$

for some integer $e \geq 0$.

Definition C.9 The number e in (C.14) is called the *multiplicity* of M with respect to $\mathfrak{q}$ and is denoted by $e(\mathfrak{q}, M)$. When $\mathfrak{q} = \mathfrak{m}$, we put $e(\mathfrak{m}, M) = e(M)$ and call it the *multiplicity* of M.

Here we summarise some basic properties of the multiplicity. The proofs of all these properties can be found in [Mat89, pp. 107–109].

Proposition C.14 *Let R be a Noetherian local ring of dimension d, $\mathfrak{q}$ an $\mathfrak{m}$-primary ideal of R and M a finite R-module. We have:*

1. $e(\mathfrak{q}, M) = \lim_{t\to\infty} \frac{d!}{t^d} \text{length}(M/\mathfrak{q}^t M)$.

2. *Additivity: let* $\{\mathfrak{p}_1, \ldots, \mathfrak{p}_r\}$ *be the minimal primes of* R *such that* $\dim R/\mathfrak{p}_i = d$, *then*

$$e(\mathfrak{q}, M) = \sum_{i=1}^{r} e(\overline{\mathfrak{q}_i}, R/\mathfrak{p}_i)\,\mathrm{length}_{R_{\mathfrak{p}_i}}(M_{\mathfrak{p}_i}),$$

where $\overline{\mathfrak{q}_i}$ *is the image of* $\mathfrak{q}$ *in* $R/\mathfrak{p}_i$.

3. *Assume* R *is an integral domain, then*

$$e(\mathfrak{q}, M) = e(\mathfrak{q}, R)\,\mathrm{rank}(M).$$

4. *If* $\mathfrak{q}$ *is a parameter ideal, then* $e(\mathfrak{q}, R) \leq \mathrm{length}(R/\mathfrak{q})$.

The following result characterises Cohen–Macaulay Noetherian local rings in terms of their multiplicity with respect to a parameter ideal.

Theorem C.9 *The following three conditions on a Noetherian local ring* R *are equivalent:*

1. R *is Cohen–Macaulay,*
2. $e(\mathfrak{q}, R) = \mathrm{length}(R/\mathfrak{q})$, *for some parameter ideal* $\mathfrak{q}$,
3. $e(\mathfrak{q}, R) = \mathrm{length}(R/\mathfrak{q})$, *for any parameter ideal* $\mathfrak{q}$.

Proof See [Mat89, Theorem 17.11]. □

We can give an analogue of Theorem C.9 for modules, in the particular case that $R = \mathcal{O}_d$ and $\mathfrak{q} = \mathfrak{m}_d$.

Proposition C.15 *For any finite* $\mathcal{O}_d$*-module* M, *we have*

$$e(M) \leq \dim_{\mathbb{C}} \frac{M}{\mathfrak{m}_d M},$$

with equality if and only if M *is Cohen–Macaulay of dimension* d.

Proof The ring $\mathcal{O}_d$ is an integral domain and by item (6) of Proposition C.14,

$$e(M) = e(\mathcal{O}_d)\,\mathrm{rank}(M) = \mathrm{rank}(M) = \dim_{Q(\mathcal{O}_d)}(M \otimes_{\mathcal{O}_d} Q(\mathcal{O}_d)),$$

where $Q(\mathcal{O}_d)$ is the field of fractions of $\mathcal{O}_d$. On the other hand, by Corollary C.2, $\dim_{\mathbb{C}} M/\mathfrak{m}_d M$ is equal to the minimal number of generators of M over $\mathcal{O}_d$. If $\{g_1, \ldots, g_r\}$ is a minimal generating system of M over $\mathcal{O}_d$, with $r = \dim_{\mathbb{C}} M/\mathfrak{m}_d M$, then $\{g_1 \otimes 1, \ldots, g_r \otimes 1\}$ is also a generating system of $M \otimes_{\mathcal{O}_d} Q(\mathcal{O}_d)$ over $Q(\mathcal{O}_d)$ and hence, $e(M) \leq r$.

The equality $e(M) = r$ holds if and only if $\{g_1 \otimes 1, \ldots, g_r \otimes 1\}$ is a basis of $M \otimes_{\mathcal{O}_d} Q(\mathcal{O}_d)$ over $Q(\mathcal{O}_d)$, which is equivalent to $\{g_1, \ldots, g_r\}$ being a basis of M over $\mathcal{O}_d$. This shows that $e(M) = r$ if and only if M is free over $\mathcal{O}_d$. By the

Auslander–Buchsbaum formula Theorem C.4, this is equivalent to M being Cohen–Macaulay of dimension d. □

We have seen that the multiplicity of a ring or module with respect to a parameter ideal has very nice properties. In a certain sense, the general case of an $\mathfrak{m}$-primary ideal can be reduced to this one by taking a reduction of the ideal. This method was introduced by Northcott and Rees in[NR54]. Given any ring R and an ideal $I \subset R$, a *reduction* of I is another ideal J such that $J \subset I$ and $I^{k+1} = JI^k$, for some $k \geq 1$.

Lemma C.7 *Let R be a Noetherian local ring and $\mathfrak{q}$ an $\mathfrak{m}$-primary ideal in R. If $\mathfrak{b}$ is a reduction of $\mathfrak{q}$, then $\mathfrak{b}$ is also $\mathfrak{m}$-primary and for any finite R-module M, $e(\mathfrak{q}, M) = e(\mathfrak{b}, M)$.*

Proof If $\mathfrak{q}^{k+1} = \mathfrak{b}\mathfrak{q}^k$ for some k, then $\mathfrak{q}^{k+1} \subset \mathfrak{b} \subset \mathfrak{q}$, so $\mathfrak{b}$ is also $\mathfrak{m}$-primary. Moreover, we have that $\mathfrak{q}^{k+t} = \mathfrak{b}^t\mathfrak{q}^k$, for all $t \geq 1$, hence

$$\text{length}\left(\frac{M}{\mathfrak{b}^{k+t}M}\right) \geq \text{length}\left(\frac{M}{\mathfrak{q}^{k+t}M}\right) = \text{length}\left(\frac{M}{\mathfrak{b}^t\mathfrak{q}^k M}\right) \geq \text{length}\left(\frac{M}{\mathfrak{b}^t M}\right).$$

The equality $e(\mathfrak{q}, M) = e(\mathfrak{b}, M)$ follows by taking limits when $t \to \infty$, by item (1) of Proposition C.14. □

Theorem C.10 *Let R be a Noetherian local ring of dimension d with infinite residue field $k = R/\mathfrak{m}$. Let $\mathfrak{q} = (u_1, \ldots, u_s)$ be an $\mathfrak{m}$-primary ideal and consider d linear combinations $y_i = \sum_{i=1}^{s} a_{ij}u_j$, with $i = 1, \ldots, d$ and $a_{ij} \in R$. If the coefficients a_{ij} are generic, then the ideal $\mathfrak{b} = (y_1, \ldots, y_d)$ is a parameter ideal which gives a reduction of $\mathfrak{q}$.*

Remark C.3

(i) The precise meaning of the term "generic" in Theorem C.10 is the following. Consider the space $M_{s\times d}(k)$ of $s \times d$ matrices with coefficients in k and let $(\overline{a_{ij}}) \in M_{s\times d}(k)$ the matrix given by the residue classes of the coefficients a_{ij}. Then, there exists a non-empty Zariski open subset $D \subset M_{s\times d}(k)$ such that $\mathfrak{b}$ is a reduction of $\mathfrak{q}$, for all a_{ij} such that $(\overline{a_{ij}}) \in D$.

(ii) In the particular case that R is a k-algebra (for instance, when $R = \mathcal{O}_{X,x_0}$), then we can choose the generic coefficients a_{ij} in k, so the reduction is generated by k-linear combinations of the generators of $\mathfrak{q}$.

Exercises for Sect. C.6

1 Show that $e(\mathcal{O}_d) = 1$.

Appendix D
Local Analytic Geometry

We define analytic spaces, and introduce notions of local analytic geometry, beginning with the Weierstrass Preparation Theorem, Malgrange's version for smooth geometry, and Mather's module-theoretic reformulations. We discuss normal rings and the normalisation of an analytic space germ.

D.1 The Preparation Theorem

The preparation theorem as used in this book has rather an algebraic appearance, but is in fact a theorem of analysis. The classical Weierstrass preparation theorem on which it is based concerns division of analytic functions, and is more evidently "analytic". To state it we introduce the following notation: if $x = (x_1, \ldots, x_n)$ is a point of $\mathbb{C}^n$, then we write $x' = (x_1, \ldots, x_{n-1})$.

Theorem D.1 (Weierstrass Preparation Theorem) *Let $f \in \mathcal{O}_n$ and suppose that $f(0, \ldots, 0, x_n) = c_k x_n^k +$ higher order terms, where $c_k \neq 0$. Then there exist a unique unit $u \in \mathcal{O}_n$ and unique functions $a_1, \ldots, a_k \in \mathcal{O}_{n-1}$ such that*

$$f(x) = u(x)\left(x_n^k + a_1(x')x_n^{k-1} + \cdots + a_{k-1}(x')x_n + a_k(x')\right)$$

and $a_i(0) = 0$ for all $i = 1, \ldots, k$.

Proof See, e.g., [dJP00, Theorem 3.2.4] □

Thus, up to multiplication by a unit, f is a polynomial in x_n with coefficients in $\mathcal{O}_{n-1}$. The theorem cannot be immediately applied to every $f \in \mathcal{O}_n$: for example, $f(x_1, x_2) = x_1 x_2$ is identically zero on each coordinate axis. But a linear change of coordinates suffices to make the theorem applicable to any given non-zero function.

D. Mond, J. J. Nuño-Ballesteros, *Singularities of Mappings*,
Grundlehren der mathematischen Wissenschaften 357,
https://doi.org/10.1007/978-3-030-34440-5

The preparation theorem is the basis of many results in several complex variables and analytic geometry. Its geometrical import is the fact that it shows that for any non-zero germ of function, after a linear coordinate change the projection of its zero set to the hyperplane $\{x_n = 0\}$ is finite-to-one.

A version of the theorem for C^∞ functions was proved by Bernard Malgrange in 1962–64 (see [Mal67]), at the urging of René Thom, and made possible Thom's Catastrophe Theory, and Mather's celebrated series of papers on the stability of C^∞ mappings, [Mat68a, Mat69a, Mat68b, Mat69b, Mat70, Mat71]. Mather himself [Mat68a], Nirenberg [Nir71] and Łojasiewicz [Loj71] published alternative proofs. Now we look at Mather's "algebraic" formulation [Mat68a]. When $\mathbb{F} = \mathbb{C}$ we suppose that (X, x_0) and (Y, y_0) are germs of analytic spaces, and that $f: (X, x_0) \to (Y, y_0)$ is an analytic map-germ. In the real case $\mathbb{F} = \mathbb{R}$, we suppose that $f: (X, x_0) \to (Y, y_0)$ is a smooth map-germ between manifolds. Then any $\mathscr{O}_{X,x_0}$-module M acquires the structure of $\mathscr{O}_{Y,y_0}$-module "via f": for $h \in \mathscr{O}_{Y,y_0}$ and $m \in M$, we define

$$h \cdot m = (h \circ f)m.$$

In general, even when M is finitely generated over $\mathscr{O}_{X,x_0}$, it may not be finitely generated over $\mathscr{O}_{Y,y_0}$—an example is given below. The preparation theorem gives a simple condition guaranteeing finiteness over $\mathscr{O}_{Y,y_0}$.

Theorem D.2 *Let $f: (X, x_0) \to (Y, y_0)$ be an analytic map-germ between analytic spaces when $\mathbb{F} = \mathbb{C}$ or a smooth map-germ between manifolds when $\mathbb{F} = \mathbb{R}$. Let M be a finitely generated $\mathscr{O}_{X,x_0}$-module. The following statements are equivalent.*

1. *M is finitely generated over $\mathscr{O}_{Y,y_0}$ via f.*
2. $\dim_{\mathbb{F}} M/f^* \mathfrak{m}_{Y,y_0} M < \infty$.

The deduction of Theorem D.2 from Theorem D.1 can be found in [Mat68a], in [Mar82, Chapter], and in [GG73, Chapter IV] for C^∞ mappings between smooth manifolds. The case of analytic map-germs between analytic set-germs can be found in [dJP00, Theorem 3.2.10].

From the conclusion that M is finitely generated over $\mathscr{O}_{Y,y_0}$, one can easily go on to find a set of generators:

Corollary D.1 *Suppose, with the hypotheses of Theorem D.2, that $M/f^* \mathfrak{m}_{Y,y_0} M$ is generated over $\mathbb{F}$ by the classes in $M/f^* \mathfrak{m}_{Y,y_0} M$ of the elements $m_1, \ldots, m_n \in M$. Then M is generated over $\mathscr{O}_{Y,y_0}$ by the m_i, $i = 1, \ldots, n$, themselves.*

Proof For brevity, write $\mathscr{O}_Y$ and $\mathfrak{m}_Y$ in place of $\mathscr{O}_{Y,y_0}$ and $\mathfrak{m}_{Y,y_0}$. From the hypothesis we deduce immediately that

$$M = f^* \mathfrak{m}_Y M + \mathrm{Sp}_{\mathbb{F}}\{m_1, \ldots, m_n\}.$$

As we are thinking of M as an $\mathcal{O}_Y$-module, we can rewrite this as

$$M = \mathfrak{m}_Y\, M + \mathrm{Sp}_{\mathbb{F}}\{m_1, \ldots, m_n\}.$$

Let N be the $\mathcal{O}_Y$-submodule of M generated by $m_1, \ldots, m_n$. Then the last equality clearly implies that

$$M = \mathfrak{m}_Y\, M + N.$$

Now because the preparation theorem tells us that M is a finitely generated $\mathcal{O}_Y$-module, we can conclude, by Nakayama's lemma C.1, that $M = N$. □

Example D.1 Let $f : (\mathbb{F}^3, 0) \to (\mathbb{F}^3, 0)$ be defined by $f(x, y, z) = (x^2, y^2, z^2)$. Then

$$\frac{\mathcal{O}_3}{f^*\,\mathfrak{m}_3\,\mathcal{O}_3} = \frac{\mathcal{O}_3}{(x^2, y^2, z^2)} = \mathrm{Sp}_{\mathbb{C}}\{1, x, y, z, xy, yz, zx\}$$

so every $h \in \mathcal{O}_3$ can be written in the form

$$\begin{aligned} h &= (c_{000} \circ f) + (c_{100} \circ f)x + (c_{010} \circ f)y + (c_{001} \circ f)z \\ &\quad + (c_{011} \circ f)yz + (c_{101} \circ f)xz + (c_{110} \circ f)xy. \end{aligned}$$

◊

We include here a slightly more general version of the preparation theorem which can be useful when we consider multi-germs of analytic mappings instead of just mono-germs. We recall that for a finite point set $S = \{x_1, \ldots, x_r\}$, a multi-germ of analytic spaces is a finite disjoint union $(X, S) = (X_1, x_1) \cup \cdots \cup (X_r, x_r)$ of germs of analytic spaces. The ring $\mathcal{O}_{X,S}$ is easily seen to be equal to $\oplus_{i=1}^{r} \mathcal{O}_{X_i, x_i}$.

A multi-germ of analytic mapping $f : (X, S) \to (Y, y_0)$ is given by a system of analytic map germs $f_i : (X_i, x_i) \to (Y, y_0)$, $i = 1, \ldots, r$. Such a map induces a morphism of $\mathbb{C}$-algebras $f^* : \mathcal{O}_{Y,y_0} \to \mathcal{O}_{X,S}$ in the obvious way.

Corollary D.2 *Let $f : (X, S) \to (Y, y_0)$ be an analytic multi-germ between analytic spaces when $\mathbb{F} = \mathbb{C}$ or a smooth multi-germ between manifolds when $\mathbb{F} = \mathbb{R}$. Let M be a finitely generated module over $\mathcal{O}_{X,S}$. The following statements are equivalent:*

1. *M is also finitely generated over $\mathcal{O}_{Y,y_0}$ via f.*
2. $\dim_{\mathbb{F}} M/f^*\,\mathfrak{m}_{Y,y_0}\, M < \infty$.

More precisely, M is generated over $\mathcal{O}_{Y,y_0}$ by $m_1, \ldots, m_k \in M$ if and only if $M/f^*\,\mathfrak{m}_{Y,y_0}\, M$ is generated over $\mathbb{F}$ by the classes $\overline{m}_1, \ldots, \overline{m}_k$ in $M/f^*\,\mathfrak{m}_{Y,y_0}\, M$.

Proof For each $i = 1, \ldots, r$ we denote by $u_i \in \mathcal{O}_{X,x}$ the germ which is the constant function 1 in $\mathcal{O}_{X_j,x_j}$ and 0 in $\mathcal{O}_{X_i,x_i}$ for $i \neq j$. Then $M_i := \{u_i\} \cdot M$ is an $\mathcal{O}_{X_i,x_i}$-

module as well as an $\mathcal{O}_{X,S}$-module, and $M = M_1 \oplus \cdots \oplus M_r$. Moreover, we have the following obvious equivalences:

1. M is generated over $\mathcal{O}_{X,S}$ by $\{m_1, \ldots, m_k\}$ if and only if for each $i = 1, \ldots, r$, M_i is generated over $\mathcal{O}_{X_i,x_i}$ by $\{u_i m_1, \ldots, u_i m_k\}$.
2. M is generated over $\mathcal{O}_{Y,y_0}$ via f by $\{m_1, \ldots, m_k\}$ if and only if for each $i = 1, \ldots, r$, M_i is generated over $\mathcal{O}_{Y,y_0}$ via f_i by $\{u_i m_1, \ldots, u_i m_k\}$.
3. $M/f^* \mathfrak{m}_{Y,y_0} M$ is generated over $\mathbb{C}$ by $\{\overline{m}_1, \ldots, \overline{m}_k\}$ if and only if for each $i = 1, \ldots, r$, $M_i/f_i^* \mathfrak{m}_{Y,y_0} M_i$ is generated over $\mathbb{C}$ by $\{\overline{u_i m_1}, \ldots, \overline{u_i m_k}\}$.

Now the corollary is an easy consequence of Theorem D.2 and Corollary D.1. □

The following is also a parameterised version of the above corollary:

Corollary D.3 *Let F be a d-parameter unfolding of germ $f \colon (\mathbb{F}^n, S) \to (\mathbb{F}^p, 0)$. Let M be a finitely generated module over $\mathcal{O}_{n+d}$ and let $M_0 = M/(u_1, \ldots, u_d)M$. The following statements are equivalent:*

1. *M is finitely generated over $\mathcal{O}_{p+d}$ via F^*.*
2. *M_0 is finitely generated over $\mathcal{O}_p$ via f^*.*

More precisely, M is generated over $\mathcal{O}_{p+d}$ by $m_1, \ldots, m_k \in M$ if and only if M_0 is generated over $\mathcal{O}_p$ by the residue classes $\overline{m}_1, \ldots, \overline{m}_k \in M_0$.

Proof Since $F(x, u) = (f_u(x), u)$ with $f_0 = f$ we have an isomorphism

$$\frac{M}{F^*\mathfrak{m}_{p+d}M} \cong \frac{M_0}{f^*\mathfrak{m}_p M_0}$$

which sends the residue classes of $m_1, \ldots, m_k$ in $\frac{M}{F^*\mathfrak{m}_{p+d}M}$ to the residue classes of $\overline{m}_1, \ldots, \overline{m}_k$ in $\frac{M_0}{f^*\mathfrak{m}_p M_0}$. Now just apply Corollary D.2 simultaneously to M and M_0. □

Exercises for Sect. D.1

1

(i) Let $f \colon (\mathbb{F}^3, 0) \to (\mathbb{F}, 0)$ be defined by $f(x_1, x_2, x_3) = x_1^2 + x_2^3 + x_3^4$, and let $\nabla f \colon (\mathbb{F}^3, 0) \to (\mathbb{F}^3, 0)$ be the map with components $\partial f/\partial x_1, \partial f/\partial x_2, \partial f/\partial x_3$. Find a minimal set of generators for $\mathcal{O}_3$ (source) as module over $\mathcal{O}_3$ (target).

(ii) Let $f : S = (\mathbb{F}^2, 0) \to (\mathbb{F}^2, 0) = T$ be defined by $f(x, y) = (x, xy)$, and consider, as $\mathcal{O}_S$-module, $\mathcal{O}_S$ itself. Show that $\mathcal{O}_S$ is *not* finitely generated over $\mathcal{O}_T$ by showing that the germs $y^k, k \in \mathbb{N}$ are linearly independent in $\mathcal{O}_S / \mathfrak{m}_T \mathcal{O}_S$.

D.2 Local Properties of Analytic Sets and Finite Mappings

One of the key applications of the Weierstrass preparation theorem is the Noether normalisation theorem [1]. For each $k \leq n$, we consider $\mathcal{O}_k$, with variables $x_1, \ldots, x_k$ as a subring of $\mathcal{O}_n$, with variables $x_1, \ldots, x_n$.

Theorem D.3 (Noether Normalisation) *Let $I \subset \mathcal{O}_n$ be an ideal. Then, after a generic linear change of coordinates, there exists $k \leq n$ such that $\mathcal{O}_k \subset \mathcal{O}_n/I$ and such that $\mathcal{O}_n/I$ is finitely generated as $\mathcal{O}_k$-module. Moreover, if $I \neq 0$, then $k < n$.*

Proof See, e.g., [dJP00, Corollary 3.3.19] □

Noether normalisation is used in the proof of Rückert's analytic version of the Hilbert Nullstellensatz. If I is any ideal of $\mathcal{O}_{\mathbb{C}^n,S}$, then the (analytic) zero locus of I is denoted by $V(I)$ and is defined as the set-germ of points x in a neighbourhood of S in $\mathbb{C}^n$ such that $f(x) = 0$, for all $f \in I$. By definition, (X, S) is a germ of analytic subset of $(\mathbb{C}^n, S)$ if $(X, S) = V(I)$, for some ideal I in $\mathcal{O}_{\mathbb{C}^n,S}$. On the other hand, given any analytic set-germ (X, S) in $(\mathbb{C}^n, S)$, the subset of $\mathcal{O}_{\mathbb{C}^n,S}$ of functions f such that $f(x) = 0$, for all $x \in X$ forms an ideal, which is denoted by $I(X, S)$. As in the algebraic case, the Rückert Nullstellensatz gives the relationship between the two constructions.

Theorem D.4 (Rückert Nullstellensatz) *For any ideal $I \subset \mathcal{O}_{\mathbb{C}^n,S}$, we have*

$$I(V(I)) = \sqrt{I}.$$

Proof See [dJP00, Theorem 3.4.4] for a proof in the mono-germ case. The multi-germ case follows easily. □

By construction, $V(I(X, S)) = (X, S)$ for any analytic set-germ (X, S). So, the Rückert Nullstellensatz says us that there is one-to-one correspondence between the analytic set-germs (X, S) in $(\mathbb{C}^n, S)$ and the radical ideals of $\mathcal{O}_{\mathbb{C}^n,S}$. Obviously, if instead of analytic set-germs we consider complex space germs (so that $\mathcal{O}_{X,S} = \mathcal{O}_{\mathbb{C}^n,S}/I$, with I a not necessarily radical ideal), then we have a one-to-one correspondence between the complex space germs in $(\mathbb{C}^n, S)$ and all the ideals of $\mathcal{O}_{\mathbb{C}^n,S}$.

We note that the Nullstellensatz does not hold for C^∞ functions or real polynomials; it requires functions taking values in an algebraically closed field. Indeed, the equivalence of geometrical finiteness and algebraic finiteness fails for smooth real-valued functions. The function $f\colon (\mathbb{R}^2, 0) \to (\mathbb{R}, 0)$, $f(x_1, x_2) = x_1^2 + x_2^2$, satisfies $f^{-1}(0) = \{(0, 0)\}$, but $\dim_{\mathbb{R}} \mathcal{E}_2/f^*(\mathfrak{m}_1)\mathcal{E}_2 = \infty$ and the ring $\mathcal{E}_2$ is not a finite module over $\mathcal{E}_1$.

[1] The use of "normalisation" here conflicts with its use to describe the space associated with the integral closure of the ring $\mathcal{O}_X$ in its total ring of fractions (see Definition D.5). The two have nothing to do with one another.

Recall from Definition 4.5 that a continuous map $f : X \to Y$ is *finite* if it is closed ($f(C)$ is closed in Y whenever C is closed in X) and has finite fibres. A continuous map-germ $f : (X, S) \to (Y, y_0)$ is *finite* if it has a representative $f : X \to Y$ which is finite.

Theorem D.5 *Let $f : (X, S) \to (Y, y_0)$ be an analytic map-germ. The following statements are equivalent:*

1. *$\mathcal{O}_{X,S}$ is finitely generated over $\mathcal{O}_{Y,y_0}$ via f.*
2. $\dim_{\mathbb{C}} \mathcal{O}_{X,S}/f^* \mathfrak{m}_{Y,y_0} < \infty$.
3. *As germs at S, $f^{-1}(y_0) = S$.*
4. *f is finite.*

Proof The equivalence between the two first statements is just a particular case of Corollary D.2. The equivalence between (2) and (3) is a consequence of the Nullstellensatz. Let $(X, S) = (X_1, x_1) \cup \cdots \cup (X_r, x_r)$ and let f_i be the restriction of f to each (X_i, x_i). Then $\dim_{\mathbb{C}} \mathcal{O}_{X,S}/f^* \mathfrak{m}_{Y,y_0} < \infty$ if and only if $\dim_{\mathbb{C}} \mathcal{O}_{X,x_i}/f_i^* \mathfrak{m}_{Y,y_0} < \infty$ for each $i = 1, \ldots, r$. By Proposition C.4, this is equivalent to $\sqrt{f_i^* \mathfrak{m}_{Y,y_0}} = \mathfrak{m}_{X_i,x_i}$. By the Nullstellenstaz, this is equivalent to

$$f_i^{-1}(x_0) = V(f_i^* \mathfrak{m}_{Y,y_0}) = V(\sqrt{f_i^* \mathfrak{m}_{Y,y_0}}) = V(\mathfrak{m}_{X_i,x_i}) = \{x_i\}.$$

The implication (4) $\Longrightarrow$ (3) is clear, since if f has finite fibres, then each x_i is isolated in $f^{-1}(y_0)$ and hence we may take a small enough representative such that $f^{-1}(y_0) = S$. The implication (3) $\Longrightarrow$ (4) is not so obvious; it is based on another important result in analytic geometry, the local parametrisation theorem (see [dJP00, Theorem 3.4.14]). □

The Noether normalisation theorem can be now restated in geometrical terms as follows:

Corollary D.4 *Let $(X.x_0)$ be an analytic set-germ in $(\mathbb{C}^n, x_0)$. There exists $k \leq n$ such that for any generic linear projection $\pi : \mathbb{C}^n \to \mathbb{C}^k$, the restriction $\pi : (X, x_0) \to (\mathbb{C}^k, y_0)$ is finite and surjective. Moreover, if (X, x_0) is a proper subset, then $k < n$.*

The finiteness of $\pi : (X, x_0) \to (\mathbb{C}^k, y_0)$ is equivalent to $\mathcal{O}_{X,x_0}$ being finitely generated over $\mathcal{O}_{\mathbb{C}^k,y_0}$ and the surjectivity is equivalent to $\pi^* : \mathcal{O}_{\mathbb{C}^k,y_0} \to \mathcal{O}_{X,x_0}$ being a monomorphism. Moreover, the number k is precisely the Weierstrass dimension of (X, x_0) (see [dJP00, Definition 4.1.3]). In fact, we have:

Lemma D.1 *If $f : (X, x_0) \to (Y, y_0)$ is any finite surjective map-germ, then* $\dim(X, x_0) = \dim(Y, y_0)$.

Proof See [dJP00, Lemma 4.1.4]. □

Finite mappings are central in the theory of singularities of mappings, since $\mathscr{A}$-finite map germs are finite on their critical set. Finite mappings have the fortunate property that the image of a finite analytic map is analytic. We recall from Proposition 2.1 that if $f\colon (X,S) \to (Y, y_0)$ is such that $f^{-1}(y_0) = \{x_0\}$, then f has a well-defined image $(f(X), y_0)$ as a set-germ. The following result is known as Remmert's proper mapping theorem.

Theorem D.6 *The image $(f(X), y_0)$ of any finite analytic map-germ $f\colon (X,S) \to (Y, y_0)$ is analytic of the same dimension as (X,S).*

Proof A proof that the image is analytic can be found in [Ło91, §5, Chapter V]. To show that $\dim(f(X), y_0) = \dim(X,S)$, we apply Lemma D.1 to each restriction $f\colon (X_i, x_i) \to (f(X_i), y_0)$. □

An alternative proof that $(f(X), y_0)$ is analytic can be given by means of sheaf theory. Consider $\mathscr{O}_{X,x_0}$ as a finitely generated module over $\mathscr{O}_{Y,y_0}$, then its support is precisely $(f(X), y_0)$ and hence $f(X)$ is analytic (see Corollary E.1). If $Y \subset \mathbb{C}^m$, we take $f\colon (X,S) \to (\mathbb{C}^m, y_0)$, which is also finite and has the same image. It follows that $(f(X), y_0) = V(\ker f^*)$.

Corollary D.5 *Let $f\colon (X,S) \to (Y, y_0)$ be a finite analytic map-germ with (Y, y_0) irreducible and such that* $\dim(X,S) = \dim(Y, y_0)$. *Then f is surjective.*

In the last part of this section, we give the definition of singular locus of a complex space-germ (X, x_0) (not necessarily reduced).

Definition D.1 We say that (X, x_0) is *smooth* or that x_0 is a *regular* point of X if (X, x_0) is isomorphic (as a complex space-germ) to $(\mathbb{C}^d, 0)$, for some d. Otherwise, we say that (X, x_0) is singular.

The above definition of regular point means that $\mathscr{O}_{X,x_0} \cong \mathscr{O}_d$, so we necessarily must have $d = \dim(X, x_0)$. Assume that $(X, x_0) \subset (\mathbb{C}^n, x_0)$ and $\mathscr{O}_{X,x_0} = \mathscr{O}_{\mathbb{C}^n,x_0}/I$, where I is generated by $f_1, \dots, f_r$. We fix representatives of X and the functions f_i on some open subset $U \subset \mathbb{C}^n$ such that X is given by the vanishing of f_i on U. Then, it makes sense to consider the set Σ of points $x \in X$ such that X is not regular at x. The set-germ (Σ, x_0) is called the *singular locus* of (X, x_0) and is denoted by $\mathrm{Sing}(X, x_0)$. The fact that $\mathrm{Sing}(X, x_0)$ is well defined and is analytic is a consequence of the so-called Jacobian criterion and will be sketched in Exercise D.2.1. What is clear from the definitions that (X, x_0) is smooth if and only if $\mathrm{Sing}(X, x_0) = \emptyset$.

Example D.2

1. The complement of $\mathrm{Sing}(X, x_0)$ in (X, x_0) is a locally a complex submanifold at each point x, but this set may have several connected components, and hence, it may have different dimensions on each component. For instance, if we consider X as the set in $\mathbb{C}^3$ given by the union of the plane $x = 0$ and the line $y = z = 0$, then $\mathrm{Sing}(X, 0)$ is only the origin. On $X \setminus \{0\}$ we find points of dimension 2 on the plane and points of dimension 1 on the line. In general, the dimension

of X at a point x in the complement of $\mathrm{Sing}(X, x_0)$ is equal to the dimension of the irreducible component which contains x. Hence, the dimension at x is always the same if and only if (X, x_0) is equidimensional (i.e., all the irreducible components have the same dimension).

2. Let $(X, 0)$ be complex space-germ of $(\mathbb{C}^2, 0)$ given by $x^2 = 0$. Then $\mathrm{Sing}(X, 0)$ is the whole space $(X, 0)$, although the underlying analytic set-germ, $(X^{red}, 0)$ is smooth.

◊

When (X, x_0) is an analytic set-germ (that is, when it is reduced), then one expects that almost all of the points of X are smooth. That is what next proposition says.

Proposition D.1 *Let* (X, x_0) *be an analytic set-germ of dimension* d*. Then its singular locus* $\mathrm{Sing}(X, x_0)$ *is contained in some analytic subset* $(D, x_0) \subset (X, x_0)$ *of dimension* $< d$*.*

Proof See [dJP00, 4.3.2] . □

By definition, (X, x_0) is smooth if its local ring $\mathcal{O}_{X,x_0}$ is isomorphic to $\mathcal{O}_d$. The algebraic counterpart of this notion is the concept of regular local ring. We say that a Noetherian local ring R is *regular* if its maximal ideal $\mathfrak{m}$ is a parameter ideal, that is, is generated by d elements, where $d = \dim R$ (see Sect. C.6).

Exercises for Sect. D.2

1 Let $(X, x_0) \subset (\mathbb{C}^N, x_0)$ be a complex space-germ of dimension d whose defining ideal is generated by $f_1, \ldots, f_r$ and let $f = (f_1, \ldots, f_r)$.

(i) Show that the minimal number of generators of $\mathfrak{m}_{X,x_0}$ is equal to the minimal number n such that there is an embedding $(X, x_0) \hookrightarrow (\mathbb{C}^n, 0)$ (this number is called the *embedding dimension* of (X, x_0)). Hint: Use Nakayama's lemma to deduce that the minimal number of generators of $\mathfrak{m}_{X,x_0}$ is equal to $N - \mathrm{rank}\, d_{x_0} f$.

(ii) Show that (X, x_0) is smooth if and only if $\mathcal{O}_{X,x_0}$ is regular.

(iii) Suppose (X, x_0) is equidimensional. Show that $\mathrm{Sing}(X, x_0)$ is given by the vanishing of the f_j and the $N - d$ minors of the Jacobian matrix of f.

D.3 Degree and Multiplicity

The following lemmas are necessary in order to define the degree of a surjective finite map-germ $f\colon (X, S) \to (Y, y_0)$. Roughly speaking, it is the number of points in a generic fibre; it is well defined when (Y, y_0) is irreducible. We consider first the case that $S = \{x_0\}$ and (X, x_0) is also irreducible. Since f is finite and surjective, $f^*\colon \mathscr{O}_{Y,y_0} \to \mathscr{O}_{X,x_0}$ is a monomorphism and $\mathscr{O}_{X,x_0}$ is finitely generated over $\mathscr{O}_{Y,y_0}$. Moreover, because (X, x_0) and (Y, y_0) are irreducible, their local rings are integral domains, so we can pass to their total fraction fields to get a finite field extension $Q(\mathscr{O}_{Y,y_0}) \subset Q(\mathscr{O}_{X,x_0})$. Recall that the *degree* of this field extension, $[Q(\mathscr{O}_{X,x_0}) : Q(\mathscr{O}_{Y,y_0})]$, is the dimension of $Q(\mathscr{O}_{X,x_0})$ as a vector space over $Q(\mathscr{O}_{Y,y_0})$.

Lemma D.2 *Let $f\colon (X, x_0) \to (Y, y_0)$ be a finite surjective analytic map-germ with (X, x_0) and (Y, y_0) irreducible and let $k = [Q(\mathscr{O}_{X,x_0}) : Q(\mathscr{O}_{Y,y_0})]$. There exists a hypersurface $(D, y_0) \subset (Y, y_0)$ such that for all small enough representatives Y of (Y, y_0) and D of (D, y_0), the fibre $f^{-1}(y)$ has cardinality k, for all $y \in Y \setminus D$.*

Proof A proof, based on the local parametrisation theorem [dJP00, Theorem 3.4.14], is sketched in [dJP00, 4.4.16]. □

The case where (X, S) is reducible follows.

Lemma D.3 *Let $f\colon (X, S) \to (Y, y_0)$ be a finite surjective analytic map-germ with (Y, y_0) irreducible of dimension d. Let $(X, S) = \bigcup_{j=1}^{\ell}(X_j, x_j)$ be the decomposition of (X, S) into irreducible components and let*

$$k = \sum_{\dim(X_j,x_j)=d} [Q(\mathscr{O}_{X_j,x_j}) : Q(\mathscr{O}_{Y,y_0})].$$

There exists a hypersurface $(D, y_0) \subset (Y, y_0)$ such that for all small enough representatives Y of (Y, y_0) and D of (D, y_0), the fibre $f^{-1}(y)$ has cardinality k, for all $y \in Y \setminus D$.

Proof For each j such that $\dim(X_j, x_j) = d$, the restriction $f_j\colon (X_j, x_j) \to (Y, y_0)$ must be finite and surjective by Corollary D.5. Hence, there exists a hypersurface (D_j, y_0) in (Y, y_0) such that $f_j^{-1}(y)$ has cardinality k_j, for all $y \in Y \setminus D_j$, where $k_j = [Q(\mathscr{O}_{Z_j,p_j}) : Q(\mathscr{O}_{Y,y_0})]$. If $\dim(X_j, x_j) < d$,then $\dim(f(X_j), y_0) < \dim(Y, y_0)$ by Theorem D.6 and hence there exist a hypersurface (D_j, y_0) in (Y, y_0) which contains $(f(X_j), y_0)$. We take (D, y_0) to be the hypersurface $\cup_{j=1}^{\ell}(D_j, y_0)$. □

Definition D.2 With the hypothesis of Lemma D.3, the integer k is called the *degree* of f and is denoted by $\deg(f)$. In this case we say that f is *generically k-to-1*.

If $f\colon (X, S) \to (Y, y_0)$ is finite with irreducible image, then this definition applies to the restriction of f to its image and the degree is called *degree onto the image*. If the image of f is not irreducible, then the definition makes sense over each irreducible component of the image, although the degree may vary from one component to another. Of course, if $S = \{x_0\}$ is a single point and (X, x_0) is irreducible, then the image of f is also irreducible.

Example D.3 If $f : (\mathbb{C}^n, S) \to (\mathbb{C}^p, 0)$ is $\mathscr{A}$-finite and $n < p$, then it is finite and has degree 1 onto its image. By the Mather–Gaffney criterion Theorem 4.5, there exists a representative $f : U \to V$ which is finite, $f^{-1}(0) = S$ and f is locally stable on $V \setminus \{0\}$. In particular, its double point set $D^2(f)$ has dimension $2n - p < n$. Take $D \subset f(U)$ any hypersurface containing $f(D^2(f))$, so that $f^{-1}(y)$ has cardinality 1, for all $y \in f(U) \setminus D$. ♢

Remark D.1 Suppose M and N are C^∞ manifolds of the same dimension with N connected and $f : M \to N$ is a proper C^∞ mapping. The degree of f is defined as the number of preimages of a regular value $y \in N$, counted with sign, depending on if they preserve or reverse the orientations (see [GP10]). There is a local version as follows: if $f : (M, x_0) \to (N, y_0)$ is a C^∞ map and $f^{-1}(y_0) = \{x_0\}$ (as set-germs), then we can fix a proper representative $f : U \to V$, with V connected and such that $f^{-1}(y_0) = \{x_0\}$ (as sets). By definition, the degree of the germ f is the degree of the proper map $f : U \to V$ (which is independent of the representative, as one can easily check). Both the global and local degrees can even be defined for continuous mappings between oriented topological manifolds, by using homology theory, so these notions are topological in nature. ♢

When (Y, y_0) is smooth, the degree of f can be computed algebraically, by means of the Samuel multiplicity. Let $f : (X, x_0) \to (\mathbb{C}^d, 0)$ be a finite surjective map-germ. Since f is finite, the ideal $f^*\mathfrak{m}_d$ is $\mathfrak{m}_{X,x_0}$-primary in $\mathscr{O}_{X,x_0}$. Moreover, $f^*\mathfrak{m}_d$ is generated by the components $f_1, \dots, f_d$ and $d = \dim(X, x_0)$ by Lemma D.1. So, $f^*\mathfrak{m}_d$ is in fact a parameter ideal.

Theorem D.7 *Let $f : (X, x_0) \to (\mathbb{C}^d, 0)$ be a finite surjective map-germ. Then,*

$$\deg(f) = e(f^*\mathfrak{m}_d, \mathscr{O}_{X,x_0}).$$

Proof Since both the degree and the multiplicity are additive, we can assume without loss of generality that (X, x_0) is irreducible. We know that $f^* : \mathscr{O}_d \to \mathscr{O}_{X,x_0}$ is a monomorphism and that $\mathscr{O}_{X,x_0}$ is finitely generated over $\mathscr{O}_d$. To compute the multiplicity, we can consider $\mathscr{O}_{X,x_0}$ both as a ring and as an $\mathscr{O}_d$-module as follows:

$$\begin{aligned} e_{\mathscr{O}_{X,x_0}}(f^*\mathfrak{m}_d, \mathscr{O}_{X,x_0}) &= \lim_k \frac{d!}{k^d} \dim_{\mathbb{C}} \frac{\mathscr{O}_{X,x_0}}{(f^*\mathfrak{m}_d)^k} \\ &= \lim_k \frac{d!}{k^d} \dim_{\mathbb{C}} \frac{\mathscr{O}_{X,x_0}}{\mathfrak{m}_d^k \mathscr{O}_{X,x_0}} = e_{\mathscr{O}_d}(\mathfrak{m}_d, \mathscr{O}_{X,x_0}). \end{aligned}$$

Now $e_{\mathcal{O}_d}(\mathfrak{m}_d, \mathcal{O}_d) = 1$ and hence, by Proposition C.14, $e_{\mathcal{O}_d}(\mathfrak{m}_d, \mathcal{O}_{X,x_0})$ is equal to the rank of $\mathcal{O}_{X,x_0}$ over $\mathcal{O}_d$. But this is obviously equal to $[Q(\mathcal{O}_{X,x_0}) : Q(\mathcal{O}_d)]$, which is equal to $\deg(f)$, by Lemma D.2. □

Recall that if $f\colon (X, x_0) \to (\mathbb{C}^d, 0)$ is finite, then its *multiplicity* is $\mathrm{mult}(f) = \dim_{\mathbb{C}} \mathcal{O}_{X,x_0}/f^*\mathfrak{m}_d$. As a consequence of Proposition C.15 and Theorem D.7 we have:

Corollary D.6 *Let $f\colon (X, x_0) \to (\mathbb{C}^d, 0)$ be a finite surjective map-germ. Then,*

$$\deg(f) \leq \mathrm{mult}(f),$$

with equality if and only if (X, x_0) is Cohen–Macaulay.

Corollary D.7 *Let $f\colon (\mathbb{C}^n, 0) \to (\mathbb{C}^n, 0)$ be analytic such that $f^{-1}(0) = \{0\}$. Then,*

$$\deg(f) = \mathrm{mult}(f) = \dim_{\mathbb{C}} \frac{\mathcal{O}_n}{(f_1, \ldots, f_n)}.$$

For analytic set-germs (X, x_0), we also have a notion of multiplicity defined in geometrical terms, which turns out to be equal to the Samuel multiplicity of its local ring $e(\mathcal{O}_{X,x_0})$.

Definition D.3 Let (X, x_0) be an analytic set-germ in $\mathbb{C}^n$, let $\pi\colon (X, x_0) \to (\mathbb{C}^d, y_0)$ be the restriction to X of a generic linear projection from $\mathbb{C}^n$ to $\mathbb{C}^d$, and suppose that π is finite and surjective. The *multiplicity* of (X, x_0) is defined as the degree of π, $\deg(\pi)$.

By construction, $L_{y_0} := \pi^{-1}(y_0)$ is a generic $(n-d)$-linear subspace in $\mathbb{C}^n$ such that $L_{y_0} \cap X = \{x_0\}$. For a generic y close enough to y_0, $L_y := \pi^{-1}(y)$ is a subspace parallel to L_{y_0} which intersects X in a finite number of points; this number is precisely $\mathrm{mult}(X, x_0)$. In other words, the multiplicity can be seen as the local intersection number at x_0 of X with a generic plane of complementary dimension. The fact that $\mathrm{mult}(X, x_0)$ is independent of the choice of the projection will be clear from the next corollary.

Corollary D.8 *For any analytic set-germ (X, x_0), we have*

$$\mathrm{mult}(X, x_0) = e(\mathcal{O}_{X,x_0}).$$

Proof Assume for simplicity that $x_0 = 0$. The maximal ideal $m_{X,0}$ is generated by the classes $\overline{x_1}, \ldots, \overline{x_n}$ of the coordinate functions in $\mathbb{C}^n$. The components of $\pi\colon (X, 0) \to (\mathbb{C}^d, 0)$ are obtained by taking $\mathbb{C}$-linear combinations of $\overline{x_1}, \ldots, \overline{x_n}$. By Theorem C.10, we may take these $\mathbb{C}$-linear combinations generic enough in such a way that $\pi^*\mathfrak{m}_d$ defines a reduction of m_{X,x_0}. Thus,

$$e(\mathcal{O}_{X,x_0}) = e(\mathfrak{m}_{X,x_0}, \mathcal{O}_{X,x_0}) = e(\pi^*\mathfrak{m}_d, \mathcal{O}_{X,x_0}) = \deg(\pi) = \mathrm{mult}(X, x_0),$$

by Lemma C.7. □

Exercises for Sect. D.3

1 Let $f : (\mathbb{C}^n, 0) \to (\mathbb{C}^n, 0)$ be analytic, and such that $f^{-1}(0) = \{0\}$. Show that in Lemma D.2, the hypersurface $(D, 0)$ is equal to the discriminant $\Delta(f(C))$, where C is the critical locus of f. In particular, k is equal to the number of preimages of a regular value.

2 Let $f : (\mathbb{C}^n, 0) \to (\mathbb{C}^n, 0)$ be analytic such that $f^{-1}(0) = \{0\}$. Show $\deg(f)$ is equal to the degree of f, when considered as a C^∞ map-germ between oriented C^∞ manifolds (as in Remark D.1).

3 Let $f : (X, S) \to (Y, y_0)$ and $g : (Y, y_0) \to (Z, z_0)$ be finite surjective analytic map-germs with (Y, y_0) and (Z, z_0) irreducible. Show that

$$\deg(g \circ f) = \deg(g) \deg(f).$$

4 Prove the following version of the inverse function theorem: if $f : (\mathbb{C}^n, 0) \to (\mathbb{C}^n, 0)$ is analytic and injective, then f is a biholomorphism.

5 Let $(X, 0)$ be a hypersurface in $(\mathbb{C}^n, 0)$ such that $I(X, 0) = (g)$, for some $g \in \mathcal{O}_n$. Show that $\mathrm{mult}(X, 0) = \mathrm{ord}(g)$, the order of g (i.e., the minimum k such that $g \in \mathfrak{m}_n^k$). *Hint:* After a linear change of coordinates, we can assume that the projection onto the first $n - 1$ coordinates is generic and that g has minimum order in the variable x_n.

6 Let $(X, 0)$ be a curve in $(\mathbb{C}^n, 0)$ defined as the image of an analytic map-germ $\gamma : (\mathbb{C}, 0) \to (\mathbb{C}^n, 0)$ which is finite and has degree 1 onto its image. Show that $\mathrm{mult}(X, 0) = \inf\{\mathrm{ord}(\gamma_i)\}_{i=1}^n$, where γ_i are the components of γ.

D.4 Normalisation of Analytic Set-Germs

Let R be a reduced ring (that is, a ring with no nilpotent elements) and let $Q(R)$ be its total ring of fractions. The *integral closure* of R is the subset $\overline{R} \subseteq Q(R)$ of elements $a \in Q(R)$ which satisfy a relation

$$a^n + b_1 a^{n-1} + \cdots + b_n = 0,$$

for some $n \geq 1$ and $b_i \in R$, for all $i = 1, \ldots, n$. It is not difficult to prove that $\overline{R}$ is a subring of $Q(R)$ which contains R (see, e.g., [GP08, Proposition 3.2.2]).

Definition D.4 A reduced ring R is *normal* if $\overline{R} = R$. A germ of analytic space (X, S) is said to be *normal* if the ring $\mathcal{O}_{X,S}$ is normal.

Normality is a local property for Noetherian rings. In fact, if R is Noetherian, the following three statements are equivalent:

1. R is normal,
2. $R_{\mathfrak{p}}$ is normal, for every prime ideal $\mathfrak{p} \subset R$,
3. $R_{\mathfrak{m}}$ is normal, for every maximal ideal $\mathfrak{m} \subset R$,

(see [GP08, Proposition 3.2.6]). In particular, if $S = \{x_1, \ldots, x_r\}$, then (X, S) is normal if and only if each mono-germ (X, x_i) is normal for all $i = 1, \ldots, r$ (Exercise D.4.1 below).

A Noetherian normal local ring is a domain [GP08, Theorem 1.5.7] (see also Exercise D.4.2 below). So a normal space germ is irreducible.

In general, a necessary and sufficient condition for a ring to be normal is given by Serre's conditions R1 and S2:

Proposition D.2 *A reduced ring R is normal if and only if the two following conditions are satisfied:*

(R1) *$R_{\mathfrak{p}}$ is regular, for all $\mathfrak{p} \in \operatorname{Spec}(R)$ such that* $\operatorname{height} \mathfrak{p} = 1$;
(S2) *the ideal (a) in R has no embedded primes, for every non-zero divisor $a \in R$.*

Proof See [GP08, Theorem 4.4.11]. □

Thus, smooth germs are normal. We will use the following slightly weaker version of Serre's criterion, which gives a sufficient but not necessary condition for normality:

Proposition D.3 *Let (X, x_0) be a germ of analytic space. If (X, x_0) is Cohen–Macaulay and* $\operatorname{Sing}(X, x_0)$ *has codimension at least 2, then (X, x_0) is normal.*

Proof The local ring $R = \mathscr{O}_{X,x_0}$ is Cohen–Macaulay and for every non-zero divisor $a \in R$, $R/(a)$ is also Cohen–Macaulay. Hence (a) has no embedded primes. This gives condition S2.

We now prove condition R1. Let $\mathfrak{p} \subset R$ be a prime ideal of height one. Assume that $(X, x_0) \subset (\mathbb{C}^n, 0)$ and that the ideal $I = I(X, x_0)$ is generated by $f_1, \ldots, f_r$ in $\mathscr{O}_n$, so that $R = \mathscr{O}_n / I$. Let $\mathfrak{q}$ be the inverse image of $\mathfrak{p}$ by the quotient map $\mathscr{O}_n \to R$. Then $\mathfrak{q}$ is a prime ideal in $\mathscr{O}_n$ which contains I and we have an isomorphism

$$R_{\mathfrak{p}} \cong \frac{(\mathscr{O}_n)_{\mathfrak{q}}}{I(\mathscr{O}_n)_{\mathfrak{q}}}.$$

Moreover, if $\dim(X, x_0) = d$, then

$$\dim V(\mathfrak{q}) = \dim \frac{\mathscr{O}_n}{\mathfrak{q}} = \dim \frac{R}{\mathfrak{p}} = d - 1,$$

since R is Cohen–Macaulay and $\operatorname{height} \mathfrak{p} = 1$.

Let J be the ideal in $\mathscr{O}_n$ generated by I and by the minors of order $n-d$ of the Jacobian matrix of $f = (f_1, \dots, f_r)$. By the Jacobian criterion Exercise D.2.1, $V(J) = \mathrm{Sing}(X, x_0)$, which has dimension $\le d-2$ by hypothesis. Thus, $J \not\subseteq \mathfrak{q}$, that is, at least one of the $(n-d)$-minors of the Jacobian matrix is not in $\mathfrak{q}$. Without loss of generality, we can suppose that

$$\frac{\partial(f_1, \dots, f_{n-d})}{\partial(x_1, \dots, x_{n-d})} \notin \mathfrak{q}. \tag{D.1}$$

For simplicity write S and $\mathfrak{m}$ for the local ring $(\mathscr{O}_n)_\mathfrak{q}$ and its maximal ideal $\mathfrak{q}(\mathscr{O}_n)_\mathfrak{q}$. We claim that the residue classes of $f_1, \dots, f_{n-d}$ in $\mathfrak{m}/\mathfrak{m}^2$ are linearly independent over $\mathbb{C}$. In fact, suppose that $\sum_{i=1}^{n-d} \lambda_i f_i \in \mathfrak{m}^2$, for some $\lambda_i \in \mathbb{C}$. This implies that $h \sum_{i=1}^{n-d} \lambda_i f_i \in \mathfrak{q}^2$, for some $h \in \mathscr{O}_n \setminus \mathfrak{q}$. Differentiating with respect to x_j, we get

$$\frac{\partial h}{\partial x_j} \sum_{i=1}^{n-d} \lambda_i f_i + h \sum_{i=1}^{n-d} \lambda_i \frac{\partial f_i}{\partial x_j} \in \mathfrak{q},$$

and hence, since $h \sum_i \lambda_i f_i \in \mathfrak{q}$,

$$h^2 \sum_{i=1}^{n-d} \lambda_i \frac{\partial f_i}{\partial x_j} \in \mathfrak{q}.$$

But $h \notin \mathfrak{q}$ and $\mathfrak{q}$ is prime, so

$$\sum_{i=1}^{n-d} \lambda_i \frac{\partial f_i}{\partial x_j} \in \mathfrak{q}, \text{ for } j = 1, \dots, n-d.$$

From this it follows that $\partial(f_1, \dots, f_{n-d})/\partial(x_1, \dots, x_{n-d}) \in \mathfrak{q}$, contradicting (D.1). This proves the claim.

Finally, as a localisation of a regular local ring, the ring S is regular, and, by the claim, $f_1, \dots, f_{n-d}$ are part of a regular system of parameters of S. Thus, $(f_1, \dots, f_{n-d})S$ is a prime ideal in S of height $n-d$. But IS also has height $n-d$ and contains $(f_1, \dots, f_{n-d})S$, so necessarily $IS = (f_1, \dots, f_{n-d})S$. By the Jacobian criterion, $S/IS = R_\mathfrak{p}$ is regular. □

From the proposition, it follows immediately that

(i) a curve germ is normal if and only if it is smooth;
(ii) An isolated hypersurface singularity (IHS) is normal provided its dimension is at least 2, since hypersurfaces are Cohen–Macaulay.
(iii) More generally, an isolated complete intersection singularity (ICIS) of dimension ≥ 2 is normal, since complete intersections are also Cohen–Macaulay.

D.4.1 Extension Theorems

Geometrically, the key property of normal spaces is that they satisfy the extension theorems of Riemann and Hartogs:

Theorem D.8 *Let (X, x_0) be a germ of analytic space.*

1. *(X, x_0) is normal if and only if any function $f\colon X \smallsetminus \operatorname{Sing}(X) \to \mathbb{C}$ which is holomorphic and locally bounded has a holomorphic extension on (X, x_0).*
2. *If (X, x_0) is normal and (Z, x_0) is an analytic subset of codimension 2, then any holomorphic function $f\colon X \smallsetminus Z \to \mathbb{C}$ has a holomorphic extension to (Z, x_0).*

Proof See [dJP00, Theorem 4.4.15] for item 1 and [Ło91, pp. 341] for item 2. □

We need another extension theorem which seems harder to pin down in the literature:

Theorem D.9 *If (X, x_0) is Cohen–Macaulay and (Z, x_0) is an analytic subset of codimension ≥ 2, then a holomorphic function $f : X \smallsetminus Z \to \mathbb{C}$ has a holomorphic extension to (X, x_0).*

Proof The proof uses the local cohomology sheaves $\mathscr{H}^i_Z(\mathscr{S})$ associated to a sheaf $\mathscr{S}$ on X, for which the best reference for our purposes seems still to be Hartshorne's notes of Grothendieck's course, [Har67]. Let $j : X \smallsetminus Z \to X$ be the inclusion. From corollary 1.9 of [Har67] we get an exact sequence

$$0 \to \mathscr{H}^0_Z(\mathscr{O}_X) \to \mathscr{O}_X \to j_*j^{-1}\mathscr{O}_X \to \mathscr{H}^1_Z(\mathscr{O}_X) \to 0. \tag{D.2}$$

Our statement is thus equivalent to the vanishing of $\mathscr{H}^1_Z(\mathscr{O}_X)$. By Theorem 3.8 of [Har67], we have

$$\mathscr{H}^i_Z(\mathscr{O}_X) = 0 \ \text{ for } i < n \quad \text{iff} \quad I - \operatorname{depth}(\mathscr{O}_X) \geq n,$$

where I is an ideal defining Z in X. The I-depth of $\mathscr{O}_X$ is defined above, in Definition C.5: it is the length of the longest $\mathscr{O}_X$-regular sequence in I. Finally, for any Cohen–Macaulay ring R and ideal $I \subset R$, the I-depth of R is equal to the height of the ideal I. Geometrically, this means that the I-depth of $\mathscr{O}_X$ is equal to the codimension of $Z = V(I)$ in X. Thus, provided the codimension of Z in X is at least 2, $\mathscr{H}^1_Z(\mathscr{O}_X) = 0$, from which it follows, by the exactness of (D.2), that the natural morphism $\mathscr{O}_X \to j_*j^{-1}\mathscr{O}_X$ is surjective—in other words, that every function holomorphic on $X \smallsetminus Z$ is the restriction of a function holomorphic on X.

□

D.4.2 Normalisation

Definition D.5 If an analytic map-germ $f\colon (X, S) \to (Y, y_0)$ is finite, surjective and generically 1-1, and (X, S) itself is normal, then we say that $f\colon (X, S) \to (Y, y_0)$ is a *normalisation* [2] of (Y, y_0).

If $f\colon (X, S) \to (Y, y_0)$ is a normalisation of (Y, y_0), then $f^* : \mathcal{O}_{Y,y_0} \to \mathcal{O}_{X,S}$ is a monomorphism which identifies $\mathcal{O}_{X,S}$ with the integral closure of $\mathcal{O}_{Y,y_0}$ in its total ring of fractions. Finiteness is required here because the integral closure of an analytic ring R in its total ring of fractions is a finite R-module. A good reference for this is [Gun70, pages 153–163]. The situation in the algebraic category is more obscure. A theorem of Emmy Noether (see [Eis95, Theorem 4.14]) establishes this for integral domains, but for more general rings an extensive list of conditions is needed.

The next theorem ensures the existence of a normalisation of any analytic set-germ:

Theorem D.10 *Any analytic set-germ* (Y, y_0) *admits a normalisation* $f\colon (X, S) \to (Y, y_0)$.

Proof Assume first that (Y, y_0) is irreducible. Then $R = \mathcal{O}_{Y,y_0}$ is a domain and we denote by $\overline{R}$ the integral closure of R in its field of fractions $Q(R)$. It follows that $\overline{R}$ is finitely generated over R and by [dJP00, Corollary 3.3.26],

$$\overline{R} \cong \bigoplus_{i=1}^{k} S_i,$$

where the S_i are analytic $\mathbb{C}$-algebras. But $Q(\overline{R}) = Q(R)$ is a field, so necessarily $k = 1$. Suppose that $\overline{R} = \mathcal{O}_n/I$, for some ideal $I \subset \mathcal{O}_n$. Then $(X, x_0) = V(I)$ is a normal analytic set-germ such that $\mathcal{O}_{X,x_0} \cong \overline{R}$. Moreover, the morphism of analytic $\mathbb{C}$-algebras $\varphi\colon R = \mathcal{O}_{Y,y_0} \to \mathcal{O}_{X,x_0}$ induces an analytic map-germ $f\colon (X, x_0) \to (Y, y_0)$ which is a normalisation. In fact, f is surjective, finite and generically 1-1 because φ is injective, $\overline{R}$ is finitely generated over R and $Q(\overline{R}) = Q(R)$, respectively.

The general case follows easily. Assume that $(Y_1, y_0), \ldots, (Y_r, y_0)$ are the irreducible components of (Y, y_0). For each $i = 1, \ldots, r$, we have a normalisation $f_i\colon (X_i, x_i) \to (Y_i, y_0)$. As (X, S) we take the multi-germ $(X_1, x_1) \sqcup \cdots \sqcup (X_r, x_r)$ and as f the map induced by the f_i. □

It follows from the proof of the theorem that when (Y, y_0) is irreducible, then a normalisation is a mono-germ $f\colon (X, x_0) \to (Y, y_0)$. An interesting particular case is when (Y, y_0) has dimension one: if (Y, y_0) is an analytic curve-germ, then (X, S)

[2] Again we remark that normalisation in this sense is not related to the Noether normalisation of Theorem D.3.

has to be smooth and hence, we find a normalisation of the form $\gamma\colon (\mathbb{C}, S) \to (Y, y_0)$. This map is usually called a *parametrisation* of the curve (Y, y_0).

The normalisation of an analytic germ is unique up to isomorphism: if (X_1, S_1) and (X_2, S_2) are normal (multi-)germs and there are finite generically 1-1 maps $f_i\colon (X_i, S_i) \to (Y, y_0)$ for $i = 1, 2$ then there is a unique isomorphism $\varphi\colon (X_1, S_1) \to (X_2, S_2)$ such that $f_1 = f_2 \circ \varphi$. This fact is based on the universal property of the normalisation, whose proof is left as an exercise (see Exercise 3 below).

An important theorem of Seidenberg [Sei66] states that all vector fields on a space lift to vector fields on the normalisation. Recall from Definition 8.4 that if $(X, S) \subseteq (\mathbb{C}^n, S)$ any analytic germ, then $\mathrm{Der}(-\log(X))_S$ is the submodule of $\theta_{\mathbb{C}^n,S}$ of vector fields tangent to X at its smooth points. By definition, a vector field on (X, S) is an analytic map-germ $\xi\colon (X, S) \to T\mathbb{C}^n$ which can be extended to a vector field $\overline{\xi} \in \mathrm{Der}(-\log(X))_S$. The module of all vector fields on (X, S) is denoted by $\theta_{X,S}$. Seidenberg's original formulation was algebraic: if R is a domain containing $\mathbb{Q}$ and ϑ is a derivation on R, then the derivation $\bar{\vartheta}$ defined on the field of fractions $\mathbb{Q}(R)$ by means of the quotient rule satisfies $\bar{\vartheta}(\bar{R}) \subset \bar{R}$, where $\bar{R}$ is the integral closure of R in $\mathbb{Q}(R)$. We give here a version adapted to our purposes.

Theorem D.11 *Let $f\colon (X, S) \to (Y, y_0)$ be the normalisation of (Y, y_0). For all $\xi \in \theta_{Y,y_0}$ there exists $\eta \in \theta_{X,S}$ such that $df \circ \eta = \xi \circ f$.*

Proof We follow Looijenga's proof in [Loo84]. Assume that $(Y, y_0) \subset (\mathbb{C}^p, y_0)$ and $(X, S) \subset (\mathbb{C}^n, S)$. Let $\overline{\xi} \in \mathrm{Der}(-\log Y)_{y_0}$ be an extension of ξ. By integrating $\overline{\xi}$ we get a flow $\overline{\phi}\colon (\mathbb{C}^p \times \mathbb{C}, (y_0, 0)) \to (\mathbb{C}^p, y_0)$ which restricts to map $\phi\colon (Y \times \mathbb{C}, (y_0, 0)) \to (Y, y_0)$, since $\overline{\xi}$ is tangent to Y. Now consider the diagram:

$$\begin{array}{ccc}
(X, S) & \xrightarrow{\quad f \quad} & (Y, y_0) \\
\Big\uparrow{\scriptstyle \psi} & & \Big\uparrow{\scriptstyle \phi} \\
(X \times \mathbb{C}, S \times \{0\}) & \xrightarrow{f \times \mathrm{id}} & (Y \times \mathbb{C}, (y_0, 0))
\end{array}$$

The germ $(X \times \mathbb{C}, S \times \{0\})$ is also normal and by the universal property of the normalisation, there exists a unique analytic map $\psi\colon (X \times \mathbb{C}, S \times \{0\}) \to (X, S)$ such that $\phi \circ (f \times \mathrm{id}) = f \circ \psi$. Let $\overline{\psi}\colon (\mathbb{C}^n \times \mathbb{C}, S \times \{0\}) \to (\mathbb{C}^n, S)$ be any analytic extension of ψ and define $\overline{\eta} \in \theta_{\mathbb{C}^n,S}$ as the vector field given by $\overline{\psi}(x) = \frac{\partial \overline{\psi}}{\partial t}(x, 0)$ and $\eta\colon (X, S) \to T\mathbb{C}^p$ as the restriction of $\overline{\eta}$.

For general $x \in X$, X is smooth at x, Y is smooth at $f(x)$, and the germ $f\colon (X, x) \to (Y, f(x))$ is a diffeomorphism. By construction, $\eta = (f^{-1})_*\xi$ in a neighbourhood of x, so η is tangent to X and η is a lift of ξ. By continuity, η is tangent at all the smooth points of X, that is, $\eta \in \theta_{X,S}$ and moreover, $df \circ \eta = \xi \circ f$. □

We will make use of the following result about $\mathscr{A}$-finite germs.

Proposition D.4 *Let $f\colon (\mathbb{C}^n, S) \to (\mathbb{C}^p, 0)$ be $\mathscr{A}$-finite, let Σ be the set-germ of critical points of f and let $D = f(\Sigma)$.*

1. *If $n < p$, $f\colon (\mathbb{C}^n, S) \to (D, 0)$ is the normalisation of its image $(D, 0)$.*
2. *If $n \geq p$, Σ is Cohen–Macaulay of dimension $p - 1$ (with the analytic structure given by the $p \times p$ minors of the Jacobian matrix J_f). If also $p \geq 2$, Σ is reduced and the restriction $f : \Sigma \to D$ is finite and generically one-to-one.*
3. *If $n \geq p$, the restriction $f : \Sigma \to D$ is the normalisation of D provided that either f is stable or $p \geq 3$.*

Proof By the Mather–Gaffney criterion Theorem 4.5, there exists a representative $f : U \to V$ such that $f^{-1}(0) \cap \Sigma = S$ and the restriction $f : U \setminus f^{-1}(0) \to V \setminus \{0\}$ is locally stable. The first condition implies that $f : \Sigma \to D$ is always finite, by Theorem D.5.

1. Let $n < p$. Then $\Sigma = U$. By obvious dimensional inequalities, for a general point $y \in D = f(U)$, $f^{-1}(y)$ is a single point x. Thus, f is finite and generically one-to-one, so $f\colon (\mathbb{C}^n, S) \to D$ is the normalisation of D.
2. Let $n \geq p$. We have $\Sigma = \Sigma^{n-p+1} \cup \cdots \cup \Sigma^n$, where each Σ^i is the set of points $x \in U$ such that f has corank i at x. Also $\Sigma^i = (j^1 f)^{-1}(\tilde{\Sigma}^i)$, where $j^1 f : U \to J^1(U, V)$ is the 1-jet extension of f and $\tilde{\Sigma}^i$ is the submanifold of 1-jets of corank i. Since f is stable at each $x \in \Sigma^i \setminus S$, $j^1 f$ is transverse to $\tilde{\Sigma}^i$ at x and Σ^i is smooth of codimension $i(n - p + i)$ at x. As a consequence, Σ has dimension $p - 1$. But Σ is defined by the vanishing of the $p \times p$ minors of the Jacobian matrix. By Theorem C.6, Σ is determinantal and hence Cohen–Macaulay.

   If $p \geq 2$, then $D \neq \{0\}$. At a generic point $y \in D = f(\Sigma)$, the multi-germ of f at y is stable of $\mathscr{A}$-codimension $n - p + 1$ (the minimum possible $\mathscr{A}$-codimension of points in D). Now this means that $f^{-1}(y) \cap \Sigma = \{x\}$, for some $x \in U$ and f has a fold singularity at x. In particular, $f : \Sigma \to D$ is generically one-to-one. Moreover, we also deduce that Σ is smooth at a generic point x. Thus, Σ is generically reduced and hence reduced, by Proposition C.5.
3. By item 2, Σ is Cohen–Macaulay of dimension $p - 1$, and we have seen in the proof that Σ is smooth along $\Sigma^{n-p+1} \setminus S$. The singular locus of Σ is contained in $\Sigma' = S \cup \Sigma^{n-p+2} \cup \cdots \cup \Sigma^n$, which has dimension equal to $\max\{n - 2(n - p + 2), 0\}$. We have

$$p - 3 - (n - 2(n - p + 2)) = n - p + 1 \geq 0,$$

   so $n - 2(n - p + 2) \leq p - 3$. If f is stable, then $\dim \Sigma' = n - 2(n - p + 2)$, if Σ' is not empty, so the codimension of Σ' in Σ is ≥ 2. If $p \geq 3$, we have $\max\{n - 2(n - p + 2), 0\} \leq p - 3$ and again Σ' has codimension ≥ 2 in Σ. In both cases Σ is normal by Proposition D.3, and $f : \Sigma \to D$ is the normalisation of D.

□

Exercises for Sect. D.4

1 Suppose the ring R is a direct sum of rings, $R = R_1 \oplus R_2$. Show, directly from the definition of normality, that

(i) if R is normal, then so are R_1 and R_2;
(ii) if R_1 and R_2 are normal, then so is R.

2 Consider the node singularity $(X, 0)$ defined in $\mathbb{C}^2$ by the equation $x_1x_2 = 0$.

(i) Show that $(X, 0)$ is not normal directly from the definition: prove that the quotient $(x_1 - x_2)/(x_1 + x_2)$ lies in the integral closure of $\mathscr{O}_{X,0}$ in its total ring of fractions, but it is not in $\mathscr{O}_{X,0}$.
(ii) What is the normalisation of $(X, 0)$? Find a function on the normalisation of $(X, 0)$ which corresponds to $(x_1 - x_2)/(x_1 + x_2)$.

3 The universal property of the normalisation:

(i) Let $f: (X, S) \to (Y, y_0)$ be a normalisation of (Y, y_0) and let $g: (Z, S') \to (Y, y_0)$ be any analytic map-germ, with (Z, S') normal. Show that there exists a unique analytic map-germ $\varphi: (Z, S') \to (X, S)$ such that $g = f \circ \varphi$. (Hint: Prove a corresponding statement for rings).
(ii) Use (i) to prove the uniqueness of the normalisation, as stated above.

Appendix E
Sheaves

We introduce coherent analytic sheaves, and use coherence, and the material from Appendices C and D, to prove conservation of multiplicity in the cases we need.

E.1 Presheaves and Sheaves

If X and Y are spaces, and we select some class of germs of maps $X \to Y$—e.g., germs of continuous maps, or germs of complex analytic maps in case X and Y are complex analytic spaces—then we can put together all of the germs into a global object, a *sheaf*. In the most important cases, $Y = \mathbb{R}$ or $\mathbb{C}$. For example, we can associate to each open $U \subset X$ the set

$$\mathscr{O}_X(U) := \{f : U \to \mathbb{C} : f \text{ is complex analytic}\}$$

and make it into a $\mathbb{C}$-algebra by defining the operations pointwise

$$(f+g)(x) = f(x)+g(x), \quad (fg)(x) = f(x)g(x), \quad (\lambda f)(x) = \lambda f(x) \text{ for } \lambda \in \mathbb{C}.$$

If $U \subset V$ there is a restriction map $\rho_{U,V} : \mathscr{O}(V) \to \mathscr{O}(U)$ which is a homomorphism of $\mathbb{C}$-algebras, and if $U \subset V \subset W$ then evidently

$$\rho_{U,V} \circ \rho_{V,W} = \rho_{U,W}. \tag{E.1}$$

D. Mond, J. J. Nuño-Ballesteros, *Singularities of Mappings*,
Grundlehren der mathematischen Wissenschaften 357,
https://doi.org/10.1007/978-3-030-34440-5

Let $\mathcal{U}_x$ be the collection of all neighbourhoods of a point x. The equivalence relation by which we arrived at the notion of germ of function or mapping becomes a relation on the disjoint union $\coprod_{U \in \mathcal{U}_x} \mathcal{O}(U)$: let $f \in \mathcal{O}(U)$ and $g \in \mathcal{O}(V)$: then

$$f \sim g \text{ if there exists open } W \text{ with } x \in W \subset U \cap V \text{ such that } \rho_{W,U}(f) = \rho_{W,V}(g). \tag{E.2}$$

The set of equivalence classes, $\mathcal{O}_{X,x_0}$, is in a natural way a $\mathbb{C}$-algebra: if $f, g \in \mathcal{O}_{X,x_0}$, then they can be represented by some $f_1 \in \mathcal{O}(U)$ and $g_1 \in \mathcal{O}(V)$, for some open neighbourhoods U, V of x_0, and then the restrictions $\rho_{U\cap V,U}(f_1)$ and $\rho_{U\cap V,V}(g_1)$ in $U \cap V$ can be added or multiplied in the usual way. The sum and product of these restrictions then determine germs at x_0, which, as one can easily check, are independent of the choices of representative f_1, g_1.

The map $\rho_{x_0,U} : \mathcal{O}(U) \to \mathcal{O}_{X,x_0}$ defined by sending $f \in \mathcal{O}(U)$ to its germ at x_0 is a $\mathbb{C}$-algebra homomorphism. Evidently

$$\rho_{x_0,V} = \rho_{x_0,U} \circ \rho_{U,V}.$$

The procedure outlined here can be applied equally well to functions of other types: continuous, C^∞, real analytic, etc.

Definition E.1

(i) A *presheaf* $\mathcal{S}$ of rings/groups/vector spaces, etc. on a topological space X is a rule assigning to each open set U an object $\mathcal{S}(U)$ of the cited category, and to each inclusion of open sets $U \subset V$ a morphism $\rho_{U,V} : \mathcal{S}(V) \to \mathcal{S}(U)$ in the category, such that $\mathcal{S}(\emptyset) = 0$, and whenever $U \subset V \subset W$ are open then

$$\rho_{U,W} = \rho_{U,V} \circ \rho_{V,W}.$$

The elements of $\mathcal{S}$ are generally known as *sections* of $\mathcal{S}$ over U, and the morphism ρ_{UV} is the *restriction morphism*, since this is what it is in the examples described above.

(ii) A presheaf $\mathcal{S}$ is a *sheaf* if it has the following two properties. Let $\{U_\alpha : \alpha \in A\}$ be an open covering of V.

S1 If the restrictions $\rho_{U_\alpha,V}$ of the sections s_1 and s_2 of $\mathcal{S}(V)$ are the same for all $\alpha \in A$, then $s_1 = s_2$.

S2 If the sections $s_\alpha \in (U_\alpha), \alpha \in A$ agree under restriction, i.e. $\rho_{U_\alpha \cap U_\beta, U_\alpha}(s_\alpha) = \rho_{U_\alpha \cap U_\beta, U_\beta}(s_\beta)$ for all $\alpha, \beta \in A$, then there is a section $s \in \mathcal{S}(V)$ such that for all α, $s_\alpha = \rho_{U_\alpha,V}(s)$.

It is easy to see that the examples at the start of this section are indeed sheaves. The construction in the Exercise E.1.4 below is particularly important.

The restriction morphisms of a sheaf allow one to define the notion of *germ at x of a section*, by subjecting the disjoint union of all $\mathcal{S}(U)$, where U ranges over all open neighbourhoods of x, to the equivalence relation (E.2). The set of all germs at x of sections of $\mathscr{S}$ is denoted $\mathscr{S}_x$, and called the *stalk of* $\mathscr{S}$ *at* x.

One then defines a *sheaf space* $\tilde{\mathscr{S}}$ by

$$\tilde{\mathscr{S}} := \coprod_{x \in X} \mathscr{S}_x,$$

and a *sheaf projection* $\pi : \tilde{\mathscr{S}} \to X$ sending all of $\mathscr{S}_x$ to x. If U is open, then a section $s \in \mathscr{S}(U)$ determines a germ $s_x \in \mathscr{S}_x$ for each $x \in X$, and thus a map $U \to \tilde{\mathscr{S}}$; the set $\tilde{\mathscr{S}}$ is endowed with the topology generated by all the sets $\tilde{s}(U)$, so that this map is automatically continuous, and, indeed, the procedure just described determines a bijection

$$\mathscr{S}(U) \to \{\tilde{s} : U \to \tilde{\mathscr{S}} : \tilde{s} \text{ is continuous, and } \pi \circ \tilde{s} = \mathrm{id}_U\}.$$

For more details, see, e.g., [GR04, Chapter A §0]. It is worth noting that some accounts take the sheaf space $\tilde{\mathscr{S}}$ as the starting point of the definition of sheaves.

If $\mathscr{S}$ is a presheaf on X and $f : X \to Y$ is a continuous map, then $f_*\mathscr{S}$ (the *push-forward of* $\mathscr{S}$ *by* f) is the presheaf on Y defined by

$$(f_*\mathscr{S})(U) := \mathscr{S}(f^{-1}(U)),$$

with $\rho_{V,U} : (f_*\mathscr{S})(U) \to (f_*\mathscr{S})(V)$ defined to be $\rho_{V,U} : \mathscr{S}(U) \to \mathscr{S}(V)$. If $\mathscr{S}$ is a sheaf, then so is $f_*\mathscr{S}$ (Exercise E.1.4).

The notion of sheaf on a topological space allows us to pass from local to global when working with an analytic set-germ or, more generally, a complex space germ (X, x_0). The global version of these objects are called analytic or complex spaces, respectively, and they play the role of schemes in algebraic geometry.

Definition E.2 A *ringed space* is a pair $(X, \mathscr{O}_X)$, where X is Hausdorff topological space and $\mathscr{O}_X$ is a sheaf of rings on X. It is called a *locally ringed space* when the stalk $\mathscr{O}_{X,x}$ is a local ring, for all $x \in X$. A *sheaf of* $\mathscr{O}_X$*-modules* (or sometimes just "$\mathscr{O}_X$*-module*") is a sheaf $\mathscr{S}$ on X such that for all open $U \subset X$, $\mathscr{S}(U)$ is an $\mathscr{O}_X(U)$-module. If an $\mathscr{O}_X$-module is a subsheaf of $\mathscr{O}_X$, then it is called an *ideal sheaf.*

A *morphism* of locally ringed spaces is a pair $(f, f^*)\colon (X, \mathscr{O}_X) \to (Y, \mathscr{O}_Y)$, where $f : X \to Y$ is a continuous map and $f^* : \mathscr{O}_Y \to f_*\mathscr{O}_X$ is a morphism of sheaves of $\mathscr{O}_Y$-modules such that for all $x \in X$, the induced map on stalks $f_x^* : \mathscr{O}_{Y,f(x)} \to \mathscr{O}_{X,x}$ is a morphism of local rings.

The only sheaves of rings we consider are sheaves of functions, and the ring morphism f^* will always be given by composition with f.

Example E.1

1. If $U \subset \mathbb{C}^n$ is open and $\mathcal{O}_U$ is the sheaf of holomorphic functions on U, then $(U, \mathcal{O}_U)$ is a locally ringed space. Analogously, any complex manifold X can be seen as a locally ringed space $(X, \mathcal{O}_X)$, where $\mathcal{O}_X$ is the sheaf of holomorphic functions on X. It has the property that any point $x \in X$ has a neighbourhood V such that $(V, \mathcal{O}_X|_V)$ is isomorphic to $(U, \mathcal{O}_U)$ for some $U \subset \mathbb{C}^n$. The definitions which follow imitate this property to define abstract spaces which are locally isomorphic to some models of locally ringed spaces embedded in affine spaces.
2. Let (X, x_0) be an analytic set-germ and take a representative X which is a closed analytic subset of some open subset $U \subset \mathbb{C}^n$. Consider the pair $(X, \mathcal{O}_X)$, where $\mathcal{O}_X = (\mathcal{O}_U/\mathcal{I}(X))|_X$ and $\mathcal{I}(X)$ is the ideal sheaf on U of functions vanishing on X. Then $(X, \mathcal{O}_X)$ is a locally ringed space, which we call an *analytic model space*.
3. Let (X, x_0) be a complex space-germ with local ring $\mathcal{O}_{X,x_0} = \mathcal{O}_{\mathbb{C}^n,x_0}/I$ with I an ideal generated by $f_1, \ldots, f_r$. We take representatives of X and of the functions f_i on some open subset $U \subset \mathbb{C}^n$, such that X is defined by the vanishing of f_i on U. In this case, we consider $(X, \mathcal{O}_X)$, where now $\mathcal{O}_X = (\mathcal{O}_U/\mathcal{I})|_X$ and $\mathcal{I}$ is the ideal sheaf on U generated by $f_1, \ldots, f_r$. Again $(X, \mathcal{O}_X)$ is a locally ringed space, which is called a *complex model space*.

◊

Definition E.3 A locally ringed space $(X, \mathcal{O}_X)$ is called an *analytic space* (resp. *complex space*) if every point $x \in X$ has a neighbourhood V such that $(V, \mathcal{O}_X|_V)$ is isomorphic to some analytic model space (resp. complex model space).

For simplicity, we will denote an analytic or complex space just by X instead of $(X, \mathcal{O}_X)$, whenever there is no risk of confusion. It is clear from the definitions that an analytic space is a particular case of a complex space, so, from now on we will speak only of complex spaces, unless otherwise specified. We refer the reader to [dJP00] for other basic definitions (like complex subspaces, analytic subsets, etc.) and properties.

Exercises for Sect. E.1

1 Let X be a complex space. Show that the presheaf $\mathcal{O}_X$ is in fact a sheaf.

2 Let U be a connected neighbourhood of x_0. Is $\rho_{x_0,U}$ surjective? Injective?

3 Let $f : X \to Y$ be a map of topological spaces. For $U \subset Y$ define $\mathcal{H}^p(U) := H^p(f^{-1}(U))$ (the p-th topological cohomology of $f^{-1}(U)$).

(i) Given $U \subset V \subset X$, show how to define $\rho_{U,V} : \mathcal{H}^p(V) \to \mathcal{H}^p(U)$ so that (E.1) holds.
(ii) Do properties S1 and S2 hold here?

4 Suppose that $f : X \to Y$ is a continuous map and that $\mathscr{S}$ is a pre-sheaf on X.

(i) Show that if $\mathscr{S}$ is a sheaf then so is $f_*\mathscr{S}$.
(ii) Show that if f is a finite map then the stalk $(f_*\mathscr{S})_y$ is equal to $\oplus_{x\in f^{-1}(y)}\mathscr{S}_x$ (see [dJP00, Theorem 6.1.18]).

E.2 Coherence

In analytic geometry, everything we have said about commutative algebra so far applies only to the local ring $\mathscr{O}_{X,x_0}$. To move from there to a neighbourhood of x_0 we need the notion of *coherence*.

Example E.2

(1) How does the definition of the dimension of an analytic space-germ (X, x_0), as the dimension of the local ring $\mathscr{O}_{X,x_0}$, studied in Appendix C, relate to the dimension of a representative of (X, x_0) at smooth points, where it is a complex manifold?
(2) In Chapter 11 we consider analytic germs $f\colon (X^n, x_0) \to (\mathbb{C}^{n+1}, 0)$ and discuss the structure of $\mathscr{O}_{X,x_0}$ as $\mathscr{O}_{\mathbb{C}^{n+1},0}$ module. We know, by the preparation Theorem D.2, that if $f^{-1}(0) = \{x_0\}$, then $\mathscr{O}_{X,x_0}$ is a finitely generated module over $\mathscr{O}_{\mathbb{C}^{n+1},0}$. We focus on constructing and interpreting a presentation of $\mathscr{O}_{X,x_0}$ as $\mathscr{O}_{\mathbb{C}^{n+1},0}$ module—that is, finding an exact sequence

$$(\mathscr{O}_{\mathbb{C}^{n+1},0})^k \xrightarrow{\lambda} (\mathscr{O}_{\mathbb{C}^{n+1},0})^\ell \longrightarrow \mathscr{O}_{X,x_0} \longrightarrow 0$$

where λ is an $\ell \times k$ matrix over $\mathscr{O}_{\mathbb{C}^{n+1},0}$. We use standard arguments from commutative algebra to show that in a minimal presentation, $\ell = k$ and λ is then injective. But what does a presentation tell us about the geometry of the image of f? The Fitting ideal $\mathrm{Fitt}_k(\mathscr{O}_{X,x_0})$ determines the locus of prime ideals $p \subset \mathscr{O}_{\mathbb{C}^{n+1},0}$ such that the localisation of $\mathscr{O}_{X,x_0}$ at p requires at least $k+1$-generators over $\mathscr{O}_{\mathbb{C}^{n_1},0}$: it is the set $V\left(\mathrm{Fitt}_k(\mathscr{O}_{X,x_0})\right)$ of prime ideals containing $\mathrm{Fitt}_k(\mathscr{O}_{X,x_0})$. But what does it tell us about the preimages of points in $\mathbb{C}^{n+1}$ *outside* 0?

◊

These questions, and others like them, are answered by the notion of coherence, as we will shortly see. All that we need is contained in the book of de Jong and Pfister [dJP00] so we give only the bare bones of an account here.

Definition E.4

(i) Let X be a complex space. The sheaf $\mathscr{S}$ of $\mathscr{O}_X$-modules is *of finite type* if for each point $x_0 \in X$ there is a neighbourhood U of x_0 and sections $s_1, \ldots, s_q$

of $\mathscr{S}$ over U whose germs at each point $x \in U$ generate $\mathscr{S}_x$ over $\mathscr{O}_{X,x}$. Equivalently, there is a surjective homomorphism of sheaves $\mathscr{O}^q_{X|U} \to \mathscr{S}|U$.

(ii) $\mathscr{S}$ is *of finite relation type* if for every open set U and surjective homomorphism of sheaves $\alpha : \mathscr{O}^q_{X|U} \to \mathscr{S}|U$, the kernel of α is of finite type.

(iii) $\mathscr{S}$ is a *coherent sheaf of $\mathscr{O}_X$-modules* if it is of finite type and finite relation type.

Theorem E.1 *$\mathscr{O}_{\mathbb{C}^n}$ is a coherent sheaf of $\mathscr{O}_{\mathbb{C}^n}$-modules.* □

This celebrated theorem of Oka is proved in, e.g., [dJP00, Theorem 6.3.1], as are the following properties of coherent sheaves.

Proposition E.1

(i) If X is any complex space, then $\mathscr{O}_X$ is a coherent sheaf of $\mathscr{O}_X$-modules.

(ii) The support supp$\mathscr{S}$ of a coherent sheaf $\mathscr{S}$ on a complex space X is a closed analytic set in X (supp$\mathscr{S}$ is defined as the set of points $x \in X$ such that $\mathscr{S}_x \neq 0$).

(iii) The kernel and image of any homomorphism of coherent sheaves of $\mathscr{O}_X$-modules are coherent.

(iv) If $\varphi : \mathscr{S} \to \mathscr{S}'$ is a homomorphism of coherent sheaves, and for some $x \in X$ its restriction $\varphi_x : \mathscr{S}_x \to \mathscr{S}'_x$ is an isomorphism, then there is an open neighbourhood U of x such that $\varphi| : \mathscr{S}|_U \to \mathscr{S}'|_U$ is an isomorphism.

(v) If the stalk at x_0 of a coherent sheaf $\mathscr{S}$ of $\mathscr{O}_X$-modules is free, then there is a neighbourhood U of x_0 such that $\mathscr{S}_x$ is free for all $x \in U$.

(vi) If any two of the three sheaves in a short exact sequence of sheaves and sheaf morphisms are coherent, then so is the third. □

Theorem E.2 (Finite Coherence Theorem [GR04, I.3.3]) *Let $f : X \to Y$ be a finite mapping between complex spaces. If $\mathscr{S}$ is a coherent sheaf on X, then $f_*\mathscr{S}$ is a coherent sheaf on Y.* □

As a consequence, the image of a finite mapping is analytic. This gives a proof (not Remmert's original proof) of Remmert's Proper Mapping Theorem D.6.

Corollary E.1 *Let $f : X \to Y$ be a finite mapping between complex spaces. Then $f(X)$ is an analytic subset of Y.*

Now we return to the questions raised in the examples at the start of this Appendix.

Example E.2 (Continued)

(1) First we make sense of the Krull dimension of a local ring $\mathscr{O}_{X,x_0}$. If X is a complex space, then the dimension of X at x is defined as the dimension of $\mathscr{O}_{X,x}$ and the dimension of X is the maximum of the local dimensions at all the points (or ∞, in case their are not bounded). It follows from the results of Sect. D.2 that $\dim(X, x)$ is upper semicontinuous, that is, any $x \in X$ has a neighbourhood U such that $\dim(X, x') \leq \dim(X, x)$, for all $x' \in U$.

Analogously, the singular locus of X, denoted by Sing(X) is the set of points $x \in X$ such that $\mathscr{O}_{X,x}$ is regular. The characterisation of Sing(X, x) in the local case given by the Jacobian criterion (see Exercise D.2.1), together with the coherence theorems, can be used to show that Sing(X) is an analytic subset of X (see [dJP00, Chapter 6] for details). Moreover, when X is reduced, Sing(X) is nowhere dense in X. This is a consequence of Proposition D.1.

(2) Proposition 4.7 and the finite coherence theorem of Grauert, E.2, tell us that if a germ $f\colon (X, x_0) \to (Y, y_0)$ satisfies $\dim_{\mathbb{C}} \mathscr{O}_{X,x_0}/f^*\mathfrak{m}_{Y,y_0}\mathscr{O}_{X,x_0} < \infty$, then for a suitable representative, which we designate by $f : U \to V$, $f^{-1}(0) = \{x_0\}$ and the sheaf $f_*\mathscr{O}_U$ is a coherent sheaf of $\mathscr{O}_V$-modules. Since $f^{-1}(0) = \{x_0\}$, $(f_*\mathscr{O}_U)_x = \mathscr{O}_{X,x_0}$. Given a presentation

$$\mathscr{O}^p_{Y,y_0} \xrightarrow{\lambda} \mathscr{O}^q_{Y,y_0} \longrightarrow \mathscr{O}_{X,x_0} \longrightarrow 0 \tag{E.3}$$

the entries in the matrix λ are germs in $\mathscr{O}_{Y,y_0}$, and all can be represented on some neighbourhood V of y_0. By Proposition E.1, the sequence of *sheaves*

$$\mathscr{O}^k_V \xrightarrow{\lambda} \mathscr{O}^k_V \longrightarrow f_*(\mathscr{O}_U)|_V \longrightarrow 0 \tag{E.4}$$

is exact on some possibly smaller neighbourhood of y_0. Shrinking V and then U appropriately, we can suppose that (E.4) is exact. Thus, *the germ at y of the matrix λ presents $f_*(\mathscr{O}_U)_y$ at each point $y \in V$*. The $(q-k)\times(q-k)$-minors of λ generate a sheaf of ideals in $\mathscr{O}_V$; we denote it by $\mathscr{F}_k$ to distinguish it from the ideal Fitt$_k$ of $\mathscr{O}_{V,0}$ (though $(\mathscr{F}_k)_0 = \text{Fitt}_k$). This presentation at y is no longer minimal for most points y, but that does not matter; the Fitting ideals are independent of the choice of presentation.

Lemma E.1 *$V(\mathscr{F}_k)$ is the set of points $y \in V$ such that $f_*(\mathscr{O}_U)_y$ requires at least $k+1$ generators over $\mathscr{O}_{V,y}$.*

Proof The germ of the matrix λ at y gives a presentation of $f_*(\mathscr{O}_U)_y$. If the rank of the numerical matrix $\lambda(y)$ is $q-k$, then column operations on the matrix λ allow us to express $q-k$ of the generators of $f_*(\mathscr{O}_U)_y$ in terms of the others, modulo $\mathfrak{m}_{Y,y}$. Nakayama's lemma shows that one can do this without the help of the maximal ideal $\mathfrak{m}_{Y,y}$. □

Of course, the ideal $(\mathscr{F}_k)_y$ is generated by the germs at y of the representatives on V of the generators of Fitt$_k$.

The point of this example is simply to illustrate how coherence makes possible the meaningful passage from a module over a local ring at a point to a sheaf of modules over a neighbourhood of that point, and how the properties of the module over the local ring propagate to properties at nearby points.

(3) Let M be an $\mathscr{O}_{X,x_0}$-module. In Appendix C we defined the support of M as the set of prime ideals $\{\mathfrak{p} \in \text{Spec}(R) : M_{\mathfrak{p}} \neq 0\}$. For a prime ideal $\mathfrak{p}$, $M_{\mathfrak{p}} \neq 0$ if and only if $\mathfrak{p} \supset \text{Ann}\, M$. The analytic version of the support is defined by taking

the analytic zero locus of the ideal $\operatorname{Ann} M \subset \mathscr{O}_{X,x_0}$. In fact, if $\mathscr{O}_{X,x_0} = \mathscr{O}_n/I$ for some ideal I, then $\operatorname{Ann}(M) = J/I$, for some ideal $J \supset I$ in $\mathscr{O}_n$. Then the *analytic support* $\operatorname{supp}(M)$ is the complex set-germ $V(J)$ in $(\mathbb{C}^n, x_0)$. When M is finitely generated, we can make use of the coherence of sheaves to give a geometrical interpretation of $\operatorname{supp}(M)$ as follows: take a presentation

$$\mathscr{O}^p_{X,x_0} \xrightarrow{\lambda} \mathscr{O}^q_{X,x_0} \longrightarrow M \longrightarrow 0\,,$$

and consider representatives of the entries of λ in some open neighbourhood U of x_0 in X. Then $\lambda\colon\ \mathscr{O}^p_U \to \mathscr{O}^q_U$ is a morphism of sheaves and its cokernel defines a coherent sheaf $\mathscr{M}$ on U, whose stalk at x_0 is isomorphic to M. We say that $\mathscr{M}$ is a *representative* of M on U. The coherence of $\mathscr{M}$ implies that the (analytic) support of M coincides with the germ at x_0 of the support of $\mathscr{M}$, which is equal to the set of points $x \in U$ such that $\mathscr{M}_x \neq 0$. The definitions of algebraic and analytic support resemble one another more in the global case; $\operatorname{Ann} M \subset \mathscr{O}_{X,x_0}$ extends to a coherent sheaf of ideals $\mathscr{A}$ on U, and

$$\operatorname{supp}_{\mathrm{an}}\mathscr{M} = \{x \in U : \mathscr{A}_x \subset \mathfrak{m}_{X,x}\}.$$

Thus, blurring the distinction between (closed) point and maximal ideal, $\operatorname{supp}_{\mathrm{an}}$, like $\operatorname{supp}_{\mathrm{alg}}$, is a set of ideals containing the annihilator.

The relation between geometrical and algebraic finiteness is given by the following consequence of the Rückert Nullstellensatz .

Theorem E.3 *Let $\mathscr{S}$ be a coherent sheaf on a complex space X. Given $x \in X$, the following statements are equivalent:*

1. $\dim_{\mathbb{C}} \mathscr{S}_x < \infty$,
2. there exists an open neighbourhood U of x in X such that $\operatorname{supp}\mathscr{S} \cap U \subset \{x\}$.

Proof If $\dim_{\mathbb{C}} \mathscr{S}_x < \infty$, then some power $\mathfrak{m}_x^k$ of the maximal ideal $\mathfrak{m}_x$ annihilates $\mathscr{S}_x$, and hence annihilates each member of a set of sections which generate $\mathscr{S}_x$. By coherence, these sections generate $\mathscr{S}$ on some neighbourhood U of x. At each point $x' \in U \setminus \{x\}$, the stalk of $\mathfrak{m}_x^k$ is equal to $\mathscr{O}_{U,x'}$, and thus $\mathscr{S}_{x'} = 0$.

Conversely, let $s_1, \ldots, s_m$ generate $\mathscr{S}$ in some neighbourhood U of x in which the support of $\mathscr{S}$ consists just of $\{x\}$. For each s_i, let φ_i be the morphism of coherent sheaves $\mathscr{O}_U \to \mathscr{S}$ sending g to gs_i, and let $\varphi : \oplus_1^m \mathscr{O}_U^m \to \mathscr{S}|_U$ be the epimorphism $\varphi(g_1, \ldots, g_m) = \sum g_i s_i$. The zero-locus of the ideal sheaf $\ker(\varphi_i)$ consists just of x, and so $\operatorname{rad}\ker(\varphi_i) = \mathfrak{m}_x$ by the usual Hilbert-Rückert Nullstellensatz (Theorem D.4). Hence $\ker(\varphi_i) \supset \mathfrak{m}_x^{k(i)}$ for some finite index $k(i)$, for $i = 1, \ldots, m$. Then φ gives rise to an epimorphism $\oplus_{i=1}^m (\mathscr{O}_{U,x} \,/\, \mathfrak{m}_x^{k(i)}) \to \mathscr{S}_x$.

□

Exercises for Sect. E.2

1 Deduce (v) of Proposition E.1 from the preceding statements.

2 Let $f : X \to Y$ be a morphism of complex spaces and let $\mathscr{S}$ be a coherent sheaf on X such that the restriction $f : \mathrm{supp}(\mathscr{S}) \to Y$ is finite. Show that $f_*\mathscr{S}$ is a coherent sheaf on Y.

E.3 Conservation of Multiplicity

Our point of departure for this series of appendices on commutative algebra was the following fact. Let f be an analytic function on $\mathbb{C}^n$ with isolated critical point at x_0. When we deform f, the critical point at x_0 will in general split into several new critical points. Then the sum of the Milnor numbers of the new critical points is equal to the Milnor number of the old critical point.

This equality is known as *conservation of the Milnor number* under deformation. It is an attractive statement, with its resemblance to physical principles such as the conservation of mass in a chemical reaction. But it is very vaguely framed. Which critical points of the deformed function do we count, when we say that the sum of their Milnor numbers is equal to the Milnor number of f at x_0? How far can the parameter t stray from 0 before the statement fails?

Example E.3 Let $f_t(x) = x^3 - t(x^3 + x^4)$. What is the sum of the Milnor numbers of the critical points for different values of t? ◊

In this section we develop techniques for making such statements precise, and proving them, and we give several more examples of conservation. The first technique is the simplest, and relies on a topological interpretation of the multiplicity of a finite map germ $(\mathbb{C}^n, x_0) \to (\mathbb{C}^n, y_0)$.

E.3.1 Representatives

The two questions asked above are answered simultaneously. Let $F(x,t) = f_t(x)$ be the deformation of f, and suppose that it is defined on an open set of the form $U \times V$ with $x_0 \in U \subset \mathbb{C}^n$ and $0 \in V \subset \mathbb{C}^d$. As the critical point of $f = f_0$ at x_0 is isolated, there is a compact neighbourhood $U_0 \subset U$ of x_0 in which there is no other critical point. We will count the critical points of f_t in U_0, but only for t so small that "no critical point of f_t has moved out of U_0 as t has moved from 0". We rephrase this last clause as follows: since f_0 has no critical point on the (compact) set ∂U_0, there is an $\eta > 0$ such that for $|t| < \eta$, f_t has no critical point on ∂U_0. Then our statement becomes:

Proposition E.2 (Conservation of the Milnor Number) *Under these circumstances,* $\sum_{x \in U_0} \mu(f_t, x) = \mu(f, x_0)$ *provided* $|t| < \eta$.

Our first proof of this is based on a topological principle:

Theorem E.4 *Let* M *be a smooth compact oriented* $n+1$*-dimensional manifold with boundary, and let* $f : M \to S^n$ *be a smooth map. Then with respect to the boundary orientation on* ∂M*,* $\deg(f| : \partial M \to S^n) = 0$. □

For a proof using the differential-topological definition of degree, see, e.g., [GP10, page 110] or [Mil97, §5 Lemma 1]. The result is immediate in homology: $[\partial M] = 0$ in $H_n(M)$, so the commutative diagram

$$\begin{array}{ccccc} & & H_n(M) & & \\ & \nearrow & & \searrow^{f_*} & \\ H_n(\partial M) & & \xrightarrow{\quad f|_* \quad} & & H_n(S^n) \end{array}$$

shows that the image of $[\partial M]$ under the horizontal morphism is 0.

Definition E.5 Let $U \subset \mathbb{R}^n$ be open, and $f : U \to \mathbb{R}^n$ be a smooth map with an isolated zero at x_0. The *index* of the zero of f at x_0, $\operatorname{ind}_{x_0}(f)$, is defined to be the degree of the smooth map

$$\frac{f}{\|f\|} : S_\varepsilon^{n-1} \to S^{n-1}$$

where ε is so small that $\overline{B}_\varepsilon(x_0) \subset U$ and f has no zero in $\overline{B}_\varepsilon(x_0) \smallsetminus \{x_0\}$.

The index is independent of the choice of ε: $f/\|f\|$ extends to the region sandwiched between concentric spheres, and the orientation of the inner sphere as boundary of this region is opposite to its usual orientation, so that the degree of $f/\|f\|$ with respect to the boundary orientation is the negative of the usual degree. On the other hand, the orientation of the outer sphere as boundary of the region is its usual orientation.

Lemma E.2 *Let* $U \subset \mathbb{R}^n$ *be open and bounded, with smooth boundary* ∂U*, and suppose that the smooth map* $f : \bar{U} \to \mathbb{R}^n$ *has no zeros on* ∂U*, and only isolated zeros inside* U*. Then*

$$\sum_{x \in U \cap f^{-1}(0)} \operatorname{ind}_x(f) = \deg\left(\frac{f}{\|f\|} : \partial U \to S^{n-1}\right).$$

Proof The map $f/\|f\|$ extends to a map $\bar{U}' := \bar{U} \smallsetminus \bigcup_{x \in U \cap f^{-1}(0)} B_\varepsilon(x) \to S^{n-1}$, where ε is chosen as in the definition of the index. Therefore by Theorem E.4 its degree on $\partial U'$, with its boundary orientation, is zero. The lemma follows because

$\partial U' = \partial U \cup \bigcup_{x \in f^{-1}(0)} S_\varepsilon^{n-1}(x)$, but where each $S_\varepsilon^{n-1}(x)$ has orientation opposite to its orientation as $\partial B_\varepsilon(x)$. Thus,

$$0 = \deg\left(\frac{f}{\|f\|} : \partial U' \to S^{n-1}\right) = \deg\left(\frac{f}{\|f\|} : \partial U \to S^{n-1}\right) - \sum_{x \in f^{-1}(0)} \operatorname{ind}_x(f).$$

□

Corollary E.2 *In the circumstances of Lemma E.2, if f_t is a smooth deformation of f, and for all t f_t has no zero on ∂U and only isolated zeros in U, then*

$$\sum_{x \in U \cap f_t^{-1}(0)} \operatorname{ind}_x(f_t) = \sum_{x \in U \cap f^{-1}(0)} \operatorname{ind}_x(f).$$

Proof The left and right sides are equal, respectively, to the degree of $f_t/\|f_t\|$ and of $f/\|f\|$ on ∂U. The deformation f_t defines a homotopy $f_t/\|f_t\| : \partial U \to S^{n-1}$, whose degree is therefore constant. □

The conservation of the index under deformations can be used now to show that the index of a smooth mapping coincides with its local degree (see Remark D.1).

Proposition E.3 *Let $f : (\mathbb{R}^n, x_0) \to (\mathbb{R}^n, 0)$ be a smooth map-germ such that $f^{-1}(x_0) = \{0\}$. Then,*

$$\operatorname{ind}(f) = \deg(f).$$

Proof Take a proper representative $f : U \to V$ with V a convex neighbourhood of the origin and such that $f^{-1}(0) = \{x_0\}$. First we assume that 0 is a regular value of f, so that $\deg(f) = \pm 1$, according to the sign of the Jacobian determinant $\det(d_{x_0} f))$. Without loss of generality, we can assume that $\det(d_{x_0} f) > 0$. By the inverse function theorem, there exists an open neighbourhood $U' \subset U$ of x_0 such that $f(U') \subset V$ is open and $f : U' \to f(U')$ is a diffeomorphism which preserves orientation. Now take $\varepsilon > 0$ such that $S_\varepsilon^{n-1}(x_0) \subset f(U')$ and let $U'' = f^{-1}(B_\varepsilon(x_0)) \subset U'$. By construction, U'' is open and bounded, with boundary $\partial U'' = f^{-1}(S_\varepsilon^{n-1}(x_0))$. Moreover, f has no zeros on $\partial U''$ and the only zero on U'' is x_0. By Lemma E.2,

$$\operatorname{ind}_{x_0}(f) = \deg(f/\|f\| : \partial U'' \to S^{n-1}).$$

But $\|f\| = \varepsilon$ on $\partial U''$, so $f/\|f\| = f/\varepsilon$ is a diffeomorphism which preserves orientation. Therefore, $\operatorname{ind}_{x_0}(f) = \deg(f/\varepsilon) = 1$.

When 0 is not a regular value, we take $y \in V$ any regular value. Consider the deformation $f_t = f - ty$, with $t \in [0, 1]$. Since f is proper, f_t has no zero on ∂U and only isolated zeros in U. By Corollary E.2,

$$\operatorname{ind}_{x_0}(f) = \sum_{x \in U \cap f_t^{-1}(0)} \operatorname{ind}_x(f_t),$$

for all $t \in [0, 1]$. For $t = 1$, 0 is a regular value of f_1, so for each $x \in f_1^{-1}(0)$, we have $\operatorname{ind}_x(f_t) = \deg_x(f_1) = \pm 1$, according to the sign of $\det(d_x f_1) = \det(d_x f)$. Hence, $\operatorname{ind}_{x_0}(f) = \deg_{x_0}(f)$. □

When f is a holomorphic map on $\mathbb{C}^n$, we can regard it as a smooth map on the underlying $\mathbb{R}^{2n}$. In this case, the index also coincides with the multiplicity of f, by Corollary D.7.

Corollary E.3 *Let* $f\colon (\mathbb{C}^n, x_0) \to (\mathbb{C}^n, 0)$ *be a holomorphic map-germ with isolated zero. Then*

$$\operatorname{ind}(f) = \deg(f) = \operatorname{mult}(f) = \dim_{\mathbb{C}} \mathcal{O}_n/(f_1, \ldots, f_n). \tag{E.5}$$

Since the index is conserved under deformation, (E.5) implies also conservation of multiplicity in the holomorphic case.

Corollary E.4 *Let* $U \subset \mathbb{C}^n$ *be open and bounded, with smooth boundary* ∂U, *and* $\{f_t\}_{t \in T}$ *a smooth deformation of a holomorphic map* $f : \bar{U} \to \mathbb{C}^n$, *with* T *connected. If, for all* $t \in T$, f_t *has no zero on* ∂U *and only isolated zeros in* U, *then*

$$\sum_{x \in U \cap f_t^{-1}(0)} \operatorname{mult}_x(f_t) = \sum_{x \in U \cap f^{-1}(0)} \operatorname{mult}_x f.$$

Proof of Proposition E.2 The Milnor number of f at x is the multiplicity of the map-germ $\partial f := (\partial f/\partial x_1, \ldots, \partial f/\partial x_n)$. So by (E.5), $\mu(f, x) = \operatorname{ind}_x(\partial f)$. The proposition now follows from Corollary E.4, except for one detail: we have made no explicit assumption that all of the critical points of f_t are isolated. In fact this follows from the assumption that f_t has no critical point on ∂U_0: because $\{\partial f = 0\}$ is an analytic set, no irreducible component of dimension greater than 0 can be compact, and therefore any non-isolated singularity would meet ∂U_0. □

Exercises for Sect. E.3

1 The intersection number of plane curves $D_i = \{f_i = 0\}$ at an isolated intersection point x_0 is defined by $(D_1, D_2)_{x_0} := \dim_{\mathbb{C}} \mathcal{O}_{\mathbb{C}^2, x_0}/(f_1, f_2)$.

(i) Show that $(D_1, D_2)_{x_0} = 1$ if and only if both curves are smooth at x_0, and meet transversely there.
(ii) State and prove a theorem on "conservation of intersection number" following the example of Proposition E.2.

2 Suppose that X and Y are complete intersections in $\mathbb{C}^n$ of dimensions k and $n-k$, defined by ideals I_X and I_Y, respectively. Their intersection number at an isolated intersection point x_0 may be defined as $\dim_{\mathbb{C}} \mathscr{O}_{\mathbb{C}^n,x_0}/(I_X + I_Y)$. Make and prove statements analogous to those of the previous exercise.

3 Suppose that $f: (\mathbb{C}^2, x_0) \to (\mathbb{C}^3, y_0)$ has corank 1.

(i) Show that the ideal $\mathscr{R}_f$ of 2×2 minors of its jacobian matrix is generated by two of the three 2×2 minors. Conclude that its non-immersive locus is a complete intersection.
(ii) Use this to show that the *Whitney umbrella number* of f, defined as

$$\dim_{\mathbb{C}} \frac{\mathscr{O}_{\mathbb{C}^2,x_0}}{\mathscr{R}_f},$$

is conserved under deformation of f, and that this number, as its name suggests, is equal to the number of Whitney umbrellas on the image of a stable perturbation of f (see Sect. 5.5).

In the last exercise, we need the assumption that f have corank 1 to guarantee that its non-immersive locus is a complete intersection. Nevertheless, even if f has higher corank, the Whitney umbrella number, defined as $\dim_{\mathbb{C}} \mathscr{O}_{\mathbb{C}^2,x_0}/\mathscr{R}_f$, is still conserved as f is deformed. The ket to proving this is the fact that in an unfolding F of f, the parametrised version of the ramification algebra, $\mathscr{O}_{\mathbb{C}^2\times\mathbb{C}^d,(x_0,0)}/\mathscr{R}_F$, is Cohen–Macaulay—see Exercise E.4.2.

E.4 Conservation of Multiplicity II

In the previous section, we proved the formula of conservation of multiplicity for holomorphic maps from $\mathbb{C}^n$ to $\mathbb{C}^n$ by means only of elementary topological arguments. In this section, we develop an approach to conservation of multiplicity which shows a more general statement, using the commutative algebra and the sheaf theory outlined in Appendices C–D and the first sections of this appendix.

Let M be a finitely generated $\mathscr{O}_{X,x_0}$-module and consider a representative $\mathscr{M}$ of M as in Example E.2. We recall that to construct $\mathscr{M}$ we took a presentation

$$\mathscr{O}^p_{X,x_0} \xrightarrow{\lambda} \mathscr{O}^q_{X,x_0} \longrightarrow M \longrightarrow 0,$$

and choose representatives of the entries of λ in some open neighbourhood U of 0 in X. Then $\mathcal{M}$ is defined as the cokernel of $\lambda\colon\ \mathcal{O}_U^p \to \mathcal{O}_U^q$.

Proposition E.4 *Let M be a finitely generated $\mathcal{O}_d$-module. There exists an open neighbourhood U of the origin in $\mathbb{C}^d$ such that for all $u \in U$,*

$$e(M) = e(\mathcal{M}_u).$$

Proof Take a free resolution of M over $\mathcal{O}_d$:

$$0 \longrightarrow \mathcal{O}_d^{n_k} \longrightarrow \dots \longrightarrow \mathcal{O}_d^{n_0} \longrightarrow M \longrightarrow 0\,.$$

By taking representatives on some open neighbourhood U of 0 in $\mathbb{C}^d$, we get an exact sequence of $\mathcal{O}_U$-modules:

$$0 \longrightarrow \mathcal{O}_U^{n_k} \longrightarrow \dots \longrightarrow \mathcal{O}_U^{n_0} \longrightarrow \mathcal{M} \longrightarrow 0\,.$$

In particular, for each $u \in U$, we have a free resolution of $\mathcal{M}_u$ over $\mathcal{O}_{\mathbb{C}^d,u}$:

$$0 \longrightarrow \mathcal{O}_{\mathbb{C}^d,u}^{n_k} \longrightarrow \dots \longrightarrow \mathcal{O}_{\mathbb{C}^d,u}^{n_0} \longrightarrow \mathcal{M}_u \longrightarrow 0\,.$$

Now we tensor with the field $\mathcal{M}_{\mathbb{C}^d,u} = Q(\mathcal{O}_{\mathbb{C}^d,u})$ of meromorphic function germs on $(\mathbb{C}^d, u)$ and get again an exact sequence:

$$0 \longrightarrow \mathcal{M}_{\mathbb{C}^d,u}^{n_k} \longrightarrow \dots \longrightarrow \mathcal{M}_{\mathbb{C}^d,u}^{n_0} \longrightarrow \mathcal{M} \otimes \mathcal{M}_{\mathbb{C}^d,u} \longrightarrow 0\,.$$

By Proposition C.14,

$$e(\mathcal{M}_u) = \operatorname{rank}(\mathcal{M}_u) = \dim_{\mathcal{M}_{\mathbb{C}^d,u}}(\mathcal{M}_u \otimes \mathcal{M}_{\mathbb{C}^d,u}) = \sum_{i=0}^{k} (-1)^i n_i,$$

which is independent of u. □

Corollary E.5 *Let M be a finitely generated $\mathcal{O}_{X,x_0}$-module and let $f\colon (X, x_0) \to (\mathbb{C}^d, 0)$ be analytic such that $\dim_{\mathbb{C}} M/f^*\mathfrak{m}_d M < \infty$. Then*

1. *For every representative $\mathcal{M}$ of M, there is a possibly smaller representative $\mathcal{M}'$ such that f is finite on supp($\mathcal{M}'$).*
2. *There exists a representative $f\colon U \to W$, such that for all $u \in W$,*

$$e(f^*\mathfrak{m}_d, M) = \sum_{supp(\mathcal{M})\cap\{f(x)=u\}} e(f^*\mathfrak{m}_{\mathbb{C}^d,u}, \mathcal{M}_x).$$

Proof

1. It is enough to show that

$$\dim_{\mathbb{C}} \frac{\mathscr{O}_{\mathrm{supp}(\mathscr{M}),x_0}}{f^* \mathfrak{m}_d \mathscr{O}_{\mathrm{supp}(\mathscr{M}),x_0}} < \infty.$$

   As supp$\mathscr{M}$ is defined by the coherent sheaf of ideals extending $A := \mathrm{Ann}_{\mathscr{O}_{X,x_0}} M$, it is enough to show that

$$\dim_{\mathbb{C}} \frac{\mathscr{O}_{X,x_0}}{A + f^* \mathfrak{m}_d \mathscr{O}_{X,x_0}} < \infty.$$

   Now $A + f^* \mathfrak{m}_d \mathscr{O}_{X,x_0}$ contains some power of $\mathrm{Ann}_{\mathscr{O}_{X,x_0}}(M/f^* \mathfrak{m}_d M)$, by Exercise C.3.1(ii). Because $\dim_{\mathbb{C}} M/f^* \mathfrak{m}_d M < \infty$, $\mathrm{Ann}_{\mathscr{O}_{X,x_0}}(M/f^* \mathfrak{m}_d M)$ contains some power of $\mathfrak{m}_{X,x_0}$ (here we use the Nullstellensatz). Hence so does $A + f^* \mathfrak{m}_d \mathscr{O}_{X,x_0}$. Thus, by Theorem D.5, f is finite on supp($\mathscr{M}$).
2. By item 1, we can choose representatives $f: U \to W$ of the map-germ and $\mathscr{M}$ of the module such that the restriction $f: \mathrm{supp}\mathscr{M} \to W$ is a finite mapping. By Exercise E.2.2, $f_*\mathscr{M}$ is a coherent sheaf on W and we only have to apply Proposition E.4 to $f_*\mathscr{M}$. By construction, at a point $u \in W$,

$$(f_*\mathscr{M})_u = \bigoplus_{\mathrm{supp}(\mathscr{M})\cap\{f(x)=u\}} \mathscr{M}_x$$

   and the multiplicity of each $\mathscr{M}_x$ as $\mathscr{O}_{\mathbb{C}^d,u}$-module is equal to its multiplicity as $\mathscr{O}_{U,x}$-module with respect to the ideal $f^*\mathfrak{m}_{\mathbb{C}^d,u}$.

□

If M is a Cohen–Macaulay $\mathscr{O}_{X,x_0}$-module, then $\mathscr{M}_x$ is also Cohen–Macaulay, for all x in some neighbourhood of x_0 in X. In that case, we can use Proposition C.15 and rewrite Corollary E.5 as follows:

Corollary E.6 *Let M be a Cohen–Macaulay $\mathscr{O}_{X,x_0}$-module of dimension d and let $f: (X, x_0) \to (\mathbb{C}^d, 0)$ be analytic, and such that $\dim_{\mathbb{C}} M/f^*\mathfrak{m}_d M < \infty$. There exists a representative $f: U \to W$, such that for all $u \in W$,*

$$\dim_{\mathbb{C}} \frac{M}{f^*\mathfrak{m}_d M} = \sum_{\mathrm{supp}(\mathscr{M})\cap\{f(x)=u\}} \dim_{\mathbb{C}} \frac{\mathscr{M}_x}{f^*\mathfrak{m}_{\mathbb{C}^d,u}\mathscr{M}_x}.$$

We can now recover the conservation of multiplicity of Corollary E.4 as a particular case of Corollary E.6. In fact, let $f: (\mathbb{C}^n, x_0) \to (\mathbb{C}^n, 0)$ be a finite map-germ and consider a deformation $F(x,t) = f_t(x)$, with $f_0 = f$. We take $(X, \tilde{x}_0)$ as

$F^{-1}(0)$, with $\tilde{x}_0 = (x_0, 0)$, $\pi\colon (X, \tilde{x}_0) \to (\mathbb{C}, 0)$ as the projection $\pi(x,t) = t$ and $M = \mathcal{O}_{\mathbb{C}^n\times\mathbb{C},\tilde{x}_0}/F^*\mathfrak{m}_n$. Then, we get

$$\operatorname{mult}(f) = \dim_{\mathbb{C}} \frac{M}{\pi^*\mathfrak{m}_1 M} = \sum_{\operatorname{supp}(\mathcal{M})\cap\{\pi(x,t)=t\}} \dim_{\mathbb{C}} \frac{\mathcal{M}_{(x,t)}}{\pi^*\mathfrak{m}_{\mathbb{C},t}\mathcal{M}_{(x,t)}}$$

$$= \sum_{f_t(x)=0} \operatorname{mult}_x(f_t).$$

Exercises for Sect. E.4

1 Let $f\colon (\mathbb{C}^n, 0) \to \mathbb{C}$ have an isolated singularity at 0. Show that the number of non-degenerate critical points of a stable perturbation of f is equal to the Milnor number $\dim_{\mathbb{C}} \mathcal{O}_n / J_f$.

2 Let $f\colon (\mathbb{C}^2, 0) \to (\mathbb{C}^3, 0)$ have an isolated non-immersive point at 0. Show that the number of Whitney umbrellas on the image of a stable perturbation of f is equal to $\dim_{\mathbb{C}} \mathcal{O}_2 / \mathcal{R}_f$, where $\mathcal{R}_f$ is the ramification ideal of f, the ideal generated by the 2×2 minors of the matrix of df. Hint: use the Hilbert–Burch theorem, Theorem C.5.

3 Generalise the result of the previous exercise to the case of map-germs $(\mathbb{C}^n, 0) \to (\mathbb{C}^{2n-1}, 0)$ (see Exercise 7.2.4).

4 Define $f\colon (\mathbb{C}^6, 0) \to (\mathbb{C}^7, 0)$ by $f(x, y, u) = (x^k + u_1 y, xy + u_2 x + u_3 y, y^\ell + u_4 x, u)$, and let f_t be a generic perturbation. At how many points does f_t have corank 2?

5 Let $f\colon (\mathbb{C}^2, 0) \to (\mathbb{C}^3, 0)$ be defined by

$$f(x_1, x_2) = (x_1^2, x_2^2, x_1(x_1^2 + x_2^2) + x_2(x_1^2 - x_2^2)))$$

(see Example 8.5). How many triple points are there on the image of a stable perturbation of f? Suggestion: use Corollary 11.12.

6 What does proposition Proposition E.4 tell us when $\operatorname{supp}(\mathcal{M})$ is a proper analytic subset of U?

References

[A'C75] Norbert A'Campo. Le groupe de monodromie du déploiement des singularités isolées de courbes planes. I. *Math. Ann.*, 213:1–32, 1975.

[AGZV12a] V. I. Arnold, S. M. Gusein-Zade, and A. N. Varchenko. *Singularities of differentiable maps. Volume 1*. Modern Birkhäuser Classics. Birkhäuser/Springer, New York, 2012. Classification of critical points, caustics and wave fronts, Translated from the Russian by Ian Porteous based on a previous translation by Mark Reynolds, Reprint of the 1985 edition.

[AGZV12b] V. I. Arnold, S. M. Gusein-Zade, and A. N. Varchenko. *Singularities of differentiable maps. Volume 2*. Modern Birkhäuser Classics. Birkhäuser/Springer, New York, 2012. Monodromy and asymptotics of integrals, Translated from the Russian by Hugh Porteous and revised by the authors and James Montaldi, Reprint of the 1988 translation.

[AN72] M. Artin and M. Nagata. Residual intersections in Cohen-Macaulay rings. *J. Math. Kyoto Univ.*, 12:307–323, 1972.

[Arn72] V. I. Arnold. Normal forms of functions near degenerate critical points, the Weyl groups A_k, D_k, E_k and Lagrangian singularities. *Funkcional. Anal. i Priložen.*, 6(4):3–25, 1972.

[Ati00] R. Wik Atique. On the classification of multi-germs of maps from $\mathbf{C}^2$ to $\mathbf{C}^3$ under $\mathscr{A}$-equivalence. In *Real and complex singularities (São Carlos, 1998)*, volume 412 of *Chapman & Hall/CRC Res. Notes Math.*, pages 119–133. Chapman & Hall/CRC, Boca Raton, FL, 2000.

[BC13] Ragnar-Olaf Buchweitz and Aldo Conca. New free divisors from old. *J. Commut. Algebra*, 5(1):17–47, 2013.

[BdPW87] J. W. Bruce, A. A. du Plessis, and C. T. C. Wall. Determinacy and unipotency. *Invent. Math.*, 88(3):521–554, 1987.

[BE77a] David A. Buchsbaum and David Eisenbud. Algebra structures for finite free resolutions, and some structure theorems for ideals of codimension 3. *Amer. J. Math.*, 99(3):447–485, 1977.

[BE77b] David A. Buchsbaum and David Eisenbud. What annihilates a module? *J. Algebra*, 47(2):231–243, 1977.

[BEGvB09] Ragnar-Olaf Buchweitz, Wolfgang Ebeling, and Hans-Christian Graf von Bothmer. Low-dimensional singularities with free divisors as discriminants. *J. Algebraic Geom.*, 18(2):371–406, 2009.

D. Mond, J. J. Nuño-Ballesteros, *Singularities of Mappings*,
Grundlehren der mathematischen Wissenschaften 357,
https://doi.org/10.1007/978-3-030-34440-5

[BG82] J. W. Bruce and T. J. Gaffney. Simple singularities of mappings $\mathbf{C}, 0 \to \mathbf{C}^2, 0$. *J. London Math. Soc. (2)*, 26(3):465–474, 1982.

[BKdP97] J. W. Bruce, N. P. Kirk, and A. A. du Plessis. Complete transversals and the classification of singularities. *Nonlinearity*, 10(1):253–275, 1997.

[BM88] Edward Bierstone and Pierre D. Milman. Semianalytic and subanalytic sets. *Inst. Hautes Études Sci. Publ. Math.*, 67:5–42, 1988.

[BM96] J. W. Bruce and W. L. Marar. Images and varieties. *J. Math. Sci.*, 82(5):3633–3641, 1996. Topology, 3.

[BM06] Ragnar-Olaf Buchweitz and David Mond. Linear free divisors and quiver representations. In *Singularities and computer algebra*, volume 324 of *London Math. Soc. Lecture Note Ser.*, pages 41–77. Cambridge Univ. Press, Cambridge, 2006.

[Boa67] J. M. Boardman. Singularities of differentiable maps. *Inst. Hautes Études Sci. Publ. Math.*, 33:21–57, 1967.

[BR64] David A. Buchsbaum and Dock S. Rim. A generalized Koszul complex. II. Depth and multiplicity. *Trans. Amer. Math. Soc.*, 111:197–224, 1964.

[Brö75] Th. Bröcker. *Differentiable germs and catastrophes.* Cambridge University Press, Cambridge-New York-Melbourne, 1975. Translated from the German, last chapter and bibliography by L. Lander, London Mathematical Society Lecture Note Series, No. 17.

[BT82] Raoul Bott and Loring W. Tu. *Differential forms in algebraic topology*, volume 82 of *Graduate Texts in Mathematics*. Springer-Verlag, New York-Berlin, 1982.

[BV86] Winfried Bruns and Udo Vetter. Length formulas for the local cohomology of exterior powers. *Math. Z.*, 191(1):145–158, 1986.

[Car95] Henri Cartan. *Elementary theory of analytic functions of one or several complex variables.* Dover Publications, Inc., New York, 1995. Translated from the French, Reprint of the 1973 edition.

[Cat84] Fabrizio Catanese. Commutative algebra methods and equations of regular surfaces. In *Algebraic geometry, Bucharest 1982 (Bucharest, 1982)*, volume 1056 of *Lecture Notes in Math.*, pages 68–111. Springer, Berlin, 1984.

[CLO15] David A. Cox, John Little, and Donal O'Shea. *Ideals, varieties, and algorithms.* Undergraduate Texts in Mathematics. Springer, Cham, fourth edition, 2015. An introduction to computational algebraic geometry and commutative algebra.

[CMWA02] T. Cooper, D. Mond, and R. Wik Atique. Vanishing topology of codimension 1 multi-germs over $\mathbb{R}$ and $\mathbb{C}$. *Compositio Math.*, 131(2):121–160, 2002.

[Coo94] T.. Cooper. *Map-germs of $\mathscr{A}_e$ codimension 1*. PhD thesis, University of Warwick, 1994.

[CS81] C.de Concini and E. Strickland. On the variety of complexes. *Adv. Math.*, 41:57–77, 1981.

[Dam84] James Damon. The unfolding and determinacy theorems for subgroups of $\mathscr{A}$ and $\mathscr{K}$. *Mem. Amer. Math. Soc.*, 50(306):x+88, 1984.

[Dam87] James Damon. Deformations of sections of singularities and Gorenstein surface singularities. *Amer. J. Math.*, 109(4):695–721, 1987.

[Dam88] James N. Damon. Topological triviality and versality for subgroups of A and K. *Mem. Amer. Math. Soc.*, 75(389):x+106, 1988.

[Dam91] James Damon. $\mathscr{A}$-equivalence and the equivalence of sections of images and discriminants. In *Singularity theory and its applications, Part I (Coventry, 1988/1989)*, volume 1462 of *Lecture Notes in Math.*, pages 93–121. Springer, Berlin, 1991.

[Dam96] James Damon. Higher multiplicities and almost free divisors and complete intersections. *Mem. Amer. Math. Soc.*, 123(589):x+113, 1996.

[Dam98] James Damon. On the legacy of free divisors: discriminants and Morse-type singularities. *Amer. J. Math.*, 120(3):453–492, 1998.

[Dam01] James Damon. Nonlinear sections of nonisolated complete intersections. In *New developments in singularity theory (Cambridge, 2000)*, volume 21 of *NATO Sci. Ser. II Math. Phys. Chem.*, pages 405–445. Kluwer Acad. Publ., Dordrecht, 2001.

[Dam02] James Damon. On the freeness of equisingular deformations of plane curve singularities. *Topology Appl.*, 118(1-2):31–43, 2002. Arrangements in Boston: a Conference on Hyperplane Arrangements (1999).

[Dav83] J. M. S. David. Projection-generic curves. *J. London Math. Soc. (2)*, 27(3):552–562, 1983.

[Dem12] Jean-Pierre Demailly. *Complex Analytic and Differential Geometry*. Université de Grenoble I. Institut Fourier, https://www-fourier.ujf-grenoble.fr/~demailly/manuscripts/agbook.pdf, 2012.

[DG76] James Damon and Andre Galligo. A topological invariant for stable map-germs. *Invent. Math.*, 32:103–132, 1976.

[DGH11] James Damon, Peter Giblin, and Gareth Haslinger. *Local features in natural images via singularity theory*. Lecture Notes in Mathematics, vol. 2165, pages x + 255. Springer, Cham, 2016. https://0-doi-org.pugwash.lib.warwick.ac.uk/10.1007/978-3-319-41471-3.

[Dia15] Fabio Scalco Dias. Projection of generic 1 and 2-parameter families of space curves. *Hokkaido Math. J.*, 44(2):221–250, 2015.

[Dim92] Alexandru Dimca. *Singularities and topology of hypersurfaces*. Universitext. Springer-Verlag, Berlin, 1992.

[dJP00] Theo de Jong and Gerhard Pfister. *Local analytic geometry*. Advanced Lectures in Mathematics. Friedr. Vieweg & Sohn, Braunschweig, 2000. Basic theory and applications.

[dJvS90] T. de Jong and D. van Straten. Deformations of the normalization of hypersurfaces. *Math. Ann.*, 288(3):527–547, 1990.

[dJvS91] T. de Jong and D. van Straten. Disentanglements. In *Singularity theory and its applications, Part I (Coventry, 1988/1989)*, volume 1462 of *Lecture Notes in Math.*, pages 199–211. Springer, Berlin, 1991.

[DM91] James Damon and David Mond. $\mathscr{A}$-codimension and the vanishing topology of discriminants. *Invent. Math.*, 106(2):217–242, 1991.

[DNnB08] F. S. Dias and J. J. Nuño Ballesteros. Plane curve diagrams and geometrical applications. *Q. J. Math.*, 59(3):287–310, 2008.

[dP80] Andrew du Plessis. On the determinacy of smooth map-germs. *Invent. Math.*, 58(2):107–160, 1980.

[DP14] James Damon and Brian Pike. Solvable groups, free divisors and nonisolated matrix singularities II: Vanishing topology. *Geom. Topol.*, 18(2):911–962, 2014.

[DP15] James Damon and Brian Pike. Solvable groups, free divisors and nonisolated matrix singularities I: Towers of free divisors. *Ann. Inst. Fourier (Grenoble)*, 65(3):1251–1300, 2015.

[dPW95] Andrew du Plessis and Terry Wall. *The geometry of topological stability*, volume 9 of *London Mathematical Society Monographs. New Series*. The Clarendon Press, Oxford University Press, New York, 1995. Oxford Science Publications.

[Eis95] David Eisenbud. *Commutative algebra*, volume 150 of *Graduate Texts in Mathematics*. Springer-Verlag, New York, 1995. With a view toward algebraic geometry.

[Eph78] Robert Ephraim. Isosingular loci and the Cartesian product structure of complex analytic singularities. *Trans. Amer. Math. Soc.*, 241:357–371, 1978.

[FdBNnBPnS19] J. Fernández de Bobadilla, J. J. Nuño Ballesteros, and G. Peñafort Sanchis. A Jacobian module for disentanglements and applications to Mond's conjecture. *Rev. Mat. Complut.*, 32(2):395–418, 2019.

[FdBP08] Javier Fernández de Bobadilla and María Pe Pereira. Equisingularity at the normalisation. *J. Topol.*, 1(4):879–909, 2008.

[FK99] Anne Frühbis-Krüger. Classification of simple space curve singularities. *Comm. Algebra*, 27(8):3993–4013, 1999.

[FKN10] Anne Frühbis-Krüger and Alexander Neumer. Simple Cohen-Macaulay codimension 2 singularities. *Comm. Algebra*, 38(2):454–495, 2010.

[FL77] William Fulton and Dan Laksov. Residual intersections and the double point formula. In *Real and complex singularities (Proc. Ninth Nordic Summer School/NAVF Sympos. Math., Oslo, 1976)*, pages 171–177. Sijthoff and Noordhoff, Alphen aan den Rijn, 1977.

[Fuk99] Toshizumi Fukui. Butterflies and umbilics of stable perturbations of analytic map-germs $(\mathbf{C}^5, 0) \to (\mathbf{C}^4, 0)$. In *Singularity theory (Liverpool, 1996)*, volume 263 of *London Math. Soc. Lecture Note Ser.*, pages xxi, 369–378. Cambridge Univ. Press, Cambridge, 1999.

[Ful97] William Fulton. *Young tableaux*, volume 35 of *London Mathematical Society Student Texts*. Cambridge University Press, Cambridge, 1997. With applications to representation theory and geometry.

[Gaf79] Terence Gaffney. A note on the order of determination of a finitely determined germ. *Invent. Math.*, 52(2):127–130, 1979.

[Gaf83] Terence Gaffney. Multiple points and associated ramification loci. In *Singularities, Part 1 (Arcata, Calif., 1981)*, volume 40 of *Proc. Sympos. Pure Math.*, pages 429–437. Amer. Math. Soc., Providence, RI, 1983.

[Gaf18] Terence Gaffney. Equisingularity and the theory of integral closure. In *Singularities and foliations. geometry, topology and applications*, volume 222 of *Springer Proc. Math. Stat.*, pages 89–131. Springer, Cham, 2018.

[GG73] M. Golubitsky and V. Guillemin. *Stable mappings and their singularities*. Springer-Verlag, New York-Heidelberg, 1973. Graduate Texts in Mathematics, Vol. 14.

[GH93] C. G. Gibson and C. A. Hobbs. Simple singularities of space curves. *Math. Proc. Cambridge Philos. Soc.*, 113(2):297–310, 1993.

[Giu83] Marc Giusti. Classification des singularités isolées simples d'intersections complètes. In *Singularities, Part 1 (Arcata, Calif., 1981)*, volume 40 of *Proc. Sympos. Pure Math.*, pages 457–494. Amer. Math. Soc., Providence, RI, 1983.

[G.K75] G.Kempf. Images of homogeneous vector bundles and varieties of complexes. *Bull. Amer. Math. Soc.*, pages 900–901, 1975.

[GLS07] G.-M. Greuel, C. Lossen, and E. Shustin. *Introduction to singularities and deformations*. Springer Monographs in Mathematics. Springer, Berlin, 2007.

[GM93] Victor V. Goryunov and David Mond. Vanishing cohomology of singularities of mappings. *Compositio Math.*, 89(1):45–80, 1993.

[GM05] Victor V. Goryunov and David Mond. Tjurina and Milnor numbers of matrix singularities. *J. London Math. Soc. (2)*, 72(1):205–224, 2005.

[GMNRS09] Michel Granger, David Mond, Alicia Nieto-Reyes, and Mathias Schulze. Linear free divisors and the global logarithmic comparison theorem. *Ann. Inst. Fourier (Grenoble)*, 59(2):811–850, 2009.

[Gor84] V. V. Goryunov. Simple projections. *Sibirsk. Mat. Zh.*, 25(1):61–68, 1984.

[Gor91] V. V. Goryunov. Monodromy of the image of the mapping $\mathbf{C}^2 \to \mathbf{C}^3$. *Funktsional. Anal. i Prilozhen.*, 25(3):12–18, 95, 1991.

[Gor95] Victor V. Goryunov. Semi-simplicial resolutions and homology of images and discriminants of mappings. *Proc. London Math. Soc. (3)*, 70(2):363–385, 1995.

[GP82] Laurent Gruson and Christian Peskine. Courbes de l'espace projectif: variétés de sécantes. In *Enumerative geometry and classical algebraic geometry (Nice, 1981)*, volume 24 of *Progr. Math.*, pages 1–31. Birkhäuser, Boston, Mass., 1982.

[GP08] Gert-Martin Greuel and Gerhard Pfister. *A **Singular** introduction to commutative algebra*. Springer, Berlin, extended edition, 2008. With contributions by Olaf Bachmann, Christoph Lossen and Hans Schönemann, With 1 CD-ROM (Windows, Macintosh and UNIX).

[GP10] Victor Guillemin and Alan Pollack. *Differential topology*. AMS Chelsea Publishing, Providence, RI, 2010. Reprint of the 1974 original.

[GR04] Hans Grauert and Reinhold Remmert. *Theory of Stein spaces*. Classics in Mathematics. Springer-Verlag, Berlin, 2004. Translated from the German by Alan Huckleberry, Reprint of the 1979 translation.

[Gre80] Gert-Martin Greuel. Dualität in der lokalen Kohomologie isolierter Singularitäten. *Math. Ann.*, 250(2):157–173, 1980.

[Gun70] R. C. Gunning. *Lectures on complex analytic varieties: The local parametrization theorem*. Mathematical Notes. Princeton University Press, Princeton, N.J.; University of Tokyo Press, Tokyo, 1970.

[GWdPL76] Christopher G. Gibson, Klaus Wirthmüller, Andrew A. du Plessis, and Eduard J. N. Looijenga. *Topological stability of smooth mappings*. Lecture Notes in Mathematics, Vol. 552. Springer-Verlag, Berlin-New York, 1976.

[GZ74] S. M. Guseĭn-Zade. Dynkin diagrams of the singularities of functions of two variables. *Funkcional. Anal. i Priložen.*, 8(4):23–30, 1974.

[Har67] Robin Hartshorne. *Local cohomology*, volume 1961 of *A seminar given by A. Grothendieck, Harvard University, Fall*. Springer-Verlag, Berlin-New York, 1967.

[Har77] Robert M. Hardt. Triangulation of subanalytic sets and proper light subanalytic maps. *Invent. Math.*, 38(3):207–217, 1976/77.

[Has12] Seyed Hamid Hassanzadeh. Cohen-Macaulay residual intersections and their Castelnuovo-Mumford regularity. *Trans. Amer. Math. Soc.*, 364(12):6371–6394, 2012.

[Hat02] Allen Hatcher. *Algebraic topology*. Cambridge University Press, Cambridge, 2002.

[HE71] M. Hochster and John A. Eagon. Cohen-Macaulay rings, invariant theory, and the generic perfection of determinantal loci. *Amer. J. Math.*, 93:1020–1058, 1971.

[HK99] Kevin Houston and Neil Kirk. On the classification and geometry of corank 1 map-germs from three-space to four-space. In *Singularity theory (Liverpool, 1996)*, volume 263 of *London Math. Soc. Lecture Note Ser.*, pages xxii, 325–351. Cambridge Univ. Press, Cambridge, 1999.

[HMPnS18] M. E. Hernandes, A. J. Miranda, and G. Peñafort Sanchis. An algorithm to compute a presentation of pushforward modules. *Topology Appl.*, 234:440–451, 2018.

[Hoc81] Gerhard P. Hochschild. *Basic theory of algebraic groups and Lie algebras*, volume 75 of *Graduate Texts in Mathematics*. Springer-Verlag, New York-Berlin, 1981.

[Hou97] Kevin Houston. Local topology of images of finite complex analytic maps. *Topology*, 36(5):1077–1121, 1997.

[Hou98] Kevin Houston. On singularities of folding maps and augmentations. *Math. Scand.*, 82(2):191–206, 1998.

[Hou99] Kevin Houston. An introduction to the image computing spectral sequence. In *Singularity theory (Liverpool, 1996)*, volume 263 of *London Math. Soc. Lecture Note Ser.*, pages xxi–xxii, 305–324. Cambridge Univ. Press, Cambridge, 1999.

[Hou02] Kevin Houston. Bouquet and join theorems for disentanglements. *Invent. Math.*, 147(3):471–485, 2002.

[Hus94] Dale Husemoller. *Fibre bundles*, volume 20 of *Graduate Texts in Mathematics*. Springer-Verlag, New York, third edition, 1994.

[HVV85] J. Herzog, W. V. Vasconcelos, and R. Villarreal. Ideals with sliding depth. *Nagoya Math. J.*, 99:159–172, 1985.

[Joz78] T. Jozefiak. Ideals generated by minors of a symmetric matrix. *Comment. Math. Helv.*, 53:594–607, 1978.

[JPNnB09] V. H. Jorge Pérez and J. J. Nuño Ballesteros. Finite determinacy and Whitney equisingularity of map germs from $\mathbb{C}^n$ to $\mathbb{C}^{2n-1}$. *Manuscripta Math.*, 128(3):389–410, 2009.

[Kem76] G. Kempf. On the collapsing of homogeneous bundles. *Invent. Math.*, 37:229–239, 1976.

[Kir93] N. P. Kirk. *Computational aspects of singularity theory*. PhD thesis, University of Liverpool, 1993.

[Kir96] N. P. Kirk. *Transversal: a Maple package for singularity theory, user manual.* University of Liverpool, 1996.

[KL80] H. Kleppe and D. Laksov. The algebraic structure and deformation of Pfaffian schemes. *J. Algebra*, 64(1):167–189, 1980.

[Kle81] Steven L. Kleiman. Multiple-point formulas. I. Iteration. *Acta Math.*, 147(1-2):13–49, 1981.

[KPR07] Carsten Klotz, Otilia Pop, and Joachim H. Rieger. Real double-points of deformations of $\mathscr{A}$-simple map-germs from $\mathbb{R}^n$ to $\mathbb{R}^{2n}$. *Math. Proc. Cambridge Philos. Soc.*, 142(2):341–363, 2007.

[Kra01] Steven G. Krantz. *Function theory of several complex variables.* AMS Chelsea Publishing, Providence, RI, 2001. Reprint of the 1992 edition.

[KU97] Steven Kleiman and Bernd Ulrich. Gorenstein algebras, symmetric matrices, self-linked ideals, and symbolic powers. *Trans. Amer. Math. Soc.*, 349(12):4973–5000, 1997.

[Kut74] Ronald E. Kutz. Cohen-Macaulay rings and ideal theory in rings of invariants of algebraic groups. *Trans. Amer. Math. Soc.*, 194:115–129, 1974.

[Lê87] Lê Dũng Tráng. Le concept de singularité isolée de fonction analytique. In *Complex analytic singularities*, volume 8 of *Adv. Stud. Pure Math.*, pages 215–227. North-Holland, Amsterdam, 1987.

[Lee13] John M. Lee. *Introduction to smooth manifolds*, volume 218 of *Graduate Texts in Mathematics*. Springer, New York, second edition, 2013.

[LêR76] Lê Dũng Tráng and C. P. Ramanujam. The invariance of Milnor's number implies the invariance of the topological type. *Amer. J. Math.*, 98(1):67–78, 1976.

[LêT83] Lê Dũng Tráng and B. Teissier. Cycles evanescents, sections planes et conditions de Whitney. II. In *Singularities, Part 2 (Arcata, Calif., 1981)*, volume 40 of *Proc. Sympos. Pure Math.*, pages 65–103. Amer. Math. Soc., Providence, RI, 1983.

[Lev71] Harold I. Levine. Singularities of differentiable mappings. In *Proceedings of Liverpool Singularities Symposium, II (1969/1970)*, volume 209 of *Lecture Notes in Math.*, pages 1–89, Berlin, 1971. Springer.

[Ło65] S. Ł ojasiewicz. Ensembles semi-analytiques. IHES Notes, 1965.

[Ło91] S. Ł ojasiewicz. *Introduction to complex analytic geometry*. Birkhäuser Verlag, Basel, 1991. Translated from the Polish by Maciej Klimek.

[Loj71] S. Lojasiewicz. Whitney fields and the Malgrange-Mather preparation theorem. In *Proceedings of Liverpool Singularities—Symposium, I (1969/70)*, volume 192 of *Lecture Notes in Math.*, pages 106–115, Berlin, 1971. Springer.

[Loo84] E. J. N. Looijenga. *Isolated singular points on complete intersections*, volume 77 of *London Mathematical Society Lecture Note Series*. Cambridge University Press, Cambridge, 1984.

[LS85] Eduard Looijenga and Joseph Steenbrink. Milnor number and Tjurina number of complete intersections. *Math. Ann.*, 271(1):121–124, 1985.

[Mal67] B. Malgrange. *Ideals of differentiable functions*. Tata Institute of Fundamental Research Studies in Mathematics, No. 3. Tata Institute of Fundamental Research, Bombay; Oxford University Press, London, 1967.

[Mar76] Jean Martinet. Déploiements versels des applications différentiables et classification des applications stables. In *Singularités d'applications différentiables (Sém., Plans-sur-Bex, 1975)*, volume 535 of *Lecture Notes in Math.*, pages 1–44. Springer, Berlin, 1976.

[Mar82] Jean Martinet. *Singularities of smooth functions and maps*, volume 58 of *London Mathematical Society Lecture Note Series*. Cambridge University Press, Cambridge, 1982. Translated from the French by Carl P. Simon.

[Mat68a] J. N. Mather. Stability of C^∞ mappings. I. The division theorem. *Ann. of Math. (2)*, 87:89–104, 1968.

[Mat68b] J. N. Mather. Stability of C^∞ mappings. III. Finitely determined mapgerms. *Inst. Hautes Études Sci. Publ. Math.*, 35:279–308, 1968.

[Mat69a] J. N. Mather. Stability of C^∞ mappings. II. Infinitesimal stability implies stability. *Ann. of Math. (2)*, 89:254–291, 1969.

[Mat69b] J. N. Mather. Stability of C^∞ mappings. IV. Classification of stable germs by R-algebras. *Inst. Hautes Études Sci. Publ. Math.*, 37:223–248, 1969.

[Mat70] J. N. Mather. Stability of C^∞ mappings. V. Transversality. *Advances in Math.*, 4:301–336 (1970), 1970.

[Mat71] J. N. Mather. Stability of C^∞ mappings. VI: The nice dimensions. In *Proceedings of Liverpool Singularities-Symposium, I (1969/70)*, volume 192 of *Lecture Notes in Math.*, pages 207–253, Berlin, 1971. Springer.

[Mat73] John N. Mather. On Thom-Boardman singularities. In *Dynamical systems (Proc. Sympos., Univ. Bahia, Salvador, 1971)*, pages 233–248. Academic Press, New York, 1973.

[Mat76] John N. Mather. How to stratify mappings and jet spaces. In *Singularités d'applications différentiables (Sém., Plans-sur-Bex, 1975)*, volume 535 of *Lecture Notes in Math.*, pages 128–176. Springer, Berlin, 1976.

[Mat89] Hideyuki Matsumura. *Commutative ring theory*, volume 8 of *Cambridge Studies in Advanced Mathematics*. Cambridge University Press, Cambridge, second edition, 1989. Translated from the Japanese by M. Reid.

[McC01] John McCleary. *A user's guide to spectral sequences*, volume 58 of *Cambridge Studies in Advanced Mathematics*. Cambridge University Press, Cambridge, second edition, 2001.

[Mil63] J. Milnor. *Morse theory*. Based on lecture notes by M. Spivak and R. Wells. Annals of Mathematics Studies, No. 51. Princeton University Press, Princeton, N.J., 1963.

[Mil68] J. Milnor. *Singular points of complex hypersurfaces*, volume 61 of *Ann. of Math. Studies*. Princeton University Press, Princeton, 1968.

[Mil97] John W. Milnor. *Topology from the differentiable viewpoint*. Princeton Landmarks in Mathematics. Princeton University Press, Princeton, NJ, 1997. Based on notes by David W. Weaver, Revised reprint of the 1965 original.

[MM96] Washington Luiz Marar and David Mond. Real map-germs with good perturbations. *Topology*, 35(1):157–165, 1996.

[MNnB14] W. L. Marar and J. J. Nuño Ballesteros. Slicing corank 1 map germs from $\mathbb{C}^2$ to $\mathbb{C}^3$. *Q. J. Math.*, 65(4):1375–1395, 2014.

[Mon85] David Mond. On the classification of germs of maps from $\mathbf{R}^2$ to $\mathbf{R}^3$. *Proc. London Math. Soc. (3)*, 50(2):333–369, 1985.

[Mon86] David Mond. Singularities of the exponential map of the tangent bundle associated with an immersion. *Proc. London Math. Soc. (3)*, 53(2):357–384, 1986.

[Mon87] David Mond. Some remarks on the geometry and classification of germs of maps from surfaces to 3-space. *Topology*, 26(3):361–383, 1987.

[Mon91] David Mond. Vanishing cycles for analytic maps. In *Singularity theory and its applications, Part I (Coventry, 1988/1989)*, volume 1462 of *Lecture Notes in Math.*, pages 221–234. Springer, Berlin, 1991.

[Mon95] David Mond. Looking at bent wires—$\mathscr{A}_e$-codimension and the vanishing topology of parametrized curve singularities. *Math. Proc. Cambridge Philos. Soc.*, 117(2):213–222, 1995.

[Mon96] David Mond. How good are real pictures? In *Algebraic geometry and singularities (La Rábida, 1991)*, volume 134 of *Progr. Math.*, pages 259–276. Birkhäuser, Basel, 1996.

[Mon18] David Mond. Disentanglements of corank 2 map-germs: two examples. In *Singularities and Foliations. Geometry, Topology and Applications*, volume 222 of *Proceedings in Mathematics and Statistics*, pages 229–258. Springer, Berlin, 2018.

[Mor65] Bernard Morin. Formes canoniques des singularités d'une application différentiable. *C. R. Acad. Sci. Paris*, 260:5662–5665, 1965.

[Mor75] Bernard Morin. Calcul jacobien. *Ann. Sci. École Norm. Sup. (4)*, 8:1–98, 1975.

[Mor96] Marston Morse. *The calculus of variations in the large*, volume 18 of *American Mathematical Society Colloquium Publications*. American Mathematical Society, Providence, RI, 1996. Reprint of the 1932 original.

[MP89] David Mond and Ruud Pellikaan. Fitting ideals and multiple points of analytic mappings. In *Algebraic geometry and complex analysis (Pátzcuaro, 1987)*, volume 1414 of *Lecture Notes in Math.*, pages 107–161. Springer, Berlin, 1989.

[MT96] W. L. Marar and F. Tari. On the geometry of simple germs of co-rank 1 maps from $\mathbf{R}^3$ to $\mathbf{R}^3$. *Math. Proc. Cambridge Philos. Soc.*, 119(3):469–481, 1996.

[NBPS18] J. J. Nuño-Ballesteros and G. Peñafort-Sanchis. Iterated multiple points I: functoriality, defining equations and pathologies. Preprint, 2018.

[NBPS19] J. J. Nuño-Ballesteros and G. Peñafort-Sanchis. Iterated multiple points II: local properties and finite determinacy. Preprint, 2019.

[Nir71] L. Nirenberg. A proof of the Malgrange preparation theorem. In *Proceedings of Liverpool Singularities—Symposium, I (1969/70)*, volume 192 of *Lecture Notes in Math.*, pages 97–105, Berlin, 1971. Springer.

[NnBOOT18] J. J. Nuño Ballesteros, B. Oréfice-Okamoto, and J. N. Tomazella. Equisingularity of map germs from a surface to the plane. *Collect. Math.*, 69(1):65–81, 2018.

[NnBPnS17] Juan J. Nuño Ballesteros and Guillermo Peñafort Sanchis. Multiple point spaces of finite holomorphic maps. *Q. J. Math.*, 68(2):369–390, 2017.

[NnBT12] J. J. Nuño Ballesteros and J. N. Tomazella. Equisingularity of families of map germs between curves. *Math. Z.*, 272(1-2):349–360, 2012.

[NP15] András Némethi and Gergő Pintér. Immersions associated with holomorphic germs. *Comment. Math. Helv.*, 90(3):513–541, 2015.

[NR54] D. G. Northcott and D. Rees. A note on reductions of ideals with an application to the generalized Hilbert function. *Proc. Cambridge Philos. Soc.*, 50:353–359, 1954.

[Ohm16] Toru Ohmoto. Singularities of maps and characteristic classes. In *School on real and complex singularities in São Carlos, 2012*, volume 68 of *Adv. Stud. Pure Math.*, pages 191–265. Math. Soc. Japan, [Tokyo], 2016.

[Pel89] Ruud Pellikaan. On hypersurface singularities which are stems. *Compositio Math.*, 71(2):229–240, 1989.

[Pie79] Ragni Piene. Ideals associated to a desingularization. In *Algebraic geometry (Proc. Summer Meeting, Univ. Copenhagen, Copenhagen, 1978)*, volume 732 of *Lecture Notes in Math.*, pages 503–517. Springer, Berlin, 1979.

[Poé76] Valentin Poénaru. *Singularités C^∞ en présence de symétrie.* Lecture Notes in Mathematics, Vol. 510. Springer-Verlag, Berlin, 1976. En particulier en présence de la symétrie d'un groupe de Lie compact.

[PTPS18] I Pallarés-Torres and G. Peñafort-Sanchis. Image milnor number for weighted homogeneous map-germs. Preprint, 2018.

[Rei27a] Kurt Reidemeister. Elementare Begründung der Knotentheorie. *Abh. Math. Sem. Univ. Hamburg*, 5(1):24–32, 1927.

[Rei27b] Kurt Reidemeister. Knoten und Gruppen. *Abh. Math. Sem. Univ. Hamburg*, 5(1):7–23, 1927.

[Rie87] J. H. Rieger. Families of maps from the plane to the plane. *J. London Math. Soc. (2)*, 36(2):351–369, 1987.

[Ron73] F. Ronga. “La classe duale aux points doubles” d’une application. *Compositio Math.*, 27:223–232, 1973.

[Rot09] Joseph J. Rotman. *An introduction to homological algebra.* Universitext. Springer, New York, second edition, 2009.

[RR91] J. H. Rieger and M. A. S. Ruas. Classification of $\mathscr{A}$-simple germs from k^n to k^2. *Compositio Math.*, 79(1):99–108, 1991.

[Sai71] Kyoji Saito. Quasihomogene isolierte Singularitäten von Hyperflächen. *Invent. Math.*, 14:123–142, 1971.

[Sai80] Kyoji Saito. Theory of logarithmic differential forms and logarithmic vector fields. *J. Fac. Sci. Univ. Tokyo Sect. IA Math.*, 27(2):265–291, 1980.

[Sei66] A. Seidenberg. Derivations and integral closure. *Pacific J. Math.*, 16:167–173, 1966.

[Sha14] Ayşe Altınta¸s Sharland. Examples of finitely determined map-germs of corank 2 from n-space to $(n + 1)$-space. *Internat. J. Math.*, 25(5):1450044, 17, 2014.

[Sha17] Ayşe Sharland. Examples of finitely determined map-germs of corank 3 supporting Mond’s $\mu \geq \tau$-type conjecture. *Experimental Mathematics*, 0(0):1–6, 2017.

[Sie73] Dirk Siersma. The singularities of C^∞-functions of right-codimension smaller or equal than eight. *Nederl. Akad. Wetensch. Proc. Ser. A* ***76****=Indag. Math.*, 35:31–37, 1973.

[Sie91] Dirk Siersma. Vanishing cycles and special fibres. In *Singularity theory and its applications, Part I (Coventry, 1988/1989)*, volume 1462 of *Lecture Notes in Math.*, pages 292–301. Springer, Berlin, 1991.

[Spr09] T. A. Springer. *Linear algebraic groups.* Modern Birkhäuser Classics. Birkhäuser Boston, Inc., Boston, MA, second edition, 2009.

[SS75] Günter Scheja and Uwe Storch. Über Spurfunktionen bei vollständigen Durchschnitten. *J. Reine Angew. Math.*, 278/279:174–190, 1975.

[Tei76] B. Teissier. The hunting of invariants in the geometry of discriminants. In *Nordic Summer School/NAVF*, Symposium in Mathematics, pages 565–677, Oslo, August 5-25, 1976.

[Tei82] Bernard Teissier. Variétés polaires. II. Multiplicités polaires, sections planes, et conditions de Whitney. In *Algebraic geometry (La Rábida, 1981)*, volume 961 of *Lecture Notes in Math.*, pages 314–491. Springer, Berlin, 1982.

[Tho56] R. Thom. Les singularités des applications différentiables. *Ann. Inst. Fourier, Grenoble*, 6:43–87, 1955–1956.

[Tho76] René Thom. *Structural stability and morphogenesis.* W. A. Benjamin, Inc., Reading, Mass.-London-Amsterdam, 1976. An outline of a general theory of models, Translated from the French by D. H. Fowler, With a foreword by C. H. Waddington, Second printing.

[Tou68] Jean-Claude Tougeron. Idéaux de fonctions différentiables. I. *Ann. Inst. Fourier (Grenoble)*, 18(fasc., fasc. 1):177–240, 1968.

[Vak06] Ravi Vakil. Murphy’s law in algebraic geometry: badly-behaved deformation spaces. *Invent. Math.*, 164(3):569–590, 2006.

[Vas01] Victor Vassiliev. Resolutions of discriminants and topology of their complements. In *New developments in singularity theory (Cambridge, 2000)*, volume 21 of *NATO Sci. Ser. II Math. Phys. Chem.*, pages 87–115. Kluwer Acad. Publ., Dordrecht, 2001.

[Ver84] Andrei Verona. *Stratified mappings—structure and triangulability*, volume 1102 of *Lecture Notes in Mathematics*. Springer-Verlag, Berlin, 1984.

[Vos02] Henrik Vosegaard. A characterization of quasi-homogeneous complete intersection singularities. *J. Algebraic Geom.*, 11(3):581–597, 2002.

[vS95] D. van Straten. A note on the discriminant of a space curve. *Manuscripta Math.*, 87(2):167–177, 1995.

[Wal77] C. T. C. Wall. Geometric properties of generic differentiable manifolds. In *Geometry and topology (Proc. III Latin Amer. School of Math., Inst. Mat. Pura Aplicada CNPq, Rio de Janeiro, 1976)*, pages 707–774. Lecture Notes in Math., Vol. 597. Springer, Berlin, 1977.

[Wal81] C. T. C. Wall. Finite determinacy of smooth map-germs. *Bull. London Math. Soc.*, 13(6):481–539, 1981.

[Wal95] C. T. C. Wall. Classification and stability of singularities of smooth maps. In *Singularity theory (Trieste, 1991)*, pages 920–952. World Sci. Publ., River Edge, NJ, 1995.

[Wal09] C. T. C. Wall. Geometry of projection-generic space curves. *Math. Proc. Cambridge Philos. Soc.*, 147(1):115–142, 2009.

[Wei94] Charles A. Weibel. *An introduction to homological algebra*, volume 38 of *Cambridge Studies in Advanced Mathematics*. Cambridge University Press, Cambridge, 1994.

[Whi44] Hassler Whitney. The singularities of a smooth n-manifold in $(2n-1)$-space. *Ann. of Math. (2)*, 45:247–293, 1944.

[Whi55] Hassler Whitney. On singularities of mappings of Euclidean spaces. I. Mappings of the plane into the plane. *Ann. of Math. (2)*, 62:374–410, 1955.

[Zhi08] M. Zhitomirskii. Fully simple singularities of plane and space curves. *Proc. Lond. Math. Soc. (3)*, 96(3):792–812, 2008.

Index

D. Mond, J. J. Nuño-Ballesteros, *Singularities of Mappings*,
Grundlehren der mathematischen Wissenschaften 357,
https://doi.org/10.1007/978-3-030-34440-5

MIX
Papier aus verantwortungsvollen Quellen
Paper from responsible sources
FSC® C105338
www.fsc.org

If you have any concerns about our products,
you can contact us on
ProductSafety@springernature.com

In case Publisher is established outside the EU,
the EU authorized representative is:
Springer Nature Customer Service Center GmbH
Europaplatz 3, 69115 Heidelberg, Germany

Printed by Libri Plureos GmbH
in Hamburg, Germany